PREDICTION AND SIMULATION METHODS FOR GEOHAZARD MITIGATION

PROCEEDINGS OF THE INTERNATIONAL SYMPOSIUM ON PREDICTION AND SIMULATION METHODS FOR GEOHAZARD MITIGATION (IS-KYOTO2009), KYOTO, JAPAN, 25–27 MAY 2009

Prediction and Simulation Methods for Geohazard Mitigation

Editors

Fusao Oka
Department of Civil and Earth Resources Engineering,
Kyoto University, Japan

Akira Murakami
Graduate School of Environmental Science,
Okayama University, Japan

Sayuri Kimoto
Department of Civil and Earth Resources Engineering,
Kyoto University, Japan

CRC Press
Taylor & Francis Group
Boca Raton   London   New York   Leiden

CRC Press is an imprint of the
Taylor & Francis Group, an **informa** business

A BALKEMA BOOK

First issued in paperback 2017

CRC Press/Balkema is an imprint of the Taylor & Francis Group, an informa business

© 2009 Taylor & Francis Group, London, UK

Typeset by Vikatan Publishing Solutions (P) Ltd., Chennai, India

Published by: CRC Press/Balkema
P.O. Box 447, 2300 AK Leiden, The Netherlands
e-mail: Pub.NL@taylorandfrancis.com
www.crcpress.com – www.taylorandfrancis.co.uk – www.balkema.nl

ISBN 13: 978-1-138-11807-2 (pbk)
ISBN 13: 978-0-415-80482-0 (hbk)

Table of contents

Numerical and analytical simulation methods for geohazards

Advanced constitutive modeling of geomaterials and laboratory and field testing

Thermo-hydro-mechanical instabilities

Monitoring and non-destructive investigation methods

*Evaluation of existing prediction methods, performance-based design methods,
risk analysis and the management of mitigation programs*

Case records of geohazards and mitigation projects

Preface

The mitigation of geohazards is an important problem in geotechnical engineering. Heavy rains, typhoons and earthquakes are the main causes of geohazards. Due to changes in climate and extreme weather, geohazards are found all over the world. The Kansai branch of the Japanese Geotechnical Society (JGS) established a technical committee on the Mitigation of Geohazards in River Basins in 2006, and it has been doing site investigations on the geohazards brought about by heavy rains and typhoons. On the other hand, Technical Committee 34 (TC34) of the ISSMGE on Prediction and Simulation Methods in Geomechanics has been working on prediction and simulation methods for geomechanics. In particular, TC34 focuses on analyzing unstable behavior of the ground, such as strain localization, which is a precursor of the failure of the ground, liquefaction, landslides, seepage failure, etc. TC34 of ISSMGE on Prediction Methods in Large Strain Geomechanics was established in 2001 after the Istanbul Conference. Since then, TC34 had been successfully managed by Chairperson Professor Vardoulakis of Athens, Greece under the support of JGS, and a portion of their goals has been achieved through the promotion of exchanges among academic and practicing members. After the 16th ISSMGE in Osaka, TC34 was reestablished as Prediction and Simulation Methods in Geomechanics.

Members of the Kansai branch of JGS and TC34 of ISSMGE decided to organize an international symposium on Prediction and Simulation Methods for Geohazard Mitigation. The symposium provides a forum for discussing new prediction and simulation methods for geohazards and for exchanging ideas and information on topics of mutual interest. This symposium will mark the 60th anniversary of JGS and the 50th anniversary of the Kansai branch of JGS. The symposium is sponsored by the JGS, the Kansai branch of the JGS, TC34 of ISSMGE, TC34's supporting committees of JGS and the Commemorative Organization for the Japan World Exposition('70)

The themes of prediction and simulation methods for geohazard mitigation include:

1. Mechanisms of geohazards, namely, heavy rains, floods, typhoons, earthquakes, landslides, slope and snow slides, tsunamis, land subsidence, coastal erosion, etc.
2. Numerical and analytical simulation methods for geohazards, including conventional and advanced methods, FDM, FEM, Extended FEM, DEM, SPH and MPM.
3. Advanced constitutive modeling of geomaterials and numerical implementations and constitutive parameter determination using laboratory and field test results.
4. Thermo-hydro-mechanical instabilities, namely, large deformations, strain localization, progressive failure, liquefaction, ground water flow analysis, the rapid flow of complex geofluids such as mud flow, etc.
5. Monitoring and non-destructive investigative methods for geostructures during/after floods, earthquakes, heavy rains, etc. and design methods.
6. Evaluation of existing prediction methods, performance-based design methods aided by advanced numerical modeling, risk analysis, and the management of mitigation programs.
7. Case records of geohazards and mitigation projects.

A total of 89 papers on the above topics have been contributed from 18 countries. The members of the Organizing Committee and the International Advisory Committee reviewed 116 papers. The editors believe that all of the papers, presentations, and discussions during the symposium will open the door to new areas of research and engineering for the mitigation of geohazards. We wish to express our sincere thanks to the authors of the contributed papers for their resourceful papers and to the members of the International Advisory Committee and the Organizing Committee for their valuable support and review of the manuscripts. Acknowledgements are also given to the Japanese Geotechnical Society, the Kansai branch of the Japanese Geotechnical Society, and ISSMGE.

Editors

Fusao Oka, *Kyoto University*
Akira Murakami, *Okayama University*
Sayuri Kimoto, *Kyoto University*

Prediction and Simulation Methods for Geohazard Mitigation – Oka, Murakami & Kimoto (eds)
© 2009 Taylor & Francis Group, London, ISBN 978-0-415-80482-0

Organisation

Organizing committee

F. Oka, Chair, Kyoto University, Japan
K. Tokida, Vice Chair, Osaka University, Japan
A. Murakami, Vice Chair, Okayama University, Japan
H. Kusumi, Vice Chair, Kansai University, Japan
N. Nakanishi, Fukken Co., Ltd., Japan
T. Katayama, The General Environmental Technos Co., Ltd., Japan
N. Torii, Kobe University, Japan
K. Hayashi, Forest Engineering, Inc., Japan
S. Fukushima, Fudo Tetra Corporation, Japan
T. Yoden, NEWJEC Inc., Japan
T. Emura, Kansai Int. Airport, Co., Ltd., Japan
M. Kimura, Kyoto University, Japan
K. Kishida, Kyoto University, Japan
A. Kobayashi, Kyoto University, Japan
T. Katsumi, Kyoto University, Japan
S. Kimoto, Kyoto University, Japan
T. Konda, Geo-reserch Institute, Japan
H. Saito, JR West Japan Consultants Co., Japan
H. Yoshidu, Ministry of Land, Infrastructure and Transport, Japan
T. Nakai, TODA Corporation, Japan
K. Lee, CTI Engineering Co., Ltd., Japan
T. Kodaka, Meijo University, Japan
A. Iizuka, Kobe University, Japan
Y. Kohgo, Tokyo University of Agriculture and Technology, Japan
S. Shibuya, Kobe University, Japan
S. Sunami, Nikken Sekkei Civil Engineering Ltd., Japan
T. Noda, Nagoya University, Japan
F. Zhang, Nagoya Institute of Tech., Japan
R. Uzuoka, Tohoku University, Japan
K. Maeda, Nagoya Institute of Tech., Japan

International advisory committee

I.G. Vardoulakis, Technical University of Athens, Greece
F. Darve, Grenoble, L3S, France
D. Muir Wood, Bristol University, UK
K.T. Chau, Hong Kong Polytechnic University, Hong Kong
P. van den Berg, Delft Geotechnics, The Netherlands
R. Nova, Politecnico di Milano, Italy
P.V. Lade, The Catholic University of America, USA
M. Muniz de Farias, Federal University of Viçosa, Brazil
R. Wan, University of Calgary, Canada
M. Pastor, Centro de Estudios de Tecnicas Aplicadas, Spain
A. Cividini, Politecnico di Milano, Italy
A. Gens, Universitat Politecnica de Catalunya, Spain
S.-R. Lee, KAIST, Korea

D. Kolymbas, University of Innsbruck, Austria
J.F. Labuz, University of Minnesota, USA
D. Masin, Charles University, Czech Republic
D. Sheng, The University of Newcastle, Australia
G. Zhang, Tsinghua University, China
T. Länsivaara, Tampere University of Technology, Finland
L. Michalowski, University of Michigan, USA
H.-B. Mühlhaus, University of Queensland, Australia
J. Sulem, ENPC, France
C. Tamagnini, Università di Perugia, Italy
Y.K. Chow, National University of Singapore, Singapore
T. Wanchai, Chulalongkorn University, Thailand
K. Arai, University of Fukui, Japan
H. Sekiguchi, Kyoto University, Japan
H. Nakagawa, Kyoto University, Japan
A. Asaoka, Nagoya University, Japan
T. Nakai, Nagoya Institute of Technology, Japan
M. Hori, The University of Tokyo, Japan
J. Otani, Kumamoto University, Japan
A. Yashima, Gifu University, Japan

Member of TC34 of ISSMGE (2005–2009)

Chairman
F. Oka, Japan

Secretary
A. Murakami, Japan

Core Members
K.-T. Chau, Hong Kong
F. Darve, France
M. Muniz de Farias, Brazil
P.V. Lade, USA
D. Muir Wood, UK
R. Nova, Italy
P. van den Berg, The Netherlands
I.G. Vardoulakis, Greece

Members
R. Francisco de Azevedo, Brazil
D. Chan, Canada
R. Charlier, Belgium
J. Chuk, Hong Kong
A. Cividini, Italy
G. Exadaktylos, Greece
A.V. Filatov, Kazakhstan
J. Gaszynski, Poland
A. Gens, Spain
Y. Hu, Australia
S. Ryull Kim, Korea
Y.S. Kim, Korea
T. Kodaka, Japan
D. Kolymbas, Austria
J.F. Labuz, USA
G. Lamer, Hungary
T. Lansivaara, Finland

X.S. Li, Hong Kong
J. Maranha, Portugal
D. Masin, Czech Republic
R.L. Michalowski, U.S.A.
H.-B. Mühlhaus, Australia
A. Noorzad, Iran
V.N. Paramonov, Russia
M. Pastor, Spain
K. Rajagopal, India
J. Paulo Bile Serra, Portugal
D. Sheng, Australia
J. Sulem, France
C. Tamagnini, Italy
R. Uzuoka, Japan
S.-N. Vlasta, Croatia
R. Wan, Canada
F. Zhang, Japan
G. Zhang, China

Member of the research committee on the Mitigation of Geohazards in River Basins, Kansai branch of JGS

F. Oka, Chairperson, Kyoto University
Y. Ikeda, OYO Corporation
H. Uemoto, Soil Engineering Institute Co. Ltd.
M. Okuno, NTT Infranet
R. Kato, Nikken Sekkei Civil Engineering Ltd.
K. Shibata, Ministry of Land, Infrastructure and Transport, Kinki Regional Development Bureau
R. Azuma, Kyoto University
H. Takamori, WASC Kisojiban Institute
K. Tokida, Vice Chair, Osaka University
N. Nakanishi, Fukken Co., Ltd.
T. Katayama, The General Environmental Technos Co., Ltd.
N. Torii, Kobe University
K. Hayashi, Forest Engineering, Inc.
S. Fukushima, Fudo Tetra Corporation
T. Yoden, NEWJEC Inc.
S. Kimoto, Kyoto University
K. Lee, CTI Engineering Co., Ltd.
T. Kodaka, Meijo University
S. Sunami, Nikken Sekkei Civil Engineering Ltd.
K. Arai, University of Fukui
H. Sekiguchi, Kyoto University
H. Nakagawa, Kyoto University
S. Baba, TOYO Construction Co. Ltd.
K. Fujisawa, Okayama University
A. Fujisawa, Kawasaki Geological Engineering Col. Ltd.

JGS TC34 supporting committee

Y. Kohgo, Tokyo University of Agriculture and Technology
T. Kodaka, Meijo University
K. Komiya, Chiba Institute of Technology
H. Sakaguchi, JAMSTEC
S. Sunami, Nikken Sekkei Civil Engineering Ltd., Japan
K. Sekiguchi, National Research Institute for Earth Science and Disaster Prevention
A. Takahashi, Public Works Research Institute
F. Tatsuoka, Tokyo University of Science

T. Tamura, Kyoto University
F. Zhang, Nagoya Institute of Technology
Y. Tobita, Tohoku Gakuin University
T. Nakai, Nagya Institute of Technology
M. Nakano, Nagoya University
T. Noda, Nagoya University
Y. Higo, Kyoto University
M. Hori, The University of Tokyo
K. Maeda, Nagoya Institute of Technology
T. Matsushima, Tsukuba University
T. Miyake, TOYO Construction Co. Ltd.
A. Murakami, Okayama University
A. Yashima, Gifu University
Y. Yamakawa, Tohoku University
N. Yoshida, Kobe University

Mechanisms of geohazards

The LIQSEDFLOW: Role of two-phase physics in subaqueous sediment gravity flows

S. Sassa
Port and Airport Research Institute, Yokosuka, Japan

H. Sekiguchi
Kyoto University, Kyoto, Japan

ABSTRACT: The paper describes an extension of the computational code LIQSEDFLOW proposed by the authors. The salient features of the code lie in the capabilities to describe the multi-phased physics of subaqueous sediment gravity flows. Specifically, it combines Navier-Stokes/continuity equations and equations for advection and hindered settling of grains for a liquefied soil domain, with a consolidation equation for the underlying, progressively solidifying soil domain, via a transition layer that is characterized by zero effective stress and a small yet discernable stiffness. Evolutions of the flow and solidification surfaces are traced as part of solution by using a volume-of-fluid (VOF) technique. The predicted features of gravity flows of initially fluidized sediments with different concentrations conform to the observed performances in two-dimensional flume tests. The present results demonstrate the crucial role of two-phase physics, particularly solidification, in re-producing the concurrent processes of flow stratification, deceleration, and redeposition in subaqueous sediment gravity flows.

1 INTRODUCTION

Sediment gravity flows under water have become an increasingly important subject for research in relation to geomorphodynamics of sediment routing systems that connect river basins, estuaries and coastal oceans. Also, submarine landslides and flow slides have received considerable practical attention in view of their destructive power and associated consequences in nearshore and offshore facilities (Hampton et al. 1996, UNESCO 2009). Fluid-sediment interactions are a key process that features any of subaqueous sediment gravity flows. However, current flow models are mostly depth-averaged and/or rheologically based. Thus they cannot adequately describe the multi-phased nature such as pore fluid migration and associated solidification that should occur in the flowing sediments leading to redeposition. Integration of fluid-dynamics and soil mechanics approaches is indispensable in advancing the multi-phased physics of subaqueous sediment gravity flows and thereby facilitating a rational analysis framework pertaining to geohazard mitigation.

The authors have proposed a theoretical framework called "LIQSEDFLOW" (Sassa et al. 2003) to predict the flow dynamics of hyperconcentrated sediment-water mixtures that may result from liquefaction or fluidization under dynamic environmental loading. The emphasis of the analysis procedure is placed on considering the multi-phased nature of sediment gravity flows. Notably, it accounts for the occurrence of progressive solidification due to pore fluid migration in the flowing fluidized sediment.

The present paper describes an extension of the above-mentioned LIQSEDFLOW, with due consideration of the effect of hindered settling and advection of grains in the course of flowage. The predicted performances of subaqueous sediment gravity flows are discussed in light of the characteristics of flow stratification, deceleration, and redeposition. Comparison is made between the predictions and the observations made in the flume experiments of Amiruddin et al. (2006).

2 THEORY FOR SUBAQUEOUS SEDIMENT GRAVITY FLOWS

2.1 *Problem definition*

Consider a body of submerged granular soil that has just undergone liquefaction or fluidization under the action of storm waves, earthquakes or excessive seepage forces. The liquefied sediment, with a mass density ρ_2, will start collapsing under gravity into an ambient fluid with a mass density ρ_1 (Fig. 1). The depth of the ambient fluid is assumed here to be constant in the course of flowage. The liquefied flow will undergo hindered settling and advection of grains while undergoing progressive solidification. Progressive solidification is a sort of phase-change process that allows transitory fluid-like particulate sediment to reestablish a grain-supported framework during continued disturbances and has been demonstrated theoretically as well as experimentally in gravity flows (Sassa et al. 2003; Amiruddin et al. 2006) and under wave loading (Miyamoto et al. 2004).

The pore fluid pressure p at a generic point in the sediment may be divided into two components. Namely,

$$p = p_e + p_s \tag{1}$$

where p_s is the hydrostatic pressure which is expressed as $p_s = \rho_1 g(h - z)$, and p_e represents the excess pressure due to contractancy of the sediment.

2.2 Formulation for the domain of liquefied flow

For the purpose of non-dimensional formulation, let us denote a reference length as a and a reference velocity as $U_r = \sqrt{ga}$. The non-dimensional time T may then be expressed as $U_r t/a$ and the non-dimensional excess pore pressure P_e may be expressed as $p_e/(\rho_2 - \rho_1)U_r^2$. In view of Figure 1(a), it is important here to note that the boundary conditions on the flow surface can be described by $P_e = 0$, with reference to equation (1).

The two-dimensional system of Navier-Stokes equations, considering the effect of the excess pore pressure P_e, for describing the dynamics of the subaqueous liquefied flow may then be expressed as (Sassa et al. 2003):

$$\frac{\partial U}{\partial X} + \frac{\partial V}{\partial Z} = 0 \tag{2}$$

$$\frac{\partial U}{\partial T} + U\frac{\partial U}{\partial X} + V\frac{\partial U}{\partial Z} = -\frac{\rho_2 - \rho_1}{\rho_2}\frac{\partial P_e}{\partial X}$$

$$+ \frac{1}{R_e}\left(\frac{\partial^2 U}{\partial X^2} + \frac{\partial^2 U}{\partial Z^2}\right) \tag{3}$$

$$\frac{\partial V}{\partial T} + U\frac{\partial V}{\partial X} + V\frac{\partial V}{\partial Z} = -\frac{\rho_2 - \rho_1}{\rho_2}\frac{\partial P_e}{\partial Z}$$

$$+ \frac{1}{R_e}\left(\frac{\partial^2 V}{\partial X^2} + \frac{\partial^2 V}{\partial Z^2}\right) - \frac{\rho_2 - \rho_1}{\rho_2} \tag{4}$$

Here $U = u/U_r$ is the non-dimensional velocity in the X-direction, $V = v/U_r$ is the non-dimensional velocity in the Z-direction and R_e is the Reynolds number as expressed by HU_r/ν where ν is the kinematic viscosity of the flowing liquefied soil.

The hindered settling and advection of grains during the course of liquefied flow may be effected by the following equation in terms of sediment concentration C:

$$-\frac{\partial (C \cdot U)}{\partial X} - \frac{\partial (C \cdot V)}{\partial Z} + \frac{\partial (C \cdot W_s)}{\partial Z} = \frac{\partial C}{\partial T} \tag{5}$$

where $W_s = w_s/U_r$ is the non-dimensional particles settling velocity. The particles settling velocity depends on sediment concentrations C. Here, we adopt the following equation proposed by Richardson & Zaki (1954):

$$W_s = W_0 (1 - C)^n \quad \text{with} \quad n = 4.65 \tag{6}$$

where $W_0 = w_0/U_r$ stands for the non-dimensional terminal free settling velocity (Stokes settling velocity).

2.3 Formulation for the domain undergoing solidification

The process of solidification in the course of the liquefied flow may be described by the two-dimensional equation of consolidation. Let K_W be the Darcy permeability coefficient, which is non-dimensionalized by dividing by U_r. Let M be the constrained modulus of the soil skeleton, which is non-dimensionalized by dividing by $(\rho_2 - \rho_1)U_r^2$. The consolidation equation then reads

$$\frac{\partial (\sigma_m - P_e)}{\partial T} = -\frac{\rho_2 - \rho_1}{\rho_1} M \cdot K_w \left(\frac{\partial^2 P_e}{\partial X^2} + \frac{\partial^2 P_e}{\partial Z^2}\right) \tag{7}$$

where σ_m is the non-dimensional mean total stress of the solidifying particulate sediment. Note here that the value of the constrained modulus M should increase with increasing effective confining pressure. With reference to Sassa et al. (2001), we adopt the following simple relationship:

$$M = (Z_s - Z) \cdot M_r \quad \text{for} \quad 0 \le Z \le Z_s \tag{8}$$

where M_r is the reference value of M for the sediment having a non-dimensional thickness of unity.

3 NUMERICAL SOLUTION PROCEDURES

We will subsequently describe the numerical solution procedure pertaining to the entire system under consideration. Note that the locations of the flow surface as well as the solidification surface are unknowns that should be worked out in the solution procedure for the process of flowage (Fig.1).

The governing equations as described above were discretized in a non-uniform Eulerian mesh by using the MAC finite difference method (Amsden & Harlow, 1970). For tracking the flow surface the volume-of-fluid (VOF) technique (Hirt & Nichols, 1981) was applied, along with an efficient volume-advection scheme (Hamzah, 2001) so as to ensure the conservation of mass in the course of flowage. The evolution of the interface between the domains of liquefied and solidified soil was also effectively traced using the VOF technique. Specifically, the soil undergoing solidification was treated as being an obstacle to the flowing liquefied soil, such that the velocities

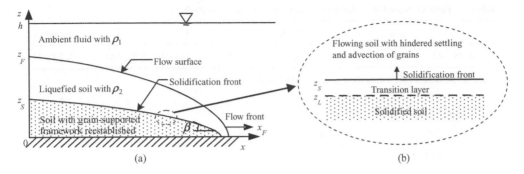

Figure 1. (a) Subaqueous sediment gravity flows undergoing progressive solidification; (b) a transition layer incorporated between flowing liquefied soil and solidified soil.

in the solidified soil become zero. This assumption is justifiable since the solidified zone should have a much higher stiffness and frictional resistance than the liquefied soil.

In view of Figure 1(b), a transition layer with zero effective stress yet having marginally discernable stiffness is introduced in such a way that it occupies the lowermost part of the liquefied soil domain and immediately overlies the solidified soil domain. By doing so, one can realize the phase change that may occur in accordance with the advance of the solidification front. In a computational step, the solidification front may be judged to be an active one if the effective stress increment in the transition layer becomes positive. Also, in order to address the effect of hindered settling upon solidification, we make the solidification front active if the sediment concentration there exceeds by a certain amount, 1%, the mass concentration of the initially fluidized sediment. Then, the solidification front can move upwards by an amount equal to the prescribed thickness of the transition layer. Concurrently, the liquefied soil domain retreats by the same amount, and the transition layer assumes a new (higher) location. The slope of the solidification surface, β, may be modified, if necessary, so as not to exceed a critical angle β_{cr} in view of the frictional resistance of the soil.

4 ANALYSES OF SUBAQUEOUS SEDIMENT GRAVITY FLOWS

A series of analyses of subaqueous sediment gravity flows were performed in conjunction with the two-dimensional flume tests conducted by Amiruddin et al. (2006). Namely, the problem was concerned with the two-dimensional dam break problem as shown in Figure 2. At the beginning of the calculation, the release gate was instantaneously removed, and the rectangular column of fluidized sediment was allowed to flow out over a horizontal floor in the channel with

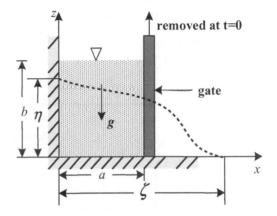

Figure 2. Two-dimensional dam break problem.

length of 1.5 m. The sand used was silica No. 6 with an average grain size of 0.32 mm. In Figure 2, the width of the column a, which is a reference length for the non-dimensional formulations as described above, was equal to 0.25 m. The computational domain with length of 1.5 m and height of 0.75 m was discretized into a total of 1500 rectangular elements. The time increment of the calculation was set at 0.00015 s.

The parameters used in the analyses are summarized in Table 1. It is instructive to note that the prescribed four different sediment concentrations $C = $ 30%, 34%, 38% and 42% were relatively high, but smaller than the concentration at loosest packing of this sand, namely 46%. This means that we targeted hyperconcentrated sand-water mixtures which could allow for hindered settling of grains during the course of flowage. Indeed, the Stokes setting velocity $w_0 = $ 20 mm/s, that was used in the standard analysis shown in Table 1, corresponds to the experimentally determined velocity, via equation (6) above, from the results of a series of sedimentation tests of Amiruddin (2005). The critical angle β_{cr} in view of frictional

Table 1. Parameters used in analyses for subaqueous sediment gravity flows.

c: %	b/a	R_e	k_w: mm/s	M	$\tan \beta_{cr}$	w_0: mm/s
30, 34, 38*, 42	0.7	100000	0.15	400	0.35, 0.4, 0.5*, 0.55	0, 20*, 40

The asterisk denotes the values used in the standard analysis.

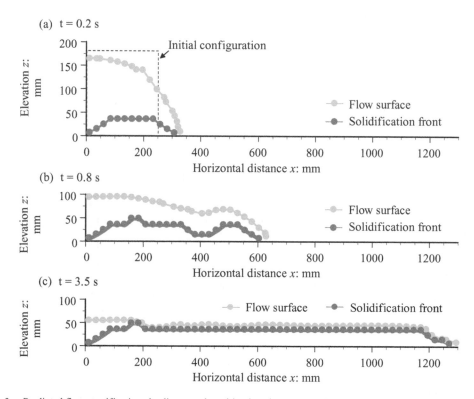

Figure 3. Predicted flow stratifications leading to redeposition in subaqueous sediment gravity flow ($c = 38\%$).

resistance of solidified soil was varied depending on the four different sediment concentrations as shown in Table 1.

The calibration of the numerical code developed was performed and verified against a dam-break problem of fluid flow (Sassa et al. 2003). Thus, in what follows, we will focus on the results of analyses of the hyperconcentrated sediment gravity flows as stated above.

4.1 Flow stratifications leading to redeposition

The predicted changes in the configuration of the subaqueous sediment gravity flow at sediment concentration $C = 38\%$ are illustrated in Figure 3 for three different times. Note that at $t = 0$, the solidification front coincides with the bottom of the liquefied sediment. It is seen that at $t = 0.2$ s the liquefied sediment has been collapsing and the solidification front has progressed upwards. The liquefied sediment then undergoes a significant flow deformation at $t = 0.8$ s. The solidified zone has developed further upwards and laterally in a wavy fashion. At this stage, a marked decelerating flow regime has taken place due to interactions between the flowing fluidized sediment and the accreting solidified soil with a grain-supported framework reestablished. Eventually, the liquefied flow becomes markedly elongated at $t = 3.5$ s, and comes to stop since the solidification front reaches the flow surface at essentially the same instant of time. This indicates the occurrence of "freezing" of the main body of the sediment gravity flow.

The above-described predicted features of flow stratifications leading to redeposition are found to be well consistent with what has been observed in the flume experiments on subaqueous sediment gravity flows as reported by Amiruddin et al. (2006).

6

4.2 Effects of solidification and hindered settling of grains on sediment gravity flows

In the course of the subaqueous sediment gravity flow, the sediment concentration changes due to advection and hindered settling of grains as well as due to the occurrence of progressive solidification. One such example is shown in Figure 4, where the predicted profiles of sediment concentrations with elevation, at $t = 0.8$ s, at three different stations are plotted for $C = 38\%$. It is seen that at given elevation z, the sediment concentration decreases with increasing x, namely in the direction of flow-out, due to horizontal advection of grains. Also, at given station x, the sediment concentration increases downward due to vertical advection and hindered settling of grains.

It is also interesting to note the occurrence of upward advection of grains just above the solidification front, at $x = 360$ mm, which stems from the clockwise fluid motions in the depression of the solidified sediment.

The effects of solidification and hindered settling of grains on sediment gravity flows can be more clearly seen in the form of Figure 5. In this figure, the predicted flow-out distances with and without the effects of progressive solidification, hindered settling and advection of grains are plotted for $C = 38\%$.

In the case without solidification, a simple form of accelerating flow takes place until the flow head hit the downstream end of the channel. By contrast, in the liquefied sediment flow with solidification, the accelerating flow occurs only at the initial stage of the flow, and deceleration becomes noticeable at and after $t = 0.4$ s. In fact, the subsequent flow process is characterized by marked decelerating flow regime resulting in the eventual stoppage of the flow at $t = 4$ s. It is also seen that the hindered settling of grains has only a marginal influence on the flow deceleration in the later process

of flowage, say after $t = 1.5$ s, with the increase in the Stokes settling velocity concerned. This means that the marked flow deceleration is effected essentially due to the occurrence of progressive solidification.

4.3 Comparison between predicted and observed performances

We now compare the predicted and observed time histories of flow-out distance for four different sediment concentrations, as shown in Figure 6. In this figure, the predicted and measured performances of the gravity flow of water are also plotted. Both of them exhibit a rapid rate of flowage, and agree with each other well. By contrast, all of the predicted sediment gravity flows exhibit decelerating flowage, except for the early stage of flow initiation. Indeed, the flow potential of subaqueous sediment gravity flows decreases markedly with the increases in sediment concentrations, together with the increases in the critical angle β_{cr} in view of frictional resistance of solidified soil. These predicted flow-out characteristics compare favorably with the observed flow-out characteristics.

In summary, all the results described above demonstrate the predictive capabilities of the computational code LIQSEDFLOW in realistically simulating the dynamics of subaqueous sediment gravity flows. This also indicate that the marked flow deceleration leading to "freezing" of the flow is indeed predictable, without introducing any artificial viscosity or yield stress with questionable physical significance, but, on the basis of the two-phase physics with solidification.

5 CONCLUSIONS

The computational code LIQSEDFLOW proposed by the authors has been extended here so as to incorporate the effect of hindered settling and advection of grains on the processes of subaqueous sediment gravity flows. The principal findings and conclusions obtained using the LIQSEDFLOW may be summarized as follows.

a. The predicted features of flow stratification, deceleration and redeposition of subaqueous sediment gravity flows following fluidization conform well to the experimentally observed features of the gravity flows as reported by Amiruddin et al. (2006).

b. The sediment concentrations of the hyperconcentrated sand-water mixtures vary significantly through the occurrence of advection and hindered settling of grains in the course of flowage. However, the effect of the hindered settling upon flow deceleration appears only marginal, and the occurrence of progressive solidification is essentially responsible for the marked flow deceleration leading to redeposition.

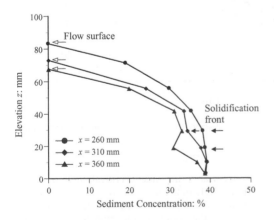

Figure 4. Predicted profiles of sediment concentrations with elevation, at $t = 0.8$ s, at three different stations in subaqueous sediment gravity flow ($c = 38\%$).

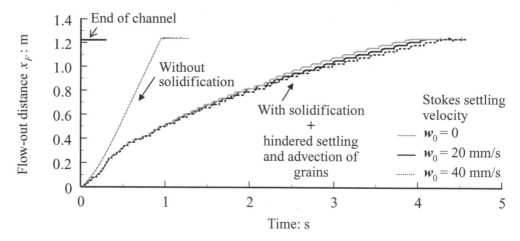

Figure 5. Predicted flow-out distances with and without the effects of progressive solidification and hindered settling and advection of grains in subaqueous sediment gravity flow ($c = 38\%$).

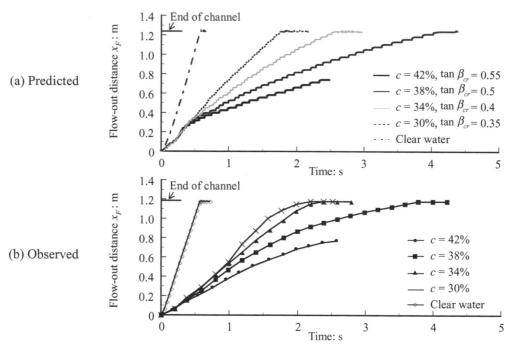

Figure 6. Comparison of predicted and observed time histories of flow-out distance for clear water and four different sediment concentrations.

c. The flow-out potentials of the subaqueous gravity flow decrease considerably with increasing sediment concentrations. Notably, we have shown that the effect of progressive solidification upon flowage at given sediment concentration can be reasonably reproduced by introducing the concept of a

concentration-dependent, critical angle of deposition in light of the frictional resistance of sediment with grain-supported framework reestablished.

d. Overall, the present results emphasize the crucial role of the two-phase physics, particularly solidification in reproducing the sequence of processes of subaqueous sediment gravity flows, thereby warranting wider applications of LIQSEDFLOW to geohazard mitigation.

REFERENCES

Amiruddin. 2005. *The dynamics of subaqueous sediment gravity flows and redepositional processes*. Doctoral thesis, Kyoto University.

Amiruddin, Sekiguchi, H. & Sassa, S. 2006. Subaqueous sediment gravity flows undergoing progressive solidification. *Norwegian Journal of Geology* 86: 285–293.

Amsden, A.A. & Harlow, F.H. 1970. A simplified MAC technique for incompressible fluid flow calculations. *J. Comput. Phys.* 6: 322–325.

Hampton, M.A., Lee, H.J. & Locat J. 1996. Submarine landslides. *Reviews of Geophysics* 34: 33–60.

Hamzah, M.A. 2001. *Numerical simulations of tsunami pressure acting upon coastal barriers on wet and dry lands*. Doctoral thesis, Kyoto University.

Hirt, C.W. & Nichols, B.D. 1981. Volume of fluid (VOF) method for the dynamics of free boundaries. *J. Comput. Phys.* 39: 201–225.

Miyamoto, J., Sassa, S. & Sekiguchi, H. 2004. Progressive solidification of a liquefied sand layer during continued wave loading. *Geotechnique* 54 (10): 617–629.

Richardson, J.F. & Zaki, W.N. 1954. Sedimentation and fluidization: Part 1. *Trans. Inst. Chem. Eng.* 32: 35–53.

Sassa, S., Sekiguchi, H. & Miyamoto, J., 2001. Analysis of progressive liquefaction as a moving boundary problem. *Géotechnique* 51(10): 847–857.

Sassa, S., Miyamoto, J. & Sekiguchi, H. 2003. The dynamics of liquefied sediment flow undergoing progressive solidification. In *Submarine Mass Movements and Their Consequences (eds. J. Locat and J. Mienert), Advances in Natural and Technological Hazards Research* 19, Kluwer Academic Publishers: 95–102.

UNESCO. 2009. International Geoscience Programme IGCP 511: Submarine Mass Movements and Their Consequences (2005–2009), http://www.geohazards.no/IGCP511

Prediction and Simulation Methods for Geohazard Mitigation – Oka, Murakami & Kimoto (eds)
© 2009 Taylor & Francis Group, London, ISBN 978-0-415-80482-0

Effort of submergence to decrement of shear strength for relatively a desiccated compacted soil

T. Nishimura

Ashikaga Institute of Technology, Tochigi, Japan

ABSTRACT: Rainfall event often cause major damages to embankments. It is known that wetting due to rainfall causes a decrease in shear strength in unsaturated soils. Changes of soil moisture alter soil suction potential. There are few studies to investigate the influence of submergence on compacted soil with relatively high soil suctions. This study focuses on shear strength of compacted, unsaturated soils. This study used a direct shear apparatus to measure shear strength of compacted soil specimens at high soil suction, at reduced soil suction, and when submerged. Considerable reductions in shear strength were measured in this test program.

1 INTRODUCTION

Rainfall event often cause damages to embankments. Failures of embankments result in the destruction of many structures and great loss of human life. It is known that wetting due to rainfall causes decrease in shear strength in unsaturated soils. Soil moisture corresponds directly to soil suction potential. Soil suctions influence shear resistance and deformations at contact point of soil particles in response to external loads. There are reports regarding reduction of shear strength in compacted soils due to decrease of soil suction as a result of wetting or submergence. There are few studies to investigate the influence of submergence on compacted soil with relatively high soil suctions. Soils with high soil suction, such as desiccated soils, are found at ground surface which is a boundary zone expose to the environment. Climatic conditions have direct impact for example; torrential rain can cause disaster for compacted soils such as embankments.

2 PURPOSE OF THIS STUDY

This study focuses on the shear strength of compacted, unsaturated soils. Techniques are available to control soil suction in laboratory test. Two techniques, namely, pressure plate technique and vapor pressure technique are widely used in unsaturated soil tests. These techniques have been used in triaxial tests and direct shear tests on unsaturated soils.

This study used a direct shear apparatus to measure shear strength of compacted soil specimens at high suction, reduced suction when submerged in water. The compacted soil specimens were maintained in a constant relative humidity environment till equilibrium in order to achieve high soil suction.

Subsequently, reduction in soil suction was achieved by increment of relative humidity. Suction was further reduced by submergence in water. Considerable reductions in shear strength were measured in this test. Experimental data sets are presented to compare the variations in shear strength before and after submergence.

3 BACK GROUND

3.1 *Shear strength of unsaturated soils and soil-water characteristic curve*

Several empirical or semi-empirical models have been proposed to predict the shear strength of unsaturated soils at limited soil suction ranges using the soil-water characteristic curve (SWCC) and the saturated shear strength parameters (Vanapalli et al. 1996, Fredlund et al. 1996, Oberg & Sallfours 1997, Khalili & Khabbaz 1998). The soil-water characteristic curve is defined as the relationship between soil suction and soil moisture (either gravimetric water content or volumetric water content or degree of saturation). The soil-water characteristic curve is consisted of three identifiable stages of unsaturation condition, namely, boundary effect and transition zone and residual zone of unsaturation (Vanapalli et al. 1998). Suggested prediction models are useful in reducing not only time consuming experimental studies but also specialized testing equipments necessary for the determination of the shear strength of unsaturated soils.

Conventional triaxial apparatus and direct shear apparatus have been modified with capability for controlling and measuring of matric suction. Determination of strength parameters includes net normal stress and matric suction as two stress state variables. Most of the experimental test results published in the literature on shear strength of unsaturated soils are limited to a low suction range of 0 to 500 kPa, which is also the common range of interest in geotechnical engineering practice. This suction range typically constitutes the boundary effect and transition zone in the soil-water characteristic curve for many fine-grained soils where the shear strength of unsaturated soils increase nonlinearly with an increase in the matric suction (Gan et al. 1988, Escario & Juca 1989, Vanapalli et al. 1996). There are limited studies however undertaken in the residual zone of unsaturation (RZU), but it is also of interest in some practical applications of geotechnical engineering.

3.2 Shear strength of unsaturated soils in residual zone of unsaturation

The residual zone of unsaturation is conventionally measured using a vapor pressure technique (Fredlund & Rahardjo 1993). The pressure plate technique is suitable for measurement in low suction (i.e., matric suction ranges from 0 to 500 kPa). Nishimura (2003) measured relationship between gravimetric water content and soil suction at residual zone of unsaturation using vapor pressure technique following drying and wetting path in the high soil suction. The soil samples were placed in glass desiccators, each desiccator containing a different salt solution. A range of high suction values can be achieved by controlling relative humidity (RH) using several different salt solutions.

There is evidence in the literature of shear strength measurement of unsaturated soils at high suction values corresponding to the residual zone of unsaturation. The shear strength of unsaturated soils at high soil suction can be lower in comparison to shear strength at suction values in boundary effect zone (Vanapalli and Fredlund 2000, Vanapalli et al. 2000). The residual zone of unsaturation for typically soils such as silt and clays occur when soil suction values are greater than 1,500 kPa. The soil suction of 1500 kPa is transition between high soil suction and low suction. The high soil suction relates with vapor, and low suction is defined as difference between pore-air pressure and pore-water pressure. The value of soil suction at a temperature of 20 degrees can be calculated using the relationship (Kelvin's equation) below by knowing the relative humidity.

$$\psi = -135022 \ln (RH) \qquad (1)$$

where: ψ = soil suction (kPa), RH = relative humidity (%).

The matric suction is defined as equation (2).

$$\text{Matric suction} = (u_a - u_w) \qquad (2)$$

where: u_a = pore-air pressure (kPa), u_w = pore-water pressure (kPa).

A study by Vanapalli and Fredlund (2000) on six compacted fine-grained soils shows that the variation of

shear strength with respect to high soil suction following drying path can be reasonably predicted using several semi-empirical equations. Several investigators have used this technique to achieve high suction values in the specimens and studied the mechanical behavior of unsaturated soils (Cui & Delage 1996, Blatz & Graham 2000, Nishimura & Fredlund 2003, Nishimura & Vanapalli 2005).

Unconfined compression tests were conducted on a compacted, unsaturated silty soil subjected to high suctions (Nishimura & Fredlund 2001). In the residual zone of unsaturation, even if a high total suction is applied following a drying path, the shear resistance between the soil particles changes very little. Beyond the residual suction, the shear strength of an unsaturated soil remains relatively constant.

Nishimura et al. (2007) presented the mechanical behavior of non-plastic silty soil under both saturated and unsaturated conditions using critical state soil mechanics concept. The triaxial test was conducted for saturated soil and unsaturated soil at low suction and high soil suction. The result of study shows that the slope of the critical state line, M(s) for unsaturated soils was equal to the slope of the critical state line for saturated soils. The unconfined compressive strength of the compacted unsaturated soil was determined using specially designed triaxial equipment (Nishimura et al. 2008(a)).

The variation of shear strength with respect to high soil suction values both for drying and wetting paths exhibited essentially horizontal shear strength envelope. A series of direct shear tests were conducted on statically compacted silty soil at high suction ranges (from 2,000 kPa to 300,000 kPa) following drying and wetting path (Nishimura et al. 2008(b)). The shear strength behavior was studied to interpret the influence of hysteresis. The influence of hysteresis on shear strength is negligible for the soil tested in direct shear at high soil suction.

4 TEST PROGRAM

4.1 Soil material

A non-plastic silty soil with a uniform grain size distribution was used in the test program (Fig. 1). Table 1 shows properties of the soil material used in test program. The optimum moisture content of the soil from the Proctor's compaction curve is equal to 17%. The test program involved the determination of unsaturated shear strength behavior of the silty soil with an initial water content of 20%, corresponding to wet side of optimum condition. The initial void ratio of the statically compacted specimen was 0.89.

Figure 2 shows the soil-water characteristic curve over the entire range of suction (i.e. 0 to 1,000,000 kPa) of the soil material following the drying path. The

SWCC was obtained using different suction control techniques (i.e., pressure plate technique and vapor pressure technique), depending on magnitude of suction values.

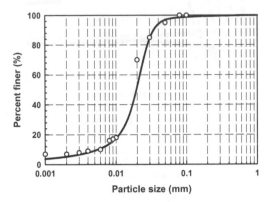

Figure 1.   Grain size distribution curve.

Table 1.   Properties of the soil used in the present study.

Sand (%)	1.6
Silt (%)	89.1
Clay (%)	9.2
Liquid Limit, w_L (%)	24.7
Plastic Limit, w_P (%)	22.8
Plasticity Index, I_p	1.9
Specific gravity, G_s	2.65
Max. dry density, γ_d (max) (kN/m^3)	15.1
Initial void ratio	0.89
Optimum moisture content, OMC (%)	17
Effective cohesion (c′) (kPa)	0
Effective internal friction angle (ϕ') (degrees)	32.3

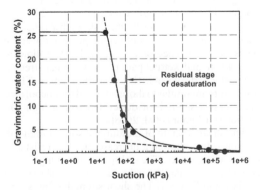

Figure 2.   Soil-water characteristic curve.

Table 2.   The relationship between RH and suction values for different salt solutions at 20°C.

Salt solutions (Chemical symbol)	Relative humidity (%)	Suction (kPa)
K_2SO_4	98	2,830
KNO_3	95	6,940
$NH_4H_2PO_4$	93.1	9,800
NaCl	75	39,000
$Mg(NO_3)_2 \cdot 6H_2O$	54	83,400
$MgCl_2 \cdot 6H_2O$	33	148,000
LiCl	11	296,000

4.2   Specimens with high soil suction values

The soil specimens that were statically compacted at a water content of 20% were placed in glass desiccators, each containing a different salt solution. These salt solutions will create each an environment with a different relative humidity (RH) condition. The seven salt solutions used in the present study to achieve high suction values are summarized in Table 2. These salt solutions are capable of achieving RH in the range from 98% to 11% in the soil specimens inside the glass desiccators. The suction values that can be achieved in the specimens range from 2,830 kPa to 26,000 kPa. The glass desiccators are then placed in a temperature controlled chamber at 20 degrees for a least 30 days to allow soil specimen to achieve equilibrium condition with respect to suction values.

4.3   Constant vertical stress direct shear test

Three different series of direct shear tests were carried out on soil specimens with an initial water content of 20%. In the first series of tests, direct shear tests were performed on soil specimens to determine the shear strength behavior following the drying paths. The direct shear box with the statically compacted specimen was placed in the glass desiccator to achieve desired suction value in the soil specimen by controlling relative humidity as shown in Figure 3. The soil specimens are placed in the glass desiccators for at least 30 days for the soil specimens in the direct shear box assembly to attain a constant suction value. The direct shear box was then transferred to the modified direct shear test apparatus. All soil specimens were subjected to a vertical stress of 3.5 kPa and sheared at a rate of 0.25 mm/min.

In the second series of direct shear tests, soil specimens were subjected to high soil suction corresponding to a relative humidity of 11%. These tests were conducted to determine shear strength of compacted unsaturated soil following wetting paths. Subsequently, the soil specimen in equilibrium with RH of 11% was removed to another glass desiccator containing a salt

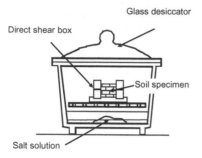

Figure 3. Direct shear box along with soil specimen in the desiccator to achieve target value of high suction.

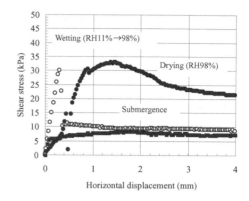

Figure 4. Shear stress versus horizontal displacement.

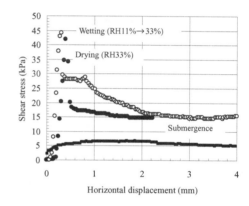

Figure 5. Shear stress versus horizontal displacement.

solution with a higher equilibrium RH. After a period of one month, the direct shear box assembly with the soil specimens was set up in the modified direct test apparatus. The shear strength of soil specimens following wetting paths was measured at a rate of 0.25 mm/min with vertical stress of 3.5 kPa.

In the third series of direct shear tests, shear strength of unsaturated soil subjected to submergence was measured. Before submergence, the soil specimens have high soil suction values following drying path and wetting path inside the glass desiccators containing various salt solutions. The soil specimen with high soil suction and a vertical stress of 3.5 kPa was placed in the direct shear apparatus and was submerged in order to decrease soil suction. The soil specimen was allowed to absorb water for at least 24 hours in the direct shear apparatus. The shear strength of saturated soil specimen was measured with same procedure above mentioned.

5 TEST RESULTS

5.1 Stress versus horizontal displacement relationships

This test program performed shear test with compacted unsaturated soils at seven different high soil suction values. Figures 4 and 5 show the shear stress versus horizontal displacement from direct shear test with constant vertical stress of 3.5 kPa, following drying path, wetting path and submergence. The maximum shear stress was reached at a horizontal displacement of less than 1.0 mm, with the exception of the specimen following a drying path at RH 98%. Due to the effect of desiccation, increment of shear stress occurs rapidly with shear strain. The shear strength of specimens following a drying path is similar to those following a wetting path. The shear strength show a large decrease due to submergence as result of increment of soil moisture.

5.2 Shear strength of unsaturated soil after submergence

Table 3 summarized the data obtained from three series of this test program. Figure 6 show the variation of shear strength with respect to suction for all these tests. The contribution of suction to shear strength continues at high soil suction ranges beyond the residual soil suction. The shear strength at a soil suction of 2830 kPa was relatively lower than the shear strength at other soil suction values regardless of drying paths and wetting paths. The influence of hysteresis due to following drying path and wetting path is less for shear strength of unsaturated soils at high soil suction ranges. These test results are similar to those obtained for unconfined compressive strength of desiccated soil at high soil suction (Nishimura et al. 2008(a)).

The shear strength of unsaturated soils at high soil suction values showed a decrease of shear resistance due to submergence. The soil specimen having a shear strength of 40 kPa was reduced to lower than 10 kPa as shown in Figure 7. The shear strength of all specimens subjected to submergence was 10 kPa regardless of soil suction values or hysteresis following drying path and wetting path.

This study focuses on the submerged shear strength in the experimental research. Nishimura (2005) conducted direct shear test at constant vertical stress for non-plastic silty soil. The same soil material was used in this test program. Measurement of shear strength was performed for saturated silty soil and unsaturated silty soil at a vertical stress of 3.5 kPa. Applied matric suction is lower than 20 kPa. Failure line for both saturated and unsaturated conditions is unique, as shown in Figure 7. The shear strength of the soil specimen subjected to submergence is slightly larger in compare to the effective cohesion shown in Figure 7. Also, the shear strength of submerged specimens corresponds to about 10 kPa to 15 kPa of consolidation pressure or matric suction.

Table 3. Data sets obtained from direct shear tests with regard to three test series.

Suction (MPa)	Shear strength (Drying path) (kPa)	Shear strength (Wetting path) (kPa)	Shear strength (Submergence) (kPa)
2.83	32.9	30.0	8.4
6.94	39.1	37.8	4.9
9.8	38.1	37.8	7.9
39	37.8	41.9	7.3
83.4	37.1	42.3	9.7
148	41.9	44.1	6.8
296	40.9	40.9	6.2

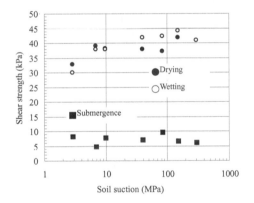

Figure 6. The variation of shear strength with respect to soil suction.

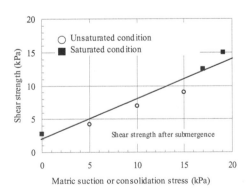

Figure 7. The shear strength for saturated or unsaturated silty soil with low matric suction.

6 CONCLUSIONS

This study focuses on shear strength of compacted, unsaturated soils having high soil suction values. Vapor pressure technique was used to apply high soil suction corresponding to the residual zone of unsaturation. Constant vertical stress direct shear tests were used to measure shear strength in the high soil suction range; following both drying path and wetting path. In addition, the desiccated soil specimens were submerged in water, and the change in shear strength was measured.

The following conclusions can be drawn from this study:

1. The influence of hysteresis with respect to drying paths and wetting paths on shear strength is negligible for non-plastic silty soil at high soil suction ranges (2,830 kPa ≤ soil suction ≤ 296,000 kPa).
2. The shear strength of soil with high soil suction shows large decrease due to submergence in water (i.e. increase of soil moisture and lost soil suction).
3. After submergence, remaining shear strength was larger than the effective cohesion.

ACKNOWLEDGEMENTS

This research work was supported by the Grants-in-Aid for Science Research (No. 19656117) from Ministry of Education, Culture, Slope, Science and Technology, Japan) and Department of Civil Engineering in Ashikaga Institute of Technology.

REFERENCES

Blatz, J.A. and Graham, J. 2000. A system for controlled suction in triaxial tests. *Géotechnique* 50(4): 465–478.
Cui, Y.J. and Delage, P. 1996. Yielding and plastic behavior of an unsaturated silt. *Géotechnique* 46(2): 291–311.
Escario, V., and Juca, J. 1989. Strength and deformation of partly saturated soils. *Proc. 2nd Int. Conf. on Soil Mechanics and Foundation Engineering*, Rio de Janerio, 2: 43–46.
Fredlund, D.G. and Rahardjo, H. 1993. Soil mechanics for unsaturated soils. John Wiley and Sons, INC., New York.

Fredlund, D.G., Xing, A., Fredlund, M.D., Barbour, S.L. 1996. The relationship of the unsaturated soil shear strength to the soil-water characteristic curve. Canadian Geotechnical Journal, 32: 440–448.

Gan, J.K.M., Rahardjo, H., and Fredlund, D.G. 1988. Determination of the shear strength parameters of an unsaturated soil using the direct shear test. Canadian Geotechnical Journal, 25: 500–510.

Khalili, N., and Khabbaz, M.H. 1998. Unique relationship for the determination of the shear strength of unsaturated soils. *Geotechnique* 48(5): 681–687.

Nishimura, T. and Fredlund, D.G. 2001. Failure envelope of a desiccated, unsaturated silty soil. *Proceedings of the 15th International Conference on Soil Mechanics and Geotechnical Engineering*, Vol.1, Istanbul 27–31 August: 615–618.

Nishimura, T. 2003. Highly soil suction portion of the soil-water characteristic curve. International Conference on Problem-atic Soils, UK, July.

Nishimura, T. and Fredlund, D.G. 2003. A new triaxial apparatus for high total suction using relative humidity control. 12th Asian Regional Conference on Soil Mechanics and Geotechnical Engineering: 65–68.

Nishimura. T. 2005. Evaluation of the resistance of an unsaturated silty soil under different shear test process. 58th Canadian Geotechnical Conference, GeoSask 2005, Saskatoon, September.

Nishimura, T, Vanapalli, S.K. 2005. Volume change and shear strength behavior of an unsaturated soil with high soil suction. 16th International Conference on Soil Mechanics and Geotechnical Engineering: 563–566.

Nishimura, T., Toyota, H., Vanapalli, S. and Won, O. 2007. Evaluation of Shear Strength Parameters of an Unsaturated Non-Plastic Silty Soil. 60th Canadian Geotechnical Conference: 1029–1036.

Nishimura, T., Toyota, H., Vanapalli, S. and Won, O. 2008(a). Determination of the shear strength behavior of an unsaturated soil in the high suction range using the vapor pressure technique. *Proceedings of the First European Conference on Unsaturated Soils*, E-UNSAT 2008: 441–447.

Nishimura, T., Toyota, H., Vanapalli, S. and Won, O. 2008(b). The shear strength behavior of a silty soil in the residual zone of unsaturation. The 12th International Conference of International Association for Computer Methods and Advances in Geomechanics (IACMAG): 2213–2221.

Oberg, A., and Sallfours, G. 1997. Determination of shear strength parameters of unsaturated silts and sands based on the water retention curve. Geotechnical Testing Journal, 20: 40–48.

Vanapalli, S.K., Fredlund, D.G., Pufahl, D.E., Clifton, A.W. 1996. Model for the prediction of shear strength with respect to soil suction. *Canadian Geotechnical Journal*, 33: 379–392.

Vanapalli, S.K., Sillers, W.S., Fredlund, M.D. 1998. The meaning and relevance of residual state to unsaturated soils. 51st Canadian Geotechnical Conference: 101–108.

Vanapalli, S.K., Wright, A., and Fredlund, D.G. 2000. Shear strength of two unsaturated silty soils over the suction range from 0 to 1,000,000 kPa, *Proc. 53rd Canadian Geotechnical Conference*, Montreal: 1161–1168.

Vanapalli, S.K., and Fredlund, D.G. 2000. Comparison of empirical procedures to predict the shear strength of unsaturated soils uses the soil-water characteristic curve. Geo-Denver 2000, ASCE, Special Publication, 99: 195–209.

Prediction and Simulation Methods for Geohazard Mitigation – Oka, Murakami & Kimoto (eds)
© 2009 Taylor & Francis Group, London, ISBN 978-0-415-80482-0

Mechanical behavior of anisotropic sand ground bearing coastal structure and its evaluation

S. Takimoto
Fudo Tetra Corporation, Tokyo, Japan

S. Miura
Hokkaido University, Sapporo, Japan

S. Kawamura
Muroran Institute of Technology, Muroran, Japan

S. Yokohama
Hokkaido University, Sapporo, Japan

ABSTRACT: This paper presents the fundamentals of mechanical behavior of anisotropic sand beds subjected to wave-structure interactions. A series of model tests was conducted in 1 g field, and the stress conditions in the prototype were replicated by means of both the cyclic loading system and the oscillating water pressure loading system. Similarly, the feature of failure of caisson-breakwaters in Japan due to ocean wave loadings was investigated in detail. Based on the experimental results in which the shape of plastic wedge zone of anisotropic ground at the peak of bearing capacity was strongly depending on the initial sand deposition condition, an evaluation method for the bearing capacity was proposed according to the upper bound solution in the limit analysis.

1 INTRODUCTION

Composite breakwaters, covered with wave dissipating blocks, are widely employed in Japan. However, a large amount of failures have occurred in many ports in Japan. Wave-induced failures of caisson-breakwaters by Typhoon 18, 2004 have been recently reported for several ports in Hokkaido, Japan (Ministry of Land, Infrastructure and Transport Hokkaido Regional Development Bureau, 2008).

For such damages of caisson-breakwaters, it has been known that many cases are induced due to the sliding failure. However, ground failure due to over-turning or tilting of caisson is not rare case. Kawamura et al. (1997) has summarized the feature of break-waters after disaster based on field data reported by Port and Airport Research Institute. Figure 1 shows the relationship between wave force P and moment M about the center on bottom of structure (Kawamura et al., 1997). Their parameters are normalized to effec-tive vertical force (W–U, W: structure weight, U: uplift force) and structure breadth B, respectively. As can be seen in the figure, it is found that around 30% of all failures seem to be induced less than $P/(W-U) = 0.6$. This fact means that a large amount of failures are induced mainly due to sliding. It should be noted

that there are considerably the overturning, tilting and settlement failures, too. Therefore, it is important to accurately evaluate bearing capacity of foundations (including rubble mound) in coastal region.

On the other hand, it has been well known that the anisotropy of sand deposits plays an important role controlling bearing capacity in foundation engineering (e.g. Oda and Koishikawa, 1979).

In this study, therefore, mechanical behavior of anisotropic ground bearing structure subjected to wave loading was clarified to obtain fundamentals for eval-uation of the stability in structure-ground the system.

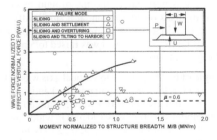

Figure 1. Relationship between wave force and moment of breakwaters after disaster.

A series of model tests was conducted in 1 g field, and the stress conditions seen in the field were simulated by means of both the cyclic loading systems and the oscillating water pressure loading system.

Based on the experimental results in which the shape of plastic wedge zone of anisotropic ground at the peak of bearing capacity was strongly depending on the initial sand deposition condition, an evaluation method for the bearing capacity was proposed according to the upper bound solution in the limit analysis.

2 TEST APPARATUS AND TEST PROCEDURE

Figure 2 shows the whole view of apparatus developed by Miura et al. (1995). This setup does not require a wave channel to simulate various stress states induced by both of wave force and oscillation of structure. Cyclic loads and oscillating water pressure can be given sinusoidally on a model structure through the vertical and horizontal rams and the loading device of oscillating water pressure.

The soil container was 2000 mm in length, 700 mm in depth and 600 mm in width, and its front wall was made of a reinforced glass to observe deformation of sand beds with motion of model structure. To examine the deformation-pore water pressure behavior in ground, transducers for pore water pressure and displacement were also set up, as illustrated in Figure 2.

Model grounds were constructed by using a sand hopper with various slits that can easily control its density (Miura et al., 1984). In this study, several anisotropic grounds were prepared to clarify effect of fabric anisotropy on deformation behavior of sand beds bearing structure. The construction procedure of anisotropic grounds is as follows (Kawamura et al., 2007);

a. The soil container was inclined at an angle to the horizontal.
b. Toyoura sand (ρ_s = 2.65 g/cm^3, ρ_{dmax} = 1.648 g/cm^3, ρ_{dmin} = 1.354 g/cm^3 and D$_{50}$ = 0.18 mm) was pluviated through air into the inclined soil box to a depth of 400 mm.
c. After the soil box was returned to the level state, water was permeated into the ground from eight porous disks on the bottom at a small differential head (4.9 kPa) so as to free from disturbance of initial fabric. Water level was raised to 5 mm from the ground surface.

Angle of bedding plane to vertical axis was defined as β (counterclockwise being positive) and was taken as 45–90°. Relative densities were 50 and 80%.

The model structure was 100 mm in width, 580 mm in length, 100 mm in height and 0.127 kN in weight, and its base surface was made rough by attaching the sand paper (G120).

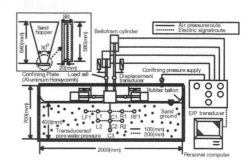

Figure 2. Test apparatus.

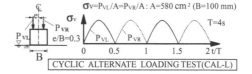

Figure 3. Test procedure for CAL-L.

A series of cyclic loading tests was carried out to examine the fundamentals of mechanical behavior of structure-ground system subjected to cyclic loads. In cyclic alternate loading test (referred to CAL), cyclic load (P_{VL}, P_{VR}) was given alternately to the model structure with a period of 4 sec (see Figure 3). For CAL test, the first loading was given to the structure through the right ram (CAL-R) or the left ram (CAL-L).

In order to simulate stress state at an element of sand ground bearing coastal structure in maritime field, a series of model test was carried out (see Figure 4). In-situ stress condition may be reproduced by appropriately combining vertical load (P_{VL}, P_{VR}), horizontal load (P_{HL}, P_{HR}) and an oscillating water pressure σ_c based on the following relationship (Kawamura et al., 1999);

$$\left\{ \frac{\sigma_z}{\sigma_{z\,max}}, \frac{\sigma_x}{\sigma_{z\,max}}, \frac{\tau_{xz}}{\sigma_{z\,max}} \right\} = \left\{ \frac{\sigma_{zm}}{\sigma_{zm\,max}}, \frac{\sigma_{xm}}{\sigma_{zm\,max}}, \frac{\tau_{xzm}}{\sigma_{zm\,max}} \right\}$$

(1)

where $\{\sigma_z, \sigma_x, \tau_{xz}\}$ are vertical stress, horizontal stress and shear stress, respectively. $\sigma_{z\,max}$ also denotes the maximum value of vertical stress induced during one period of wave propagation. The suffix m indicates in the model test. These stresses induced by external force (moment M, vertical V and horizontal loads H) that was transmitted to the structure were derived based on the Cerrutti and Boussinesq solution for the two-dimensional plane strain problem, if the ground was a homogeneous elastic body.

Settlement (S_{VL}, S_{VR}) and horizontal displacement (S_{HL}, S_{HR}) illustrated schematically in Figure 5 were derived geometrically from measurement (Y_L, Y_R)

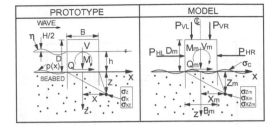

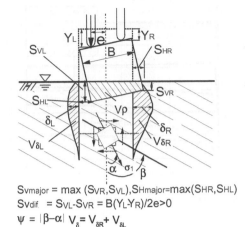

Figure 4. Test procedure for WRT.

on the model structure. The major values between S_{VL} and S_{VR} or S_{HL} and S_{HR} were defined as S_{Vmajor} and S_{Hmajor}, respectively. Differential settlement $S_{Vdif.}$ was the difference value between S_{VL} and S_{VR} (e.g. if $S_{VL} > S_{VR}$, $S_{Vdif} > 0$). Lateral deformation (δ_L, δ_R) in the ground was measured by using eight strands of spaghetti with the diameter of 1.9 mm vertically inserted at 25 mm intervals in the ground (Kawamura et al., 1999).

On the basis of the above measurement, the volumes of deformation V_δ and V_ρ were calculated. V_δ and V_ρ are the lateral deformation area of spaghetti deformed and the settlement area of model structure, respectively (see the shaded area in Figure 5).

Angle of vertical axis to maximum principal stress σ_1 was defined as α (counterclockwise being positive), and the difference between β and α was used as ψ. α at the depth of 0.1 B in ground beneath structure was regarded as a typical value, which was derived from the Boussinesq solution. This reason is that the top of plastic wedge zone in the ground appears at the depth of about 0.1 B when ultimate bearing capacity is mobilized (Kawamura et al., 2003). These parameters were conveniently used to evaluate of the mechanical behavior of ground.

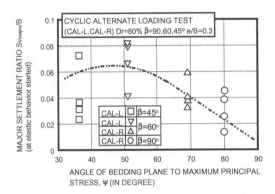

$$S_{Vmajor} = \max(S_{VR}, S_{VL}), S_{Hmajor} = \max(S_{HR}, S_{HL})$$
$$S_{Vdif} = S_{VL} - S_{VR} = B(Y_L - Y_R)/2e > 0$$
$$\psi = |\beta - \alpha| \quad V_\delta = V_{\delta R} + V_{\delta L}$$

Figure 5. Definition of deformation

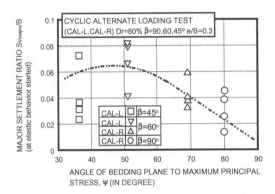

Figure 6. Relationship between Ψ and major settlement ratio at hardening behavior.

3 TEST RESULTS AND DISCUSSIONS

3.1 *Feature of cyclic deformation behavior of anisotropic ground bearing structure*

Before commencing discussions on deformation behavior of ground bearing coastal structure, the feature of deformation of several anisotropic grounds was described based on the test data obtained from a series of CAL tests (Kawamura et al., 2007).

Figure 6 shows the relationship between the angle of bedding plane to maximum principal stress ψ gnd major settlement S_{Vmajor} at the hardening behavior of ground for cyclic loading test (CAL test). From the figure, it is pointed out that the major settlement increases until about $\psi = 45°$ and thereafter decreases with the decrease of ψ, although variations in settlement exist for each ground. As described in the

previous research (Kawamura and Miura, 2003), the trend is similar to that of variation in strength behavior with the change in fabric anisotropy. Park and Tatsuoka (1994) have confirmed a similar tendency for several kinds of sands in plane strain compression test.

Therefore, deformation direction in the ground-structure system depends strongly on ψ and develops in the direction of bedding plane where ψ becomes the smallest in the ground beneath the structure, provided that ψ was more than about 45°.

3.2 *Deformation behavior of anisotropic ground for wave reproduction test (WRT)*

A typical relationship among displacement of the model structure (S_{Vmajor}, S_{Hmajor}), deformation volume (V_ρ, V_δ) in the ground having Dr = 50% and $\beta = 90°$, the normalized excess pore water pressure $\Delta u/\sigma'_{vo}$ and the number of loading cycles Nc was

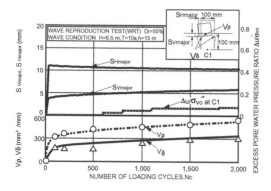

Figure 7. Deformation behavior and pore water pressure for WRT.

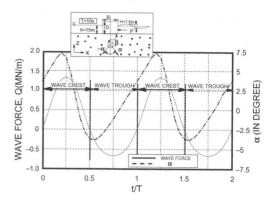

Figure 8. Changes in wave force Q and α for $\beta = 90°$ during wave propagation.

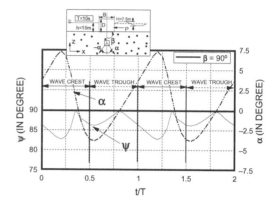

Figure 9. Changes in α and ψ for $\beta = 90°$ during wave propagation.

shown in Figure 7. In the figure, σ'_{vo} denotes effective overburden pressure at the measured point C1 (a depth of 100 mm) (see Figure 2). The displacement due to sliding of structure becomes bigger than that due to settlement in the model test. This indicates that the failure is induced mainly attributed to the sliding mode. On the other hand, V_δ related to lateral deformation increases gradually even if the horizontal displacement of structure becomes a steady state. Furthermore, the pore water pressure does not accumulate remarkably. The similar tendency was also obtained for the case for Dr = 80%. Therefore, it can be said that there is the possibility that the ground bearing coastal structure becomes a progressive failure induced not only by liquefaction of sand ground but also by lateral deformation due to the increase of settlement.

Based on the above results, the changes in ψ in ground beneath a coastal structure was revealed using elastic solution. Figures 8 and 9 show the change in horizontal force (wave force) P and α, change in ψ and α for $\beta = 90°$ during a wave propagation, respectively. Wave force P was calculated by the Sainflou formula under wave and structure conditions (Wave height 7.5 m, wave period 10 sec, the depth of water 15 m, structure breadth B and height D, 20 m respectively) as shown the inserted figure. Assuming that the top of plastic wedge in the ground appears at the depth of 0.1 B (structure width: B = 20 m), the direction of maximum principal stress α was calculated at a depth $z = 20$ m according to the Cerrutti and the Boussinesq solution (Kawamura et al., 1999). ψ ranges from 0° to 90° is calculated as 180°—$(\beta - \alpha)$ if more than 90°.

From Figure 8, it turns out that α rotates from seaside (wave crest) to harbor side (wave trough) during a wave propagation, and its phase is approximately of the same as that of wave force. On the other hand, from the Figure 9, it is noted that the change in ψ occurs remarkably, especially ψ at wave crest

is smaller than that at wave trough. As the results, it can be seen that the development direction of deformation develops toward seaside and occurs mainly during wave crest for this condition. Therefore, the experimental data seems to explain well the deformation behavior of coastal structure such as caisson-breakwater.

3.3 An evaluation method for the bearing capacity based on the upper bound solution

It was found that deformation behavior of structure-grand system subjected to cyclic loading varied with depending strongly on depositional condition (fabric anisotropy effect). An evaluation method for the bearing capacity based on the upper bound theorem of the limited analysis was proposed herein.

To define critical failure mechanism, the shape of the plastic wedge zone of anisotropic ground at the peak of bearing capacity was revealed on the basis

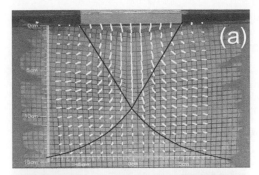

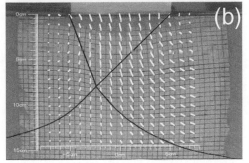

Figure 10. Deformation behavior in ground for SCL: (a) $\beta = 90°$, (b) $\beta = 45°$.

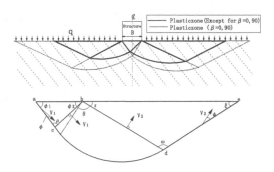

Figure 11. Failure mechanism of anisotropic ground.

In this study, critical failure mechanism was considered based on the experimental results, as shown in Figure 11.

The work dissipation rate is expressed as follows;

$$D_{int} = cr_0 \frac{\sin \psi_2}{\sin \psi_1} V_1 \cos \phi$$

$$+ cr_0 \frac{\sin \varepsilon}{\sin \xi} V_1 \exp(2\theta \tan \phi) \cos \phi$$

$$+ \frac{1}{2} c \cot \phi r_0 V_1 \{\exp(2\theta \tan \phi) - 1\} \qquad (3)$$

where, c^2 is cohesion, V_0 and V_1 are velocity components, ψ_1, ψ_2, ξ and ε is angle of wedge, respectively. θ is the angle of log spiral area in Figure 10. ϕ is angle of internal friction.

On the other hand, the rate of work of the external forces can be written as;

$$D_{ext} = D_{ext}^w + D_{ext}^{p,q} \qquad (4)$$

$$D_{ext}^{p,q} = p \frac{\sin \rho}{\psi_1} r_0 V_1 \sin(\psi_1 - \phi) - q \frac{\sin \omega \cos \varepsilon}{\sin \xi}$$

$$\times r_0 V_1 \exp(2\theta \tan \phi) \qquad (5)$$

where γ_t is unit weight of soil, $D_{ext}^{p,q}$ is the rates of work on vertical force and surcharge, D_{ext}^w is the rate of work on weight of soil and can be calculated based on the area of each wedge, for instance, the component on the wedge of $\triangle$abc is as follows;

$$D_{ext}(\triangle abc)^w = \triangle abc \gamma_t V_1 \sin(\psi_1 - \phi)$$

$$= \frac{1}{2} \gamma_t r_0^2 \frac{\sin \psi_2 \sin \rho}{\sin \psi_1} V_1 \sin(\psi_1 - \phi) \qquad (6)$$

Consequently, the bearing capacity factors N_q, N_c, and N_γ can be derived based on equation (2). For example, the factor N_γ is expressed as follows;

of test data obtained from static loading test (SCL) (Kawamura et al., 2003).

Figure 10 shows the changes in deformation in the ground until S_{Vmajor} becomes 10 mm. They were taken using a rubber membrane. From the figures, it is noted that the difference in formation of the plastic wedge attributed to the effect of anisotropy exists. For instance, the formation of wedge is almost symmetrical for $\beta = 90°$, whereas the top of wedge moves to the left side for $\beta = 45°$. Furthermore, variation of the depth of slip surface due to its formation can be also confirmed. This indicates that variation in bearing capacity mobilization is mainly induced by change in development of plastic wedge.

In general, the upper bound theorem of limit analysis can be mathematically represented by the following relation,

$$\int_v \dot{D}(\varepsilon_{ij}) dV \geq \int_{sv} T_i v_i dS_v + \int_{st} T_i v dS_t + \int_V \gamma_i v_i dV \qquad (2)$$

The left-hand side of expression (2) is the rate of work dissipation during incipient failure of the ground, on the other hand, the right hand side is the rate of works of all the external forces. T_i is the stress vectors on boundaries Si and St. v_i is the velocity vector in the kinematically admissible mechanism, γ_i is the unit weight vector, and V is the volume of the mechanism (e.g. Radoslaw, 1997).

$$N\gamma = \frac{\left\langle \begin{array}{l} \dfrac{\sin\psi_2}{\sin\rho}\left[\dfrac{\dfrac{\sin\omega\sin\varepsilon\cos\varepsilon}{\sin\xi}\exp(3\theta\tan\phi)-\dfrac{\sin\psi_1\sin\rho\sin(\psi_2-\phi)}{\sin\psi_2}}{\dfrac{\exp(3\theta\tan\phi)\{\sin(\psi_1+\theta)+3\tan\phi\cos(\psi_1+\theta)\}-\sin\psi_1-3\tan\phi\cos\psi_1}{9\tan^2\phi+1}}\right] \\ +\dfrac{\sin\psi_1}{\sin\rho}\left[\dfrac{\dfrac{\sin\omega\sin\varepsilon\cos\varepsilon}{\sin\xi}\exp(3\theta\tan\phi)-\dfrac{\sin\psi_2\sin\rho\sin(\psi_1-\phi)}{\sin\psi_1}}{\dfrac{\exp(3\theta\tan\phi)\{\sin(\psi_2+\theta)+3\tan\phi\cos(\psi_2+\theta)\}-\sin\psi_2-3\tan\phi\cos\psi_2}{9\tan^2\phi+1}}\right] \end{array}\right\rangle}{\left\{\dfrac{\sin\rho}{\sin\psi_2}\sin(\psi_2-\phi)+\dfrac{\sin\rho}{\sin\psi_1}\sin(\psi_1-\phi)\right\}} \tag{7}$$

In order to confirm the validity of equation (7), the factors N_γ was compared to those reported by Meryhof (1963), Hansen (1970), Vesic (1973). For this analysis, wedge angle of plastic zone was defined as $\psi_1 = \psi_2 = \pi/4 + \phi2$, $\varepsilon = \xi = \pi/4 - \phi/2$. Figure 12 depicts that the relationship between the calculated N_γ and ϕ. The test data in static loading test (SCL test) is also plotted in the Figure (Kawamura et al., 2003). From the figure, it is found that the value has good agreement with those suggested by them. Therefore, this method is valid for evaluating the bearing capacity.

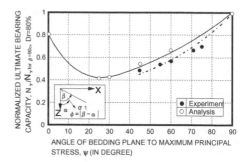

Figure 14. Changes in bearing capacity factor due to the difference in ψ.

Changes in the bearing capacity due to fabric anisotropy of sand ground were indicated herein.

Figure 13 shows the trace of the top position of plastic wedge due to the difference in fabric anisotropy, which were derived based on the shape of plastic wedge obtained from a series of static loading tests, as shown in Figure 10. If this fact is reliable, bearing capacity of anisotropic ground can be derived using upper bound solution.

Figure 14 shows the changes in the calculated N_γ due to the difference in fabric anisotropy in terms of N_γ versus angle of bedding plane to maximum principal stress, ψ. N_γ is normalized to that of $\beta = 90°$ The data of static loading test was similarly shown in the figure. As can be seen in the figure, the normalized N_γ decreases until around $\psi = 30°$ and thereafter increases with the decrease of ψ, and has good agreement with the experimental data.

Therefore, it is found that the proposed expressions to predict the bearing capacity of anisotropic sand ground can explain well the data obtained from a series of model tests.

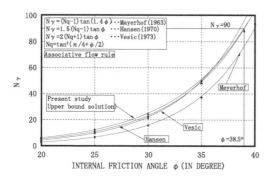

Figure 12. Relationship between the calculated N_γ and ϕ.

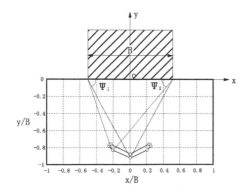

Figure 13. Trace of the top position of plastic wedge due to the difference in fabric anisotropy.

4 CONCLUSIONS

The following conclusions are drawn from the study:

1. For the disaster of caisson-breakwaters, its failure mode was mainly the sliding mode, however

there were considerably the overturning, tilting and settlement failures.

2. In the model test, the failure of grounds beneath structure due to cyclic loading such as wave force became the progressive failure with lateral deformation.

3. The predominant direction of deformation in the ground-structure system depended strongly on the deposition condition of the ground.

4. The proposed expressions to predict the bearing capacity of anisotropic sand ground can explain well the data obtained from a series of model tests.

REFERENCES

Hansen, J.B. 1970. A revised and extended formula for bearing capacity. *Geoteknisk Inst.*, Bulletin 28: 5–11.

Kawamura, S., Miura, S. and Yokohama, S. 1997. Experiments on wave-induced flow failure of seabed bearing dissipating structures. *Proc. of Coastal Engineering*, JSCE, 44 (2): 936–940. (in Japanese)

Kawamura, S., Miura, S., Yokohama, S. and Miyaura, M. 1999. Model experiments on failure of sand bed beneath a structure subjected to cyclic loading and its countermeasure. *Journal of Geotechnical Engineering*, JSCE, III-47/ No.624: 77–89. (in Japanese)

Kawamura, S. and Miura, S. 2003. Bearing capacity-lateral deformation behavior of anisotropic ground beneath structure under various loading conditions. *Journal of Geotechnical Engineering*, JSCE, III-63/No.736: 115–128. (in Japanese)

Kawamura, S., Miura, S. and Yokohama, S. 2007. Effect of fabric anisotropy on deformation behavior of ground bearing structure subjected to cyclic loading. *Journal of Geotechnical Engineering*, JSCE, 63/C(1): 81–92. (in Japanese)

Meyerhof, G. G. 1963. Some recent research on the bearing capacity of foundations. *Canadian Geotechnical Journal* (1): 16–31.

Ministry of Land, Infrastructure and Transport Hokkaido Regional Development Bureau 2008. http:// www.hk.hkd. mlit.go.jp/ port/hisai.html

Miura, S., Toki, S. and Tanizawa, F. 1984. Cone penetration characteristics and its correlation to static and cyclic deformation-strength behaviors of anisotropic sand. *Soils and Foundations* 24(2): 58–74.

Miura, S., Tanaka, N., Kondo, H., Sato, K. and Kawamura, S. 1995. Sand flow failure induced by ocean wave and oscillation of coastal structures. Proc., *First International Conference on Earthquake Geotechnical Engineering* 2: 743–748.

Oda, M. and Koishikawa, I. 1979. Effect of strength anisotropy on bearing capacity of shallow footing in a dense sand. *Soils and Foundations* 19(3): 16–28.

Park, C. S. and Tatsuoka, F. 1994. Anisotropic strength and deformation of sand in plain strain compression. *Proc., XIII International Conference on Soil Mechanics and Foundation Engineering* (1): 1–4.

Radoslaw, L. M. 1997. An estimate of influence of soil weight on bearing capacity using limit analysis. *Soils and Foundations* 37(4): 57–64.

Vesic, A.S. 1973. Analysis of ultimate loads of shallow foundations. *Journal of Soil Mechanics and Foundation Engineering* 99(1): 45–76.

Prediction and Simulation Methods for Geohazard Mitigation – Oka, Murakami & Kimoto (eds)
© 2009 Taylor & Francis Group, London, ISBN 978-0-415-80482-0

Failure mechanism of volcanic slope due to rainfall and freeze-thaw action

S. Kawamura
Muroran Institute of Technology, Muroran, Japan

S. Miura & T. Ishikawa
Hokkaido University, Sapporo, Japan

H. Ino
Kajima Co. Ltd., Sapporo, Japan

ABSTRACT: Collapse of slope formed from volcanic soils has been frequently caused in Hokkaido, Japan. This study aims at clarifying failure mechanism of volcanic slope caused by both rainfall and freeze-thaw action. In particular, the effects of freeze-thaw action, slope angle, density and friction of impermeable layer in the slope on mechanical behavior at failure were detailedly investigated on volcanic slopes having several shapes.

1 INTRODUCTION

In Hokkaido, there are over 40 Quaternary volcanoes, and pyroclastic materials cover over 40% of its area. The big volcanic activities were generated from the Neogene over the Quaternary period, and various pyroclastic materials such as volcanic ash, pumice and scoria have been formed during those eruption.

Such volcanic soils have been used as a useful construction material, especially foundations or embankment structures. However, the research on volcanic coarse-grained soils from the engineering standpoint is extremely superficial in comparison with cohesionless soils.

Recent the earthquakes and severe rainfalls in Hokk-aido generated the most serious damage in the grounds, the cut slopes and the residential embankment slopes, which composed of volcanic soils. For example, the slope failure of residential embankment due to the 1991 Kushiro-oki earthquake and the surface failure of cut slope in the Hokkaido expressway due to the severe rainfall (1986). In particular, serious damages due to rainfall with freeze-thaw action have increased.

Figure 1 shows two types of failure mode for volcanic cut slopes observed in the fields. The surface slope failure is due to forming a frozen layer such as an impermeable layer in spring season and is due to softening of the slope by freeze-thaw action in summer season.

Since weather change is thought likely to result from global warming from now, it is anticipated that such failures due to rainfall with freeze-thaw action more increase. Therefore, it will be required to accurately

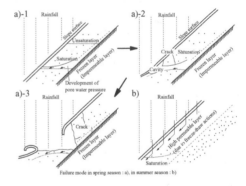

Figure 1. Failure modes of volcanic slope in spring and summer seasons.

grasp the failure mechanism of slope-foundations formed from such volcanic soils which have been classified as problematic soils.

The purposes in this paper are to reveal the mechanism of surface failure of slopes by model rainfall tests, and to elucidate the effects of freeze-thaw action on its mechanical behavior during rainfall.

2 TEST MATERIAL AND TEST PROCEDURE

Volcanic coarse-grained soil which was sampled from the ejecta of Shikotsu caldera in Hokkaido was used in this study. Sampling site is shown in Figure 2. This sample is hereafter referred to as Kashiwabara volcanic soil. The index property of sample is shown in Table 1, compared to that of Toyoura sand. As shown in

Table 1, a finer content is less than 1.3%. A low value of dry density is also shown in the sample, because constituent particle is very porous and high crushable. The details of mechanical behavior of Kashiwabara volcanic soil have been also reported by Miura et al. (1996).

Figure 3 depicts the whole view of apparatus used in rainfall testing. The soil container was 2,000 mm in length, 700 mm in depth and 600 mm in width, and its front wall was made of a reinforced glass to observe deformation with failure.

Model slopes were constructed by tamping the volcanic soil or by pluviating so as to be the desired density (ρ_d = 0.48 g/cm^3 or 0.45 g/cm^3, variations

Figure 2. Location of sampling site.

Table 1. Index property of Kashiwabara volcanic soil.

Sample name	ρ_s(g/cm^3)	$\rho_{d\,in\,situ}$ (g/cm^3)	D_{50} (mm)	U_c	F_c(%)
Kashiwabara	2.34	0.53	1.3	3.1	1.3
Toyoura sand	2.68	–	0.18	1.5	0

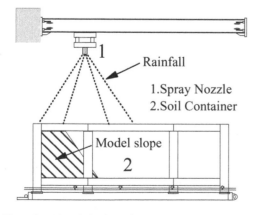

Rainfall

1. Spray Nozzle
2. Soil Container

Model slope

Figure 3. The whole view of apparatus.

in density are within 2.5%, respectively). In the case of rainfall test with freeze-thaw action, after constructed the model slope by the similar method, the surface of slope was made to freeze up by dry ice during 8 hours and was thawed in 20°C.

Rainfall intensities were 60 and 100 mm/hr and were accurately simulated by using several spray-nozzles. During rainfall testing, the changes in deformation behavior, saturation degree and temperature were monitored using digital video camera, soil moisture meters and thermocouple sensors, respectively. Particularly, the deformation behavior was estimated according to the particle image velocimetry (PIV) analysis (White et al. 2003). On the other hand, pore water pressure was similarly monitored in this study, however, it was difficult to accurately grasp its behavior, because that its value was very small. Moreover, it was clarified that pore water pressure behavior at the slope failure was not sensitive as compared with soil moisture behavior (Kawamura et al. 2007). For the latter discussions, therefore, the behavior of deformation and saturation was mainly described.

Figure 4 or Table 2 show test conditions and typical slope shapes and the setting positions of measurement devices, respectively. It has been revealed that the existence of an impermeable layer in the slope was important for evaluating such failures of volcanic slopes (Kawamura et al. 2007). In order to obtain a better understanding of the mechanism, a series of model test on volcanic slopes having an impermeable layer was carried out herein. In particular, for rainfall test, the effects of slope angle, friction of impermeable layer and slope density on the failure were investigated detailedly. The slope angle ranged from 45 degree to 65 degree, and the coefficient of base friction μ measured in this study was 0.33 under smooth condition or 2.28–4.31 under rough condition. On the other hand, for rainfall test with freeze-thaw action, the effect of freeze-thaw action on the failure mechanism was discussed compared to that without its action. A series of rainfall model tests was performed until 3 hours or slope failure. According to the preliminary test (Kawamura et al. 2007), since slope failure was rapidly developed after shear strain of 4–6% was induced at the peak of saturation degree, the mechanical behavior at shear strain of 4–6% was regarded as that at failure.

3 TEST RESULTS AND DISCUSSIONS

3.1 *Surface failure of volcanic slope due to rainfall*

Before commencing discussions on failure mechanism of volcanic slope subjected to freeze-thaw action, the feature of failure of volcanic slope caused by rainfall was investigated.

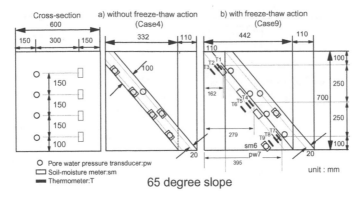

Figure 4. Typical model shapes (65 degree slope) and the setting positions of measurement devices.

Table 2. Test conditions.

	Case 1	Case 2	Case 3	Case 4	Case 5	Case 6	Case 7	Case 8	Case 9	Case 10
Slope condition	Without freeze-thaw action					Without freeze-thaw action				Without freeze-thaw action
Slope angle(°)	45	50	55	65	50	50	55	65		65
Length of base B (mm)	696	636	572	442	603, 636, 669, 702, 767	767	702	552	442	442
Rainfall intensity R (mm/h)	100, 60			100	100			100		
Layer thickness (mm)	100				75, 100, 125, 150, 200	100	50	100		
Density (g/cm³)	0.45, 0.48			0.45, 0.48	0.45			0.45		
Friction μ	0.33, 2.81			4.31	0.33	4.31				

Figure 5 shows a typical deformation of failed slope. The slope has an angle of 50 degree, and an impermeable layer with the rough condition of base friction. Rainfall intensity is 100 mm/hr. In this Figure, the slip line induced in the slope is illustrated as a solid line. From the figure, it is apparent that the surface slip failure occurs with large deformation. Figure 6 shows the development of saturation degree under the same condition. The saturation degree is gradually increasing for each position. In particular, for the saturation behavior of sm4 which is placed around the slip line, the degree is suddenly decreasing after failure in comparison with that of sm3 which is placed beneath the slip line. The similar tendency was obtained for the other cases except for those of 45 degree slope. From the results, it can be seen that the rise of saturation degree around the slip line becomes a factor to control the collapse of slope and the changes in soil moisture can explain dailatancy behavior due to shear deformation. Therefore, it is important to trace the changes in soil moisture for the prediction of instability.

3.2 Key factors to cause surface failure of volcanic slope due to rainfall

Based on the data of a series of model test, key factors to cause surface failure of volcanic slope due to rainfall were discussed herein.

Figures 7 (a), (b) and (c) show the relationships between slope angle and amount of rainfall on density, rainfall intensity and base friction, respectively.

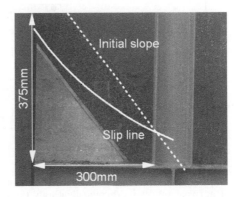

Figure 5. Typical deformation of volcanic slope after failure (50 degree slope).

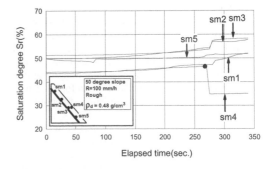

Figure 6. Typical saturation behavior in the slope.

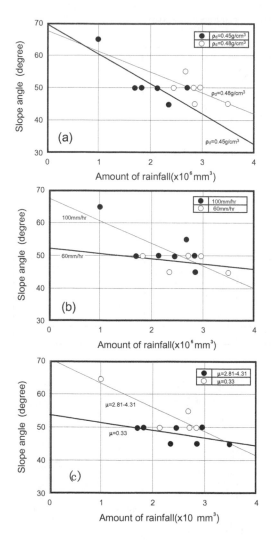

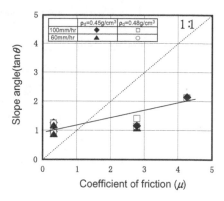

Figure 8. Relationship between base friction and slope angle.

Figure 7. Relationship between slope angle and amount of rainfall at failure; (a) density, (b) rainfall intensity, (c) base friction.

each density, irrespective of the difference of rainfall intensity. For example, the effect of slope angle on the failure became higher than that of friction until around $\mu = 1.2$, if they can be considered as the same physical index. This fact explains that the slope failure can be quantitatively evaluated by using slope angle or friction if their relative relation is estimated. For the above reason, it is pointed out that surface slope failure depends strongly on slope angle and friction of impermeable layer, however is not so affected by slope density.

The cause of failure is not only the rise of water level from the base of impermeable layer but also the difference in development of saturation (the difference in the water retention ability of volcanic slope), as shown in Figure 6. Namely, surface slope failure seems to be induced by the expansion of some area having high retention ability of water. However the reason of the differences of formation of area having high retention of water or of the development of saturation degree in the slope was not made clear. Kitamura et al. (2007) has pointed out the importance of unsaturated behavior of slopes. Therefore, it is significant to detailedly grasp unsaturated behavior of volcanic slope for evaluating the stability on the cases of surface slope failure.

3.3 Effect of freeze-thaw action on mechanical behavior of volcanic slope during rainfall

Based on the above test results, a series of rainfall model test with freeze-thaw action was performed.

For the model test with freeze-thaw action, an outer layer of thickness 100 mm was formed on the frozen layer (see Figure 4). Figure 9 shows the changes in temperatures during freeze action. It is apparent that all areas around the setting positions of measurement devices freeze after about 7 hours (25,000 sec.).

Figures 10 (a) and (b) depict the deformation behavior of volcanic slope at the peak of saturation degree with freeze-thaw action or without, respectively. The

Amount of rainfall to cause failure is increasing with the decrease of slope angle for each case, as shown in the figures. The effects of friction and rainfall intensity on the failure became relatively higher than that of density. Yagi et al. (1999) have indicated that there is a relative relationship between total amount of rainfall and rainfall (mm/day) in the field. Therefore, it can be said that the test data well explains such a phenomenon in the field.

Figure 8 illustrates the relationship between base friction and slope angle of failed slope in terms of the coefficient of friction μ versus slope angle ($= \tan \theta$). It is evident from the figure that slope angle to cause failure is increasing with the increase of friction μ for slope angle and rainfall intensity are 65 degree and 100 mm/hr. As can be seen in these figures, there seems to basically be no difference in deformation behavior at the peak of saturation degree between the slope with frozen layer and with impermeable layer. Additionally, these failures were caused in the same deformation pattern. This fact denotes that the formation of frozen layer in spring season is a key factor for evaluating the stability of volcanic slopes, similarly that with impermeable layer.

Figure 11 shows the changes in saturation degree during freeze action. In the figure, the degree is normalized by its initial value Sr_0. From the figure, it should be noted that these tendencies become inverse attributed to the difference of the setting position. For example, sm6 increases gradually, in contrast, sm1, sm2, sm3 and sm5 decrease. This depends on the difference of supply sources of water which needs for freezing. During rainfall testing, the saturation degree of sm4 is gradually increasing with the increase of rainfall, thereafter is suddenly decreasing (see Figure 12).

From the results, it can be said that one of the causes of slope failure is the increase of saturation degree, similarly that with impermeable layer. In particular, it is anticipated that saturation degree in the frozen layer suddenly increases with thaw action from low condition to high condition in spring season. Furthermore, it can be pointed out that there is the possibility that the process of thaw action has an influence on the failure mechanism.

Based on the above results, a series of rainfall tests was similarly performed on volcanic slopes of which its surface was directly subjected to freeze-thaw action.

Figures 13 (a) and (b) show the changes in the normalized saturation degree during rainfall testing after freeze-thaw action, compared with that no freeze-thaw action. Each saturation degree is suddenly decreasing after failure (see black symbol), however there are the differences of elapsed time until failure between both cases. For example, the elapsed time subjected to freeze-thaw action is 9 times faster than that without its action.

In order to clarify the influence of freeze-thaw action on slope failure, Figure 14 illustrates deformation behavior and change in density after freeze action. It is apparent that shear strain increases due to freezing, and its vector is approximately perpendicular to the surface. The change in density can be

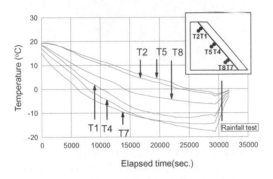

Figure 9.   Changes in temperature during freeze action.

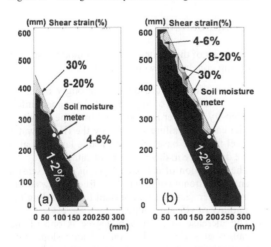

Figure 10.   Difference in deformation behavior of volcanic slope at the peak of saturation; (a) with freeze-thaw, (b) without freeze-thaw action.

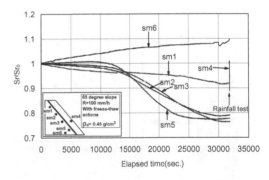

Figure 11.   Changes in saturation degree with freeze action.

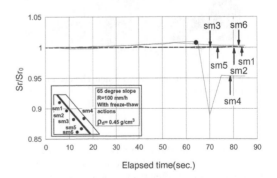

Figure 12.   Changes in saturation degree during rainfall test.

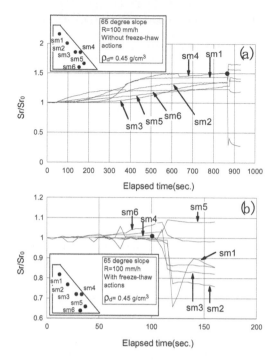

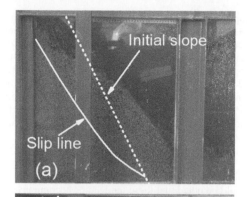

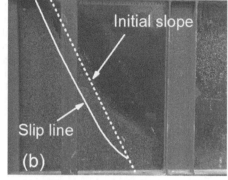

Figure 13. Changes in saturation degree; (a) with freeze thaw action, (b) without freeze thaw action.

Figure 15. Difference in deformation behavior due to freeze-thaw action or without.

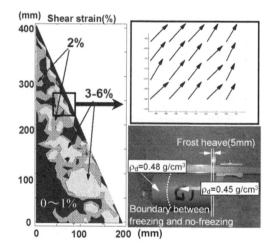

Figure 14. Deformation behavior and change in density after freeze action.

slope (see Figure 15), it is obvious that deformation behavior in the slope is affected by freeze-thaw action.

Therefore, it is important for slopes in cold region to evaluate the effect of freeze-thaw action on the failure.

4 CONCLUSIONS

On the basis of the limited number of model tests, the following conclusions were derived.

1. Surface slope failure depends strongly on slope angle, the friction of impermeable layer and rainfall intensity, however is so affected by slope density.
2. The cause of failure is not only the rise of water level from the base of impermeable layer but also the difference in development of saturation.
3. The formation of a frozen layer in spring season and the softening of the slope by freeze-thaw action in summer season are important for evaluating the stability of volcanic slopes.
4. The increase of saturation degree is one of the causes of surface failure on volcanic slope with frozen layer, similarly that with impermeable layer. However there is the possibility that the process of thaw action has an influence on the failure.

also confirmed, as shown in this figure. On the other hand, the shear strain and its vector were also almost the same as those after thaw action (Ishikawa et al. 2008). From the comparison of the shapes of failed

REFERENCES

Ishiskawa, T., Miura, S., Akagawa, S., Sato, M. and Kawamura, S. 2008. Coupled thermo-mechanical analysis for slope behavior during freezing and thawing. *Proc. of International symposium on Prediction and Simulation Methods for Geohazard Mitigation,* IS-Kyoto 2009.(in Press)

Kawamura,S., Kohata, K. and Ino, H. 2007. Rainfall-induced slope failure of volcanic coarse-grained soil in Hokkaido. *Proc. of 13th Asian regional conference on Soil Mechanics and Geotechnical Engineering:* 931–934.

Kitamura, R., Sako, K. Kato, S., Mizushima, T. and Imanishi, H. 2007. Soil tank test on seepage and failure behaviors of Shirasu slope during rainfall. *Japanese Gotechnical Journal,* JGS, 2(3): 149–168. (in Japanese)

Miura, S., Yagi, K and Kawamura, S. 1996. Static and cyclic shear behavior and particle crushing of volcanic coarse grained soils in Hokkaido. *Journal of Geotechnical Engineering,* JSCE, No.547/III-36: 159–170, 1996. (in Japanese)

White, D.G., Take, W.A. and Bolton, M.D. 2003. Soil deformation measurement using particle image velocimetry (PIV) and photogrammetry. *Geotechnique* 53(7): 619–631.

Yagi, N. Yatabe, R. and Enoki, M. 1999. Prediction of slope failure based on amount of rain fall. *Journal of Geotechnical Engineering,* JSCE, No.418/III-13: 65–73. (in Japanese)

Slope stability consisting of two different layers caused by rainfall

H. Kaneko, H. Tanaka & Y. Kudoh
Hokkaido University, Sapporo, Japan

ABSTRACT: Using centrifuge test, slope stability during rainfall was studied for grounds consisting of two layers with different permeability. All tests were conducted under 40 G centrifugal acceleration. Slope failure occurred in a restricted case where clay contents of the upper and lower layers are 0 and 10%, respectively. The thickness of the upper layer and the rainfall intensity did not significantly affect occurrence of failure. It is concluded that a key for slope failure during rainfall may be horizontal movement of water in the upper soil layer.

1 INTRODUCTION

Because of increasing abnormal climate, it is reported that a large number of slope failures are caused by heavy rain and lose valuable lives and fortune (for example, Wang and Shibata 2007). Model test may be a useful tool for studying and understanding mechanism of the slope stability during heavy rain. However, such a model test is not only costly, but also requires a lot of time to perform the test, due to its large scale. The centrifuge model test may be a powerful testing method, in which the large centrifugal acceleration is applied to the model ground so that the stress conditions are the same as those in the prototype in spite of the small scale (Soga et al. 2003). The objective of this research is to study the behavior of unsaturated ground during heavy rain by the centrifuge model test. It is considered that a slope failure may be caused by movement of water in unsaturated ground. Therefore, special attention was paid to measure pore water pressures (PWP) in the model ground by placing several pore water pressure gauges. Two types of grounds were prepared: flat and slope grounds consisting of two soil layers with different permeability.

2 OUTLINE OF EXPERIMENT

2.1 Centrifuge machine

Main features of the centrifuge machine used by this research are: the arm length is 1.5 m; the maximum centrifugal acceleration is 150 G; the maximum carrying mass is 150 kg. Water for simulating rainfall can be supplied by a rotary joint.

2.2 Pore water pressure gauge

The pore water pressure gauge used by this research is made by SSK Company. The diameter of this gauge is 8 mm, which is small enough to place in such a scaled ground. The gauges can measure negative PWP, i.e., suction, because its measurement surface is covered by a ceramic filter.

2.3 Test method

The ground for this study consists of two soil layers. These two layers were made from a mixture of Toyoura sand and Kasaoka clay. For obtaining different the coefficient of permeability (k) for these layers, Toyoura sand was mixed with Kasaoka clay by following ratios in weight: 0% (no Kasaoka clay), 5% and 10%.

The model ground is created by a following manner: The mixture soil was filled in a strong box and compacted by every 2.5 cm thickness. Water content of the mixture was determined to become 95% of the maximum dry density (see compaction curve in Fig. 1). At this water content, the degree of saturation (S_r) is about 50% in all mixtures. In the filling process, the pore water pressure gauges were placed at determined positions (see Fig. 2). Table 1 shows properties of these mixtures after compaction. From permeability test, it is found that k decreases to about 1/4 by increasing 5% of clay content (see Table 1).

In this study, two types of the model ground were prepared: i.e., flat and slope ground. The flat ground consists of the upper and the lower layers, whose thicknesses were 5 cm and 15 cm, respectively (their dimension is in the model). In the flat ground, drainage was not allowed either before or during raining.

The slope ground was made by cutting the flat ground. The slope angle was 45° and the slope height was 15 cm. The excess water caused by increasing acceleration and rainfall was drained through a hole placed in the toe side of slope of the strong box (see Figs. 3 and 4).

Tests for both flat and slope grounds were conducted under 40 G centrifugal acceleration. Rainfall

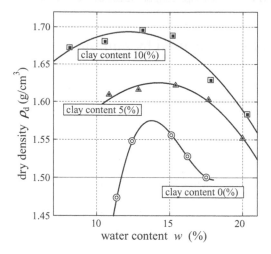

Figure 1. Compaction curve.

Figure label within graph: clay content 10(%), clay content 5(%), clay content 0(%)

dry density ρ_d (g/cm^3)

water content w (%)

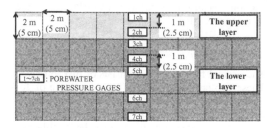

Figure 2. Location of flat ground.

Table 1. Properties of the mixtures.

Clay content	Wet density ρ_t (g/cm^3)	Coefficient of permeability k (m/sec)	Void rate e
0%	1.76	2.75×10^{-4}	0.69
5%	1.80	4.48×10^{-5}	0.63
10%	1.87	1.51×10^{-5}	0.55

Table 2. Similarity low.

	Scale effect*
Length (m)	n
Time (hr)	n^2
Stress (kPa)	1
Rainfall intensity(mm/hr)	$1/n$

* Centrifugal acceleration n (G).

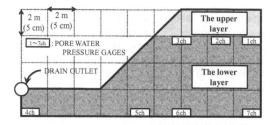

Figure 3. Location of slope ground (the thickness of the upper layer is 2 m.

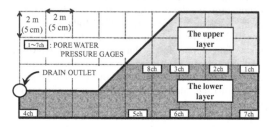

Figure 4. Location of slope ground (the thickness of the upper layer is 4 m).

slope height in the prototype becomes 6 m (15 cm × 40 = 6 m). Illustrations of the slope ground with location of pore water pressure gauges are shown in Figs. 3 and 4, where the scale is indicated in the prototype (number in the parenthesis is in the model ground). In the slope ground, the thickness of upper layer varied to 2 m (Fig. 3) and 4 m (Fig. 4).

started after observing a constant reading of pore water pressure gauges, indicating that PWP in the ground was under equilibrium between gravity force and suction acting on soil particles. Then, to simulate rainfall, water was provided by three nozzles, which were equipped on the upper flame fixed with the strong box.

According to the similarity low for centrifugal test, k apparently increases n times. Therefore, time and intensity of rainfall become n^2 and n, respectively, where n is the scale of the model (see Table 2). Hereafter, the size, time and intensity of rainfall are calculated, corresponding to the prototype. The centrifuge acceleration for all tests was fixed as 40 G, so that the

3 TEST RESULT

3.1 Flat ground

Experimental conditions of the flat ground are shown in Table 3. Figures 5, 6 and 7 show the PWP distributions in the ground before and during rainfall. Because of no drainage hole in the flat ground experiments, water filled in voids in the upper ground moves downwards due to large centrifugal force. The PWP in the lower part of the ground is distributed nearly in triangle manner and the ground water table (GWT) may be located at 4 m depth from the ground surface. The

34

Table 3. Experimental condition of the flat ground.

	Clay content		Rainfall intensity (mm/hr)	Duration of rainfall (hr)
	Upper layer	Lower layer		
No. 1	0%	5%		
No. 2	5%	10%	30	35.6
No. 3	0%	10%		

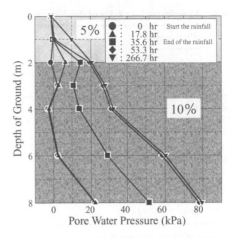

Figure 6. Distribution of pore water pressure (test No. 2).

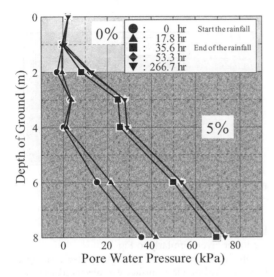

Figure 5. Distribution of pore water pressure (test No. 1).

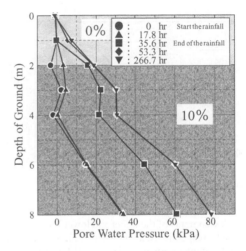

Figure 7. Distribution of pore water pressure (test No. 3).

location of GWT may be understandable, since the ground was made to be $S_r = 50\%$. In another word, most water possessed in the ground made at 1 G was expelled by acceleration of 40 G and water content of the ground became nearly zero at the start of rainfall. A small suction varying from 2 kPa to 7 kPa was measured in the upper part of the ground. However, this suction was disappeared by the rainfall, and the GWT gradually rose.

It can be seen in all experiment results (Figs. 5, 6 and 7) that the values of the pore water pressure gauges of the ground level (0 m in depth) did not changed, but indicates zero during the rainfall. From this result, it may be considered that all rain water was infiltrated into the ground without suspending at the ground surface. This inference is also supported by observation of a CCD camera equipped in the front of the strong box.

At the distribution of PWP before and 17.8 hr after rainfall (in Figs. 6 and 7), it can be seen that the PWP only on the boundary between two layers (2 m in depth from the ground surface) increased. That is, a part of water may halt on the boundary because the coefficient

of permeability of lower layer (the clay content 10%) is smaller than that of the upper layer (the clay content 0 or 5%), and it is inferred that the "parched" GWT may be temporarily formed on the boundary. As shown in Fig. 4 where the clay content of upper layer is 0% and lower layer is 5%, however, the PWP on the boundary was not as significantly changed as two previous cases did. In this case, it may be considered that k of both the upper (the clay content 0%) and lower (the clay content 5%) layer is large enough to for the rain water infiltrate, without forming a parched GWT.

3.2 Slope ground

Experimental conditions of the slope ground are indicated in Table 4. Slope failure broke out only in cases

Table 4. Experimental condition of the slope ground.

	Clay content		Rainfall intensity (mm/hr)	Duration of rainfall (hr)	Slope failure
	Upper layer	Lower layer			
No. 4	0%	5%			×**
No. 5	5%	10%	30	14.8	×
No. 6	0%	10%			○***
No. 7	0%	5%			×
No. 8	5%	10%	15	29.6	×
No. 9	0%	10%			○
No. 10*	0%	5%			×
No. 11*	5%	10%	30	14.8	×
No. 12*	0%	10%			○

* thickness of the upper layer is 4 m.
** ×: slope failure did not occur.
*** ○: slope failure occurs.

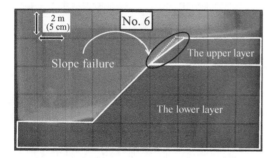

Figure 8. Slope failure (test No. 6).

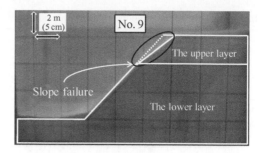

Figure 9. Slope failure (test No. 9).

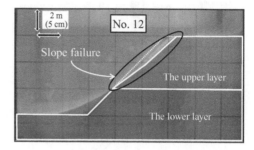

Figure 10. Slope failure (test No. 12).

that clay contents of the upper and lower layers are 0% and 10%, respectively, as shown in Table 4. Neither the rainfall intensity nor the thickness of the upper layer affected the occurrence of the slope failure. Patterns of the slope failure are shown in Figs. 8, 9 and 10. Manner of the slope failure did not depend on the thickness of the upper layer and the slope failure occurred at the upper layer. And it is found that the clay content of 0% for the upper layer is not necessary to bring about slope failure, since test conditions of test No. 4, 7 and 10 did not bring about slope failure.

Let consider mechanism of the slope failure. Changes in location of the GWT for test No. 4, 5 and 6 during the rainfall are shown in Figs. 11, 12 and 13, respectively. In these figures, GWT is calculated from measurement of pore water pressure gauges placed on the bottom of the strong box and the boundary surface between the upper and lower soil layers (see Figs. 2 and 3), assuming that the PWP is hydrostatically distributed in each layer. It can be seen that GWT was not formed on the boundary surface between two layers in Fig. 11, where the clay content of upper layer is 0% and lower

layer is 5%. That is, all rain water was infiltrated into the lower layer, similarly to Fig. 5.

On the other hand, in other test cases, rising of GWT on the boundary surface was observed, as shown in Figs. 12 and 13. It is observed in these cases that the rain water halted on the boundary surface and the parched GWT was temporarily formed, because the PWP on the boundary surface rose. It is anticipated that the lower layer was not well permeable ground, because the clay content of the lower layer is 10% in the same manner as the flat ground showing in Figs. 6 and 7. When Figs. 12 and 13 are compared, it can be recognized that the heights of the parched GWT in Fig. 13 are slightly lower than those in Fig. 12. This fact indicates that horizontal water flow may take place from the upper layer to the slope because the permeability of the upper layer in Fig. 13 is smaller than that in Fig. 12. It is considered that this movement of water brought about the slope failure.

Let look at test results from test No. 7, 8 and 9, where these ground conditions were the same as test No. 4, 5 and 6, but the rain intensity was 15 mm/hr. The slope failure also occurred even under smaller rainfall intensity of 15 mm/hr, when the clay content of the upper layer is 0% and that of the lower layer is 10%. When failure patterns in Figs. 7 and 8 are compared, it is recognized that they are almost same, in spite of different rainfall intensity. It is likely that the difference in the rainfall intensity does not strongly affect

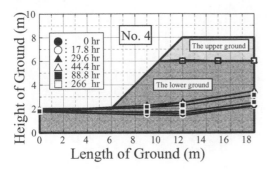

Figure 11. Distribution of groundwater table (test No. 4).

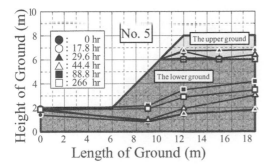

Figure 12. Distribution of groundwater table (test No. 5).

possibility of the slope failure. However, it should be kept in mind that in these experimental conditions, the rainfall intensity varies in relatively narrow range from 15 mm/hr to 30 mm/hr. More tests are necessary to confirm the effect of the rainfall intensity on the slope stability.

Finally let focus on test results from No. 10, 11 and 12, with different of the thickness of the upper layer. The slope failure occurred when the clay content of the upper and the lower layers are 0% and 10%, respectively. It may be concluded that the thickness of the upper layer does not have influence on the slope stability. As shown in Figs. 5 and 7, these patterns of the slope failure are nearly identical. However, as the thickness of the upper layer becomes larger, the range of the slope failure extends much more.

4 CONCLUSION

To simulate slope failure triggered by heavy rain, centrifuge tests were carried out for model grounds consisting of two layers with different permeability. Although several tests having different conditions

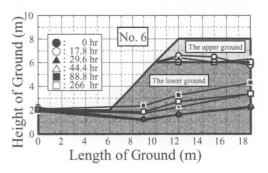

Figure 13. Distribution of groundwater table (test No. 6).

were carried out, failure was observed at restricted cases: i.e., clay contents of the upper and lower layers were 0% and 10%, respectively. Mechanism of the failure is considered as follows: Some part of the rain water is halted on the boundary surface between the two layers when the permeability of the lower layer is lower, and a parched ground water table is formed. In addition, this halted water moves horizontally to the slope surface when the permeability of the upper layer is large. This movement brought about the slope failure.

The present study indicates that possibility of the slope failure increases when the ground consists of several layers with different coefficient of permeability. Especially when difference of the permeability of upper and lower layer is large, the slope stability deteriorates.

REFERENCES

Soga, K., Kawabata, J., Kehavarzi, C., Coumolos, H. & Waduge, W.A.P. 2003. Centrifuge Model of Nonaqueous Phase Liquid Movement and Entrapment in Unsaturated Layered soils. *Journal of geotechnical and geoenvironmental engineering, February 2003*: 173–182.
Wang, F. & Shibata, H. 2007. Influence of soil permeability on rainfall-induced flowslides in laboratory flume test. *Canadian geotechnical journal* 44(9): 1128–1136.

Mechanism of soil slope reinforced using nailing during earthquake

L. Wang, G. Zhang, J.-M. Zhang & C.F. Lee
Tsinghua University, Beijing, China

ABSTRACT: Centrifuge model tests were conducted to investigate the behavior of cohesive soil slopes during an earthquake, considering both the nailing-reinforced and unreinforced slopes. The displacement exhibited an evidently irreversible accumulation that depended on the shaking magnitude. Significant deformation localization occurred mainly in the upper parts of the unreinforced slope when shaking, leading to a final landslide. The nailing significantly changed displacement distribution of the slope and arrested a landslide that occurred in the unreinforced slope during the earthquake. The strain analysis was introduced to discuss the reinforcement mechanism using the measured displacement fields. The reinforcement mechanism can be described using a basic concept: shear effect, which refers to that the nailing decreases the shear strain in comparison with the unreinforced slope. The significant shear effect was induced in the nailing reinforcement zone; preventing the possible sliding.

1 INTRODUCTION

The landslide, one of the most important geological disasters in the world, has often been triggered by earthquakes. The mechanism of earthquake-induced landslide has been of great concern in geotechnical engineering. The nailing becomes an important reinforcement structure for the unreliable slopes, and has been widely used in many natural or artificial slopes.

Centrifuge model tests can reproduce the gravity stress field and the gravity-related deformation process, therefore, they have been an important approach for the investigation of the behavior and failure mechanism of slopes. The failure and deformation mechanism of kinds of slopes during an earthquake have been analyzed using dynamic centrifuge model tests (e.g., Taboada-Urtuzuastegui et al. 2002; Thusyanthan et al. 2005; Lili and Nicholas 2006). Such a testing approach has been widely used for investigating the response of reinforced slopes with different reinforcement structures, such as geomembranes, geotextiles, and soil nails (Porbaha and Goodings 1996, Zornberg et al. 1998; Zhang et al. 2001; Thusyanthan et al. 2007).

The centrifuge model tests have barely been used for the investigation on the behavior of nailing-reinforced slopes under a strong earthquake condition. Moreover, it has not been adequately illustrated how the nailing-soil interaction affects the deformation field within the entire slope and, as a result, increases the stability level under an earthquake condition, although such an earthquake is a main cause of landslides. Therefore, further study is needed to obtain a solid basis for an effective method to evaluate the aseismic stability level of a soil slope that is reinforced by using nailing.

Dynamic centrifuge model tests were conducted to investigate the behavior of cohesive soil slopes during an earthquake, considering both the nailing-reinforced and unreinforced slopes. On the basis of the observations, we analyzed the failure process of the unreinforced slope under the earthquake. The behavior and reinforcement mechanism of the nailing-reinforced slope were further discussed.

2 DESCRIPTION OF TESTS

2.1 Device

The centrifuge model tests were conducted using the 50 g-ton geotechnical centrifuge machine of Tsinghua University. The earthquake was simulated by using a specially manufactured shake table for this centrifuge machine, which can generate arbitrary horizontal earthquake waves with maximum acceleration of 20 g via a complex hydraulic pressure servo-system.

The model container for the tests, made of aluminum alloy, is 50 cm long, 20 cm wide, and 35 cm high. A transparent lucite window was installed on one container side, through which the deformation process of the soil can be observed and recorded.

2.2 Test model

Figure 1 shows the schematic view of the model slope reinforced using the nails. The unreinforced slope was identical except for removal of the nails. The soil was retrieved directly from the soil mountain of the Beijing Olympic Forest Park. The average grain size of the soil is 0.03 mm, and the plastic limit and liquid limit are 5% and 18%, respectively.

The soil was compacted into the container by 6-cm-thickness layer, with a dry density and water content of 1.45 g/cm^3 and 17%, respectively. The slope for the tests was obtained by removing the redundant soil; the slope was 1.5:1 and 25 cm in height, respectively (Fig. 1). A 6-cm-high horizontal soil layer under the slope was set to diminish the influence of the bottom container on the deformation of the slope. In addition, silicone oil was painted on both sides of the container to decrease the friction between the slope and the sides of the container.

A thin columniform needle, made of steel with elastic modulus of 210 GPa, was used to simulate the nails of the reinforced slope. The needle is approximately 1mm in the diameter. This is equivalent to a prototype nail with a diameter of 5 cm at a centrifugal acceleration of 50 g. The nails were buried perpendicular to the slope surface of 10 cm depth in the soil and arranged in an interval of 2 cm (Fig. 1).

2.3 Measurements

An image-record and displacement measurement system was used to record the images of soil during the centrifuge model tests (Zhang et al. 2009). This system can capture about 50 image frames per second. An image-correlation-analysis algorithm was used to determine the displacement vectors of soil without disturbing the soil itself (Zhang et al. 2006). The displacement history of an arbitrary point in the soil can be measured with sub-pixel accuracy. White terrazzo particles were embedded in the lateral side of the soil for the image-based measurements. In addition, a few patterns were affixed to the container to obtain its displacement during the earthquake. It should be noted that the area within the dotted line was used for displacement measurements owing to the requirement of the measurement system (Fig. 1) and it covers the main deformation zone. Cartesian coordinates were established, with the origin as the slope foot, specifying positive as upward in the vertical direction (y-axis), and to the right in the horizontal direction (x-axis), respectively (Fig. 1). The measurement accuracy can reach 0.02 mm based on the model dimension for the centrifuge tests in this paper.

A series of accelerometers, with a measurement accuracy of 0.3%, was buried in the soil to measure the acceleration response at different altitudes during the earthquake (Fig. 1a). These transducers were linked to the automatic data-acquisition system so that real-time records can be obtained.

2.4 Test procedure

The model slope was installed on the shaking table of the centrifuge machine. The centrifugal acceleration gradually increased to 50 g and was maintained

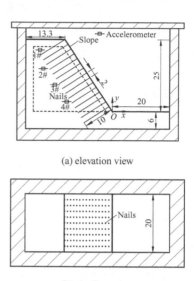

(a) elevation view

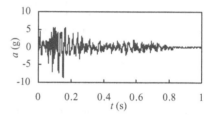

(b) planform view

Figure 1. Schematic view of nailing-reinforced slope (unit: cm).

Figure 2. Seismic wave input. t, time; a, acceleration.

during shaking tests. After the deformation of the slope became stable, an earthquake wave was input on the container bottom. This wave lasted 1 s, with a maximum acceleration of 8.7 g in the model dimension (Fig. 2), equivalent to 0.174 g in the prototype dimension at a centrifugal acceleration of 50 g. The images of slope were recorded with a rate of 48 frames/second using the image-based measurement system. The displacement field and its change can be obtained based on the images. The measured results were all based on the model dimension.

3 RESPONSE OBSERVATIONS

3.1 Failure configuration

A significant landslide occurred in the unreinforced slope during the earthquake (Fig. 3), which can also be found in the measured contours of displacements (Fig. 4). It should be noted that the borders in Fig. 4

Figure 3. Post-earthquake image of unreinforced slope.

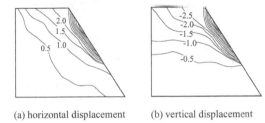

(a) horizontal displacement   (b) vertical displacement

Figure 4. Post-earthquake displacement contours of unre-inforced slope.

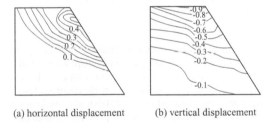

(a) horizontal displacement   (b) vertical displacement

Figure 5. Post-earthquake displacement contours of nailing-reinforced slope.

were designated as the dotted area in Fig. 1, but not the actual slope borders.

On the other hand, it can be seen from the displacement contours of the nailing-reinforced slope that such a slope exhibited only residue deformation that resulted from the earthquake (Fig. 5); a landslide was avoided. This indicated that the nailing increased the stability level of the slope significantly.

3.2 *Acceleration response*

Figure 6 shows the seismic response of a point at the 2# accelerometer location in the reinforced slope (Fig. 1a). It should be noted that the present displacement is relative to the container during the earthquake.

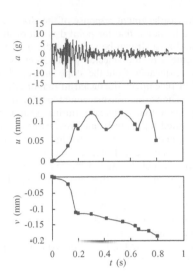

Figure 6. Seismic response of a point at the accelerometer location of reinforced slope. a, acceleration; u, horizontal displacement; v, vertical displacement; t, time.

In addition, high-frequency displacement response waves had been filtered owing to the frame-rate limit of the captured images (48 frames/s) and measurement accuracy (0.02 mm). This resulted in that a few minor displacements, especially in the horizontal direction, cannot be captured. Whereas, it is believed that the measurement results are sufficient for the rule analysis of displacement response. The horizontal displacement fluctuated in magnitude, with the tendency of accumulation being to the outer slope during shaking; the vertical displacement increased monotonically during shaking. This demonstrated that significant irreversible deformation occurred from the earthquake application. The irreversible deformation increased rapidly in the early earthquake period, accompanied with the remarkable input wave, and the rate of increase was low when the input wave dropped off. This indicated that the increase rate of the deformation significantly depended on the magnitude of the input shaking acceleration.

3.3 *Displacement response*

It can be seen from the displacement contours that the horizontal displacement of both slopes increased from the inner slope to the free surface, and the vertical displacement increased with increasing altitude (Figs. 4–5). The post-earthquake deformation of the unreinforced slope was significantly larger than that of the reinforced slope; this demonstrated that the nailing had a significant effect on the deformation of the slope.

Several points were selected at equal interval in the nail-direction of the reinforced slope to discuss the dynamic displacement response of the slope (Fig. 7). It should be noted that the point d is out of the reinforcement area. It can be seen that the peak relative displacement of the selected points decreased with increasing distance from the slope surface. The horizontal displacement fluctuated in magnitude during shaking. The vertical displacement exhibited approximately monotonic increase during shaking; this demonstrated that significant irreversible volumetric deformation occurred due to shaking application.

The displacement differences between the points a, b and c, which located in the reinforcement area, were nearly the same, demonstrating that the nailing induced a tendency of uniform deformation of the reinforcement area. Whereas, the point d, out of the reinforcement area, exhibited a different feature of displacement from former points; this indicated that the influence of the nailing is relatively insignificant in the unreinforced area.

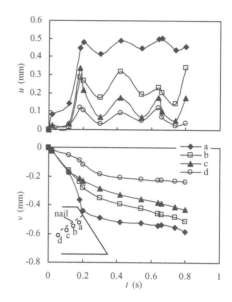

Figure 7. Displacement histories of typical points of reinforced slope. u, horizontal displacement; v, vertical displacement; t, time.

4 MECHANISM ANALYSIS

4.1 *Failure process of unreinforced slope*

According to the images and corresponding displacement contour lines, the displacement histories of typical points, before the landslide of the unreinforced slope were given to illustrate the failure process (Fig. 8).

In the upper part of the slope (Fig. 8a), the displacement was significant during shaking in both horizontal and vertical directions. The displacements of points A and B were fairly close during the shaking; however, the displacement of point C varied more significantly. Such a difference increased during shaking, indicating that there was large shear deformation between points B and C. In the middle part of the slope (Fig. 8b), the horizontal displacements of the three points became significant different during shaking (Fig. 9b). In addition, the vertical displacement of the point F, close to the slope surface, was significantly different from those of other points. This demonstrated that there was significant deformation localization in the middle of the slope, which became more evident in the region near the surface.

In the lower part of the slope (Fig. 8c), the changes of displacement were fairly smaller than those of the upper slope. Moreover, the displacement difference between different locations was similar. This indicated that there was insignificant deformation localization, which can be confirmed by the contour lines of the post-earthquake deformation (Fig. 4).

It can be concluded that the shear deformations concentrated in a narrow band, indicating a significant deformation localization due to the earthquake. Comparisons of the earthquake wave and displacement histories showed that the significant deformation localization occurred when a larger acceleration was inputted (Figs. 2, 8). This deformation localization occurred mainly in the upper and middle parts of the slope; this led to final landslide.

4.2 *Reinforcement mechanism*

The nailing reinforcement mechanism was analyzed based mainly on the comparison of displacement distribution of the lateral side of both slopes. It was discovered that significant irreversible deformation of the slope was induced by the earthquake application (Fig. 6). Thus, the distribution of post-earthquake horizontal displacement at a horizontal line was carefully analyzed to investigate the influence of the nailing on the slope's deformation resulting from the earthquake (Fig. 9).

A close examination of the displacement distribution at a horizontal line, $y = 11.5$ cm, showed a significant difference between the reinforced and unreinforced slopes (Fig. 9b). For the reinforced slope, the horizontal displacement increased from the inner slope and the increase rate became fairly small within the reinforcement area; on the contrast, the horizontal displacement increased at a rapid rate from the inner slope for the unreinforced slope. Similar rules can be found in the other altitudes (Fig. 9). It can be concluded that the nailing significantly changed the displacement distribution of the slope. In other words, the nailing significantly reduced the horizontal displacement within the reinforcement area; this decreased the deformation localization evidently.

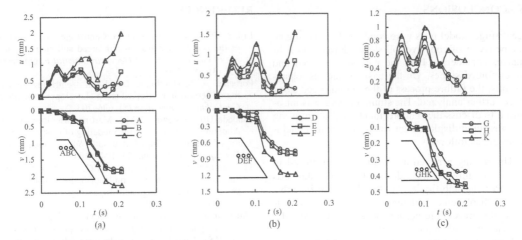

Figure 8. Displacement histories of typical points of unreinforced slope. u, horizontal displacement relative to the container; v, vertical displacement; t, time.

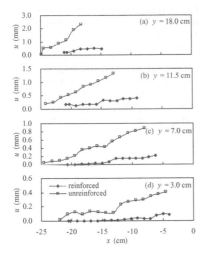

Figure 9. Post-earthquake horizontal displacement distributions. u, horizontal displacement; y, vertical coordinate; x, horizontal coordinate.

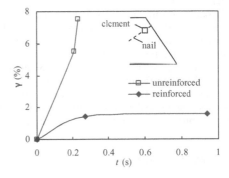

Figure 10. Strain history of typical element during earthquake. γ, slope-direction shear strain.

The strain of a soil element within the slope was introduced for a further analysis of the reinforcement mechanism. In this paper, a two-dimensional, four-node square isoparametric element, 1 cm long, was used for strain analysis. All the strain components, at an arbitrary plane of the element, can be easily derived by the measured displacement via the image-measurement system.

The observation showed that the slip surface of the unreinforced slope was approximately parallel to the slope surface (Fig. 4a); thus the slope-direction shear strain was used to describe the shear deformation of the slope. Figure 10 shows the history of shear strain of a typical element in the unreinforced and reinforced slopes. It can be seen that the shear strain of the unreinforced slope increased rapidly in the early period of the earthquake. This strain reached a significant magnitude when the landslide, just across this element, occurred. However, the shear strain increased at a smaller rate and reached a lower level far from failure if the nails were used. In other words, the nailing significantly confined the shear deformation of the slope so that the shear strain was noticeably decreased.

Therefore, the nail-reinforcement mechanism can be described using the effects of the nails on the shear. A basic concept, *shear effect*, was introduced. It refers to that the nails decrease the shear strain of the region where significant deformation concentration occurred. A significant shear effect of nailing in the slope arrested the formation of a slip surface. Accordingly, the horizontal deformation was decreased significantly.

5 CONCLUSIONS

Centrifuge model tests were conducted to investigate the behavior of cohesive soil slopes during an earthquake, considering both the nailing-reinforced and unreinforced slopes. On the basis of the test observations, the failure process of the unreinforced slope were further analyzed. The strain analysis was introduced to discuss the reinforcement mechanism using the measured displacement fields. The main conclusions are as follows:

1. The nailing arrested a landslide that occurred in the unreinforced slope during the earthquake, thus increasing stability level of the slope significantly.
2. The displacement of the slope exhibited an evidently irreversible accumulation that was significantly dependent on the magnitude of input shaking.
3. For the unreinforced slope, significant deformation localization occurred when the larger acceleration was input. This localization occurred mainly in the upper and middle parts of the slope; this led to a final landslide.
4. The nailing significantly changed displacement distribution of the slope. The reinforcement mechanism can be described using a basic concept: shear effect, which refers to that the nailing decreases the shear strain in comparison with the unreinforced slope. The significant shear effect was induced in the reinforcement zone; this effect prevented the possible sliding so that the stability level of the slope was accordingly increased during the earthquake.

ACKNOWLEDGMENTS

The authors appreciate the support of the National Basic Research Program of China (973 Program) (No. 2007CB714108) and National Natural Science Foundation of China (No. 50778105).

REFERENCES

Lili, N.-R. & Nicholas, S. 2006. Centrifuge model studies of the seismic response of reinforced soil slopes. *Journal of Geotechnical and Geoenvironmental Engineering* 132 (3): 388–400.

Porbaha, A. & Goodings, D.J. 1996. Centrifuge modeling of geotextile-reinforced steep clay slopes. *Canadian Geotechnical Journal* 33 (5): 696–703.

Taboada-Urtuzuastegui, V.M., Martinez-Ramirez, G. & Abdoun T. 2002. Centrifuge modeling of seismic behavior of a slope in liquefiable soil. *Soil Dynamics and Earthquake Engineering* 22: 1043–1049.

Thusyanthan, N.I., Madabhushi, S.P.G. & Singh, S. 2007. Tension in geomembranes on landfill slopes under static and earthquake loading- centrifuge study. *Geotextiles and Geomembranes* 25 (2): 78–95.

Thusyanthan, N.I., Madabhushi, S.P.G. & Singh, S. 2005. Seismic performance of geomembrane placed on a slope in a MSW landfill cell—A centrifuge study. *Geotechnical Special Publication* 130–142: 2841–2852.

Zhang, G., Hu Y. & Zhang J.-M. 2009. New image-analysis-based displacement-measurement system for centrifuge modeling tests. *Measurement* 42 (1): 87–96.

Zhang, G., Liang, D. & Zhang, J.-M. 2006. Image analysis measurement of soil particle movement during a soil-structure interface test. *Computers and Geotechnics* 33 (4–5): 248–259.

Zhang, J., Pu, J., Zhang, M. & Qiu, T. 2001. Model Tests by Centrifuge of Soil Nail Reinforcements. *Journal of Testing and Evaluation* 29 (4): 315–328.

Zornberg, J.G., Sitar, N. & Mitchell, J.K. 1998. Performance of geosynthetic reinforced slopes at failure. *Journal of Geotechnical and Geoenvironmental Engineering* 124 (8): 670–683.

Mechanical deterioration of ground due to microorganisms activated by eutrophication of ground ecosystems

T. Futagami & K. Terauchi
Hiroshima Institute of Technology, Hiroshima, Japan

T. Adachi
MAEDA Road Construction Co., Ltd, Tokyo, Japan

K. Ogawa
Rail Track & Structures Technology, Osaka, Japan

T. Kono
Nittoc Construction Co., Ltd., Tokyo, Japan

Y. Fujiwara
TAISEI Corporation, Tokyo, Japan

S. Sakurai
Construction Engineering Research Institute Foundation, Kobe, Japan

ABSTRACT: There are many geohazards caused by debris flows and landslides all over the world, and, especially, in Japan every year. In June 1999 and August 2005 in Hiroshima prefecture there were many large geohazards of debris flows and landslides owing to torrential downpours. Through the survey of these geohazards it seems that there are two main factors causing the increase of geohazards. The one factor is the contiguity of urbanization and hillsides and streams. The other factor is biodeterioration of mechanical structures of the ground in hillsides caused by eutrophication (nutrient enrichment) originating in the transition of ecological systems in hillsides. This paper involves the investigation of the relationships between the geohazards and biodeterioration of mechanical structures of the ground.

1 INTRODUCTION

In recent years, understanding of processes such as biodeterioration and biofoulding, in fields usually outside geotechnical engineering, has increased tremendously. Weathering of ground is caused by not only physical and chemical processes, but also biological processes (Howsam (ed.) 1990).

A lot of hazards caused by debris flows and landslides on hillsides and streams frequently happen in Japan.

In the end of June, 1999 and September 2005 in Hiroshima prefecture there were many large debris flows and landslides owing to localized torrential downpours. Through the survey of these hazards it seems that there are two main factors causing the increase of geohazards. The one factor is the contiguity of urbanization and hillsides and streams. The other

factor is biodeterioration of mechanical structures of ground in hillsides caused by eutrophication (nutrient enrichment) originating in transition of ecological systems in hillsides. The eutrophication and dampening of hillsides by the accumulation of organic detritus promote microbial growth and activities that deteriorate mechanical structures of ground. Microorganisms promote the weathering of rocks to soils (Futagami et al. 1999 & 2007).

It seems that mechanical deterioration of ground due to microorganisms activated by eutrophication of ground ecosystems has become an important theme for geotechnical engineers. A fast, quantitative analysis technique was developed to assess potential rock weathering by bacteria (Puente et al. 2006).

In this study the relationships between the debris flows and biodeterioration of mechanical structures of ground in hillside in Hiroshima are investigate.

2 DEBRIS FLOW PROMOTED BY EUTROPHICATION OF HILLSIDE ECOSYSTEMS

2.1 Eutrophication of hillside ecosystems

Hillside ecosystems are modeled as shown in Figure 1. The hillside ecosystems are composed of biotic community and hillside environment. The biotic community and the hillside environment deeply depend on each other.

The biotic community is composed of microbial community (microbes), plant community and animal community which deeply depend on each other. The plant community is composed of producers that can manufacture food from simple inorganic substances by photosynthesis. The animal community is macroconsumers that ingest other organisms and particulate organic matter. The microbial community is decomposers (bacteria, fungi, protozoa, etc.)

The eutrophication of ecosystems on hillsides and mountain slopes has been advanced by accumulation of organic detritus originating in litter (dead leaves and branches) and fallen timbers as shown in Figure 2.

Soil microorganisms live in an environment that is dominated by solid particles of various natures, sizes and shapes and having complex spatial arrangement as shown in Figure 3 (Chenu & Stotzky 2002).

Microbial plowing proceeds deep under ground. These kinds of plowing mean "The soil works itself" or "The soil lives of its own accord and plows itself", as stated by the pioneers of natural or organic farming (Fukuoka 1978 & Rodale 1942).

Microorganisms promote the weathering of rocks to soils and cause aggregation of soils by their enzymes (Nishio & Morikami 1997).

Microorganisms are influenced by the nature, properties, and arrangement of soil particles, and they also modify these particles and their arrangements. The spatial arrangement of the solid particles results in

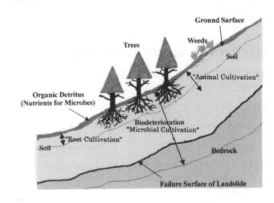

Figure 2. Debris Flows and Landslides Promoted by Eutrophication of Hillside Ecosystems.

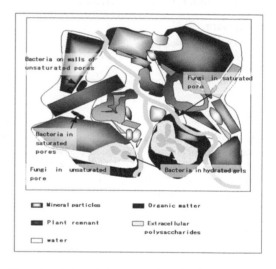

Figure 3. Schematic Representation of Various Habitats of Microorganisms in Soils (Chenu & Stotzky 2002).

a complex and discontinuous pattern of pore spaces various sizes and shapes that are more or less filled with water or air as shown in Figure 4 (Chenu, C. & Stotzky G. 2002).

2.2 Mcrobial works promoting debris flows

There seem to be three kinds of biological plowing (Figure 5). One is *kinko* (microbial plowing) by microbes (Nobunkyo 1997). Another is root plowing by plant roots. The other is animal plowing by animals such as earthworms, frogs, snakes and moles.

Aggregated structures of soils wet the ground of hillsides to decrease resistance to landslides. The landslides in upstreams cause the debris flows. It seems that microbial activities caused by eutrophication of hillsides accelerate the occurrence and hazard of debris

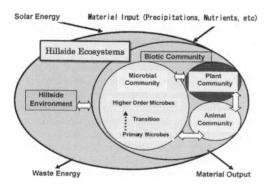

Figure 1. Complexity Model of Hillside.

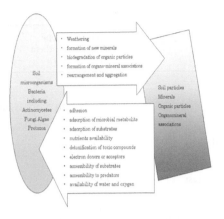

Figure 4. Possible Interaction between Microorganisms and Soil Particles (Chenu, C. & Stotzky G. 2002).

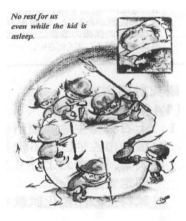

Figure 5. Microorganisms Decompose Rocks to Soils as in Decayed Teeth (Illustrated by Imai 1984).

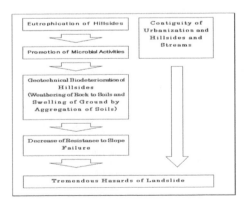

Figure 6. Occurrence Scenario of Tremendous Hazards of Debris Flows and Landslides Promoted by Eutrophication of Hillside Ecosystems.

flows and landslides. Then occurrence scenario of tremendous hazards of debris flows and landslides promoted by eutrophication of hillsides ecosystems may be obtained as shown in Figure 6.

3 INVESTIGATION ON EUTROPHICATION OF HILLSIDE

3.1 Investigation sites

The subsurface investigation on eutrophication was conducted on Sensuiminamitani River in Hatsukaichi, Hiroshima, in which a debris flow occurred in September 2005. The investigation sites are shown in Figures 7 to 8.

In Table 1 "Stable Slope" and "Unstable Slope" should be understood as follows:

Figure 7. Location Map of Subsurface Investigation of Eutrophication in Sensuiminamitani River in Hatsukaichi, Hiroshima.

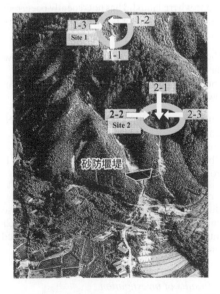

Figure 8. Photograph of Subsurface Investigation Sites of Eutrophication in Sensuiminamitani River in Hatsukaichi, Hiroshima.

Table 1. Data on Subsurface Investigation Site.

Site	No	Slope failure	Plants	Organic detritus (m)	Slope gradient (°)	Soil
1	1–1	Stable slope: Safe slope for landslide (5 m lower from center of landlide head)	Bare	0	30	Soft rock
	1–2	Unstable slope: Dangerous slope for landslide (5.0m upper from center of landlide head)	Mixed forest	0.06–0.10	20	Sandy soil
	1–3	Unstable slope: Dangerous slope for landslide (7.0m from right bank of landslide head)	Mixed forest	0.06–0.10	20	Sandy soil
2	2–1	Stable slope: Safe slope for landslide(center of landslide head)	Weeds without trees	0.01–0.03	20	Sandy soil
	2–2	Unstable slope: Dangerous slope for landslide (3.2m from right bank of landslide head)	Mixed forest with mainly pine trees	0.05–0.07	30	Sandy soil
	2–3	Unstable slope: Dangerous slope for landslide (2.8m from left bank of landslide head)	Mixed forest with mainly pine trees	0.05–0.07	30	Sandy soil

Stable Slope (Slope where landslide is not expected).

In September, 2005 during torrential rains there were landslides leaving hard bare surfaces, where landslides need not to be expected in the near future.

Unstable Slope (Slope where landslide may be expected).

Slope where landslid has not recently occurred, and where organic matter is plentiful, the ground is soft and the probability of sliding is plentiful.

The data of these investigation sites are shown in Table 1.

The items of the subsurface investigation are as follows:

1. Nutrient test (Total Carbon (TC) and Total Nitrogen (TN))
2. Simple Penetration Test
3. Observation of Microoganisms by Scanning Electron Microscope.

Observation of microbes in subsurface in hillside was conducted by using scanning electron microscope (Suzuki et al. 1995 & Futagami et al. 2003). The low-vacuum freeze drying method (Figure 9) was used for the preparation of biological specimens.

3.2 Results of investigation

The results of the subsurface investigation are shown in Figures 10–18.

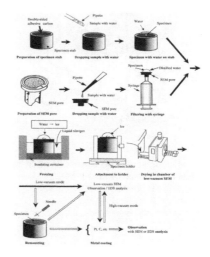

Figure 9. The low-vacuum freeze drying method.

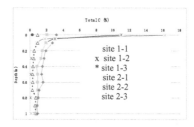

Figure 10. Vertical profiles of Total Carbon (TC).

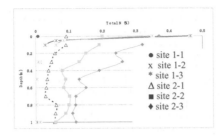

Figure 11. Vertical profiles of Total Nitrogen (TN).

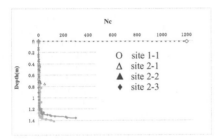

Figure 12. Vertical profiles of N_c obtained simple penetration test.

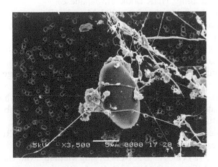

Figure 13. Spindle type microorganism.

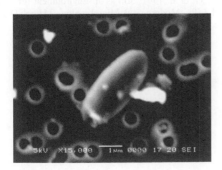

Figure 14. Semi-spindle type microorganism.

It was found that the values of eutrophication indexes (TC and TN) are low in the stable slope and high in the unstable slopes (see Figures 10–11). The values of N_c are high in the stable slopes and low in the unstable slopes (see Figure 12). Many kinds of microorganisms are found in the unstable slopes (Figure 13–18).

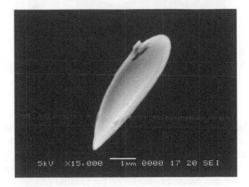

Figure 15. Sunflower-seed type microorganism.

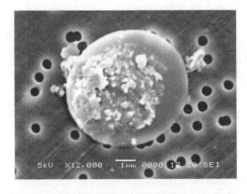

Figure 16. Spherical type microorganism.

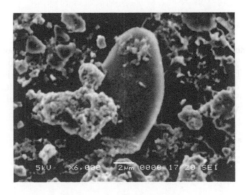

Figure 17. Spindle type microorganism.

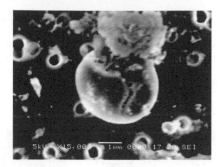

Figure 18. Semi-spheical Type Microorganism.

4 CONCLUSIONS

The relationships between debris flows and biodeterioration of mechanical structures of ground in hillsides were studied in this research. Through the results of investigation on eutrophication of hillside, it was found that biodeterioration of the ground progressed more in the unstable slopes than in the stable slopes. It seems that the eutrophication of hillsides promotes biodeterioration of mechanical structures of the ground and cause the debris flows.

It is often said that one should not dump refuse at the top of a precipice (*Gake no chikaku de gomi wo suteruna*). This is to prevent the eutrophication of place potentially dangerous for slope failure. In order to decrease disaster of debris flows and landslides promoted by eutrophication of the ecosystems on hillsides and mountain slopes, it is necessary not to nourish or wet dangerous hillsides.

ACKNOWLEDGEMENTS

The authors acknowledge the financial support of Grant-in-Aid for Scientific Research in 2005–2007 (Exploratory research, No. 17651102).

REFERENCES

Chenu, C. & Stotzky, G. 2002. Interactions between Microorganisms and Soil Particles.: An Overview, Interactions between Microorganisms and Soil Particles, IUPAC Series on Analytical and Phisical Chemistry of Environmental Systems, Vol. 8, Bollang J.-M. and Senei N. (eds.), John Wiley & Sons Ltd: 3–40.

Dixon, B. 1994. Power Unseen—How Microbes Rule the World, W.H. Freeman/Spektrum Academischer Verlag.

Esther, M.E. et al. 2006. Image Analysis for quantification of bacterial rock weathering. *Journal of Microbiological Methods,* Elsevier, 64: 275–286.

Fukuoka, M. 1978. The One Straw Revolution—An Introduction to Natural Farming. Rodale Press.

Futagami, T. et al. 1999. Heavy Landslides Caused by Transition of Hillside Ecosystems—Microbial Work Owing to Eutrophication of Slopes (Microbial Plowing), *Proceedings of Symposium on Slope Failure and Landslide Problems* (International Symposium on the Auspicious Occasion of 50th Anniversary of Ehime University), Shikoku Branch, Japanese Geotechnical Society: 89–98 (in Japanese).

Futagami, T., Terauchi, K. & Kono, T. 2007. Transportation Phenomena of Substances in Hillside Ecosystems Promoting Debris Flow, *Proceedings, 6th International Symposium on Ecohydraulics,* IAHR, Poster Session, 56.

Futagami, T., Yoshimoto, K. & Sima, S. 2003. Observation of Microbes Causing Ground Biodeterioration by Low-Vacuum SEM Freeze Drying Method, Special Supplement, Abstracts of the 13th Annual Goldschmidt Conference, Poster Session, A114.

Hattori, T. 1998. Surveying of Microbes. Shinchosha (in Japanese).

Howsam, P. (ed.) 1990. Microbiology in Civil Engineering. E. & F.N. Spon, An Imprint of Chapman and Hall, ix–x.

Likens, G.E & Bormann, F.H. 1994. Biogeochemistry of a Forested Ecosystems, Springer-Verlag.

Miyawaki, A. 1967. Plants and Man—Balances of Biotic Community. NHK Books (in Japanese).

Nishio, M. & Morikami, Y. 1997. Microbes Bring up Forests. Series of Man in Nature (Microbes and Man). No. 2 (in Japanese).

Nanao, J. & Imai, Y 1984. *Mushibakun Daisuki* (Do You Like Toothache?) (in Japanese).

Nobunkyo, ed. 1997. Bokashi Fertilizer Made by Indigenous Microbes. Video Tape (in Japanese).

Odum, E.P. 1983. Basic Ecology. CBS College Publishing.

Postgate, J. 1994. The Outer Reaches of Life. Cambridge University Press.

Rodale J.I. 1942. Organic Farming and Gardening. Rodale Press.

Suzuki, T., et al. 1995. A New Drying Method: Low-vacuum SEM Freeze Drying and Its Application to Plankton Observation. Bulletin of Plankton Society of Japan, 42 (1): 53–62 (in Japanese).

Stotzky, G. et al. 1997. Soil as an Environment for Microbial Life. Modern Soil Microbiology, Marcel Dekker. Inc.: 1–20.

Prediction and Simulation Methods for Geohazard Mitigation – Oka, Murakami & Kimoto (eds)
© 2009 Taylor & Francis Group, London, ISBN 978-0-415-80482-0

Thermal, hydraulic and mechanical stabilities of slopes covered with *Sasa nipponica*

K. Takeda
Obihiro Univeresity of Agriculture and Veterinary Medicine, Obihiro, Japan

T. Suzuki
Kitami Institute of Technology, Kitami, Japan

T. Yamada
Yamada Consultant Office, Toyonaka, Japan

ABSTRACT: Following a comprehensive field study evaluating thermal, hydraulic, and mechanical stabilities of a slope covered with *Sasa nipponica,* a type of dwarf bamboo (known as *Sasa*), against slope failure in a cold region (eastern Hokkaido in Japan), it was found that thermal stability increases due to a decrease in freezing and frost heaving of the surface layer caused by the adiabatic effects of *Sasa* litter and the snow it retained. Further, *Sasa* plays a role in reducing the effect of rainfall due to the water retention by the *Sasa* rhizome layer in the surface soil, thereby increasing hydraulic stability. Through surveys of surface failures caused by an earthquake in places on wild slopes covered with *Sasa*, the mechanical stability was evaluated using the safety factor in a static condition expressed as a function of the leaf area index (LAI), which exceeds 1.37, owing to the reinforcing effect of the rhizomes.

1 INTRODUCTION

On the artificial slopes along roads in cold regions, slope surface failure and erosion occur occasionally owing to frost heave damage to protective grids, which usually cease to function because they protrude from the ground surface due to repeated freeze–thaw cycles, as shown in Figure 1. In Japan, these grids are generally used to keep the planting ground on steep slopes stable, but the damage is mainly concentrated in eastern Hokkaido, as shown in Figure 2 (Takeda et al. 2000). On the other hand, hardly any damage occurs on native slopes covered with *Sasa nipponica* Makino et Shibata (called *Sasa*), a type of dwarf bamboo, especially to grow in such the cold region, which has less snow, illustrated in Figures 2 (Hokkaido 1983) and 3.

In general, rhizomes of *Sasa* species characteristically have strong tensile strength (Takeda et al. 2001), the same as that of a tree (O'Louglin et al. 1982), allowing it to uniformly cover the ground and form a surface layer (Karisumi 1969) like a mattress on slopes (called a *Sasa* mat). Comparing native and artificial slopes, it is expected that *Sasa* lessens the effects of frost such as frost depth and the amount of frost heave in the surface layer of the slope (Takeda et al. 1999). During rainfall, the above-ground part seems to mitigate the impact of raindrops, and the underground parts temporarily retain water and to drain it for a long time like a surface soil of forest (Ohta 1990). Further, it is anticipated that the rhizomes growing in the surface layer consolidate the soil to prevent sliding and slope failure (Karisumi 1987). Although the slope stability caused by effects of tree roots has been studied such as Wu et al. (1979) and Tsukamoto (1987), any stability by that of the rhizomes hardly has been done. They cannot be applied adequately to the design and construction of slopes to prevent disasters, because these functions of *Sasa* are not well understood quantitatively for the slope stability.

Figure 1.   Surface failure on the artificial slope and frost heave damage to protective grids, marked with a circle (left).

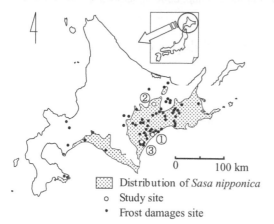

Figure 2.   Study sites and distributions of frost heave damage and *Sasa nipponica* in Hokkaido, Japan.

Figure 3.   Cross section of the ground covered with *Sasa nipponica* (left) and its rhizomes (right).

In this study, for clarifying quantitatively the functions of *Sasa* for thermal, hydraulic, and mechanical stability of slopes to prevent slope failure, a comprehensive field study and a laboratory test were conducted on flat ground and slopes covered with *Sasa*.

2   STUDY METHODS

The studies consist of a field observation, a field survey, and a laboratory test.

2.1   *Observation of the thermal effects of Sasa*

To estimate the thermal environment formed by *Sasa*, the seasonal maximum values of frost depth and heave amount were observed in November 1994–April 1995 at 10 spots on the ground where *Sasa* grows densely or sparsely, located on a hill along the coast in Urahoro in Hokkaido, marked with ① in Figure 2. These spots, including bare ground, are established on frost-susceptible soil, and are almost flat to eliminate the influence of the slope's direction. The difference in the growth density was expressed as the culm density of *Sasa*, given by the number of culms in 1 m². In this observation, the frost depth and heave amount were measured using a frost tube that detected the depth from the disturbed traces of blue-colored jelly by methylene blue in the tube due to freezing and a newly devised displacement gauge having a screw-type anchor that is screwed deeper than the maximum frost depth, respectively. Further, the air and ground temperatures were measured.

2.2   *Observation of the hydraulic functions of Sasa*

To evaluate the hydraulic functions of *Sasa*, the water balance was compared on slopes covered with and without *Sasa*. On a slope near Rikubetsu (marked ② in Fig. 2), which has a gradient of 20° and faces southwest, we partly stripped away the *Sasa*, including the rhizomes. Then, the sites covered with and without *Sasa*, called the *Sasa* site and the bare site, 5 m in width × 5 m in slope length, and 1 m × 5 m, respectively, were prepared for observation by enclosing their upper ends and sides with zinc to a depth of 60 cm to avoid inflow of water.

Because the surface of the *Sasa* site, consisting of the organic soil layer including a *Sasa* mat 25 cm thick and volcanic sand layers 1 and 2, is established as the surface level, the surface of the bare site, consisting of only the latter two, is 25 cm below the surface, as shown in Figure 4. To calculate the water balance, suction was automatically measured using a tensiometer (indicated as a circle in Fig. 4) at 10 cm, 35 cm, and 50 cm depth (D10, D35, and D50) in the *Sasa* site, and at 35 cm and 50 cm depth (D35 and D50) in the bare site, from July through September 2000. In addition, the outflow rate was measured at 25 cm and 55 cm depths at the lower end of both sites using a run 1 m wide and a rain gauge. The precipitation, air temperature, and surface temperature were also measured. Using the soil moisture characteristic curve obtained in the laboratory tests, suction during heavy rains was converted into volumetric water content, which was then used for calculating the water balance.

2.3   *Survey of the mechanical functions of Sasa*

To evaluate the stability of the surface layer on slopes reinforced by *Sasa*, field surveys and laboratory tests were conducted on the native slope covered with *Sasa* on which partial sliding surface failures occurred due to the earthquake (January 1993, M7.8) at Taiki, as indicated by ③ in Figures 2 and 5. The following procedures were carried out: (1) field survey of surface

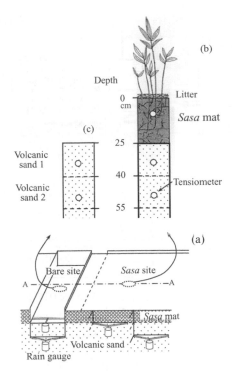

Figure 4. Outline of study sites (a), and cross sections of *Sasa* (b) and bare sites (c).

Figure 5. Surface failure on the wild slope covered with *Sasa nipponica* induced by the earthquake.

sliding on the slope; (2) vegetative investigation for *Sasa* on the slope; (3) laboratory tension test for the rhizomes; and (4) laboratory direct shear test for soil sampled at the sliding surface on the slope. In addition, by the vegetative investigation, the relationship between the above- and below-ground parts of *Sasa* on the slope was investigated in eastern Hokkaido (Takeda et al. 2003).

3 RESULTS

Based on the studies, the estimation of the thermal, hydraulic, and mechanical effects of *Sasa* are presented as follows.

3.1 *Thermal effects of Sasa*

As shown in Figures 6 and 7, both the seasonally maximum values of frost depth F_d (cm) and heave amount H_a (cm) decrease with increasing of culm density S_d (culms/m^2), given by the following equations:

$$F_d = -0.126\,S_d + 43.8 \qquad (1)$$

$$H_a = -0.0198\,S_d + 5.40 \qquad (2)$$

Further, using the daily temperature changes at 5 cm and 0 cm above the surface, the thermal conductivity of the litter layer and the snow cover retained by *Sasa* was calculated, based on the difference in oscillations observed when the change is regarded as a sine curve (Com. Meth. Phys. Prop. Soil 1978, Takeda et al. 1999). It was estimated to be 0.035 W/mK, indicating

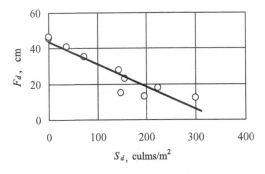

Figure 6. Relationship between the seasonally maximum frost depth F_d and the culm density of *Sasa nipponica* S_d.

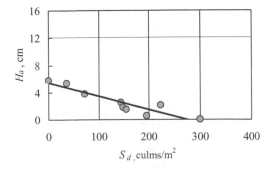

Figure 7. Relationship between the seasonally maximum frost heave amount H_a and the culm density of *Sasa nipponica* S_d.

that *Sasa* works like a thermal insulating material available in the market in extruded polystyrene forms (Jpn. Soc. Thermophysical Prop. 1990).

When frozen soil thaws, slope failure often occurs at the boundary of thawing and frozen layers. Therefore, since the frost action occurs in the *Sasa* mat on the *Sasa* slope owing to the decrease in the frost depth and heave amount, the thermal stability of the slope increases.

3.2 Observation of suction in surface layers during rainfall

During the observation, two rainfalls offered suitable opportunities to observe large changes in the suction (July 15–18 and September 24–25). Observation results for the former event are shown in Figure 8, which shows changes in precipitation, suction, and outflow. The suction responds sensitively to the two peaks of precipitation, 8.5 and 15.2 mm/h. Also, the outflow changes considerably on the surface of the bare site (25 cm in depth), while there is hardly any change at 25 cm below the *Sasa* mat. Using the soil moisture characteristic curve obtained in laboratory tests, the suction is converted into the volumetric water content, as presented in Figure 8 (g). During this observation, the total precipitation was 33.2 mm. On September 24–25, similar results were obtained when the total precipitation was 32.2 mm. Further, the amount of evaporation was calculated by the bulk method using the surface temperature of both sites and the air temperature, and confirmed to be negligible small (Kondo, 1999). Also, to estimate the capacity of rainfall intercepted by the above-ground part of *Sasa* and its litter, the water retained in a unit area 1 m² was 842 cm³ from the laboratory test. Through the calculation of water balance using the results, it was found that the average percentage of precipitation retained in the *Sasa* mat was 77%, from 79.8% and 73.8% shown in Table 1, while the infiltration at the bare site averages 71%.

Since a comparison showed that the downward force of the Sasa mat that retains rainwater is less than 10% that of the resistance to sliding, such as the tensile force of the rhizomes and the soil strength, it was verified that the hydraulic stability increases by *Sasa* coverage (Takeda et al. 2008).

3.3 Mechanical stability of the slope

To estimate the mechanical stability of the slope covered with *Sasa*, we used the survey results of the sliding surface and found that the slope length is 33 m, the width 45 m, the average gradient α 40°, the average thickness H 0.6 m, the direction 135° clockwise from the north, and the area S 1,041 m². Based on the results, a model for analysis was established, as

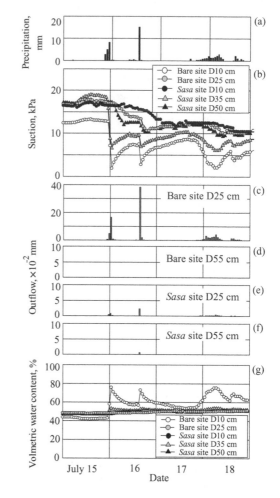

Figure 8. Changes of precipitation (a), suction (b), outflow (c)–(f), and volumetric water content (g) at the *Sasa* and bare sites on the slope.

presented in Figure 9. In this model, the safety factor is expressed as a function of the tensile force of rhizomes per unit slope length, considering the seismic acceleration K_h, as follows: where the symbols

$$F'_s = \frac{L}{SH\gamma_1(\sin\alpha + K_h\cos\alpha)}t$$
$$+ \frac{c + H\gamma_t(\cos\alpha - K_h\sin\alpha)\tan\phi}{H\gamma_t(\sin\alpha + K_h\cos\alpha)} \quad (3)$$

and values obtained from the field investigation and the laboratory tests, are summarized in Table 2, $W(=SH\gamma_t)$ is the weight of the sliding layer, and $T(=tL)$ is the tensile force of rhizomes around the sliding layer.

Table 1. Water balance at the *Sasa* and bare sites.

	July 15 22:00– 16 17:00		September 24 15:00–25 22:00	
Duration of rainfall				

(a) Water balance at the *Sasa* site

Amount & Percentage	l*	%	l*	%
Precipitation	166.0	100.0	161.0	100.0
Interception of *Sasa* Surface (D0 cm)	4.2	2.5	4.2	2.6
Sasa mat	132.5	79.8	118.8	73.8
Outflow D25 cm	0.2	0.2	0.1	0.1
Volcanic sand 1	16.5	9.9	24.8	15.4
Volcanic sand 2	11.3	6.8	12.8	7.9
Outflow D55 cm	0.0	0.0	0.1	0.1
Infiltration	1.3	0.8	0.2	0.1

(b) Water balance at the bare site

Precipitation	166.0	100.0	161.0	100.0
Outflow D25 cm	3.2	1.9	6.8	4.2
Volcanic sand 1	25.5	15.4	25.5	15.8
Volcanic sand 2	17.3	10.4	18.0	11.2
Outflow D55 cm	0.0	0.0	0.1	0.0
Infiltration	120.0	72.3	110.6	68.7

* Total amount (l) in the range of 1 m (W) × 5 m (L)

Table 2. Values used in the analysis for slope stability.

Items for analysis	Values
Unit weight of failed slope γ_t, kN/m^3	9.5
Tensile force of rhizomes per unit sectional length of slope surface layer t, kN/m	0.0, 5.8–20.3
Length of broken part around the failed slope L, m	73.1
Cohesion at the sliding surface c, kPa	2.35
Internal friction angle at the sliding surface ϕ °	27.5
Ratio of horizontal seismic acceleration to gravity, K_{h*}	0.335

* National Res. 1993.

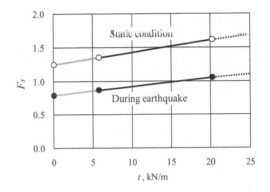

Figure 10. Relationship between safety factor F_s and tensile force of *Sasa* rhizomes per unit slope length t.

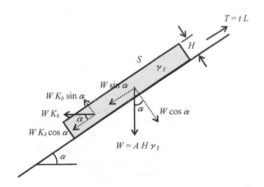

Figure 9. Analysis model for the mechanical stability of wild slope covered with *Sasa nipponica*.

As shown in Figure 10, the safety factor during the earthquake is determined to be about 1.0; thus, the result can be considered relevant, because five surface failures were found distributed on slopes in the range of 1 km. For the other two failures of them analyzed in the same manner (Takeda et. al., 2001), the factors were about 1.0. Next, assuming that we can apply these values to the estimation of the safety factor in the static condition, it was determined to be 1.26 on the slope without *Sasa*, and more than 1.37–1.65 using *t* values of 5.8–20.3 kN/m. Therefore, it was found that the increases in the factor with the tensile force indicate the reinforcing effect

of rhizomes on the surface layer of the slope, while the factor for the slope without *Sasa* is dependent on the strength parameters of soils at the sliding surface.

4 DISCUSSION

The results of the study indicate that the thermal, hydraulic, and mechanical stabilities of the slope increase due to the *Sasa* cover. The functions for evaluating each type of stability are different, but it is necessary to establish a common function when the stability is synthetically discussed. Moreover, it is more important for the function to sample easily. Thus, as the function, we used the leaf area index (LAI, m^2/m^2), which is the total leaf area in a unit area 1 m^2 to represent the activity of the above-ground part of *Sasa*.

For evaluating thermal stability, the culm density is converted into LAI by the dry weight and the dry density of leaves. From the results of 12 other investigation sites in the field and flat ground in eastern Hokkaido (Takeda et al. 2003), the mean value of

LAI, 0.60 m²/m², is modified to a September value of 0.69 m²/m² by seasonal correction, and that of culm density is obtained as 99.9 culms/m² (Agata et al. 1979). From these values, the maximum frost depth and heave amount in Equations (1) and (2) are given as a function of LAI, as follows:

$$F_d = -18.3 \text{ LAI} + 43.8 \tag{4}$$

$$H_a = -2.88 \text{ LAI} + 5.40 \tag{5}$$

When the hydraulic stability is discussed, the percentage of the water remaining in each layer after precipitation, the water retention rate w_r (%), is defined. The mean w_r of the values from both rainfall events, 79.8% and 73.8%, is 76.8% on the *Sasa* slope, while the mean for the bare slope is 15.6%. Further, the LAI on the *Sasa* and bare slopes were 4.35 and 0 m²/m², respectively. For about 30 mm of precipitation during a few days, when it is assumed that w_r linearly increases with LAI, the relationship is presented as the following equation:

$$w_r = 14.1 \text{ LAI} + 15.6 \tag{6}$$

Since w_r is generally considered to increase with the dry weight of the underground part and LAI increases with the weight (Takeda et al. 2003), the above equation is regarded as relevant.

To evaluate the mechanical stability of the slope covered with *Sasa*, the safety factor is presented as a function of the tensile force of rhizomes per unit sectional length of the slope surface layer. Through the field survey and the laboratory test, the force is given by the product of the sectional area of rhizomes and their tensile strength p (MPa). Since the relationship between LAI and the accumulated sectional area of rhizomes per unit sectional length of slope surface layer S_r is also obtained by the other surveys at 17 sites of the slopes covered with *Sasa*, given in rhizomes per unit sectional length of slope surface (see Eq. (7) and

Fig. 11), the tensile force t is presented by LAI and p as Equation (8),

$$S_r = 1.26 \text{ LAI} + 1.03 \tag{7}$$

$$t = (1.26 \, LAI + 1.03) \times p \tag{8}$$

where p is in the range 8.4–33.8 MPa. Moreover, as F_s' in Equation (3) was found to be a linear relationship as a function of t, the factor F_s is generalized by use of Equation (8) as a function of LAI as follows.

$$F_s = \frac{Lp \, (1.26 \text{ LAI} + 1.03)}{S H \gamma_t \, (\sin \alpha + K_h \cos \alpha)} \tag{9}$$

$$+ \frac{c + H \gamma_t \, (\cos \alpha - K_h \sin \alpha) \, \tan \phi}{H \gamma_t \, (\sin \alpha + K_h \cos \alpha)} \tag{10}$$

As these equations indicate, the stability is evaluated as a function of LAI representing the above-ground part of *Sasa*, without surveying the underground part.

5 CONCLUSION

The results of a field survey and laboratory test were analyzed and discussed. The following conclusions could be made:

1. *Thermal stability*: The annual maximum values of frost depth and frost heave amount of the surface layers on the slope are significantly suppressed by the *Sasa* cover, so that the thermal stability of slope increases.
2. *Hydraulic stability*: Once a *Sasa* mat has formed on the surface layer of the slope, the water retention increases remarkably, as does the hydraulic stability.
3. *Mechanical stability*: Using the surface failure on the *Sasa*-covered slope caused by the earthquake, the mechanical stability is evaluated by the safety factor in the static condition expressed as a function of leaf area index (LAI), which exceeds 1.37 owing to the reinforcing effect of rhizomes.

As *Sasa nipponica* grows not only in eastern Hokkaido, but also in the area along the Pacific Ocean from Honshu to the Kyushu Islands in Japan (Suzuki 1978), this evaluation can be applied to slope protection techniques in warmer areas, which has less snow.

ACKNOWLEDGEMENTS

We would like to thank Mr. A. Okamura (Ashimori Industry Co. Ltd.), Mr. T. Itoh (Mie Pref. Office) and Prof. S. Shibata (Kyoto Univ.) for their support and valuable advice.

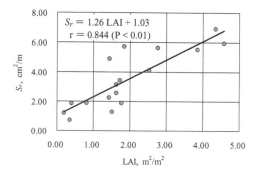

Figure 11. Relationship between the leaf area index (LAI) and the accumulated cross sectional area of rhizomes per unit sectional length of slope surface layer S_r.

REFERENCES

Karisumi, N. 1969. Structure of under ground part of Sasa species. *The reports of the Fuji Bamboo Garden* 14: 27–40. (in Japanese)

Suzuki, S. 1978. *Index to Japanese Bambusaceae*: 206–207. Tokyo: Gakken Co. Ltd. (in Japanese)

Committee on Method to Measure the Physical Properties of Soil (ed.), 1978. *Method to measure the physical properties of soil*: 287–289. Tokyo: Yokendo Co. Ltd. (in Japanese)

Agata, W. & Kamata, E. 1979. Ecological characteristics and dry matter production of some native grasses in Japan. *J. Jpn. Grassl. Sci.* 25 (2): 103–109. (in Japanese)

Wu, H.T., Mckinnell III, P.W. & Swanston, N.D. 1979. Strength of tree roots and landslide on Prince of Wales Island, Alaska. *Can. Geotech. J.* 16: 19–33.

O'Louglin, C. & Ziemer, R.R. 1982. The importance of root strength and deterioration rates upon edaphic stability in steepland forests. *Proceedings of an I.U.F.R.O. workshop, P.I.07-00 Ecology of subalpine Zones, Corvallis, OR.*: 70–78.

Hokkaido Forestry Research Institute Doto Station 1983. Distribution map of Sasa group in Hokkaido. (in Japanese)

Karisumi, N. 1987. Structure of the underground part of Sasa species and soil-straining powers. *Bamboo J.* 4: 167–174. (in Japanese)

Tsukamoto, Y. 1987. Evaluation of effect of tree roots on slope stability. *Bull. Exp. For. Tokyo Univ. Agric. Technol.* 23: 65–124. (in Japanese)

Ohta, T. 1990. A conceptual model of storm runoff on steep forested slopes. *J. Jpn. For. Soc.* 72(3): 201–207. (in Japanese)

Japan Society of Thermophysical Properties (ed.), 1990. *Thermophysical properties handbook:* 625 p. Tokyo: Yokendo Co. Ltd. (in Japanese)

National Res. Inst. For Earth Sci. Disaster Prevention Sci. and Technol. Agency,1993. The 1993 Kushiro-oki Earthquake. *Prompt Rep. on Strong-motion Accelerograms* 41: 5. (in Japanese)

Kondo, J. (Ed.) 1999. *Meteorology on water environment:* 194–198. Tokyo: Asakura Publishing Co. Ltd. (in Japanese)

Takeda, K. & T. Okamura, A. 1999. Thermal environment formed by dwarf bamboo in a cold region. *J. Jpn. Soc. Reveget. Technol.* 25(2): 91–101. (in Japanese)

Takeda, K. & Itoh, T. 2000. Development of effective slope protection technique in cold regions. In Thimus, F., J. (ed.), *Ground Freezing* 2000: 251–256. Rotterdam: Balkema.

Takeda, K., Yamada, T. Okamura, A. & Itoh, T. 2001. Effect of the rhizome of *Sasa nipponica* on the slope reinforcement for preventing surface failure. *J. Jpn. Soc. Reveget. Technol.* 26(3): 198–208. (in Japanese)

Takeda, K. & Itoh, T. 2003. Effect of the rhizomes of *Sasa nipponica* on the slope reinforcement for preventing surface failure II. *J. Jpn. Soc. Reveget. Technol.* 28(3): 431–437. (in Japanese)

Takeda, K., Suzuki, T., Itoh, Y. & Hamatsuka, T. 2008. Water retentivity effects of ground cover with *Sasa nipponica* on surface layer of slope. *Japanease Geotech. J.* 3(2): 121–132. (in Japanese)

Landslide dam failure and prediction of flood/debris flow hydrograph

H. Nakagawa, R. Awal, K. Kawaike, Y. Baba & H. Zhang
Kyoto University, Kyoto, Japan

ABSTRACT: The formation and failure of landslide dam are common geomorphic process in the mountainous area all over the world. Landslide dams may fail by erosion due to overtopping, abrupt collapse of the dam body or progressive failure. The prediction of outflow hydrograph and flood routing is essential to mitigate downstream hazards. This study is focused on prediction of landslide dam failure due to sliding and overtopping and prediction of outflow hydrograph. A stability model coupled with a transient seepage flow model is further integrated with dam surface erosion and flow model. The main advantage of an integrated model is that it can detect failure mode and flood/debris flow hydrograph due to either overtopping or sliding based on initial and boundary conditions. The proposed model is tested for three different experimental cases of landslide dam failure and reasonably reproduced the resulting flood/debris flow hydrograph.

1 INTRODUCTION

The formation and failure of landslide dam are common geomorphic process in the mountainous area all over the world. Landslide dams are also common in Japan because of widespread unstable slopes and narrow valleys exist in conjunction with frequent hydrologic, volcanic and seismic landslide triggering events. Historical documents and topography have revealed the formation of many landslide dams, some of which broke and caused major damage in Japan. The 2004 Chuetsu earthquake resulted in many landslide dams particularly in the Imo River basin. In 2005, typhoon 14 caused a large landslide dam near the Mimi-kawa river. Recently, eleven landslide dams were formed in Iwate and Miyagi prefecture due to landslides triggered by Iwate-Miyagi Inland Earthquake on 14th June 2008.

A landslide dam is made up of a heterogeneous mass of unconsolidated or poorly consolidated material. In general, shapes of landslide dams are triangular without flat crest. Sudden, rapid and uncontrolled release of water impounded in landslide dam has been responsible for some major disasters in mountainous region. To provide adequate safety measures in the event of such a catastrophic failure we have to predict resulting outflow hydrograph. It will serve as an upstream boundary condition for subsequent flood routing to predict inundation area and hazard in the downstream. Peak discharge produced by such events may be many times greater than the mean annual maximum instantaneous flood discharge.

Landslide dams may fail by the erosive destruction due to overtopping, abrupt collapse of the dam body or progressive failure (Takahashi 1991). Landslide dams most commonly fail by overtopping, followed by breaching from erosion by the overtopping water. Although abrupt collapse of the dam body is not common, the peak discharge produced by such failure is very high compared with failure due to overtopping. If the infiltration rate of the dam body is high and strength of the dam body is medium, instantaneous slip failure may occur. However, in-depth knowledge of the mechanism of the dam failures and measured data are still lacking. A simulation model of the dam failure processes by different failure modes will therefore be useful to predict flood/debris flow hydrograph.

Basically, there are two methods to predict probable peak discharge from potential failure of landslide dam (Walder & O'Connor 1997). One method relies on regression equations that relate observed peak discharge of landslide dam failure to some measure of impounded water volume: depth, volume, or some combination thereof (Costa 1985, Costa & Schuster 1988, Walder & O'Connor 1997) and regression equations that relate experimental peak discharge to some measure of impounded water volume: depth, torrent bed gradient and inflow discharge (Tabata et al. 2001). The other method employs computer implementation of a physically based mathematical model. Several researchers have developed physically based model such as Takahashi & Kuang (1988), Fread (1991), Takahashi & Nakagawa (1994), Mizuyama (2006) and Satofuka et al. (2007). Although, landslide dam failure is frequently studied as an earthen dam failure, very few models are developed for landslide dam failure that can treat the flow as both sediment flow and debris flow. If the concentration of sediment is above 10%, non-Newtonian viscous flow has to be taken into account. During surface erosion of landslide dam, sediment concentration increased more than 10%, so the model to predict the flood/debris flow hydrograph due to landslide dam failure should be capable to treat all types of flow based on sediment concentration.

The main objective of this study is to predict flood/debris flow hydrograph due to sudden sliding and overtopping failure of landslide dam through flume experiments and numerical simulations. A stability model coupled with a seepage flow model was used to determine moisture movement in the dam body, time to failure and geometry of failure surface for transient slope stability analysis (Awal et al. 2007). The model is further integrated with model of dam surface erosion and flow to predict the outflow hydrograph resulted from failure of landslide dam by overtopping and sudden sliding. The main advantage of an integrated model is that it can detect failure mode due to either overtopping or sliding based on initial and boundary conditions.

2 NUMERICAL MODEL

The model of the landslide dam failure to predict flood/debris flow hydrograph consists of three models. The seepage flow model calculates pore water pressure and moisture content inside the dam body. The model of slope stability calculates the factor of safety and the geometry of critical slip surface according to pore water pressure, moisture movement in the dam body and water level in the upstream reservoir. The model of dam surface erosion and flow calculates dam surface erosion due to overflowing water. General outline of proposed integrated model is shown in Figure 1. A brief description of each model is given below.

2.1 Model of seepage flow

The seepage flow in the dam body is caused by the blocked water stage behind the dam. The transient flow in the dam body after formation of landslide dam can be analyzed by Richards' equation. To evaluate the change in pore water pressure in variably saturated soil, pressure based modified Richards' equation is used (Freeze 1971, Awal et al. 2007).

$$\frac{\partial}{\partial x}\left(K_x(h)\frac{\partial h}{\partial x}\right) + \frac{\partial}{\partial z}\left(K_z(h)\left(\frac{\partial h}{\partial z}+1\right)\right) = C\frac{\partial h}{\partial t} \quad (1)$$

where h = water pressure head; $K_x(h)$ and $K_z(h)$ = hydraulic conductivity in x and z direction; C = specific moisture capacity ($\partial\theta/\partial h$); θ = soil volumetric water content; t = time; x = horizontal spatial coordinate; and z = vertical spatial coordinate taken as positive upwards. Equation 1 represents flow in both the unsaturated domain as well as in the saturated domain. For saturated domain, $K_x(h) = K_z(h) = K_s$; $\theta = \theta_s$; and $C = 0$, where K_s = saturated hydraulic conductivity; and θ_s = saturated water content. Line-successive over-relaxation (LSOR) is often

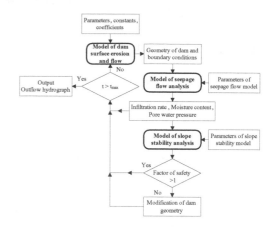

Figure 1. General flow chart of an integrated model to predict flood/debris flow hydrograph due to landslide dam failure.

a very effective method of treating cross-sectional problem grids. LSOR scheme is used in this study for the numerical solution of Richards' equation (Freeze 1971).

In order to solve Richards' equation, the constitutive equations, which relate the pressure head to the moisture content and the relative hydraulic conductivity, are required. In this study, constitutive relationships proposed by van Genuchten (1980) are used for establishing relationship of K–h and θ–h, with $m = 1-(1/\eta)$.

$$S_e = \frac{\theta-\theta_r}{\theta_s-\theta_r} = \begin{cases} \frac{1}{(1+|\alpha h|^\eta)^m} & \text{for} \quad h \geq 0 \\ & \text{for} \quad h < 0 \end{cases} \quad (2)$$

$$K = \begin{cases} K_s S_e^{0.5}\left[1-\left(1-S_e^{1/m}\right)^m\right]^2 & \text{for} \quad h < 0 \\ & \text{for} \quad h \geq 0 \end{cases} \quad (3)$$

where, α and η = van Genuchten parameters; S_e = effective saturation; θ_s and θ_r = saturated and residual moisture content respectively.

2.2 Model of slope stability

The evaluation of transient slope stability of landslide dam by the limit equilibrium method involves calculating the factor of safety and searching for the critical slip surface that has the lowest factor of safety. Many attempts have been made to locate the position of critical slip surface by using general noncircular slip surface theory coupled with different non-linear programming methods. The numerical procedure behind the identification of critical noncircular slip surface with the minimum factor of safety based on dynamic programming and the Janbu's simplified method is mainly based on the research of Yamagami & Ueta (1986). The algorithm combines the Janbu's simplified method with dynamic programming on the basis of Baker's successful procedure (1980).

Janbu's simplified method can be used to calculate the factor of safety for slip surfaces of any shape. The sliding mass is divided into vertical slices and the static equilibrium conditions of each slice are considered as sum of the vertical forces equal to zero and sum of the forces parallel to failure surface equal to zero. For the soil mass as a whole, sum of the vertical forces $\sum F_y = 0$ and sum of the horizontal forces $\sum F_x = 0$ are considered as equilibrium condition.

Based on the above considerations the factor of safety, F_s for Janbu's simplified method is defined as:

$$F_s = \frac{1}{\sum_{i=1}^{n} W_i \tan \alpha_i}$$

$$\times \sum_{i=1}^{n} \left\{ \frac{c_i l_i \cos \alpha_i + (W_i - u_i l_i \cos \alpha_i) \tan \phi}{\cos^2 \alpha_i \left(1 + \frac{1}{F_s} \tan \alpha_i \tan \phi\right)} \right\} \quad (4)$$

where W_i = weight of each slice including surface water; l_i = length of the base of each slice; u_i = average pore water pressure on the base of the slice; α_i = inclination of the base to the horizontal; n = total number of slices; and c_i and ϕ = Mohr − Coulomb strength parameters.

The details of transient slope stability analysis of landslide dam by using dynamic programming and Janbu's simplified method can be found in Awal et al. (2007).

2.3 Model of dam surface erosion and flow

The mathematical model developed by (Takahashi & Nakagawa 1994) was used for the modeling of surface erosion and flow. The model was capable to analyse the whole phenomena from the beginning of overtopping to the complete failure of the dam as well as to predict flood/debris flow hydrograph in the downstream. The infiltration in the dam body was not considered in the model; therefore, time to overflow after formation of landslide dam can not be predicted from previous model. In this study, infiltration in the dam body is also incorporated.

The model is two-dimensional and it can also collapse to treat one-dimensional for overtopping from full channel width. In case of sudden sliding failure, simplified assumption is made for initial transformation of the dam body after the slip failure. Based on many experiments the slipped mass is assumed to stop at the sliding surface where slope is less than angle of repose and the shape of the slipped

mass is assumed as trapezium. There is some time lag between slip failure and movement of the slipped soil mass but in the model, the time necessary for such a deformation is assumed as nil. The erosion process by the overtopping water is analysed for the modified dam shape.

The erosive action of the overtopping flow removes material from the top part of the dam. The overtopped flow grows to debris flow by adding the eroded dam material to it, if the slope and length of dam body satisfy the critical condition for the occurrence of a debris flow.

The main governing equations are briefly discussed here. The depth-wise averaged two-dimensional momentum conservation equation for the x-wise (down valley) direction and for the y-wise (lateral) direction are

$$\frac{\partial M}{\partial t} + \beta' \frac{\partial (uM)}{\partial x} + \beta' \frac{\partial (vM)}{\partial y} = gh \sin \theta_{bxo}$$

$$- gh \cos \theta_{bxo} \frac{\partial (h + z_b)}{\partial x} - \frac{\tau_{bx}}{\rho_T} \quad (5)$$

$$\frac{\partial N}{\partial t} + \beta' \frac{\partial (uN)}{\partial x} + \beta' \frac{\partial (vN)}{\partial y} = gh \sin \theta_{byo}$$

$$- gh \cos \theta_{byo} \frac{\partial (h + z_b)}{\partial y} - \frac{\tau_{by}}{\rho_T} \quad (6)$$

The continuity of the total volume is

$$\frac{\partial h}{\partial t} + \frac{\partial M}{\partial x} + \frac{\partial N}{\partial y} = i \{c_* + (1 - c_*) s_b\} - q \quad (7)$$

The continuity equation of the particle fraction is

$$\frac{\partial (ch)}{\partial t} + \frac{\partial (cM)}{\partial x} + \frac{\partial (cN)}{\partial y} = ic_* \quad (8)$$

The equation for the change of bed surface elevation is

$$\frac{\partial z_b}{\partial t} + i = i_{sml} + i_{smr} \quad (9)$$

where $M = uh$ and $N = vh$ are the x and y components of flow flux; u and v are the x and y components of mean velocity; h = flow depth; z_b = elevation; ρ_T = apparent density of the flow; $\rho_T = c(\sigma \rho) + \rho$; c = volume concentration of the solids fraction in the flow; σ = density of the solids, ρ = density of water; β' = momentum correction coefficient; τ_{bx} and τ_{by} = x and y components of resistance to flow; i = erosion or deposition velocity; c_* = solids fraction in the bed; s_b = degree of saturation in the bed (applicable only in cases of erosion, when deposition takes place substitute $s_b = 1$); i_{sml} and i_{smr} = mean recessing velocity of the left and right hand side banks of the incised channel respectively; t = time; g = acceleration due to gravity; and q = infiltration rate.

Shear stress, erosion or deposition velocity and channel enlargement for overtopping from partial channel width were evaluated using the model presented in Takahashi & Nakagawa (1994).

3 EXPERIMENTAL STUDY

A rectangular flume of length 5 m, width 20 cm and depth 21 cm was used. The slope of the flume was set at 17 degree. Mixed silica sand of mean diameter 1 mm was used to prepare triangular dam in the flume. The height of the dam was 20 cm and the longitudinal base length was 84 cm. The schematic diagram of the flume is shown in Figure 2. van Genuchten parameters (including θ_r) of sediment mixture were estimated by non-linear regression analysis of soil moisture retention data obtained by pF meter experiment. Water content reflectometers (WCRs) were used to measure the temporal variation of moisture content during seepage process. The arrangements of WCRs are shown in Figure 2. To measure the movement of dam slope during sliding, red colored sediment strip was placed in the dam body at the face of flume wall. A digital video camera was placed on the side of the flume to capture the shape of slip surface due to sudden sliding. The shape of the dam body at different time step due to surface erosion after overtopping and the shape of slip surface during sliding were measured by analyses of video taken from the flume side. Load cell and servo type water gauge were used to measure sediment and total flow in the downstream end of the flume.

4 RESULTS AND DISCUSSIONS

Numerical simulations and flume experiments were performed to investigate the mechanism of landslide

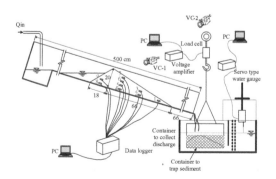

Figure 2. Experimental setup.

dam failure and resulting hydrograph due to overtopping and sudden sliding. Experimental conditions and parameters used for simulations in different cases are shown in Table 1. K and δ_d are the parameters of erosion and deposition velocity respectively. The other parameters used in the calculations are $\theta_s = 0.287$; $\theta_r = 0.045$; $\alpha = 5.5$ m^{-1}; η 3.2; $c_* = 0.655$; $\sigma t = 2.65$ g/cm^3; $d = 1$ mm; and $\Delta t = 0.002$ sec. Following three cases are considered:

4.1 Overtopping (from full channel width)

Steady discharge of 550 cm^3/sec was supplied from the upstream part of the flume. The model started simulation after the start of inflow. Overtopping occurred after the filling of the reservoir. Overtopped water proceeds downstream eroding the crest as well as the downstream slope of the dam body.

The simulated and experimental outflow hydrograph at 66 cm downstream of the dam are represented in Figure 3. Transformation of the dam body with time is shown in Figure 4. The shape of the simulated surface of the dam body at each time steps are similar to observed. The simulated outflow hydrograph is not matching perfectly due to difference in time to overspill the reservoir and rate of dam surface erosion between simulation and experiment.

4.2 Overtopping and channel breach (from partial channel width)

Notch of the width 5 cm and depth 0.5 cm was incised at the crest and downstream face of the dam in the left side of the dam body so that the erosion of the surface of dam body can be observed from left side of the flume. Steady discharge of 49.0 cm^3/sec was supplied from the upstream part of the flume, after the filling of the reservoir, it overflowed from the notch at the crest of the dam. The overtopping flow incised a channel on the slope of the dam and that channel increased its cross-sectional area with time caused by the erosion of released water. The simulated and experimental outflow hydrograph are represented in Figure 5. Figure 6 shows the comparison of the simulated and experimental shapes of dam surface at different time steps. In both experiment and simulation the channel incised almost vertically that may be due to rapid drawdown of reservoir and small inflow rate. The overflowing

Table 1. Experimental conditions and parameters for simulation.

Case	Q (cm^3/sec)	Water content	Permeability K_s (m/sec)	K	δ_d
I	550	50%	0.00018	0.11	0.005
II	49	50%	0.00018	0.11	0.005
III	30.5	20%	0.00030	0.11	0.005

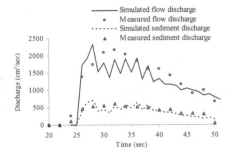

Figure 3.   Outflow hydrograph.

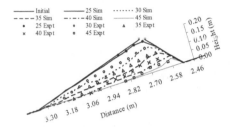

Figure 4.   Comparison of dam surface erosion.

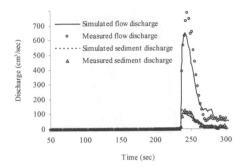

Figure 5.   Outflow hydrograph.

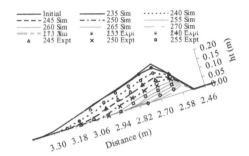

Figure 6.   Comparison of dam surface erosion.

water depth was very small so the shear stress due to flowing water in the side wall of incised channel was also small and above the water level there was some apparent cohesion added by water content and adhesion so the side wall is very steep. Armouring effect is also negligible due to small particle size of the dam body.

4.3   Sudden sliding

Steady discharge of 30.5 cm³/sec was supplied from the upstream part of the flume. A gradual rise of water level in the reservoir caused water to infiltrate into the dam body and it increased mobilized shear stress and dam was failed by sudden collapse when it became larger than resisting shear stress. The sudden sliding of the dam body was observed at 447 sec in the experiment whereas in the simulation it was observed at 410 sec. The simulated time was slightly earlier than the experimentally observed time that may be due to the assumption of immobile air phase in unsaturated flow and variation of saturated hydraulic conductivity. Moreover, the effects of interslice forces are ignored in Janbu's simplified method. Increase in shear strength due to the negative pore-water pressures are not considered in the formulation of factor of safety. Figure 7 shows the comparison of simulated and experimental slip surface. For the same experimental conditions, moisture content in the dam body was measured by using WCRs. Figure 8

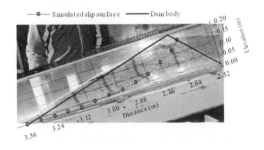

Figure 7.   Comparison of simulated and experimental slip surface.

shows the simulated and experimental results of moisture profile at WCR-4, WCR-5, WCR-6, WCR-8, and WCR-9 which are in good agreement. The geometry of predicted critical slip surface was also similar to that observed in the experiment.

Figure 9 shows the simulated and experimental results of outflow hydrograph. There is some time lag between failure of dam and movement of the slipped soil mass but in the model, the time necessary for such a deformation is assumed as nil so the simulated peak is earlier than experimental peak. Peak discharge depends on the shape of the dam body assumed after sliding and parameters of erosion and deposition velocity.

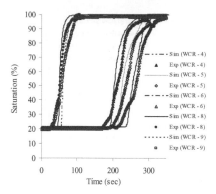

Figure 8. Simulated and experimental results of water content profile for different WCRs.

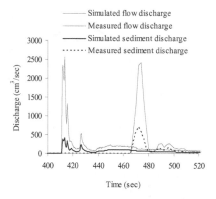

Figure 9. Outflow hydrograph.

The movement of moisture in the dam body measured by using WCRs, critical slip surface observed in the experiment and predicted outflow hydrograph are close to the result of numerical simulation.

5 CONCLUSIONS

An integrated model is developed by combining slope stability model, transient seepage flow model and dam surface erosion and flow model for the simulation of outflow hydrograph due to landslide dam failure by overtopping and sliding. The proposed model is tested for three different experimental cases of landslide dam failure due to overtopping and sliding and reasonably reproduced the resulting hydrograph. The numerical simulation and experimental results of movement of moisture in the dam body, predicted critical slip surface and time to failure of the dam body are also in good agreement. The predicted hydrograph can be used for flood disaster mitigation in the downstream. The model can be further extended to three-dimensions for the better representation of failure process of landslide dam.

REFERENCES

Awal, R., Nakagawa, H., Baba, Y. & Sharma, R.H. 2007. Numerical and experimental study on landslide dam failure by sliding. *Annual J. of Hydraulic Engineering,* JSCE 51: 7–12.

Baker, R. 1980. Determination of the critical slip surface in slope stability computations. *International Journal for Numerical and Analytical Methods in Geomechanic* 4: 333–359.

Costa, J.E. 1985. Floods from dam failures, U.S. Geological Survey, Open-File Rep. No. 85-560, Denver, 54.

Costa, J.E. & Schuster, R.L. 1988. The formation and failure of natural dams, Geological Society of America Bulletin 100: 1054–106.

Fread, D.L. 1991. BREACH: an erosion model for earthen dam failures, U.S. National Weather Service, Office of Hydrology, Silver Spring, Maryland

Freeze, R.A. 1971. Influence of the unsaturated flow domain on seepage through earth dams. *Water Resources Research* 7(4): 929–941.

Mizuyama, T. 2006. Countermeasures to cope with landslide dams—prediction of the outburst discharge, *Proc. Of 6th Japan-Taiwan Join Seminar on Natural Disaster Mitigation.*

Satofuka, Y., Yoshino, K., Mizuyama, T., Ogawa, K., Uchikawa, T. & Mori, T. 2007. Prediction of floods caused by landslide dam collapse. *Annual J. of Hydraulic Engineering,* JSCE 51: 901–906 (in Japanese).

Tabata, S., Ikeshima, T., Inoue, K. & Mizuyama, T. 2001. Study on prediction of peak discharge in floods caused by landslide dam failure. *Jour. of JSECE,* 54(4): 73–76 (in Japanese).

Takahashi T. & Kuang, S.F. 1988. Hydrograph prediction of debris flow due to failure of landslide dam, *Annuals, Disas. Prev. Res. Inst.,* Kyoto Univ. 31(B-2): 601–615.

Takahashi T. & Nakagawa, H. 1994. Flood/debris flow hydrograph due to collapse of a natural dam by overtopping. *Journal of Hydroscience and Hydraulic Engineering,* JSCE, 12(2): 41–49.

Takahashi T. 1991. *Debris flow, Monograph Series of IAHR,* Balkema.

van Genuchten, M. Th. 1980. A closed-form equation for predicting the hydraulic conductivity of unsaturated soils. *Soil Sci. Soc. Am. J,* 44: 892–898.

Walder, J.S. & O'Connor, J.E. 1997. Methods for predicting peak discharge of floods caused by failure of natural and constructed earthen dams. *Water Resources Research* 33(10): 2337–2348.

Yamagami, T. & Ueta, Y. 1986. Noncircular slip surface analysis of the stability of slopes: An application of dynamic programming to the Janbu method. *Journal of Japan Landslide Society* 22(4): 8–16.

Prediction and Simulation Methods for Geohazard Mitigation – Oka, Murakami & Kimoto (eds)
© 2009 Taylor & Francis Group, London, ISBN 978-0-415-80482-0

Some geohazards associated with the 8.0 Wenchuan earthquake on May 12, 2008

K.T. Chau

The Hong Kong Polytechnic University, Hong Kong, China

ABSTRACT: The May 12, 2008 Great Wenchuan Earthquake has resulted in more than 69,227 deaths, 17,923 people are listed as missing, and 374,643 were injured (up to September 22, 2008). Along the Central Longmenshan fault, which generated this great earthquake, the earthquake intensity (also called macrointensity) at the towns of Yingxiu and Beichuan County (Qushan Town) is XI on Chinese Intensity Scale (similar to the MMI scale). The author of this paper made two separate excursions to the affected areas in Sichuan in May and in July and August. In this paper, some earthquake-induced geohazards will be summarized, including the huge Niu-juan Gully rock avalanche at the epicenter, the Wangjiayan landslide and Jingjiashan rock avalanche at Qushan Town, the surface ruptures at Bailu, and Xiaoyudong. Discussions will be made on the geohazard-associated fatality and damages. The paper concludes with lessons learnt from these earthquake-induced-geohazards, including proper site-assessment before town planning.

1 INTRODUCTION

1.1 *May 12, 2008 Wenchuan Earthquake*

The May 12, 2008 Great Wenchuan Earthquake has resulted in more than 69,227 deaths, 17,923 missing, 374,643 injured (up to September 22, 2008), and at least 4.8 million people became homeless. Unofficial estimation of total economic loss may be as high as US$75 billion dollars. A total of 391 dams were damaged by the earthquake, but luckily no failure of dam was reported. It occurred at about 14:28:01 pm local time and China Earthquake Administration estimated the surface magnitude is 8.0 whereas the moment magnitude is 7.9 (assigned by USGS). However, some Chinese seismologists estimated the moment magnitude may be as high as 8.3 based on seismic moment calibrated from seismic stations all over the world. The ground shaking lasted for about 3 minutes, depending on the distance from the fault zone. The peak ground acceleration (PGA) recorded at various seismic stations had been exceeding or close to 1g (i.e. 9.81 m/s^2). It was believed that at the epicentral areas, the PGA is much larger than 1g. Aftershocks continued to shake the area months after the earthquake, eight of them exceeding a magnitude of 6.0. The largest one was magnitude 6.4 on May 25, 2009 when the author had just finished his first excursion to Sichuan. The maximum slip on the fault surface was estimated from 9 m to 15 m with an average of exceeding 2 m, that seems agree with the slip deformation observed at surface ruptures (maximum vertical and horizontal displacement are 6.5 m and 4.9 m respectively) (Xu et al. 2008).

It was generally believed that rupture fault is the Longmenshan Central Fault (there are also upper and lower faults in Longmenshan). Historically, the largest historical earthquake Longmenshan Fault is the 6.5 Wenchuan earthquake occurred in 1657. The epicentral intensity for this earthquake in two stripe areas running north-east direction covering 2419 km^2 are XI on Chinese Intensity Scale (similar to the MMI scale), including Yingxiu and Beichuan County (Qushan Town) (see Fig. 1). The area of intensity VI or above covered 440,000 km^2. The epicenter is very close to Yingxiu Town within the Wenchuan County, and therefore, it was officially called Wenchuan Earthquake.

Tectonically, Longmenshan located at the most eastern part of the Qinghai-Tibet plateau, which is resulted from the push of the Indian plate into the Eurasian plate (about 4 cm/year). The plateau is pushed

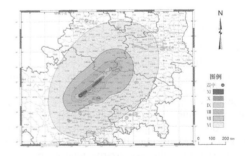

Figure 1. The official isoseismal map of the May 12 2008 Wenchuan earthquake. Two stripe-shaped zones of intensity XI were shown.

northward at Himalayas but pushed eastward in the Longmenshan area. Longmenshan Fault is a reverse fault or thrust fault that makes Longmenshan mountain ranges rising suddenly from the Sichuan Basin. From GPS measurements, the relative slip across Longmenshan is only 1–3 mm/year, which is substantially smaller than that from the Indian plate pushing. The largest historical earthquake occurred at Longmenshan fault is the 6.5 Wenchuan Earthquake occurred on April 21, 1657. In terms of seismic hazard level, both Yingxiu and Beichuan are within intensity VII zone (a probability of exceedance of once every 475 years). Clearly, such seismic hazard level is highly underestimated, apparently because of the lack of big historical earthquakes along the Longmenshan Fault. Even structures built according to the most updated seismic code of China, they should not survive the Wenchuan earthquake without severe damages in the region of intensity of XI. In fact, much effort of earthquake monitoring and studies has been paid to the Xianshuihe fault zone which is on the northwest of the Longmenshan Fault, instead of to Longmenshan Fault. In this sense, the May 12 Wenchaun earthquake is a big surprise to most Chinese seismologists.

Back analysis of the focal mechanism also suggested a strike-slip component in addition to the dip slip component. The strong ground motion records of the Wolong station (home of giant pandas) suggested that there are two sequences of fault slip event at the fault plane. At this moment, the details of these two events have not been worked out completely. The earthquake was felt in nearly the whole Mainland China, except Heilongjiang, Jinlin, and Xinjiang. It was felt in Macau, Hong Kong, Taiwan, Vietnam, Thailand, Nepal, Mongolia, Bangladesh, India, Pakistan, and Russia. For the first time since the establishment of the People's Republic of China, a national mourning was enforced for a disaster. A three-day period of national mourning for the quake victims starting from May 19, 2008, and the whole nation stopped for 3 minutes at 14:28 on May 19, a week after the disaster. The author of this paper witnessed that historical moment at Taiyuan Airport.

1.2 Associated geohazards

The author have spent more than two weeks in the severely affected areas, including Pingtong, Leigu, Qushan (Beichuan County), Anchang, An County, Fuxin, Jinhua, Hanwang, Hongbai, Yinghua, Luoshui, Mianzhu City, Shifang City, Deyang City, Bailu, Hongyan, Guixi, Xiaoyudong, Tongji, Longmenshan, Yong'an, Zipingpu, Juyuan, Yingxiu, Pengzhou, Dujiangyan, Mianyang, Tongkou, and Danjingshan. It was observed the geohazards inudced by earthquake are sometimes more deadly than the ground shaking itself. They include landslides, rockfall, rock avalanche, liquefaction, quake lake, and debris flows.

This paper will summarize some of the earthquake-induced-geohazards observed in these places, including the rock avalanche at the epicenter near Yingxiu, the landslides and rockfall at Qushan, the surface ruptures at Bailu and Xiaoyudong.

2 ROCK AVALANCHE AND LANDSLIDES

2.1 Rock avalanche at epicenter near Yingxiu

About 5 kilometers southwest of Yingxiu is the geographical epicenter, and the most affected gullies in this area is the Niu-juan Gully. The entrance of this gully is near the famous collapsed Baihua Bridge (about 2.3 km from Yingxiu). In fact, we can see the gully entrance from the town of Yingxiu as shown in Fig. 2.

The gully was filled with fragments or stones of granite (whereas the rock in the gully is sandstone). Apparently, a huge rock avalanche was caused by the sudden collapse of a granite mountain on the upstream of a major branch of the Niu-juan Gully. The highest point of this mountain is 2610 m (see Figs. 3 & 4). It was estimated that more than 3 millions cubic meters of rocks and stones collapsed and smashed into the Niu-juan Gully at a junction point between this branch and the primary gully of Niu-juan Gully. The stone deposits are as thick as 40 m, and about 20 people were buried under the stone deposits. Along the gully, there are signs of glacier-type of erosion. The trees on two sides were swept away along the gully, and the super-elevation of the stone deposits at the junction point and curving point is as high as 60 m. According to survivors living in the small villages in the gully (actually retold from the major of the town of Yingxiu): "crushed stones flied from the ground into the sky after a huge sound of explosion, trees were upright in one moment and down for another moment, and the ground opened up and closed again within split of seconds". These descriptions from survivors

Figure 2. The entrance of Niu-juan Gully (arrow) viewed from the temporary primary school of Yingxiu.

Figure 3. The plan view of Niu-juan Gully rock avalanche.

Figure 4. Terrain plot of the rock avalanche and the location of quake lake formed by this avalanche.

Figure 5. The entrance point of Niu-juan Gully. Boulders of granites were found at the front end.

Figure 6. The junction of the avalanche entry point. The camera are facing uphill and southwest while the avalanche is jumping over the gully from the right side of the photo (from northwest).

Figure 7. The camera is facing northwest, and viewing the source of the avalanche. The erosion of both sides of the waterfall is very severe, but trees above erosion line are intact.

are very difficult for normal people to comprehend. I attempt to offer my speculation and explanation here. The high speed avalanche contained mainly flying stones with chopped trees. As the trees flying by with stones, some may appear upright at one moment, and others may appear down at the other moment. The rock avalanche may come into surges that the width of the gully was appearing changing size, and it might create an image of ground opening up and closing down. The area must be clouded by dust and mud when the mountain collapsed, and the visibility must be extremely poor. Therefore, it is not a surprise that survivor did not see the mountain collapse, and the scenes perceived by survivors are indeed possible as a result of a huge rock avalanche. In fact, earthquake-induced rock avalanche is not uncommon in other part of the world (Jibson et al. 2006).

Some photographs taken at various parts of the gully are shown in Figs. 5–7. In particular, Fig. 5 showed the entrance to the Niu-juan Gully, and stone deposits can be seen in the front end. Fig. 6 showed the location of the main impact junction where the rock avalanche came from. A quake lake was formed further uphill in the direction that the camera was pointing (southwest). The stone deposit resting on the other side of the slope is estimated up to 60 m. When we went further up and turned to the northwest side, a waterfall was seen (Fig. 7). The saddle is about 50 m above the gully while the erosion on both sides of the waterfall may be up to 100 m high.

Photo in Fig. 8 was taken from the top of the slope adjacent to the quake lake. The projectile of the rock avalanche impact can be seen clearly, indicating by the arrow. The deposit on the other side is up to 60 m high. Figure 9 shows the upper valley above the secondary gully above the waterfall. The photograph was taken at about 1160 m. The distance from the entrance to the impact junction in front of the quake lake is about 1 km with a vertical rise of about 120 m, and thus an average gradient for this part of the gully is about 0.12 or 6.8 degrees (after the gully is filled up with stones and debris). The angle of projectile from the top of the eroded slopes (right of Fig. 6) to the top of the deposit on the other slope (left of Fig. 6) is about 24 degrees (see the projectile in Fig. 8 as well). The rock fragments and boulders appeared to be air-borne (at least at this section of the avalanche). For the upper section about the waterfall, the average gradient from the top of the scar of the big collapse at the middle of the gully

Figure 8.   Turning point of the rock avalanche from a nearby mountain, and arrow showing the direction of avalanche. The camera is facing northeast (downhill), and the quake lake is on the lower right corner.

Figure 9.   The source zone is in the background, which is about 2610 m high. Clearly, there is a big collapse of mountain range in the middle. Surprisingly, there is not much chopping of trees in the middle valley.

shown in Fig. 9 to the point where the photo was taken is about 26 degree. The vertical drop is about 930 m with a horizontal distance of about 1.9 km. The total debris is estimated as 3.2 million cubic meters, calculating from the size of the debris in filling up the lower gully (160 m wide, 20 m deep and 1 km long). The total vertical drop from the scar area to the entrance of gully is about 1280 m and a horizontal travel distance of about 3.4 km. The rock avalanche velocity must be very high for such a travel distance.

For open channel flow, there is a standard formula to estimate the average flow velocity from the superelevation on the flow rise up on a curved channel (Chow, 1959):

$$V_{ave} = \left(\frac{\Delta h}{w} r_c g \right)^{1/2} \tag{1}$$

where $\Delta h, w, r_c$, and g denote the height of superelevation at a bend, the flow width, the radius of curvature at the bend, and the gravitational constant. Using the estimated data of 60 m superelevation, 100 m wide gully, 380 m of radius of curvature, and 9.81 m/s^2, we obtain an average flow velocity of 47 m/s or 169 km/hr. The maximum flow velocity may then be up to 75 m/s or 271 km/hr (using a factor of 1.6 between average and maximum). The total duration of the rock avalanche may last for about 45 seconds.

These values are comparable to other famous rock avalanches observed in other parts of the world. For example, the 1881 Elm rock avalanche in Switzerland (11 million m^3) traveled at a maximum speed of 70 m/s (Hsu, 1975), the 1903 Frank rock avalanche in Canada (3 million m^3) traveled at a maximum speed of 40 m/s (Sosio et al, 2008), and the earthquake-triggered 1959 Madison Canyon rock avalanche in USA (30 million m^3) traveled at a maximum speed of 50 m/s (Hadley, 1964). Therefore, although our estimation given here is extremely rough (in the sense that all parameters of depth of deposit, height of superelevation, and radius of curvature were estimated approximately), the results appear to be in the right order of magnitude.

However, the failure and travel mechanisms of such fast moving rock avalanche are still an unresolved problems despite numerous models have been proposed. For example, it has been suggested that immense frictional heating has produced steam trapped in the sliding plane that allow the rock fragments to cruise as air-borne. This idea should be seriously challenged, as the tumbling rock mountain is highly permeable, and steam (even generated) cannot be trapped to support the flowing rock fragments. The idea of dispersive pressure by Hsu (1975) may be feasible, but yet no experiments have been done to verify the idea and there is no field evidence to support it either. Such endeavor is clearly out of the scope of the present paper.

2.2 Landslide in old Qushan Town

Apart from Yingxiu, the other intensity XI region is around Beichuan County or the Qushan Town. In this section, we will summarize the major landslide at Qushan. Figure 10 showed the satellite photos of the Qushan Town as well as the terrain plot showing the locations of the Wangjiayan landslide and Jingjiashan rock avalanche.

The major landslide destroyed the old twon of Qushan is called Wangjiayan landslide. The geolgical formation for the slope is mainly slate, pyllite and metamorphized sandstones. The original slope is about 37 degrees. The landslide started from 1020 m down to 640 m. The runout distance is about 500 m long, involving a total of 7 millions of earth materials. This slide alone caused about 100 house collapse and 1600 death. This probably responsible to half of the death toll in Qushan Town. The origin Beichuan County is located in a place further north and was formed 566 BC in Zhou Dynasty, which was moved to the current Qushan Town in 1952. The selection of this site is clearly a bad choice. The geohazard is clearly not assessed properly when the town was proposed. Wangjiayan location is actually a reoccurrence site of landslide. In a sense, the geohazard associated with the earthquake is more destructive than the collapses of individual buildings.

2.3 Rock avalanche in new Qushan Town

Another major geohazard at Qushan is the major rock avalanche occurred at Jingjiashan. This rock avalanche dropped from 1020 m down to about 680 m, and total length of collapsed slope is about 570 m. A total of 5 millions cubic meters of rocks involved in the event. The Beichuan School Secondary School (New Town Campus) was completedly covered by huge boulders. Only the flagstaff and entrance stairway still can be seen (Fig. 13), and the rest of the campus was under boulders. The bodies of 900 students and teachers were never recovered from the deposit of huge boulders. The average slope anlge before the avalanche is about 30 degrees. The geological formation for the Jingjiashan is mainly limestone formation.

Again, this rock avalanche is probably not the first event at this location. There is strong evidence that similar event has happened in the past. Fig. 15 shows a vertical cliff with vegetations just above the top scar

Figure 12.   Another photo showing the portion of old town covered by the Wangjiayan landslide.

Figure 10.   Qushan Town and the Wangjiayan landslide and Jianjiashan rock avalanche.

Figure 11.   The Wangjiayan landslide in old Qushan Town.

Figure 13.   Jingjiashan rock avalanche, covering the Beichuan Secondary School.

Figure 14. Jingjiashan rock avalanche, viewed from the top of the mountain.

Old scar of previous historical slide

Figure 15. Show the top cliff just above the scar of the Jingjiashan rock avalanche.

Figure 16. Boulder field found in a construction site just next to the Jingjiashan rock avalanche.

Figure 17. Surface rupture passing through two school buildings at Bailu.

of this event. Evidently, this cliff was created as previous historical slip (probably before the formation of town in 1952). From the size of the trees on the slope, the last major event may happen 60 to 100 years ago. Fig. 16 showed a boulder field in a construction site in the town just north of the Jingjiashan collapse. One of this boulder is huge and with heavy vegetations on it, and according to local people this boulder has been there since the development of the town. We speculate that this is the remains of a former rock fall event happened at Jingjiashan. The death of these 900 students and teachers could have been avoided by proper site selection. Clearly, buildings standing on a steep cliff enjoy a spectacular view both in the front and behind. But, rock avalanche like Jingjiashan event remind us there is a high price to pay for such view. And, similar event will happen again somewhere along this mountain range, just a matter of time.

The Qushan Town was evacuated and sealed up on May 23 when the author first visited this area 10 days after the earthquake. The new Beichuan County will be re-built in another location, but, the new site is yet to

be determined. The Qushan Town will be preserved as an "Earthquake Monument" for our next generation to remember this tragedy and memorize those lost their lives here.

3 SURFACE RUPTURES

3.1 Bailu

In addition to strong ground shaking exceeding 1 g, two very distinct ground ruptures running northeast direction was observed after the earthquake. One running primarily follows the Central Longmenshan Fault for 240 km whereas another one runs parallel to the Lower Longmenshan Fault for 72 km. The surface ruptures have resulted in collapsed of houses, bridges and infrastructures, either directly or indirectly.

One of the most visited sites is Bailu because a ground rupture passed right through the playground between two school buildings of the Bailu Central School, which consists of 2 classes of kindergarten, 19 classes of primary and secondary studies. The total number of students is around 950, and luckily no students die. The vertical offset is from 2 to 3 m, and is parallel to the Longmenshan Fault direction of northeast-trending. The surface rupture, however, passed right through buildings of staff quarters, and the wives of the Vice-Headmaster and of a few teachers did not survive the earthquake. The 3-story school building on the upper block did not suffer any damage, whereas the 4-story lower school building was damaged severely. Normally, in the case of reverse fault setting upper block buildings suffer more damages comparing to those on the lower block in an earthquake. This peculiar observation at Bailu Central School deserves more detailed investigation.

The surface rupture continued to the northeast passing the Bailu River into the old Bailu Town, where a very spectacular rupture passed a local street with a sudden rise of 3 m across an originally flat street. Fig. 18 shows the photo taken at that location. All houses located along the ruptures collapsed. The rupture then passed the major road of Bailu, and kind of disappeared in the hills. When we went all the way up to the top of the hills, we found the surface rupture continued into a resort village. Very spectacular rupture passed through two wooden structures and ripped a swimming pool apart in the resort village (see Fig. 19).

In Bailu Town, a total of 9851 houses collapsed with 68 death and 1123 injured. Most of the causalities were related to rupture-induced collapses of buildings. In this sense, surface rupture is a very damaging geohazard associated with earthquake. Regulation has been imposed in Southern California to specify the distance of buildings away from potential surface ruptures. But, there is major problem for such regulation because the separation distance is given as a single value, whereas in reality the location of surface rupture depends on the dip angle, thickness of soil, mechanical properties of soil, amount of slippage on the fault, the type fault rupture, and the rate of slip imposed. However, existing regulation did not take that into consideration. Using the existing regulation, the towns of Yingxiu and Bailu should be abandoned permanently. However, both

towns are being rebuilt at the moment. Clearly, more scientifically-based and realistic recommendations on separation distance must be derived urgently.

3.2 Xiaoyudong

Another spectacular structural collapse is the Xiaoyudong bridge. Fig. 20 shows two collapsed sections of the Xiaoyudong bridge, with the surface rupture right next to the foundation of the bridge. A vertical offset of 1.2 m was observed. This surface rupture was found running continuously from southeast to northwest (perpendicular to the major Longmenshan Fault

Figure 19. A swimming pool is ripped apart by surface rupture at the top of a mountain in Bailu.

Figure 18. Surface rupture passing through a street in the old Bailu Town.

Figure 20. Surface rupture-induced collapse at Xiaoyudong bridge, and associated ground rupture.

direction) for 6 km. There are both reverse dip slip as well as strike slip components for this surface rupture. It is a typical secondary type of surface rupture running away from the primary fault that generated the earthquake. In a sense, this type of surface rupture is much more dangerous than the primary surface rupture running parallel to the earthquake-generating fault, because of their unpredictable path of occurrence. The next major earthquake from Longmenshan Fault may or may not generate secondary rupture. Even in the case of it does, the repeatability of the same route of surface rupture is very poor. Therefore, the safe separation distance from surface rupture is very difficult to be specified.

4 CONCLUSIONS

In this paper, we have summarized some earthquake-induced geohazards that we observed at the affected sites in Sichuan after the Wenchuan earthquake, including high-speed rock avalanche of Niu-juan Gully eanr epicenter, Wangjiayan landslide and Jingjiashan rock avalanche, and surface rupture at Bailu and Xiaoyudong. We can conclude that the consequences of geohazard are much more devastating than the building collapse caused by ground shaking alone, because it normally involves hundreds of buildings (e.g. the Wangjiayan landslide) and rescuing of people trapped by these rock avalanches and landslides are extremely difficult (e.g. both Wangjiayan and Jingjiashan). Therefore, more much effort should be paid to mitigate or avoid such earthquake-induced geohazards.

ACKNOWLEDGEMENTS

The paper was fully supported by PolyU area of strategic development and by PolyU Chair Professor fund (Project No. 1-ZZBF). The authors want to express the help of Profs. Guo X., Wen Z.P. and Cui P. in helping me to travel to various sites in Sichuan after the Wenchuan earthquake.

REFERENCES

Chow, V. T. 1959. *Open-Channel Hydraulics*. McGraw-Hill, Inc.: NY.
Hadley, J.B. 1964. Landslides and related phenomena accompanying the Hebgen Lake earthquake of August 17, 1959. *U.S. Geological Survey Professional Paper* 435: 107–138.
Hsü, K.J. 1975. Catastrophic debris streams (Sturzstroms) generated by rockfalls. GSA. Bull. 86: 129–140.
Jibson, R.W., Harp, E.L., Schulz, W. & Keefer, D.W., 2006. Large rock avalanches triggered by the M 7.9 Denali Fault, Alaska, earthquake of 3 November 2002. *Engineering Geology* 83 (1–3): 144–160.
Sosio, R., Crosta, G.B. & Hungr, O. 2008. Complete dynamic modeling calibration for the Thurwieser rock avalanche (Italian Central Alps). *Engineering Geology* 100 (1–2): 11–26.
Xu, X.W., Wen, X.Z, Ye, J.Q., Ma, B.Q., Chen, J., Zhou, R.J., He, H.L., Tian, Q.J., He, Y.L., Wang, Z.C., Sun, Z.M., Feng, X.J., Yu, G.H., Chen, L.C., Chen, G.H., Yu, S.E., Ran, Y.K., Li, X.G., Li, C.X. & An, Y.F., 2008. The Ms 8.0 Wenchuan earthquake surface ruptures and its seismogenic structure. *Seismology and Geology* 30 (3): 597–629.

Prediction and Simulation Methods for Geohazard Mitigation – Oka, Murakami & Kimoto (eds)
© 2009 Taylor & Francis Group, London, ISBN 978-0-415-80482-0

Use of LIDAR and DEM in the study of the massive February 17, 2006, Leyte, Philippines, Rockslide

M. Gutierrez

Colorado School of Mines, Golden, USA

ABSTRACT: This paper presents the results of a study of the February 17, 2006 rockslide which occurred in Guinsaugon, Southern Leyte, Philippines. The rockslide created a large scarp on the 800-m high Mt. Canabag and involved a large amount of debris consisting of mud and boulders. The debris flow resulting from the slide had a volume of about 25 million m³ and completely inundated the village of Guinsaugon located at the foot of Mt. Canabag. The study used LIDAR to establish the geometry of the scarp created by the slide, and to obtain data on fracturing at the side. Three-dimensional distinct element modeling (DEM) was used to determine the cause of the triggering of the slide and the subsequent long-running flow of the debris materials. Different loading mechanisms were tested in the simulations, and the results were compared with accounts made by slide survivors and observations made during post-slide reconnaissance surveys of the rockslide site.

1 INTRODUCTION

This paper presents the results of a study of the February 17, 2006 rockslide which occurred in Guinsaugon, Leyte, Philippines. Guinsaugon (population: 1,857) is a small village in the town of Saint Bernard located in the southern part of the island of Leyte (Fig. 1). The Guinsaugon rockslide occurred between 10:30 and 10:45 AM on February 17, 2006. It involved the movement of an extremely large piece of rock on the eastern face of the 800-m high Mt. Canabag, a very steep mountain with slopes steeper than 50° on its eastern face. The scarp created by the slide is about 600 m high, 200 m at its deepest part and possibly about 600 m wide at its base (Fig. 2). Prior to the slide, there was an overhanging rock formation at the location of scarp, as identified in pre-slide topographic maps and old photographs of the area.

The rockslide created a large amount of debris, consisting of mud and boulders, which were as thick as 30 m in some areas. The debris completely inundated the village of Guinsaugon located at the foot of Mt. Canabag. In the aftermath of the slide, at least 1,328 persons were reported missing and presumed dead. Numerous houses and buildings, including an elementary school with about 300 students and teachers in attendance, were completely buried.

The rockslide followed extensive rain which fell on the area since February 1, 2006. The amount of rain is much higher than normal, due to the weather phenomenon in the Pacific Ocean known as La Niña, which causes cooling of the water surface temperature in the Pacific Ocean as opposed to the warming

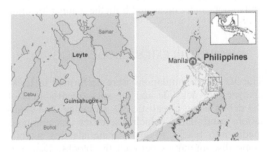

Figure 1. Map showing Guinsaugon, Leyte, Philippines—the site of February 17,2006 landslide (from CNN).

Figure 2. Scarp on Mt. Canabag and debris from the rockslide (from Getty Images).

of water surface temperature during El Niño years. A rainfall metering station 7 km southwest of Guinsaugon recorded an accumulated rainfall of 780 mm from February 1 to 17, 2006, which is about five times higher than during normal rainy seasons (Inter-Agency Committee 2006). In addition to the rainfall, PHIVOLCS (the Philippine Institute of Volcanology and Seismology) recorded two shallow earthquakes with different magnitudes about the time of the slide. However, it is not clear whether these earthquakes had any impact on the triggering of the slides or whether they were generated as a result of the slide.

The nature of the Guinsaugon rockslide is not fully understood in terms of geological, geomechanical and hydrological processes, and no conclusive triggering mechanisms have so far been proposed. To investigate the underlying mechanism of the rockslide, numerical distinct element simulations using 3DEC (Three-Dimensional Distinct Element Code) developed by HItasca (2006) are performed. LIDAR imaging was used to establish the geometry of the scarp created by the slide and to obtain geological data on fracturing at the site of the slide. Other model parameters required in the DEM modeling were established from in situ and laboratory testing of materials taken from the site (Gutierez 2007). Different triggering mechanisms were tested in the simulations, and the results of the numerical modeling were compared with accounts made by slide survivors and observations made during a post-slide field reconnaissance of the rockslide site.

2 LIDAR IMAGING OF THE ROCKSLIDE SITE

LIDAR imaging was used to obtain accurate data on the geometry of the escarpment and fracturing in rockslide site. LIDAR (Light Detection and Ranging) uses a laser to generate an image of a 3D surface. LIDAR systems emit rapid pulses of laser light to precisely measure distances from a sensor mounted on a fixed based. The laser pulse is bounced off a surface and the distance between the scanner and the surface is equal to the speed of light multiplied by the travel time between the laser source and target. Commercially available LIDARs can collect as many as 20,000 points per second with an accuracy of a few millimeters. The immediate result of LIDAR imaging is the 3D point cloud data, which can be used to build a 3D surface model.

For the post-slide LIDAR scanning of the slide, a Riegl LMS-Z620, Laser Profile Measuring System with a range of 2 km and an angular resolution of 0.0025° was used. A comprehensive 3D data acquisition software package operable from any standard laptop or PC is included. The scanner comprises state-of-the-art digital signal processing and echo waveform analysis, enabling precise distance measurements even under bad visibility conditions. The laser scanner is

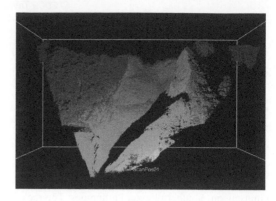

Figure 3.   3D point cloud image of the scarp from LIDAR survey.

combined with an oriented and calibrated high resolution digital camera for a hybrid digital scanning and imaging system. This hybrid system provides data which lend itself to automatic or semi-automatic processing of scan data and image data to generate products such as colored pointclouds, textured triangulated surfaces or orthophotos with depth information.

To georeference the points in the point cloud, reflector targets were placed at several locations on the ground surface and the GPS coordinates of the locations were taken before the LIDAR imaging is performed. The GPS coordinates of the reflector targets are then used to establish a suitable transformation matrix from the local LIDAR coordinates to a global coordinate system. Two or more LIDAR scans can be performed, and the point clouds from each scan can be matched via the known locations of the reflector targets. As a check, orientations of the slip planes calculated from the LIDAR survey are compared with those obtained from manual compass surveys. In addition to establishing the geometry of slide surfaces, the LIDAR surveys are also used to establish morphology and surface roughness of the main slipping plane, and establish statistical data on fracture orientations, spacings, lengths and persistence. Figure 3 shows the 3D point cloud image taken from the LIDAR scanning. Each point from the point cloud has its own unique x, y and z coordinates measured from the scanner location.

3 DISCRETE ELEMENT MODELING

The 3D DEM simulations were performed to back-analyze the rockslide and the subsequent spread of the debris using a realistic 3D geometry. The simulations used 3DEC, which discretizes discontinuous media into blocks of intact materials that interact along discontinuities between the blocks. Discontinuities, which can be joints, faults, fractures, cracks and bedding planes, are represented by spring and dashpot contact elements to characterize the stiffness and damping properties of the discontinuity. Contact stiffness have linear normal and shear components, and for elastoplastic contacts, frictional, cohesive, tensile and dilation can, in addition, be prescribed. Intact blocks can be rigid, elastic or elastoplastic with pre-scribed yield/failure criteria and flow rules. 3DEC is based on the wave equation which is solved numeri-cally using explicit time integration.

3.1 Objectives of the numerical simulations

The main objective of the numerical simulations is to obtain a reliable understanding of the mechanism(s) responsible for the slide. In particular, the simulations will try to investigate: 1) the mode of failure, that is whether the overhang failed by overturning or top-pling, or by sliding along failure surfaces, 2) the effects of hydrologic conditions on the slide, whether the increased pore pressure in the fault and the increased weight of the overhanging block triggered the slide, and 3) the effects of the small earthquakes in triggering of the slide and in initiating the debris flow.

3.2 Distinct element models

The first step in establishing the 3DEC models is to develop a Digital Terrain Model (DTM) of the site. This is done by digitizing a topographic map of the site prior to the slide. Figure 4 shows the DTM of the site on a 100 m × 100 m grid covering an area 2.9 km in the east-west direction and 1.7 m in the north-south direction. Because of the coarse grid used, some details of the topography cannot be accurately rep-resented. However, a finer grid would require more computational resources (in terms of computer mem-ory and computing time), and it may not be possible to obtain results in a reasonable amount of computing time.

Once the DTM has been established, discontinu-ities representing the failure surfaces are introduced. Three major failure surfaces have been identified cor-responding to: (1) a fault which is part of the PFZ—the Philippine Fault Zone (Besana & Ando 2005), (2) a vertical shear failure surface, and (3) the bedding plane (Fig. 5). The model of the scarp and the overhanging rock removed from the face of the mountain are shown in Fig. 6.

The failure surfaces are given normal and shear stiffness, and shear strength (friction angle, cohesion, tensile strength, and dilation angle) properties accord-ing to a Mohr-Coulomb fracture model. These prop-erties are estimated from fracture surface roughness profilometry, Schmidt hammer tests done on exposed failure surfaces in the field, and simple laboratory tests (unconfined compression and Brazilian tests) on rocks samples taken from the field. The intact rocks are modeled as rigid blocks and only their total unit

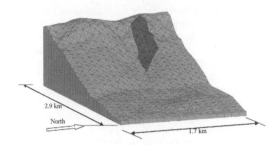

Figure 4. Digital Terrain Model (DTM) of the slide area showing the surface of the detached block.

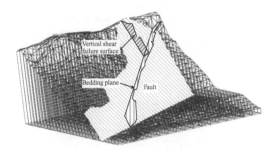

Figure 5. Failure surfaces used in the DEM model.

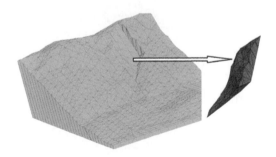

Figure 6. Modeled scarp and detached block.

weights, which are determined from laboratory tests, are required.

3.3 Analysis of the triggering mechanism

The first objective of the DEM modeling is to inves-tigate the triggering mechanism responsible for the rockslide and the type of slope failure. A 3DEC model is established where the overhanging block is assumed to be rigid, intact and contains no discon-tinuities. Quasi-static conditions are assumed and no dynamic forces are applied. In the first model, it is assumed that the triggering comes mainly from the rainfall-induced hydraulic pressurization of the fault. The fault is assumed to be pressurized with a hydrostatic pressure that increases linearly from the ground surface. 3DEC is then used to determine whether the overhanging block will move under its own weight and the increased pore pressure along the fault. Under these assumptions, the results indicate movement of overhanging block. The results of the simulation for the effect of hydraulic pressurization as a triggering load are shown in Fig.7.

As can be seen, the overhanging block moved from its original position, and the predominant mode of failure is slip along the scarp surfaces. This mode of failure is consistent with a witness account of the slide. The vectors of block displacements show that initially the slide occurred downwards along fault dip direction, and there is very little separation along the vertical shear surface. The block moved a few meters by slip along the fault after which it started to move

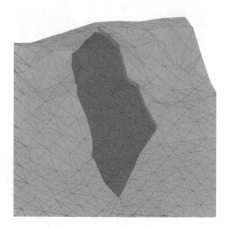

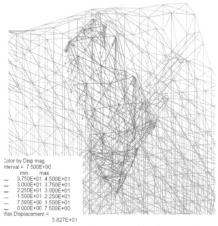

Figure 7. Movement of the overhanging block due to hydraulic pressurization of the fault. Top: deformed geometry, Bottom: displacement vectors (in m).

and slide along the lower slide plane, and the block started to separate from the vertical failure plane. The block experienced almost no rotation indicating that the block did not topple or overturn. The block continued to move downwards as a rigid body and did not get wedged in the escarpment. Although the block cannot break and disintegrate in the model, as it was assumed to be rigid, its continuous movement is consistent with the witness observation that the blocks remained intact while sliding for some distance. The results indicate that rainfall-induced increased pore pressure in the fault is sufficient to trigger the slide.

3.4 *Analysis of the debris flow*

The second objective of the DEM simulation is to model the subsequent flow of debris materials following the triggering of the slide. Debris flow simulations will be carried to determine timing, extent and nature of the debris flow. To model the debris flow, the falling block is subdivided into smaller blocks to represent the natural fracturing of the rock, and to allow the block to disintegrate into smaller pieces due to failure along natural discontinuities. Fracture geometry and distribution were determined from the LIDAR survey of the vertical shear failure surface which forms one of the scarp surfaces, and data logging for field rock mass classification of rock exposures.

Three fracture sets were identified from rock exposures. One set is parallel to the fault, the second is parallel to the vertical shear failure surface, and the third set corresponds to the bedding plane. Thus, at a much larger scale, the three surfaces forming the scarp appear to be part of the three fracture sets identified from the rock exposures. During the data logging, fracture orientation, length, spacing, persistence and fracture surface characteristics were recorded. These data are then used to generate the fracture patterns close to the sliding block. Figure 8 shows the fracture patterns for the region around and within the overhanging block that have been generated from field data.

Because of the very small spacing of the fractures, which are generally less than 1 m, it was not possible to distinctly model and include all fractures in the simulation of debris flow due to computational limitations. By trial and error, it was found that the smallest fracture spacing that can be modeled in the 3DEC, without requiring too much computational time, is 25 m. This is the fracture spacing shown in Fig. 8.

Similar to the model to investigate the triggering mechanism, the model shown in Fig. 8 is subjected to increased pressure in the fault. Results of the simulations are shown in Fig. 9 showing the disintegration of the detached blocks into smaller debris and the movements of the debris with time. Figure 10 shows the displacement vectors at two time instances. The results

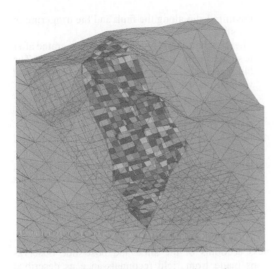

Figure 8.   Model with fractured overhanging rock.

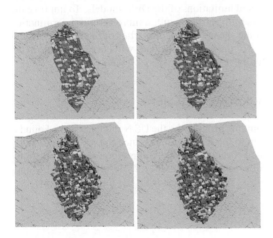

Figure 9.   Block movements at different instances of time following the triggering of the rockslide.

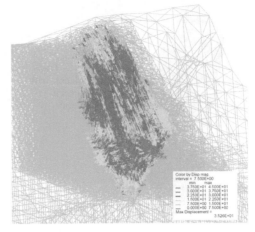

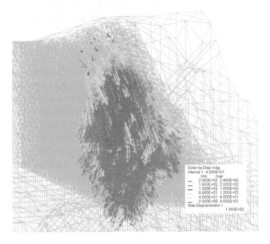

Figure 10.   Displacement vectors (in m) at two instances of time.

again confirm that increased pressure in the fault is enough to trigger both the slide and the debris flow.

The main results of the debris flow simulations can be summarized as follows: 1) the detached block disintegrated after it slipped only a few meters along the fault, 2) the blocks at the top of the scarp disintegrated first before the rest of the block, 3) initially the blocks at the top moved mainly downwards along the dip direction of the fault while those at bottom tended to move along the bedding plane, 4) with increasing time, the blocks started to spread although the main flow direction tended to follow the small valley at the foot of the mountain, and 5) some debris material moved above the lower failure plane (i.e., the bedding plane) instead of moving along the plane.

Close investigation of the falling overhanging rock, right at the point it starts to disintegrate, reveals that the rock tended to break more along the bedding and vertical planes (Fig. 11). Also, failure along the existing fractures due to large separations appears to subdivide the falling block into several clusters before further sliding and disintegrating. This result appears to be consistent with the witness observation that the overhanging rock broke into large pieces after it fell for a short distance and before sliding further down the slope.

In addition to the simulations where hydraulic pressurization was the only load applied, simulation was also carried out to investigate the effects of earthquake loads on the triggering of the slide. The earthquake load was imposed as a horizontal acceleration of 100 gals, which is the maximum acceleration recorded from the two earthquakes detected at the time of the

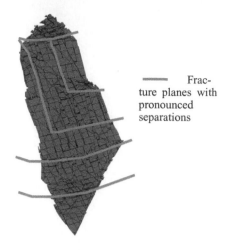

━━━ Frac-
ture planes with
pronounced
separations

Figure 11. Disintegration of the falling overhanging into large clusters due to large separations along the fracture planes.

slide. The earthquake was applied as a triggering load in addition to increased hydraulic pressure along the fault. The earthquake is applied quasi-statically along the strike of the main fault towards the southeast direction. This direction is along the potential movement of the overhanging rock and will maximize the likelihood of slope failure. The result of the simulation for this combination of triggering loads showed minimal effects on the response of the detached overhanging rock. This indicates that the earthquake was possibly triggered by the slide, and not the other way around.

4 CONCLUSIONS

LIDAR imaging and DEM simulations using 3DEC of the February 17, 2006 Guinsaugon rockslide were carried out to obtain an understanding of the mechanisms responsible for the slide. The main conclusions from the study are summarized as follows:

1. The rockslide initially occurred due to slip or activation of a fault, which is a splay of the PFZ, and the downwards movement of an overhanging rock along fault dip direction. Following slippage along the fault, a vertical shear failure plane was created causing the overhanging rock to be separated from the face of the mountain. The falling rock then slid along the bedding plane at the base of the overhanging rock, and started to disintegrate to create a rock avalanche and debris flow. The overhanging block experienced almost no rotation indicating that the block did not initially topple or overturn.
2. Most possibly, rainfall-induced increased hydraulic pressurization of the main fault is responsible for

the initial slip along the fault and the triggering of the slide.
3. The detached overhanging block disintegrated after it slipped only a few meters along the fault. The block tended to break more along the bedding and vertical planes, and failure and large separations along existing fractures appear to subdivide the falling block into several clusters before further sliding and disintegrating.
4. Initially the blocks at the top of the scarp moved mainly downwards along the dip direction of the fault while those at bottom tended to move along the bedding plane. With increasing time the blocks started to spread laterally although the main flow direction tended to be funneled and follow the small valley at the foot of the mountain.

Many of the above results are consistent with witness accounts of the slide and post-slide observations made from field reconnaissance as described in (Gutierrez 2006; and Orense & Gutierrez 2008). However, the results are highly dependent on the input to and limitations of the DEM models. To improve the reliability of the result, a more careful determination of the shear strength properties of the failure surfaces and of the earthquake loads will have to be done in the future.

ACKNOWLEDGEMENT

Funding provided by the US National Science Foundation to the US Reconnaissance Survey Team to study the February 17, 2006 Guinsaugon rockslide is gratefully acknowledged (Myra McAulife is NSF's cognizant program director for the funding support).

REFERENCES

Besana, G.M. & Ando, M. 2005. The central Philippine Fault Zone: Location of great earthquakes, slow events, and creep activity. *Earth, Planets and Space* 57: 987–994.
Gutierrez, M. 2006. DEM Simulation of a Massive Rockslide. In *Proc. Intl. Symp. Geomechanics and Geotechnics of Particulate Media*, Kyushu, Japan, Sept. 12–14, 2006.
HItasca, 2008. *3DEC—Three-dimensional Distinct Element Code*, HItasca Consulting Group, Minneapolis, MN.
Inter-agency Committee on the Guinsaugon, Southern Leyte, Leyte Tragedy, 2006. *The 17 February Bgry. Guinsaugon, Southern Leyte, Landslide*, Government of the Philippines, February 28, 2006, 18 pp.
Orense, R. & Gutierrez, M. 2008. 2006 Large-Scale Rockslide-Debris Avalanche in Leyte Island, Philippines. In *"Earthquake Geotechnical Case Histories for Performance-Based Design," Intl. Soc. Soil Mech. Geotech. Eng., Tech. Comm. 24 Special Case Histories Vol.*, to appear.

Prediction and Simulation Methods for Geohazard Mitigation – Oka, Murakami & Kimoto (eds)
© 2009 Taylor & Francis Group, London, ISBN 978-0-415-80482-0

Mechanism of failure of irrigation tank due to earthquake

A. Kobayashi, T. Hayashi & K. Yamamoto
Kyoto University, Kyoto, Japan

ABSTRACT: To clarify the mechanism of seismic damage of irrigation tanks, the irrigation tanks damaged by Mid Niigata prefecture earthquake in 2004 were analyzed. The irrigation tanks which were inspected for damaged or non-damaged were classified according to some estimations statistically. Conditional damage probability of the irrigation tanks when Mid Niigata prefecture earthquake occurred was calculated by using Monte Carlo Simulation. It was found that the conditional damage probability was harmony with the actual damaged situation at Mid Niigata prefecture earthquake.

1 INTRODUCTION

There are about 210 thousands irrigation tanks in Japan, which have been the important sources of water supply for farming since ancient times. But most of the tanks were constructed many years ago and damaged so much. Kato (2005) reported that about 20 thousands irrigation tanks needed to be repaired. Due to the decrease in the population of farming areas and the increase of aging farmers, many tanks are not well controlled. Under such a circumstance, if a big earthquake or a heavy rain by typhoon is encountered, the damage probability of embankments of irrigation tanks would be very high. In past years, many irrigation tanks were damaged by earthquakes. Tani (2002) summarized the number of the irrigation tanks damaged by past each earthquake. In 1995, 1222 irrigation tanks were damaged by South Hyogo prefecture earthquake. The maintenance and the repair of irrigation tanks are the important to prevent the damage from such a disaster. It is important for the maintenance and the repair to analyze the actual some damages of the irrigation tanks by an earthquake. Some studies have been made on the analysis of the seismic damage of irrigation tanks in Japan. Yamazaki et al. (1989) analyzed the seismic damage of irrigation tanks for Nihonkai-chubu earthquake in 1983 in statistical approach. In statistical approach, it is easy to deal with many targets. On the other hand, many studies have been made on the seismic damage of the individual irrigation tanks or the earth dams by numerical analysis which includes finite element model (FEM). These analyses need much time, therefore it is difficult to analyze many targets.

In 2004, Mid Niigata prefecture earthquake occurred. 561 irrigation tanks were damaged by the earthquake. An amount of the damage of the irrigation tanks was 7.6 billion yen. Mohri et al. (2004) examined the damages of the irrigation tanks by Mid Niigata prefecture earthquake in detail. A few irrigation tanks were failed and most of the damaged irrigation tanks had crack or settlement in the embankments. These tanks did not cause the serious disaster, however there was a possibility to reduce agricultural activity. At Mid Niigata prefecture earthquake, the area of agricultural fields where the farming became impossible was about 900 ha in 2005 and 340 ha in 2006 due to the damaged irrigation facilities in spite of rapid rehabilitation (Misawa et al. 2007).

In this study, the seismic damage of the irrigation tanks at Mid Niigata prefecture earthquake is analyzed. The irrigation tanks which were inspected for damaged or non-damaged are classified according to some estimations statistically. Circular slip method by considering seismic coefficient and liquefaction is applied for the embankments of the irrigation tanks. The conditional damage probability of the irrigation tanks when Mid Niigata prefecture earthquake occurred is calculated by using Monte Carlo Simulation. Sensitivity analysis is made in Circular slip method, and the degree of influence of parameters to the safety factor in Circular slip method is estimated.

2 STATISTICAL ANALYSIS

2.1 Data of irrigation tank

Mid Niigata prefecture earthquake occurred on October 23rd, 2004. Shortly after, 240 irrigation tanks were inspected for damaged condition by Niigata prefectural government temporarily. Table 1 shows the number of the irrigation tanks classified into the damaged conditions. The conditions are classified into Unknown, Undamaged, Lightly damaged, Heavily damaged and Failed. The Unknown condition means that the damaged conditions could not be examined due to the blocked roads. It is found that much of the

Table 1. The number of irrigation tanks with damage condition.

Damage condition	Unknown	Un-damaged	1	2	3
The number of irrigation tank	33	86	57	59	5

* 1: Lightly damaged 2: Heavily damaged 3: Failed.
Source; Niigata prefecture.

Table 2. Year of construction.

	The number of irrigation tanks (Ratio (The number of tanks/Total number))			
	Total	Failed	Damaged	Undamaged
Over 60 years	100	2	68 (68.0%)	32 (32.0%)
Within 60 years	18	0	15 (83.3%)	3 (16.7%)

irrigation tanks are classified into Lightly damaged or Heavily damaged and a few irrigation tanks are classified into Failed.

2.2 Statistical analysis

The irrigation tanks which were inspected for damaged condition are analyzed according to distance from the seismic center, year of construction, height of embankment and length of embankment. Figures 1–3 show the distribution of the density of the damaged and undamaged tanks as functions of the distance from the seismic center, the height of the embankment, the length of the embankment and the width of the embankment. The damaged tanks include lightly, heavily damaged ones and failed ones. Table 2 shows the number and the ratio of the damaged and undamaged irrigation tanks of which construction year is 60 years ago or not. The design criterion of agricultural engineering facility was established in 1953. Thus, the irrigation tanks constructed within 60 years ago fulfill the design criterion.

It can be found from Figure 1 that the irrigation tanks near the seismic center were more vulnerable to damage. In Figures 2–3 and Table 2, the tanks located over 20km from the seismic center are neglected. It can be found from Figures 2–3 that the large embankments were more vulnerable to damage. It can be found from Table 2 that the irrigation tanks constructed within 60 years ago were more vulnerable to damage. On the other hand, All failed irrigation tanks were constructed over 60 years ago. Even if the tank was constructed according to the design criterion, the damage could not be avoided in the case that the tank located near the seismic center. However, the failure was avoided if the tanks were

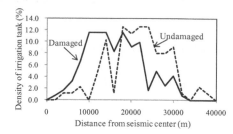

Figure 1. Distribution of density as a function of distance from seismic center.

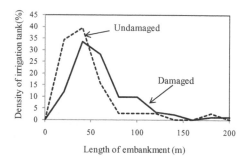

Figure 2. Distribution of density as a function of height of embankment.

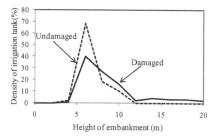

Figure 3. Distribution of density as a function of length of embankment.

constructed according to the design criterion. There are many irrigation tanks that year of construction is unknown, and the irrigation tanks constructed within 60 years ago are less than ones constructed over 60 years ago.

3 CIRCULAR SLIP METHOD

3.1 Circular slip method (CSM)

Circular slip method (CSM) by considering seismic coefficient is applied for upstream and downstream slopes of irrigation tanks according to the design criterion. CSM by considering seismic coefficient is applied for the 240 inspected irrigation tanks.

Equation (1) shows safety factor of CSM based on seismic coefficient.

$$F = \frac{\sum (c \times sl + (W \cos\theta - k_h W \sin\theta) \times \tan\phi)}{\sum (W \sin\theta + k_h W \cos\theta)} \quad (1)$$

where c = cohesion; sl = length of slope; W = unit weight; θ = slope angle; k_h = horizontal seismic coefficient; ϕ = friction angle at interface.

Δu method based on liquefaction is also applied for the 240 inspected irrigation tanks. Equation (2) shows safety factor of Δu method based on liquefaction.

$$F = \frac{\sum (c \times sl + (W \cos\theta - u \times sl - \Delta u \times sl) \times \tan\phi)}{\sum W \sin\theta}$$

$$L_u = \Delta u/\sigma_v = \begin{cases} F_L^{-7}(F_L \geq 1) \\ 1(F_l < 1) \end{cases} \quad (2)$$

where u = pore pressure; Δu = excess pore pressure; σ_v = effective load pressure; F_L = ratio of liquefaction resistance.

Tables 3–4 show the values, the standard deviation and the probability distribution of the parameters used in these methods. The precise size data are used from the database of the irrigation tanks which are controlled by Niigata prefectural government. The gradient of upstream and downstream slope is calculated by using the multiple regression equations as shown Table 3 which arc derived from other tanks of same prefecture because the database does not have the data of the slope gradient. The reservoir water level changes every month and it is difficult to find the precise data. The reservoir water level is assumed to locate at 1m lower than the crest of the embankment. Cohesion and friction angle at interface are used as the average data of irrigation tanks in several areas because these data of each inspected irrigation tank can not be found and it is difficult to investigate all the inspected tanks. Gradient of upstream slope, gradient of downstream slope and reservoir water level as Table 3 vary according to each irrigation tank. On the other hand, unit weight, cohesion and friction angle at interface as Table 3 are the same values for all the irrigation tanks. The standard deviation and the probability distribution are also configured from the actual data.

3.2 Peak ground acceleration

Using CSM, it is important to estimate a horizontal seismic coefficient. A horizontal seismic coefficient is shown as equation (3) by using peak ground acceleration (Noda 1975).

$$k_h = \begin{cases} ll\frac{A}{g} & (A < 200 \text{ gal}) \\ \frac{1}{3} \times \left(\frac{A}{g}\right)^{\frac{1}{3}} & (A > 200 \text{ gal}) \end{cases} \quad (3)$$

where A = peak ground acceleration; g = gravitational acceleration.

In general, peak ground acceleration is calculated by using an attenuation relationship as equation (4) (Fukushima 1996).

$$\log A = 0.51 M - \log(L + 0.006 \times 10^{0.51M}) - 0.0033L + 0.59 \quad (4)$$

where M = magnitude; L = distance from earthquake center.

The peak ground acceleration by an attenuation relationship is calculated by considering only the distance from the

Table 3. Parameters used in the examination.

Parameter	Value
α: Gradient of upstream slope	$1.26 + 0.06 \times H$
β: Gradient of downstream slope	$1.37 + 0.032 \times H$ $+0.00034 \times TL$
l: Reservoir water level (m)	H−1.0
W: Unit weight (kN/m^3)	18.0
c: Cohesion (kN)	27.5
ϕ: Friction angle at interface (°)	25.9

* H: Height of embankment (m), TL: Length of embankment (m).

Table 4. Probability distribution and standard deviation of parameters.

Parameter	Standard deviation	Probability distribution
α: Gradient of upstream slope	0.46	Normal
β: Gradient of downstream slope	0.4	Normal
l: Reservoir water level (m)	0.3	Normal
W: Unit weight (kN/m^3)	1.0	Normal
c: Cohesion (kN)		Uniform (10~30)
ϕ: Friction anglc at interface (°)	8.37	Normal

seismic center. However, the peak ground acceleration is heavily influenced by the geological and geographical condition. Suetomi et al. (2005 & 2007) developed the method of calculating the peak ground acceleration by considering the geological and geographical condition and make the distribution (250 ×250 m) of the peak ground accelerations at Mid Niigata prefecture earthquake. Figure 4 shows the distribution of the peak ground acceleration at Mid Niigata prefecture earthquake. In this study, this distribution is used and peak ground acceleration which is the closest to each irrigation tank is selected. Table 5 shows the averaged value of the peak ground acceleration of the irrigation tank classified into the damaged conditions.

3.3 Safety factor

Safety factor of an embankment of upstream and downstream slopes of each irrigation tank is calculated using the equation (1) of CSM, and safety factor of an embankment of upstream slope of each irrigation tank is calculated using the equation (2) of Δu method. The values of parameters are used as Table 3. Table 5 shows the averaged safety factor of the irrigation tanks classified into damaged conditions. In all cases, it is found that the smaller the averaged safety factor becomes, the more heavily the irrigation tanks were damaged. The results can explain the actual behavior observed in Mid Niigata prefecture earthquake. It is predicted that circular slip on the upstream slope occurred because the safety factor of the embankment of upstream slope is the smallest. It is found that liquefaction seldom occurred because the safety factor is larger than other cases. The averaged value of peak ground acceleration which is applied for the irrigation tanks classified into Failed is smaller than Heavily damaged. It is

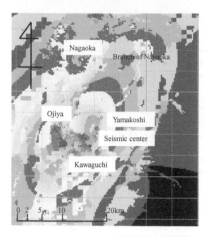

Figure 4. Distribution of peak ground acceleration at Mid Niigata prefecture earthquake.
*Circular point: Observation point, Triangular point: Seismic point.

Table 5. Safety factor and peak ground acceleration.

Damaged condition	Un damaged	Lightly damaged	Heavily damaged	Failed
CSM(up)	0.74	0.73	0.72	0.64
CSM(down)	1.77	1.74	1.62	1.55
Δu(up)	2.44	2.44	2.46	2.08
Peak ground acceleration(gal)	670	656	757	706

* up: upstream slope, down: downstream slope.

found that seismic damage of irrigation tanks is influenced by not only peak ground acceleration but also other factors.

4 CONDITIONAL DAMAGE PROBABILITY

Conditional damage probability of the inspected irrigation tanks when Mid Niigata prefecture earthquake occurred is calculated by using Monte Carlo Simulation. For probability variables, each random number based on each averaged value as shown in Table 3 and each standard deviation and probability distribution as shown in Table 4 is generated and equation (1) or (2) is calculated. The conditional damage probability is calculated as Equation (5).

Conditional damage probability = (The times when the safety factor in equation (1) or (2) is less than 1)/(Trial times (1000)). (5)

Table 6 shows the conditional damage probability of the irrigation tanks classified into damaged conditions in CSM (upstream and downstream slope) and Δu method (downstream slope). When the damaged condition is worse, the averaged value of the conditional damage probability is higher. The conditional damage probability is found to be harmony with the actual damaged situation at Mid Niigata prefecture earthquake. The conditional damage probability which is calculated by this method has an advantage

Table 6. Conditional damaged probability.

Damaged condition	Undamaged	Lightly damaged	Heavily damaged	Failed
CSM(up)	0.212	0.225	0.263	0.292
CSM(down)	0.081	0.110	0.121	0.143
Δu(up)	0.014	0.025	0.027	0.034

* up: upstream slope, down: downstream slope.

Table 7. Sensitivity analysis (CSM (upstream slope)).

Damaged condition	Undamaged	Lightly damaged	Heavily damaged	Failed
$\sigma_\alpha \partial F/\partial \alpha$	0.15	0.25	0.18	0.29
$\sigma_l \partial F/\partial l$	−0.10	−0.021	−0.028	−0.26
$\sigma_W \partial F/\partial W$	−0.045	−0.074	−0.063	−0.039
$\sigma_c \partial F/\partial c$	0.79	1.15	1.01	0.70
$\sigma_\phi \partial F/\partial \phi$	0.12	0.12	0.11	0.11
$\sigma_{kh} \partial F/\partial k_h$	−0.14	−0.18	−0.15	−0.12

* σ_A: Standard deviation, α: Gradient of downstream slope, l: Reservoir water level, W: Unit weight, c: Cohesion, ϕ: Friction angle at interface, k_h: Horizontal seismic coefficient.

Table 8. Sensitivity analysis (CSM (downstream slope)).

Damaged condition	Undamaged	Lightly damaged	Heavily damaged	x damaged
$\sigma_\beta \partial F/\partial \beta$	0.22	0.23	0.23	0.25
$\sigma_W \partial F/\partial W$	−0.060	−0.074	−0.071	−0.052
$\sigma_c \partial F/\partial c$	4.1	3.9	4.0	4.3
$\sigma_\phi \partial F/\partial \phi$	0.23	0.23	0.22	0.22
$\sigma_{kh} \partial F/\partial k_h$	−0.17	−0.18	−0.17	−0.15

Table 9. Sensitivity analysis (Δu method (upstream slope)).

Damaged condition	Undamaged	Lightly damaged	Heavily damaged	Failed
$\sigma_\alpha \partial F/\partial \alpha$	0.41	0.42	0.42	0.55
$\sigma_l \partial F/\partial l$	−0.041	−0.018	−0.018	−0.012
$\sigma_W \partial F/\partial W$	−0.076	−0.11	−0.11	−0.066
$\sigma_c \partial F/\partial c$	1.3	1.7	1.6	1.2
$\sigma_\phi \partial F/\partial \phi$	0.21	0.23	0.28	0.20

that unknown parameters are effectively controlled by using the probability variables. Annual damage probability by considering an earthquake can be calculated by using this conditional damage probability and annual probability density function of an earthquake. It is effective to estimate the damaged probability of irrigation tanks in other areas by using this method.

5 SENSITIVITY ANALYSIS

In general, a parameter which heavily influences an equation can be found by using sensitivity analysis. Sensitivity analysis of the parameters as shown in Table 3 and the horizontal seismic coefficient are carried out. The parameter which influences the safety factor as equation (1) or (2) can be found by this analysis. Tables 7–9 show the results of sensitivity analysis in CSM (upstream slope), CSM (downstream slope) and Δu method (upstream). In sensitivity analysis, a partial differential value with a standard deviation is not influenced by the unit of the parameters. As shown in Tables 7–9, the reservoir water level, the unit weight and the horizontal seismic coefficient are negatively correlated to the safety factor, and the gradient of upstream and downstream slopes, the cohesion and the friction angle at interface are positively correlated to the safety factor. The cohesion most heavily influences the safety factor. The information of the sensitivity analysis is useful when irrigation tanks are repaired. The lower slope gradient, the larger cohesion and the larger friction angle at interface are set to prevent the damage. Especially, the cohesion needed to be paid the most attention to because the cohesion influences the safety factor. On the other hand, the unit weight does not influence the safety factor so much.

6 CONCLUSION

To clarify the mechanism of seismic damage of irrigation tanks, the irrigation tanks damaged by Mid Niigata prefecture earthquake in 2004 were analyzed. The irrigation tanks inspected for damaged or non-damaged were classified according to some estimations statistically. Circular slip method by considering seismic coefficient is applied for the embankments of upstream and downstream slope of the irrigation tanks. Δu method by considering liquefaction is applied for the embankments of upstream slope of the irrigation tanks. Conditional damage probability of the irrigation tanks when Mid Niigata prefecture earthquake occurred was calculated by using Monte Carlo Simulation. Sensitivity analysis was made in Circular slip method, and the degree of influence of parameters to the safety factor in Circular slip method was estimated. The conclusions can be summarized as follows;

1. The irrigation tanks near the seismic center were more vulnerable to damage. The large embankments were more vulnerable to damage. Even if the tank was constructed according to the design criterion, the damage could not be avoided in the case that the tank located near the seismic center. However, the failure was avoided, if the tanks were constructed according to the design criterion.
2. By using Circular Slip Method, the smaller the safety factor becomes, the more heavily the irrigation tanks were damaged by Mid Niigata prefecture earthquake. It is predicted that circular slip on the embankment of upstream slope occurred because the safety factor of the embankment of upstream slope is the smallest. It is predicted that liquefaction seldom occurred because the safety factor is larger than other cases.
3. When the damaged condition is worse, the averaged value of the conditional damage probability is higher. The conditional damaged probability is found to be harmony with the actual damaged situation at Mid Niigata prefecture earthquake.
4. The reservoir water level, the unit weight and the horizontal seismic coefficient are negatively correlated to the safety factor, and the gradient of upstream and downstream slopes, the cohesion and the friction angle at interface are

positively correlated to the safety factor. The cohesion most heavily influences the safety factor.

To obtain the precise safety factor, the database of irrigation tanks needs to be improved. By using this conditional damage probability and the probability density function of an earthquake, the damage probability by considering an earthquake can be calculated. The damage probability is very useful to maintain and repair irrigation tanks in other areas.

ACKNOWLEDGMENT

Niigata prefecture gave us the data related to the disaster very kindly. Authors are so grateful for their corporation.

REFERENCES

Fukushima, Y. 1996. Derivation and revision of attenuation relation for peak horizontal acceleration applicable to the near source region. *Research Report of Shimizu Corporation*. No.63. (in Japanese).

Kato, T. 2005. Flood mitigation function and its stochastic evaluation of irrigation ponds. *Research Report of National Institute for Rural Engineering in Japan*. No.44. (in Japanese).

Misawa, S & Yoshikawa, N. & Takimoto, H. & Hashimoto, S. 2007. Damages to irrigation and drainage canals in Niigata Chuetsu Earthquake and their restorations. *Journal of the Japanese Society of Irrigation, Drainage and Reclamation Engineering* 75(3): 197–200. (in Japanese).

Mohri, Y. & Hori, T. & Matsushima, K. & Ariyoshi, M. 2006. Damage to small earthdam and pipeline by the Mid Niigata Prefecture Earthquake in 2004. *Engineering Report of National Institute for Rural Engineering in Japan*: No.205: 61–76. (in Japanese).

Noda, S. & Jobe, T. & Chiba, T. 1975 Intensity and ground acceleration of gravity quay. *Report of the Port and Harbour Research Institute*. 14(4): 67–111. (in Japanese).

Suetomi, I & Iwata, E. & Isoyama, R. 2005. A procedure for high-accuracy map of peak ground motion using observed records. *Proceedings of 28th JSCE Earthquake Engineering Symposium*. (in Japanese).

Suetomi, I. & Ishida, E. & Fukushima, Y. & Isoyama, R. & Swada, S. 2007. Mixing method of geomorphologic classification and borehole data for estimation of average shear-wave velocity and distribution of peak ground motion during the 2004 Niigata-Chuetsu Earthquake. *Journal of Japan Association for Earthquake Engineering*. 7(3): (in Japanese).

Tani, S. 2002. Earthquake-proof design method to improve the safety of small earth dams. *Soil mechanics and foundation engineering* 50(1): 16–18. (in Japanese).

Yamazaki, A. & Miyake, K. & Nakamura M. & Ikemi, H. 1989. A statistical analysis of seismic damage of small dams for irrigation tank. *Journal of Japan Society of Civil Engineers*. No.404: 361–366. (in Japanese).

Yamazaki, A. & Miyake, K. & Nakamura M. & Ikemi, H. 1989. Seismic damage ratio based on questionnaires of small earth dams for irrigation. *Journal of Japan Society of Civil Engineers*. No.404: 367–374. (in Japanese).

Prediction and Simulation Methods for Geohazard Mitigation – Oka, Murakami & Kimoto (eds)
© 2009 Taylor & Francis Group, London, ISBN 978-0-415-80482-0

Simplified dynamic solution of the shear band propagation in submerged landslides

A.M. Puzrin & E. Saurer
Swiss Federal Institute of Technology, Zurich, Switzerland

L.N. Germanovich
Georgia Institute of Technology, Atlanta, USA

ABSTRACT: Using the energy balance approach, a failure mechanism of submarine landslides based on the phenomenon of shear band propagation has been investigated. Dynamic analysis includes inertia effects in the sliding layer and viscous resistance of the water, but ignores elastic and plastic wave propagation in the sliding layer. The solution allows assessing the velocity and acceleration of both the landslide and the shear band at the moment when the slide fails due to the limiting equilibrium (i.e., the initial post-failure velocity). The effects of the initial landslide velocity on the tsunami wave height are discussed and validated for the Storegga slide example.

1 INTRODUCTION

Tsunami waves represent a serious hazard for the world coastlines. Understanding the mechanisms of tsunamis and their sources is a key task for the tsunami hazard assessment and mitigation. Recent devastating tsunami events, such as the 1998 Papua New Guinea tsunami and the event in 2004 in the Indian Ocean have aroused the public and scientific interest on an improved understanding of the triggering mechanisms and tsunami hazard assessment (e.g. Liam Finn 2003, Okal & Synolakis 2003, Ioualalen et al. 2007). Although tsunamis often occur directly due to normal faulting of earth plates, it has been shown that submarine landslides, triggered by earthquakes, significantly affect the tsunami wave height (Bardet et al. 2003). Wright & Rathije (2003) provided an overview on earthquake related triggering mechanisms of submarine and shoreline slope instabilities. The authors distinguish between direct, such as acceleration- or liquefaction-induced sliding, and indirect triggering mechanisms, such as a delayed failure mechanism due to excess pore water pressure. The general tendency, however, is to assume that the landslide fails simultaneously along the entire sliding surface, which can be tens and hundreds kilometres long. This unrealistic assumption is also behind the fact that numerical simulations of landslide induced tsunamis (e.g. Harbitz 1992, Murty 2003, Bondevik et al. 2005) tend to underestimate the tsunami wave height.

These limitations can be only overcome, if failure of the landslide is considered as a dynamic process. Such an approach has been proposed by Puzrin &

Germanovich (2003, 2005). The basic idea of the mechanism is that a shear band emerges along a certain length of the potential failure surface. Within this shear band the shear strength drops due to the softening behaviour of the material. Therefore, the soil above this weakened zone starts moving downwards, causing the shear band to propagate further along the potential failure surface. This produces an initial landslide velocity already before the slide reached the state of the global limiting equilibrium; i.e. the post failure stage. Analysis of the mechanism is based on the energy balance approach of Palmer & Rice (1973): for the shear band to propagate the energy surplus produced in the body by an incremental propagation of the shear band should exceed the energy required for this propagation. The main advantage of this model is that it allows distinguishing between progressive and catastrophic shear band propagation and treats the shear band as a true physical process and not just as a mathematical bifurcation problem (Rudnicky & Rice 1975).

Analysis of the catastrophic shear band propagation in an infinite submerged slope built of normally consolidated clays has shown that relatively short initial failure zones are sufficient to cause a full-scale landslide (Puzrin et al. 2004). An attempt to assess the landslide velocity at failure was also made (Puzrin & Germanovich 2003), based on a quasi-static approach, neglecting the fundamental dynamic terms.

As will be shown below, this landslide velocity at failure plays an important role for the tsunami height assessment. Therefore, in spite of the complexity of the dynamic problem, it is worth exploring a possibility of producing a better estimate of this velocity.

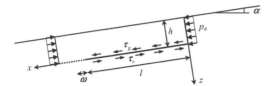

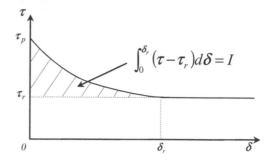

$$\int_0^{\delta_r} (\tau - \tau_r)d\delta = I$$

Figure 1. Propagation of the shear band in an infinite slope.

Figure 2. Strain softening behaviour in the shear band process zone.

This paper briefly outlines an attempt to provide an improved approximation of the true dynamic solution (for the details, the readers are referred to Puzrin et al. 2009). In this simplified approach, the stress distribution in the sliding layer is calculated using inertia terms and the viscous resistance of the water, but excluding propagation and reflection of P-waves. In spite of this simplification, the energy balance, which includes the kinetic energy of the moving landslide, leads to a non-linear differential equation. While this equation can be solved numerically, for large lengths of the shear band the landslide velocity asymptotically approaches a closed form solution. This allows for estimation of the initial landslide velocity at the moment of failure.

2 DYNAMIC SHEAR BAND PROPAGATION IN AN INFINITE SLOPE

2.1 Geometry and soil behaviour

Consider an infinite slope inclined by angle α to the horizontal with a discontinuity zone at the depth h parallel to the slope (Fig. 1). Starting from the initial weak zone of the length larger than critical $l > l_{cr}$ (Puzrin & Germanovich 2004), a shear band propagates down the slope parallel to the surface. At the top of this zone the soil fails in active failure with the active pressure p_a. It is assumed that the length of the discontinuity l is sufficiently larger than its depth and the length of the process zone ω: $l > h >> \omega$. Within this small process zone, the shear resistance τ gradually drops from the peak τ_p to the residual value τ_r, as a function of the relative displacement (Fig. 2). Within the rest of the shear band the shear resistance is constant and equal to τ_r. Outside the shear band and at the tip of the process zone, the shear resistance is equal to the peak value τ_p. If the gravitational shear stress τ_g above the shear band exceeds the residual shear strength τ_r, the soil above the shear band starts moving downwards, driving the shear band to propagate along the slope, until it comes to the surface and the slope fails (Fig. 1). We are interested in the velocity of the shear band propagation and of the landslide at the moment when the slide fails.

Although the normal stress in the x-direction is a function of depth z, in this derivation we only are interested in the average value of this stress across

the sliding layer $\bar\sigma_x(x)$. Before the shear band propagation, the average normal stress in the intact slope is $\bar\sigma_x = p_0$. As the shear band propagates, it starts growing. We assume linear elasto-plastic behaviour in the sliding layer $\bar\sigma_x = p_0 + \varepsilon_x/E$ where, E is the secant Young's modulus.

2.2 Equation of motion

Considering the sliding layer in the dynamic case, when all the points above the band are moving with the same velocity v and acceleration $\dot v$, from the equation of motion we obtain the average normal stress above the tip of the shear band (Fig. 3):

$$h\bar\sigma_x(l) = h\bar\sigma_l$$
$$= (\tau_g - \tau_r)l + p_a h - \rho h l\dot v - \rho h\dot l v - \mu v l \quad (1)$$

where ρ is the density of the soil, μ is the viscosity of the water. The fourth term on the right side reflects the fact that the mass of the moving body is increasing during the shear band propagation (the additional mass accelerates from zero velocity to v).

2.3 Energy balance approach

The energy balance criterion for an incremental dynamic propagation of the shear band can be expressed in the following equation:

$$\Delta W_e - \Delta W_i - \Delta D_l - \Delta D_\mu - \Delta K = \Delta D_\omega \quad (2)$$

where W_e is the external work made by gravitational forces on downhill movements of the layer; W_i denotes the internal work of the normal stress acting parallel to the slope surface on the change of strains in the layer; D_l is the dissipated energy due to plastic work along the shear band; D_μ is the dissipated energy at the soil-water interface; K is the kinetic energy and D_ω is the plastic work required to overcome the peak shear resistance at the tip of the band, i.e. the softening in the process zone.

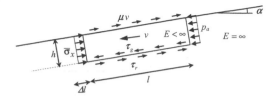

Figure 3. Motion of the sliding layer.

Incremental propagation of the shear band by Δl (Fig. 3) over the time increment Δt produces displacement of the entire sliding layer, proportional to the strain ε_l in the portion of the sliding layer above Δl:

$$\Delta \delta = \varepsilon_l \Delta l \tag{3}$$

Velocity of the sliding layer is then given by

$$v = \frac{\Delta \delta}{\Delta t} = \varepsilon_l \frac{\Delta l}{\Delta t} = \varepsilon_l \dot{l} \tag{4}$$

where $\dot{l} = \Delta l / \Delta t$ is the velocity of the shear band propagation.

Substituting the corresponding work increments (Puzrin el al. 2009) into the energy balance and dividing each term by the time increment Δt, after certain manipulation we obtain

$$[(\tau_g - \tau_r)l + p_a h - \mu v l - \rho h l \dot{v} - \rho h l v]\varepsilon_l \\ - h \int_0^{\varepsilon_l} \sigma_x d\varepsilon_x = I \tag{5}$$

where

$$I = \int_0^{\delta_r} (\tau - \tau_r)d\delta \approx \frac{1}{2}(\tau_p - \tau_r)\delta_r \tag{6}$$

The term in the square brackets in equation (5) can be recognised from equation (1), leading to

$$\bar{\sigma}_l \varepsilon_l - \int_0^{\varepsilon_l} \bar{\sigma}_x d\varepsilon_x = \frac{I}{h} \tag{7}$$

2.4 Assessment of the landslide velocity

The terms on the left side of the equation (7) appear to be equal to the complimentary strain energy:

$$\bar{\sigma}_l \varepsilon_l - \int_0^{\varepsilon_l} \bar{\sigma}_x d\varepsilon_x = \int_{p_0}^{\bar{\sigma}_l} \varepsilon_x d\bar{\sigma}_x \tag{8}$$

The average linear strain ε_x can be related to the average normal stress $\bar{\sigma}_x$ in the layer along the shear band:

$$\varepsilon_x = \frac{\bar{\sigma}_x - p_0}{E} \tag{9}$$

Equation (9) can be then substituted into (8), integrated, and the result substituted into (7):

$$\bar{\sigma}_l \varepsilon_l - \int_0^{\varepsilon_l} \bar{\sigma}_x d\varepsilon_x = \frac{(\bar{\sigma}_l - p_0)^2}{2E} = \frac{I}{h} \tag{10}$$

which gives

$$\bar{\sigma}_l = p_0 + \sqrt{\frac{2IE}{h}} \quad \varepsilon_l = \frac{\bar{\sigma}_l - p_0}{E} = \sqrt{\frac{2I}{hE}} \tag{11}$$

i.e., the shear band propagates at the constant normal lateral stress in the sliding layer above the band tip.

Equation of motion (1) can be then rewritten as:

$$\rho h l \dot{v} + \rho h \dot{l} v + \mu v l - (\tau_g - \tau_r)(l - l_{cr}) = 0 \tag{12}$$

where

$$l_{cr} = \frac{h\bar{\sigma}_l - p_a h}{\tau_g - \tau_r} = \frac{\sqrt{2IEh} - (p_a - p_0)h}{\tau_g - \tau_r} \tag{13}$$

is the critical length of the initial shear band beyond which it starts propagating. Substitution of the Equation (4) into (12) gives the following non-linear second order differential equation:

$$(y + l_{cr})\ddot{y} + (\dot{y})^2 + a(y + l_{cr})\dot{y} - by = 0 \tag{14}$$

where

$$y = l - l_{cr} \tag{15}$$

$$a = \frac{\mu}{\rho h} \quad b = \frac{(\tau_g - \tau_r)}{\varepsilon_l \rho h} = \frac{(\tau_g - \tau_r)}{\rho h}\sqrt{\frac{hE}{2I}} \tag{16}$$

with initial conditions:

$$y(0) = \dot{y}(0) = 0 \tag{17}$$

Equation (14) can be solved numerically. However, for large lengths of the shear band $y \gg l_{cr}$ and zero viscosity ($a = 0$ for sub-aerial slides), it can be simplified:

$$y\ddot{y} + (\dot{y})^2 - by = 0 \tag{18}$$

and solved with initial conditions (17) analytically:

$$y = \frac{b}{6}t^2 \tag{19}$$

so that

$$l = y + l_{cr} = \frac{b}{6}t^2 + l_{cr} \quad \dot{l} = \dot{y} = \frac{b}{3}t \tag{20}$$

The dependency of the landslide velocity on the shear band length follows:

$$v = \varepsilon_l \dot{l} = \varepsilon_l \sqrt{\frac{2b}{3}}\sqrt{l - l_{cr}} \tag{21}$$

or after substituting (11) and (16) into it:

$$v = \sqrt{\sqrt{\frac{8I}{9hE}}\frac{\tau_g - \tau_r}{\rho h}}\sqrt{l - l_{cr}} \tag{22}$$

At large $y \gg l_{cr}$, solution of Equation (14) will always approach approximation (22) from below, because propagation of the shear band in the approximate Equation (18) starts with non-zero acceleration $\ddot{y} = b/3$, while in the full Equation (14) with initial conditions (17), it begins with the zero acceleration $\ddot{y} = 0$.

Approximations (20)–(22) are strictly speaking only valid for sub-aerial landslides. For submarine landslides, viscosity cannot be neglected ($a \neq 0$) and another approximation has to be made. Introducing dimensionless length and time:

$$Y = y/l_{cr} \quad T = at \tag{23}$$

into equation (14) and dividing both parts by $Y+1$ we obtain:

$$\ddot{Y} + \frac{(\dot{Y})^2}{Y+1} + \dot{Y} - c\frac{Y}{Y+1} = 0, \quad \text{where } c = \frac{b}{a^2 l_{cr}} \tag{24}$$

We are looking for a limiting condition for the shear band propagation velocity. If for large $Y \gg 1$ the velocity stabilizes, the second term in the above equation becomes small and the forth approaches c, leading to

$$\ddot{Y} + \dot{Y} - c = 0 \tag{25}$$

which can be solved with initial conditions (17) analytically:

$$\dot{Y} = c(1 - e^{-T}) \quad \text{or} \quad \dot{y} = \frac{b}{a}(1 - e^{-at}) \tag{26}$$

confirming that the shear band velocity cannot grow infinitely and stabilizes at larger y, limiting the initial landslide velocity to the maximum value of:

$$v = \varepsilon_l \dot{l} = \varepsilon_l \frac{b}{a} = \frac{\tau_g - \tau_r}{\mu} \tag{27}$$

Note, that the shear band propagation velocity is not limited by the shear wave velocity. This does not represent a problem however, because the proposed formulation is analogous to the shock wave propagation in a finite layer, and does not require energy delivered to the shear band tip. For realistic water resistance, velocity (27) can become rather large in an infinite slope case. In reality, however, no infinite slopes exist in the nature, and even if they did exist, the shear band would sooner or later propagate to the surface, causing the slope failure at a finite velocity.

2.5 Dependency of tsunami wave height on landslide velocity

In general, wave height η and landslide velocity v_{max} are correlated via the Froude number (e.g. Harbitz et al. 2006). For tsunamigenic landslides this number relates the linear long-wave velocity c_0 at a water depth of H; $c_0 = \sqrt{gH}$ (where g is the gravity acceleration) to the maximum landslide velocity v_{max} and is defined as

$$Fr = \frac{v_{max}}{\sqrt{gH}} \tag{28}$$

Sub-critical, critical and super-critical landslide motions are defined as $Fr < 1$, $Fr = 1$ and $Fr1$, respectively. Critical landslide motion produces tsunami waves several times higher than the thickness of the landslide h (Fig. 4). For larger water depths, most of the landslides

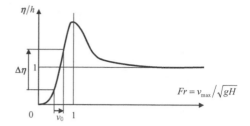

Figure 4. Typical dependency of the tsunami height on the landslide velocity (after Ward 2001).

are going to be sub-critical, therefore, a substantial initial velocity of the landslide would result in significantly larger increase in the tsunami wave height.

2.6 Example: The Storegga slide

The Storegga slide, situated on the continental slope off the western coast of Norway, is one of the largest and best-studied submarine landslides in the world (e.g. Bugge 1988, Haflidason et al. 2004, Kvalstad et al. 2005). Mainly three separate slides occurred in the area. The first slide occurred about 30,000–50,000 years before present involving a volume of 3880 km^3 and a run-out distance of 350–400 km from the headwall. The average thickness was about 114 m (Bugge et al. 1988).

The second slide occurred about 8,200 years ago. It has been found that the second Storegga slide consisted of one giant slide, with a volume of about 3100 km^3 and a runout of about 750 km, followed by a multitude of smaller events. The average thickness was about 88 m (Bugge et al. 1988). The third event was limited to the upper part of the second slide scar.

Numerous numerical simulations of the tsunami wave heights for the first and second Storegga slides (Harbitz 1992, Bondevik et al. 2005, Grilli & Watts 2005 a, b, De Blasio et al. 2005) indicated that rather high maximum velocities of up to 35–60 m/s are required to get a correct correlation to the run-out distances. Such high velocities are difficult to explain without including the initial landslide due to shear band propagation.

The average height of the first slide was $h = 114$ m, the average inclination $\alpha = 0.5$ deg; the total unit weight of the soil $\gamma = 17$ kN/m^3; i.e. $\rho = 1.7$ t/m^3. Therefore, the gravitational shear stress:

$$\tau_g = \gamma \cdot h \cdot \sin \alpha = 17 \cdot 114 \cdot \sin 0.5° = 16.9 \text{ kPa} \tag{29}$$

The undrained shear strength for the calculation of the τ_p in normally consolidated clays may be approximated using the formula $\tau_p = s_u = 1/4\gamma' z$; assuming total static liquefaction: $\tau_r = 0$; $\delta_r = 0.1$m, so that:

$$I \approx \frac{1}{2}(\tau_p - \tau_r)\delta_r = 9.98 \text{ kN/m} \tag{30}$$

The at rest earth pressure coefficient $K_0 = 0.5$. For the calculation of the earth pressure at the top of the sliding layer we assume a gap filled with water:

$$p_a = \gamma_w z \tag{31}$$

The initial average stress in x-direction:

$$p_0 = K_0 \gamma' z + \gamma_w z \tag{32}$$

Therefore at the average height of $z = h/2 = 57$ m:
$p_a - p_0 = -K_0 \gamma' z = -199.5$ kPa.
The secant Young's modulus of $E = 1.0$ MPa; viscosity of water $\mu = 0.001\ Pa \cdot s \cdot m^{-1}$. Substitution of these parameters into the Equations (13) and (16) gives the critical length:

$$l_{cr} = \frac{\sqrt{2IEh} - (p_a - p_0)\, h}{l_g - l_r} \approx 1434 \text{ m} \tag{33}$$

and parameters:

$$a = \frac{\mu}{\rho h} = 5.16 \cdot 10^{-6}\ s^{-1} \tag{34}$$

$$b = \frac{(\tau_g - \tau_r)}{\rho h}\sqrt{\frac{hE}{2I}} = 6.60\ \text{ms}^{-2} \tag{35}$$

for the first Storegga slide.

Equation (24) can be now solved together with the boundary conditions (17) numerically. However, the dimensionless parameter c in this case appears to be very large:

$$c = 1.43 \cdot 10^8 \tag{36}$$

which could lead to numerical instabilities. In order to stabilize the solution, the initial condition has to be applied in the form:

$$Y(0) = \varepsilon << 1 \tag{37}$$

The resulting numerical solution is shown in Figure 5, and compared with the analytical approximation given by Equation (22), i.e. for sub-aerial landslides. As is seen, the approximation is very precise at high values of the shear band length. This is hardly surprising because of the very small value of parameter a (Equation 34). For the failure length $l = 150$ km, which was suggested by Harbitz (1992), numerical solution predicted the landslide velocity of $v_0 = 10.66$ m/s, while the analytical approximation was $v_0 = 10.69$ m/s.

Figure 6 shows the solution and its approximation in the vicinity of $l = l_{cr}$. It demonstrates why the approximation limits the velocity from above, namely due to the non-zero acceleration at $l = l_{cr}$. Very soon, however, accelerations in both solutions become very close.

For the second Storegga slide, the only difference is the landslide thickness $h = 88$ m (Harbitz, 1992), which in our case does not effect the initial landslide velocity. The Storegga landslides are the longest landslides discovered so far, and their initial velocities in

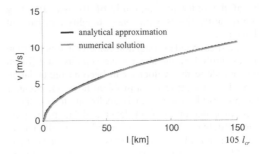

Figure 5. Landslide velocity as a function of the shear band length.

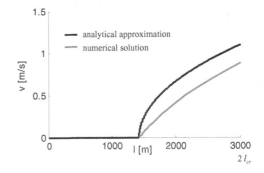

Figure 6. Landslide velocity in the beginning of the shear band propagation.

the order of 10 m/s are, probably, the largest initial velocities that could develop in realistic environment.

Nevertheless, these initial velocities represent 17–30% of the estimated maximum velocities of 35–60 m/s. Assuming an average water depth of the landslide of 1500 m results in a tsunami wave velocity $c_0 = \sqrt{gH}$ of about 120 m/s, i.e. in the $Fr = 0.3$–$0.5 < 1$. For this sub-critical landslide motion, increase in the Froude number by 17–30% would result in significant increase in tsunami wave height (Fig. 4).

3 SUMMARY AND CONCLUSIONS

A stable numerical solution and its closed form approximation have been obtained for the velocity of the submarine tsunamigenic landslides at failure. The landslide mechanism is based on the phenomenon of the dynamic shear band propagation, analyzed using the energy balance approach. Inertia effects and viscous water resistance have been included into the analysis, while the propagation and reflection of the P-waves within the sliding layer have been neglected. For the Storegga slides these velocities appeared to be of the order of magnitude of 10 m/s, affecting significantly the tsunami wave heights and run-out distances.

The shear band propagation velocity in this solution is not limited by the shear wave velocity, and for an infinite slope this velocity can become rather large. This does not represent a problem however, because the proposed formulation is analogous to the shock wave propagation in a finite layer, and does not require energy delivered to the shear band tip. In the nature however, no infinite slopes exist and the shear band would sooner or later propagate to the surface, causing the slope failure at a finite velocity.

The Storegga landslides are the longest landslides discovered so far, and their initial velocities in the order of 10 m/s are, probably, the largest initial velocities that could develop in realistic environment.

ACKNOWLEDGEMENTS

The project has been supported by Swiss National Science Foundation (SNF) (Grant No. 200021-109195) and by the US NSF grants OCE-0242163 and CMC-0421090.

REFERENCES

Bardet, J.-P., Synolakis, C.E., Davies, H.L., Imamura, F. & Okal, E.A. 2003. Landslide tsunamis: recent findings and research directions. *Pure and Applied Geophysics* 160: 1793–1809.

Bondevik, S., Løvholt, F., Harbitz, C., Mangerud, J., Dawson, A. & Svendsen, J.I. 2005. The Storegga Slide tsunami—comparing field observations with numerical simulations. *Marine and Petroleum Geology* 22: 195–208.

Bugge, T., Belderson, R.H. & Kenyon, N.H. 1988. The Storegga slide. *Phil. Trans. R. Soc. Lond. A* 325: 357–388.

De Blasio, F.V., Elverhøi, A., Issler, D., Harbitz, C.B., Bryn, P. & Lien, R. 2005. On the dynamics of subaqueous clay rich gravity mass flows-the giant Storegga slide, Norway. *Marine and Petroleum Geology* 22: 179–186.

Grilli, S.T. & Watts, P. 2005. Tsunami generation by submarine mass failure. I: Modeling, experimental validation, and sensitivity analyses. *Journal of Waterway, Port, Coastal and Ocean Engineering* 131 No. 6: 283–297.

Grilli, S.T. & Watts, P. 2005. Tsunami generation by submarine mass failure. II: Predictive Equations and case studies.. *Journal of Waterway, Port, Coastal and Ocean Engineering* 131 No. 6: 298–310.

Haflidason, H., Sejrup, H.P., Nygard, A., Mienert, J., Bryn, P., Lien, R., Forsberg, C.F., Berg, K. & Masson, D. 2004.

The Storegga slide: architecture, geometry and slide development. *Marine Geology* 213: 201–234.

Harbitz, C.B. 1992. Model simulations of tsunamis generated by the Storegga slides. *Marine Geology* 105: 1–21.

Harbitz, C.B., Løvholt, F., Pedersen, G. & Masson, D.G. 2006. Mechanisms of tsunami generation by submarine landslides: a short review. *Norwegian Journal of Geology* 86: 255–264.

Ioualalen, M., Asavanant, J., Kaewbanjak, N., Grilli, S.T., Kirby, J.T. & Watts, P. 2007. Modeling the 26 December 2004 Indian Ocean Tsunami: Case study of impact in Thailand. *Journal of Geophysical Research* 112, C07024: 1–21.

Kvalstad, T.J., Andresen, L., Forsberg, C.F., Berg, K., Bryn, P. & Wangen, M. 2005. The Storegga slide: evaluation of triggering sources and slide mechanics. *Marine and Petroleum Geology* 22: 245–256.

Liam Finn, W.D. 2003. Landslide-generated tsunamis: geotechnical considerations. *Pure and Applied Geophysics* 160: 1879–1894.

Murty, T.S. 2003. Tsunami wave height dependence on landslide volume. *Pure and Applied Geophysics* 160: 2147–2153.

Okal, E.A. & Synolakis C.E. 2003. A theoretical comparison of tsunamis from dislocations and landslides. *Pure and Applied Geophysics* 160: 2177–2188.

Palmer, A.C. & Rice J.R. 1973. The growth of slip surfaces in the progressive failure of over-consolidated clay. *Proceedings of the Royal Society A* 332'F 527–548.

Puzrin, A.M. & Germanovich, L.N. 2003. The mechanism of tsunamigenic landslides. Geomechanics: *Testing, Modeling and Simulation* (GSP 143): 421–428.

Puzrin, A.M. & Germanovich, L.N. 2005. The growth of shear bands in the catastrophic failure of soils. *Proceedings of the Royal Society A* 461: 1199–1228.

Puzrin, A.M., Germanovich, L.N. & Kim, S. 2004. Catastrophic failure of submerged slopes in normally consolidated sediments. *Géotechnique* 54 No. 10: 631–643.

Puzrin, A.M., Saurer, E., Germanovich, L.N. 2009. Dynamic shear band propagation mechanism for tsunamigenic landslides. *(in preparation)*

Ramberg, W. & Osgood W.R. 1943. Describtion of stress-strain curve by three parameters. *NACA TN-902, National Advisory Committee for Aeronautics Tech Note* 902.

Rudnicky, J.W. & Rice, J.R. 1975. Conditions for the localization of deformation in pressure-sensitive dilatant materials. *J. Mech. Phys. Solids* 23: 371–394.

Ward, S.N. 2001. Landslide tsunami. *Journal of Geophysical Research* 106, No. 11: 201–215.

Wright, S.G. & Rathije, E.M. 2003. Triggering Mechanisms of Slope Instabilities and their Relation to Earthquakes and Tsunamis. *Pure and Applied Geophysics* 160: 1865–1877.

Prediction and Simulation Methods for Geohazard Mitigation – Oka, Murakami & Kimoto (eds)
© 2009 Taylor & Francis Group, London, ISBN 978-0-415-80482-0

Simulation of wave generated by landslides in Maku dam reservoir

S. Yavari-Ramshe & B. Ataie-Ashtiani
Sharif University of Technology, Tehran, Iran

ABSTRACT: In this work, impulsive wave generation and propagation generated by landslides are studied numerically for a real case. Maku dam reservoir, in the northwestern of Iran is considered as the case study. Generated wave heights, wave run-up, maximum wave height above the dam crest and the probable overtopping volume have been evaluated, using a two-dimensional numerical model (LS3D). This model is validated using available three-dimensional experimental data for simulating impulsive wave caused by sub-aerial landslides. Based on the results, the generated wave height for first and second scenarios are 12 m and 18 m respectively. The wave height of 8 m is observed close to dam body. Because of the freeboard of 9 m, no dam overtopping happen in normal condition. In rainy seasons water level is close to spillway level, therefore the maximum wave height above dam crest is 4 m and the volume of flood overtopping is about 20000 m^3.

1 INTRODUCTION

Landslides in dam reservoirs initiate horrible water waves that cause probable damages to reservoir side-walls, dam body, adjacent hydraulic structures and agricultural or residential areas (Semenza 2000). If they have huge size, the reservoir capacity decrease and subsequent economical damage is expected. So it is important to estimate the characteristics of impulsive wave generated by landslides in dam reservoirs. Ataie-Ashtiani and Malek-Mohammadi (2007a) studied the landslide generated wave in Shafaroud dam reservoir, in the north of Iran. They used FUNWAVE numerical model (Kirby et al. 1998) for simulating wave propagation. They have also provided empirical equations for the first wave characteristics (Ataie-Ashtiani & Malek-Mohammadi 2007b).

A large number of analytical, experimental and numerical studies have been performed about the landslide generated waves. Ataie-Ashtiani and Najafi-Jilani (2006) have performed a comprehensive review on the experimental and numerical studies.

Four distinct stages can be distinguished in simulating landslide waves in dam reservoirs: Generation, Propagation, Overtopping and Inundation (Figure 1). In the present study, a forth order Boussinesq type model with moving bottom boundary (LS3D), developed by Ataie-Ashtiani and Najafi-Jilani (2007) has been applied to Muko dam reservoir in Iran. Generated wave height, wave run-up, maximum wave height above the dam crest and the probable overtopping volume are estimated for two scenarios.

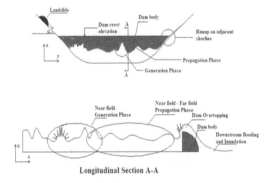

Figure 1. Sketch illustrating separation of Generation, Propagation, Inundation and Overtopping.

2 LS3D MODEL VALIDATION

LS3D model has been extended for simulating impulsive waves caused by sub-aerial landslides, according to assumptions of Lynnet and Liu (2002).

Experimental data of Ataie-Ashtiani and Nikkhah (2008) are used to validate LS3D model. A schematic of wave tank is shown in figure 2. Eight Validyne DP15 differential pressure transducers were used as wave gages. They were located at the central axes of the tank.

For one of these experiments, numerical and experimental results are compared in Figure 3. This figure shows time series of wave amplitude at first (ST1)

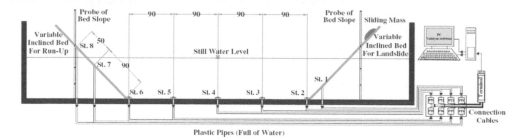

Figure 2. Schematics of experimental set up for sub-aerial landslide generated waves, all dimensions are in centimeter.

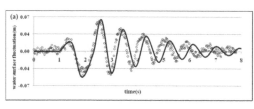

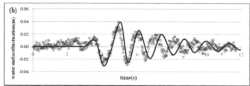

Figure 3. Comparison of LS3D numerical model (-) and Ataie-Ashtiani & Nikkhah(2008) experimental(o) results for subaerial impulsive waves in (a): generation and (b): propagation stages.

and sixth (ST6) gages location. Based on the results, numerical and experimental data are properly matched. Error of wave amplitude estimation is less than 5%. A time phase difference of 10–15% is observed between numerical and experimental results.

3 MAKU DAM: SETTING, GEOLOGIC SITUATIONS AND PROBABLE SLIDES

Maku(Barun) dam is located in the north part of Maku town, west Azarbayjan province, Iran, in 11°39'17" north latitude and 44°28'55" east longitude, on the Zangmar river. The Zangmar river originates in the mountains above Maku, along the Turkish-Iranian border, not far from Mount Ararat and flows south and east into the Araxes at the town of Pol Dasht (Fig. 4).

Dam is located in a seismic region (Fig. 5). "Badavli" fault is near the dam site. Hence, seismic conditions are intensified its crucial landslide-susceptibility status (Mahab Ghods 1999).

Maku dam is 75 m high storage earth dam, with a reservoir capacity of 135 Mm^3 (Fig. 6). Length and

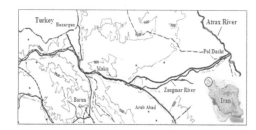

Figure 4. Geographical situation of "Maku" dam ("Niro ministry).

Figure 5. Satellite picture of Maku damsite and existing faults.

width of dam are 350 m and 10 m respectively. The dam crest level is 1690 m from sea level (Mahab Ghods 1999).

According to geological investigation, the most probable slide is located in left beach of dam reservoir near the dam axes. The formation and extension of tensile cracks in left beach is the main cause of slide forming (Figure 7(a)). This slide is shown in Figure 8. There is another probable slide in right beach (figure 8) that is a ring-shaped slide. The location of these two slides is shown in Figure 9 (Mahab Ghods 1999).

According to the three-dimensional geologic maps, the first scenario, located in left beach, is submarine landslide and the second scenario, located in right beach, is sub-aerial landslide. The characteristics and properties of the scenarios can be observed in Table 1. Parameters in this table are B: slide length, T: slide thickness, γ: relative slide density, d: initial vertical distance of slide mass from water surface and α: angle of slide surface.

4 SIMULATION STEPS

The LS3D model received AutoCAD file of dam-site topographic map. Time step and grid dimensions in x and y directions, are chosen as 0.05 seconds and 5×5 meters respectively. The normal water level of 1680 m is also given to program. According to three dimensional topographic map, the model identify reservoir extent and computational borders intelligently. After computational borders correction to prevent model divergence, below generated mesh is obtained (Figure 10).

Effect of reservoir sides in plan, drying/wetting succession of borders, on water body is exerted by definition of a reflection factor (Grilli et al. 2002). This factor is determined according to material and vegetation types of reservoir sides. It is estimated about 0.6 for Maku dam reservoir.

Table 1. Probable landslide properties of Maku dam site.

scenario	B(m)	T(m)	γ	d(m)	(deg)
1	80	15	1.9	−15	30
2	60	10	1.9	5	32

5 RESULT DISCUSSION

According to final corrected mesh of Maku dam reservoir, average length and width of dam reservoir are

Figure 6. Maku earth dam in west Azerbaijan Province, Iran.

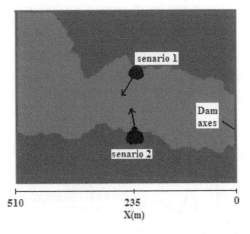

510   235   0

X(m)

Figure 9. Location of probable landslides due to dam axes.

Figure 7. Slide patterns in Maku reservoir beaches.

Figure 8. (a) close vision and (b) landscape of probable landslide of left beach.

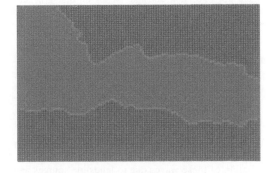

Figure 10. Generated mesh after correction of computational borders for Maku dam reservoir.

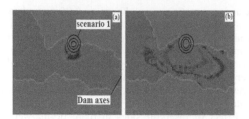

Figure 11. Propagation of generated impulsive waves caused by first scenario in (a) near field and (b) far field.

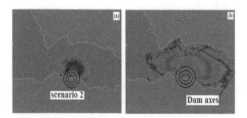

Figure 12. Propagation of generated impulsive waves caused by second scenario in (a) near field and (b) far field.

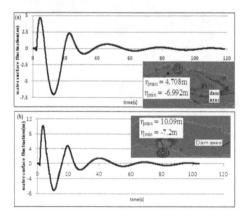

Figure 13. Location and time series of first generated impulsive wave in near field for (a) first and (b) second scenario.

about 510 m and 185 m respectively. Generated impulsive waves propagation in near field and far field are shown in Figures 11 and 12 for two scenarios in plan.

5.1 Generated wave amplitude

The heights of first impulsive wave in near field, for two scenarios are about 12 m and 18 m respectively. Location of these waves due to landslide sources and time series of water surface fluctuations are shown in figure 13 for two scenarios.

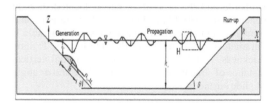

Figure 14. Generation, Propagation and Run-up stages of impulsive waves.

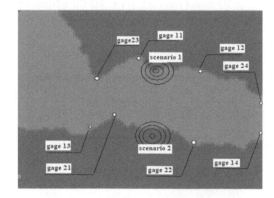

Figure 15. Location of gages for estimating wave run-up at beaches and near Maku dam body.

5.2 Wave run-up

In LS3D model, simulation of wave run-up at reservoir sidewalls is according to numerical method of Synolakis (1987). He gave a simple equation for estimating wave run-up (Equation (1)) as results of experiment on non-breaking positive solitary wave with non-permeable flat sidewall. This equation is:

$$\frac{R}{h_0} = 2.831 \left(\frac{H}{h_0} \right)^{1.25} \sqrt{\cot \beta} \tag{1}$$

The parameters in this equation are β: slope angle of run-up surface, R: run-up height, h_0: normal water depth of reservoir and H: the height of first wave closes to sidewall. These parameters are shown in figure 14.

The wave run-up is estimated in several gages near sidewalls and dam body, for two scenarios. The location of these gages can be observed in figure 15. In this figure, the first number of each gage name is the number of scenario that making run-up and the second number is the number of gage. The amounts of wave run-up at each gage with distance from landslide source and slope angle of sidewall are brought in table 2. In this table, parameters are X_p(m): distance from slide source, h_0(m): average depth, a_c(m): positive wave amplitude and a_t(m): negative wave amplitude.

Table 2. Wave run-up at Maku reservoir sidewalls.

Gage number	X_p(m)	$h_{0(m)}$	β(o)	a_c(m)	a_t(m)	H(m)	R(m)
11	70	50	23	2.19	5.71	7.9	22.2
12	90	50	35	1.35	4.07	5.42	10.5
13	168	50	48	1.75	4.23	5.98	9.5
14	280	50	25	2.42	4.56	6.98	17
21	90	50	20	4.7	5.22	9.9	30.7
22	100	50	70	3.8	2.6	6.38	6.5
23	135	50	65	4.25	2.5	6.74	7.8
24	245	50	25	5.62	2.9	8.5	22

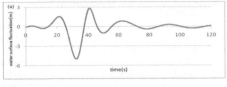

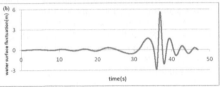

Figure 16. time series of water surface fluctuations close to dam body caused by (a) first and (b) second scenarios.

It is observed from table that near to landslide sources, height of run-up reach about 30 m and this show that if landslides take place, water covers extensive arias of reservoir beaches.

Table 3. Summary of Maku dam numerical simulation results.

Scenario	V_s (Mm^3)	D (m)	H_{max} (m)	V_{max} (m^3)	V_{min} (m^3)	h_{max} (m)	h_{min} (m)
1	0.6	375	12	13000	0	3	0
2	0.15	230	18	20000	0	4	0

5.3 Estimation of probable overtopping

The levels of Maku dam crest and normal water are 1690 m and 1680 m, respectively. The height of 1 m is designed for waves caused by winds and earthquakes, so the dam freeboard is 9 m. In rainy seasons, water level can rise close to spillway level. The spillway of Maku dam is a tunnel spillway in left dam sidewall with diagonal shaft and is located at level of 1685 m. As it is more probable that landslides happen in rainy seasons, we supposed the best (water level of 1680 m) and the worst (water level of 1685 m) states to estimate probable dam overtopping.The volume of overtopping water is calculated with below equation:

$$V = b \int \eta dt \tag{2}$$

The parameters are V: volume of overtopping water, b: the dam length, η: water surface height due to normal water surface and t: time. Time series of reaching wave to dam body are shown in figure 16 for two scenarios.

Maximum wave heights close to dam body are 7.8 m and 8.5 m for first and second scenarios, respectively. So in the best state, where the freeboard is 9 m, no overtopping happens. In rainy seasons, when the freeboard is 4 m, for first scenario about 3 m and for second scenario about 4 m of reaching wave height will be upper than the dam crest. According to equation (2) in this state volumes of probable overtopping are about 13000 Mm^3 and 20000 Mm^3 for two scenarios, respectively.

6 CONCLUSIONS

Landslide wave generation in the Maku dam reservoir in the West Azerbaijan province in Iran, was studied using a two-dimensional forth-order boussinesq type numerical model (LS3D). Impulsive wave generation and propagation caused by two scenarios of probable landslides, one as submarine at left reservoir sidewall, and another as sub-aerial at right reservoir sidewall, were simulated with LS3D model. Summary of results are observed in table 3.

As the Muko dam freeboard is about 9 m, the probability of overtopping with high volumes at normal water level (1680 m) condition is low. If landslides occur at rainy seasons, in the maximum water level (1685 m) condition, overtopping can reach up to 20000 m^3.

REFERENCES

Ataie-Ashtiani, B. & Malek-Mohammadi, S. 2007a. Mapping Impulsive Waves due to Subaerial Landslides into a Dam Reservoir: Case Study of Shafa-Roud Dam. Dam Engineering, March2008, XVIII(3): 1–25

Ataie-Ashtiani, B. & Malek-Mohammadi, S. 2007b. Near field Amplitude of Sub-aerial Landslide Generated Waves in Dam Reservoirs. Dam Engineering, February/March XVII (4): 197–222.

Ataie-Ashtiani, B. & Najafi-Jilani, A. 2006. Prediction of Submerged Landslide Generated Waves in Dam Reservoirs: An Applied Approach. Dam Engineering, XVII(3): 135–155.

Ataie-Ashtiani, B. & Najafi-Jilani, A. 2007. A Higher-order Boussinesq-type Model with Moving Bottom Boundary: Applications to Submarine Landslide Tsunami Waves.

International Journal for Numerical Methods in Fluids, 35 (6): 1019–1048.

Ataie-Ashtiani, B. & Nik-khah, A. 2008. Impulsive Waves Caused by Subaerial Landslide. Environmental Fluid Mechanics(2008) 8: 263–280.

Grilli, S.T., Vogelmann, S. & Watts, P. 2002. Development of a 3D Numerical Wave Tank for Modeling Tsunami Generation by Underwater Landslides. Journal of Engineering Analysis with Boundary Elements 26: 301–313.

Kirby, J.T., Wei, G., Chen, Q., Kennedy, A.B. & Dalrymple, R.A. 1998. FUNWAVE 1.0 Fully Nonlinear Boussinesq wave model Documentation and user's Manual. Research Report No. CACR-98-06/September1998.

Lynett, P. & Liu, P.L. 2002. A Numerical Study of Submarine-Landslide-Generated Waves and Run-up. Philosophical Trans. Royal society, London, U.K..A458, pp.2885–2910.

Najafi-Jilani, A. & Ataie-Ashtiani, B. 2007. Estimation of Near-Field Characteristics of Tsunami Generation by Submarine Landslide. Ocean Engineering (2007), doi:10.1016/ j. oceaneng.2007: 11.006.

Semenza, E. 2000. La storia del Vajont, raccontata dal geologo che ha scoperto la frana. Tecomproject, Ferrara, 2002.

Synolakis C.E. 1987. The run-up of solitary waves. Journal of Fluid Mechanics 185: 523–545.

"Niro" ministry. Local water organization of west Azerbaijan province of Iran. possibility studies of border rivers of Iran and Turkey. Drainage network of "Bazargan" plain. (in Persian).

"Mahab Ghods" Inc. 1999. Final report of Maku Dam reservoir project (in Persian).

Prediction and Simulation Methods for Geohazard Mitigation – Oka, Murakami & Kimoto (eds)
© 2009 Taylor & Francis Group, London, ISBN 978-0-415-80482-0

Relationship between climate change and landslide hazard in alpine areas interested by thawing permafrost

V. Francani & P. Gattinoni
Politecnico di Milano, Milano, Italy

ABSTRACT: The paper deals with the landslide hazard assessment with regard to the thaw of mountain permafrost, that in the alpine region interests chiefly moraine and debris deposits situated at an altitude higher than 2300 m a.s.l. The aim of the study was the understanding of the causes of the presence of geomorphologic and climatic critical characteristics, through a detailed hydrogeological and geomechanical modeling of the mountain slopes interested by thawing permafrost. The modeling results showed that, because of the high water content, the kinematisms triggered by thawing permafrost are mostly debris flows and they are often concomitant with piping phenomena and liquefaction.

1 INTRODUCTION

The global warming related to the climate change of the recent decades is one of the main causes of both glacier retreat and thawing permafrost (Bateau et al., 2005), that is the permanently frozen ground. Whereas the first effect is well evident in the eyes of the world, the second one is more hidden even if it is much more critical, because of its influence on slopes stability (Davies et al. 2001, Fischer et al. 2006, Gudem & Barsch 2005, Kääb et al. 2005).

This paper deals with the landslide hazard assessment with regard to the thaw of mountain permafrost. The aim of the study was the understanding of the causes of the presence of geomorphologic and climatic critical characteristics, through a detailed hydrogeological and geomechanical modeling.

To this aim, the study was subdivided in different phases. First, the areas subjected to thawing permafrost and the related landslide kinematics were identified both according to previous Authors (Chiarle et al. 2004, Noetzli et al. 2006) and on the basis of landslide events that interested the alpine region in the last decades (Table 1). This analysis allowed to point out the typical features of the areas exposed to this kind of risk and to reconstruct the conceptual model of the phenomenon, useful in the following modeling analysis.

Then, some case histories in Valmalenco (Sondrio District, Northern Italy) were modelled. In particular, the ground temperature profiles were obtained, corresponding to different climate change scenarios. On the base of these results, the tenso-deformative response of the slope was simulated, pointing out the effects of thawing permafrost on stability.

The applicative interest on the topic is connected to the risk arising from thawing permafrost in alpine region and especially from the landslides triggered by it, that can damage not only touristic facilities at high

Table 1. Landslides occurred during the last century in periglacial areas of the Alps, possibly related to the thawing permafrost.

Year Place	Rainfall	Altitude (m a.s.l.)	Vol. (10^3 m^3)	Length (km)
1993 Levanne (To)	heavy/long	2525	800	5.7
2001 Alphubel, (CH)	no	3060	25–40	5.1
1950 Sissone V. (So)	1 d	2380		5.0
1987 Presanella (Tn)	heavy	2200		4.6
1996 Rutor (Ao)	rainstorm	2500	300	4.0
1991 M. Blanc (Ao)	3 h	2600		3.4
1979 Belvedere gl. (Vb)	no	2200		3.0
1987 M. Leone (Vb)	heavy/long	2350		3.0
1990 Dolent moraine (CH)	250 mm/ 40 d	2610	30	2.4
1997 Chambeyron (FR)	no	2700	10(?)	2.3
1999 Montasio (Ud)	rainstorm	2150		2.0
1994 M. Pelmo (Bl)	115 mm/ 48 h	2185	200	1.7
2003 M. Blanc (Ao)	no	2080		1.3
1998 M. Blanc (Ao)	20 min	2020		1.2
1994 Fletschorn (CH)	heavy/long	2300	100	

altitude (i.e. cableways), but also towns located in the valleys below.

2 CONCEPTUAL MODEL

2.1 Geomorphological characteristics

The conceptual model was implemented with reference to a specific area of the Italian Alpine Region, the Valmalenco (Sondrio District), where there are evidences of the phenomenon being studied:

– active rock glaciers,
– debris flow occurrence with thermal zero at high altitude (Sissone Valley, Table 1 and Fig. 1).

From a geological point of view, the studied area is characterized by the outcropping of metamorphic and magmatic formations (Trommsdorff et al. 2005). Glacial deposits cover a wide surface of steep slopes, at an altitude between 2100 and 2400 m a.s.l.

Based on the geological and geomorphological setting of the area, the geometrical scheme of the slope was pointed out (Fig. 2), considering also that the face of the active rock glaciers, often considered evidence of permafrost (Haeberli 2005), are located at an altitude between 2300 and 2700 m a.s.l. and that the thermal zero is at 2550 m a.s.l.

2.2 Thermal profile

The thermal profile of the slope was obtained with reference to different climate scenarios (IPCC 2001).

According to the temperature series of the study area, at the altitude of interest (2750 m a.s.l.) the length of the ablation season, having an average value equal to 103 days, was correlated to the average annual temperature. For an annual temperature rise of 0.035°C, an increase of 1 day in the ablation season was assessed.

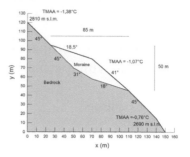

Figure 2. Geometrical scheme of the slope.

Table 2. Thermal and geomechanical properties.

	Bedrock	Moraine	Ice
Thermal capacity kJ/(m³°C)	2490	1875	1880
Thermal conduct. J/(sm°C)	2,5	2,0	2,23
Density (kg/m³)	2700	1800	920
Friction angle (°)	55	30–43	$\phi(T)$
Cohesion (kPa)	1.4E2	1–13	c(T)
Bulk modulus (Pa)	2.3E9	2.4E8	2E9
Shear modulus (Pa)	1.25E8	2.5E7	9.2E8
Tensile strength (Pa)	8E6	–	2E5
Porosity (%)	0.1	30–60	–
Hydraulic conductivity (m/s)	1E-11	1E-4	1E-6

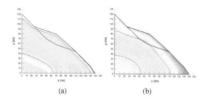

(a)   (b)

Figure 3. Permafrost active layer (represented by the more superficial dotted line): (a) at present day (depth between 3.2 and 3.4 m from land surface); (b) after an average temperature rise equal to 2.5°C.

Then, considering the average values of the thermal properties of the materials (Table 2), the lowering of the permafrost active layer (layer interested by yearly cycle of frost-thaw) was pointed out as a function of the temperature rise (Figs. 3–4).

2.3 Type of permafrost

In the conceptual model different kind of permafrost were considered to be present in the moraine below the active layer:

– saturated, that is present in the moraine as a cement of continuous ice, having a volume depending on the moraine porosity (Fig. 5a);

Figure 1. Sissone Valley (Valmalenco, Northern Italy). There are evidences of the debris flows occurred in 1950 and 1987.

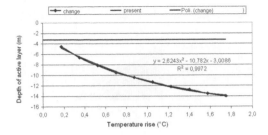

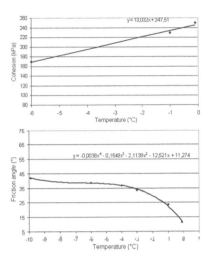

Figure 4. Permafrost active layer lowering versus temperature rise.

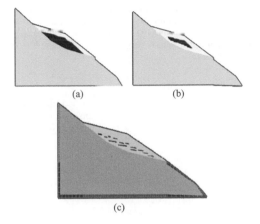

Figure 5. Permafrost (dark colour into the moraine, see Fig. 2) typology: (a) continuous permafrost, (b) single lens of massive ice and (c) sporadic permafrost.

Figure 6. Ice cohesion (above) and friction angle (below) versus temperature.

As far as the thawing permafrost is concerned, it was modelled as an internal boundary condition, considering a water flow melting from the ice during the ablation season.

Form a mechanical point of view, even if the creep deformations are generally expressed by the Glen's low, the permafrost behaviour in the slope failure can be described through the Mohr-Coulomb criterion (Fortt & Schulson 2007), considering both the cohesion and the friction angle depending on the temperature (Fig. 6).

3.1 *Parametrical long time simulations*

First, some parametrical simulations were carried out (Table 2) to identify the critical geomorphologic and climatic conditions.

The results show the remarkable importance of the permafrost typology (Fig. 7). In particular, in presence of massive ice lenses, the critical temperature rise is equal to 0.3–0.5°C; whereas, for continuous and sporadic permafrost, the critical temperature rise is equal respectively to 1.2 and 0.8°C. Also, in the case of sporadic permafrost, the location of the lenses (in terms of distance from the moraine foot and depth) has great effects on the stability (Fig. 8), whereas, in presence of continuous permafrost, the most affecting parameter is the moraine porosity (Fig. 9).

3.2 *Kinematisms and failures mechanisms*

The extension of the areas potentially interested by the phenomenon depends on its kinematism. According to the results obtained, the failure mechanism is typically rotational (Fig. 10). Moreover, the simulations show

– discontinuous, with massive ice lenses in which the ice fills a volume greater than the pore volume of the moraine ((Fig. 5b);
– sporadic, that appears like ice lenses widespread in the moraine (Fig. 5c).

3 SIMULATIONS AND RESULTS

Even if the problem being studied is evidently coupled thermal-hydraulic and mechanical, the simulations were carried out with an uncoupled approach.

First the thermal profile was simulated through the software Geoslope, then it was inserted as input in the software FLAC (Fast Lagrangian Analysis of Continua, Itasca 1998) for the solution of the coupled hydro-mechanical problem. Having no laboratory tests, the mechanical properties were assigned on the base of literature values, considering a range for those parameters thought of great importance for the following simulations (Table 2).

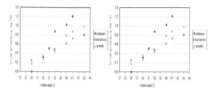

Figure 7. Critical temperature rise versus friction angle for the different permafrost typology (Fig. 4).

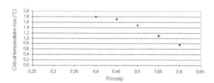

Figure 8. Critical temperature rise versus horizontal distance of the sporadic permafrost from the moraine foot.

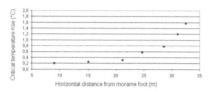

Figure 9. Critical temperature rise versus moraine porosity in presence of continuous permafrost.

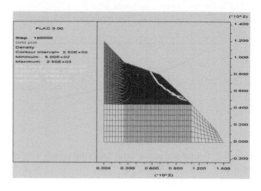

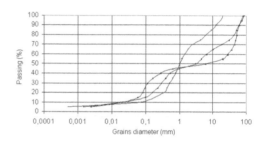

Figure 10. Example of simulations results: max shear stress increment (in white) for the case of sporadic permafrost (in black are the lenses and the contact moraine-bedrock). The grid used for modeling is very fine (min 0.3 m) in the most interesting area.

Figure 11. Grains size distributions obtained for the Sissone Valley moraines.

that the failure generally come with a sudden emptying out of the aquifers. Therefore, a significant water release takes place, bringing about the evolution of the phenomenon in debris flow, especially in presence of massive ice.

If a continuous permafrost is present, the locally high porosity can bring about a groundwater flow concentration along main directions having high permeability, according to the piezometrical gradient that often achieves its highest value near the foot moraine (groundwater exit point). If the piezometrical gradient achieves a critical threshold, localized piping phenomena can initiate, with the removal of finest materials and the consequent further increase of permeability. Evidently, the process grows by itself: once soil particles are removed by erosion, the magnitude of the erosive forces increases due to the increased concentration of the flow.

Actually, the slope moraines are a typical geological situations where piping phenomena can be significant, because of the concurrence of several adverse conditions, such as:

– the grains size, characterized by high heterogeneity, with glacial clays (Fig. 11) prone to piping (Fig. 12);

– the high slope dip and moraine porosity, that bring about a critical piezometrical gradient for piping initiation; in the case study, the critical threshold is equal to 0.52, according to the Zaslavsky & Kassiff (1965) method;
– the high piezometrical gradient, localized at the moraine foot, that often exceeds the critical threshold.

The pipe evolution was simulated iteratively (Bonomi et al. 2005), starting from the identification of those cells where processes of pipe initiation take place. In these cells, pipes were inserted in the model, as an equivalent material having higher permeability and porosity and lower density (Watanabe & Imai 1984), and then a new simulation was carried out to identify those cells interested by pipes growing (Fig. 13).

If, during the evolution of the phenomenon, the pipes clog, they may behave as closed pipes (not connected to the land surface); then, the porewater pressures at the pipe outlet became higher and higher with the increase of the pipe length. In the latter case,

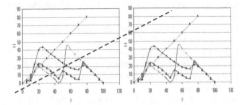

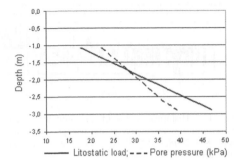

Figure 12. Application of the Kenney and Lau's method (1985) to the 3 grain size distribution curves in Figure 11. F is the percentage of grains having diameter D; H is the percentage increment of grains having diameter 4D with reference to the ones having diameter D. The dotted line divides the stability zone (above) from the instability zone (below); the 3 curves are the H(F) curves for the case examined.

Figure 14. Lithostatic load and pore pressure for different depth along a vertical section at the moraine foot.

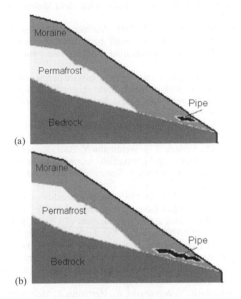

Figure 13. Piping evolution. The cells having piezometrical gradient equal or higher then j_{cr} are showed at the foot of the slope in a deep colour: a) starting phase; b) pipes widening.

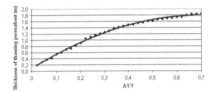

Figure 15. Thickness of the thawing permafrost versus the temperature deviation with respect to its average value.

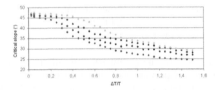

Figure 16. Critical slope versus the temperature deviation with respect to its average value, for different moraine porosity.

the more the porewater pressure increases, the more the shear strength decreases, thus leading to the failure. Also, in undrained conditions, if the increment in porewater pressures is very high, it can even set the effective stresses to zero with the consequent soil liquefaction (Fig. 14). The simulations also showed that the depth to which the liquefaction can take place increases with the increase of the temperature, involving larger and larger volumes of the moraine.

3.3 Short time simulations

Finally, some short time simulations were carried out, studying the effects on the stability of a very hot

summer. To this aim, the thermal profiles were simulated, pointing out the correlation between the thawing permafrost and the temperature deviation with respect to its average summer value (Fig. 15), that for the study area is equal to 6°C at 2700 m a.s.l. in the period 1864–2008.

Then, some simulations allowed to identify the critical temperature deviation for different moraine porosity and slope dip (Fig. 16). The results show that instability can occur for $\Delta T/T$ higher than 0.3, that corresponds to summer temperature deviation at an altitude of 2700 m a.s.l. higher than 1.8°C. Considering the temperature series of the studied area (Fig. 17), such a summer temperature deviation has became more and more frequent in the last few decades, with a max value equal to +4.03°C at an altitude of 1800 m a.s.l.

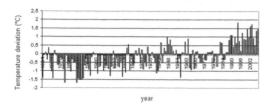

year

Figure 17. Historical series of the summer temperature deviation, with respect to its average value of the period 1864–2008, for the studied area, at an altitude of 1800 m a.s.l..

4 CONCLUSIONS

The study pointed out the critical conditions, both climatic and geomorphologic, for the moraine landslide in permafrost area.

It is important to highlight that, even if the modelling was carried out with reference to the Valmalenco area, the results can be considered of more general validity. Actually, the conceptual model used for the simulations is typical of a large amount of alpine slope at altitudes above 2500 m a.s.l.

As far as the critical climatic conditions are concerned, the temperature increase able to trigger the instability (at about 1°C for long period stability and a summer deviation at about 2°C for the short period stability) has a high probability to occur in the future, according to the more recent scenarios of the climate change.

Furthermore, the parametrical simulations emphasized that the permafrost typology is one of the most important elements affecting the moraine stability in consequence of global warming. In particular, the melting of massive ice and sporadic permafrost is the more probable triggering cause of the instability phenomena in permafrost area.

Finally, the modeling results showed that, because of the high water content, the kinematisms triggered by thawing permafrost are mostly debris flows and they are often concomitant with piping phenomena and liquefaction.

As regards to the precautionary measures, besides the obvious necessity to prevent the global warming, the systematic monitoring of the periglacial areas is very important, both as an alarm device and to improve the present knowledge concerning the thawing permafrost consequent to climate change.

REFERENCES

Bonomi, C., Francani, V., Gattinoni, P. & Villa, M. 2005. Il piping come fattore d'innesco del franamento: il caso di Stava, *Quaderni di Geologia Applicata* 12(2): 41–56.

Buteau, S., Fortier, R., Delisle, G. & Allard, M. 2004. Numerical simulation of the impacts of climate warming on a permafrost mound. *Permafrost and Periglacial Processes* 15: 41–57.

Chiarle, M., Iannotti, S., Mortara, G. & Deline, P. 2007. Recent debris flow occurrences associated with glaciers in the Alps. *Science Direct-Global and Planetary Change* 56: 123–136.

Davies, M.C.R., Hamaza, O. & Harris, C. 2001. The effect of rice in mean annual temperature on the stability of rock slopes containing ice-filled discontinuities. *Permafrost and Periglacial Processes* 12: 137–144.

Fischer, L., Kaab, A., Huhhel, C. & Noetzli, J. 2006. Geology glacier retreat and permafrost degradation as controlling factors of slope instabilities in high- mountains rock wall: the Monte Rosa east face, *Nat.Hazard Earth Syst. Sci* 6: 761–772.

Fortt, A.L. & Schulson, E.M. 2007. The resistence to sliding along Coulombic shear faults in ice. *Acta Materialia* 55: 2253–2264.

Gude, M. & Barsch, D. 2005. Assessment of geomorphic hazards in connection with permafrost occurrence in the Zugspitze area (Bavarian Alps). *Geomorphology* 66: 85–93.

Haeberli, W. 2005. Investigating glacier-permafrost relationship in high-mountain areas: historical background, selected examples and research needs, *Geological Society, London,* Special Publications 242: 29–37

Kääb, A., Huggel, C., Fischer, L., Guex, S., Paul, F., Roer, I., Salzmann, N., Schlaefli, S., Schmutz, K., Schneider, D., Strozzi, T. & Weidmann Y. 2005. Remote sensing of glacier- and permafrost- related hazards in high mountains: an overview, *Natural Hazards Earth System Sciences* 5: 527–554.

Kenney, T.C. & Lau, D. 1985. Internal stability of granular filters. *Canadian Geotechnical Journal,* 22(2): 215–225.

IPCC 2001. Third Assessment Report (TAR) of Working Group I, Cambridge (UK): Cambridge University Press,.

Noetzli, J., Huggel, C., Hoelzle, M. & Haeberli, W. 2006. GIS-based modelling of rock-ice avalanches from Alpine permafrost areas. *Computational Geosciences* 10(2006): 161–178.

Trommsdorff, V., Montrasio, A., Hermann, J., Müntener, O., Spillmann, P. & Gieré, R. 2005. The Geological Map of Valmalenco. *Schweiz. Mineral. Petrogr.Mitt.* 85, 1–13.

Zaslavsky, D. & Kassiff, G. 1965. Theoretical formulation of piping mechanism in cohesive soil. *Géotechnique,* XV: 305–314.

Watanabe, K. & Imai, H. 1984. Transient behaviour of groundwater flow in idealized slope having a soil pipe in it - numerical analysis of the saturated-unsaturated groundwater flow by the use of finite element method of three dimensional form. *Journal of the Japan Society of Engineering Geology* 25: 1–8.

Prediction and Simulation Methods for Geohazard Mitigation – Oka, Murakami & Kimoto (eds)
© 2009 Taylor & Francis Group, London, ISBN 978-0-415-80482-0

Numerical analysis on deformation/failure patterns of embankments observed in 2004 Chuetsu earthquake

T. Noda, M. Nakano, E. Yamada, A. Asaoka & K. Itabashi
Nagoya University, Nagoya, Japan

M. Inagaki
Central Nippon Expressway Company Limited, Kanazawa, Japan

ABSTRACT: Soil-water coupled finite deformation analysis without distinguishing between dynamic and static analyses was carried out for three damage patterns in embankments with the ground stiffness and slope conditions seen in the 2004 Niigata Chuetsu earthquake, and the damage mechanisms were investigated. The elasto-plastic constitutive equation for the soil is capable of describing the behavior of a wide range of soils, from sand to intermediate soil to clay, etc., by focusing on the soil skeleton structure (structure, over consolidation, and anisotropy). The main conclusions from the results of the analysis were as follows. (1) In ground with laminating layers of weak sand and clay, disturbance of the structure of the clay layers occurs, particularly directly below the embankment during and after an earthquake, which causes settlement. (2) In the case of stiff ground, differential settlement of the crown of the embankment and bulging of the bottom of the slope are caused by localized liquefaction within the embankment and shear deformation of the slope surface.

1 INTRODUCTION

In the Niigata Chuetsu earthquake of October 2004, liquefaction and ground lateral flow occurred in the thin sand and clay layers of the foundation soils of the Kanetsu Expressway. Additionally, shaking down settlement accompanied by up to 8% compression strain occurred in the expressway's embankments and foundation soils that contained clay soils, and this resulted in the road being completely severed. Even allowing for the fact that the seismic motions in the vicinity of Ojiya IC were unexpectedly strong, at 1314 gal, the highway engineers, who expected at most 1% compression strain in the embankment during an earthquake, were surprised by this shaking down settlement. Fig. 1 shows the three typical damage patterns that occur in embankment-foundation soil systems, as summarized by the former Japan Highways Public Corporation (JH). According to JH, they are classified as follows: (a) comparatively good foundation soils, and so the embankment material flows; (b) settlement occurs in the embankment itself with the shape of the embankment being maintained to a certain extent; and (c) poor foundation soils, and so, as they soften, settlement occurs in them and the embankment (Okubo et al., 2005). In this paper, the various types of foundation soil and embankment conditions corresponding to these damage patterns were simplified, dynamic/static soil–water coupled finite deformation calculations (Asaoka and Noda, 2007;

Noda et al., 2008) were carried out using the SYS Cam-clay model (Asaoka et al., 2002) as the soil constitutive model, and the various types of seismic behavior of the embankment-foundation soil system were investigated.

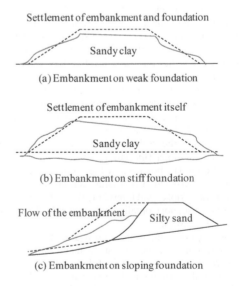

(a) Embankment on weak foundation

(b) Embankment on stiff foundation

(c) Embankment on sloping foundation

Figure 1. Patterns of embankment damage (Okubo et al., 2005).

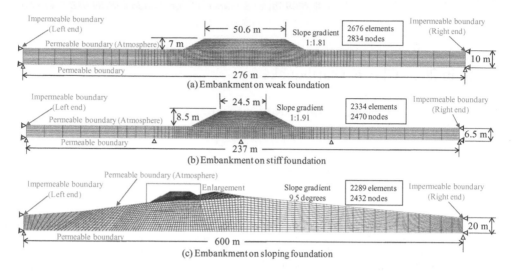

Figure 2. Finite element mesh (after construction of embankment).

2 CALCULATION CONDITIONS (FOUNDATION SOIL AND EMBANKMENT PARAMETERS AND SEISMIC FORCE)

Figs. 2 and 3 show the finite element meshes at the three positions that were subjected to analysis. The calculations were carried out under the conditions of plane strain over the entire cross-section. The bottom surface was engineering bedrock with a viscous boundary at its bottom surface and was considered to be a stratum with Vs > 300 m/s. In addition, the distance to the side surfaces was considered to be sufficient to prevent them from being affected by embankment loading, and they were modeled as periodic boundaries (constant displacement boundaries). Furthermore, the hydraulic boundaries were assumed such that the water level surface coincided with the ground surface at which the water pressure was zero, the left and right end surfaces were assumed to be undrained boundaries, and the bottom surface was assumed to be a drained boundary, taking the sand and gravel strata into consideration. Table 1 show the soil stratigraphy at the positions of the calculations (the stratigraphy of the sloping foundation soil is shown within Fig. 3). The weak foundation soil consisted of alternating clay layers with low N-values and loose sand layers, and the stiff foundation soil was dense, horizontally- and uniformly-deposited sand/gravel layers.

The sloping foundation soil was a thick deposit of very hard sandstone below a stepped cut sandy gravel layer, which was below a thin silt layer. Table 1 shows the material constants and initial values used in the analyses. For positions where undisturbed or disturbed samples could be obtained by sampling or boring, the

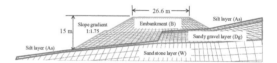

Figure 3. Finite element mesh of sloping foundation (enlargement).

density (initial void ratio) of the in-situ initial conditions was determined by reverse analysis, based on laboratory test results, using a constitutive equation. For strata for which there were no test samples, the soil material was specified based on the physical characteristics of similar materials previously tested by the authors. The initial values of the specific volume and structure within a stratum were assumed to be uniform and homogeneous, and only an over consolidation ratio was applied corresponding to the soil overburden pressure.

In the calculations, the embankment was first built onto the foundation soil at a loading rate of 0.08 m/day up to the specific height in order to take into consideration the construction process (Takeuchi et al., 2006). Next, consolidation was applied for the number of years' service of the road (25 years), and the embankment was left to allow the excess pore water pressure, due to embankment loading, to dissipate. Then, half the acceleration of the free ground surface, calculated from the measured ground surface wave at K-NET Ojiya, shown in Fig. 4, was applied to the bottom surface of the foundation soil, and the consolidation calculations were continued until the excess pore water pressure in the foundation soil dissipated and the foundation soil stabilized after the earthquake.

Table 1. Material constants and initial conditions.

(a) Weak foundation soils

(m)	B GL+7.0~GL+1.0	Ts GL+1.0~GL	Ac1 GL~GL-2.0	Ac2-1 GL-4.0~GL-6.0	Ac2-2 GL-6.0~GL-7.0	Ac2-3 GL-7.0~GL-10.0
Elasto-plastic parameters						
λ	0.128	0.050	0.185	0.260	0.193	0.193
$\tilde{\kappa}$	0.020	0.012	0.010	0.022	0.030	0.030
M	1.30	1.00	1.40	1.70	1.35	1.35
N	1.91	1.98	2.05	2.63	2.405	2.405
ν	0.20	0.30	0.35	0.25	0.30	0.30
Evolution parameters						
m	0.25	0.06	2.8	3.0	0.8	0.8
a	3.4	2.2	0.01	0.3	0.6	0.6
b	0.4	1.0	1.0	1.0	1.0	1.0
c	1.5	1.0	1.0	1.0	1.0	1.0
c_s	0.92	1.0	0.1	0.25	0.3	0.3
b_r	0.7	3.5	0.001	0.01	0.01	0.01
m_r	0.6	0.6	1.0	1.0	1.0	1.0
Initial conditions						
η_0	0.545	0.545	0.545	0.545	0.545	0.545
ν_0	1.94	1.79	2.20	3.67	2.37	2.58
$1/R_0^*$	6.0	12.5	3.5	43.0	2.0	5.3
$1/R_0$				0.270	0.250	0.250
k	1.75×10^{-4}	2.80×10^{-4}	1.00×10^{-6}	1.00×10^{-7}	1.00×10^{-7}	1.00×10^{-7}
ρ_s	2.641	2.650	2.631	2.559	2.648	2.469

(b) Stiff foundation soils

(m)	B GL+8.0	As GL~GL-1.0	Ag GL-1.0~GL-4.5	As1 GL-4.5~GL-6.0	Ac2 GL-6.0~GL-6.5
Elasto-plastic parameters					
λ	0.128	0.220	0.050	0.050	0.063
$\tilde{\kappa}$	0.020	0.055	0.012	0.012	0.012
M	1.30	1.30	1.00	1.00	1.45
N	1.91	2.50	1.98	1.98	1.51
ν	0.20	0.20	0.30	0.30	0.30
Evolution parameters					
m	0.25	1.00	0.06	0.06	0.15
a	3.4	0.3	2.2	2.2	10.0
b	0.4	0.5	1.0	1.0	1.0
c	1.5	0.4	1.0	1.0	1.0
c_s	0.92	0.001	3.5	3.5	0.3
b_r	0.7	0.7	0.6	0.6	0.65
m_r	0.6				
Initial conditions					
η_0	0.545	0.545	0.545	0.545	0.545
ν_0	1.94	2.15	1.89	1.90	1.57
$1/R_0^*$	6.0	1.5	1.25	1.25	1.37
$1/R_0$		0.100	0.545	0.545	0.344
k	1.75×10^{-4}	1.00×10^{-7}	2.80×10^{-7}	2.80×10^{-5}	1.80×10^{-7}
ρ_s	2.641	2.640	2.650	2.650	2.650

(c) Sloping foundation soils

(m)	B	As1	Dg	W
Elasto-plastic parameters				
λ	0.083	0.063	0.050	0.050
$\tilde{\kappa}$	0.008	0.012	0.012	0.012
M	1.10	1.51	1.00	1.00
N	1.91	1.51	1.98	1.98
ν	0.20	0.30	0.30	0.30
Evolution parameters				
m	0.20	0.15	0.06	0.06
a	4.0	10.0	2.2	2.2
b	0.4	1.0	1.0	1.0
c	1.0	1.0	1.0	1.0
c_s	0.8	0.3	1.0	1.0
b_r	0.5	0.3	3.5	3.5
m_r	1.1	0.65	0.7	0.7
Initial conditions				
η_0	0.545	0.545	0.545	0.545
ν_0	1.94	1.56	1.79	1.79
$1/R_0^*$	20.0	12.0	1.25	1.25
$1/R_0$	0.545	0.545		
k	2.80×10^{-4}	1.00×10^{-7}	1.00×10^{-5}	1.00×10^{-5}
ρ_s	2.650	2.650	2.650	2.650

λ: Compression index, $\tilde{\kappa}$: Swelling index, M: Critical state constant, N: NCL intercept (at $p' = 98.1\text{kPa}$), ν: Poisson's ration, m: Degradation parameter of overconsolidated state, a, b, c and c_s: Degradation parameter of structure, b_r: Evolution parameter of β, m_r: Limit of rotation, η_0: Initial stress ratio, ν_0: Initial specific volume, $1/R_0^*$: Initial degree of structure, $1/R_0$: Initial degree of overconsolidated state, ς_0: Initial degree of anisotropy, k: Coefficient of permeability (cm/s), ρ_s: soil particle density (g/cm³)

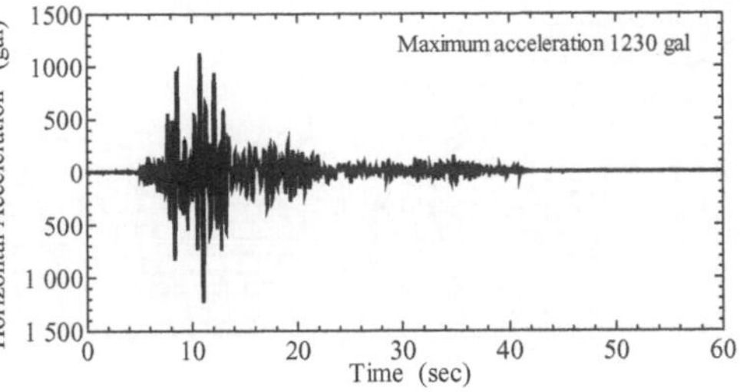

Figure 4. Input seismic wave acceleration (K-NET Ojiya; EW direction).

The expressway embankment had been constructed to satisfy the construction specification of 90% consolidation, and it would normally be considered to be an embankment stabilized in the unsaturated condition. However, prior to the earthquake, a typhoon had passed through and subjected the area to five consecutive days of rain (corresponding to about 1/15 the annual rainfall amount), and so, the embankment may have been close to being in the saturated condition. Therefore, in the calculations, both the embankment and the foundation soil were assumed to be in the saturated condition.

3 CALCLATION RESULTS (COUPLED BEHAVIOR OF THE FOUNDATION-EMBANKMENT SYSTEM BEFORE, DURING, AND AFTER EARTHQUAKE)

3.1 Embankment on weak foundation soils

Fig. 5 shows a comparison of the deformations near the slope surface of weak foundation soil (light color: before the earthquake) before and after the earthquake. During the earthquake, there was conspicuous settlement of the foundation soil at the center of the embankment and upheaval at the base of the slope, and the embankment itself deformed in a shape that followed the deformation of the foundation soil. Upheaval of the foundation soil at the base of the slope was confirmed at the site of the damage (Photo 1). Fig. 6 shows the variation with time of the amount of settlement of the foundation soil surface. At the time of construction of the embankment, 0.5 m of settlement occurred below the embankment. During the earthquake, the maximum upheaval of the base of the slope and maximum settlement of the shoulders of the slope occurred, and this produced a "W-shaped" deformation. Furthermore, there was about 0.3 m of settlement overall from the end of the earthquake until the end of consolidation. Fig. 7 shows the amount of settlement, obtained from the time-settlement relationship, for the different strata beginning at the start of the

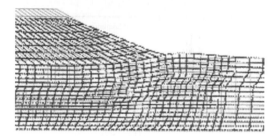

Figure 5. Deformation of weak foundation before and after the earthquake.

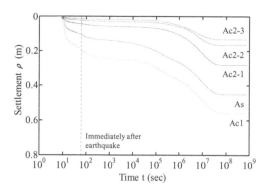

Figure 7. Time-settlement relationship below the center of the embankment.

Photo 1. Upheaval of foundation near the slope base.

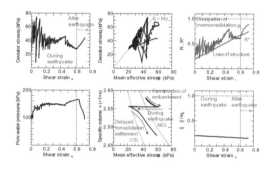

Figure 8. Shear strain distribution 20 years after earthquake.

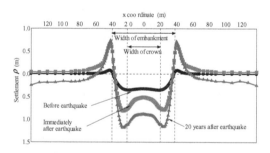

Figure 6. Amount of settlement of foundation (except for embankment).

Figure 9. Element behavior of the clay lower layer.

earthquake. In the sand (As) and silt (Ac1) layers, volumetric compression occurred immediately after the earthquake as a result of shaking down. On the other hand, in the upper and lower layers of clay 2 (Ac2), which were highly structured, settlement occurred slowly over a period of years after the earthquake. Fig. 8 shows the shear strain distribution 20 years after the earthquake. There is significant strain in the foundation soil below the base of the slope that is equal to that occurring in the layers with delayed consolidation.

Next, Fig. 9 shows the seismic behavior of the highly-structured clay layer as the element behavior of the lower clay layer below the base of the slope. When constructing the embankment, although

volumetric compaction was produced in the consolidation process, the stress state was above the normal consolidation line (NCL). However, when the earthquake occurred, there was an increase in excess pore water pressure, decay of structure (in other words, disturbance), and a reduction in the average effective stress p'. After the earthquake, decay of structure and volumetric compaction occurred in conjunction with recovery of p'. As a result, in the case of the embankment on the weak foundation soil, the decay of structure of the loosely deposited, highly-structure clay layer during the earthquake is believed to have contributed greatly to deformation of the embankment-foundation soil system.

3.2 Embankment on stiff foundation soil

Fig. 10 shows a comparison of the deformation of the stiff foundation soil (light color: before the earthquake) before and after the earthquake. Immediately after the earthquake, virtually no settlement or upheaval was seen in the foundation soil. On the other hand, there was lateral bulging of the slope surface and base of the slope and differential settlement of the crown of the embankment. This embankment deformation is in agreement with the site survey (Photo 2). Fig. 11 shows the variation with time of the shear stress distribution in the embankment. Significant strain occurred at the base of the slope of the embankment beginning immediately after the earthquake. The shear strain below the center of the embankment was small, but delayed settlement after the earthquake (omitted from the figure).

Figs. 12 and 13 show the element behavior at the base of the slope and immediately below the center of the embankment, respectively. The soil elements at the base of the embankment slope had almost no rise in water pressure during the earthquake and sheared almost in a drained manner, with the final strain reaching 50%. During consolidation after the earthquake, the specific volume increased due to water absorption expansion from near the slope surface. In contrast, the soil elements immediately below the center of the embankment showed behavior close to undrained shear during the earthquake, and p' rapidly decreased, although liquefaction did not occur (in other words, p' did not completely approach zero). Furthermore, during the earthquake, accumulation of over consolidation occurred in conjunction with decay

of structure, and thereafter, volumetric compression and dissipation of the accumulated excess pore water pressure occurred during the consolidation process.

3.3 Embankment on sloping foundation soil

Fig. 14 shows a comparison of the deformation of the sloping foundation soil (light color: before the earthquake) before and after the earthquake. As with the stiff foundation, there was almost no deformation

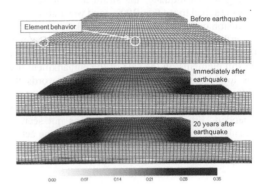

Figure 11. Deformation of stiff foundation after earthquake.

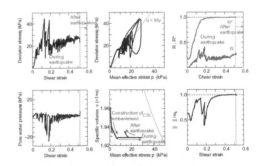

Figure 12. Element behavior at the base of the slope.

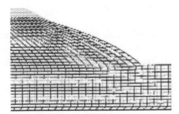

Figure 10. Deformation of stiff foundation before and after the earthquake.

Photo 2. Bulging on the slope surface.

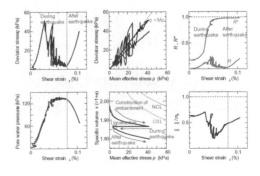

Figure 13. Element behavior at the immediately below the center of the embankment.

of the foundation immediately after the earthquake, but there was differential settlement of the crown of the embankment and bulging of the slope surface on the left-hand side. Fig. 15 shows the shear strain before and after the earthquake and at the time of completion of consolidation. Immediately after the earthquake, the shear strain suddenly increased on the side of the embankment that was lower than the step cut surface. Fig. 16 shows the behavior of the soil elements in the base of the slope on the left side of the embankment, where the strain was particularly significant. During the earthquake, the strain in these soil elements reached more than 100%, and water absorption expansion occurred after the earthquake.

Furthermore, although not shown in the figures, the strain gradually increased after the earthquake in the soil elements on the side of the embankment that was lower than the small step (in other words, the part where excavation was carried out to expose the sandy gravel layer [Dg]), and the extent of the strain expanded. From this, although it was not seen in the calculations, the reason for collapse of the actual embankment is believed to be the occurrence of water absorption expansion in the embankment due to inflow of water into it, as shown in Fig. 17.

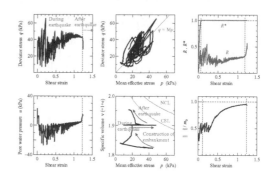
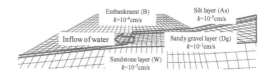

Figure 16. Element behavior in the base of the slope on the left side of the embankment.

Figure 17. Inflow of water into embankment on the sloping foundation.

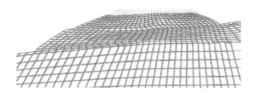

Figure 14. Deformation of sloping foundation before and after the earthquake.

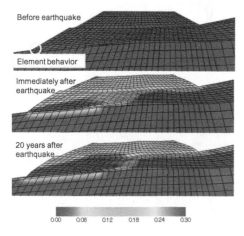

(Values exceeding 0.3 are shown as 0.3)

Figure 15. Shear strain distribution.

4 CONCLUSIONS

In this paper, different deformation and collapse patterns for three locations in the 2004 Niigata Prefecture Chuetsu earthquake were considered, and the deformation behavior during and after the earthquake was investigated using *GEOASIA*. The following is a summary of the calculation results, the main knowledge gained, and our speculations.

1. In the case of weak foundation soils containing loosely deposited sand and clay, decay of structure (in other words, disturbance) occurred, particularly in the clay layers of the foundation during and after the earthquake, due to the local loading of the embankment, and this resulted in settlement. There was also upheaval in the foundation soils during the earthquake near the base of the slope.
2. In the case of comparatively stiff foundation soils, differential settlement of the crown and bulging of the base of the slope occurred due to the combination of local liquefaction within the embankment and shear deformation of the slope surface.
3. In the case of the embankment on sloping foundation soils, bulging occurred on the slope surface on the valley side during the earthquake, and the slope is believed to have collapsed after the earthquake due to water absorption expansion and shear deformation in the embankment resulting from inflow of water from the mountain side as a result of the sloping terrain.

We have also demonstrated that, in the case of earthquake damage due to complex causes in embankment-foundation soil systems, such as that discussed in this paper, *GEOASIA* can seamlessly analyze the various mechanical phenomena (shear deformation/collapse, delayed consolidation settlement, shaking down settlement, and water absorption expansion) intrinsic to soils without switching programs for pre-, co- and post-seismic behavior.

REFERENCES

Asaoka, A., Noda, T., Yamada, E., Kaneda, K. and Nakano, M. 2002. An elasto-plastic description of two distinct volume change mechanisms of soils, *Soils and Foundations*, 42(5): 47–57.

Asaoka, A. and Noda, T. 2007. All soils all states all round geo-analysis integration, *Int. Workshop on Constitutive Modelling—Development, Implementation, Evaluation, and Application*, Hong Kong, China: 11–27.

Noda, T., Asaoka, A. and Nakano, M. 2008. Soil-water coupled finite deformation analysis based on a rate-type equation of motion incorporating the SYS Cam-clay model, *Soils and Foundations*, 48(6): 771–790.

Okubo, K., Hamasaki, T. and Yokoyama, Y. 2005. Damage States in Soil Structures, *Highway Technology*, 177: 27–33 (in Japanese).

Takeuchi, H., Takaine, T. and Noda, T. 2006. Nonlinearity in geometry on the consolidation deformation of saturated clay soil, *Journal of Applied Mechanics, JSCE*, 9: 539–550 (in Japanese).

Numerical and analytical simulation methods for geohazards

A proposal of 3D liquefaction damage map of Kochi harbor

M. Tanaka & Y. Watabe
Port and Airport Research Institute, Yokosuka, Japan

S. Sakajo
Geospheree Environmental Technology Corp., Tokyo, Japan

T. Sadamura
Kiso-Jiban Consultants Co., Ltd., Tokyo, Japan

ABSTRACT: In the present paper, the authors made a new liquefaction risk map at Kochi harbor. Conventional methods could show only 2D map, which is not convenient for the cost estimation of countermeasures, because it is not accurate to the depth. Herein, one dimensional vertical ground was cut from 3D ground model by the 3D soil layer estimation proposed by the authors. Liquefaction risk was calculated realistically, based on the technical standard of harbor with SHAKE analysis.

1 GENERAL INSTRUCTIONS

1.1 Procedure of damage map

Figure 1 show the procedure of making liquefaction map. This procedure consists of 1) soil layer estimation and 2) making liquefaction map. At the first step, a 3D ground model is made by the authors (Tanaka et al., 2007) Then precise liquefaction analysis in one dimension is conducted using a soil column cut out from the 3D ground model, where the depths of sandy and clay sub soils and bed rock layers are modeled precisely. Only this method could evaluate the liquefaction damages to the depth precisely, which will lead to the risk management regarding to liquefaction. This paper presents the abstract of method and an example of results of liquefaction map at Kochi harbor. Particularly, for the evaluation of liquefaction, a program code of SHAKE is used to compute one dimensional seismic response against the input strong motions in Japan.

1.2 The used database

For soil layer estimation, data base from PARI and geological information in Shikoku district and Kochi geological database are used. Table 1 shows the major data in the database from PARI, including standard penetration test (SPT), laboratory tests, in-situ tests. This database will be open to public in the near future, which will be very useful for many purposes in many projects of harbor and airports. This importance and effectiveness was confirmed by the application to the project of Haneda airport expansion by the authors.

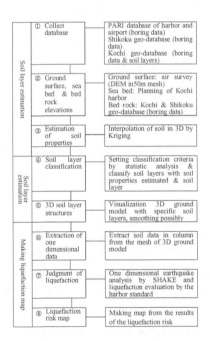

Figure 1. Procedure of making liquefaction risk map.

2 GEOLOGICAL CONDITIONS AT KOCHI HARBOR

2.1 Range of risk map with boring locations

Figure 2 shows the risk map area to consider by 3.7 km in East-west by 2.8 km in South-north including Kochi

Table 1. Major Data in database.

Items	Major data	
Fundamental	Longitude, latitude, elevation	
SPT	Depth, number of blows, length of penetration	
Physical properties	Density	Specific gravity, water content, submerged unit weight, void ratio, saturation degree
	Soil particle size	Gravel, sand, silt, clay (%)
	Consistency	Liquid limit and plasticity limit
	Soil classification	Japanese unified classification, Name of soil layer
Oedometer test	Compression index Cc, Pre-consolidation stress Pc′	
Un-confined test	$q_u (kN/m^2)$	
Tri-axial compression test	Cohesion c (kN/m^2), inner frictional angle $\phi(°)$	
Lateral loading test in the bore hole	Coefficient of reaction (K/m^3)	
PS exploration	Wave velocity Vs &Vp (m/sec)	

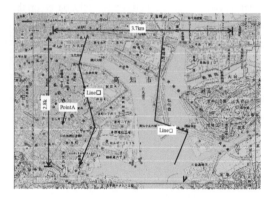

Figure 2. Target area and boring locations in Kochi harbor.

Table 2. Geological age in Kochi harbor.

Age	Geological age	Layer name	Remarks
Alluvium	Gen Urado harbor age	G I	River deposit
	Later Urado harbor age	S I a	
		M I	Sea deposit
		S I v	Sea deposit and partially river deposit
	Early Urado harbor age	S I b	ditto
		M II	Sea deposit
Last Ice Age (Gravel Layer)	Akatsuki Urado age	G II a	Sea deposit
Diluvium	Early Kochi basin age	M III a	ditto
	Old Urado harbor age	G III a	ditto

harbor, in the red color square on PARI database system. The blue dots are locations of boring, where are the locations of boring data for 79 data in the target area. Among them, 10 boring data has some test data but 69 has only SPT data. Therefore, there were some problems on the procedure of making liquefaction map as follows.

- Soil classification has to use the soil properties for the regression analysis. Because it is not easy to do so if the data are limited.
- Furthermore, if data is only limited except SPT, Kriging may not be enough to do accurately.

Then, in this study, other data of soil properties was imported out of this target area, in the Kochi city, to avoid this problem. And the area is referred without analysis where soil layer constitutions are already defined.

2.2 Soil layers in Shikoku

The order of soil layers are preliminary studied based on Kochi geological map as shown in Table 2. Alluvium layers were deposited around the age of Urado harbor age and diluvium layers were deposited around before that. G IIa is a gravel layer at the last ice age to separate Alluvium layers from Diluvium layers but it was classified as a Diluvium.

3 3D LAYER ESTIMATIONS

3.1 Soil characteristics from database

By all data using database, soil characteristics were analyzed to obtain criterion to classify soil to be a specific layer. Pre analytical results are shown Figures 3–4, which are classification of soil geological types, respectively. In these figures, criterion to separate soils are sown by solid lines, which is regression analysis among soil properties, N-value, natural water contents and Fine content (Fc) from database, although

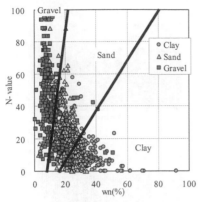

1) Sand and clay separation from N-values and W_c

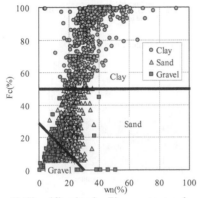

2) Classification by Fine content and w_n

Figure 3. Classification among soil types (Sand, Clay and Gravel).

PI is not used because of few data in the database. To classify soils with these criterion, at the unknown portions, soil properties would be interpolated by Kriging using those known at boring points.

3.2 Spatial interpolation of soil properties

Figure 5 shows the results of interpolation by Kriging for N-value, water content plasticity index PI and Fc (fine content of soil) on the target line as seen in Figure 2. Figure 6 shows N-value in 3D space interpolated by Kriging.

3.3 Soil layer estimation results comparing with Kochi geological map

Soil layer could be estimated by the author's method as shown in Figures 7 and 8, which s are at two cross sections on the lines of ① and ② in Figure 2. Also, these results were compared with those of Kochi

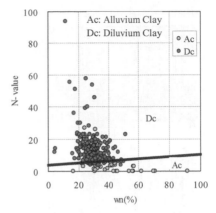

1) Classification between Ac and Dc layers

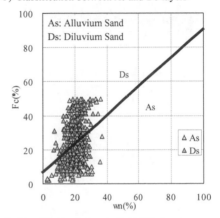

2) Classification between As and Ds layers

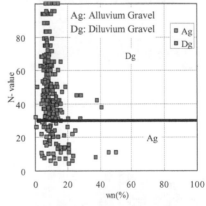

3) Classification between Ag and Dg layers

Figure 4. Classification between deposit ages.

geological map. The estimated ones coincide roughly well with Kochi geological maps. Alluvium sand layer, As is deposited near the ground surface, which could be easily liquefied by the earthquake. On the other

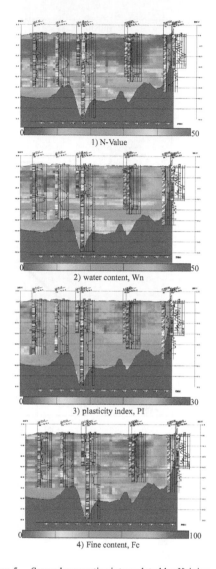

1) N-Value

2) water content, Wn

3) plasticity index, PI

4) Fine content, Fc

Figure 5. Several properties interpolated by Kriging.

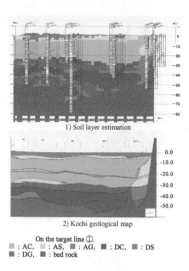

1) Soil layer estimation

2) Kochi geological map

On the target line ①.
■ : AC, ■ : AS, ■ : AG, ■ : DC, ■ : DS
■ : DG, ■ : bed rock

Figure 7. On the target line ①.

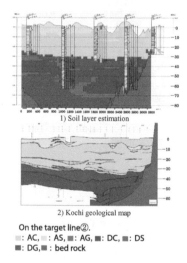

1) Soil layer estimation

2) Kochi geological map

On the target line②.
■ : AC, ■ : AS, ■ : AG, ■ : DC, ■ : DS
■ : DG, ■ : bed rock

Figure 8. On the target ②.

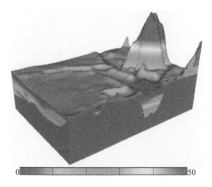

Figure 6. N-value in 3D space interpolated by Kriging.

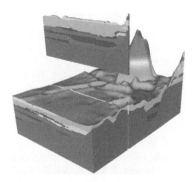

Figure 9. Soil layer estimation in 3D space.

hand, a thick alluvium clay layer, Ac is deposited beneath that, which could not be liquefied.

Furthermore, Figure 9 shows the 3D model image by this analysis. A given cross section could be cut from 3D body, which is very useful.

4 LIQUEFACTION RISK MAP

4.1 Seismic wave and SHAKE input

Figure 10 shows the seismic waves for SHAKE (GES, 1994). It was originally recorded by Tokachi-oki earthquake in 1968, at Hachinohe (Ministry of Transportation). Maximum acceleration was adjusted for 250 gal and 400 gal for this calculation.

4.2 Liquefaction judgment by harbor standard

At the each mesh, liquefaction judgment can be conducted by the harbor standard, where N_{65} and α_{eq} are compared in the given table in the standard to define liquefaction degree. These could be calculated from

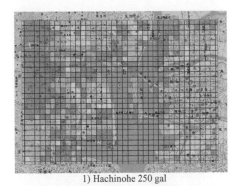

1) Hachinohe 250 gal

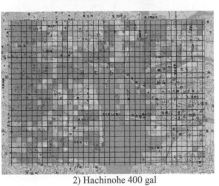

2) Hachinohe 400 gal

Liquefaction risk map by harbor Standard.
■ : class I :liquefied, ■ : class II :could be liquefied
▨ : class III: my not be liquefied, ▨ : class IV:not liquefied

Figure 12.   Liquefaction risk map by harbor Standard.

the following equations. SHAKE is required to obtain $\tau_{\max}$ to the depth to define α_{eq}.

$$N_{65} = \frac{N - 0.019\left(\sigma_V' - 65\right)}{0.0041\left(\sigma_V' - 65\right) + 1.0} \tag{1}$$

where, N_{65} is equivalent N-value, N is N-value, σ_v' is effective vertical pressure (in kN/m^2).

$$\alpha_{eq} = 0.7\frac{\tau_{\max}}{\sigma_V'}g \tag{2}$$

where, α_{eq} is equivalent acceleration (in gal) and $\tau_{\max}$ is maximum acceleration (kN/m^2).

4.3 Liquefaction risk maps at Kochi harbor

Figure 12 shows two liquefaction risk maps based on the above harbor Standard, where liquefaction occurred a little more widely in 400 gal than 250 gal, because Ac layer is very thick. These will be very helpful for the future planning.

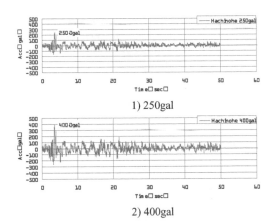

1) 250gal

2) 400gal

Figure 10.   Seismic wave of Hachinohe.

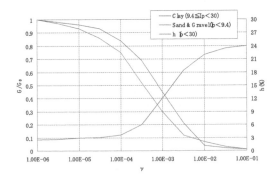

Figure 11.   Non-linearity for SHAKE.

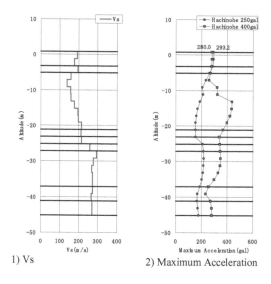

Figure 13. Information extracted from 3D soil layer estimation results (water table was assumed to be on the ground surface).

Altitude	SoilLayer	N-value	γt (kN/m³)	Vs (m/s)
0.84	AC	7.1	17.0	192.3
		5.7	17.0	178.3
-3.16	AS	11.8	20.0	196.6
-6.16	AC	4.0	17.0	159.3
		2.8	17.0	140.4
		4.0	17.0	158.9
		4.1	17.0	159.7
		5.7	17.0	178.5
		7.7	17.0	197.0
		7.9	17.0	199.5
		9.9	17.0	214.8
-21.16	DC	9.8	17.0	213.7
-23.16	DS	15.3	20.0	214.6
-25.16	DC	17.1	17.0	257.7
-27.16	DG	38.5	20.0	293.7
		32.2	20.0	276.1
		30.0	20.0	269.7
		30.8	20.0	272.0
		31.2	20.0	273.4
-37.16	DS	31.1	20.0	273.1
		28.9	20.0	266.2
-41.16	DC	20.0	17.0	271.4
-45.16		20.0	17.0	271.4

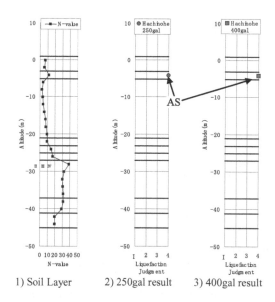

1) Soil Layer   2) 250gal result   3) 400gal result

Figure 15. Liquefaction results at Point A.

1) Vs   2) Maximum Acceleration

Figure 14. SHAKE result at Point A (Estimation result).

5 REMARKS ON LIQUEFACTION RISK IN THE ONE DIMENSION

Figure 13 shows a data of one dimensional liquefaction analysis of SHAKE at Point A, extracted from the 3D soil layer result by the size of 50 m by 100 m.

Figure 14 shows the detailed results of SHAKE with the used shear wave velocities. It was found that the amplification of acceleration decreased in Ac layer. Therefore maximum acceleration in As layer is similar for 250 gal and 400 gal. Therefore, as shown in Figure 15, liquefaction judgments at Point A became not liquefied as Class IV for the both cases. This implies that the system of As seen in this, the procedure in this paper could give a very precise evaluation for the liquefaction risk to the depth.

6 CONCLUSIONS

Main conclusions are developed as follows.

1. General procedure of making liquefaction map was shown.
2. Geological characteristics were studied at Kochi harbor.
3. Classification of soil types could be conducted based on the soil properties using the database at PARI.
4. 3D soil layer estimation was successfully obtained.
5. Then, liquefaction map at Kochi harbor was obtained realistically.

Tonankai and Nankai earthquakes are expected as Level 2 earthquake to attack Shikoku Island in the near future, individually and at the same time. Kochi harbor will be faced to very strong motions, which has never experienced to occur sever liquefactions and sinking down of ground.

The above results showed the importance of soil layers for the liquefaction evaluation, which could not be realized by the conventional liquefaction map in the wide area such a Kochi harbor. As the 3D ground model system has been proposed to clarify the soil layer constitutions using the soil data-base by PARI, which has SPT and physical properties. Through this study, the system was found to contribute making the precise liquefaction map. This system also can be used for the settlement analysis by earthquake and flood analysis by Tsunami. Therefore, it is highly recommended to apply to other harbors and airports in Japan for various natural disasters including liquefaction. On the basis of the system, the rational and effective countermeasures and also mitigations of people hopefully would be planed and designed.

REFERENCES

Maze et al. 2001. *Spatial modeling*. Kyoritsu Publisher.
Shiono et al. 1986. Geoinfomatics. *Estimation of cross section of soil layers based on optimization*. No. 11: 197–236.
Shiono et al. 2000. Theory of soil investigation and, Geoinfomatics. *Estimation of cross section of soil layers based on optimization* 11(4): 241–252.
Committee of geological information application at Shikoku. Shikoku geo-database.
Kochi Association of Architectural Firms. 1992. Kochi geo-database.
Office of Kochi Harbor and Airport, Planning of Kochi harbor.
Tanaka, Watabe, Miyata & Sakajo. 2007. 3D-VISUALIZ-ATION OG GROUND OF NEW RUNWAY AT HANEDA AIRPORT. *Soils and Foundations* 47(1): 131–139.
GES. 1994. applicability of SHAKE. *symposium on seismic motions of soft ground*.

Numerical simulations of progressive/multi-stages slope failure due to the rapid increase in ground water table level by SPH

Ha H. Bui, R. Fukagawa & K. Sako
Ritsumeikan University, Kusatsu, Shiga, Japan

ABSTRACT: In this paper, an attempt is make to simulate progressive and multi-stage slope failures induced by rapid increase in ground water table level. Herein, the SPH framework (Bui et al., 2008) for solving large deformation and failure of geomaterials is adopted where soil was modeled using an elasto-plastic constitutive model based on the Drucker-Prager yield criterion. For shake of simplicity, the rapid increase in water table level is modeled by step loading rather than solving seepage flow equation. Several smoothed particle slope failures are presented to investigate failure mechanism under different conditions, including low water table level, high water table level, and gradually increase in underground water table level. It is argued the SPH method is more powerful alternative to the traditional methods especially in handling large deformation and slope failure problems.

1 INTRODUCTION

Rainfall, especially heavy storms, has caused many landslides and slope failures. Such rain-induced landslides and slopes failures have posed serious threats and caused severe damage to many such countries as Japan, Hong Kong, Brazil, etc., over the years. As a result, studies on slope stability and slope failure under rainfall are attracting increasing attention in many countries especially in Japan.

As part of that effort, our recent experimental study of slope failure due to rainfall noticed that slope failure was initiated from the toe of the slope. Experimental data showed that rise in ground water level due to rainfall speeded up seepage flow downstream to the slope toe and then causing soil erosion at the slope toe, which induced slope failure. Furthermore, multi-stages failure process was also observed in these experiments if the slope was subjected to a long heavy rainfall condition. To have an insight about the above failure mechanisms under heavy rainfall condition, there is a need to develop a numerical method that can simulate gross discontinuous failure of a slope during heavy rainfall.

In the field of computational geomechanics, finite element method (FEM) has been found to be the most attracting method for studying slope failure due to heavy rainfall. This is because FEM can be employed either to solve unsaturated/saturated seepage flows or to model soil deformation. However, FEM often encounters grid distortion problem when attempting to simulate large deformation and failure flows of soil such as the erosion or multi-stages slope failure mentioned above. On the other hand, smoothed particle hydrodynamics (SPH), proposed by Gingold &

Monaghan (1977) and Lucy (1997), has been recently developed for solving large deformation and failure flow of geomaterials (Bui et al. 2007, 2008). The SPH method represents a powerful alternative approach for slope stability analysis and discontinuous slope failure simulation (Bui et al. 2009) which offers similar advantages to FEM. Furthermore, SPH provides physical insight into the failure mechanisms of the slope due to its capability to simulate gross discontinuous failure. In this paper, the above SPH framework is adopted in an attempt to simulate progressive and multi-stages slope failures induced by the rapid increase in water table level. For shake of simplicity, the rapid increase in water table level is modeled herein by step loading procedure rather than solving seepage flow equation, which is postponed to a near future publication. Three examples of slope failure are presented to investigate failure mechanism under different conditions, including low water table level, high water table level, and progressive increase in water table level. As for the slope with low water table level, failure is not occurred. To trigger slope collapse, the shear strength reduction method (Griffiths et al. 1999) is applied. Results obtained from this paper argued that the SPH method is a powerful alternative to the traditional methods especially in handling large deformation and failure of slope during heavy rainfall condition.

2 MOTIONS OF SOIL IN SPH

An equilibrium equation of a continuum media, such as soil, can be described using following equation,

$$\rho \ddot{u}^\alpha = \nabla_\beta \sigma'^{\alpha\beta +} + \rho g^\alpha \tag{1}$$

where α and β denote Cartesian components x, y, z with the Einstein convention applied to repeated indices; ρ is the density; u is the displacement; σ' is the effective stress tensor; and g is the acceleration due to gravity.

In the SPH framework, the above partial differential equation can be approximated through the use of a kernel interpolation, leading to

$$\ddot{u}_i^\alpha = \sum_{j=1}^{N} m_j \left(\frac{\sigma_i'^{\alpha\beta} + \sigma_j'^{\alpha\beta}}{\rho_i \rho_j} + \Pi_{ij}\delta^{\alpha\beta} \right) \frac{\partial W_{ij}}{\partial x^\beta} + g^\alpha \quad (2)$$

where i is the particle under consideration; N is the number of neighbouring particles, i.e. those in the support domain of particle i; m is the mass of particle; Π_{ij} is the artificial viscosity, which is employed to damp out unphysical stress fluctuation; and W_{ij} is the smoothing function, which is chosen to be the cubic spline function (Monaghan 1985).

The completed artificial viscosity appeared in equation (2) can be defined using the formulation proposed by Monaghan (1985). Herein, only an effective term of this formulation is adopted,

$$\Pi_{ij} = \begin{cases} \frac{-\alpha_\Pi c_{ij}\phi_{ij}}{\rho_{ij}} & (\dot{u}_i - \dot{u}_j) \cdot (x_i - x_j) < 0 \\ 0 & (\dot{u}_i - \dot{u}_j) \cdot (x_i - x_j) \geq 0 \end{cases} \quad (3)$$

in which

$$\phi_{ij} = \frac{h(\dot{u}_i - \dot{u}_j) \cdot (x_i - x_j)}{|x_i - x_j|^2 + 0.01h^2}, \quad c_{ij} = \frac{c_i + c_j}{2} \quad (4)$$

where α_Π is a constant parameter which is set to be 0.1; h is the smoothing length; and c is the sound speed in soil, which should be chosen by using the following equation (Bui et al. 2009),

$$c_i = \sqrt{\frac{E_i}{2(1 + \upsilon_i)\rho_i}} \quad (5)$$

where E is the elastic Young's modulus; and υ is Poisson's ratio. It is worth to know that the use of equation (5) to calculate the sound speed in soil considerably reduces computational time for the current SPH application.

As discussed by Bui et al. (2008), the SPH method encounters the so-called "tensile instability" problem when applying to cohesive soil. To remove this problem, they have proposed to use an artificial stress term, originally invented by Gray et al. (2001). Accordingly, equation (2) is modified to,

$$\ddot{u}_i^\alpha = \sum_{j=1}^{N} m_j \left(\frac{\sigma_i'^{\alpha\beta} + \sigma_j'^{\alpha\beta}}{\rho_i \rho_j} + F_{ij}^n R_{ij}^{\alpha\beta} + \Pi_{ij}\delta^{\alpha\beta} \right)$$
$$\times \frac{\partial W_{ij}}{\partial x^\beta} + g^\alpha \quad (6)$$

where

$$F_{ij}^n R_{ij}^{\alpha\beta} = \left[\frac{W_{ij}}{W(d_o, h)} \right]^n (R_i^{\alpha\beta} + R_j^{\alpha\beta}) \quad (7)$$

where n is a parameter; d_o is the initial distance between particles; h is the smoothing length, which specified the non-zero region of the smoothing function; and R_{ij} is obtained as follows. For each particle the effective stress tensor $\sigma'^{\alpha\beta}$ is diagonalised. Then an artificial stress term is evaluated for any of the diagonal components $\bar{\sigma}'^{\alpha\beta}$ which are positive,

$$\bar{R}_i^{\gamma\gamma} = -\varepsilon_0 \frac{\bar{\sigma}_i'^{\gamma\gamma}}{\rho_i^2} \quad (8)$$

where ε_0 is a small parameter ranging from 0 to 1. The artificial stress in the original coordinates system R_{ij} is then calculated by reversing coordinates transformation. Gray et al. (2001) derived optimal values from the dispersion analyses, and suggested to use $\varepsilon_0 = 0.3$ and $n = 4$ when applied SPH to solid. However, Bui et al. (2008) showed that these selections can not remove the tensile instability problem found when simulating cohesive soil. Instead, they suggested to use $\varepsilon_0 = 0.5$ and $n = 2.55$, and showed that these values have no effect on the modeled soil behaviour. In this paper, the same values are applied.

When considering the presence of water in soil, it is necessary to take into account the pore-water pressure (p_w) into the momentum equation. Directly replacing the total stress tensor, which consists of effective stress and pore-water pressure, into equation (6) will result in numerical instability for soil particles near the submerged soil surface. To remove this problem, Bui et al. (2009) has derived a new momentum equation where the pore-water pressure is implemented in the following way,

$$\ddot{u}_i^\alpha = \sum_{j=1}^{N} m_j \left(\frac{\sigma_i'^{\alpha\beta} + \sigma_j'^{\alpha\beta}}{\rho_i \rho_j} + F_{ij}^n R_{ij}^{\alpha\beta} + \Pi_{ij}\delta^{\alpha\beta} \right) \frac{\partial W_{ij}}{\partial x^\beta}$$
$$+ \sum_{j=1}^{N} \frac{m_j}{\rho_i \rho_j} (p_{wj} - p_{wi}) \frac{\partial W_{ij}}{\partial x^\alpha} + g^\alpha \quad (9)$$

It is easy to see that this equation ensures that the gradient of a constant pore-water pressure field vanishes. Furthermore, the above expression of pore-water pressure automatically imposes the dynamic boundary condition at the free surface. For details of driving this equation, we refer the readers to our coming publication (Bui et al. 2009). Finally, the above motion equation (8) can be solved directly using the standard Leapfrog algorithm if the effective stress tensor is known. Thus, it is necessary to derive constitutive equations for the effective stress tensor that are applicable in the SPH framework.

3 SOIL CONSTITUTIVE MODEL

According to the classical plasticity theory, the total strain-rate tensor of an elasto-plastic material $\dot{\varepsilon}$ can be decomposed into two parts: an elastic strain rate tensor $\dot{\varepsilon}_e$ and a plastic strain rate tensor $\dot{\varepsilon}_p$,

$$\dot{\varepsilon}^{\alpha\beta} = \dot{\varepsilon}_e^{\alpha\beta} + \dot{\varepsilon}_p^{\alpha\beta} \tag{10}$$

The elastic strain rate tensor $\dot{\varepsilon}_e^{\alpha\beta}$ is given by a generalized Hooke's law, i.e.,

$$\dot{\varepsilon}_e^{\alpha\beta} = \frac{\dot{s}^{\alpha\beta}}{2G} + \frac{1-2\upsilon}{E}\dot{\sigma}'^m\delta^{\alpha\beta} \tag{11}$$

where $s'^{\alpha\beta}$ is the deviatoric effective shear stress tensor; υ is Poisson's ratio; E is the elastic Young's modulus; G is the shear modulus and σ'^m is the mean effective stress.

The plastic strain rate tensor is calculated by the plastic flow rule, which is given by

$$\dot{\varepsilon}_p^{\alpha\beta} = \lambda\frac{\partial g}{\partial\sigma'^{\alpha\beta}} \tag{12}$$

where λ is the rate of change of plastic multiplier, and g is the plastic potential function.

In the current study, the Drucker-Prager model with non-associated flow rule is applied. In addition, this study assumes that the yield surface is fixed in the stress space. Accordingly, plastic deformation will occur only if the following yield criterion is satisfied,

$$f(I_1, J_2) = \sqrt{J_2} + \alpha_\phi I_1 - k_c = 0 \tag{13}$$

where I_1 and J_2 are, respectively, the first and second invariants of stress tensor; α_ϕ and k_c are the Drucker-Prager constants, which are calculated by,

$$\alpha_\phi = \frac{\tan\phi}{\sqrt{9 + 12\tan^2\phi}}, \text{ and } k_c = \frac{3c}{\sqrt{9 + 12\tan^2\phi}} \tag{14}$$

where c is the cohesion and ϕ is the friction angle.

For the non-associated plastic flow rule, the plastic potential function is given by,

$$g = 3I_1\sin\psi + \sqrt{J_2} \tag{15}$$

Substituting equations (11), (12) into (10), and adopting the Jaumman stress rate for large deformation treatment, the stress-strain relation for the current soil model at particle i becomes,

$$\begin{aligned}\dot{\sigma}_i'^{\alpha\beta} = {}&\sigma_i'^{\alpha\gamma}\dot{\omega}_i^{\beta\gamma} + \sigma_i'^{\gamma\beta}\dot{\omega}_i^{\alpha\gamma} + 2G_i\dot{e}_i^{\alpha\beta}\\&+K_i\dot{\varepsilon}_i^{\gamma\gamma}\delta_i^{\alpha\beta} - \dot{\lambda}_i\left[9K_i\sin\psi_i\,\delta^{\alpha\beta}\right.\\&\left.+(G/\sqrt{J_2})_i s_i'^{\alpha\beta}\right]\end{aligned} \tag{16}$$

where $\dot{e}_i^{\alpha\beta}$ is the deviatoric shear strain rate tensor; ψ is the dilatancy angle; and $\dot{\lambda}_i$ is the rate of change of plastic multiplier, which in SPH is specified by,

$$\dot{\lambda}_i = \frac{3\alpha_\phi K\dot{\varepsilon}_i^{\gamma\gamma} + (G/\sqrt{J_2})_i s_i'^{\alpha\beta}\dot{\varepsilon}_i^{\alpha\beta}}{27\alpha_\phi K_i\sin\psi_i + G_i} \tag{17}$$

and $\dot{\varepsilon}_i^{\alpha\beta}$, $\dot{\omega}_i^{\alpha\beta}$ are the strain rate and spin rate tensors, which are respectively defined by,

$$\dot{\varepsilon}_i^{\alpha\beta} = \frac{1}{2}\left(\frac{\partial\dot{u}^\alpha}{\partial x^\beta} + \frac{\partial\dot{u}^\beta}{\partial x^\alpha}\right)_i, \quad \dot{\omega}_i^{\alpha\beta} = \frac{1}{2}\left(\frac{\partial\dot{u}^\alpha}{\partial x^\beta} - \frac{\partial\dot{u}^\beta}{\partial x^\alpha}\right)_i \tag{18}$$

The above soil constitutive model requires six soil parameters, including Young's modulus (E), cohesion (c), friction angle (ϕ), dialatancy angle (ψ), specific unit weight, and Poisson's ratio (υ).

4 NUMERICAL MODELS AND BOUNDARY CONDITIONS

4.1 Slope model and setting conditions

Figure 1a shows the slope model used for numerical investigations. This slope is assumed to be made of homogeneous soil with $E = 10^5$ kN/m^2, $\upsilon = 0.3$, $\phi = 20°$, $c = 20.6$ kN/m^2, and $\gamma = 21$ kN/m^3 (below and above the water table). Initially, the water table level is set at 10 m high, and then it is enforced to increase by step loading, which is controlled using two parameters θ and h. The parameter θ controls the gradient of pore-water pressure downstream to the slope toe, while the parameter h controls the water level. Each parameter has a maximum value, for an instant $\theta_{max} = 15°$ and $h_{max} = 6$ m are employed in the current paper. When the loading starts, the water level, left side from the slope toe, is pushed up via θ and h, while the water level in the remaining slope is kept constant. The loading is performed in 8s and simulation is carried out until the completion of slope collapse. The typical development of hydrostatic pore-water pressure at the base of the slope foundation (left hand side from the crest of the slope) and corresponding pore-water distribution in the slope is showed in Figure 1b. In addition to the case of loading water table, two numerical simulations employed constant water table are also performed to verify the difference and to validate the SPH method. The first case is associated with the low water table level ($\theta_{min} = 0°$ and $h_{min} = 0$ m), while the second one is concerned with the high water table level ($\theta_{max} = 15°$ and $h_{max} = 6$ m).

5 NUMERICAL APPLICATIONS AND DISCUSSION

5.1 Validation of new SPH equation for slope failure simulations

In this section, mechanism of slope failure via new SPH equation is examined. Firstly, two numerical examples with low and high water table levels are performed and then comparing to the limit equilibrium method (Bishop's method) in term of safety factor and critical slip surface.

123

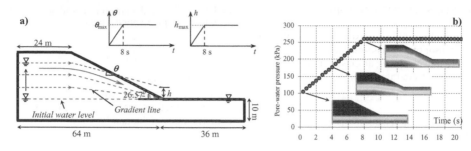

Figure 1.   Slope model geometry and initial setting condition.

Similar to FEM, a serial of SPH slopes are performed using the shear strength method (Griffths et al. 1999) in association with the new convergent and unconvergent analyses (Bui et al. 2009) to find factor of safety (*FOS*) of the current slopes. On the other hand, the critical slip surfaces are automatically predicted via the SPH contour plot of accumulated plastic shear strain.

Figure 2 shows the SPH result for slope stability analysis and slope failure simulation, which corresponds to the unconvergent solution. Herein, the SPH method predicts $FOS = 1.32$ for the slope with ($\theta_{min} = 0°$ and $h_{min} = 0$ m) and $FOS = 0.91$ for the slope with ($\theta_{max} = 15°$ and $h_{max} = 6$ m). These results are in very close agreement with the limit equilibrium method, which gave $FOS = 1.34$ for the former case and $FOS = 0.92$ for the latter case.

Regarding the critical slip surface, Figure 2a shows the contour plot of accumulated plastic strain corresponding to unconvergent SPH solutions as compared to the circular slip surface obtained from the limit equilibrium method. It can be seen that SPH predicts very clear critical slip surface via thin layer of plastic strain localization, which is normally difficult to model by FEM. Again, close agreements are obtained between SPH and limit equilibrium method. The slight difference between two methods can be explained due to the assumption of circular slip surface employed in the limit equilibrium method. The toe failure mechanism is also observed in both cases with a deeper mechanism extending into the foundation layer for the case with lower water table level. Furthermore, two slip surfaces were observed for the case with high water table level, suggesting that multi-stage failure process may be simulated by SPH.

Figure 2b shows the final configuration of the slopes after collapse, via contour plot of total displacements. The result shows that gross discontinuities of the slope along the potential slip surface, which is unable to model by FEM, can be described very well by SPH. Numerical simulation can be performed as long as desired without encountering any problems. Furthermore, comparison between two cases shows

that the failure zone is enlarged for the case with lower water table level. However, the lager maximum displacement was observed for the case with higher water table level.

5.2   Simulation of slope failure process due to the increase in ground water level

The above numerical validations of SPH for slope stability analyses and slope failure simulations show good agreement with the traditional approaches. Thus, the SPH method is now extended to simulate slope failure process due to the rapid increase in underground water table, which is obtained using the step loading procedure as described in section 4.1.

Figure 3 shows the formation of critical slip surface via contour plot of accumulated plastic shear strain. From this figure, it can be seen that failure process was initiated from the toe of slope and then propagating to the crest of the slope. The plastic shear strain developed very slowly at the beginning of the failure process, and then turned to develop rapidly until the total collapse of the slope. We expected that by adopting the powerful SPH technique in handling large deformation and failure, erosion mechanism at the slope toe can be reproduced. However, because of the regardless of seepage behavior, and because the soil cohesion was kept constant in the current simulations, such erosion process hasn't yet been reproduced through the current SPH model. We suggested that by carefully solving unsaturated-saturated seepage flow equation, and taking into account change of soil strength due to rainfall, the above erosion mechanism can be simulated by SPH.

Figure 4 shows the failure process of the slope after collapse via contour plots of accumulated plastic shear strain and corresponding total displacements. Again, the gross discontinuous failure of the slope along the critical slip surface can be simulated very well by SPH, and numerical simulation can be performed as long as desired without encountering any problems. In comparison to the previous slopes, in which the water table level of ($\theta_{max} = 15°$ and $h_{max} = 6$ m) was employed,

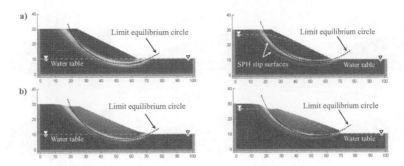

Figure 2. Slope failure simulation by SPH: (a) Contour plot of accumulated plastic shear strain; (b) Contour plot of total displacement at final deformation. Color is auto adaptive.

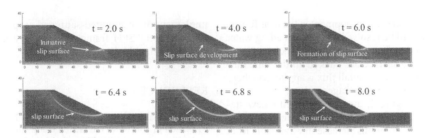

Figure 3. Formation of critical slip surface during the step loading of water table level via contour plot of accumulated plastic shear strain.

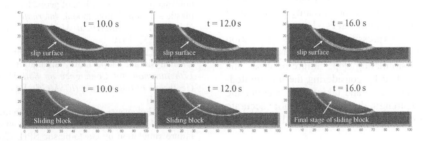

Figure 4. Failure mechanism of the slope due to the rapid increase in water table via contour plots of plastic shear strain (upper images) and total displacement (lower images).

there is a slight difference in term of total displacement of the crest of the sliding block. The current case shows higher displacement, suggesting that there is the effect of inertial force. On the other hand, the critical slip surface is almost the same between two cases. Regarding the slope failure process, it is well-known that the slope failure due to heavy rainfall is often associated with multi-stages failure. Herein, only the first stage of slope failure was observed since the water table level in the current simulation was constrained. However, if we remove the constraint of the water table level, the multi-stages failure process of the current slope can be reproduced.

Figure 5 shows the second failure stage of the current slope obtained when the constraint of water table level was removed. Herein, only h is permitted to increase gradually without limitation, while the maximum value of θ was constrained to $\theta = 15°$. Therefore, when the gradient line moves beyond the crest of the slope, the step loading of water table level only affects the left hand side section of the slope counting from the crest of the slope. The current assumption of pore-water pressure increment is not practical; however, the author's purpose is to demonstrate the ability of SPH in simulating multi-stages slope failure. Therefore, such fictitious assumption is accepted in the current

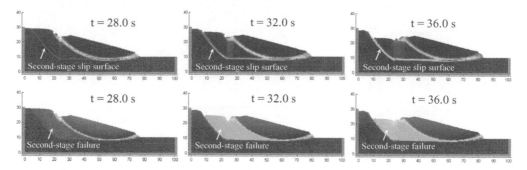

Figure 5. Multi-stage failure process (second stage failure) of the slope via contour plots of plastic shear strain (upper images) and total displacement (lower images).

simulation. For further development in the near future, seepage flow will be resolved instead of loading water level employed in the current paper. In such case, soil particles on free surface are tracked all the time during failure process and rainfall flux is applied via these particles to simulate multi-stages failure.

As seen in Figure 4, the slope reached a new stable position after the first stage failure. However, because of the continuous increase in water table level the second failure stage occurred at the deeper section to the left as shown in Figure 5 via contour plots of shear plastic strain and total displacement. The process will be continued if the water table level is kept increasing. This phenomenon is reality as compared to the well-known multi-stages slope failure at the Tokai-Hokuriku Expressway. From this simulation result, it confirmed that the SPH method is a powerful alternative for simulating slope failure process. The authors suggest that by considering the unsaturated-saturated seepage flow and taking into account the change of soil strength during rainfall, more accurate slope failure simulation can be performed by the SPH method.

6 CONCLUDING REMARKS

The smoothed particle method in conjunction with an elasto-plastic (Drucker-Prager) stress-strain model has been shown to be a reliable and robust method for simulating the progressive and multi-stages slope failures due to the rapid increase in ground water table level. One of the main advantages of SPH is that it can handle the gross discontinuous failure of soil, thereby applicable for simulating the multi-stages slope failure. Furthermore, by using the proposed method, it is also possible to predict safety factor of a slope and maximum displacement of soil after a slope failure. The authors are encouraged by the current simulation results but recognized the need for further

implement of unsaturated-saturated seepage behavior to the current SPH framework. Such works will be soon published in the near future.

REFERENCES

Bui, H.H., Sako, K. & Fukagawa, R. 2007. Numerical simulation of soil-water interaction using smoothed particle hydrodynamics (SPH) Method, *Journal of Terramechanics* 44(5): 339–346.

Bui, H.H., Fukagawa, R., Sako, K. & Ohno, S. 2008. Lagrangian mesh-free particles method (SPH) for large deformation and post-failure of geomaterial using elastic-plastic soil constitutive model, *International Journal for Numerical and Analytical Methods in Geomechanics* 32(12): 1537–1570.

Bui, H.H., Fukagawa, R. & Sako K. 2009. Slope stability analysis and slope failure simulation by SPH, *The 17th International Conference on Soil Mechanics and Geotechnical Engineering (ICSMGE)*, Egypt, Oct. 2009. (Accepted)

Bui, H.H., Fukagawa, R. & Sako, K. Innovative solution for slope stability analysis and discontinuous slope failure simulation by elasto-plastic SPH, *Geotechnique*, (submitted).

Gingold, R.A. & Monaghan, J.J. 1977. Smoothed particle hydrodynamics: Theory and application to non spherical stars. *Mon. Not. Roy. Astron. Soc.,* 181: 375–389.

Gray, J.P., Monaghan, J.J., & Swift, R.P. 2001. SPH elastic dynamics. *Computer Methods in Applied Mechanics and Engineering* 190: 6641–6662.

Griffths, D.V. & Lane, P.A. 1999. Slope stability analysis by finite elements. *Geotechnique* 49(3): 387–403.

Monaghan, J.J. & Lattanzio J.C. 1985. A refined particle method for astrophysical problems. *Astronomic and Astrophysics* 149: 135–143.

Libersky, L.D., Petschek, A.G., Carney, T.C., Hipp, J.R. & Allahdadi, F.A. 1993. High strain lagrangian hydrodynamics: A three dimensional SPH code for dynamic material response, *Journal of Computational Physics* 109: 67–75.

Lucy L. 1977. A numerical approach to testing the fission hypothesis. *Astronomical Journal* 82: 1013–1024.

Prediction and Simulation Methods for Geohazard Mitigation – Oka, Murakami & Kimoto (eds)
© 2009 Taylor & Francis Group, London, ISBN 978-0-415-80482-0

A multiphase elasto-viscoplastic analysis of an unsaturated river embankment associated with seepage flow

F. Oka, S. Kimoto, N. Takada & Y. Higo
Kyoto University, Kyoto, Japan

ABSTRACT: In the present study, a multiphase deformation analysis of a river embankment has been carried out using an air-soil-water coupled finite element method considering the unsaturated seepage flow. A numerical model for unsaturated soil is constructed based on the mixture theory and an elasto-viscoplastic constitutive model. The theory used in the analysis is a generalization of Biot's two-phase mixture theory for saturated soil. An air-soil-water coupled finite element method is developed using the governing equations for three phase soil based on the nonlinear finite deformation theory, i.e., the updated Lagrangian method. Two-dimensional numerical analyses of the river embankment under seepage conditions have been conducted, and the deformation associated with the seepage flow has been studied. From the numerica/11 methods, it has been found that the seepage-deformation coupled three-phase behavior can be simulated well with by the proposed method.

1 INTRODUCTION

In recent years, many natural disasters have occurred in the world due to floods have occurred in the world associated with torrential rains, typhoons and hurricanes etc. In many cases, river embankments have been failed due to seepage and overflow. In the present study, a multiphase deformation analysis of a river embankment has been carried out using an air-soil-water coupled finite element method considering the unsaturated seepage flow. A numerical model for unsaturated soil is constructed based on the mixture theory and an elasto-viscoplastic constitutive model. As for the stress variables in the formulation of unsaturated soil, we use the skeleton stresses and suction simultaneously. The skeleton stresses are used in the constitutive model instead of the Terzaghi's effective stress for the saturated soil, and the suction is incorporated through the constitutive parameters of the model. An air-soil-water coupled finite element method is developed using the governing equations for the three-phase soil based on the nonlinear finite deformation theory, i.e., the updated Lagrangian method. Two-dimensional numerical analyses of the river embankment under seepage conditions have been conducted, and the strain localization associated with the seepage flow has been studied. From the numerical methods, it has been found that seepage-deformation coupled behavior can be simulated well with the proposed method.

2 GOVERNING EQUATIONS

2.1 Partial stress for the mixture

The total stress tensor is assumed to be composed of three partial stress values for each phase.

$$\sigma_{ij} = \sigma_{ij}^s + \sigma_{ij}^f + \sigma_{ij}^a \tag{1}$$

where σ_{ij} is the total stress tensor, and σ_{ij}^s σ_{ij}^f, and σ_{ij}^a are the partial stress tensors for solid, liquid, and air, respectively.

The partial stress tensors for unsaturated soil can be given by

$$\sigma_{ij}^f = -nS_r p^f \delta_{ij} \tag{2}$$

$$\sigma_{ij}^a = -n(1 - S_r)p^a \delta_{ij} \tag{3}$$

$$\sigma_{ij}^s = \sigma_{ij}' - (1 - n)S_r p^f \delta_{ij} - (1 - n)(1 - S_r)p^a \delta_{ij} \tag{4}$$

where σ_{ij}' is the skeleton stress, P^f and P^a are the partial stress levels for the pore water pressure and the pore air pressure, respectively and, n is the porosity, and S_r is the degree of saturation.

The skeleton stress is used as the basic stress variable in the model for unsaturated soil. Definitions for the skeleton stress and the average fluid pressure (Oka et al. 2008) are given as follows:

$$\sigma'_{ij} = \sigma_{ij} + P^F \delta_{ij} \tag{5}$$

$$P^F = S_r p^f + (1 - S_r) p^a \tag{6}$$

where P^F is the average pore pressure.

Adopting the skeleton stress provides a natural application of the mixture theory to unsaturated soil. The definition in Equation (5) is similar to Bishop's definition for the effective stress of unsaturated soil. In addition to Equation (5), the effect of suction on the constitutive model should always be taken into account. This assumption leads to a reasonable consideration of the collapse behavior of unsaturated soil, which has been known as a behavior that cannot be described by Bishop's definition for the effective stress of unsaturated soil. Introducing suction into the model, however, makes it is possible to formulate a model for unsaturated soil, starting from a model for saturated soil, by using the skeleton stress instead of the effective stress.

2.2 Mass conservation law

The mass conservation law for the three phases is given by

$$\frac{\partial \bar{\rho}^J}{\partial t} + \frac{\partial (\bar{\rho}^J \dot{u}_i^J)}{\partial x_i} = 0 \tag{7}$$

where $\bar{\rho}^J$ is the average density for the J phase, and $\dot{u}_i^J$ is the velocity vector of the J phase.

$$\bar{\rho}^s = (1 - n)\rho^s \tag{8}$$

$$\bar{\rho}^f = nS_r \rho^f \tag{9}$$

$$\bar{\rho}^a = n(1 - S_r)\rho^a \tag{10}$$

where $J = s, f$, and a in which super indices s, f, and a indicate the solid, the liquid, and the air phases, respectively, n is the porosity, and S_r is the saturation. ρ^J is the mass bulk density of the solid, the liquid, and the gas.

2.3 Conservation laws of linear momentum for the three phases

The conservation laws of linear momentum for the three phases are given by

$$\ddot{u}_i^s - Q_i - R_i = \frac{\partial \sigma_{ij}^s}{\partial x_j} + \bar{\rho}^s b_i \tag{11}$$

$$\bar{\rho}^f \ddot{u}_i^f + R_i = \frac{\partial \sigma_{ij}^f}{\partial x_j} + \bar{\rho}^f b_i \tag{12}$$

$$\bar{\rho}^a \ddot{u}_i^a + Q_i = \frac{\partial \sigma_{ij}^a}{\partial x_j} + \bar{\rho}^a b_i \tag{13}$$

where b_i is a body force, Q_i denotes the interaction between solid and gas phases, and R_i denotes the interaction between solid and liquid phases.

These interaction terms Q_i and R_i can be described as

$$R_i = nS_r \frac{\gamma_w}{k^f} \dot{w}_i^f \tag{14}$$

$$Q_i = n(1 - S_r) \frac{\rho^a g}{k^a} \dot{w}_i^a \tag{15}$$

where k^f is the water permeability coefficient, k^a is the air permeability, $\dot{w}_i^f$ is the average relative velocity vector of water with respect to the solid skeleton, and $\dot{w}_i^a$ is the average relative velocity vector of air to the solid skeleton. The relative velocity vectors are defined by

$$\dot{w}_i^f = nS_r(\dot{u}_i^f - \dot{u}_i^s) \tag{16}$$

$$\dot{w}_i^a = n(1 - S_r)(\dot{u}_i^a - \dot{u}_i^s) \tag{17}$$

Using Equation (16), Equation (12) becomes

$$\bar{\rho}^f \left(\ddot{u}_i^s + \frac{1}{nS_r} \ddot{w}_i^f \right) + R_i = \frac{\partial \sigma_{ij}^f}{\partial x_j} + \bar{\rho}^f b_i \tag{18}$$

When we assume that $\ddot{w}_i^f \cong 0$, and use Equations (3), (7), and (16), Equation (18) becomes

$$nS_r \rho^f \ddot{u}_i^s + nS_r \frac{\gamma_w}{k^f} \dot{w}_i^f = -nS_r \frac{\partial p^f}{\partial x_i} + nS_r \rho^f b_i \tag{19}$$

After manipulation, the average relative velocity vector of water to the solid skeleton and the average relative velocity vector of air to the solid skeleton are shown as

$$\dot{w}_i^f = -\frac{k^f}{\gamma_w} \left(\frac{\partial p^f}{\partial x_i} + \rho^f \ddot{u}_i^s - \rho^f b_i \right) \tag{20}$$

$$\dot{w}_i^a = -\frac{k^a}{\rho^a g} \left(\frac{\partial p^a}{\partial x_i} + \rho^a \ddot{u}_i^s - \rho^a b_i \right) \tag{21}$$

Based on the above fundamental conservation laws, we can derive equations of motion for the whole mixture. Substituting Equations (8), (9), and (10) into the given equation and adding Equations (11)–(13), we have

$$\rho \ddot{u}_i^s + nS_r \rho^f (\ddot{u}_i^f - \ddot{u}_i^s) + n(1 - S_r)\rho^a (\ddot{u}_i^a - \ddot{u}_i^s)$$

$$= \frac{\partial \sigma_{ij}}{\partial x_j} + \rho b_i \tag{22}$$

where ρ is the mass density of the mixture as $\rho = \bar{\rho}^f + \bar{\rho}^a + \bar{\rho}^s$, $\ddot{u}_i^s$ is the acceleration vector of the solid phase, b_i is the body force vector, and σ_{ij} is the total stress tensor.

From the following assumptions,

$$\ddot{u}_i^s \cong (\ddot{u}_i^f - \ddot{u}_i^s) \qquad (23)$$

$$\ddot{u}_i^s \cong (\ddot{u}_i^a - \ddot{u}_i^s) \qquad (24)$$

Hence, the equations of motion for the whole mixture are derived as

$$\rho \ddot{u}_i^s = \frac{\partial \sigma_{ij}}{\partial x_j} + \rho b_i \qquad (25)$$

2.4 Continuity equations for the fluid phase

Using the mass conservation law for the solid and the liquid phases, Equation (7), and with the assumption of the incompressibility of soil particles, we obtain

$$\frac{\partial \{nS_r(\dot{u}_i^f - \dot{u}_i^s)\}}{\partial x_i} + S_r \dot{\varepsilon}_{ii}^s + nS_r \frac{\dot{\rho}^f}{\rho^f} + n\dot{S}_r = 0 \qquad (26)$$

Incorporating Equation (20) and $p = -K^f \varepsilon_{ii}^f$ (K^f: volumetric elastic coefficient) into the above equation leads to the following continuity equation for the liquid phase:

$$-\frac{\partial}{\partial x_i}\left[\frac{k^f}{\gamma_w}\left(\rho^f \ddot{u}_i^s + \frac{\partial p^f}{\partial x_i} - \rho^f b_i\right)\right]$$

$$+ S_r \dot{\varepsilon}_{ii}^s + n\dot{S}_r + nS_r \frac{p^f}{K^f} = 0 \qquad (27)$$

Similarly, we can derive the continuity equation with the assumption that the spatial gradients of porosity and saturation are sufficiently small.

$$-\frac{\partial}{\partial x_i}\left[\frac{k^a}{\gamma_w}\left(\rho^a \ddot{u}_i^s + \frac{\partial p^a}{\partial x_i} - \rho^a b_i\right)\right]$$

$$+ (1 - S_r) \dot{\varepsilon}_{ii}^s - n\dot{S}_r + n(1 - S_r)\frac{\dot{\rho}^a}{\rho^a} = 0 \qquad (28)$$

Since saturation is a function of the suction, i.e., the pressure head, the time rate for saturation is given by

$$n\dot{S}_r = n\frac{dS_r}{d\theta}\frac{d\theta}{d\psi}\frac{d\psi}{dp}\dot{p}^c = \frac{C}{\gamma_w}\dot{p}^c \qquad (29)$$

where $\theta = \frac{V_w}{V}$ is the volumetric water content, p^c is the matrix suction $p^c = -(p^a - p^f)$, $\psi = p^c/\gamma_w$ is the pressure head for the suction, and $C = \frac{d\theta}{d\psi}$ is the specific water content.

2.5 Constitutive equations

In this study, the saturated elasto-viscoplastic model for the overstress-type of viscoplasticity with soil structure degradation proposed by Kimoto & Oka (2005) has been extended to unsaturated soils using the

skeleton stress and including the effects of suction. The collapse behavior of unsaturated soils is macroscopic evidence of the structural instability of the soil skeleton and it is totally independent of the chosen stress variables (Oka et al. 2008). In the present model, the collapse behavior is described by the shrinkage of the overconsolidated boundary surface, the static yield surface, and the viscoplastic surface due to the decrease in suction.

It is assumed that the strain rate tensor consists of the elastic stretching tensor D_{ij}^e and the viscoplastic stretching tensor D_{ij}^{vp} as

$$D_{ij} = D_{ij}^e + D_{ij}^{vp} \qquad (30)$$

The elastic stretching tensor is given by a generalized Hooke type of law, namely,

$$D_{ij}^e = \frac{1}{2G}\dot{S}_{ij} + \frac{\kappa}{3(1+e)}\frac{\dot{\sigma}_m'}{\sigma_m'}\delta_{ij} \qquad (31)$$

where S_{ij} is the deviatoric stress tensor, σ_m' is the mean skeleton stress, G is the elastic shear modulus, e is the initial void ratio, κ is the swelling index, and the superimposed dot denotes the time differentiation.

2.5.1 Overconsolidation boundary surface

The overconsolidated boundary surface separates the normally consolidated (NC) region, $f_b \geq 0$, from the overconsolidated region, $f_b < 0$, as follows:

$$f_b = \bar{\eta}_{(0)}^* + M_m^* \ln \frac{\sigma_m'}{\sigma_{mb}'} = 0 \qquad (32)$$

$$\bar{\eta}_{(0)}^* = \{(\eta_{ij}^* - \eta_{ij(0)}^*)(\eta_{ij}^* - \eta_{ij(0)}^*)\}^{\frac{1}{2}} \qquad (33)$$

where η_{ij}^* is the stress ratio tensor ($\eta_{ij}^* = S_{ij}/\sigma_m'$), and (0) denotes the state at the end of the consolidation, in other words, the initial state before the shear test. M_m^* is the value of $\eta^* = \sqrt{\eta_{ij}^* \eta_{ij}^*}$ when the volumetric strain increment changes from negative to positive dilatancy, which is equal to ratio M_f^* at the critical state. σ_{mb} is the strain-hardening parameter, which controls the size of the boundary surface. The suction effect is introduced into the value of σ_{mb} as

$$\sigma_{mb}' = \sigma_{ma}' \exp\left(\frac{1+e}{\lambda - \kappa}\varepsilon_{kk}^{vp}\right)$$

$$\times \left[1 + S_I \exp\left\{-S_d\left(\frac{P^c}{P_i^c} - 1\right)\right\}\right] \qquad (34)$$

where ε_{kk}^{vp} is the viscoplastic volumetric strain, P^c is the present suction value, P_i^c is a reference suction, S_I denotes the increase in yield stress when suction increases from zero to the reference value $P_i^c \cdot S_d$ controls the rate of increase or decrease in σ_{mb} with suction and σ_{ma} is a strain-softening parameter used to

129

describe the degradation caused by structural changes, namely,

$$\sigma'_{ma} = \sigma'_{maf} + (\sigma'_{mai} - \sigma'_{maf})\exp(-\beta z) \tag{35}$$

$$z = \int_0^t \dot{z}\,dt \quad \text{with} \quad \dot{z} = \sqrt{\dot{\varepsilon}_{ij}^{vp}\dot{\varepsilon}_{ij}^{vp}} \tag{36}$$

in which σ'_{mai} and σ'_{maf} are the initial and the final values of σ'_{ma}, respectively, while β controls the rate of degradation with viscoplastic strain, and $\dot{\varepsilon}_{ij}^{vp}$ is the viscoplasitic strain rate.

2.5.2 Static yield function

To describe the mechanical behavior of the soil at its static equilibrium state, a Cam-clay type of static yield function is assumed:

$$f_y = \bar{\eta}_{(0)}^* + \tilde{M}^* \ln \frac{\sigma'_m}{\sigma'^{(s)}_{my}} = 0 \tag{37}$$

where $\tilde{M}^*$ is assumed to be constant in the NC region and varies with the current stress in the OC region (Kimoto & Oka 2005).

The static strain hardening parameter $\sigma'^{(s)}_{my}$ controls the size of the static yield surface. In the same way as for the over consolidation boundary surface, the parameter $\sigma'^{(s)}_{my}$ varies with the changes in suction as well as with the changes in viscoplastic volumetric strain and structural degradation:

$$\sigma'^{(s)}_{my} = \frac{\sigma'^{(s)}_{myi}}{\sigma'_{mai}}\sigma'_{ma}\exp\left(\frac{1+e}{\lambda-\kappa}\varepsilon_{kk}^{vp}\right)$$

$$\times \left[1 + S_I \exp\left\{-S_d\left(\frac{P_i^c}{P^c} - 1\right)\right\}\right] \tag{38}$$

where $\sigma'^{(s)}_{myi}$ is the initial value of $\sigma'^{(s)}_{my}$.

2.5.3 Viscoplastic potential function

The viscoplastic potential function is given by

$$f_p = \bar{\eta}_{(0)}^* + \tilde{M}^* \ln \frac{\sigma'_m}{\sigma'_{mp}} = 0 \tag{39}$$

where σ'_{mp} denotes the mean skeleton stress at the intersection of the viscoplastic potential function surface and the σ'_m axis.

2.5.4 Viscoplastic flow rule

Finally, the viscoplastic stretching tensor is based on a Perzyna's type of viscoplastic theory and is given as

$$D_{ij}^{vp} = \gamma\langle\Phi_1(f_y)\rangle\frac{\partial f_p}{\partial \sigma'_{ij}} \tag{40}$$

where the symbol $\langle\rangle$ is defined as

$$\langle\Phi_1(f_y)\rangle = \begin{cases} \Phi_1(f_y) & ; \ f_y > 0 \\ 0 & ; \ f_y \le 0 \end{cases} \tag{41}$$

in which Φ_1 denotes a material function for rate sensitivity. Herein, the value of f_y is assumed to be positive for any stress state in this model, in other words, the stress state always exists outside of the static yield function, so that viscoplastic deformation always occurs. Based on the experimental results of constant strain-rate triaxial tests, the material function Φ_1 is defined by an exponential function (Kimoto & Oka 2005).

$$\gamma\Phi_1(f_y) = C_{ijkl}\sigma'_m \exp\left\{m'\left(\bar{\eta}_{(0)}^*\tilde{M}^*\ln\frac{\sigma'_m}{\sigma'^{(s)}_{my}}\right)\right\} \tag{42}$$

where m' is the viscoplastic parameter that controls rate sensitivity and the viscoplastic parameter C_{ijkl} is a fourth rank isotropic tensor given by

$$C_{ijkl} = a\delta_{ij}\delta_{kl} + b(\delta_{ik}\delta_{jl} + \delta_{il}\delta_{jk}),$$

$$C_1 = 2b, \quad C_2 = 3a + 2b \tag{43}$$

where a and b are material parameters, which have a relation with the deviatoric component C_1 and volumetric component C_2 of the viscoplastic parameter.

2.6 Soil water characteristic curve

The soil-water characteristic model proposed by Van Genuchten (1980) is usedto describe the unsaturated seepage characteristics for which effective saturation S_e is adopted as

$$S_e = \frac{\theta - \theta_r}{\theta_s - \theta_r} = \frac{nS_r - \theta_r}{\theta_s - \theta_r} \tag{44}$$

where θ is the volumetric water content, θ_s is the volumetric water content in the saturated state, which is equal to porosity n, and θ_r is the residual volumetric water content retained by the soil at a large value of suction head which is a disconnected pendular water meniscus. For relatively large and uniform sand particles, such as those of Toyoura sand, θ_r becomes zero which is equal to common saturation.

In order to determine the soil-water characteristics, effective saturation S_e can be related to negative pressure head ψ through the following relation:

$$S_e = (1 + |\alpha\psi|^{n'})^{-m} \tag{45}$$

where α is a scaling parameter which has the dimensions of the inverse of ψ, and n' and m determine the

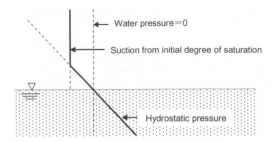

Figure 1. Initial pressure head.

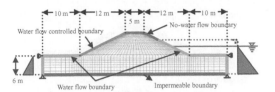

Figure 2. Finite element mesh and boundary conditions.

Table 1. Material parameters for viscoplastic model.

Parameters	Value
Compression Index λ	0.136
Swelling Index κ	0.0175
Initial void ratio e_0	1.05
Initial elastic shear modulus G_0	32400 (kPa)
Initial mean skeleton stress σ'_{mi}	205 (kPa)
Failure stress ratio M^*	1.01
Viscoplastic parameter m'	23.0
Viscoplastic parameter C_1	1.0×10^{-11} (1/s)
Viscoplastic parameter C_2	1.5×10^{-11} (1/s)
Structural parameter β	0.0
Suction parameter s_I	0.2
Suction parameter s_d	0.25
Water Permeability for full saturation k_s^W	1.0×10^{-5} (m/s)
Air Permeability for full saturation k_s^G	1.0×10^{-3} (m/s)

shape of the soil-water characteristic curve. The relation between n' and m leads to an S-shaped type of soil-water characteristic curve, namely,

$$m = 1 - \frac{1}{n'} \qquad (46)$$

In the present analysis, we follow the Guide for Structural Investigations of River Embankments (2002), and set $\alpha = 2$ and $n' = 4$.

Specific water content C, used in Equation (5), can be calculated as

$$C \left(\equiv \frac{d\theta}{d\psi} \right) = \alpha(n' - 1)(\theta_s - \theta_r)S_e^{1/m}(1 - S_e^{1/m})^m \qquad (47)$$

Specific permeability coefficient k_r, which is a ratio of the unsaturated to the saturated permeability, is defined by

$$k_r = S_e^{1/2}\{1 - (1 - S_e^{1/m})^m\}^2 \qquad (48)$$

Applying the above-mentioned relations, we can describe the unsaturated seepage characteristics.

In the analysis, the unsaturated region is treated in the following manner: In the embankment, the initial suction, i.e., the initial negative pore water pressure is assumed to be constant. Below the water level, the pore water pressure is given by the hydrostatic pressure. In the transition region between the water level and the suction constant region, we assumed that the pore water pressure is linearly interpolated (Fig. 1).

When the pressure head is negative, the increase in the soil modulus due to suction is considered. The effective saturation and the saturation are calculated with Equations (44) and (45) using the negative pressure head, i.e., suction. Applying the obtained effective saturation, specific water content C is then calculated by Equation (30) and specific permeability k_r is calculated by Equation (49).

3 NUMERICAL RESULTS

Figure 2 shows the model of the river embankment and the finite element mesh used in the analysis with boundary conditions. The soil parameters for DL clay

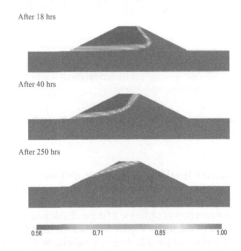

Figure 3. Distribution of saturation with time.

are used and listed in Table 1. The water level of the river has been increased for 18 hours and then remains constant. Figure 3 shows the distribution of the saturation with different times. The phreatic surface proceeds from the river side to the land side of the embankment. Figure 4 indicates the development of the air pressure with time. The magnitude of the

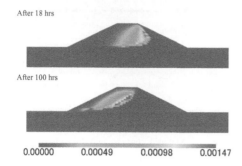

Figure 4. Distribution of air pressure with time.

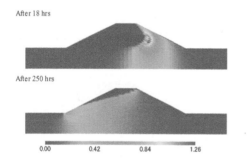

Figure 5. Distribution of accumulated shear strain with time.

Figure 6. Distribution of horizontal hydraulic gradient with time.

Table 2. Material parameters for hydraulic properties.

Parameters	Value
Shape parameter for water permeability a	3.0
Shape parameter for air permeability b	2.3
Maximum saturation $S_{r\,max}$	1.0
Minimum saturation $S_{r\,min}$	0.0
Van Genuchten parameter α	0.2 (1/kPa)
Van Genuchten parameter n	0.0

Parameters for Initial stress analysis	Value
Young's modulus E	7900 (kPa)
Poisson ratio ν	0.33
Internal friction angle ϕ'	30.0
Cohesion	0.0

4 CONCLUSION

Two-dimensional numerical analyses of a river embankment under seepage conditions have been conducted based on the mixture theory. Using the proposed seepage-deformation coupled three-phase analysis method, we have predicted the behavior of levees during the increase in the river water level.

REFERENCES

Ehlers, W., Graf, T. & Ammann, M. 2004. Deformation and localization analysis of partially saturated soil. *Comp. Meth.in Appl. Mech. and Eng.* 193: 2885–2910.

Japan Institute of Construction Engineering 2002. Guide for structure investigations of river embankments (in Japanese).

Kimoto, S. & Oka, F. 2005. An elasto-viscoplastic model for clay considering destructuralization and consolidation analysis of unstable behavior. *Soils and Foundations* 45(2): 29–42.

Oka, F., Kodaka, T.,S. Kimoto, S., Kato, R. & Sunami, S. 2007. Hydro-Mechanical Coupled Analysis of Unsaturated River Embankment due to Seepage Flow. *Key Engineering Materials* 340–341: 1223–1230.

Van Genuchten, M.T. 1980. A closed-form equation for predicting the hydraulic conductivity of unsaturated soils. *Soil Science Society of America Journal* 44: 892–898.

Oka, F., Feng, H. & Kimoto, S. 2008. A numerical simulation of triaxial tests of unsaturated soil at constant water and constant air content by using an elasto-viscoplastic model. Proc. *1st Europ. Conf. on Unsaturated Soils*, D. Toll and Wheeler, S.J. eds., Taylor and Francis Gr.: 735–741.

air pressure is relatively small but well simulated in the analysis. The development of the strain during the seepage flow is presented in Figure 5, in which a larger strain occurs at the toe of the embankment on the land side of the levee. Figure 6 shows the distribution of the horizontal local hydraulic gradient. In this simulation, the maximum value of 1.26 has been reached in the embankment after 18 hours. After 250 hours, the maximum value of 0.66 of the horizontal gradient is observed at the toe of the embankment. According to the Japanese design guide for river levees (2002), the levees becomes unstable if the horizontal hydraulic gradient is more than 0.5. It is worth noting that the numerically obtained the gradient, 0.66, is similar to the limit given in the design guide.

Prediction and Simulation Methods for Geohazard Mitigation – Oka, Murakami & Kimoto (eds)
© 2009 Taylor & Francis Group, London, ISBN 978-0-415-80482-0

Low energy rockfall protection fences in forested areas: Experiments and numerical modeling

S. Lambert, D. Bertrand, F. Berger & C. Bigot
Cemagref, St-Martin-d'Hères, France

ABSTRACT: This paper presents the first steps towards the development of an innovative type of rockfall protection fence on forested slopes specially designed for low energy events. The fence is made of a hexagonal wire mesh and it is tightened between two supporting trees. The behavior of the fence has been evaluated thanks to real scale experiments performed in situ. Rocks with a mean mass of 200 kg have been thrown on the fence, with a velocity of 17 m/s. Based on a discrete element model, numerical simulations has allowed the estimation of the influence of parameters related to the block trajectory. The results obtained are encouraging and should lead to the development of efficient low energy fences to be used on forested slopes.

1 INTRODUCTION

Forest management in mountainous regions implies frequent sylvicultural interventions on slopes. These human interventions mainly consist of wood felling and can also lead to the opening of forest roads. But in such forested contexts, engines, workers or falling trees can initiate rock falls. Indeed, any block in a previous stable position on the slope (i.e. near the soil surface or even resting on a tree trunk) can be moved and in some cases pushed down the slope. These blocks can vary in weight from one kilogramme to few tons. As a consequence, a site where rocks used not to be a threat for the elements at risk down the slope can temporarily turn to a potentially dangerous site. In addition, in these areas there are no protection structures except the forest stands.

Thus there is a real need for the development of specific rock fall protection structures adapted to this context. The main feature for this type of structure is that it must be a temporary structure. Indeed, protection is required for only the duration of forestry works. In some cases, the life time of this protection can be extended not more to 10 years. This can be the case where the forest previously offered an efficient protection against rock falls. This kind of protective structure must counterbalance the lack of protection within the growing period of the trees.

The other feature is that the structure must be compatible with the forested context. This means that it should not require heavy interventions neither than heavy engines such as for soil-moving or even nailing.

Considering both of these features, it appeared that the technical solution has to be rapid and light to install, requiring no specific engine. The innovative proposed solution consists of a rockfall catchment fence made of a cable supported wire mesh tightened between trees.

Obviously, using trees as support instead of the commonly used poles simplify the installation. But it introduces new consideration in the design of the structure. The first one is that the length of the fence depends on the distance between trees. Trees to be considered here must be strong enough to support the loads applied. Trees must then be safe and have a large enough diameter. In the absence of liable data concerning the response of tree trunks under dynamic loadings, this leads to consider the biggest trees as possible support. The other point to consider is the connection between the supporting cable and the trees. This connection should not irreversibly hurt the trees. Specific structural disposition must be considered in this way.

In addition, it is recommended that the fences are installed as close as possible of the probable rock departure points. Indeed, in such a case the rock velocity remains smaller than 25 m/s which is the maximum velocity of rocks on forested slopes reported in the literature (Azzoni et al. 1995, Dorren et al. 2005). Thus, considering the range of mass of the rocks to be stopped, the kinetic energy is expected to be less than 200 kJ, which is considered as a low value for rockfall protection structures.

A part from considerations relating to the trees, the development of this type of fence is investigated based on real scale impact experiments performed in parallel to discrete element modelling.

The aim of this paper is to present some of the first results obtained during both experimental and numerical studies.

2 EXPERIMENTS

2.1 Experimental site

The experimental site is located in the Northern French Alps and had previously been used for real size rock fall experiments on a forested slope (Dorren et al. 2006). On this site, the forest stand is mainly constituted of fir, spruce, beech and maple. The average slope angle; in the area where the tested fence has been implement is of 32°.

In order to build a fence perpendicular to the slope two trees located at the same altitude have been chosen as support. The distance between these trees is of 22 m. The diameter at breath height of the trees is 60 cm for one and 45 cm for the other.

2.2 Material and methods

2.2.1 Fence description

The fence is composed of a wire mesh, three cables and two rigid rods (Figs. 1, 2).

For the sake of simplicity and practicality a commercially available product has been considered as wire mesh. It consisted of a 3 meters large hexagonal wire mesh (or double twisted). The mesh height and width were 60 mm and 80 mm, respectively and was made up of a wire having a diameter of 2.7 mm. The weight of the wire mesh roll used was less than 100 kg which is thought to be possible to move on slopes without engines.

The upper cable, or supporting cable, is a 12 mm in diameter steel cable. The lower cable had the same diameter. The mid height cable had a diameter of 8 mm. The connection between the cables and the fence was performed using cable clips, placed each 20 cm.

On each side, the wire mesh was connected to two vertical steel rods. The aim of these rods is to maintain at the appropriate distance the upper and lower cables, in this case 3 m. It reduces any mesh necking resulting from the impact. In fact, this function can't be taken over by the trees unless tightening firmly the cable around the tree trunk, with a risk of damage for the tree.

The three cables were rolled around the trees and the dead extremities were tightened to the cable using three cable clips. The tree was protected from damage by the cable loop using woody elements.

The height between the soil and the lower cable as well as the distance between the rods and the trees

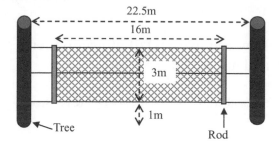

Figure 1. Sketch of the fence, with typical dimensions in the context of the experimentations.

Figure 2. View of a fence before impact.

during the experiments are rather high. Indeed, these large voids would obviously reduce the ability of the fence to intercept rock falls. At this stage of the study, it is thought not to be of great importance. Moreover, this distance was necessary to place a force transducer to measure the tension in the supporting cable. This sensor is a 200 kN load transducer. It was connected to an acquisition system adapted to dynamics loading measurement.

2.2.2 Experimental procedure

The experiments consisted of impacting the fence with boulders of varying mass with the aim of evaluating its capacities. A total of 17 impact experiments were performed on four different fences, using rocks of mass varying from 30 to 500 kg.

For this purpose, a cable was tightened between two trees (Fig. 3). This cable allowed conveying a trolley supporting the rock. The rock was dropped just before the impact (Fig. 4). The location of the impact was depending on the moment of the rock's dropping: It was then possible to impact the fence from the lower to the upper part.

The difference of altitude between the departure point of the trolley and the impact point allowed reaching velocities from 14 up to 17 m/s. This velocity was calculated thanks to image analysis. On the site two

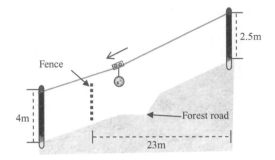

Figure 3. Sketch of the experimental device.

numerical cameras have been used to film the experiments. This value is satisfactory considering the aim of this study.

The tension in the upper cable and the tree acceleration were continuously measured during the impact.

The accelerometer was nailed on one of the supporting tree at distance of 2 m from the ground, close to the mid-height cable. The axis of measurement was aligned with the cable. It was a ± 200 g capacitive accelerometer with a 0–2.5 kHz frequency response. This accelerometer has been used to estimate the tree displacement during the dynamic loading of the fence.

The tests were filmed with a numerical video camera and a numerical high speed camera (up to 500 frames/second). Knowing the travel distance of the rock, these cameras allowed to calculate the velocity of the boulder and to visualize its trajectory during the impact.

The deformation of the mesh was quantified measuring its height in the centre. The maximum displacement of the centre of the fence was estimated thanks to the movies.

The test sequence consisted in impacting successively a same fence increasing the weight of the rock. This has been done until the rupture has been reached. In some cases, the impact led to local damage of the fence. When this damage only consisted in wire cut, with a small induced hole in the net, the fence was tested again without any reparation. In case of a larger hole, obtained after many impacts on a damaged net, simple reparation where made prior the next impact in order to evaluate the strength of a repaired structure.

2.3 Experimental results

The results presented concern the fence described on Figure 1 with indicated dimensions. This fence was impacted several times. The results concern the two first impacts.

A same block 200 kg in mass and angular in shape was launched at a 17 m/s velocity for both tests. In terms of energy, this corresponds to a 30 kJ impact.

Figure 4. Rock before the dropping, at the beginning of impact and after falling in the net.

During the first experiment the block impacted the fence in its centre while the impact location for the second experiment was just beneath the supporting cable (upper cable).

The reference in terms of time considered for the plotting of the curves is defined as the moment when the tension in the cable starts increasing. It is thought

135

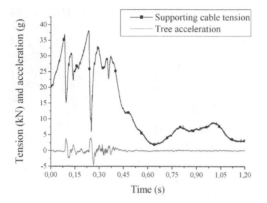

Figure 5. Tension in the cable and acceleration of the tree measured during the first impact.

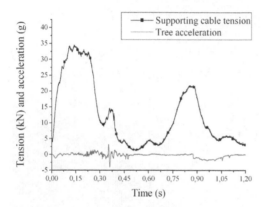

Figure 6. Tension in the cable and acceleration of the tree measured during the second impact.

to be a few milliseconds after the impact of the rock on the net. Indeed, in an attempt to measure the acceleration in the centre of the fence, a delay of about 7 ms was observed between the impact by the block and the increase in tension. As this acceleration measurement is generally not reliable it is not presented here.

Figures 5 and 6 present the tension in the cable and the acceleration of the tree in both cases.

Initially, the tension in the cable is about 20 kN. During the first impact (Fig. 5), the maximum tension is about 38 kN. Sudden tension drops are also observed (mainly at about 120 ms and 220 ms). After 60 ms, the tension fluctuates slightly around a value of about 4.3 kN. This value corresponds to the tension at rest after impact.

The value of the impact duration is evaluated to 500 ms.

In parallel, the acceleration curve has peaks immediately after the tension drops. No other general trend can be deduced form this curve.

The displacement of the centre of the fence reached the maximum value of 1.6 m. The height of the fence, measured in its centre, reduced of about 5 cm. The impact led to the cutting of one wire, in the middle of the impacted zone. This cutting off is certainly due to the angular shape of the block.

The tension in the cable was not modified and the mesh was not repaired before performing the second test. During this second impact, the tension in the cable reached a maximum value of 35 kN 15 ms after contact. 75 ms after this peak, a second peak appears.

The displacement of the centre of the fence reached the maximum value of 2.4 m.

After the impact, the mesh was deformed in the impacted zone, two wires were cut and the fence height was reduced by about 10 cm.

2.4 Interpretation

During the impact the tension in the cable increases rapidly. As a consequence, the cable clips placed on each side of the fence, on the cables, are solicited by increasing loads. If the load exceeds a threshold value the cables clips slid. This seems to be the origin of the sudden drops observed mainly during test 1.

Due to the sliding of the cable clips, the fence tends to sag and the tension at rest is reduces from 20 to less than 5 kN.

During the second test, the block impacted the fence near the upper cable, about 1 meter above the previous impact zone. As a consequence the rock was pushed down during the impact. As a consequence, during the rebound, the rock felt on the lower part of the net, leading to an increase in tension in the supporting cable, as observed at 870 ms.

Despite the impacts were of same energy the maximum displacement in the centre of the fence is very different from one test to the other (1.6 m/2.4 m). This is due to the increase of length of the supporting cable. On the other hand, the maximum tension in the cable doesn't seem to be affected.

Concerning the acceleration of the trees, the interpretation of the data provided is not easy. Indeed, the main variations seem to correspond to the only brutal variations in cable tension. Some low frequency vibrations were expected to occur but the data reveal no such trend. Specific investigations are necessary to validate and analyse the data.

2.5 Trends from other experiments

The previous results concern 2 tests out of 17. The others allowed investigating the influence on the net behaviour of the impact location, of the mesh repairing and of the rock mass increase. The main conclusions drawn from these are:

- Angular blocks can cut wires even for low mass rocks (observed from 200 kg blocks with a 17 m/s velocity);
- In case of an impact on a cable, this cable can be cut as well;
- In the presence of a cut wire the rupture hardly propagates in the mesh during a following impact. This is due to the fact that the mesh is a double-twisted mesh;
- When net perforation occurs, it generally occurs after the first contact with the mesh and results from the free fall of the rock in the mesh near the lower support cable, after the initial impact (see last picture on Fig. 4);
- Reparation of a perforated mesh consisting in placing a 3 m* 3 m patch is perfectly efficient.

3 NUMERICAL SIMULATIONS

3.1 *Method*

Bertrand et al. (2008) have previously developed a numerical model based on the discrete element method (DEM) to simulate the behaviour of double-twisted wire meshes used in various fields of civil engineering. These authors also investigated the behaviour of rockfall protection fences made of such meshes. The fence considered was vertical and was hanged between two poles.

The model developed was used as a base of the numerical simulations. It was necessary to adapt it to account for the specificities of the fence considered. These concern the geometry of the fence, the presence of the cables and rods and the boundary conditions.

Nevertheless, using this adapted model to proceed simulations at the real scale appeared to require very long calculation times. Thus, the numerical model was used to investigate the influence of some parameters related to the rock kinematics (velocity, impact point, orientation). The influence of some structural choices has been also investigated in an optimization process.

The simulations concerned a fence made of a 4 m* 1.86 m net. The mesh height and width were 80 mm and 100 mm, respectively and was made up of a wire having a diameter of 2.7 mm. The upper and lower cables have a diameter of 5 mm. The distance between the supports was 7 m.

In the following, the results concern the investigation of the influence of the initial velocity of the rock, the orientation of the rock before impact and the location of the impact. The rock had a mass of 50 kg.

3.2 *Some results*

3.2.1 *Varying the incident velocity*
The incident velocity (*Vi*) of the block before impact was varied from 5 to 25 m/s.

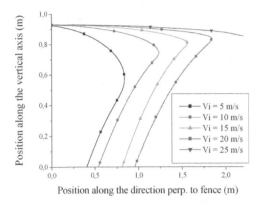

Figure 7. Trajectory of the rock after contact with the fence for different incident velocities.

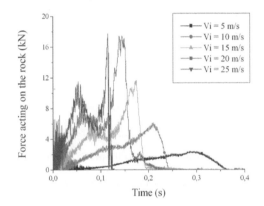

Figure 8. Force acting on the rock during impact for different incident velocities.

Figure 7 shows the trajectory of the block during the impact, in a vertical plane. The 25 m/s velocity block perforates the mesh. Figure 8 shows the evolution of the response of the mesh depending on the incident velocity of the rock. The impact duration reduces with *Vi*, from 36 to 23 ms over the 5–20 m/s *Vi* range. The maximum force applied on the boulder during the 20 m/s impact is of about 17 kN. The corresponding tension in the upper cable was of about 10 kN (Fig. 9).

3.2.2 *Varying the impact location*
Impacts on 3 different locations are illustrated on Figures 10–12. The 50 kg boulder having a velocity of 20 m/s either hit the fence in its centre (1), at a mid distance between the fence centre and the upper cable (2) and in the left lower corner of the fence (3).

Both the penetration of the fence and the force acting on the boulder are significantly modified by the impact location (Figs. 10, 11). In case of the impact in the corner, the rock endures a vertical and ascendant

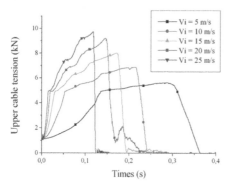

Figure 9. Tension in the upper cable during the impact for different incident velocities.

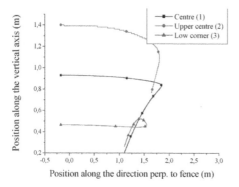

Figure 10. Trajectory of the rock after impact on different points of the wire mesh.

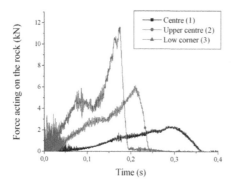

Figure 11. Force acting on the rock during impact for different incident velocities.

force due to the proximity of the lower cable. Moreover, the rock trajectory is also deviated out of the vertical plane during the impact. The tension in the upper cable is less in this case.

The maximum tension in the upper cable in the case of an impact close to the upper cable is less than during a centred impact what is counter-intuitive. This is

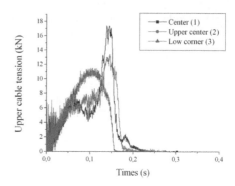

Figure 12. Tension in the upper cable during the impact for different incident velocities.

the consequence of the fact that a block impacting the fence close to the upper cable is pushed downward by the mesh.

The conclusion is that the centred impact is the most critical for the upper cable and that the fence behaves satisfactorily whatever the impact point in case of a 50 kg boulder with an incident velocity of 20 m/s.

4   CONCLUSION

The real size impact tests show that the tested fences have an unexpectedly good behaviour to rock impacts due to their high deformability and due to the mechanical behaviours of the trees. Repeated impacts often led to only minor damages to the net and to the trees. Besides, the numerical study allowed investigating the influence of some parameters providing essential complementary data.

These results constitute the first and encouraging step towards the development of efficient low energy fences fixed on trees.

REFERENCES

Azzoni, A. & de Freitas, M.H. 1995. Experimentally gained parameters, decisive for rock fall analysis. *Rock mechanics and rock engineering* 28: 111–124.
Bertrand, D., Nicot, F., Gotteland P. & Lambert S. 2008. DEM numerical modeling of double-twisted hexagonal wire mesh *Canadian Geotechnical journal* 45: 1104–1117.
Dorren, L.K.A., Berger, F., Le Hir, C., Mermin, E. & Tardif, P. 2005. Mechanics, effects and management implications of rockfall in forests. *Forest ecology and management* 215: 183–195.
Dorren, L.K.A., Berger, F., & Putters U.S. 2006. Real size experiments and 3D simulation of rockfall on forest slopes. *Natural Hazards and Earth Systems Science* 6: 145–153.

Prediction and Simulation Methods for Geohazard Mitigation – Oka, Murakami & Kimoto (eds)
© 2009 Taylor & Francis Group, London, ISBN 978-0-415-80482-0

Numerical analysis of erosion and migration of soil particles within soil mass

K. Fujisawa, A. Murakami & S. Nishimura
Okayama University, Okayama, Japan

ABSTRACT: Internal Erosion caused within soil structures, such as earth dams and irrigation ponds, lead to the piping failures of the structures. Most of studies on the internal erosion of soils have been experimental and the computational method to analyze the phenomenon is currently limited. The purpose of this study is to propose the numerical method for solving the internal erosion and the transport of eroded soil particles and to verify the applicability of the method. In this paper, three equations concerning the seepage flow, the change of porosity due to internal erosion and the transport of the eroded soil particles are introduced and computationally solved. The equation of the seepage flow is solved with Finite Element Method, and Finite Volume Method is applied to the equation of the transport of the soil particles. It is shown that the numerical approach can reproduce the experimental date of the previous study.

1   INTRODUCTION

Piping is the phenomenon that a flow path, where the seepage flow concentrates, appears within soil structures. Usually the flow path is created due to the erosion and the migration of soil particles. Piping is the primary cause of dam breaks. Actually, Foster et al. (2000 a,b) investigated world-wide embankment dam failures and accidents, and reported that 46% of them were triggered by piping. The water leakage from aging irrigation ponds, which are small embankment dams to store irrigation water, has been frequently reported in Japan. This phenomenon is considered to result from piping. Therefore, piping is a serious problem for soil structures subjected to seepage flow.

Up to now, there have been a dozen studies on piping and these works can be divided into roughly two categories. One is so-called "boiling", which is the failure of a soil skeleton induced by the inter-granular seepage force greater than the resistible force the soil mass has due to its weight. The studies on boiling have been focused on the determination of the critical hydraulic head or gradient (e.g., Meyer et al. 1994, Ojha et al. 2000). The other is detachment of fine particles from the soil fabric due to the seepage flow and the transport of them out of the soil mass, which increases the porosity and leads to piping. In this paper, this phenomenon is called "internal erosion" (e.g. Khilar et al. 1985, Reddi et al. 2000, Wan & Fell 2004). Although internal erosion does not cause the failure of the soil fabric as boiling does, these two phenomena are related with each other, as shown by Skempton & Brogan (1994). They carried out the experiments of boiling using the mixtures of gravels and sands and observed that the critical hydraulic gradient became even smaller than the conventional estimate when the sands of the mixture subjected to erosion within the sample. Their results have shown that internal erosion remains important even when boiling is considered.

Richards & Reddy (2007) have indicated that the computational methods for evaluation of piping potential are currently limited. As mentioned above, piping is a phenomenon deeply related with internal erosion. However, soil mechanics, at present, cannot deal with the internal erosion of soils and the transport of eroded soil particles. The objective of this study is to propose the numerical method for analyzing the internal erosion and the transport of eroded soil particles and to verify the applicability of the method in light of the previous experimental studies.

In order to solve the internal erosion and the transport of eroded fine particles within soils, the followings have to be considered: (1) the velocity field of intergranular seepage flow (2) the increase of the porosity of soils (3) the transportation of eroded fine soil particles. In this paper, governing equations for the above three items are introduced with Eulerian formulation and the numerical method to solve the equations are proposed using finite element method and finite volume approach. After that, the proposed numerical model is applied to the results of the internal erosion tests conducted by Reddi et al. (2000) and the applicability of the model is verified.

2   GOVERNING EQUATIONS

2.1   *Erosion rates and four phases of soils*

Erosion rates of soils are introduced and soil phases are considered here in advance of deriving the governing

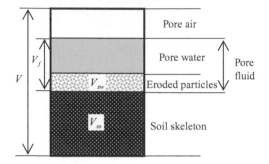

Figure 1. Four phases of soils (soil skeleton, eroded soil particles, pore water and pore air).

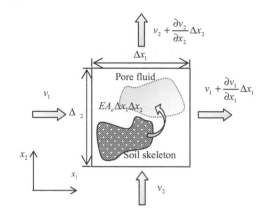

Figure 2. Balance of pore fluid within an infinitesimal soil element.

equations. Erosion rates are defined as the volume of eroded soil particles from unit surface area of erodible region within unit time, which have the dimensions of velocity. The previous empirical studies have adopted the following form of erosion rates as a function of the shear stress exerted onto erodible soil particles (e.g., Khilar et al. 1985, Reddi et al. 2000):

$$E = \alpha(\tau - \tau_c) \qquad (1)$$

where E, α, τ and τ_c denote the erosion rates, the erodibility coefficient, the shear stress and the critical shear stress. If the shear stress exerted by fluid τ is smaller than the critical shear stress τ_c, erosion does not happen.

Figure 1 shows the four phases of soils in order to deal with the internal erosion and the transport of detached soil particles from the soil skeleton. Usually, soils are divided into three phases of pore air, pore water and soil particles. However, when the internal erosion of soils is considered, there exist two types of soil particles, i.e., the particles of the soil skeleton and those eroded, or detached from the soil fabric.

As shown in Figure 1, the mixture of pore water and eroded soil particles is defined as the pore fluid in this paper. V, V_{ss}, V_{sw} and V_f in the figure denote the volume of the soil mass, soil skeleton, eroded soil particles and pore fluid, and respectively. Using these definitions, the soil properties are introduced as follows:

$$\theta = \frac{V_f}{V}, \quad n = 1 - \frac{V_{ss}}{V}, \quad C = \frac{V_{sw}}{V_f} \qquad (2)$$

where θ, n and C denote the volumetric fluid content, the porosity and the concentration of detached soil particles from the soil fabric contained in the pore water, respectively.

2.2 Derivation of governing equations

Two dimensional problem is treated in this study. Figure 2 shows the volumetric balance of pore fluid within an infinitesimal element of Δx_1 in width and Δx_2 in height. x_1, x_2, v_1, v_2 and A_e in the figure denote the horizontal and vertical axis, the horizontal and vertical component of the pore fluid velocity and the surface area of the erodible region per unit volume, respectively.

Considering the direction of x_1, the pore fluid flows into the small soil element at the rate of $v_1 \Delta x_2$ and flows out of it at the rate of $v_1 \Delta x_2 + \partial v_1/\partial x_1 \, \Delta x_1 \Delta x_2$, so that the pore fluid within the element decreases by $\partial v_1/\partial x_1 \, \Delta x_1 \Delta x_2$ due to the flow in the horizontal direction. Similarly, the pore fluid decreases by $\partial v_2/\partial x_2 \, \Delta x_1 \Delta x_2$ due to that in vertical direction. In addition to the inflow and outflow, the pore fluid increases by the volume of eroded soil particles since the pore fluid is the mixture of the pore water and eroded particles. The volume of eroded soil particles per unit time is given by $E A_e \Delta x_1 \Delta x_2$. From the above discussion, the rate of the change of volumetric fluid content $\partial\theta/\partial t$ has the following relation:

$$\frac{\partial \theta}{\partial t} \Delta x_1 \Delta x_2 = (E A_e) \Delta x_1 \Delta x_2 - \frac{\partial v_1}{\partial x_1} \Delta x_1 \Delta x_2$$
$$- \frac{\partial v_2}{\partial x_2} \Delta x_1 \Delta x_2 \qquad (3)$$

The internal erosion decreases the volume of soil skeleton by that of eroded particles, which leads to the following equation:

$$\frac{\partial (1-n)}{\partial t} \Delta x_1 \Delta x_2 = -E A_e \Delta x_1 \Delta x_2 \qquad (4)$$

The volume of eroded soil particles within the pore fluid also changes because of the inflow and outflow and the detachment of the soil particles from the fabric, which can be treated in the same manner of deriving equation (3).

140

Due to the horizontal and vertical flow of the pore fluid, the volume of the detached particles within the small element decreases by $\partial Cv_1/\partial x_1\,\Delta x_1\Delta x_2$ and $\partial Cv_2/\partial x_2\,\Delta x_1\Delta x_2$. Additionally, the soil particles are detached at the rate of $E\,A_e\Delta x_1\Delta x_2$. Therefore, the increase rate of eroded soil particles within the small soil mass $\partial C\theta/\partial t$ has the following relation:

$$\frac{\partial C\theta}{\partial t}\Delta x_1\Delta x_2 = (E\,A_e)\Delta x_1\Delta x_2 \\ -\frac{\partial Cv_1}{\partial x_1}\Delta x_1\Delta x_2 - \frac{\partial Cv_2}{\partial x_2}\Delta x_1\Delta x_2 \quad (5)$$

Divided by $\Delta x_1\Delta x_2$, equations (3) to (5) are reduced to the following three partial differential equations:

$$\frac{\partial\theta}{\partial t} + \frac{\partial v_i}{\partial x_i} = E\,A_e \qquad (6)$$

$$\frac{\partial(1-n)}{\partial t} = -E\,A_e \qquad (7)$$

$$\frac{\partial C\theta}{\partial t} + \frac{\partial Cv_i}{\partial x_i} = E\,A_e \qquad (8)$$

where Einstein summation convention is applied (the summation convention will be applied hereafter). A similar equation to equation (8) was adopted by Khilar et al. (1985) and Reddi & Bonala (1997) to analyze the migration of soil particles. When the soil is saturated, the following relation is satisfied:

$$\theta = n\,Sr = n \qquad (9)$$

where Sr is the degree of saturation. Substituting equation (9) to equations (6) and (8) and rewriting equation (7), these three equations can be reduced to as follows:

$$\frac{\partial n}{\partial t} + \frac{\partial v_i}{\partial x_i} = E\,A_e \qquad (10)$$

$$\frac{\partial n}{\partial t} = E\,A_e \qquad (11)$$

$$\frac{\partial nC}{\partial t} + \frac{\partial Cv_i}{\partial x_i} = E\,A_e \qquad (12)$$

Usually, the internal erosion happens when the soil is saturated and the seepage flow velocity is sufficiently rapid. For this reason, the internal erosion of soils is analyzed using equations (10) to (12) in this paper. Darcy's law is applied to the intergranular seepage flow:

$$v_i = k_s\frac{\partial h}{\partial x_i}\,, \quad h = z + \frac{u_w}{\rho\,g} \qquad (13)$$

where k_s, h, z, u_w, ρ, g denote the permeability, the hydraulic head, the elevation head, the pressure and the density of pore fluid and the gravitational acceleration,

respectively. With the aid of equations (11) and (13), equation (10) can be reduced into

$$\frac{\partial}{\partial x_i}\left(k_s\frac{\partial}{\partial x_i}\left(z + \frac{u_w}{\rho\,g}\right)\right) = 0 \qquad (14)$$

which is the same as the equation conventionally used for the analysis of seepage flow. From equations (10) and (11), these equations seem to be coupled with each other. However, equation (14) shows that it is not need to be solved simultaneously with equation (11). The unknown variables of equations (11), (12) and (13) are the pressure of the pore fluid, the porosity and the concentration of the eroded soil particles.

2.3 Estimate of shear stress

The shear stress exerted to erodible soil particles need to be estimated in order to determine the values of the erosion rates of soils, as seen from equation (1). Efforts to estimate the shear stress in the interior of soils have required the idealization of soil pores as an ensemble of pore tubes. However, the pore tube dimension distributes within soils and it is quite difficult to determine the spatially distributed value of the shear stress. To overcome this difficulty, the representative pore tube dimension has been frequently used. If the permeability and the porosity of a soil material are given, the representative pore diameter $\hat{D}$ is obtained as

$$\hat{D} = 4\sqrt{\frac{2K}{n}} \qquad (15)$$

where K denotes the intrinsic permeability, defined as

$$K = \frac{k_s\,\mu}{\rho\,g} \qquad (16)$$

where μ is the viscosity of pore fluid. Assuming Hagen-Poiseulle flow in the pore tube, the shear stress exerted onto the pore wall is estimated by the following equation (e.g., Reddi et al. 2000):

$$\tau = \rho\,gI\sqrt{2K/n} \qquad (17)$$

where I stands for the hydraulic gradient. With equation (17), the erosion rates of soils can be known.

3 NUMERICAL METHOD

The numerical method to solve the governing equations (11), (12) and (14) is explained in this section. Equation (14) describing the seepage flow within soils is solved with respect to the pressure of pore fluid u_w by FEM (finite element method). In this analysis, 4-node isoparametric element is adopted. FVM (finite volume method) is applied for solving equation (12) with respect to the concentration of eroded soil particles C. The installation of FVM is easy especially when advection equations, such as equation (12), are treated because the method does not require to the inverse matrix calculation. The details of the application of FVM to equation (12) are accounted for later in this section.

Figure 3 shows the flow chart of this computational analysis. First, equation (14) is solved using FEM and the results gives the spatial distribution of the pressure of the pore fluid u_w. From equation (13), or Darcy's law, the value of the seepage flow velocity is also available anywhere in the calculation region, using the obtained pressure distribution. The pressure of the pore fluid given, the hydraulic gradient I can be obtained. The determination of the hydraulic gradient allows the evaluation of the shear stress using equation (17). With the aid of equation (1), the erosion rate of the material E can be estimated. The surface area of the erodible region, which is equivalent to that of movable soil particles, is evaluated by the particle distribution curve or the test of the specific surface area. Giving the erosion rate E and the erodible surface area per unit volume A_e, the increment of the porosity is calculated from equation (11) as follows:

$$n^{m+1} = n^m + \Delta t E^m A_e^m \qquad (18)$$

where Δt is the time increment for one time step and the superscript m denotes the time step number.

After these above process, equation (12), describing the alteration of the particle concentration within the pore fluid, is solved with FVM. As finite volume cells, the finite elements of the above analysis are used. To apply FVM, equation (12) is spatially integrated in the cell:

$$\int_{\Omega^l} \frac{\partial nC}{\partial t} d\Omega^l + \int_{\Omega^l} \frac{\partial v_i C}{\partial x_i} d\Omega^l = \int_{\Omega^l} E A_e d\Omega^l \qquad (19)$$

where Ω^l denotes the area of the lth cell. Applying Green's formula to equation (19), it reduces to

$$\int_{\Omega^l} \frac{\partial nC}{\partial t} d\Omega^l + \int_{s^l} C v_i t_i^l ds^l = \int_{\Omega^l} E A_e d\Omega^l \qquad (20)$$

where t_i^l is the unit normal outward vector at the boundary of the lth cell. Introducing the representative porosity and particle concentration in lth cell at mth time step, $n^{l,m}$ and $C^{l,m}$, equation (20) is discretized into

$$\Omega^l \frac{n^{l,m+1} C^{l,m+1} - n^{l,m} C^{l,m}}{\Delta t} + q_i^m t_{ij}^l \Delta s_j^l \qquad (21)$$

$$= \Omega^l \frac{n^{l,m+1} - n^{l,m}}{\Delta t}$$

where t_{ij}^l and Δs_j^l mean the unit normal outward vector and the length of jth boundary segment of lth cell, as shown in Figure 4. q_i^m in equation (21) denotes

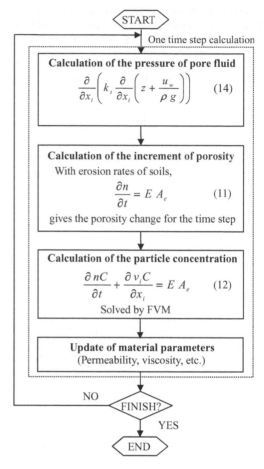

One time step calculation

Calculation of the pressure of pore fluid

$$\frac{\partial}{\partial x_i}\left(k_s \frac{\partial}{\partial x_i}\left(z + \frac{u_w}{\rho g}\right)\right) \qquad (14)$$

Calculation of the increment of porosity

With erosion rates of soils,

$$\frac{\partial n}{\partial t} = E A_e \qquad (11)$$

gives the porosity change for the time step

Calculation of the particle concentration

$$\frac{\partial nC}{\partial t} + \frac{\partial v_i C}{\partial x_i} = E A_e \qquad (12)$$

Solved by FVM

Update of material parameters
(Permeability, viscosity, etc.)

NO

FINISH?

YES

END

Figure 3. Flow chart of numerical analysis of internal erosion of soils.

the flux of the eroded soil particles passing through the boundary at mth time step. The value of the flux is given as follows, using FVS (flux vector splitting) approach (Toro, 1999)

$$q_i^m t_{ij}^l = C^{l,m} v_+^{l,m} + C^{r,m} v_-^{r,m} \qquad (22)$$

$$v_+^{l,m} = \frac{1}{2}\left(\bar{v}_i^{l,m} t_{ij}^l + \left| \bar{v}_i^{l,m} t_{ij}^l \right| \right) \qquad (23)$$

$$v_-^{r,m} = \frac{1}{2}\left(\bar{v}_i^{r,m} t_{ij}^l - \left| \bar{v}_i^{r,m} t_{ij}^l \right| \right) \qquad (24)$$

Here, the subscript r denotes the number of the neighbor cell shearing the boundary of lth cell, as shown in Figure 4.

Establishing equation (21) and solving it with respect to $C^{l,m+1}$ for all the cells, the distribution of the particle concentration is calculated. Finally, the

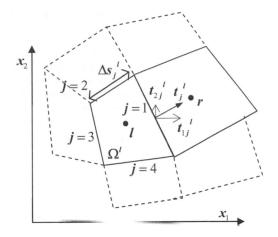

Figure 4.   Geometry of finite volume cell.

density and viscosity of the pore fluid is updated with the following equations (Julien, 1998):

$$\rho = C\rho_s + (1 - C)\rho_w \qquad (25)$$

$$\mu = \eta(1 + 2.5C) \qquad (26)$$

where ρ_s, ρ_w and η denote the density of soil particles and pure water, and the viscosity of pure water. The permeability alters because of the change of the fluid viscosity, so that it is also updated as follows (e.g. Lambe & Whitman, 1979; Budhu, 2007):

$$k_s = D_r^2 \frac{\rho g}{\mu} \frac{C_T n^2}{(1 - n)^3} \qquad (27)$$

where C_T and D_r are the material constant and the representative particle diameter, respectively.

4   VERIFICATION

The applicability of the numerical model presented in the previous section is verified here. Reddi et al. (2000) conducted the internal erosion test using the mixture of Ottawa sand (70%) and kaolinite clay (30%). The details of the experiments are referred to their pape and the brief explanation is made here.

They compacted the mixture in the mold and prepared the cylindrical sample with 101.6 mm in diameter and 50 mm in thickness. The sample was permeated by distilled water and the seepage flow in the thickness direction was generated, which caused the erosion in the interior of the samples. They measured the turbidity of the effluent, which was converted to the kaolinite particle concentration or the discharge rate of kaolinite. During the experiments, they controlled the flow rate and recorded the water pressure at the inlet of the samples.

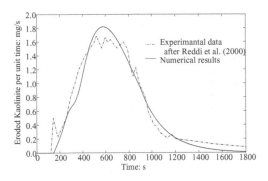

Figure 5.   Relationship between discharge rate of kaolinite and elapsed time after Reddi et al. (2000).

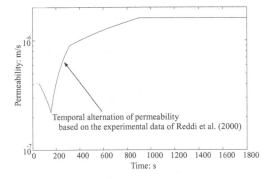

Figure 6.   Relationship between recorded permeability and elapsed time during the internal erosion test of Reddi et al. (2000).

The dash line in Figure 5 shows the recorded discharge rate of kaolinite. During the test, the flow rate was increased linearly from 0 to 200 ml/min till 900 minutes and kept constant at 200 ml/min after that. Since the pore water pressure was continuously measured, the variation of the permeability with respect to time was determined. Figure 6 shows the temporal alteration of the permeability during the test, which was drawn on the basis of their figure.

In order to verify the proposed numerical model, the reproduction of the experimental date was carried out. The numerical result of the kaolinite discharge rate is shown as the solid line in Figure 5. It can be seen that the numerical result provides good agreement with the experimental data.

In this numerical analysis, the finite element mesh shown in Figure 7 was used. The width of the mesh could be arbitrarily chosen because the experiment of Reddi et al. (2000) was a one-dimensional problem. As the boundary condition, the flow rate equivalent to their control was imposed on the top and the air pressure was applied as the pore fluid pressure at the bottom. Integrating the kaolinite discharge rate

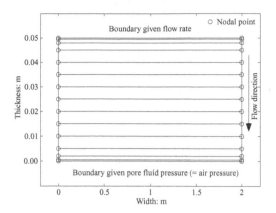

Figure 7. Finite element mesh and boundary conditions used in the numerical analysis.

in Figure 5 with respect to time to obtain the total eroded mass ($= 1.1 \times 10^{-3}$ kg), and multiplying it by the specific surface area of kaolinite ($= 20 \times 10^3$ m²/kg), the initial surface area of erodible soil particles was determined. Dividing it by the volume of the test sample ($= 4.05 \times 10^{-4}$ m³), the surface area per unit volume A_e was obtained. Other parameters of the gravitational acceleration ($= 9.8$ m/s²), the density of water ($= 1000$ kg/m³) and soil particles ($= 2600$ kg/m³) and the viscosity of pure water ($= 1.005 \times 10^{-3}$ kg/m·s) were given, and the critical shear stress ($= 1.1$ Pa) and the initial porosity ($= 0.27$) were provided by the results of Reddi et al. (2000). During the numerical analysis, the permeability was updated according to its change shown in Figure 6. Only the erodibility coefficient α could not be decided, so that its value was arbitrarily given to meet the experimental data shown in Figure 5. Up to now, the direct determination of the erodibility coefficient for internal erosion has not been done for its great difficulty.

This numerical analysis has shown that the proposed numerical model can reproduce the experimental data sufficiently well and that the method may allow the indirect determination of the erodibility coefficient for internal erosion by the measurement of the effluent particle concentration when its value is unknown.

5 CONCLUSIONS

This paper has presented the numerical method to analyze the internal erosion of soils and its verification. In order to describe the internal erosion, the concept of the four phases of soils is important. The usage of erosion rates of soils has enabled the Eulerian derivation of governing equations for the seepage flow, the change of the porosity and the transport of detached soil particles from the soil skeleton.

As for the numerical analysis, finite element method realized the accurate solutions of the seepage flow field, while finite volume method allowed the easy installation and the low computational load because the method does not need to solve matrices. From the verification of the proposed numerical model, it has been shown that the model can reproduce the data of the internal erosion test sufficiently well and suggested the possibility of the indirect determination of erodibility coefficients.

If this method of internal erosion of soils is coupled with the deformation analysis of soils, the applicability to realistic problems, such as piping failure and hollowing of soil structures, is considerably wide.

REFERENCES

Budhu, M. 2007. *Soil Mechanics and Foundation, 2nd Edition,* John Wiley & Sons: 83.

Foster, M., Fell, R. and Spannagle, M. 2000a. The statistics of embankment dam failures and accidents. *Can. Geotech. J.* 37: 1000–1024.

Foster, M., Fell, R. and Spannagle, M. 2000b. A method for assessing the relative likelihood of failure of embankment dams by piping. *Can. Geotech. J.* 37: 1025–1061.

Julien, P.Y. 1998. *Erosion and Sedimentation,* Cambridge University Press: 15.

Khilar, K.C., Fogler, H.S. and Gray, D.H. 1985. Model for piping-plugging in earthen structures. *Journal of Geotechnical Engineering, ASCE,* 111(7): 833–846.

Lambe, T.W. and Whitman, R.V. 1979. *Soil Mechanics, SI version,* John Wiley & Sons: 283.

Meyer, W., Schuster, R.L. and Sabol, M.A. 1994. Potential for seepage erosion of landslide dam. *Journal of Geotechnical Engineering* 120(7): 1211–1229.

Ojha, C.S.P., Sigh, V.P. and Adrian, D.D. 2003. Determination of critical head in soil piping. *Journal of Hydraulic Engineering* 129(7): 511–518.

Reddi, L.N. and Bonala, M.V.S. 1997. Analytical solution for fine particle accumulation in soil filters, *Journal of Geotechnical and Geoenvironmental Engineering* 123(12): 1143–1152.

Reddi, L.N., Lee, In-M and Bonala, M.V.S. 2000. Comparison of internal and surface erosion using flow pump test on a sand-kaolinite mixture, *Geotechnical Testing Journal* 23(1): 116–122.

Richards, K.S. and Reddy, K.R. 2007. Critical appraisal of piping phenomenon in earth dams, *Bull. Eng. Geol. Environ.* 66: 381–402.

Skempton, A.W. and Brogan, J.M. 1994. Experiments on piping in sandy gravels, *Geotechnique* 44(3): 449–460.

Toro, E.F. 1999. *Riemann Solvers and Numerical Method for Fluid Dynamics, 2nd edition,* Springer.

Wan, C.F. and Fell, R. 2004. Investigation of rate of erosion of soils in embankment dams, *Journal of Geotechnical and Geoenvironmental Engineering* 130(4): 373–380.

Prediction and Simulation Methods for Geohazard Mitigation – Oka, Murakami & Kimoto (eds)
© 2009 Taylor & Francis Group, London, ISBN 978-0-415-80482-0

Assessment of well capture zone using particle tracking

K. Inoue & T. Tanaka
Kobe University, Kobe, Japan

G.J.M. Uffink
Delft University of Technology, Delft, The Netherlands

ABSTRACT: The aim of this study is to assess time-related capture zones relevant to the activity of extraction wells with different pumping rates. The flow is steady and two-dimensional in a homogeneous domain with isotropic and anisotropic transmissivities. Spatial locations of time-related well capture zones for nonreactive solute are delineated using backward particle tracking approach. The results showed the effects of well config-uration, pumping rate and interaction between individual capture zones belonging to each extraction well on the variation of second order spatial moments as well as the form of capture zones. Moreover, total probability that initial particle distributions associated with time-related capture zones were extracted by the well within a given time was identified through random walk particle tracking. Extraction well comprising the outer capture zone displayed a substantial increase of total probability due to the particle pass through a boundary between the inner and outer capture zones.

1 INTRODUCTION

Pump-and-tread is probably the oldest and most reliable of all technologies available for groundwater remediation and groundwater quality management. Although the method has been criticized recently for being incapable of cleaning up a contaminated aquifer completely, there is no doubt that this method alone can remove a large portion of the contaminants from the aquifer and prevent the contamination from further spreading (Kunstmann & Kinzelbach 2000). Capture zones delineate the area around a pumping well from which groundwater is captured a specified time period. They play an important role in the pro-tection of groundwater supply wells, which are used in the definition of source protection zones. The accu-rate assessment of a capture zone is therefore of great importance, not only from an environmental point of view, but in terms of public health.

Vassolo et al. (1998) developed the method of inverse modeling for delineating the capture zone distribution of a pair of drinking water supply wells under the conditions of uncertainty of groundwater recharge in Monte Carlo simulations. Additionally, van Leeuwen et al. (2000) determined capture zone probability distributions and introduced the measures of conditioning. Feyen et al. (2001) assessed the predictive uncertainty in well capture zones in het-erogeneous aquifers with uncertain parameters. These methods only take into account the advective transport of a contaminant.

Another aspect of concern associated with the capture zone delineation and groundwater quality management is the total probability of well by which a particle injected at a point in an aquifer is extracted under the effect of dispersion. Although a few authors modeled the transport by advective dispersion and proposed analytical solutions for the probability that a solute particle is extracted by a well (e.g., van Kooten 1995) little attempt has been made to investigate the effect of random dispersion on the probability distribu-tion of contaminant arrival to the well. The objectives of this study are to assess time-related capture zones relevant to the activity of extraction wells with dif-ferent pumping rates and to elucidate the effect of dispersion on the total probability of particles released from capture zones under the conditions of steady and two-dimensional flow in a homogeneous domain with isotropic and anisotropic transmissivities. Particle tracking approaches are used to illustrate the tran-sient variation of capture zones and make inferences on the effect of dispersion on the capture probability of wells.

2 METHODOLOGY

2.1 *Groundwater flow model*

A two-dimensional steady state saturated groundwater flow in isotropic and anisotropic homogeneous porous formations is assumed in this study. The flow of fluid in porous media is described by the following equation (Bear, 1972).

$$\nabla \cdot (\mathbf{K} \nabla h) = Q \tag{1}$$

where h is the piezometric head, $\mathbf{K}$ is the hydraulic conductivity tensor and Q is the sink/source term. Equation (1) is solved numerically by finite element method. Then, the average groundwater flow velocity in each direction can be calculated by dividing the Darcy velocities by the porosity of the medium.

$$\mathbf{v} = \frac{\mathbf{K}}{n} \nabla h \tag{2}$$

where $\mathbf{v}$ is the pore water velocity vector, n is the porosity. In this study, two-dimensional groundwater flow is of interest so that the magnitude of anisotropy β is defined by K_x/K_y.

2.2 Forward and backward particle tracking

In porous media, the mass transport equation is described by the classical advection and dispersion equation which can be written as (Bear, 1972)

$$R\frac{\partial c}{\partial t} = -\nabla(\mathbf{v}c) + \nabla(\mathbf{D}\nabla c) \tag{3}$$

where c is the concentration and R is the retardation factor. In this equation, $\mathbf{D}$ represents the local hydrodynamic dispersion tensor and can be written as

$$D_{ij} = (\alpha_T |v| + D_d) I_{ij} + (\alpha_L - \alpha_T) \frac{v_i v_j}{|v|} \tag{4}$$

where α_L is the longitudinal dispersivity, α_T is the transverse dispersivity, D_d is the effective diffusion coefficient and I_{ij} is the unit matrix.

Particle tracking method is a lagrangian approach in which a large number of particles are tracked to simulate transport phenomena. The evolution in time of a particle is driven by a drift term that relates to the advective movement and a superposed Brownian motion responsible for dispersion.

$$X_{p,i}(t + \Delta t) = X_{p,i}(t) + A_i(X_p(t))\Delta t \tag{5}$$

$$+ \sum_{j=1}^{3} B_{ij}(X_p(t)) \Xi_i \sqrt{\Delta t}, \quad i = 1, 2, 3$$

where $X_{p,i}(t)$ is the i-component of the particle location at time t, Δt is the time increment, and Ξ_i is a vector which contains three normally distributed random numbers with zero mean and unit variance. The drift vector of the random walk A_i is expressed as follows (Uffink, 1990)

$$A_i = \frac{v_i(X_p(t)) + \sum_{j=1}^{3} \frac{\partial D_{ij}}{\partial x_j}(X_p(t))}{R(X_p(t))}, \quad i = 1, 2, 3 \tag{6}$$

The displacement matrix B_{ij} represents the dispersion and has the form (Burnett and Frind, 1987)

$$B_{ij} = \begin{pmatrix} \frac{v_1}{|v|}\sqrt{\frac{2\alpha_L|v|}{R}} & \frac{-v_1v_3\sqrt{\frac{2\alpha_T|v|}{R}}}{|v|\sqrt{v_1^2+v_2^2}} & \frac{-v_2\sqrt{\frac{2\alpha_T|v|}{R}}}{\sqrt{v_1^2+v_2^2}} \\ \frac{v_2}{|v|}\sqrt{\frac{2\alpha_L|v|}{R}} & \frac{-v_2v_3\sqrt{\frac{2\alpha_T|v|}{R}}}{|v|\sqrt{v_1^2+v_2^2}} & \frac{v_1\sqrt{\frac{2\alpha_T|v|}{R}}}{\sqrt{v_1^2+v_2^2}} \\ \frac{v_3}{|v|}\sqrt{\frac{2\alpha_L|v|}{R}} & \frac{\sqrt{v_1^2+v_2^2}}{|v|}\sqrt{\frac{2\alpha_T|v|}{R}} & 0 \end{pmatrix} \tag{7}$$

In this study, solute of concern is assumed to be conservative and to have no interaction with the solid matrix because of the simplification of problem. In backward tracking, particles are initialized near the sinks and are moved backward toward the source in the reverse flow direction. Backward particle tracking is implemented by

$$X_{p,i}(t+\Delta t) = X_{p,i}(t) - \frac{v_i(X_p(t))}{R(X_p(t))}\Delta t, \quad i = 1, 2, 3 \tag{8}$$

Equation (8) is solved numerically using the fourth order Runge-Kutta scheme.

3 TIME-RELATED CAPTURE ZONES

3.1 Capture zones delineation

Model domain is assumed to be a horizontal two-dimensional aquifer with the porosity of 0.2. As shown in Figure 1, the overall dimensions of the model domain are 400 m in the x-direction and 400 m in the y-direction. The boundary condition of the first type, or Dirichlet boundary condition, for the steady state groundwater flow model is assigned to all boundaries in order to consider a uniform regional flow with a hydraulic gradient of 0.005 from the origin to the point of (400, 400). Three extraction wells are located at point PW1, PW2 and PW3 with the coordinates of (150, 250), (250, 250) and (250, 150), respectively, and are completely penetrating homogeneous isotropic and anisotropic aquifers with a thickness of 10 m. Three types of the magnitude of anisotropy with 0.5, 1.0 and 2.0 are of concern in this study. Three extraction wells are being continuously pumped at the same prescribed rates, Q, of 8, 20, 40 m^3/day to investigate the effects of not only the magnitude of anisotropy but the pumping rate on the extension of the time-related capture zones.

The ratio of longitudinal to transverse dispersivity is assumed to be 0.1 (Bear, 1972) while longitudinal dispersivity is set to another value of 2.0 m. Retardation factor is assumed to be 1.0 in order to simplify the problem. In backward particle tracking, 5000 particles of 50 g of total mass are released to each well under

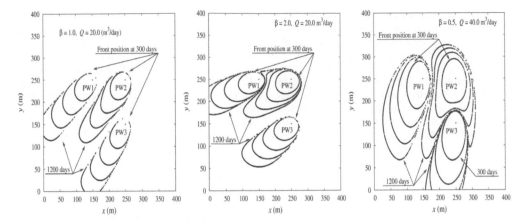

Figure 1. Model domain and time-related capture zones derived from backward particle tracking: (left) $\beta = 1$ and $Q = 20$ m³/day, (middle) $\beta = 2$ and $Q = 20$ m³/day, and (right) $\beta = 0.5$ and $Q = 40$ m³/day.

the time increment of 2 days. Parameters used in the analysis are listed in Table 1.

As representative cases of time-related capture zones derived from backward particle tracking of Equation (8), particle distributions, or isochrones, released from each well under a given hydrogeological condition are illustrated in Figure 1. Each well has own time-related capture zones whose spatial location depends highly on the magnitude of anisotropy and the pumping rate. As shown in the left and middle in Figure 1, a part of particle distribution front extends in the reverse direction of flow with time, while all particles under the pumping rate of 40 m³/day vary those locations during the period of time of concern, except for the capture zones related to the well PW3 as seen in the right of Figure 1. Additionally, isochrones derived from the well PW2 comprise the outer capture zone of the wells PW1 and PW3 due to the well location of PW2 and vary substantially with time since the inner capture zones restrict the inside groundwater to flow to the inside wells. This point indicates that the arrival point of a particle located at a specific point depends on the hydrogeological conditions such as the magnitude of anisotropy and the pumping rate.

3.2 Spatial moments

The overall development of a solute plume can be measured in terms of spatial moments of its mass distribution. In order to assess the variation of time-related capture zones for each well, spatial moments are introduced and are calculated from snapshots of particles in a space at given times as follows:

$$X_{G,i} = \frac{1}{m(t)} \sum_{k=1}^{NP_t} \frac{m_p^k X_{p,i}^k(t)}{R(X_p^k(t))}, \quad m(t) = \sum_{k=1}^{NP_t} \frac{m_p^k}{R(X_p^k(t))} \quad (9)$$

Table 1. Parameters used in the analysis.

Parameter	Value
Hydraulic gradient	0.005
Longitudinal dispersivity	2.0, 4.0, 8.0 m
Dispersivity ratio	0.2
Porosity	0.2
Aquifer thickness	10 m
Hydraulic conductivity in x and y directions	2.5, 5.0 m/day
Pumping rate	0, 8.0, 20.0, 40.0 m³/day
Magnitude of anisotropy	0.5, 1.0, 2.0
Number of the particles	15000
Time step	2 day
Total simulation time	30 years

$$S_{ij}(t) = \frac{1}{m(t)} \sum_{k=1}^{NP_t} \frac{m_p^k X_{p,i}^k(t) X_{p,j}^k(t)}{R(X_p^k(t))} - X_{G,i}(t) X_{G,j}(t) \quad (10)$$

where $m(t)$ is the total liquid phase solute mass, m_p^k is the mass assigned to the k-th particle, $X_{G,i}$ is the center of mass, S_{ij} is the liquid phase second order spatial moments associated with the distribution of particles at a given time, NP_t is the number of particles in the system at time t, and $X_{p,i}^k$ is the i-component of the k-th particle location.

In the left of Figure 2, the results of second order spatial moments in the y direction are shown as a function of displacement distance from the pumping well to the centroid of particles, which is calculated using Equation (9). This figure includes the case (cross symbol) where only the well PW2 is active with the pumping rate of 40 m³/day as a comparison. The second order spatial moments increase with the increase of displacement and approach a constant value, indicating that the extension rate of isochrones becomes constant. Under the pumping rate of 8 m³/day, all second order spatial moments show the same behavior despite the location of wells. According to the increase of pumping rate, the second spatial moments increase nonlinearly with displacement and approach an asymptotic value, except for the well PW2. Under the pumping rate of 40 m³/day, isochrones corresponding to the well PW2 extend irregularly to form the outer capture zones of the wells PW1 and PW3, leading to the marked increase of the second order spatial moment and the difference from the behavior of the case where the wells PW1 and PW3 are inactive. Conversely, the second order spatial moments for the wells PW1 and PW3 approach an asymptotic value in a relatively short displacement because the extension of the capture zones is restricted by the pumping at the well PW2 to form inner and outer isochrones in the domain.

In order to investigate the behavior of the second spatial moments related to the well PW2, the second spatial moments in x and y directions as a function of time are plotted in the middle and right of Figure 2, respectively. It is seen that asymptotic constant values depend on the magnitude of anisotropy. The largest value of the second spatial moment in x direction, S_{11}, is in the case of $\beta = 2$, while the second spatial moment in y direction, S_{22}, displays the largest values under the magnitude of anisotropy of 0.5. This reflects the difference between the velocity components in x and y directions according to the magnitude of anisotropy, which influences to the extension of time-related capture zone.

3.3 Capture probability of well

Backward particle tracking along the streamlines but the reverse direction of inherent flow, as mentioned above, has been conducted under pure advection without dispersion. Therefore, if the direction of flow velocity is switched to the ordinal direction and forward particle tracking without dispersion is carried out, all particles comprising an isochrone at a given time are extracted by the well simultaneously. Inherently, however, porous formations provide the dispersive spreading of particles (Kunstmann & Kinzelbach 2000). Consequently, the total probability that particle enters the pumping well varies according to the degree of dispersion and the location of particle within the largest capture zone, which may be identified using backward particle tracking.

Under the initial locations of particles, which are obtained in a certain time using backward particle tracking, random walk particle tracking is implemented under the longitudinal and transverse dispersivities of 2.0 m and 0.4 m, respectively, in order to elucidate the relation between the release location of particles and the capture probability of well. Some arrival time distributions at each extraction well

are displayed in Figure 3. In this figure, the ratio of cumulative mass captured by each well to total mass released on an isochrone is plotted. In other words, if all particles distributed initially are captured by a corresponding well, which originates an isochrone, the relative mass approaches 1 with time. Figure 3 indicates that relative mass, or capture probability of well approaches to a certain value with time according to the pumping rate. Despite of the magnitude of anisotropy, as the pumping rate becomes larger, the total capture probability increases. As seen in Figure 1, extension of front positions of capture zones concerned with small pumping rate is restricted to a portion of the isochrones, leading to relatively narrow time-related capture zones. The probability that a particle passes over the boundary of time-related capture zones and moves to the downgradient without extraction increases relative to the high extraction case where almost all fronts of the isochrones are located without overlapping within a certain time. This point corresponds to the findings of van Kooten (1995).

Among three well of concern, the well PW2 has the significant variation of the relative mass and shows the value beyond 1.0, especially for the case under the high pumping condition. This means that some released particles located within the capture zones of PW1 and PW3 move into the capture zones of PW2 and enter the well PW2 due to dispersion. Thus, a well comprising the outer capture zones is expected to extract a larger number of particles than a well generating the inner capture zones.

On the other hand, as seen in the right of Figure 1, the isochrones associated with travel time of 900 and 1200 days under the pumping condition of 40 m³/day are not closed due to the movement of particles over the domain during backward particle tracing. Moreover, a large portion of the isochrones of PW3 may locate on a separating streamline with the capture zones of the well PW2. A few studies pointed out that near a separating streamline the dispersivity has a large impact on the arrival of a particle at the extraction well, where as far away from a separating streamline its effect may be neglected (Uffink 1990, van Kooten 1995). Particle tracking results indicate that approximate 32% of particles comprising the capture zones in 900 days for PW3 enters the capture zones for PW2, whereas about 10% of particles forming the isochrone of PW2 moves into the capture zone of PW3. As for the case in 600 days where the isochrone is closed, less part of which shares the boundary between the capture zones of PW2 and PW3. About 12% of particles enters the well PW2 from the region of PW3, while reverse movement of particle is less than 4%. Hence, the difference of the number of particles moving to another capture zone possibly induces earlier grow of relative mass as shown in the right of Figure 3. Since initial particles are located at the same travel time the arrival

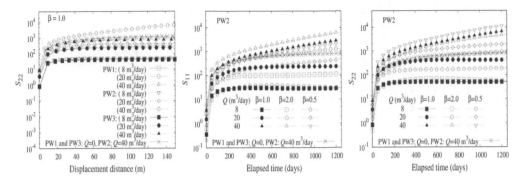

Figure 2. Spatial moment variations: (left) displacement distance vs S_{22} for $\beta = 1$ and (middle) and (right) S_{11} and S_{22} variations as a function of time, respectively.

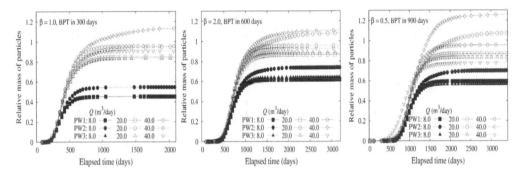

Figure 3. Variation of relative mass of particles captured by the wells: (left) initial particle locations correspond to the travel distance in 300 days under pure advection and $\beta = 1$, (middle) initial particle in 600 days under pure advection and $\beta = 2$ and (right) initial particle locations in 900 days under pure advection and $\beta = 0.5$.

time distribution depends on the isochrone location rather than the form of isochrone.

For the purpose of comparison of capture properties for each well, Figure 4 depicts the total capture probability of well, which is the asymptotic value, as a function of travel time under pure advection. In this figure, PWs means the number of particles, which form the isochrones of the well and arrive at the corresponding well, divided by the initial number of particles, while PWt is the total number of particles entering the well in spite of the attribute of initial particle location. Therefore, the value of PWt is equal to or larger than the value of PWs. Figures confirm the decrease of capture probability of well as the travel time under pure advection becomes longer, whereas the capture probability of well increases with the increase of pumping rate. Far away from the pumping well the probability moving into another capture zone increases. Total capture probability is influenced on the number of particles entering from another capture zone. In the right of Figure 4, the value of PWt of PW2 corresponding to the travel time in 600 days is slightly larger than that in 900 days, indicating that the increase of the

number of particles moving into the capture zone of PW3 from that of PW2. Thus, initial location of particles has an effect on the capture probability not only of PW2 comprising the outer capture zone but of PW1 or PW3 forming the inner capture zone.

3.4 Effect of microdispersivity

Major parameter controlling the capture probability of well is the dispersivity of the medium, which is in accordance with the studies by Uffink (1990) and van der Hoek (1992). The longitudinal dispersivity controls the elongation of the particles with, whereas transverse dispersivity dominates the spreading of particles in the direction perpendicular to the flow direction. While the ratio of transverse to longitudinal dispersivity is taken constant at a value of 0.2, evaluation of the capture probability of well as a function of longitudinal dispersivity is performed.

Capture probabilities of particles of well are shown in Figure 5 as a function of travel time under pure advection from initial location of particles to the extraction well. In this figure, the ratio of PWt to PWs,

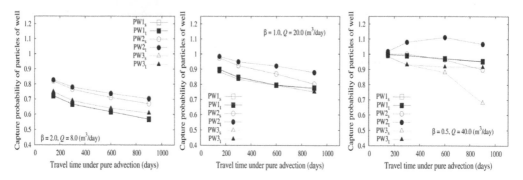

Figure 4. Total capture probability of wells: (left) $\beta = 2$ and $Q = 8$ m³/day, (middle) $\beta = 1.0$ and $Q = 20$ m³/day and (right) $\beta = 0.5$ and $Q = 40$ m³/day.

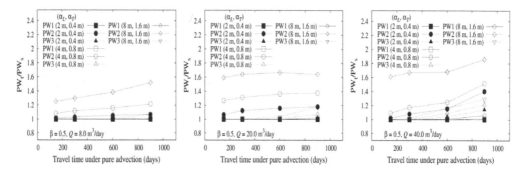

Figure 5. Variation of capture probabilities of well for three types of microdispersivity: (left) $Q = 8$ m³/day, (middle) $Q = 20$ m³/day and (right) $Q = 40$ m³/day.

PWt/PWs is plotted to clarify the effect of microdispersivity on the capture probability. As for the results of the well PW1 and PW3, almost all values of PWt/PWs become approximately 1. In the case of the pumping rate of 40 m³/day, as initial particles locate farther from the pumping well and as the microdispersivity increases, the values of PWt/PWs for PW3 slightly increase. This situation indicates that some particles, which are released at the well PW2 in backward particle tracking and comprise the isochrones for PW2, pass through the boundaries between capture zones of PW2 and PW3 due to the effect of dispersion and are captured by the well PW3. As for the results of PW2, moreover, the values of PWt/PWs range from 1 to 1.8 and are influenced on microdispersivity and initial particle location as well as the pumping rate. The increase of pumping rate may bring the distance between isochrones for each well close to each other as seen in Figure 1. In addition, microdispersivity has an effect on the change of probability that particles move between the capture zones belonging to different wells, resulting in the increase of the number of particles captured by the well PW2. Consequently, in the prediction of capture probability of well,

dispersion property in an aquifer is one of the uncertain factors and affects the total mass withdrawn by an extraction well.

4 CONCLUSIONS

The conclusions drawn from this study are the following:

1. Particle tracking approach demonstrated the effects of well configuration, pumping rate and interaction between individual capture zones belonging to each extraction well on the variation of second order spatial moments as well as the form of capture zones were clarified.
2. Total probability that initial particle distributions associated with time-related capture zones were extracted by the well within a given time was identified through random walk particle tracking.
3. Extraction well comprising the outer capture zone displayed a substantial increase of total probability due to the particle passing through the inner and outer capture zones.

4. The increase of microdispersivity resulted in the increase of the number of particles moving across the boundary between the capture zones derived from different wells.

REFERENCES

Bear, J. 1972. *Dynamics of fluids in porous media*. Dover Publications: 764.

Burnett, R.D. & Frind, E.O. 1987. Simulation of contaminant transport in three dimensions, 2. Dimensionality effects. *Water Resour. Res.* 23(2): 695–705.

Feven, L., Beven, K.J. de Smedt, F. & Freer, J. 2001. Stochastic capture zone delineation within the generalized likelihood uncertainty estimation methodology: conditioning on head observations. *Water Resour. Res.* 37(3): 625–638.

Kunstmann, H. & Kinzelbach, W. 2000. Computation of stochastic wellhead protection zones by combining the first-order second-moment method and Kolmogorov backward equation analysis. *J. Hydrol.* 237: 127–146.

Tompson, A.F.B. & Gelhar, L.W. 1990. Numerical simulation of solute transport in three-dimensional, randomly heterogeneous porous media. *Water Resour. Res.* 26(10): 2541–2562.

Uffink, G.J.M. 1990. *Analysis of dispersion by the random walk method*. Ph.D. Dissertation. Delft University of Technology. Netherlands: 150.

van der Hoek, C.J. 1992. Contamination of a well in a uniform background flow. *Stocha Hydrol. Hydrau.* 6: 191–207.

van Kooten, J.J.A. 1995. An asymptotic method for predicting the contamination of a pumping well. *Adv. Water Resour.* 18(5): 295–313.

van Leeuwen, M. Butler, A.P. te Stroet, C.B.M. & Tompkins, J.A. 2000. Stochastic determination of well capture zones conditioned on regular grids of transmissivity measurements. *Water Resour. Res* 36(4): 949–957.

Vassolo, S., Kinzelhach, W. & Schafer, W. 1998. Determination of a well head protection zone by stochastic inverse modeling. *J. Hydrol* 206: 268–280.

Zheng, C. & Bennett, G.D. 2002. *Applied contaminant transport modeling second edition*, Wiley Interscience: 621.

Prediction and Simulation Methods for Geohazard Mitigation – Oka, Murakami & Kimoto (eds)
© 2009 Taylor & Francis Group, London, ISBN 978-0-415-80482-0

2D simulation analysis of failure of rock slope with bedding planes by DEM

K. Ise & H. Kusumi
Graduate school of Kansai University, Osaka, Japan

S. Otsuki
Graduate school of Kyoto University, Kyoto, Japan

ABSTRACT: As is known, there are many fractures and discontinuities in rock slope, and these might be often occurred the slope failure. In this paper, we try to clarify the mechanism of slope failure in modeling of rock slope with discontinuities using two dimensional DEM (distinct element method). However, it is difficult for DEM to express both of the continuum and discontinuities. So, we introduce the concept of bonding force, and it is made to be an applicable analysis method for the continuum. In the rock slope model in this simulation, the slope shape and the location of discontinuities can be arbitrarily set. The simulated rock slope has bedding planes, and we are reflected these in the model. Using this model, we try to simulate a failure rock slope, and to visible progress of fractures. As the results of this analysis, it is recognized that the discontinuities are formed surface and internally the rock slope. Moreover, the factors of failure can be visualized.

1 INTRODUCTION

Finite element method and boundary element method are numerical method for continuum. These methods are widely used for rock mass and ground. But, it is difficult for these methods to handle large-scale deformation and destruction. On the other hand, DEM (distinct element method) devised by P.Cundall is useful for discontinuity body analysis. Especially this method got a lot of attention as a solution for a large deformation problem involved with destruction. Researches on DEM have been carried out by many researchers such as P.Cundall, M.Hakuno and M.Hisatake.

However, the force between elements in this method was limited to the repulsive force, and it was difficult to apply the method to continuum such as rock mass and concrete. Then, Hakuno proposes EDEM (extended distinct element method) which can consider filling materials between elements. In this method, an element spring and a pore spring exist. And, this method can express dilatancy effect observed in the aggregate of granular matter such as ground and concrete. In addition, the approach of DEM was developed by Hisatake and others into CEM (contact element method). This method assumed the application to viscous ground. Like this, DEM was more refined, and it was possible to analyze enormous number of element by the development of the recent computer technology.

In this study, we try to clarify the mechanism of slope failure in modeling of rock slope with discontinuities using DEM. However, it is difficult for DEM to express both continuum and discontinuities. So,

we introduce the concept of the bonding force into DEM, and it is made to be an applicable analysis method for the continuum. By carrying out the simulation using this method, it has become possible to examine the progress of the fractures in the rock slope.

2 ANARYSIS METHOD

2.1 DEM

DEM is an analysis method devised by P.Cundall, and the analysis object is mainly discontinuous body of rock mass and ground. This method analyzes the dynamic behavior of rock mass considering the simulation object as an aggregate of the minute particles. Interparticle force is generated by setting a virtual spring, making it possible to calculate acceleration, velocity and displacement with the use of the force and to track the behavior of particles. The microscopic relationship between the particles is shown in Figure 1. In this analysis method, interparticle force is calculated by multiplying the contact distance (Δn) by spring stiffness.

2.2 Bonding force

Interparticle force is not only the repulsive force, when the model of granular material is applied to the solid like rock mass. Then, the tensile force is expressed by introducing the bonding force in this study.

Figure 2 shows two kind of bonding radii of r_{b1} and r_{b2}. r_{b1} shows the distance in which the bonding force comes to the yield, and r_{b2} shows the distance in

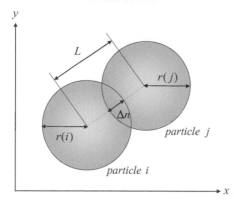

Figure 1. The relationship between the particles.

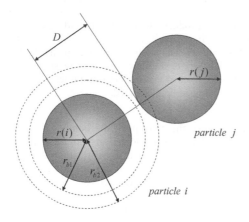

Figure 2. The region where the bonding force acts.

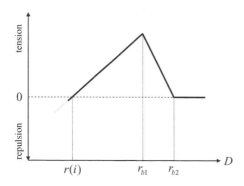

Figure 3. The force between the particles.

which the binding force breaks. In short, the bonding force increases from contact point r to r_{b1}, and it decreases from r_{b1} to r_{b2}. In addition, the bonding force is broken at r_{b2}. At this time, the value of the tensile force is zero (see Figure 3). The repulsive

Figure 4. The circumstances of failure.

force and the bonding force can be formulated as follows.

$$F_{ij} = \begin{cases} K \cdot \Delta n & (D < r(i)) \\ K \cdot (D - r(i)) & (r(i) < D \leq r_{b1}) \\ K \cdot (r_{b2} - D) & (r_{b1} < D \leq r_{b2}) \\ 0 & (D > r_{b2}) \end{cases} \quad (1)$$

In the equation (1), F_{ij} shows the interparticle force between particle i and particle j, K shows the spring stiffness, and D shows the distance from center of particle i to particle j.

3 OUTLINE OF ROCK SLOPE FAILURE

An analysis object in this study is a rock slope failure arose in Nara Prefecture, which broke down on the 31st January, 2007. The circumstances after it failed are shown in Figure 4. This slope is mainly composed of sandstones and mudstones. The slope has bedding planes with the gradient of about 20~40 degrees. The failure scale was about 35 m in height and 30 m in width, and the volume of failed rock mass was about 1,100 m³.

4 ESTABLISHMENT OF SIMULATION MODEL

In this study, to examine the effect that bedding planes exercise on rock slope failure, we established two simulation models. Figure 5 shows the geological cross

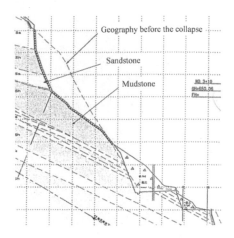

Figure 5. Geological cross section diagram.

(a)   (b)

Figure 6. Simulation models.

Table 1. Data of simulation model.

Number of particles	8303
Particles size (maximum)	0.25 [m]
Particles size (minimum)	0.11 [m]
X direction	47.387 [m]
Y direction	47.892 [m]

section diagram, and Figure 6 shows two simulation models.

These simulation models are composed of random-sized particles, and the number of the particles is about 8,000. The packing which packed the random-sized particles was carried out under the gravity. The aggregate of particles was cut off in a slope shape, after all movements of particles stopped. Furthermore, the geological structures of the slope were set in this simulation model. In Figure 6, white particles represent the sandstones, and blue particles show the mudstones.

In short, (a) is the simulation model composed only of sandstones, and (b) is the simulation model that assumes the actual rock slope, and is composed of sandstones and mudstones. The Data of these simulation models are shown in Table 1.

5 DECISION OF MECHANICAL CONSTANTS OF ROCK MATERIAL

5.1 Simulation analysis of uniaxial compression test

In DEM, value of physical properties of simulation object is controlled by interparticle parameters. However, decision technique of parameters in this analysis method has not been established. Then, in advance of failure simulation, we tried the simulation analysis of uniaxial compression test by DEM and examined the parameters that are expressible of simulated rock slope.

5.2 Analytical conditions

It is estimated that there are some correlations between bonding force and physical properties of rock mass, especially uniaxial compressive strength. Then, we examined the effect that bonding radius exercises on uniaxial compressive strength. Table 2 and Table 3 represent the analytical conditions and the results of simulation analysis. We tried the simulation analysis using 6 cases that were changed only the bonding radius.

5.3 Analytical results

Figure 7 shows the result obtained from simulation analysis of uniaxial compression test. As the result of this analysis, it is recognized that there is proportional connection between the uniaxial compressive strength and the bonding radius.

Table 2. Analytical conditions and results.

Case	$\left\{\frac{r_{b1}}{r_{(i)}}-1\right\} \times 100$ [%]	$\left\{\frac{r_{b2}}{r_{(i)}}-1\right\} \times 100$ [%]	Uniaxial compressive strength [MPa]
1	0.001	0.002	28.61
2	0.010	0.020	31.17
3	0.100	0.200	28.59
4	1.000	2.000	44.56
5	5.000	10.000	125.14
6	10.000	20.000	270.10

Table 3. Analytical conditions.

time interval	1.0×10^{-5} [s]
damping coefficient	100.0 [N · s/m]
friction coefficient	0.5
density	2500 [kg/m³]
spring stiffness (normal)	2.5×10^7 [N/m]
spring stiffness (shear)	1.0×10^7 [N/m]

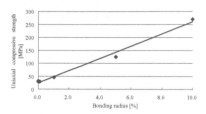

Figure 7. Relationship between uniaxial compressive strength and bonding radius.

5.4 Decision of the parameters

In this study, the uniaxial compressive strength of sandstones is set 100 MPa, and that of mudstones is set 50 MPa. Therefore, from analytical result, we decided that the bonding radii of the sandstones are $r_{b1} = 3.0\%$, $r_{b2} = 6.0\%$ and those of the mudstones are $r_{b1} = 1.0\%$, $r_{b2} = 2.0\%$. In addition, the ratio of r_{b1} to r_{b2} was fixed at twice.

6 EXPRESSION OF FRACTURES

In this analysis, the number of ruptures of interparticle bonding force increases in the failure process of the slope. By visually expressing it, the fracture propagation under failure was visualized. In short, the color of particle is changed every time-step, and by seeing it in broad perspective, we regard its particles as the development part of fracture.

Figure 8 is pattern diagrams which show that the color of particle changes by breaking bonding force. (a) shows the initial state, and the color of particle is white. (b), (c) and (d) show the state that interparticle bonding force is broken at 1 part, 2 parts and 3 or more parts, and the color of particle changes from blue into red and green.

7 ANALYTICAL RESULTS

7.1 Ruptures of interparticle bonding force

Using two simulation models and analytical conditions decided by simulation analysis of uniaxial compression test, we tried to simulate a rock slope failure, and to visible progress of fractures. Case01 is the simulation analysis used model (a), and Case02 is it used model (b).

Figure 9 shows the number of ruptures interparticle bonding force obtained by the analysis. As seen in this result, there are differences in the number of ruptures bonding force of Case01 and those of Case02. And in Case02, it increases exponentially at about 3,500,000 step. Therefore, it is estimated that large fractures are formed at this time.

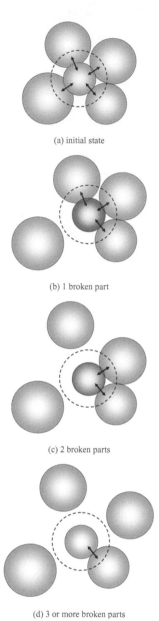

(a) initial state

(b) 1 broken part

(c) 2 broken parts

(d) 3 or more broken parts

Figure 8. Rupture number of the bonding force.

7.2 Fracture propagation

Figure 10 shows the fracture propagation in this simulation analysis. In this figure, white particles show the initial state, and blue particles, red particles and green particles show the state that interparticle bonding force is broken at 1 part, 2 parts and 3 or more parts. From this results, in Case02, it is recognized that ruptures of interparticle bonding force concentrate in

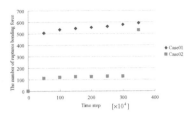

Figure 9. The number of ruptures interparticle bonding force.

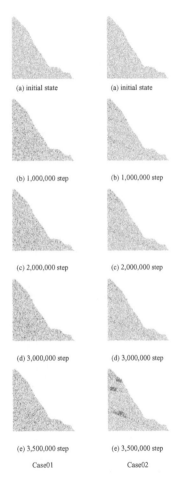

(a) initial state   (a) initial state

(b) 1,000,000 step   (b) 1,000,000 step

(c) 2,000,000 step   (c) 2,000,000 step

(d) 3,000,000 step   (d) 3,000,000 step

(e) 3,500,000 step   (e) 3,500,000 step

Case01   Case02

Figure 10. The development of the fracture.

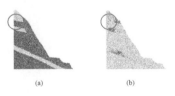

(a)   (b)

Figure 11. Comparisons with the actual phenomenon.

the vicinity of the bedding planes and that the large fractures are formed on the surface and inside of the rock slope at 3,500,000 step.

7.3 Comparisons with the actual phenomenon

Red particles of Figure 11(a) show the failed rock mass of the actual phenomenon. The upper part of these corresponds with the fracture formed slope surface in this simulation analysis. From this result, it is said the simulation analysis were attempted more accurately, and it is supposed that this fracture is one of the factors in the failure.

8 CONCLUSIONS

In this study, the rock slope failure was simulated by DEM using the bonding force, and its failure process was supposed. Knowledge obtained by this study is stated as follows.

Modeling the rock slope composed of the sandstones and the mudstones, the bonding radii led by the result of laboratory test were able to be acted on each layer.

Furthermore, as an estimation of the failure process, it is supposed that fractures have progressed by the bedding planes and a weight of the rock slope, and that the failure arose because of the formed fractures and pre-exiting fractures.

REFERENCES

Cundall, P.A. 1971. A computer model for simulating progressive, large scale movement in blocky rock system, Symp. ISRM Nancy France Proc. 2: 129–136.

Cundall, P.A. & Strack, O.D.L. 1979. A discrete numerical model for granular assemblies. *Geotechnique* 29(1): 47–65.

Donze, F., Mora, P. & Magnier, S. 1979. Numerical simulation of faults and shear zones. Geophys.J.Int. 116: 46–52.

Iwashita, K. & Hakuno, M. 1990. A modified distinct element analysis for progressive failure of a cliff, Journal of the Japanese Society of Soil Mechanics and Foundation Engineering 30(3): 197–208.

Numerical prediction of seepage and seismic behavior of unsaturated fill slope

R. Uzuoka, T. Mori, T. Chiba, K. Kamiya & M. Kazama
Tohoku University, Sendai, Japan

ABSTRACT: A dynamic three-phase coupled analysis is newly proposed. The equations governing the dynamic deformation of unsaturated soil are derived here based on porous media theory and constitutive models. The weak forms of the momentum balance equations of the overall three-phase material and the continuity equations (mass and momentum balance equations) of the pore fluids (water and air) at the current configuration are implemented in a finite element model. The discretized equations are solved by fully implicit method and the skeleton stress is also implicitly integrated. Predictions of the seismic responses of an actual fill slope during a future earthquake are performed under different moisture conditions of unsaturated fill.

1 INTRODUCTION

Fill slopes located on an old valley have been damaged because the ground water table was high after rainfall. Moreover the capillary zone in an artificial ground with a volcanic soil with high water retention is usually thick; therefore it possibly liquefies during earthquake (Uzuoka et al. 2005).

Cyclic triaxial tests with unsaturated soil have been performed by many researchers (e.g. Yoshimi et al. 1989, Tsukamoto et al. 2002, Selim & Burak 2006). Recently liquefaction mechanism of unsaturated soil has been discussed (e.g. Okamura & Soga 2006, Unno et al. 2008) and it is suggested that the behaviors of pore air and suction play an important role during liquefaction of unsaturated soil.

Liquefaction analyses of saturated ground based on Biot's porous media theory (Biot 1962) have been studied since 1970s (e.g. Ghaboussi & Dikmen 1978, Zienkiewicz & Shiomi 1984, Oka et al. 1994). Liquefaction analyses of unsaturated ground have been performed using porous media theory (e.g. Meroi & Schrefler 1995). Most liquefaction analyses, however, assumed that pore air pressure was zero and the behavior of pore air was not directly treated. Therefore, liquefaction analysis without considering suction cannot precisely predict the seismic behavior of unsaturated fill slope.

In this study, a dynamic three-phase coupled analysis is newly proposed. The equations governing the dynamic and finite deformation of unsaturated soil are derived here based on porous media theory. The soil water characteristic curve (SWCC) modeling is one of the most important issues to reproduce the change in suction and water saturation during cyclic loading. A simplified SWCC model with logistic function is proposed to reproduce the wetting process during cyclic loading.

Numerical simulations of seismic behavior of an actual fill slope are performed with the proposed numerical method. The material parameters of the constitutive models are determined through the calibration of the laboratory tests. Predictions of the seismic responses of the fill slope during a future earthquake are performed under different moisture conditions of unsaturated fill.

2 NUMERICAL METHOD

2.1 Balance and constitutive equations

Firstly the basic equations are derived based on porous media theory (e.g. de Boer 2000, Schrefler 2002). The partial densities of soil skeleton, pore water and air are defined as follows,

$$\begin{aligned}
\rho^s &= n^s \rho^{sR} = (1-n)\rho^{sR} \\
\rho^w &= n^w \rho^{wR} = ns^w \rho^{wR} \\
\rho^a &= n^a \rho^{aR} = ns^a \rho^{aR} = n(1-s^w)\rho^{aR}
\end{aligned} \tag{1}$$

where ρ^s, ρ^w and ρ^a are the partial densities of soil skeleton, pore water and air respectively. ρ^{sR}, ρ^{wR} and ρ^{aR} are the real densities of each phase, n^s, n^w and n^a are the volume fractions of each phase. n is the porosity, s^w is the degree of water saturation and s^a is the degree of air saturation.

Mass balance equation for α phase ($\alpha = s, w, a$) is

$$\frac{D^\alpha \rho^\alpha}{Dt} + \rho^\alpha \operatorname{div} \mathbf{v}^\alpha = 0 \tag{2}$$

where $D^\alpha \rho/Dt$ is the material time derivative with respect to α phase, v^α is the velocity vector of α phase. The mass exchange among three phases is ignored here. The linear momentum balance equation of α phase is

$$\rho^\alpha \frac{D^\alpha \mathbf{v}^\alpha}{Dt} = \rho^\alpha \mathbf{a}^\alpha = \operatorname{div}\sigma^\alpha + \rho^\alpha \mathbf{b} + \hat{\mathbf{p}}^\alpha \tag{3}$$

where $\mathbf{a}^\alpha$ is the acceleration vector of α phase, σ^α is the Cauchy stress tensor of α phase, $\mathbf{b}$ is the body force vector, $\hat{\mathbf{p}}^\alpha$ is the interaction vector of α phase against other phases.

Constitutive equations are the followings. The partial Cauchy stress of each phase is assumed as

$$\begin{aligned}
\sigma^s &= \sigma' - (1-n)(s^w p^w + s^a p^a)\mathbf{I} \\
\sigma^w &= -ns^w p^w \mathbf{I}, \quad \sigma^a = -ns^a p^a \mathbf{I}
\end{aligned} \tag{4}$$

where σ' is the skeleton stress tensor (e.g. Gallipoli et al. 2003), p^w is the pore water pressure and p^a is the pore air pressure. These pressures are defined as positive in compression. The interaction vector for each phase is assumed as

$$\begin{aligned}
\hat{\mathbf{p}}^s &= -\hat{\mathbf{p}}^w - \hat{\mathbf{p}}^a \\
\hat{\mathbf{p}}^w &= p^w \operatorname{grad} n^w - \frac{n^w \rho^{wR} g}{k^{ws}} n^w \mathbf{v}^{ws} \\
\hat{\mathbf{p}}^a &= p^a \operatorname{grad} n^a - \frac{n^a \rho^{aR} g}{k^{as}} n^a \mathbf{v}^{as}
\end{aligned} \tag{5}$$

where g is the gravity acceleration, k^{ws} and k^{as} is the permeability coefficient of water and air respectively. The compressibility of pore water under an isothermal condition is assumed as

$$\frac{D^s \rho^{wR}}{Dt} = \frac{\rho^{wR}}{K^w} \frac{D^s p^w}{Dt} \tag{6}$$

where K^w is the bulk modulus of pore water. The compressibility of pore air under an isothermal condition assumed as

$$\frac{D^s \rho^{aR}}{Dt} = \frac{1}{\Theta \overline{R}} \frac{D^s p^a}{Dt} \tag{7}$$

where Θ is the absolute temperature, $\overline{R}$ is the specific gas constant of air. The constitutive relation between water saturation and suction is assumed as

$$\frac{D^s s^w}{Dt} = c\frac{D^s p^c}{Dt} = c\frac{D^s(p^a - p^w)}{Dt} \tag{8}$$

where c is the specific water capacity. The specific water capacity is calculated from the soil water characteristic curve (SWCC). The SWCC is assumed as

$$\begin{aligned}
s^w &= (s_s^w - s_r^w)s_e^w + s_r^w \\
s_e^w &= \{1 + \exp(a_{lg}p^c + b_{lg})\}^{-c_{lg}}, \quad p^c = p^a - p^w
\end{aligned} \tag{9}$$

where s_s^w is the saturated (maximum) degree of saturation, s_r^w is the residual (minimum) degree of saturation and s_e^w is the effective water saturation. The relationship between s_e^w and suction p^c is assumed as a logistic function with the material parameters a_{lg}, b_{lg} and c_{lg}. The logistic SWCC is continuous function at $p^c = 0$; therefore the convergence in the iterative numerical scheme can be achieved. The SWCC during undrained cyclic shear has a similar shape as "wetting" curve

(Unno et al. 2008). The above SWCC is modified to fit "wetting" curve during undrained cyclic shear as shown in later. The permeability coefficient of water and air are assumed to be dependent on the effective water saturation as

$$k^{ws} = k_s^w (s_e^w)^{\xi_k}, \quad k^{as} = k_s^a (1 - s_e^w)^{\eta_k} \tag{10}$$

where k_s^w is the saturated (maximum) coefficient of water permeability, k_s^a is the dry (maximum) coefficient of air permeability, ξ_k and η_k are the material parameters.

Combining these equations, we derive the governing equations which include the momentum balance equations of the overall three-phase material and the mass and momentum balance equations (continuity equations) of the pore water and air with the following assumptions. 1) The soil particle is incompressible, 2) the mass exchange among phases is neglected, 3) The material time derivative of relative velocities and advection terms of pore fluids to the soil skeleton are neglected, 4) an isothermal condition are assumed. The momentum balance equations of the overall three-phase material is derived as

$$\rho \mathbf{a}^s = \operatorname{div}\{\sigma' - (s^w p^w + s^a p^a)\mathbf{I}\} + \rho \mathbf{b} \tag{11}$$

where ρ is the overall density of three-phase material. The mass and momentum balance equations of the pore water and air are derived as

$$\begin{aligned}
&\left(\frac{ns^w \rho^{wR}}{K^w} - n\rho^{wR}c\right)\frac{D^s p^w}{Dt} + n\rho^{wR}c\frac{D^s p^a}{Dt} \\
&+ s^w \rho^{wR}\operatorname{div}\mathbf{v}^s \\
&+ \operatorname{div}\left\{\frac{k^{ws}}{g}\left(-\operatorname{grad}p^w + \rho^{wR}\mathbf{b} - \rho^{wR}\mathbf{a}^s\right)\right\} = 0
\end{aligned} \tag{12}$$

$$\begin{aligned}
&\left(\frac{ns^a}{\Theta\overline{R}} - n\rho^{aR}c\right)\frac{D^s p^a}{Dt} + n\rho^{aR}c\frac{D^s p^w}{Dt} \\
&+ s^a \rho^{aR}\operatorname{div}\mathbf{v}^s \\
&+ \operatorname{div}\left\{\frac{k^{as}}{g}\left(-\operatorname{grad}p^a + \rho^{aR}\mathbf{b} - \rho^{aR}\mathbf{a}^s\right)\right\} = 0
\end{aligned} \tag{13}$$

This simplified formulation is called $u - p^w - p^a$ formulation. Although the governing equations are derived in the regime of finite strain, we assume infinitesimal strain in the following study for simplicity.

2.2 Constitutive equation for skeleton stress

A simplified elasto-plastic constitutive equation for skeleton stress is used here. Assuming that plastic deformation occurs only when the deviatoric stress ratio changes, the yield function is assumed as

$$f = \|\eta - \alpha\| - k = \|\mathbf{s}/p' - \alpha\| - k = 0 \tag{14}$$

where p' is the mean skeleton stress, $\mathbf{s}$ is the deviatoric stress tensor, k is the material parameter which defines the elastic region. α is the kinematic hardening parameter (back stress) and its nonlinear evolution rule (Armstrong & Frederick 1966) is assumed as

$$\dot{\boldsymbol{\alpha}} = a\left(b\dot{\boldsymbol{e}}^p - \boldsymbol{\alpha}\dot{\varepsilon}_s^p\right)$$
$$\dot{\varepsilon}_s^p = \left\|\dot{\boldsymbol{e}}^p\right\|$$

(15)

where a, b are the material parameters, $\dot{\boldsymbol{e}}^p$ is the plastic deviatoric strain rate tensor. With non-associated flow rule, the plastic potential function is assumed as

$$g = \|\boldsymbol{\eta} - \boldsymbol{\alpha}\| + M_m \ln\left(p'/p_a'\right) = 0$$

(16)

where M_m is the material parameter which defines the critical state ratio, p_a' is p' when $\|\boldsymbol{\eta} - \boldsymbol{\alpha}\| = 0$. Finally the elastic module are assumed as

$$K^e = -K^* p' \quad G^e = -G^* p'$$

(17)

where K^e is the elastic bulk modulus, G^e is the elastic shear modulus, K^* and G^* are the dimensionless elastic module respectively.

2.3 Finite element formulation and time integration

Weak forms of the equations (11)–(13) are implemented in a finite element formulation. Newmark implicit scheme is used for time integration. The primary variables are the second-order material time derivative of displacement of soil skeleton $\mathbf{a}^s$, pore water pressure $\ddot{p}^w$ and pore air pressure $\ddot{p}^a$. The weak forms are linearized and solved by Newton–Raphson method iteratively at each time step. The linearized forms of the weak forms are derived as

$$D\delta w^s[\Delta \mathbf{a}^s] + D\delta w^s[\Delta \ddot{p}^w] + D\delta w^s[\Delta \ddot{p}^a] = -\delta w_{(k)}^s$$
$$D\delta w^w[\Delta \mathbf{a}^s] + D\delta w^w[\Delta \ddot{p}^w] + D\delta w^w[\Delta \ddot{p}^a] = -\delta w_{(k)}^w$$
$$D\delta w^a[\Delta \mathbf{a}^s] + D\delta w^a[\Delta \ddot{p}^w] + D\delta w^a[\Delta \ddot{p}^a] = -\delta w_{(k)}^a$$

(18)

where $D\delta w^s[\Delta \mathbf{a}^s]$ is directional derivative of δw^s with respect to $\Delta \mathbf{a}^s$, the $\delta w_{(k)}^s$ is the residual at the iteration step of (k). The iteration is continued until the norm of the residual vectors becomes less than the convergence tolerance of 1.0×10^{-7}. In the finite element formulation, Galerkin method and isoparametric 8-node elements are used. The soil skeleton displacement and the fluid pressures are approximated at 8 nodes and 4 nodes respectively to avoid a volumetric locking.

2.4 Implicit stress integration

Implicit stress integration and consistent tangent modulus at infinitesimal strain (e.g. Simo & Taylor 1985) are used to achieve the convergence of global iteration

of (18). The return mapping algorithm in stress space is used as

$$r_1 = \boldsymbol{\sigma}' - \boldsymbol{\sigma}'^{(\text{tr})} + \Delta\gamma\, \mathbf{c}^e \frac{\partial g}{\partial \boldsymbol{\sigma}'}$$
$$r_2 = \boldsymbol{\alpha} - \boldsymbol{\alpha}_n + \Delta\boldsymbol{\alpha}$$
$$r_3 = f$$

(19)

where $\boldsymbol{\sigma}'^{(tr)}$ is the trial skeleton stress for a given strain increment at the global iteration step, $\Delta\gamma$ is the plastic multiplier, $\mathbf{c}^e$ is the elastic tensor, $\boldsymbol{\alpha}_n$ is $\boldsymbol{\alpha}$ at the previous time step, $\Delta\boldsymbol{\alpha}$ is the increment of $\boldsymbol{\alpha}$. Until the norm of left-handed residual vector r of (19) becomes less than the convergence tolerance of 1.0×10^{-10}, the nonlinear equations of (19) are solved iteratively by Newton-Rapshon method with respect to $\boldsymbol{\sigma}'$, $\boldsymbol{\alpha}$ and $\Delta\gamma$. The skeleton stress and back stress for a given strain increment at the global iteration step are obtained by local iteration of (19) at each stress integration point. The consistent tangent modulus is obtained as

$$\mathbf{c}^{ep} = \frac{\partial \boldsymbol{\sigma}'}{\partial \boldsymbol{\varepsilon}^{e(tr)}} = \frac{\partial \boldsymbol{\sigma}'}{\partial \boldsymbol{\varepsilon}}$$

(20)

where $\mathbf{c}^{ep}$ is the elasto-plastic tensor, $\boldsymbol{\varepsilon}^{e(tr)}$ is the trial elastic strain (given strain). The differentiation of (20) is carried out at each converged stress derived from the local iteration of (19).

3 NUMERICAL DATA FOR PREDICTION

Numerical predictions of seismic behavior of an actual fill slope are performed with the proposed numerical method. The fill slope is located at a residential area in Sendai city, Japan. The seismic motions on the fill and moisture content in the fill have been measured by the authors since 2006 (Mori et al. 2008).

3.1 Material parameters

The fill material is silty sand. The physical and mechanical properties are investigated with in-situ and laboratory tests. The material parameters of the constitutive models for skeleton stress and SWCC are determined through the calibration of the undrained cyclic triaxial tests with unsaturated soil. The detailed description of the testing method is referred to Unno et al. (2008).

The specimen was made of silty sand obtained from the in-situ fill. The initial dry density of the specimen was about 1.2 g/cm^3 which roughly agreed with the dry density of the in-situ fill. The effective degree of water saturation was from about 23% to 60% by controlling air pressure during the isotropic consolidation process. The pore water pressure was almost zero after the consolidation and the pore air pressure increased with the decrease in water saturation. The

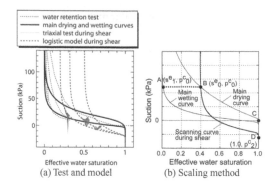

Figure 1. Soil water characteristic curves (SWCC).

Table 1. Material parameters of silty sand from the fill.

Elasto-plastic model parameters	
Dimensionless shear modulus, G^*	140
Dimensionless bulk modulus, K^*	600
Nonlinear hardening parameter, a	5512
Nonlinear hardening parameter, b	−1.8
Critical state stress ratio, M_m	1.8
Yield function parameter, k	0.0245
SWCC parameters	
Maximum degree of saturation, s_s^w	0.99
Minimum degree of saturation, s_r^w	0.35
Main drying curve, a_{lg}, b_{lg}, c_{lg}	1.5, 3.0, 0.05
Main wetting curve, a_{lg}, b_{lg}, c_{lg}	0.25, −2.5, 0.2
Scaling parameter, p_2^c (kPa)	−15
Physical parameters of water and air	
Bulk modulus of water, K^w (kPa)	1.0×10^6
Real density of air, ρ^{aR} (t/m³)	1.23×10^{-3}
Gas parameter, $1/(\bar{R}\Theta)$ (s²/m²)	$\times 1.25 \times 10^{-5}$

net stress was about 45 kPa for all specimens and the mean skeleton stress varied with the initial suction dependent on initial water saturation.

The cyclic shear was applied to the specimen under undrained air and water conditions. The input axial strain was the sinusoidal wave with multi step amplitudes whose single amplitudes were 0.2, 0.4, 0.8, 1.2, 1.6, and 2.0 with every ten cycles. The frequency of the sinusoidal wave was 0.005 Hz. This loading rate is slow enough to achieve an equilibrium condition between air and water pressure.

Figure 1 (a) shows SWCC during the undrained cyclic triaxial tests. The measured SWCC are not on the main drying and wetting curves obtained from water retention tests; therefore the scanning curves should be used to reproduce the SWCC during undrained cyclic shear. The logistic SWCC function of (9) is modified to reproduce the scanning curves

with simple scaling method in Figure 1 (b).

$$s_e^w = \frac{1 - s_{e0}^w}{1 - s_{e1}^w} \left[1 + \exp(a_{lg}p_a^c + b_{lg}) \right]^{-c_{lg}}$$

$$+ \left(1 - \frac{1 - s_{e0}^w}{1 - s_{e1}^w} \right)$$

$$p_a^c = \frac{p_1^c}{p_0^c} \left\{ \frac{p_0^c}{p_0^c - p_2^c} (p^c - p_0^c) + p_0^c \right\}$$

(21)

where p_0^c and s_{e0}^w are the initial suction and effective water saturation respectively, $p_1^c(= p_0^c)$ and s_{e1}^w are the suction and effective water saturation on main wetting curve respectively and p_2^c is the suction on the scanning curve at $s_e^w = 1$. The modified logistic SWCC roughly reproduce the measured SWCC during undrained shear in Figure 1 (a). Further investigation on SWCC during undrained shear is necessary for more precise reproduction.

In the triaxial test simulations, the finite element formulation presented in the previous section is not used. Assuming that the variables in the specimen are homogeneous, only the local equilibrium is considered. Table 1 shows the calibrated material parameters of the constitutive model. Figure 2 shows the time histories of pore water pressure, pore air pressure, suction, mean skeleton stress (positive in compression) and void ratio from tests and simulations in the case with the initial effective water saturation of 46%. In the test results (denoted "Test" in the figures), the pore water and air pressure increase, while the suction and mean skeleton stress decrease during cyclic undrained shear. In this case, the suction and the mean skeleton stress do not attain zero, which means that the specimen do not liquefy completely. In the simulated results (denoted "Model" in the figures), the model well reproduces the overall tendency of the test results. Therefore, the simplified constitutive equation of soil skeleton is applicable to predict pore water and air responses of unsaturated soil in the framework of three-phase porous media theory. The modifications of the constitutive equations are necessary for more precise reproduction.

3.2 Finite element model and boundary conditions

Figure 3 shows the cross section of the in-situ fill slope. Assuming plane strain condition, the cross section is used for finite element modeling. The fill is elasto-plastic material and the air, clay and base rock are linear elastic material with Lame coefficients λ and μ. Kelvin type viscosity is assumed in all materials and its viscous coefficient is proportional to the elastic modulus with a multiplier coefficient α. Tables 1 and 2

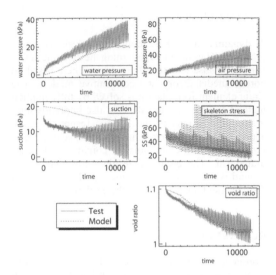

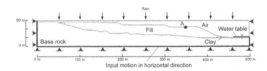

Figure 2. Test and simulation with initial effective water saturation of 46%.

Figure 3. Cross section of the fill.

Table 2. Material parameters for seismic analysis.

	Rock	Fill	Clay	Air
n	0.37	0.57	0.64	1.0
ρ^{sR}(t/m^3)	2.72	2.59	2.72	0.0
ρ^{wR}(t/m^3)	1.0			
k^w_s(m/s)	1.0×10^{-12}	1.0×10^{-7}	1.0×10^{-10}	1.0
k^a_s(m/s)	1.0×10^{-13}	1.0×10^{-8}	1.0×10^{-11}	1.0
ξ_k/η_k	–	3.0/0.05	–	–
λ (kPa)	293876	–	99915	0.01
μ (kPa)	293877	–	11102	0.001
α	0.001	0.001	0.001	0.001

show the material parameters for all layers. Smooth infiltration of rain to the fill is simulated with "aerial elements" which have a special SWCC and large permeability of water in vertical direction (Uzuoka et al. 2008). Use of the aerial elements makes no numerical treatment (e.g. switching between natural and basic boundary) on the surface.

The soil displacement at the bottom boundary is fixed in all directions and the lateral boundaries are vertical rollers. The bottom and lateral boundaries are impermeable and a part of the surface on the clay layer is permeable with zero water pressure.

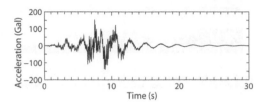

Figure 4. Time history of input acceleration (Sendai city 2002).

Figure 4 shows the time history of input acceleration. This acceleration history is a calculated wave by a numerical simulation of future earthquake (Sendai city 2002). The coefficients in Newmark implicit time integration are 0.5 and 0.25. The time increment is 0.002 seconds.

4 NUMERICAL RESULTS OF SEEPAGE AND SEISMIC ANALYSES

Predictions of the seismic responses of the fill slope during a future earthquake are performed under some different moisture conditions of unsaturated fill.

4.1 Initial conditions with seepage analyses

The moisture condition in the fill is strongly dependent on weather conditions before the earthquake. Two typical cases are considered in this study. Case 1 is the case with low ground water table after light precipitation and Case 2 is the case with high ground water table after heavy precipitation. Rainfall is applied on the top boundary in the seepage analyses without inertia terms in (11)–(13). Because of difficulties to reproduce the actual rainfall history since distant past, the precipitation in Case 1 is assumed as 175 mm/year. This amount of infiltration roughly corresponds to the amount of measured flux at the toe of the fill when it has not rained for about a week before the measurement. The precipitation in Case 2 is assumed as 263 mm/year. Figure 5 shows the distributions of water saturation after the precipitation in both cases after the steady state is achieved. The ground water table in Case 1 roughly agrees with the measured one without heavy precipitation before the measurement.

4.2 Seismic response of the fill

Following the seepage analyses, the seismic response analyses are performed with the input acceleration in Figure 4. Figure 6 shows the distributions of skeleton stress reduction ratio after 30 seconds in both cases. The skeleton stress reduction ratio is defined as $1-p'/p'_0$ where p'_0 is the initial value of p'. The skeleton stress reduction ratio in the saturated fill below

163

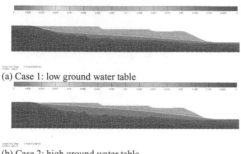

(a) Case 1: low ground water table

(b) Case 2: high ground water table

Figure 5. Distribution of water saturation before the earthquake.

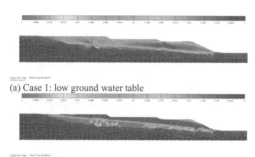

(a) Case 1: low ground water table

(b) Case 2: high ground water table

Figure 6. Distribution of skeleton stress reduction ratio after the earthquake.

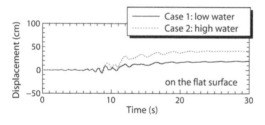

Figure 7. Time histories of horizontal displacement on the flat surface (point A Figure 3).

the ground water table reaches almost one at some elements in both cases; liquefaction occurs in some saturated area. The skeleton stress reduction ratio in the unsaturated fill above the ground water table in Case 1 is larger than that in Case 2. Figure 7 shows the time histories of horizontal displacement at the point A in Figure 3 on the flat surface. Higher ground water table causes larger residual horizontal displacement on the fill after the earthquake.

5 CONCLUSIONS

A dynamic three-phase coupled analysis is newly proposed. The equations governing the dynamic deformation of unsaturated soil are derived here based on porous media theory and constitutive models. The finite element formulation is solved by fully implicit method and the skeleton stress is also implicitly integrated. Predictions of the seismic responses of an actual fill slope during a future earthquake are performed under different moisture conditions of unsaturated fill. The predicted results show that higher ground water table causes larger residual horizontal displacement on the fill after the earthquake.

REFERENCES

Armstrong, P.J. & Frederick, C.O. 1966. A mathematical representation of the multiaxial Bauschinger effect. C.E.G.B. Report RD/B/N731, Berkeley Nuclear Laboratories, Berkeley, UK.

Biot,MA. 1962. Mechanics of deformation and acoustic propagation in porous media. *Journal of Applied Physics* 33: 1482–1492.

de Boer, R. 2000. Contemporary progress in porous media theory. *Applied Mechanics Reviews* 53(12): 323–369.

Ghaboussi, J. & Dikmen, S.U. 1978. Liquefaction analysis of horizontally layered sands. *Proc. ASCE* 104(GT3): 341–356.

Gallipoli, D., Gens, A., Sharma, R. & Vaunat, J. 2003. An elasto-plastic model for unsaturated soil incorporating the effects of suction and degree of saturation on mechanical behaviour. *Geotechnique* 53(1): 123–135.

Meroi, E.A. & Schrefler, B.A. 1995. Large strain static and dynamic semisaturated soil behavior. *Int. J. for Numerical and Analytical Methods in Geomechanics* 19(8): 1–106.

Mori, T., Kazama, M., Uzuoka, R. and Sento, N. 2008. Fill slopes: Stability assessment based on monitoring during both heavy rainfall and earthquake motion. In Chen et al. (eds), *Landslides and Engineered Slopes:* 1241–1246. London: Taylor & Francis Group.

Oka, F., Yashima, A., Shibata, T., Kato, M. & Uzuoka, R. 1994. FEM-FDM coupled liquefaction analysis of a porous soil using an elasto-plastic model. *Applied Scientific Research* 52: 209–245.

Okamura, M. & Soga, Y. 2006. Effects of pore fluid compressibility on liquefaction resistance of partially saturated sand. *Soils and Foundations* 46(5): 93–104.

Schrefler, B.A. 2002. Mechanics and thermodynamics of saturated/unsaturated porous materials and quantitative solutions. *Applied Mechanics Reviews* 55(4): 351–388.

Selim, A. & A. Burak, G. 2006. Cyclic stress-strain behavior of partially saturated soils. *Proc. 3rd Int. Conf. on Unsaturated Soils:* 497–507.

Sendai city. 2002. Report on earthquake damage investigation.

Simo, J.C. & Taylor, R.L. 1985. Consistent tangent operators for rate-independent elastoplasticity. *Computer Methods in Applied Mechanics and Engineering* 48: 101–118.

Tsukamoto, Y., Ishihara, K., Nakazawa, H., Kamada, K.Huang, Y. 2002. Resistance of partly saturated sand to liquefaction with reference to longitudinal and shear wave velocities. *Soils and Foundations* 42(6): 93–104.

Unno, T., Kazama, M., Uzuoka, R.Sento, N. 2008. Liquefaction of unsaturated sand considering the pore air pressure and volume compressibility of the soil particle skeleton. *Soils and Foundations* 48(1): 87–99.

Uzuoka, R., Sento, N., Kazama, M. and Unno, T. 2005. Landslides during the earthquake on May 26 and July 26, 2003 in Miyagi, Japan. *Soils and Foundations* 45(4): 149–163.

Uzuoka, R., Kurihara, T., Kazama, M.Sento, N. 2008. Finite element analysis of coupled system of unsaturated soil and water using aerial elements. *Proc. the Conference on Computational Engineering and Science* 13: 219–222. (in Japanese)

Yoshimi, Y., Tanaka, K. Tokimatsu, K. 1989. Liquefaction resistance of a partially saturated sand. *Soils and Foundations* 29(3): 157–162.

Zienkiewicz, O.C. & Shiomi, T. 1984. Dynamic behavior of saturated porous media: The generalized Biot formulation and its numerical solution. *Int. J. for Numerical and Analytical Method in Geomechanics* 8: 71–96.

Prediction and Simulation Methods for Geohazard Mitigation – Oka, Murakami & Kimoto (eds)
© *2009 Taylor & Francis Group, London, ISBN 978-0-415-80482-0*

Finite element simulation for earthquake-induced landslide based on strain-softening characteristics of weathered rocks

A. Wakai & K. Ugai
Gunma University, Kiryu, Japan

A. Onoue
Nagaoka National College of Technology, Nagaoka, Japan

S. Kuroda
National Institute for Rural Engineering, Tsukuba, Japan

K. Higuchi
Kuroiwa Survey and Design Office Co., Ltd., Maebashi, Japan

ABSTRACT: In this study, an effective and simple analytical method based on FEM with total stress formulations is developed to predict a catastrophic landslide induced by strong earthquake. It is well known that such a catastrophic failure would be induced by the strain-softening behavior of the weak materials around the slip surface, which can be appropriately taken into account in the proposed method. The validation of the proposed method is made through the numerical simulations for two different types of catastrophic landslides occurred during 2004 Mid Niigata Prefecture Earthquake in Japan, utilizing the measurements in the ground investigation and the results of the laboratory tests. As a result, the observed phenomena can be simulated well by the analyses, and the developed method is found to be effective for the prediction.

1 INTRODUCTION

The elasto-plastic finite element method (FEM) has been confirmed to be very effective for predicting the residual deformation of the slope due to strong earthquake. It always requires a suitable constitutive model for the analysis, which is usually based on rational mechanical theories and often has a lot of input parameters to be determined so that the model should behave just like the real soils. Here, we should notice that it might be rather difficult for us to determine all of those input parameters appropriately, because of limitation of time and cost. Wakai & Ugai (2004) proposed a simple but good constitutive model for seismic analysis of slope. They adopted the joint use of G-γ, h-γ relationships and c-ϕ strength parameterization in the model. The number of parameters required for the model is only seven, and it is advantageous for applications.

By the way, the following two kinds of slope failure can be distinguished from each other. The fist one is **"a non-catastrophic type failure"**, where the sliding mass stops after the earthquake. In the case, the residual displacement is one of the important matters for the seismic design of adjacent structures. The second one is **"a catastrophic type failure"** so-called as **"a collapse"**. In the case, the sliding mass continues moving even after the earthquake, as far as there are no obstacles on the way of moving. It means that the self weight of the sliding mass will become not to be supported at all after sliding, although it has been statically supported before the earthquake. The strain-softening characteristics of soils along the slip surface during the earthquake may cause such a long distance traveling failure. In this study, a new constitutive model of soil with such strain-softening characteristics is introduced into the finite element analysis, and an effective and simple analytical method is developed to simulate the following two different cases of **catastrophic type slope failure** induced by a strong earthquake.

2 CASE 1: YOKOWATASHI LANDSLIDE

2.1 *A rockslide along weak bedding plane*

Many landslides in mountain area occurred during The 2004 Mid Niigata Prefecture Earthquake in Japan. In the first half of this paper, a dip slope landslide along the Shinano River, Yokowatashi landslide, is simulated by the proposed analytical method. As seen in Figure 1(a), about 4 m thickness block of the upper

(a) Whole of collapse

(b) Undisturbed block sample for laboratory tests

Figure 1.   Yokowatashi landslide.

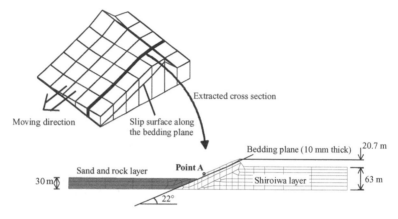

Figure 2.   Two dimensional finite element meshes for the simulation.

Shiroiwa layer of soft silt rock has slid down more than 72 m toward the Shinano River during the earthquake. The remaining upper Shiroiwa layer is laid on the planer bedding plane which inclination is approximately 22°. Figure 1(b) is a close photograph showing rock sampling for the laboratory tests. A thin seam layer of 5–10 mm thick was sandwiched between the upper and lower Shiroiwa layers. The material consisting of the seam is tuff, which is like a weathered sandy rock.

2.2   Analytical model with total stress formulations

The finite element mesh of eight-nodes elements is shown in Figure 2. The time history of the response at **Point A** in the figure will be mentioned later. The upper and lower soft rock layers were assumed to be elastic material, and the sandwiched tuff seam was assumed to be elasto-plastic material having a thickness of 10 mm taking strain softening characteristics into consideration. The surface soil at the foot of the slope is sand and gravel spreading down to the Shinano, which was modeled as elasto-plastic material with no strain-softening.

The basic concept of the newly proposed model with strain-softening is the same as the simple cyclic loading model originally proposed by Wakai and Ugai (2004). In those models, the undrained shear strength τ_f with Mohr-Coulomb's c and ϕ is specified as the upper asymptotic line of the hyperbolic stress-strain curve. In addition, in the new model, the shear strength value τ_f was modified so as to be a simple decreasing function of the accumulated plastic strain γ_p to incorporate the strain softening characteristics (Wakai et al. 2005). Thus, the shear strength value during earthquake is given as,

$$\tau_f = \tau_{f0} + \frac{\tau_{fr} - \tau_{f0}}{A + \gamma^p} \gamma^p \tag{1}$$

where the initial strength is denoted as,

$$\tau_{f0} = c \cdot \cos \phi + \left(\frac{\sigma_1 + \sigma_3}{2} \right)_{initial} \times \sin \phi \tag{2}$$

In the new model, the shear stiffness ratio, G_0, was also assumed to decrease in proportion to the decrease of shear strength.

168

Figure 3. Test specimen consisting of upper and lower silt rocks with tuff seam in between.

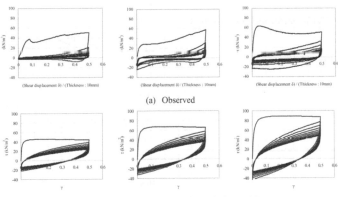

(a) Observed

(b) Simulated by the proposed constitutive model

Figure 4. Comparison between observed and simulated hysteretic loops. (from left to right; the applied initial normal stress as $\sigma = 40$ kN/m^2, 80 kN/m^2 and 120 kN/m^2)

Table 1. Material parameters used in the analysis.

Layer	Shiroiwa silt rock	Tuff seam	Sand and gravel
Young's modulus E (kN/m^2)	100000	30000	30000
Poisson's ratio ν	0.3	0.3	0.3
Cohesion c (kN/m^2)	–	24	0
Internal friction angle ϕ (deg)	–	30.9	35
Dilatancy angle ψ (deg)	–	0	0
$b \cdot \gamma_{G_0}$ (A parameter for damping)	–	8.0	18.
n (A parameter for damping)	–	1.40	1.35
Unit weight γ (kN/m^3)	20.0	18.0	18.0
Residual strength ratio τ_{fr}/τ_{f0}	–	0.30	–
A (A parameter for damping)	–	4.0	–

The constants of Rayleigh damping were assumed to be $\alpha = 0.171$ and $\beta = 0.00174$ which are almost equivalent to a damping ratio of 3% for a vibration period of 0.2 through 2.0 s. The material properties used in the analysis are summarized in Table 1.

2.3 Strain-softening modeling

Figure 3 shows an intact sample consisting of the upper and lower Shiroiwa layers and the tuff seam in between. The sample was subjected to the cyclic direct shear test under constant volume condition. Figure 4 compares the simulated hysteretic loop during cyclic loading by the proposed model to the observed one. The horizontal axis is described in strain. As for the figures of the tests, the strain value was estimated as the horizontal displacement divided by the thickness of the sandwiched layer of 10 mm. As seen in the figures, it is shown that the increase of the number of cycle gradually decreases the mobilized shear stress during cyclic loading. Although they don't perfectly agree, they are almost similar to each other in various confining pressures.

2.4 Analytical result

The acceleration record observed at Takezawa (EW) (Figure 5) was used as an input wave in the analyses. Two analytical cases were conducted to examine the

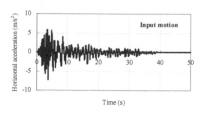

Figure 5. Input horizontal acceleration.

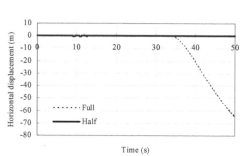

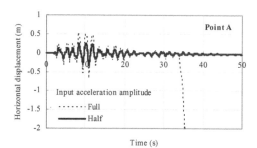

Figure 6. Histories of horizontal displacement at **Point A**.

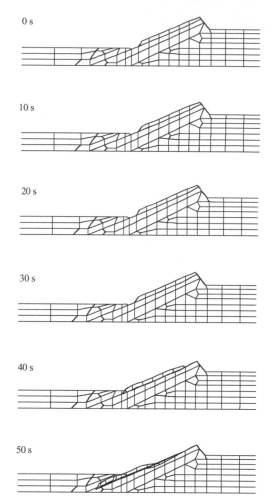

0 s

10 s

20 s

30 s

40 s

50 s

Figure 7. Deformation mode of the slope.

influence of seismic intensity on the inducement of sliding. One is the case where the observed acceleration record is input as it is at the base of the analytical area (here, denoted as the **Full** case), and the other is the case where the acceleration amplitude is compressed to one half that of the observed wave is input (the **Half** case).

Figure 6 shows the time histories of horizontal displacement at the foot of the slope, namely **Point A**, in Figure 2. As seen in the figures, the slope did not suffer any big damages after earthquake in the **Half** case, while a so-called collapse occurred in the **Full** case. The horizontal displacement is 20 m at an elapsed time t = 40 s and almost 65 m at t = 50s. The residual deformation at 0, 10, 20, 30, 40 and 50 s after the beginning of the seismic motion in the **Full** case is shown in Figure 7. The long distance movement of the upper Shiroiwa block along the bedding plane can be seen. Such a result agrees actual phenomena. However, you should notice that the formulations used in this study are based on the infinitesimal strain theory.

The predicted velocity of the sliding mass may not have sufficient accuracy, and the large deformation effect needs to be considered to obtain a more accurate result for moving body.

3 CASE 2: AMAYACHI LANDSLIDE

3.1 A catastrophic landslide caused by softening of strongly-weathered mudstone layer

In the second half of this paper, Amayachi landslide (Figure 8) is simulated numerically. At the time of the earthquake, the slope greatly deformed together with adjacent road and farmland. The approximate scale of the sliding block is 250 m in length, 150 m in width, 15 m in depth. The horizontal component of the residual displacement of the sliding block is estimated to be about 40 m. The slope is a dip slope and it has a

Figure 8. Amayachi landslide.

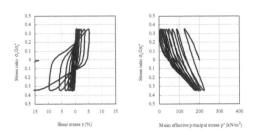

Figure 9. An example of laboratory test results.

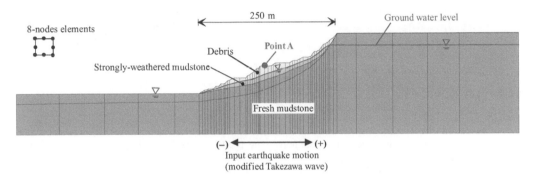

Figure 10. Ground profiles in the slope and corresponding finite element mesh for analysis.

few bedding planes whose inclination mostly accords with the slope angle. The slope mainly consists of Neogene deposits sedimented by early Diluvium. The deep part of the slope is the fresh mudstone that has not been weathered. The slid block consists of two parts; the deeper part as the strongly-weathered mudstone, and the surface part as the debris accumulated by past collapses. In this study, a few undisturbed soil block sample of the debris was sampled for the laboratory element test. The sampled block consisting of mud lumps and the filling materials seems slightly fragile.

A series of the undrained cyclic triaxial tests of saturated undisturbed specimens were performed. Not only the disappearance of existing soil cohesion but also the sudden strength degradation like liquefaction of loose sand can be seen in Figure 9, which is an example of the observed results under the initial confining pressure as 200 kPa. According to the above, the strength degradation of the strongly-weathered mudstone induced by cyclic loading is considered in the following analysis.

3.2 Analytical model with total stress formulations

A geological structure as shown in Figure 10 is assumed in the analysis. The finite element mesh (8-nodes) is also overdrawn in the same figure. As for the strongly-weathered mudstone layer, the newly proposed constitutive model already described in the previous chapter is adopted. Although the material parameters used in the analysis was omitted here because of limitation of space of writing, you should notice that most of the parameters were based on the results of the laboratory tests. For example, softening parameters were determined so that the simulated liquefaction strength curve of the soil agrees with the ones obtained by the laboratory test of the undisturbed sample of the debris, which might be similar to the strongly-weathered mudstone. The observed liquefaction strength curve could be fitted by the proposed model as shown in Figure 11.

3.3 Analytical result

By using the input wave that is the same one as in the case of Yokowatashi landslide, the dynamic response

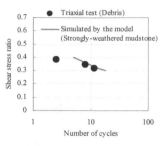

Figure. 11   Liquefaction strength curve.

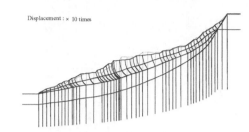

Figure. 12   Calculated deformation just after the earthquake.

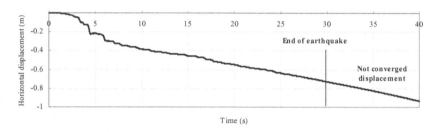

Figure 11.   Calculated time history of horizontal displacement at **Point A**.

analysis based on the finite element method was performed. Figure 12 shows the time histories of horizontal displacement at **Point A** in Figure 10. The sliding displacement in horizontal direction is about 73 cm at the end of the earthquake. However, the important thing is not the value itself, but the continuous increase of the value after the earthquake. It suggests that the movement of the sliding block does not stop even after the earthquake. This phenomenon can be explained by such a mechanism that the total of the shear strength of the softened area, the strongly-weathered mudstone, became smaller than the total of the shear stress induced only by the self weight of upper block.

Figure 13 shows the residual deformation just after the earthquake in order to evaluate the whole shape of the collapsed slope. As can be seen in the figure, the largely deformed area is concentrated on the strongly-weathered mudstone layer below the ground water level.

4   CONCLUSIONS

According to the above numerical simulations for two different types of catastrophic landslides occurred during 2004 Mid Niigata Prefecture Earthquake in Japan, the proposed analytical method with total stress formulations was proved to be very effective for prediction of such landslides triggered by the strain-softening

characteristics of weathered rocks in the slope. It will be useful for controlling the seismic hazard risk in mountainous area. It seems also very important for us to obtain more appropriate material parameters based on undisturbed specimens of various types of strain-softening materials in the slope.

ACKNOWLEDGEMENTS

This study was conducted in "Research project for utilizing advanced technologies in agriculture, forestry and fisheries" supported by the Ministry of Agriculture, Forestry and Fisheries of Japan. The authors wish to thank for the great support.

REFERENCES

Onoue, A., Wakai, A., Ugai, K., Higuchi, K., Fuku-take, K., Hotta, H. & Kuroda, S. 2006. Slope failures at Yokowatashi and Nagaoka College of Technology due to the 2004 Niigata-ken Chuetsu Earthquake and their analytical considerations. *Soils and Foundations* 46(6): 751–764.

Wakai, A. & Ugai, K. 2004. A simple constitutive model for the seismic analysis of slopes and its applications. *Soils and Foundations* 44(4): 83–97.

Wakai, A., Kamai, T. & Ugai, K. 2005. Finite element simulation of a landfill collapse in Takamachi Housing complex. *Proc. Symposium on Safeness and performance evaluation of ground for housing,* JGS: 25–30 (in Japanese).

Distinct element analysis for progressive failure in rock slope

T. Nishimura & K. Tsujino
Tottori University, Tottori, Japan

T. Fukuda
Geoscience Research Laboratory, Osaka, Japan

ABSTRACT: This paper investigates a numerical modeling of progressive failure in rock mass using the distinct element analysis. The numerical modeling consists of two analyses. One is to get the mechanical properties of synthetic specimens of circular rigid elements with the bonded effect between elements. The other is to analyze deformation of a scaled rock slope in an accelerated field where the gravitational acceleration is increased, representing the centrifugal experiment. Over several stages of the applied acceleration to the scaled slope model, evolution of displacements and the resulting initiation of failure surface are displayed.

1 INTRODUCTION

Most rock slopes are inhomogeneous structures comprising anisotropic layers of rock characterized by different material properties, and they are often discontinuous because of jointing, bedding and faults. In rock slope stability analyses, the failure surface is often assumed to be predefined as a persistent plane or series of interconnected planes, where the planes are fitted to the surfaces based on the structural observation. Such assumptions are partly due to the constraints of the analytical technique employed (e.g. limit equilibrium method, the distinct element method, etc.) and can be valid in cases in which the response of single discontinuity or a small number of discontinuities is of critical importance on the stability. However, especially on a large scale slope, it is highly unlikely that such a system of fully persistent discontinuous planes exists a priori to form the failure surface. Instead, the persistence of the key discontinuities may be limited and a complex interaction between pre-existing flaws, stress concentration and resulting crack generation, is required to bring the slope to failure. In small engineered slopes, excavation gives significant changes in stress distribution in the slopes and may generate fully persistent planes keep propagating with stress re-distribution. Larger natural rock slopes seldom experience such a disturbance and have stood in relatively stable features over the period of thousands of years. This does not imply that in natural rock slopes a system of discontinuities may not be interconnected developing the portion of where the failure surface will be formed. Strength degradation may occur in rock mass with time-dependent manner and drive the slope unstable state. Thus, rock slope instability

problem requires the progressive failure modeling to drive the slope to catastrophic events.

In this paper, a numerical modeling of progressive failure in rock mass using the distinct element method (Cundall 1971) is presented. The modeling can give a possible failure volume of rock material based on the geometrical data and strength properties, such as cohesion and internal friction angle, but can not mimic the strength degradation of rock material.

2 MODELING OF THE FAILURE SURFACE

Based on the coulomb shearing strength criterion, both the shearing resistance depending on the normal stress value (i.e. the frictional strength) and the cohesion of intact rock between discontinuous joints resist to shear failure. At the tip of the joints, stresses would increase and subsequent failure in rock would occur. Progressive failure in rock mass would involve the failure of intact rock as their strength is exceeded. There have been a number of investigations focused on the failure process of rock slope using the finite element method and the boundary element method.

Kaneko et al. (1997) used the displacement-discontinuity method (DDM) and fractures' principles to model the progressive development of shear crack in rock slope. In their analysis, rock material was assumed to be homogeneous and any pre-existing cracks were not considered. They compared the DDM results with the conventional limit equilibrium method (LEM) and discussed the allowable slope height under the given strength parameters and the slope angle. Eberhardt et al. (2004) discussed modeling of progressive failure surface development linking initiation and degradation to eventual catastrophic slope failure, using a hybrid method that combines both continuum and discontinuum numerical techniques to model fracture propagation. The continuum modeling of shear plane development is based on a simplified representation of rock mass strength degradation over an equivalent continuum that incorporates both intact and discontinuity effects. Another option must be examined through the adaptation of a discontinuum technique. Jiang et al. (2008) proposed an expanded distinct element method for simulating crack generation and propagation due mainly to the stress redistribution. In their modeling, cracks only propagate with pre-determined path in the plastic zones obtained a priori elasto-plastic analysis which has been carried out to estimate the direction of principal stresses and the extent of plastic zones.

The modeling of crack generation is done by:

i. continuous discritization of elements (or blocks) along the propagation direction of new cracks during the simulation. Crack initiation is controlled according to a facture criterion specified by a constitutive model.
ii. a priori distribution of cracks, which are bonded with the equivalent strength to the rock matrix, along the possible direction of propagation of cracks.

These typical procedures are illustrated in Figure 1. The procedure of re-discritization of elements during the simulation must be a difficult and a time consuming work, and the pre-placement of crack elements must be much influenced by stress state obtained from the prior elasto-plastic analysis. The second procedure could not represent the stress state subsequent to crack generation. These two procedures could not afford a large number of potential cracks because of limited computing time. The short coming of the distinct element method related to the modeling for crack generation is partly reduced by a modeling using circular element. A system composed of many circular elements with bonded effect at their contact point reproduces many features of rock

behavior including elasticity, fracturing and damage accumulation (Potyondy et al. 2004). The simulation of crack generation and propagation can be accomplished by continuously bond breaking under a given condition. Bond breaking occurs along element surface and then the cracks are oriented perpendicular to the line connecting the center of the two previously bonded elements. The incremental crack length and distribution can be depends upon the packing of elements and size. Particle size is not free parameter and is related to the propagation of cracks and the fracture of rock material.

3 DISTINCT ELEMENT MODEL IMPLEMENTATION

3.1 Input parameters and macro-parametres

For continuum analysis, the input parameters (such as Young's modulus and shear strength) can be given by experiment performed on laboratory size specimens, e.g. uniaxial test and Brazilian test. For the distinct element method, which mimics the macro-properties of rock material using a simple packing assembly, the input parameter such as contact stiffness are not known. The relation of the macro-properties and the micro-properties of the assembly are measured by conducting a calibration analysis. As mentioned in the previous section, particle size is not free parameter; it must dominate the mechanical parameters of the synthetic specimen composed of circular element. Potyondy et al. (2004) performed an analysis of a cubic packing of disks joined by contact bonds subjected to an extension strain. The results of the analysis have been shown to be equivalent to a material by Linear Elastic Fracture Analysis (LEFM) in terms of observable behavior and mathematical description. They also concluded that modulus and strength parameters must be independence, if the number of particles across the specimen is large enough to obtain a representative volume of the material response.

3.2 A synthetic specimen and a reduced slope model

In the numerical attempts to the process using the distinct element method, prior to the slope failure simulation, effects of micro-properties (such as particle diameter in mm or cm, contact stiffness and bond strength) on macro-properties (such as Young's modulus, unconfined compressive strength) are often analyzed using synthetic specimens. Nevertheless, the same size of elements could not be adopted because current computing power limits the number of elements to represent slope profile over hundreds meters in height. A slope modeling that reproduces the macro-properties of the synthetic specimens should be

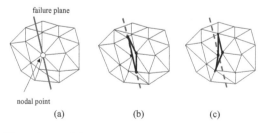

Figure 1. Modeling of crack generation: (a) element discritization and failure plane, (b) continuous discritization of elements along the direction of new cracks, (c) a priori placement of crack element along the possible direction of the crack propagation.

created and executed in reasonable times on standard desktop-type computer, providing the slope profile and the same level of stresses in the slope. Figure 2 illustrates reproduction of in-situ stress value in slope with a full-scale model and a reduced-scale model. The deformation and the failure of the reduced-size slope model will be analyzed in this procedure in which the gravitation acceleration is increased from $G(= 9.8 \text{ m/s}^2)$ to $nG(n \geq 1)$. The relation between the reduced slope model and the prototype is written by:

Length: $L_p = nL_m$

Area : $A_p = n^2 A_m$ (1)

Volume : $V_p = n^3 V_m$

The unit weight of the rock material in the slope model and the overburden pressure at depth $z_m\ (= z_p/n)$ will be given by:

$\gamma_m = n\gamma_p$

$\sigma_m = \gamma_m z_m = (n\gamma_p)(z_p/n) = \sigma_p$ (2)

The slope model could reproduce the same magnitude of the overburden pressure to the prototype. The material set-up procedure is the following stages.

1. *Particle generation*
Circular elements are placed in a rectangular area so as not to have overlap between particle-particle and particle-wall.

2. *Particle network building*
Through this stage, an assembly is made. In the assembly, all particles have contact points with neighboring elements. This is done by the introduction of acceleration for a specific direction such as the gravitational field.

3. *Trimming*
The assembly is cut into models for the specimen simulation and the slope failure simulation, etc. Throughout the assembly, the bond effect can be installed between all elements that are in contacts (if the bonded effect is needed). The contact forces due to the bonded effects or the acceleration could not be a significant effect, when the bond force is much smaller than the setting bond strength and the magnitude of applied external forces. Potyondy called the forces 'lock in force'.

The microproperties in Table 1 with three radii of 1 cm, 0.7 cm and 0.5 cm are used. Then a rectangular specimen (120×60 cm^2) for biaxial test and a circle specimen for the Brazilian test (diameter: 60 cm) were created along the stage (3), as shown in Figure 3 and 4. Table 2 lists the analytical conditions and the ratio of the number of element generated in the stage (2). In the table, the lower suffices, e.g. 20 of D_{20}, mean the diameter of circular elements.

Table 1. Micro-parameters of synthetic rock material for rock slope model.

Element		
density	ρ	2650 (kg/m^3)
diameters	D	2.0, 1.4, 1.0 (cm)
contact stiffness	k_n	100 MN/m
	k_s	25 MN/m
coefficient of friction	μ	0.577 (tan$^{-1} \mu = 30°$)
Bond		
stiffness	E_n	100 MN/m^2
	E_s	25 MN/m^2
shear strength	τ_c	1.0 MN/m^2
tensile strength	σ_c	1.0 MN/m^2

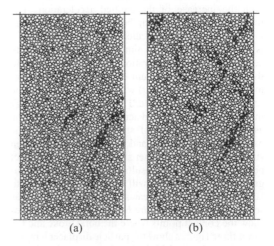

(a)   (b)

Figure 3. Specimen simulation of synthetic rock material and the damage patterns during uniaxial test (a) and biaxial test (b).

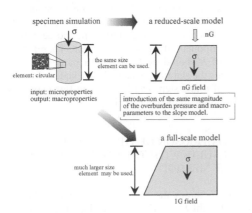

Figure 2. Reproduction of in-situ stress in slope simulation under the acceleration field.

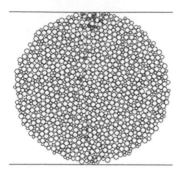

Figure 4. Specimen simulation of synthetic rock material and the damage patterns during the Brazilan test.

4 MEASURE PROPERTIES OF THE SYNTHETIC SPECIMEN AND SLOPE FAILURE SIMULATION

Near peak load in the specimen simulations, cracks initiate and propagate. The linkage of the bond braking cut across the specimens as seen in Figures 3 and 4 in which colored points represent tension-induced or shear-induced failure. The macro-parameters obtained from both tests are shown in Table 2.

Figure 5 shows the slope model which is cut from the original assembly. The depth of ground h is 200 cm and the width w_d is 200 cm. Three slopes with the toe angle β of 60°, 70° and 80° are used, and the height of the three slopes is constant and equals to 100 cm. As seen in Eq. (1), the overburden pressure σ_m in the slope model corresponds to the pressure at depth nh, e.g. for $n = 100$, the depth of 100 cm in the model would corresponds to the depth of 100 m.

Crack generation and slope failure were analyzed with the increase of the applied acceleration. The increase is explained with the value of n in nG where G is the gravitational acceleration. The increase of n is step-wise with the minimum increment equals to 5. If the applied acceleration gives little damage to the slope models, the sum of incremental displacement in one time step (Δt) will decrease and the slope reaches a quasi-static state. This means that the energy introduced by the applied acceleration is stored in the assembly in the form of strain energy in both the particle-particle contact and the bond material. When stress values reach at either strength, failure at contact point will occur and resulting failure plane propagates. This propagation will drive the slope model to large deformation and catastrophic slope failure as seen in Figure 6. In Figure 6, the upper three figures show the contact points where the bond broke and the lower three figures show the particle displacement, for $\beta = 80°$ with $n = 145$. Failure is generated at the toe of the slope and gradually spread to upward with the accumulation of time step. The value of (steps × Δt)

Table 2. Analytical condition and macro-parameters of synthetic rock material for rock slope model.

Analytical condition		
time increment:	$\Delta t = 1.0 \times 10^{-5}$ (sec)	
upper boundary displacement:	$\Delta u = 1.0 \times 10^{-6}$(cm)	
the ratio of the number		
of element: $D_{20} : D_{14} : D_{10} = 45 : 17 : 38$		
Macro-parameters		
uniaxial strength	q_u (MN/m^2)	1.440
tensile strength	σ_t(MN/m^2)	0.156
young's modulus	E (MN/m^2)	145.0
poisson's ratio	ν	0.295
cohesion	c (MN/m^2)	0.359
angle of internal friction	ϕ (°)	36.9

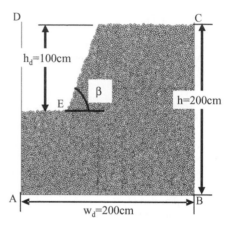

Figure 5. Rock slope model with constant inclination using the assembly of circular element.

gives the elapsed time since the slope model has been set under the given acceleration. This value could be only used as a relative rate of deformation under the given acceleration field. Needless to mention, such an increase of the gravitation never occur on the earth, and if balance of a slope disturbs, the slope will fail. Slope fails due to a triggering event and degradation of a component of strength with time. Therefore, the value of n should be treated as a parameter which indicates the stability. No conclusive statements about value of n and steps can be made in this paper. This will be done in the later work.

5 COMPARISON OF THE NUMERICAL RESULTS WITH THE CONVENTIONAL LEF METHOD

Figure 7 shows the relation between the allowable slope height $H = nh_d$ and the slope toe angle. The height for the slope of $\beta = 80°$ is the smallest. This could result in that the slope of 80° is the most unstable

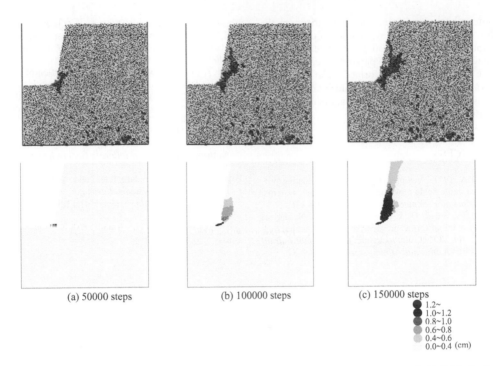

(a) 50000 steps	(b) 100000 steps	(c) 150000 steps

● 1.2~
● 1.0~1.2
● 0.8~1.0
● 0.6~0.8
○ 0.4~0.6
  0.0~0.4 (cm)

Figure 6. Propagation of failure point and displacement of elements forming under the acceleration of 145 G. The inclination of the slope β is 80°.

slope. This figure shows also the relation based on the limit equilibrium method (LEM). The LEM calculation was conducted using the mechanical properties (the cohesion and the internal friction angle) obtained from the synthetic specimen analysis described in the section 3.2, assuming that the slope ground is homogeneous and the unit weight $\gamma = 2500$ kN/m³. This value of γ is a reference value of rock and is not calculated using the value of the porosity in the assembly as seen in Figure 6.

The comparison indicates that for the angle smaller than $\beta = 70°$, the allowable height by the LEM is smaller than that by the numerical result, and for the angle greater than $\beta = 70°$, the LEM produces the greater allowable height. The numerical result may be less sensitive against the change the slope toe angle. This could be due to the differences in both methods: (1) the numerical method includes the progressive failure modeling, (2) the circular failure envelope is assumed in the LEM but in the numerical method the envelope is not circular.

6 CONCLUSION

Various numerical methods (continuum and discontinuum methods, and hybrid methods which combine

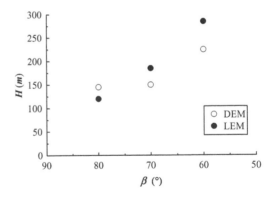

Figure 7. Comparison of the allowable slope height by the DEM analysis with the height by the limit equilibrium method.

both continuum and discontinuum techniques to simulate fracturing process) have been applied to demonstrate the evolution of failure in rock slope. In this paper, a numerical modeling of progressive failure in rock mass using the distinct element method with the bonded effect element is introduced to modeling the rock slope failure. The modeling could give a possible failure volume of rock material based on the geometrical data and strength properties, such as cohesion and internal friction angle. The comparison of the result with the LEM result was shown in terms of the allowable slope height. No decisive statement about the effect of the mechanical parameters can be made because the analysis was performed under the limited input values. On going work should be done incorporating the effects of the macro parameters and the initial stress condition.

REFERENCES

Cundall, P.A. 1971. A computer model for simulating progressive, large-scale movements in blocky rock systems. *Symposium on rock mechanics*, Nancy, 2:129–136.

Eberhardt, E., Stead, D. & Coggan, J.S. 2004. Numerical analysis of initiation and progressive failure in natural rock slopes—the 1991 Randa rockslide. *International Journal of Rock Mechanics and Mining sciences* 41: 69–87.

Einstein, H.H., Veneziano, D., Baecher, G.B. & O'reilly, K.J. 1983. The effect of discontinuity persistence on rock slope stability. *International Journal of Rock Mechanics and Mining sciences* 20–5: 227–236.

Jiang. Y, Li, B., Yamashita,Y. 2008. Simulation of cracking near a large underground cavern in a discontinuous rock mass using the expanded distinct element method. *International Journal of Rock Mechanics and Mining sciences*, Available on line, May, 2008.

Kaneko, K., Otani, J., Noguchi, Y. & Togashiki, N. 1997. Rock fracture mechanics analysis of slope failure. *Deformation and Progressive failure in Geomechanics*, Nagoya, Japan: 671–676.

Potyondy, D.O. & Cundall, P.A. 2004. A bonded-particle model for rockck systems. *International Journal of Rock Mechanics and Mining sciences* 41:1329–1364.

Backward-Euler stress update algorithm for the original Cam-clay model with the vertex singularity

T. Pipatpongsa, M.H. Khosravi & H. Ohta
Tokyo Institute of Technology, Tokyo, Japan

ABSTRACT: The original Cam-clay model (1963) is the earliest critical state soil model which can characterize stress, strain and strength of soils under the unified framework of plasticity theory. However the surface of this model in stress space is considered non-smooth. The vertex causes the discontinuity on the yield surface and is viewed as a drawback of the model. The aim of this research is to formulate a backward-Euler stress update algorithm with consideration to the vertex singularity for the original Cam-clay model. Koiter's associated flow rule was employed to both yield function and constraint function. The constraint function is defined as a plane passing the vertex. This simple method can conveniently handle the vertex without discretizing the yield surface into a finite number of yield loci. Numerical results of the proposed algorithm were illustrated in single and varying number of sub-steps. Error-maps were successfully generated to evaluate the high performance of the algorithm.

1 INTRODUCTION

In civil and geotechnical engineering, behaviors of soil materials are sometimes critical in the viewpoint of geo-hazards phenomenon like creep, localization and large deformation. Therefore, the efficient algorithm is necessary to integrate the rate constitutive models for analyzing stress-strain responses under these extreme states by numerical methods. Backward-Euler integration scheme has been extensively proved to be one of the most robust, stable and accurate integration algorithm for constitutive relations applied in the realm of plasticity. Amongst a numerous number of constitutive equations, the original Cam-clay model (Roscoe, Schofield & Thurairajah, 1963) is the earliest constitutive equations based on critical state theory that is capable to characterize stress, strain and strength of soils under the unified framework of plasticity theory. Numerical implementations of this model under backward-Euler scheme were found efficient in both isotropic version (Yatomi & Suzuki 2001) and anisotropic version (Pipatpongsa et al. 2001) in form of the Sekiguchi-Ohta model. Still, the algorithm for handling the vertex in general stress space has not been completely included.

The yield surface of the original Cam-clay model falls into a category of non-smooth yield surface. The shape of original Cam-clay yield surface is basically smooth except at the vertex. The singularity at the vertex causes numerical troubles when associated flow rule is applied because discontinuity is encountered on the gradient of yield surface.

In fact, the theoretical foundation which can solve non-smooth yield surface using backward-Euler integration scheme has been substantially developed. Simo, Kennedy & Govindjee (1988) provides the algorithms for evaluating plastic flow for constitutive laws with the intersecting yield surfaces by employing Koiter's associated flow rule (Koiter 1953).

However there is a limitation of using Koiter's flow rule because the yield criteria must be applied to a finite number of individual yield surfaces intersecting one another. But in the original Cam-clay model, there is only one smooth yield surface with the vertex, therefore uncertainty is left on how to validate the algorithm if the yield surface is dissociated to a finite number of yield loci or discretized to a pair of conjugate yield loci surrounding the vertex. Recently, Pipatpongsa & Ohta (2008) proposed the alternative method using Koiter's flow rule to determine plastic flow at the vertex singularity in rate form by additionally considering a constraint plane.

Therefore, the aim of this research is to extend the previous research by formulating a backward-Euler stress update algorithm for the original Cam-lay model with consideration of the vertex singularity. The results of this research can provide a basic background for further application to finite element analyses. Also, it is expected that the equivalent implementation to the Sekiguchi-Ohta model can be developed based on this study.

2 MODEL DESCRIPTION

2.1 Regular yield function and constraint function

The yield function of the original Cam-clay model is expressed by the following equation where M is critical state frictional parameter, p is mean stress, p_c is isotropic hardening stress parameter and q is a deviatoric stress.

$$f = Mp \ln\left(\frac{p}{p_c}\right) + q = 0 \qquad (1)$$

The yield surface of the original Cam-clay model looks like bullet-shaped surface aligned with the space diagonal in principle stress space as represented in Figure 1. The pointed vertex or the corner of the model is clearly marked by a sharp point in the three-dimensional yield surface. According to Pipatpongsa et al. (2008), it is suggested to consider the constraint function defined in addition to the regular yield surface. This simple method can facilitate a formulation of a generalized stress-strain relationship at the vertex singularity. The recommended constraint function is expressed in Eq. (2) which matches with stress dimension of Eq. (1). The plane of constraint function is shown in Figure 1 as a diagonal plane passing the corner to limit the value of p within p_c during loading process.

$$\bar{f} = p - p_c = 0 \qquad (2)$$

2.2 Plastic flow and vertex singularity

The hardening stress p_c at the singular vertex keeps the stress history of the model. Therefore, the vertex of the yield surface can move along the hydrostatic axis

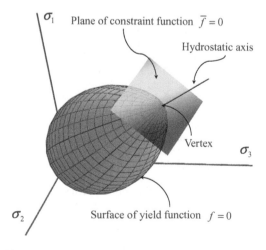

Figure 1. Representation of the original Cam-clay yield and the constraint functions in principal stress space. Hardening/softening responses move the vertex along hydrostatic axis.

to exhibit hardening/softening process. The evolution rule of p_c with rate of volumetric plastic strain $\dot{\varepsilon}_v^p$ obeys the following equation where λ is compression index, κ is swelling index and e_o is initial void ratio. Elastic and plastic parts of strains are denoted by superscript e and p respectively.

$$\dot{p}_c = p_c \frac{1 + e_o}{\lambda - \kappa} \dot{\varepsilon}_v^p \qquad (3)$$

According to the associated flow rule, plastic flow $\dot{\boldsymbol{\varepsilon}}^p$ can be evaluated by Eq. (4) where γ is a consistency parameter. Correspondingly, volumetric plastic strain can be obtained thru Eq. (5). Derivation of the gradient of yield surface with respect to generalized stress $\boldsymbol{\sigma}$ and relevant tensors are summarized in Box 1. The gradient of yield surface is discontinuous and undefined at the vertex. Hence, the normality postulate is ruled out, leading to an ambiguity of the direction of plastic flow.

$$\dot{\boldsymbol{\varepsilon}}^p = \gamma \frac{\partial f}{\partial \boldsymbol{\sigma}}, \quad \dot{\varepsilon}_v^p = \dot{\boldsymbol{\varepsilon}}^p : \mathbf{1} = \gamma \frac{\partial f}{\partial p} \qquad (4), (5)$$

In order to evaluate a plastic flow at the corner, the gradient of constraint plane which coincided with hydrostatic axis in principal stress space is used to correct the direction of plastic flow. The concept of associated flow rule proposed by Koiter (1953) is generally employed to handle non-smooth yield surfaces. Consequently, the plastic flow at the corner is evaluated by a summation of two distinct plastic

Box 1. Yield and constraint functions with derivatives.

1. Yield and constraint functions
$f = Mp \ln\left(\frac{p}{p_c}\right) + q = 0$, $\quad \bar{f} = p - p_c = 0$

2. Gradient of yield surface and constraint plane
$\frac{\partial f}{\partial \boldsymbol{\sigma}} = \frac{\partial f}{\partial p}\frac{\partial p}{\partial \boldsymbol{\sigma}} + \frac{\partial f}{\partial q}\frac{\partial q}{\partial \boldsymbol{\sigma}}$, $\quad \frac{\partial \bar{f}}{\partial \boldsymbol{\sigma}} = \frac{\partial \bar{f}}{\partial p}\frac{\partial p}{\partial \boldsymbol{\sigma}}$
where
$\boldsymbol{\sigma} = p\mathbf{1} + \mathbf{s}$ is a generalized stress in second-order tensor
$\mathbf{1} = \delta_{ij}\mathbf{e}^{<i>} \otimes \mathbf{e}^{<j>}$ is second-order identity tensor
$\delta_{ij} = \mathbf{e}^{<i>} \cdot \mathbf{e}^{<j>}$ is Kronecker delta
where $i = 1 \ldots 3\, \mathbf{e}^{<i>}$ is unit vector basis having 1 at row i and 0 for others. Single dot ($\cdot$) and double dot ($:$) denote single and double contraction of two tensors respectively. $\otimes$ denotes dyadic product of two tensors

3. Stress variables: $p = \frac{1}{3}\boldsymbol{\sigma} : \mathbf{1}$, $q = \sqrt{\frac{3}{2}\mathbf{s} : \mathbf{s}}$
where $\mathbf{s} = \boldsymbol{\sigma} - p\mathbf{1}$ is stress deviator

4. Tensor derivatives: $\frac{\partial p}{\partial \boldsymbol{\sigma}} = \frac{1}{3}\mathbf{1}$, $\frac{\partial q}{\partial \boldsymbol{\sigma}} = \sqrt{\frac{3}{2}}\mathbf{n}$, $\mathbf{n} = \frac{\mathbf{s}}{\sqrt{\mathbf{s} \cdot \mathbf{s}}}$

5. First derivative with respect to stress/hardening variables Yield function:
$\frac{\partial f}{\partial p} = \ln\left(\frac{p}{p_c}\right) + 1$, $\frac{\partial f}{\partial q} = 1$, $\frac{\partial f}{\partial p_c} = -\frac{p}{p_c}$

Constraint function: $\frac{\partial \bar{f}}{\partial p} = 1$, $\frac{\partial \bar{f}}{\partial p_c} = -1$
Note: $\partial f / \partial \boldsymbol{\sigma}$ is discontinuous because a partial derivatives $(\partial f/\partial q)(\partial q/\partial \boldsymbol{\sigma}) = \sqrt{3/2}\mathbf{n}$ in which $\mathbf{n}$ is undefined at the vertex where $\mathbf{s} = \mathbf{0}$

6. Second derivative with respect to stress variables Yield function:
$\frac{\partial^2 f}{\partial p^2} = \frac{1}{p}$, $\frac{\partial^2 f}{\partial p \partial q} = \frac{\partial^2 f}{\partial q \partial p} = 0$,
$\frac{\partial^2 f}{\partial q^2} = 0$, $\frac{\partial^2 f}{\partial p_c \partial p} = \frac{\partial^2 f}{\partial p \partial p_c} = -\frac{1}{p_c}$

Constraint function: $\frac{\partial^2 \bar{f}}{\partial p^2} = 0$, $\frac{\partial^2 \bar{f}}{\partial p_c \partial p} = \frac{\partial^2 \bar{f}}{\partial p \partial p_c} = 0$

flows contributed by both yield and constraint functions as shown by Eqs. (6)–(7). The additional consistency parameter $\bar{\gamma}$ is introduced. Plastic flow at the singular vertex is schematized in Figure 2.

$$\dot{\boldsymbol{\varepsilon}}^p = \gamma \frac{\partial f}{\partial \boldsymbol{\sigma}} + \bar{\gamma} \frac{\partial \bar{f}}{\partial \boldsymbol{\sigma}}, \quad \dot{\varepsilon}_v^p = \dot{\boldsymbol{\varepsilon}}^p : \mathbf{1} = \gamma \frac{\partial f}{\partial p} + \bar{\gamma} \frac{\partial \bar{f}}{\partial p}$$

$$(6), (7)$$

During loading condition, consistency requirement given by Eq. (8) must be generally satisfied. In particular loading condition, consistency requirement is enforced on both yield and constraint functions. Therefore, the consistency condition described by Eq. (9) is additionally required if stress state violates the constraint function at the same time with yield function.

$$\dot{f} = \frac{\partial f}{\partial \boldsymbol{\sigma}} \dot{\boldsymbol{\sigma}} + \frac{\partial f}{\partial p_c} \dot{p}_c = 0, \quad \dot{\bar{f}} = \frac{\partial \bar{f}}{\partial \boldsymbol{\sigma}} \dot{\boldsymbol{\sigma}} + \frac{\partial \bar{f}}{\partial p_c} \dot{p}_c = 0$$

$$(8), (9)$$

Eqs. (4) and (8) characterize plastic flow in regular mode while Eqs. (6), (8) and (9) characterize plastic flow in corner mode. Therefore, there are three possible modes of solution which are elastic, regular and corner modes. Following the general approach advocated in Simo, Kennedy and Govindjee (1988), the classical Kuhn-Tucker complementarily conditions are capable to appropriately characterize loading/unloading conditions using two consistency parameters as given in Table 1.

Table 1. Loading/unloading conditions.

Modes of solution	Criteria	Consistency parameters
Elastic mode	$f \leq 0$ and $\bar{f} \leq 0$	$\longrightarrow \gamma = 0$ and $\bar{\gamma} = 0$
Regular mode	$f > 0$ and $\bar{f} < 0$	$\longrightarrow \gamma > 0$ and $\bar{\gamma} = 0$
Corner mode	$f > 0$ and $\bar{f} > 0$	$\longrightarrow \gamma \geq 0$ and $\bar{\gamma} \geq 0$

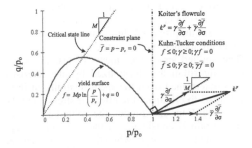

Figure 2. Schematization of plastic flow determined by Koiter's flow rule at the vertex of the original Cam-clay model is shown as the summation of two plastic flows outward surfaces.

2.3 Nonlinear stress-strain relation

According to e-log(p) relation obtained from triaxial tests, rate of change between p and volumetric strain ε_v^e is related by bulk modulus K.

$$\dot{p} = K(p)\dot{\varepsilon}_v^e \quad \text{where } K(p) = \frac{1 + e_o}{\kappa} p \tag{10}$$

Rate of change between stress deviator tensor $\mathbf{s}$ and strain deviator tensor ε_d^e is related by shear modulus G which can be related to K by Poisson's ratio ν.

$$\dot{\mathbf{s}} = 2G(p)\dot{\boldsymbol{\varepsilon}}_d^e \quad \text{where } G(p) = \frac{3(1 - 2\nu)}{2(1 + \nu)} K(p) \tag{11}$$

As a result, rate of stress is related to rate of strain by Eq. (12). Due to co-axial direction for isotropic elastic relationship from Eq. (13), rate change of q and deviatoric strain ε_s^e can be obtained by Eq. (14).

$$\dot{\boldsymbol{\sigma}} = \dot{p}\mathbf{1} + \dot{\mathbf{s}} = K\dot{\varepsilon}_v^e \mathbf{1} + 2G\dot{\boldsymbol{\varepsilon}}_d^e, \tag{12}$$

$$\mathbf{n} = \boldsymbol{\varepsilon}_d / \sqrt{\boldsymbol{\varepsilon}_d : \boldsymbol{\varepsilon}_d} \tag{13}$$

$$\dot{q} = 3G\dot{\varepsilon}_s^e \quad \text{where } \dot{q} = \sqrt{\frac{3}{2}\dot{\mathbf{s}} : \dot{\mathbf{s}}}, \dot{\varepsilon}_s^e = \sqrt{\frac{2}{3}\dot{\boldsymbol{\varepsilon}}_d^e : \dot{\boldsymbol{\varepsilon}}_d^e}$$

$$(14)$$

3 BACKWARD-EULER INTEGRATION

3.1 Integration of stresses and hardening variable

Rate form of stress is integrated over strain increment $\delta\varepsilon$ from the initial conditions denoted by subscript i to the current stresses $\boldsymbol{\sigma}$ as shown in Eqs. (13), (15)–(20). According to Borja (1991), the closed-form integration of nonlinear hypo-elasticity and related derivatives are summarized in Box 2.

$$\Delta\boldsymbol{\varepsilon} = 1/3\Delta\varepsilon_v \mathbf{1} + \Delta\boldsymbol{\varepsilon}_d, \tag{15}$$

$$\Delta\varepsilon_s = \sqrt{2/3\Delta\boldsymbol{\varepsilon}_d : \Delta\boldsymbol{\varepsilon}_d} \tag{16}$$

$$\boldsymbol{\varepsilon}_i = 1/3\varepsilon_{vi}\mathbf{1} + \boldsymbol{\varepsilon}_{di}, \tag{17}$$

$$\boldsymbol{\varepsilon}_d = \boldsymbol{\varepsilon}_{di} + \Delta\boldsymbol{\varepsilon}_d \tag{18}$$

$$\boldsymbol{\sigma}_i = p_i\mathbf{1} + \sqrt{2/3}q_i\mathbf{n}_i, \tag{19}$$

$$\boldsymbol{\sigma} = p\mathbf{1} + \sqrt{2/3}q\mathbf{n} \tag{20}$$

3.2 Integration of strains and iterative scheme

Plastic strains are updated by Newton method with restart algorithm. Boxes 3–6 summarize equations and procedures which can efficiently judge the solution for elastic, regular and corner modes. Variables of corner mode are noticed by upper bar.

181

Box 2. Updated stress and hardening variables with derivatives .

1. Mean stress, deviatoric stress
$$p = p(\varepsilon_v^e) = p_i \exp\left(\frac{\varepsilon_v^e - \varepsilon_{vi}^e}{\kappa/(1+e_o)}\right) \quad \text{where } \mu = \frac{3(1-2\nu)}{2(1+\nu)}$$

$$q = q(\varepsilon_v^e, \varepsilon_s^e) = \begin{cases} q_i + 3\mu\frac{p_i}{\kappa/(1+e_o)}(\varepsilon_s^e - \varepsilon_{si}^e) & \text{if } \varepsilon_v^e = \varepsilon_{vi}^e \\ q_i + 3\mu\frac{p-p_i}{\varepsilon_v^e - \varepsilon_{vi}^e}(\varepsilon_s^e - \varepsilon_{si}^e) & \text{otherwise} \end{cases}$$

2. Stress hardening parameter
$$p_c = p_{ci} \exp\left(\frac{\varepsilon_v^p - \varepsilon_{vi}^p}{(\lambda-\kappa)/(1+e_o)}\right) \quad \text{where } \varepsilon_v^p = \varepsilon_v - \varepsilon_v^e$$

3. Derivatives of stress and hardening variables
$$\frac{\partial p}{\partial \varepsilon_v^e} = \frac{p}{\kappa/(1+e_o)}, \quad \frac{\partial q}{\partial \varepsilon_s^e} = \begin{cases} 3\mu\frac{p_i}{\kappa/(1+e_o)} & \text{if } \varepsilon_v^e = \varepsilon_{vi}^e \\ 3\mu\frac{p-p_i}{\varepsilon_v^e - \varepsilon_{vi}^e} & \text{otherwise} \end{cases}$$

$$\frac{\partial q}{\partial \varepsilon_v^e} = \begin{cases} \frac{3}{2}\frac{\mu p_i}{(\kappa/(1+e_o))^2}(\varepsilon_s^e - \varepsilon_{si}^e) & \text{if } \varepsilon_v^e = \varepsilon_{vi}^e \\ 3\mu\left(\frac{\partial p}{\partial \varepsilon_v^e} - \frac{p-p_i}{\varepsilon_v^e - \varepsilon_{vi}^e}\right)\frac{\varepsilon_s^e - \varepsilon_{si}^e}{\varepsilon_v^e - \varepsilon_{vi}^e} & \text{otherwise} \end{cases}$$

$$\frac{\partial p_c}{\partial \varepsilon_v^p} = \frac{p_c}{(\lambda-\kappa)/(1+e_o)}, \quad \frac{\partial \varepsilon_v^p}{\partial \varepsilon_v^e} = -1$$

$$\frac{\partial p_c}{\partial \varepsilon_v^e} = \frac{\partial p_c}{\partial \varepsilon_v^p}\frac{\partial \varepsilon_v^p}{\partial \varepsilon_s^e} = -\frac{p_c}{(\lambda-\kappa)/(1+e_o)}$$

Box 3: Formulation of residual vectors and Hessian matrices.

1. State variables:
$$\mathbf{x} = \begin{Bmatrix} \varepsilon_v^e & \varepsilon_s^e & \Delta\gamma \end{Bmatrix}^T, \quad \bar{\mathbf{x}} = \begin{Bmatrix} \varepsilon_v^e & \varepsilon_s^e & \Delta\gamma & \Delta\bar{\gamma} \end{Bmatrix}^T$$

2. Residuals vectors:
$$\mathbf{r} = \begin{Bmatrix} \varepsilon_v^e - \varepsilon_v^{tr} + \Delta\gamma\frac{\partial f}{\partial p} \\ \varepsilon_s^e - \varepsilon_s^{tr} + \Delta\gamma\frac{\partial f}{\partial q} \\ f(p,q,p_c) \end{Bmatrix}, \bar{\mathbf{r}} = \begin{Bmatrix} \varepsilon_v^e - \varepsilon_v^{tr} + \Delta\gamma\frac{\partial f}{\partial p} + \Delta\bar{\gamma}\frac{\partial\bar{f}}{\partial p} \\ \varepsilon_s^e - \varepsilon_s^{tr} + \Delta\gamma\frac{\partial f}{\partial q} \\ f(p,q,p_c) \\ \bar{f}(p,p_c) \end{Bmatrix}$$

where $p = p(\varepsilon_v^e), q = q(\varepsilon_v^e, \varepsilon_s^e), p_c = p_c(\varepsilon_v^e)$ (see Box 2)

3. Hessian matrices: (refer to Box 2)
$$\frac{\partial \mathbf{r}}{\partial \mathbf{x}} = \begin{bmatrix} \partial_{x_1} r_1 & \partial_{x_2} r_1 & \partial_{x_3} r_1 \\ \partial_{x_1} r_2 & \partial_{x_2} r_2 & \partial_{x_3} r_2 \\ \partial_{x_1} r_3 & \partial_{x_2} r_3 & \partial_{x_3} r_3 \end{bmatrix}$$

where
$\partial_{x_1} r_1 = 1 + \Delta\gamma(\partial_{pp}^2 f \partial_{\varepsilon_v^e} p + \partial_{ppc}^2 f \partial_{\varepsilon_v^e} p_c)$,
$\partial_{x_2} r_1 = 0, \partial_{x_3} r_1 = \partial_p f$
$\partial_{x_1} r_2 = 0, \partial_{x_2} r_2 = 1, \partial_{x_3} r_2 = \partial_q f$
$\partial_{x_1} r_3 = \partial_p f \partial_{\varepsilon_v^e} p + \partial_q f \partial_{\varepsilon_v^e} q + \partial_{pc} f \partial_{\varepsilon_v^e} p_c$
$\partial_{x_2} r_3 = \partial_q f \partial_{\varepsilon_s^e} q, \partial_{x_3} r_3 = 0$

$$\frac{\partial \bar{\mathbf{r}}}{\partial \mathbf{x}} = \begin{bmatrix} \partial_{\bar{x}_1} \bar{r}_1 & \partial_{\bar{x}_2} \bar{r}_1 & \partial_{\bar{x}_3} \bar{r}_1 & \partial_{\bar{x}_4} \bar{r}_1 \\ \partial_{\bar{x}_1} \bar{r}_2 & \partial_{\bar{x}_2} \bar{r}_2 & \partial_{\bar{x}_3} \bar{r}_2 & \partial_{\bar{x}_4} \bar{r}_2 \\ \partial_{\bar{x}_1} \bar{r}_3 & \partial_{\bar{x}_2} \bar{r}_3 & \partial_{\bar{x}_3} \bar{r}_3 & \partial_{\bar{x}_4} \bar{r}_3 \\ \partial_{\bar{x}_1} \bar{r}_4 & \partial_{\bar{x}_2} \bar{r}_4 & \partial_{\bar{x}_3} \bar{r}_4 & \partial_{\bar{x}_4} \bar{r}_4 \end{bmatrix}$$

where
$\partial_{\bar{x}_1} \bar{r}_1 = \partial_{x_1} r_1, \partial_{\bar{x}_2} \bar{r}_1 = 0, \partial_{\bar{x}_3} \bar{r}_1 = \partial_{x_3} r_1, \partial_{\bar{x}_4} \bar{r}_1 = \partial_p \bar{f}$
$\partial_{\bar{x}_1} \bar{r}_2 = 0, \partial_{\bar{x}_2} \bar{r}_2 = \partial_{x_2} r_2, \partial_{\bar{x}_3} \bar{r}_2 = \partial_{x_3} r_2, \partial_{\bar{x}_4} \bar{r}_2 = 0$
$\partial_{\bar{x}_1} \bar{r}_3 = \partial_{x_1} r_3, \partial_{\bar{x}_2} \bar{r}_3 = \partial_{x_2} r_3, \partial_{\bar{x}_3} \bar{r}_3 = 0, \partial_{\bar{x}_4} \bar{r}_3 = 0$
$\partial_{\bar{x}_1} \bar{r}_4 = \partial_p \bar{f} \partial_{\varepsilon_v^e} p + \partial_{pc} \bar{f} \partial_{\varepsilon_v^e} p_c, \partial_{\bar{x}_2} \bar{r}_4 = 0, \partial_{\bar{x}_3} \bar{r}_4 = 0, \partial_{\bar{x}_4} \bar{r}_4 = 0$

Note: for regular mode, refer to $\mathbf{x}, \mathbf{r}$ and $\partial_{\mathbf{x}} r$; for corner mode, refer to $\bar{\mathbf{x}}, \bar{\mathbf{r}}$ and $\partial_{\bar{\mathbf{x}}} \bar{r}$

Box 4. Restart algorithm.

1. If over-loop or stack-overflow then reset $\bar{\mathbf{x}} = \bar{\mathbf{x}}_i$ and exit
2. On error in evaluating $\bar{\mathbf{r}}(\bar{\mathbf{x}})$ then restart by step 5
3. On error in evaluating $\partial_{\bar{\mathbf{x}}} \bar{\mathbf{r}}(\bar{\mathbf{x}})$ then restart by step 5
4. If $\det(\partial_{\bar{\mathbf{x}}} \bar{\mathbf{r}}(\bar{\mathbf{x}})) < \Theta$ then restart by step 5
5. Restart $\bar{\mathbf{x}} = \bar{\mathbf{x}}^{tr} rnd(1)$ and repeat step 1

Note: $rnd(1)$ is a random number between 0 and 1
Nested call in step 5 can prevent numerical error in the next iteration of Newton method due to a poor guessed value

Box 5. Newton method.

1. Select modes:
   For regular mode: assign $\bar{\mathbf{x}} = \mathbf{x}$ and $\bar{\mathbf{r}} = \mathbf{r}$
   For corner mode: assign $\bar{\mathbf{x}} = \bar{\mathbf{x}}$ and $\bar{\mathbf{r}} = \bar{\mathbf{r}}$
2. Set $\bar{\mathbf{x}} = \bar{\mathbf{x}}^{tr}$ and check restart algorithm in Box 4
3. If $\sqrt{\bar{\mathbf{r}}(\bar{\mathbf{x}}) \cdot \bar{\mathbf{r}}(\bar{\mathbf{x}})} > \Theta$ then

   3.1 Calculate $\bar{\mathbf{x}} = \bar{\mathbf{x}} - \partial_{\bar{\mathbf{x}}} \bar{\mathbf{r}}(\bar{\mathbf{x}})^{-1} \cdot \bar{\mathbf{r}}(\bar{\mathbf{x}})$
   3.2 On error then restart $\bar{\mathbf{x}}$ in Box 4 and redo step 3.1
   3.3 Repeat step 3
4. Recalculate $\bar{\mathbf{x}} = \bar{\mathbf{x}} - \partial_{\bar{\mathbf{x}}} \bar{\mathbf{r}}(\bar{\mathbf{x}})^{-1} \cdot \bar{\mathbf{r}}(\bar{\mathbf{x}})$
   Reconfirm if $\sqrt{\bar{\mathbf{r}}(\bar{\mathbf{x}}) \cdot \bar{\mathbf{r}}(\bar{\mathbf{x}})} < \Theta$ then return $\bar{\mathbf{x}}$
   Else repeat 2

Note: Θ is a predefined tolerance for numerical methods Restart algorithm provides a new guessed value

Box 6. Stress update algorithm.

1. Initialize: $p_i = p, q_i = q, p_{ci} = p_c, \varepsilon_{vi}^e = \varepsilon_v^e, \varepsilon_{si}^e = \varepsilon_s^e$
   Set: $\varepsilon_{vi}^{tr} = \varepsilon_{vi}^e + \Delta\varepsilon_v, \varepsilon_{si}^{tr} = \varepsilon_{si}^e + \Delta\varepsilon_s$
2. Elastic trial step: $p^{tr} = p(\varepsilon_v^{tr}), q^{tr} = q(\varepsilon_v^{tr}, \varepsilon_s^{tr})$
   Calculate $f^{tr} = f(p^{tr}, q^{tr}, p_{ci})$ and $\bar{f}^{tr} = \bar{f}(p^{tr}, p_{ci})$
3. Check plastic loading: If $f^{tr} > 0$
   Then check corner mode: If $\bar{f}^{tr} > 0$
      Then go to corner mode in step 4
      Else go to regular mode in step 5
   Else go to elastic mode in step 6
4. Corner mode:
   Set $\bar{\mathbf{x}}^{tr} = \begin{Bmatrix} \varepsilon_v^{tr} & \varepsilon_s^{tr} & 0 & 0 \end{Bmatrix}^T, \bar{\mathbf{x}}_i = \begin{Bmatrix} \varepsilon_{vi}^e & \varepsilon_{si}^e & 0 & 0 \end{Bmatrix}^T$
   Solve for $\bar{\mathbf{x}}$ by Newton method in Box 5
   Obtain $\begin{Bmatrix} \varepsilon_v^e & \varepsilon_s^e & \Delta\gamma & \Delta\bar{\gamma} \end{Bmatrix}^T = \bar{\mathbf{x}}$
   Check regular mode: If $\Delta\bar{\gamma} < 0$
      Then go to regular mode in step 5
      Else go to update state variables in step 7
5. Regular mode:
   Set $\mathbf{x}^{tr} = \begin{Bmatrix} \varepsilon_v^{tr} & \varepsilon_s^{tr} & 0 \end{Bmatrix}^T, \mathbf{x}_i = \begin{Bmatrix} \varepsilon_{vi}^e & \varepsilon_{si}^e & 0 \end{Bmatrix}^T$
   Solve for $\mathbf{x}$ by Newton method in Box 5
   Obtain $\begin{Bmatrix} \varepsilon_v^e & \varepsilon_s^e & \Delta\gamma \end{Bmatrix}^T = \mathbf{x}$
   Go to update stress variables in step 7
6. Elastic mode:
   Set $\varepsilon_v^e = \varepsilon_v^{tr}, \varepsilon_s^e = \varepsilon_s^{tr}$
7. Update stress variables and hardening parameter in Box 2
   $p = p(\varepsilon_v^e), \quad q = |q(\varepsilon_v^e, \varepsilon_s^e)|, \quad p_c = p_c(\varepsilon_v^e)$
8. Continue next sub-step

Note: Updated q in step 7 is taken in absolute form to prevent insignificant negative value caused by numerical tolerance

4 NUMERICAL ILLUSTRATION

Basic properties of material shown in Table 2 are used in numerical examples to explain and illustrate the performance of the proposed algorithm. Herein, the initial stresses are located on the vertex.

The applied strain increments to test the strain-driven algorithm are normalized by characteristics strains determined from the specific strains at initial yield stress to eliminate the effect of stress level. For the original Cam-clay model, the volumetric characteristic strain is considered at the yielding isotropic consolidated pressure p_{ci} while the deviatoric characteristic strain is considered at the yielding deviatoric stress at critical state representing by Mp_{ci}/e. Formula and numeric values of characteristic strains are shown in Table 3.

Table 2. Material parameters and initial stresses.

M	λ	κ	ν	e_o	p_i (kPa)	q_i (kPa)	p_{ci} (kPa)
1.00	0.10	0.02	0.20	2.00	100	0	100

Table 3. Material parameters and initial stresses.

Characteristic strains	Formula	Numeric values
Volumetric:	$\varepsilon_{vy} = \dfrac{p_{ci}}{K(p_{ci})} = \dfrac{\kappa}{1+e_o}$	0.667%
Deviatoric:	$\varepsilon_{sy} = \dfrac{Mp_{ci}/e}{3G(p_{ci})} = \dfrac{\kappa}{1+e_o}\dfrac{M}{3\mu e}$	0.109%

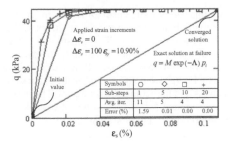

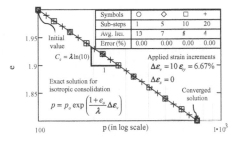

Figure 4. Deviatoric stresses under undrained shear test are computed with varying step-size of applied deviatoric strains.

Figure 5. Mean stresses under isotropic consolidation test are computed by varying step-size of volumetric strain. Slope of e-log(p) is correctly obtained with no error.

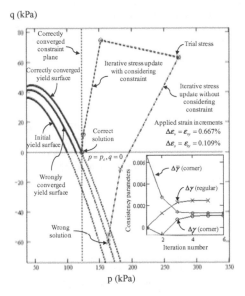

Figure 3. Return paths obtained by correct and wrong algorithms are illustrated. Two consistency parameters for yield and constraint functions can lead to the correct solution.

Discontinuity arises at the vertex of the model. However, the vertex does not remain fixed in stress space but move along hydrostatic axis to characterize hardening/softening behaviors during plastic loading steps, therefore the position of vertex is unknown. According to Box 6, trial stress is determined from trial strain initially assumed elastic state in order to find active surfaces in which yield criteria are violated. The problem is treated by initially assuming that both surfaces are active and the solution is fall into a corner mode. If $\Delta\bar{\gamma} < 0$ when the solution for a corner mode is converged, then the mode of solution is reduced to a regular mode. Without using the constraint plane, the iterative scheme in the singular region is led to a wrong solution as shown in

Figure 3 because the yield function is not delimited by p_c in the iteration. Once the constraint plane is considered, the correct return path can be obtained and $\Delta\gamma \geq 0$, $\Delta\bar{\gamma} > 0$. Figure 4 shows results of undrained shear test for 1,5,10 and 20 numbers of sub-steps to the applied strain increments. It was found a single step of $\Delta\varepsilon_s = 100\varepsilon_{sy}$ can obtain the solution at critical state by 11 iterations in regular mode with error 1.59% from the exact solution. Figure 5 shows results of isotropic consolidation test. It was found that a single step of $\Delta\varepsilon_v = 10\varepsilon_{vy}$ can obtain the exact solution in corner mode by 13 iterations. Generally, for small step size, the average number of iteration per number of sub-step is decreased while accuracy is increased.

5 ACCURACY EVALUATION

Error-map contour of the proposed algorithm are constructed to obtain the contour map generated by a series of single-step strain increments imposed to the particular stresses on yield surface. The overall accuracy is expressed relatively to the exact solutions of the updated stresses. The exact solutions are obtained numerically by taking further repeatedly sub-division until there is no significant change.

In this study, tolerance Θ is set to 10^{-6}. The error-map obtained from the initial stress at the vertex is shown in Figure 6. The considered domain of applied strain is 100 intervals of ± 2.5 of $\Delta\varepsilon_v/\varepsilon_{vy}$ and 50 intervals of 0–5 of $\Delta\varepsilon_s/\varepsilon_{sy}$. Relative error is calculated by Eq. (7) where $\boldsymbol{\sigma}^*$ and $\boldsymbol{\sigma}_c^*$ is the exact solutions for stress and hardening stress tensors. The maximum error was found 5.36% for resulted stresses located in neighborhood of the critical state.

$$Err = \sqrt{\frac{\{\boldsymbol{\sigma} - \boldsymbol{\sigma}^*\} : \{\boldsymbol{\sigma} - \boldsymbol{\sigma}^*\} + \{\boldsymbol{\sigma}_c - \boldsymbol{\sigma}_c^*\} : \{\boldsymbol{\sigma}_c - \boldsymbol{\sigma}_c^*\}}{\boldsymbol{\sigma}^* : \boldsymbol{\sigma}^* + \boldsymbol{\sigma}_c^* : \boldsymbol{\sigma}_c^*}}$$

$$= \sqrt{\frac{3(p - p^*)^2 + 2/3(q - q^*)^2 + 3(p_c - p_c^*)^2}{3p^{*2} + 2/3q^{*2} + 3p_c^{*2}}}$$

$$(21)$$

Modes of solution for single-step of applied strains are shown in Figure 7. Elastic mode was found in the resulted stress bounded within yield surface. Corner mode was limited in the boundary of $\varepsilon_s/\varepsilon_v \leq \Lambda/M$. The other region was regular mode.

$$\begin{Bmatrix} \dot{f} \\ \dot{\bar{f}} \end{Bmatrix} = \begin{Bmatrix} \dot{p}\partial_p f + \dot{q}\partial_q f + \dot{p}_c \partial_{p_c} f \\ \dot{p}\partial_p \bar{f} + \dot{q}\partial_q \bar{f} + \dot{p}_c \partial_{p_c} \bar{f} \end{Bmatrix} = \begin{Bmatrix} 0 \\ 0 \end{Bmatrix}$$

$$(22)$$

$$\begin{Bmatrix} \dot{\varepsilon}_v \\ \dot{\varepsilon}_s \end{Bmatrix} = \begin{Bmatrix} \gamma M + \bar{\gamma} \\ \Lambda \\ \gamma \end{Bmatrix}, \text{ hence } \frac{\dot{\varepsilon}_s}{\dot{\varepsilon}_v} = \frac{\gamma\Lambda}{\gamma M + \bar{\gamma}}$$

$$(23), (24)$$

The condition of corner mode can be derived thru Eqs. (8)–(24), using Eqs. (3), (10), (14) and Box 1. Eq. (24) reveals that if the yield function is inactive, then $\gamma = 0$, $\dot{\varepsilon}_s/\dot{\varepsilon}_v = 0$. Oppositely, if the constraint function is inactive, then $\bar{\gamma} = 0$, $\dot{\varepsilon}_s/\dot{\varepsilon}_v = \Lambda/M$.

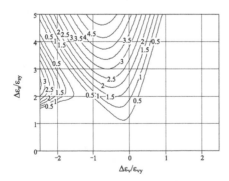

Figure 6. Percent error-map contour with loading from the vertex is constructed by applying the normalized strain increments to evaluate the stress update algorithm. The maximum error 5.36% was found in a domain of ± 2.5 $\Delta\varepsilon_v/\varepsilon_{vy}$ and 0–5.0 $\Delta\varepsilon_s/\varepsilon_{sy}$.

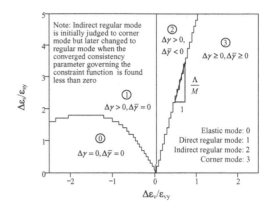

Figure 7. Modes of solution are judged from the applied volumetric and deviatoric strain increments normalized by the corresponding characteristic strains. Computed regions based on 100 intervals of $\Delta\varepsilon_v/\varepsilon_{vy}$ and 50 intervals of $\Delta\varepsilon_s/\varepsilon_{sy}$ were drawn.

6 CONCLUSION

Backward-Euler stress update algorithm for the original Cam-clay model with the vertex singularity was proposed. The scheme based on Newton method with restart algorithm was evaluated by error-map to confirm robust, accurate and stable performance.

REFERENCES

Borja, R.I., 1991. Cam-Clay plasticity, Part II: Implicit integration of constitutive equation based on a nonlinear elastic stress predictor. Comput. *Methods Appl. Mech. Engrg.* 88: 225–240

Koiter, W.T., 1953. Stress-strain relations, uniqueness and variational theorems for elastic-plastic materials with a singular yield surface. *Quart. Appl. Math.* 11: 350–354.

Pipatpongsa, T., Iizuka, A., Kobayashi, I., Ohta, H. & Suzuki, Y., 2001. Nonlinear analysis for stress-strain-strength of clays using return-mapping algorithms. *JSCE J. Appl. Mechs.* 4: 295–306.

Pipatpongsa, T. & Ohta, H., 2008. How can we describe strain changes due to isotropic consolidation by using the original cam clay model?. *The 13th National Convention on Civil Engineering of Thailand*, GTE: 96–101.

Roscoe, K.H., Schofield, A.N. & Thurairajah, A., 1963. Yielding of clays in state wetter than critical. *Geotechnique* 13, 3: 211–240.

Simo, J.C., Kennedy, J.G. & Govindjee, S., 1988. Non-smooth multisurface plasticity and viscoplasticity loading/unloading conditions and numerical algorithms. *Int. J. Numer. Methods Engrg* 26: 2161–2185.

Yatomi, C. & Suzuki, Y., 2001. Finite element method analysis of soil/water coupling problems using implicit elasto-plastic calculation algorithm. *JSCE J. Appl. Mechs.* 4: 345–356.

Erosion and seepage failure analysis of ground with evolution of bubbles using SPH

H. Sakai, K. Maeda & T. Imase
Nagoya Institute of Technology, Nagoya, Japan

ABSTRACT: Seepage failure of soil and/or ground causes important geotechnical problems such as damage of dyke under flood, erosion of soil structure nearby ocean and river and so on. Moreover, generation of gas and blow-out of air bubbles have been seen before the failure occurred in many cases. The sources of air bubbles could be thought to be air phase entrapped by seepage front and oversaturated air in pore water. In this paper, we focused the evolution effect of air bubbles in pore water on seepage failure, which the air bubbles must increase the risk of soil failure and erosion. We performed model test, and developed a new numerical simulation method accounting for flowage deformation and solid-water-air bubbles interactions using Smoothed Particle Hydrodynamics. And simulation results were verified by comparison with model test results.

1 INTRODUCTION

Large flowage deformations and the hydraulic collapse of ground piping, which induces erosion, are induced by the permeation of water through the ground. This plays an important role in the destabilization of ground during floods, liquefaction, erosion and so forth. In order to analyze these phenomena more precisely, it is necessary to model progressive seepage failure in the soil. Some reports have found that interactions between all three phases—solid, liquid and gas—play important roles. In particular, Kodaka & Asaoka (1994) may be the first instance in which the importance of the dynamics of air bubbles in geo-engineering was revealed. Furthermore, when the Tokai flooding disaster occurred in the Nagoya region on 11 September 2000, a man who witnessed the process of dike failure recounted his story in a newspaper, claiming that, after a crack generated on the dike's surface, white bubbles of water blew out of the crack, and then the dike gradually failed for about three hours. In reality, this type of phenomena has been frequently witnessed for generations. The blowing air bubbles that precede seepage failure are called "frog blows bubbles" by elderly people. Terzaghi (1942) gives a definition and discussion of hydraulic failure without air bubbles.

In this study, we conducted model tests and developed a new numerical simulation method for seepage failure with air bubbles. While discrete analysis (e.g. discrete element method (DEM)) is adapted to abruption, failure and flowage, it is an unsuitable procedure to analyze large-scale domains. Conversely, continuum analysis (e.g. conventional FEM) has the opposite properties. The *smoothed particle hydrodynamics* (SPH) method (Gingold & Monaghan 1977, Lucy 1977), a completely mesh-free technique, was used to obtain the combined benefits of both the distinct and continuum methods. In this study (Maeda & Sakai 2004, Maeda et al. 2006), we propose SPH with a new method for calculating density, surface tension and multi-phases coupling. Moreover, in this paper, the simulation results were verified by comparison with model test results, including the velocity of ground and pore water pressure at failure.

2 MODEL TEST WITH IMAGE ANALYSIS

2.1 Model test procedure

In model tests, we observed the deformation-failure around sheet-pile in sand ground submerged in two kinds of water: water with low DO (dissolved oxygen) and water with high DO and over-saturated air, (see Figure 1). In the latter, air bubbles were easily generated.

Two types of tests were conducted: a "normal piping" test with lower DO and a "holding" test. In the normal piping test, a difference in water level, h, head loss applied to the ground, was increased to a critical head loss, h_{cr}, in which piping occurred in a ground; h increased gradually within one to two hours with a set of small increments of h and five-minute holding. In the holding test, applied head loss, h, was held until piping occurred. If piping did not occur even after much time had elapsed, h was increased again to generate piping.

The velocity field of the ground can be measured using particle image velocimetry (PIV) image analysis and the strain rate fields can be calculated from

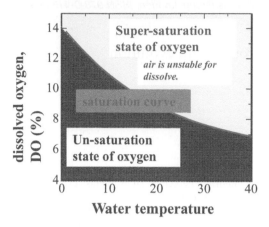

Figure 1. *DO* (dissolved oxygen) saturation curve.

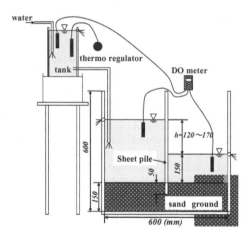

Figure 2. Model test apparatus for seepage failure.

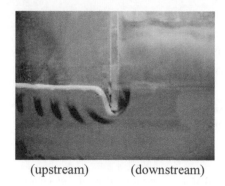

(upstream)   (downstream)

Figure 3. Piping with water of lower *DO* and without air bubbles in normal piping test: increasing head loss h to h_{cr}.

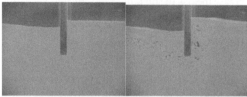

(start of holding test) (elapsed time after holding: 94 hr)

Figure 4. Deformed ground just prior to piping with water of higher *DO* in holding test: holding head loss $h = 0.8 \times h_{cr}$.

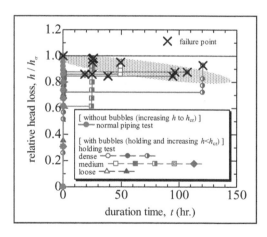

Figure 5. Decrease in safety against piping due to air bubbles.

the image analysis results. The test apparatus used is shown in Figure 2, and test conditions followed those of experiments performed by Kodaka & Asaoka (1994). Toyoura sand was employed. The dissolved oxygen (*DO*), water temperature, and permeable water volume of the ground were measured. A tensiometer pressure sensor was placed 50 mm to the right horizontally from the tip of the sheet-pile in the downstream.

2.2 *Model test results and discussions*

Figures 3–4 show the typical deformation-failure behaviours around the sheet-pile for lower *DO* and higher *DO* cases, respectively. Figure 5 shows the relative applied head loss, h/h_{cr}, indicating the safety against piping with elapsed time after h was increased and/or held. In the lower *DO* case in the normal piping test, piping occurred at $h/h_{cr} = 1$, in accordance with

the definition. However, in the higher *DO* case, even though h is less than h_{cr}, air bubbles were generated as shown in Figure 4, and, consequently, failure occurred in the holding test. When the water level was increased after holding the smaller water level difference, h, for a long time, the failure tended to occur even before the water level difference reached the critical level, h_{cr} (i.e. when $h/h_{cr} < 1$), which is similar to creep failure in a material under a constant loading condition that is less than its strength. This implies that air bubbles in the ground bring about strength degradation. The relationship between air bubble generation and DO, and the influence of air bubbles on the ground are investigated in the following sections.

In seepage failure without air bubbles, as shown in Figure 6, ground subsidence (scour) on the upstream side and ground uplift (roll-up) on the downstream side occur continuously when $h/h_{cr} > 0.90$. However, in the case with air bubbles, as shown in Figure 7, the ground surface on the downstream side was displaced intermittently. This must be related to the dynamics of the air bubbles, such as their generation, development, movement and ejection from the ground. Figure 8 shows both the change in the amount of air bubbles on the downstream side, which was calculated by image analysis, and the change in the void ratio, which is estimated by assuming that void change was only due

to air bubbles. Here, the sharp drops in the graph indicate bubble ejection with the rapid subsidence of ground. Moreover, since the accumulated amount of air bubbles was limited as shown in Figure 8, the ground surface on the downstream side was not displaced for some time following the ejection of large air bubbles. On the other hand, the ground surface on the upstream side was also not displaced for some time but was suddenly displaced after a longer period of time. This can be observed visually in Figure 9. Consequently, the displacement in the ground on the downstream side gradually propagated to the ground on the upstream side, passing below the sheet pile,

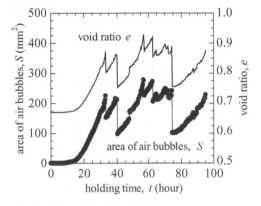

Figure 8. Changes in amount of air bubbles downstream around sheet-pile and in void ratio calculated according to Fig. 7.

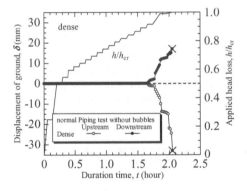

Figure 6. Displacement of ground surfaces upstream and downstream for normal piping test without air bubbles due to application head loss.

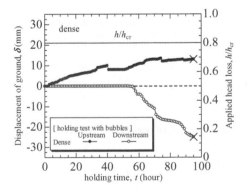

Figure 7. Displacement of ground surfaces upstream and downstream for holding test with air bubbles due to application head loss.

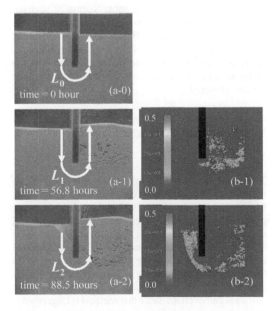

Figure 9. Variations of length of seepage around sheet-pile and maximum shear strain rate distribution (refer to Figure 7).

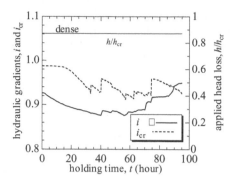

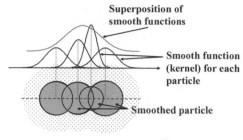

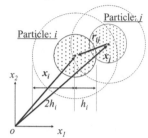

Figure 10. Changes in both hydraulic gradient and critical hydraulic gradient due to deformation of ground and increments of void ratio by evolution of air bubbles, as shown in Figure 7.

Figure 11. Smoothed particle and smoothing function in SPH.

and accompanied by the intermittent ejections of air bubbles accumulated in the ground on the downstream side. This resulted in the ground surface subsidence on the upstream side.

These air bubble dynamics bring frequent changes in the void ratio, since the critical hydraulic gradient, i_{cr}, of the ground is calculated by $i_{cr} = (G_s - 1)/(1 + e)$ where G_s is soil particle specific weight (Terzaghi 1942), as shown in Figure 10. From the photographs, seepage distances, L, around the sheet-pile can be calculated. Before ground subsidence occurs upstream, L becomes longer such that $L_0 < L_1$. Then, L becomes shorter such that $L_2 < L_1$. Consequently, the hydraulic gradient, i, increases, expressed in Figure 10, even through head loss, h, is constant. Normally, seepage failure occurs when the seepage force increases due to an increase in head loss. However, even though the head loss was held, the residence against piping, i_{cr}, decreased, and seepage load, i, increased due to the abovementioned mechanism. Consequently, seepage failure occurred.

3 NUMERICAL ANALYSIS: SPH

3.1 Analysis procedure

The SPH method is a particle-based Lagrangian method. It was originally developed by Gingold & Monaghan (1977), Lucy (1977) in the field of astrophysics to solve motions of galaxies. Then, this method was applied to viscid flows and failure of solids. The SPH method is intended not for treating the actual soil grain but for solving "particle" as soil mass (Figure 11), whose radius is h. Similarly, water "particle" is a finite volume of water, not a molecule. These particles can overlap. Since this method is Lagrangian, it can also express sliding, contact, separation, and two or three phase interactions.

The spatial averaged value $< f(\mathbf{x}) >$ of a physical quantity $f(\mathbf{x})$ at point x is given by Equation (1). Particles $\mathbf{x}'$ with $f(\mathbf{x}')$ are located within the zone of influence of the first particle $(2h)$. The physical quantity is interpolated using a smoothing function W in Figure 11,

$$f_i = < f(\mathbf{x}_i) > \cong \sum_{j=1}^{N} m_j \frac{f_j}{\rho_j} W_{ij}(\mathbf{r}_{ij}, h) \qquad (1)$$

where $\mathbf{r} = \mathbf{x}_i - \mathbf{x}_j$ and W is defined by:

$$1 = \int W(\mathbf{r}, h) d\mathbf{x}' \qquad (2)$$

In this paper, 3rd B-spline function was employed as W, and $r_{ij} = |\mathbf{r}_{ij}|$ and $S = r_{ij}/h$ are used:

$$W_{ij} = \frac{15}{7\pi h^2} \times \begin{cases} \frac{2}{3} - S^2 + \frac{1}{2}S^3 & 0 \leq S < 1 \\ \frac{1}{6}(2-S)^3 & 1 \leq S \leq 2 \\ 0 & 2 < s \end{cases} \qquad (3)$$

The density ρ_i of particle i is replaced with f_i:

$$\rho_i = \sum_{j=1}^{N} m_j \frac{\rho_i}{\rho_j} W_{ij} = \sum_{j=1}^{N} m_j W_{ij} \qquad (4)$$

However, this description shows large error in densities calculated by original theory around the interface. We improved this point by the normalization and limited summation for a focused material (e.g. material a).

$$\rho_{i\epsilon\text{material }a} = \frac{\sum\limits_{\substack{j=1\\ \epsilon\text{material }a}}^{N} m_j W_{ij}}{\sum\limits_{\substack{j=1\\ \epsilon\text{material }a}}^{N} \left(\frac{m_j}{\rho_j}\right) W_{ij}} \quad (5)$$

The SPH description for a particle i in a motion equation can be explained as follows:

$$\frac{dv_i}{dt} = -\sum_{j=1}^{N} m_j \left(\frac{\sigma_j}{\rho_j^2} + \frac{\sigma_i}{\rho_i^2} + \pi_{ij}\mathbf{I}\right) \cdot \nabla W_{ij} + \mathbf{f}_i \quad (6)$$

where σ and $\mathbf{f}$ are stress tensor and body force, respectively. The matrix $\mathbf{I}$ is the unit matrix, and Π_{ij} is the artificial viscosity.

For the purpose of coupling, the soil and the fluids of water and air were handled on different layers (see Figure 12). The frictional body forces resulting from velocity differences between two phases, $\mathbf{v}^s$ and $\mathbf{v}^f$, were employed with Biot's mixture theory (Biot 1941). The forces can be expressed as follows:

$$\mathbf{f}^{sf} = n\frac{\rho_f g}{k}(\mathbf{v}^s - \mathbf{v}^f)\mathbf{f}^{fs} = n\frac{\rho_f g}{k}(\mathbf{v}^f - \mathbf{v}^s) \quad (7)$$

Here, porosity n, permeability k, fluid density ρ_f and gravity acceleration g ($= 9.8$ m/s^2) are included.

Surface tension was introduced (Maeda et al. 2004, 2006). The boundary was reproduced by creating an array of virtual boundary particles. The leapfrog method with time step was used, holding the CFL conditions. The state equation of water and gas and the constitutive law of soil employed in detail were introduced in previous papers (Maeda & Sakai, 2004, Maeda et al. 2006).

3.2 Analysis results and discussions

Figures 13–14 show the results of the single-phase analyses for the collapse of columns of water (initial density is 1000 kg/m^3, and viscosity is 1.002×10^{-3} N · s/m^2) and frictional soil material with the internal frictional angle $\phi = 30$ (deg.), respectively. The final surface of the collapsed water column becomes flat, but one in Figure 14 shows an inclination angle that is slightly smaller than ϕ because of inertial force. These tendencies agree with actual flows.

Figure 15 shows the analysis result of *two phases*, such as the rising process of air bubbles in water: initial density of air is 1.207 kg/m^3, and viscosity is 1.810×10^{-5} N · s/m^2. The bubbles were simulated using clusters of SPH gas particles. A surface tension effect was involved by adding a force term.

Figure 16 shows a *soil-water coupling* SPH model of the experiment shown in Figure 3 without air bubbles. The elastic-perfect plastic-type model with internal friction angle and dilatancy angle was employed as the constitutive relation of soil, where elastic stiffness is described as a function of void ratio and mean effective stress (see the reference Maeda et al., 2006). This model can successfully simulate characteristics of seepage failure not only during deformation but also after the failure. We can see both the concentration of flow around the tip of the sheet-pile and the evidence of high speed flow on the downstream side due to curling induced by the erosion.

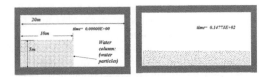

Figure 13. Single-phase analysis of water dam break.

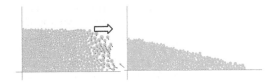

Figure 14. Single-phase analysis of frictional material column.

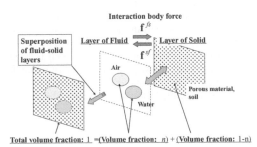

Figure 12. Interaction force to coupling between solid-liquid-gas.

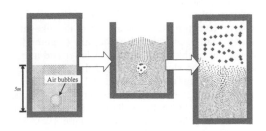

Figure 15. Two-phase analysis of bubbles in water.

189

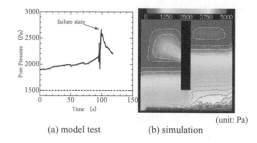

Figure 16. Seepage failure coupling analysis around sheet-pile without air bubbles; generation of failure of ground with permeability k of 1.0×10^{-5} m/s when $h = h_{cr}$.

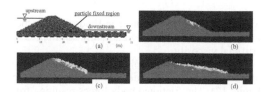

(a) model test   (b) simulation

(unit: Pa)

Figure 18. Comparison in pore water pressure around sheet-pile without air bubbles at piping; (a) from tensiometer pressure sensor in model test; (b) from SPH simulation.

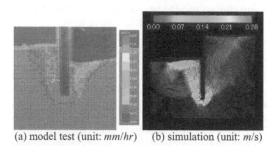

(a) model test (unit: *mm/hr*)   (b) simulation (unit: *m/s*)

Figure 17. Comparison in velocity distributions around sheet-pile without air bubbles; PIV image of ground velocity for model test; (b) water velocity for SPH simulation.

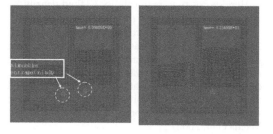

Figure 19. Seepage failure coupling analysis of dike where the colour of particle indicates its velocity (red: high; blue: low): (a) analysis configuration, (b) first local movement and erosion, (c) progression of failure, (d) washed-out after collapse attached to the crest.

Figures 17–18 show comparisons between model test results and numerical simulation results for velocity field and for pore water pressure in the ground around the sheet-pile at piping failure. The velocity of ground measured by PIV is shown in Figure 17a, and the velocity 50 mm horizontal from the sheet-pile tip, in the downstream after the failure, is about 0.3 m/s, which might be the same as the velocity of water in mixed. The SPH simulation result in Figure 17b is 0.28 m/s, which is almost identical to that of the model test. The pore water pressure at the same point as measured by the tensiometer is shown in Figure 18a, and the value at the failure state is around 2,500 Pa. The pressure distributions are influenced by the curling flow due to erosion. The pressure value analyzed in the downstream around the tip in Figure 18b is almost the same as that of the model test. The analysis results show good agreement with model test results both qualitatively and quantitatively.

Figure 19 shows the dike with water maintained at a constant level (on the upstream side of the dike) after seepage had been induced in the dike (Figure 19a). Movement of the small portion at lower and middle domains of the slope occurs, and the local erosion-triggered sliding collapses (Figure 19b). The failure gradually progresses toward the crest of the dike (Figure 19c). Once the collapse affects the upper portion, the dike is washed out (Figure 19c). This means that this proposed model can simulate a sequent collapse from small erosion to washed-out.

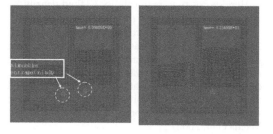

Figure 20. Seepage failure analysis with air bubbles; generation of local failure occurs at local site even when $h = 0.8 \times h_{cr}$.

Figure 20 shows a *three-phase coupling* SPH prediction of seepage failure around sheet-pile with air bubbles at below 80% of the water height difference in Figure 16. In this analysis, the air phase was replaced forcedly with another phase at initial state. The ground failure is induced by air bubbles rising as well as experiments even under a lower h than h_{cr}. We find that the movement of air bubbles induces the local deformation of the ground, bringing the degradation of bubbles.

4 CONCLUSIONS

PIV image analysis of the model test results revealed that the dynamics of air bubbles in the ground caused its degradation. Even though the head loss was held at a constant value lower than critical head loss, the residence against piping i_{cr} decreased, and seepage load i increased. In the next stage, we will reveal the interaction detail mechanism between the degradation of the ground and the evolution of air bubbles.

This paper proposed a newly developed method of SPH to solve three-phase systems (solid, liquid and gas). The analysis performance is qualitatively high. However, the quantitative verification could be seen in only some of the data. This paper showed clearly the validation and usefulness of SPH in its application to three-phase problems for flood disaster predictions and the development of disaster prevention methods. In future work, it will be possible to develop the procedure to simulate seepage failure from the generation to the evolution of bubbles and concerning entropy and enthalpy.

ACKNOWLEDGMENT

The authors are grateful to the Japan Society for the Promotion of Science for its financial support with Grants-in-Aid-for Scientific Research (B) 20360210.

REFERENCES

Biot, M.A. 1941. General theory of three dimensional consolidation. *Journal of Applied Physics* 12: 152–164.
Gingold, R.A. & Monaghan, J.J. 1977. Smoothed particle hydrodynamics: theory and application to non-spherical stars. *Monthly Notices of the Royal Astronomical Society* 181: 375–389.
Kodaka, T. & Asaoka, A. 1994. Formation of air bubbles in sandy soil during seepage process. *Journal of JSCE* 487 (III-26): 129–138. (in Japanese).
Lucy, L.B. 1977. A numerical approach to the testing of the fission hypothesis. *Astronomical Journal* 82: 1013–1024.
Maeda, K., Sakai, H. & Sakai, M. 2006. Development of seepage failure analysis method of ground with smoothed particle hydrodynamics. *Jour. of Structural and earthquake engineering, JSCE, Division A* 23(2): 307–319.
Maeda, K. & Sakai, M. 2004. Development of seepage failure analysis procedure of granular ground with Smoothed Particle Hydrodynamics (SPH) method. *Journal of Applied Mechanics, JSCE* 7: 775–786. (in Japanese).
Sakai, H. & Maeda, K. 2007. A study on seepage failure of sand ground with account for generation and development of air bubbles. *Proc. of 13th Asian Regional Conference on Soil Mechanics and Geotechnical Engineering*: 571–574.
Terzaghi, T. 1942. *Theoretical soil mechanics*, John Wiler and Sons, INC.

Prediction and Simulation Methods for Geohazard Mitigation – Oka, Murakami & Kimoto (eds)
© 2009 Taylor & Francis Group, London, ISBN 978-0-415-80482-0

Performance estimation of countermeasures for falling rock using DEM

K. Maeda & T. Yuasa

Nagoya Institute of Technology, Nagoya, Japan

ABSTRACT: This study aims to simulate the effects of non-circular shapes with rotation and breakage of falling rocks, using DEM that models a falling rock as an assembly of bonded particles. Moreover, ways to express the influence of a slope, a natural talus, and an artificial pocket are suggested. Particle properties and density of the deposits used in countermeasures are also examined, and inclined collisions with rotation are considered.

1 INTRODUCTION

Slope disasters with the destructive power of a rock fall occur due to weathering, heavy rain, and earthquakes, and cause serious damage. Existing countermeasure manuals (Japan Road Association 2000) are mainly based on a slope tens of meters high, with rocks weighing several tons. However, in practice, slopes are more commonly 50–200 meters high and rock weights are approximately 10 tons or more, and sized several meters or more. Since such slope disasters are not considered in existing countermeasure manuals, numerical simulation accounting for such factors and conditions need to be introduced.

Simplified models in the discrete element method (DEM) (Cundall & Stract 1979) or discontinuous deformation analysis (DDA) (Shi & Goodman 1984), which are usually used, treat a falling rock as a particle element and model a slope using a plate element, or simply arranged particles. In such models, several significant factors are not taken into consideration. These include: breakage of the falling rock; crush of the contact part of the rock; and the effect of weathered and deposited materials, such as soft talus or a weathered slope surface. Moreover, when a falling rock is modeled as a circle in 2D or a sphere in 3D, the influence of its shape; interaction between rotation and translational motion; and effects such as digging into the talus need to be considered. The energy dissipation effects of crush and of naturally existing talus must be considered in designing effective countermeasures. Moreover, in road construction where rock fall is a considered risk, a space is provided between the slope toe and the road, or a retaining wall is installed at the roadside; this is called a "pocket" where deposits of sand or gravel are provided to act as an artificial talus. This artificial talus is meant to act as a buffering area to absorb falling rocks. But in this test, the inclined collision between the falling rocks and the granular deposit will be challenged. This study aims to simulate the effects of non-circular, rotating shapes and

the breakage of falling rocks, using DEM that models a falling rock as an assembly of bonded particles. Moreover, ways to express the influence of a slope, a natural talus, or an artificial pocket are suggested. Particle properties and density of the deposits used in countermeasures are also examined, and inclined collisions with rotation are taken into account.

In the DEM, it is difficult to establish parameters such as particle size, rigidity, viscous constant, friction, and so forth. Therefore, the means of establishing effective parameters for a practical application of the DEM is examined in this study. Moreover, the validity of the DEM is compared with results obtained by other simple model tests.

2 DISCRETE MODELING OF FALLING ROCK

2.1 Discrete element modeling

Both the falling rock block and the two dimensional slope were modeled as assemblies of circular particle elements, where the largest and the smallest diameters are D_{max} and D_{min}, respectively and the particle size number frequency distribution is uniform. The sedimentary layer, such as natural talus and the artificial pocket, which has a shock-absorbing function, were also made by cohesion-less assembly. A conventional discrete element method (DEM) was used in which the interaction between elements was modeled by contact elements (springs k_n and k_s, dash-pots c_n and c_s, frictional slider μ, and non-extensional elements) as shown in Figure 1.

The surface of the slope and the falling rock were modeled as particle elements bonded to simulate the complex shape of rock, rolling resistance, and breakage of rock mass due to collision (refer to Figure 2). Here, a bond material element with a width of D_b, stiffness k_b^n and k_b^s, and tensile and shear strength stress s_b, was employed. Accordingly, the bending moment at the interparticle is transmitted until maximum tensile

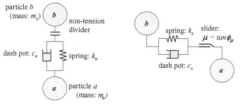

Figure 1. Contact elements with a Voigt type model in a conventional DEM: normal and tangential direction to a contact plane.

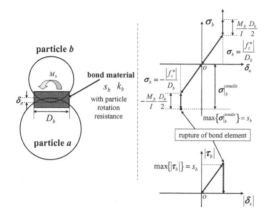

Figure 2. Bond element with flexural rigidity at interparticle.

stress with summation of normal contact stress f_c^n/D_b and tensile stress due to bending $M/I\,(D_b/2)$, attaches the bond strength s_b. Here, M is the moment at interparticle and I is the geometric moment of inertia. In this paper, it is assumed that k_b^n and k_b^s are equal to k_n and k_s, respectively and D_b is the smallest particle diameter $D_{\min}$, tentatively.

2.2 Micro and macro parameters

The micro physical parameters in the DEM must be determined to be reasonable for practicability in computation and to be consistent with the macro mechanical characteristics of the material: a set of $\mathbf{M}$ is composed of {macro stiffness; strength; dilatancy}. The macro mechanical properties $\mathbf{M}$ are determined by a function F_1 in Equation 1 (Mikasa, 1963),

$$\mathbf{M} = \{\text{Mechanical properties}\} = \{F_1(\text{particle properties; packing and fabric; stress})\} \quad (1)$$

where packing could be strongly influenced by density and fabric could be controlled by heterogeneity and anisotropy: packing and fabric make a set $\mathbf{K}$.

$$\mathbf{K} = \{\text{packing; fabric}\} \quad (2)$$

The particle properties $\mathbf{P}$ in this paper are micro parameters described by Equation 3,

$$\mathbf{P} = \{\text{Particle properties}\} = \{D_{\max}/D_{\min},\ D_{\min},\ shape,\ \rho_s,\ k_n/k_s,\ k_n,\ h,\ \mu,\ k_b,\ s_b,\ r_b\} \quad (3)$$

where ρ_s is the particle density and h is the damping factor, which is the ratio of c_n and c_s to critical damping coefficients in the normal and shear directions, respectively.

$$h = \frac{c_n}{2\sqrt{mk_n}} = \frac{c_s}{2\sqrt{mk_s}} \quad (4)$$

where m is particle mass.

Since effective stress must transmit only on particulate structures, a set of the stress states $\mathbf{S}$ as shown by Equation 5, could be determined by $\mathbf{K}$ using a function of G_1 in Equation 6.

$$\mathbf{S} = \{\text{Effective stresses}\} \quad (5)$$

$$\mathbf{S} = \{G_1\,(\mathbf{P};\ \mathbf{K})\} \quad (6)$$

Using Equations 2, 3, and 6, Equation 1 can be simplified such as Equation 7,

$$\mathbf{M} = F_1\{\mathbf{P}, \mathbf{K}, \mathbf{S}\} = F_1\{\mathbf{P}, \mathbf{K}, G_1(\mathbf{P};\mathbf{K})\} = F_2(\mathbf{P}, \mathbf{K}) \quad (7)$$

The mechanical properties are shown by a function of $\mathbf{P}$ and $\mathbf{K}$. Once the packing, fabric, and stress conditions are known, or simply assumed, the mechanical properties can be estimated according to the particle properties using the function F_2. The opposite is also possible using the inverse function F_2^{-1}.

$$\mathbf{M} = F_2\,(\mathbf{P}), \quad \mathbf{P} = F_2^{-1}\,(\mathbf{M}) \quad (8)$$

For example, considering one dimensional waves in a one dimensional DEM system with p-wave and s-wave velocities, V_p and V_s, k_n and k_s can be estimated using Equations 9 and 10.

$$k_n = \frac{1}{4}\pi\rho V_p^2, \quad k_s = \frac{1}{4}\pi\rho V_s^2 \quad (9)$$

$$V_p/V_s = \sqrt{2(1-\upsilon)/(1-2\upsilon)} \quad (10)$$

where ρ is the assembly density and υ is Poisson's ratio, which is usually around 1/3: therefore, $V_p/V_s = 2$ and $k_n/k_s = 4$. In this paper, $D_{\max}/D_{\min} = 2$ and $k_n = 5 \times 10^8$ were fixed. The diameter $D_{\min}$ was set to be 0.50 m considered to be the most frequent diameter of rock observed in situ. The representative value of friction coefficient was $\mu = 0.466$: friction angle $\phi_\mu = 25$ deg. According to the prior calculation, the kinematic behaviors of falling rock were not sensitive for k_n if it was higher than 5×10^7 N/m. In order to save computing time, k_n of 5×10^8 N/m was employed.

2.3 Strengths with different particle properties

The macro failure-strengths (peak strength) for bonded and non-bonded particle assemblies are discussed in this section.

For bonded particle assembly, the micro parameters can be determined to be suitable for macro behaviors, such as diametric loading in a Brazil cylinder test for tensile strength, shown in Figure 3, and an unconfined compression test shown in Figure 4, where the densest packing and random fabric specimen was prepared. Figure 5 shows a stress-strain relationship in the latter test. In Figure 6, the relation between q_u and s_b is summarized with different bond strengths s_b in the same packing and fabric. The peak strength q_u increases with s_b, and a strong correlation can be seen, represented by a power function; if q_u is required to be 2 MPa, the value of s_b only has to be adjusted to 10 MPa (refer to Figure 6). However, rocks usually show size effect in peak strength: the strength decreases as the specimen size increases because the existing probability of a larger defect is said to become higher since the specimen size is larger (Bieniawski 1981, Einstein et al. 1969). Experimental laws and some theories support that the strength reduces to about 30% of the strength of the smallest specimen: this effect must be considered according to the reduction of s_b.

For no-bonded particle assembly (cohesion-less material), focus is placed on the internal friction angle ϕ_f representing the macro strength index. It has been found, based on experimental results, that ϕ_f in granular material is influenced especially by surface roughness, grain shape and crushability, and

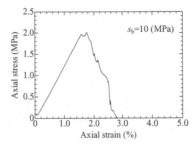

Figure 5.   Stress-strain behaviors of unconfined compression tests analyzed.

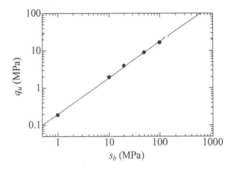

Figure 6.   Increments in unconfined compression strength with increasing bond strength.

density and confining pressure (Miura et al. 1997 & 1998; Maeda and Miura 1997); grain shape has the duality of the unstable contact condition and the mobilization of particle rotation resistance. The particle element was assumed, in this paper, to be not crushed, although falling rock and slope surface consisted of particles were allowed for breakage. It has, moreover, usually been said that ϕ_μ influenced ϕ_f strongly: $\phi_f = \phi_\mu + \upsilon_{df}$, where υ_{df} is the dilatancy angle at failure and not negative. For natural slopes and artificial fills, it was often necessary to input a cohesion-less slope with an inclination angle of over 30°. into the computer; the particle parameters, such as ϕ_μ and rotation resistance had be regulated to obtain $\phi_f > 30°$.

Bi-axial compression tests were conducted, employing both circular particles (Figure 7a) and non-circular particles. Circular (Figure 7a) and non-circular (Figures 7b-e) particle elements were considered. For example, the non-circular particles shown in Figure 7b were prepared by connecting three circular particles of the same radii; these particles were clumped. The names of circular and non-circular particles were denoted by the index "cl" added to the number of connecting particles. Simulations were performed for a material composed entirely of circular particles, and then all circular particles were replaced by non-circular particles.

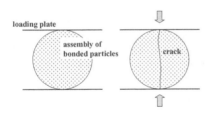

Figure 3.   Brazil cylinder test for tensile strength.

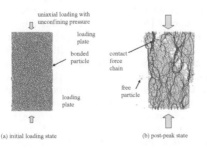

Figure 4.   Unconfined compression test analysis for rock specimen.

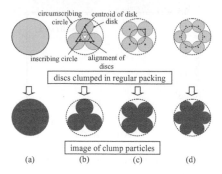

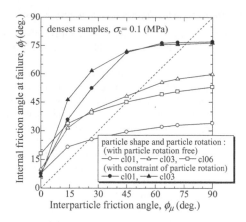

Figure 7. Particles types used in the DEM: (a) c101 (circle), (b) c103 (triangle), (c) c104 (square), (d) c106 (hexagon), where the broken line circumscribes a circle.

Figure 8. Macro internal friction angle at failure ϕ_f with different inter-particle friction angles ϕ_μ in the cases of particle rotation free and perfect particle rotation constraint.

Figure 8 shows the analyzed internal friction angle ϕ_f with a different interparticle friction angle ϕ_μ ($= \tan^{-1}\mu$) for a circular particle and non-circular particle sample, where void ratios for each sample at the initial state of shearing are the same, even with a different ϕ_μ (Maeda & Hirabayashi 2006). Furthermore, ϕ_f is plotted with perfect constraint of all particle rotations for each sample; in these cases, the extreme particle rotation resistance is mobilized and the relative displacements at the interparticle are only due to sliding.

In the all cases, ϕ_f is influenced remarkably by ϕ_μ. Although ϕ_f increases with ϕ_μ for the circular particle sample, the maximum value is around 30°. Otherwise, a high $\phi_f = 30$–40° can be obtained for non-circular samples with $\phi_\mu = 15$–25° due to particle rotation resistance by the concavity and convexity of particle, even if $\phi_\mu = 0$, ϕ_f could be induced to at least 10°, and when $\phi_\mu < 30°$, ϕ_f is greater than ϕ_μ. However, in the other range of ϕ_μ, ϕ_f is less than ϕ_μ. Besides, ϕ_f converges to a limit value when ϕ_μ exceeds 20–30°, even when ϕ_μ is close to 90°.

On the other hand, the converged limit values in the case with particle rotation constraint are higher than those without the constraint. These results reveal the failure mechanism of granular media. Even circular particles with $\phi_\mu = 15$–25° can show high ϕ_f over 40° with rotation resistance. This numerical technique might be a trick, but it is useful for control of strength with the help of non-circular, saving computing time. It is important that a circular particle model involving rotation resistance is developed.

2.4 Modeling of rock slope in-situ

The slope and the falling rock were modeled and analyzed as an assembly of particles bonded as shown in Figure 9.

First, particles were deposited in the outline domain of a slope under gravity of 9.8 m/s². Here, the surface

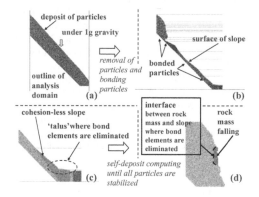

Figure 9. Calculation process for constructing a slope and falling rock mass: construction and set up (a) particle deposit in slope outline, (b) slope surface line, (c) talus zone, (d) falling rock mass.

layer thickness of the slope was made only of a few particles because the number of the particles analyzed was reduced. Second, all particles generated were bonded using the bond element illustrated in Figure 2, and then the particles outside the slope were removed along the surface lines. Since the particles inside the slope had been released from overburden pressure due to the removal and the slope swelled, self-weight analysis continued until the swelling ceased and the slope stabilized. Athough this process takes a great deal of computing time; the initial stress conditions are set inside the slope. Then, the particles run off from the slope line were removed. Thirdly, in the talus zone, bond elements were eliminated and the particle rotation was constrained to make the talus slope stable. Finally, the bond elements around the interface zone between the falling rock mass and the slope were eliminated.

The rock mass fell after all of these processes. In this paper, in-situ slope data is provided.

3 ANALYIS RESULTS AND DISCUSSIONS

3.1 Breakage effect

Figures 10(a–c) show behaviors of falling rock with different bond strengths. The breakage makes the motion of the pieces more complex due to the rotation but reduces the energies; many small pieces are scattered. The difference in colors indicates the difference in materials.

3.2 Talus effect Pocket effect and granular mat effect

Figures 11(a–c) show the influence of the presence of talus on the behaviors of falling rocks without breakage. The falling rock digs into the talus, and rolls, scooping it out.

Figure 12 shows the influence of the presence of a pocket with a granular mat on the falling rock, where the assemblies of sand or gravel particles are usually paved.

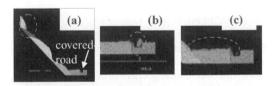

Figure 10. Falling rock behaviors with different bond strengths s_b: (a) initial state; (b) non-breakage: rock hitting road; (c) breakage: small pieces scattering.

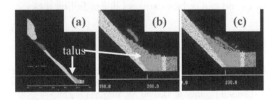

Figure 11. Falling rock behaviors with talus: (a) initial state; (b) just before collision; (c) rock into talus after collision.

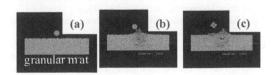

Figure 12. Falling rock behaviors with pocket: (a) before collision to pocket; (b) circular rock ball; (c) non-circular ball cl04 with rotation $\omega = 25$ rad/s.

3.3 Energy absorption and restitution performances

The falling velocity v of the largest piece of the rock was examined at the position of the road or protective barrier, such as is shown in Figure 10. Here breakage, talus, and pocket-granular mat effects were considered. Figure 13 shows the energy ratio E_k/E_i, where E_i is initial potential energy. A beneficial countermeasure can be proposed taking these effects into consideration.

Here, the energy absorption (restitution) behavior in an oblique and rotational collision of the falling mass with a slope with talus and granular mat, are examined. As shown in Figures 11 and 12, the energy absorption behaviors described as a set of **R** are controlled by a function H_1 of dynamic bearing behaviors **B** of a granular mat under low confining stress. In addition, **B** is determined not only by the strength in **M** of the mat, but also by contact conditions **C**; inclined and/or eccentric loading. The set **C** includes the surface inclination angle of the slope, the talus, the mat α, the oblique collision relative angle β, the contact area A_c, the inclination angle θ, the contact eccentricity η, the translational velocity vector $\mathbf{v} = \{v_x, v_y\}^{-1}$, and the rotary velocity ω, where the angle is determined to be positive in a counter-clockwise direction from the x-axis (horizontal axis).

$$\mathbf{R} = \{energy\ absorption;\ restitution\} = H_1\{\mathbf{B}\} \quad (11)$$

$$\begin{aligned}\mathbf{B} &= \{bearing\ properties\} = H_2\ (\mathbf{M};\ \mathbf{C}) \\ &= H_2\ (F_2\ (\mathbf{P},\ \mathbf{K})\ ;\ \mathbf{C})\end{aligned} \quad (12)$$

$$\begin{aligned}\mathbf{C} &= \{contact\ condition\} = \\ &\{\alpha,\ \beta,\ A_c,\ \theta,\ \eta,\ \mathbf{v},\ \omega\}\end{aligned} \quad (13)$$

The restitution behaviors of the rock ball were analyzed in relation to the wall element with the same friction as μ and to the granular mat, with initial rotational velocity ω_0, as shown in Figure 14. Figure 15 shows the influence of damping factor h and the restitution coefficient e_y in y-axis: $\alpha = 0°$ and $\beta = 90°$. Here e_i ($i = x, y$) is determined by the ratio of the post-collision velocity v_i' to the pre-collision velocity v_i'. For the case of rock ball-wall, e_y decreases with h: the energy absorption is determined only by vibration characteristics at contact, as in Figure 1, even with the rotation of the rock mass. On the other hand, e_y is low, around 0.15–0.20, for ball-bonded assembly and is 0 for ball-non bonded assembly because the mass is dug into the assembly.

Figure 16 shows the influence of friction on the energy ratio of translational kinetic energy $1/2 \times mv_x^2$ and rotational kinetic energy $1/2 \times I\omega^2$ after collision with both energies before collision, where m and I are the mass and the moment of inertia of the ball, respectively. The total energy is reduced by friction or rotation, and negative rotation remarkably loses $1/2 \times mv_x^2$. As the shear velocity at contact is

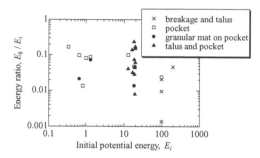

Figure 13. Energy absorption for the largest piece of falling rock with breakage, talus and pocket with granular mat.

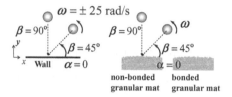

Figure 14. Analysis cases for oblique and rotational collision.

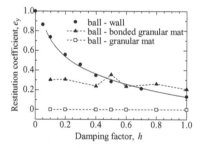

Figure 15. Influence of damping factor on normal restitution coefficient for wall, bonded, and non-bonded assemblies.

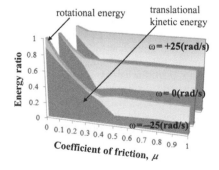

Figure 16. Change in translational and rotation energies in the shear direction for oblique collision with rotation for rock ball–wall.

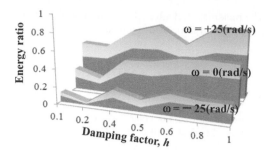

Figure 17. Change in translational and rotation energies in the shear direction for oblique collision with rotation for rock ball—granular mat.

$v_s = du_s/dt = v_x - D/2 \cdot \omega$, the shear force acting on the ball is determined not only by translational motion, but also by rotation, and the ball is accelerated for $v_x \cdot \omega < 0$ due to rotation and is decelerate for $v_x \cdot \omega > 0$. Louge and Adams (2002) also pointed out that rotation influenced strongly on kinematic restitution in oblique impacts. The same tendency was observed in the case of the ball-granular mat, as shown in Figure 17. These imply that it is necessary to focus on the rotation to understand falling rock behaviors and that it is possible to reduce the energy by controlling the rotation.

4 CONCLUSIONS

On the basis of analysis results, we clarified that DEM, along with accounting for the effects of particle properties on the mechanical properties of granular material, can simulate shape and breakage effects of falling rock, and the effect of the presence of talus and pockets on falling rock behaviors. Evaluating effects, which have not yet been considered in present designs, can assist in the development of beneficial countermeasures and can reduce damages due to falling rocks.

REFERENCES

Bieniawski, Z.T. 1981. Improved design of coal pillars for US mining conditions. *Proc of 1st Annual Conference on Ground Control in Mining,* W. Virgia: 12–22.
Cundall, P.A. & Strack, O.D.L. 1979. A Discrete Model for Granular Assemblies. *Geotechnique* 29(1): 47–65.
Einstein, H.M., Nelson, R.A. & Bruhn, R.W. 1969. Model studies of jointed rock behaviour. *Proc. 11th Symposium on Rock Mechanics,* Berkeley: 83–103.
Japan Road Association. 2000. *Manual on counter-measure for falling rock,* Maruzen.
Louge, M. & Adams, M.E. 2002. Anomalous behavior of normal kinematic restitution in the oblique impacts of a hard sphere on an elastoplastic plate. Physics Review 65(2): 021303-1-021303-6.
Maeda, K. & Hirabayashi, H. 2006. Influence of grain properties on macro mechanical behaviors of granular media by DEM. *Journal of Applied Mechanics, JSCE* 9: 623–630.

Maeda, K. & Miura, K., 1999. Confining stress dependency of mechanical properties of sands. *Soils and Foundations* 39(1): 53–68.

Mikasa, M. 1963. Soil strength and stability analysis. *Lecturetext on soil engineering for foundations, Kansai-branch, JGS*: 11–29 (in Japanese).

Miura, K., Maeda, K., Furukawa, M. & Toki, S. 1997. Physical characteristics of sands with different primary properties. *Soils and Foundations* 37(3): 53–64.

Miura, K., Maeda, K., Furukawa, M. & Toki, S. 1998. Mechanical characteristics of sands with different primary properties, *Soils and Foundations* 38(4): 159–172.

Shi, G.H. & Goodman, R.E. 1984. Discontinuous deformation analysis. *Proc. of 25th U.S. Symposium on rock Mechanics*: 269–277.

Modification of compressible smooth particles hydrodynamics for angular momentum in simulation of impulsive wave problems

S. Mansour-Rezaei & B. Ataie-Ashtiani
Sharif University of Technology, Tehran, Iran

ABSTRACT: In this work, Compressible Smooth Particles Hydrodynamics (C-SPH) is applied for numerical simulation of impulsive wave. Properties of linear and angular momentum in SPH formulation are studied. Kernel gradient of viscous term in momentum equation is corrected to ensure preservation of angular momentum. Corrected SPH method is used to simulate solitary Scott Russell wave and applied to simulate impulsive wave generated by two-dimensional under water landslide. In each of test cases, results of corrected SPH are compared with experimental results. The results of the numerical simulations and experimental works are matched and a satisfactory agreement is observed. Furthermore, vorticity counters computed by the corrected SPH are compared with uncorrected SPH so that the effect of preservation of angular momentum is illustrated. Comparison between experimental and computational results proves applicability of The C-SPH method for simulation of these kinds of problems.

1 INTRODUCTION

Impulsive waves generated by landslides and slamming on horizontal cylindrical members of a jacket are examples of phenomena which cause damage on structures and endanger human life. Importance of wave generated by land slide was a motivation for a large number of analytical, experimental and numerical studies. The generated impulsive waves due to submerge landslides has been extensively studied by numerical and experimental works (Ataie-Ashtiani & Najafi-Jilani 2007, Ataie-Ashtiani & Najafi-Jilani 2006, Ataie-Ashtiani & Najafi Jilani 2008, Najafi Jilani & Ataie-Ashtiani 2008). Ataie-Ashtiani & Nik-khah (2008) studied experimentally the impulsive waves generated by sub-aerial land slide. Ataie-Ashtiani & Malek-Mohammadi (2007) evaluated the accuracy of empirical equations to estimate generated wave amplitude in the near field.

Recently, numerical methods that do not use a grid have been used in order simulation and prediction of damage in these sophisticated hydrodynamic problems. One of these modern methods is Smoothed Particle Hydrodynamics (SPH). SPH was introduced for astrophysical applications (Lucy 1977), but rapidly extended and used in engineering fields, including free surface flows and wave prorogation (Monaghan 1994, Monaghan & Kos 1999). For example, waves overtopping offshore platform deck, the breaking of a wave on a beach and bore in-a-box problem has been simulated using compressible SPH method (Dalrymple & Rogers 2006). Gómez & Dalrymple (2004) used three dimensional SPH, and modeled the

impact of single wave to the tall structure which was located in a region with vertical boundaries. Velocity and forces computed from numerical model are compared with laboratory measurement. The results showed that SPH method can successfully be used to simulate wave problems. Oger et al. (2006) proposed new formulation of the equation based on a SPH for spatially varying resolution (variable smoothing length). They simulated wedge water entry problem by using a new method to evaluate fluid pressure on solid boundaries. Ataie-Ashtiani & Shobeyri (2008) presented an incompressible SPH (I-SPH) formulation to simulate impulsive waves generated by landslides. A new form of source term to the Poisson equation was employed, and the stability and accuracy of SPH method improved. Moreover, I-SPH method was used to simulate Dam-break flow, evolution of an elliptic water bubble, solitary wave breaking on a mild slope and run-up of non-breaking waves on steep slopes (Ataie-Ashtiani et al. 2008). Khayyer et al. (2008) corrected I-SPH method based on correction introduced by Bonet & Lok (1999), and achieved enhanced accuracy in modeling of the water surface during wave breaking and post-breaking.

A general model based on Smoothed Particle Hydrodynamics, "SPHysics", was developed jointly by researchers at the Johns Hopkins University (U.S.A.), the University of Vigo (Spain), the University of Manchester (U.K.) and the University of Rome La Sapienza (Italy) to promote the development and use of SPH within the academic and industrial communities (SPHysics user guide 2007). SPHysics is provided for two and three-dimensional simulation of dam break in a box, dam break evolution over a wet bottom, waves generated by a paddle in a beach, tsunami generated by a sliding wedge and dam-break interaction with a structure (SPHysics user guide 2007).

SPHysics model is used in this work and it is modified and improved in two aspects. First, boundary conditions in SPHysics are modified to simulate three problems about impulsive wave and impact on the water surface. Second, correction technique, which was introduced by Bonet & Lok (1999), is applied for the purpose that the accuracy of the SPH model is enhanced through preservation of angular momentum. The corrected model is used to simulate solitary wave generated by Scott Russell, two-dimensional under water landslide.

SPH formulations preserve linear momentum, but they do not usually preserve angular momentum, which plays vital role in the case of violence free surface flows (Khayyer et al. 2008). The main objective of this work is to improve precision of compressible SPH method through preservation of angular momentum for simulation of impulsive wave problems.

2 SPH FORMULATION FOR COMPRESSIBLE FLUID FLOW

2.1 Basic SPH theory

SPH is an interpolation method which allows any function to be expressed in the terms of its values at a set of disordered particles (Monaghan 1992). The fundamental principle is to approximate any function $A(\vec{r})$ by:

$$A(\vec{r}) = \int A(\vec{r}')w(\vec{r} - \vec{r}', h)d\vec{r}' \qquad (1)$$

where h is called the smoothing length and $w(\vec{r} - \vec{r}', h)$ is the weighting function or kernel. For numerical work, the integral interpolant is approximated by summation interpolant (Monaghan 1992):

$$A(\vec{r}) = \sum_b m_b \frac{A_b}{\rho_b} w_{ab} \qquad (2)$$

where the summation is over all the neighboring particles. ρ_b and m_b are mass and density, respectively, $w_{ab} = w(\vec{r}_a - \vec{r}'_b, h)$ is weighting or kernel function which is similar to the delta function.

2.2 Governing equations

Governing equations of viscous fluid which are mass and momentum conservation equations are presented following (Monaghan 1992):

$$\frac{D\vec{v}}{Dt} = -\frac{1}{\rho}\vec{\nabla}P + \vec{g} + \vec{\Theta} \qquad (3)$$

$$\frac{1}{\rho}\frac{D\rho}{Dt} + \nabla.\vec{v} = 0 \qquad (4)$$

where ρ is density, $\vec{v}$ is the velocity vector, P is the pressure, $\vec{g}$ is acceleration due to gravity and $\vec{\Theta}$ refers to the diffusion terms.

Conventional SPH notation of mass and momentum equations are that artificial viscosity has been used due to its simplicity (Monaghan 1992):

$$\frac{d\vec{v}_a}{dt} = -\sum_b m_b \left(\frac{P_b}{\rho_b^2} + \frac{P_a}{\rho_a^2} + \Pi_{ab}\right)\vec{\nabla}_a w_{ab} + \vec{g} \qquad (5)$$

$$\frac{d\rho_a}{dt} = \sum_b m_b \vec{v}_{ab}\vec{\nabla}_a w_{ab} \qquad (6)$$

In the above equations, $\vec{\nabla}_a \vec{w}_{ab}$ is gradient of the kernel with respect to the position of particle a. P_k and ρ_k are pressure and density of particle k(evaluate a a or b), m_b is mass of particle b. Π_{ab} represents the effects of viscosity (Monaghan 1992).

In this work, we used quadratic kernel function from multifarious possible kernel in SPHysics. Ataie-Ashtiani & Jalali-Farahani (2007) displayed that in simulation of impulsive waves, quadratic kernel is efficient and accurate:

$$w_{ab} = \alpha_N(q^2/4 - q - 1) \qquad 0 \le q \le 2 \qquad (7)$$

where $q = r_{ij}/h$, $\alpha_N = \frac{3}{2\pi h^2}$, $r_{ij} = |\vec{r}_i - \vec{r}_j|$ and the coefficient h is smoothing length.

2.3 Equation of state

The equation of state allows us to avoid an expensive resolution of an equation such as the Poisson's equation, but it inverts any incompressible fluid to the weakly compressible (SPHysics user guide 2007). The equation of state relates the pressure in the fluid to the local density:

$$P = B\left[\left(\frac{\rho}{\rho_0}\right)^{\gamma} - 1\right] \qquad (8)$$

where $\gamma = 7$, $B = C_0^2\rho_0/\gamma$ and C_0 is the speed of sound at reference density ($\rho_0 = 1000$ kg/m^3). We cannot use correct sound speed because we should determine the speed of sound such low that time stepping remains reasonable. On the other hands, the speed of sound should be about ten times faster that the maximum fluid velocity in order keeping changes in fluid density less than 1% (Dalrymple & Rogers 2006).

3 VISCOSITY

Artificial viscosity which originally was used in the equation of motion has few advantages and disadvantages. First of all, in free surface problems, it plays the role of stabilizer in numerical scheme. Second, Artificial viscosity prevents the particle from inter-penetrating (Dalrymple & Rogers 2006). Then, it preserves both linear and angular momentum and has acceptable manner in the case of rigid body rotations (Monaghan 1992). In contrast, the artificial viscosity has some disadvantages. It is scalar viscosity which cannot take the flow directionally into account (Khayyer et al. 2008), and it causes strong dissipation and affects shears in the fluid (Dalrymple & Rogers 2006); thus, researchers prefer to simulate viscosity in realistic manner.

One realistic expressions of viscosity is Laminar viscosity (Morris et al. 1997) and Sub-Particle Scale (SPS) technique to modeling turbulence. Researchers used SPS approach to modeling turbulence in some kind of particle methods such as MPS (Gotoh et al. 2001) and Incompressible SPH (Lo & Shao, 2002). Recently, Dalrymple & Rogers (2006) implemented this expression of viscosity in compressible SPH method. This facility is provided in SPHysics code too, and we used Laminar Viscosity and Sub-Particle Scale (SPS) Turbulence in this paper.

Implementing SPS approach in diffusion term of momentum equation (Equation 3) give:

$$\frac{d\vec{V}}{dt} = -\frac{1}{\rho}\vec{\nabla}P + \vec{g} + v_0\nabla^2\vec{V} + \frac{1}{\rho}\vec{\nabla}\tau \qquad (9)$$

where $v_0\nabla^2\vec{V}$ represents laminar viscosity term, and τ represents SPS stress tensor, which was modeled by eddy viscosity assumption very often (SPHysics user guide 2007). Dalrymple & Rogers (2006) wrote momentum equation (Equation 10) in SPH notation using laminar viscosity and SPS:

$$\frac{d\vec{V}_a}{dt} = -\sum_b m_b\left(\frac{P_b}{\rho_b^2} + \frac{P_a}{\rho_a^2}\right)\vec{\nabla}_a w_{ab} + \vec{g}$$

$$\times \sum_b m_b\left(\frac{4v_0\vec{r}_{ab}\vec{v}_{ab}}{|\vec{r}_{ab}|^2(\rho_a + \rho_b)}\right)\vec{\nabla}_a w_{ab}$$

$$+ \sum_b m_b\left(\frac{\tau_b}{\rho_b^2} + \frac{\tau_a}{\rho_a^2}\right)\vec{\nabla}_a w_{ab} \qquad (10)$$

where v_0 is the kinetic viscosity of laminar flow ($10^{-6}\text{m}^2/\text{s}$).

4 PRESERVATION OF ANGULAR MOMENTUM

In the absence of external forces, the motion of particles must be such that the total linear and angular momentum is preserved. In this section we explain that SPH formulations inherently preserve linear momentum, but generally cannot preserve angular momentum.

In the absence of external forces, combination time derivation of total linear momentum with Newton's second law gives the condition for preservation of linear momentum (Bonet & Lok 1999, Khayyer et al. 2008):

$$\sum_{i=1}^{N} A_i = 0 \qquad (11)$$

where A_i denotes the total internal force acting on particle i, and N is total number of particles.

Interaction forces between pairs of particles have two components; one part is related to the pressure gradient and another part is related to the viscose term. Khayyer et al. (2008) wrote kernel gradient as a function of r_{ij}. They showed that interaction forces between pairs of particles (due to pressure and viscosity) are exactly equal and opposite. This means that their influences are vanished, and linear momentum is preserved by SPH formulations.

Similarly, in absence of external force, combination time derivation of total angular momentum with equilibrium equation gives the condition for preservation of angular momentum. Rate of change of total angular momentum will be zero (angular momentum will preserve), provided that internal force between pairs of particles are collinear with vector of r_{ij} (Bonet & Lok 1999, Khayyer et al. 2008). Because of the fact that pressure stress tensor is isotropic, internal force due to pressure term is collinear with vector of r_{ij} (Khayyer et al. 2008). Therefore, angular momentum, which is produced by pressure force, equals zero precisely. The same is not true about internal force due to viscosity (when we use one realistic expression of viscosity) because viscous stress tensor is anisotropic, and internal viscous force which is not collinear with r_{ij} produces a momentum.

To sum up, pressure gradient in momentum equation preserves both linear and angular momentums. But, viscosity term, which is described by laminar viscosity and Sub-Particle Scale (SPS) Turbulence, can not preserve angular momentum in spite of the fact that it is good expression of realistic viscosity.

4.1 Corrective term

In previous sections, we mentioned that artificial viscosity has considerable disadvantages. Thus, applications of realistic viscosity such as laminar viscosity and Sub-Particle Scale (SPS) Turbulence are increasing. Also, we explain that viscous term in momentum equation cannot preserve angular momentum. As a result, it seems that preservation of angular momentum will help to obtain more accurate results.

Researchers have used a number of correction techniques in the hope that accuracy of the SPH method enhances through preservation of angular momentum. Some researchers correct kernel functions, while other correct gradients of kernel functions. In this work similar to the Khayyer et al. (2008), we apply correction of kernel gradients (L_i) introduced by Bonet & Lok (1999) due to its simplicity:

$$\overline{\nabla_i W_{ij}} = L_i \nabla_i W_{ij}$$

$$L_i = \left(\sum \frac{m_j}{\rho_j} \nabla_i W_{ij} \otimes (r_j - r_i) \right)^{-1} \quad (12)$$

where $\overline{\nabla_i w_{ij}}$ is corrected kernel gradient. On the ground that pressure gradient in momentum equation preserves both linear and angular momentum, this correction is only used during the calculation of viscous term in momentum equation. Using this correction technique will guarantee that the gradient of any linear velocity properly evaluated. In addition, both pressure gradient term and viscosity term preserve angular momentum. (Bonet & Lok 1999).

5 TEST CASES

In this section the results of numerical simulation for two examples of impulsive waves are given. In each of test cases, first, results of corrected SPH are compared with experimental data, and then vorticity contours of corrected and uncorrected method are compared. Comparisons show that preservation of angular momentum affects shape and intensity of vorticity.

In numerical simulations, Prediction-correction algorithm, Dalrymple boundary condition and Shephard filtering techniques are employed. These techniques are described in details in following papers (Monoghan 1989, Crespo et al 2007, Dalrymple & Knio 2000, SPHysics user guide 2007).

5.1 Scott Russell wave generator

Researchers frequently use the Scott Russell wave generator to simulate falling avalanche in dam reservoirs and to assess the behavior of waves generated by landslides near slopes. Monaghan & Kos (2000) evaluated this problem both experimentally and numerically (using SPH method). Besides, Ataie-Ashtiani & Shobeyri (2008) numerically simulated experimental data of Monaghan and Kos by employing a new form of source term to the Poisson equation.

In this section, solitary waves generated by a heavy box falling vertically into the water are considered. Also, the effect of preservation of angular momentum in generation of impulsive waves and its influence to obtain more accurate result are evaluated. Boundary conditions in SPHysics code, including shapeof tank and sliding wedge are changed according to the Monaghan & Kos (2000) experiments configuration. The experiment involved a weighted box (0.3 m × 0.4 m)

dropping vertically into a wave tank while steel water depth is 0.21 m. It should be noted that the horizontal length of the numerical tank is assumed to be 2 m, which is much shorter than the experimental tank (9 m). However, numerical results show that this assumption is not affected results even in maximum time of simulation. The bottom of the box was initially placed 0.5 cm below the water surface in the experiment to avoid splashing. Vertical velocity of the box is computed by (Monaghan & Kos 2000):

$$\frac{V}{\sqrt{gD}} = 1.03 \frac{Y}{D} \left(1 - \frac{Y}{D} \right)^{0.5} \quad (13)$$

where D is the depth of the water, Y is the height of the bottom of the box above the bottom of the tank at time t, g is the acceleration of gravity and V is the falling vertical velocity of the box at time t. Analytical shape of solitary wave generated by falling box calculated by (Lo & Shao 2002):

$$H(x, t) = a \times \sec h^2 \left[\sqrt{\frac{3a}{4d^3}} (x - ct) \right] \quad (14)$$

where H is the water surface elevation, a is wave amplitude,d is water depth and $c = \sqrt{g(d + a)}$ is solitary wave velocity.

In Figure 1 particles configurations display solitary wave formation when SPH method preserved angular momentum. Figure 2 illustrates that analytical profile of solitary wave (Equation 15) are in good agreement with numerical result.

Figure 3 judges against vorticity contour in corrected and uncorrected method at $t = 0.7$s. These pictures show that how preservation of angular momentum changes vorticity patterns in fluid motion.

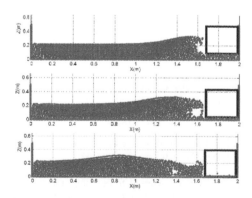

Figure 1. Particle configuration in Scott Russell wave generator computed by corrected SPH at $t = 0.28$, $t = 0.42$ and $t = 0.7$s, respectively, $l_0 = 0.015$ cm.

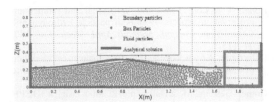

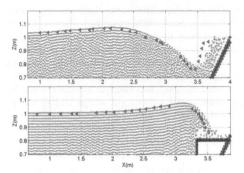

Figure 2. Comparison between analytical solution and numerical result in Scott Russell problem, $t = 0.7$s, $l0 = 0.015$ cm.

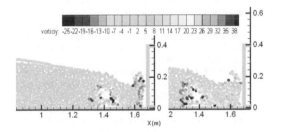

Figure 3. Vorticity contour computed by corrected (Right) and uncorrected (left) SPH at $t = 0.7$.

5.2 Under water rigid landslide

In this section, the corrected SPH is used to simulate wave generated by two-dimensional under water landslide, based on a laboratory experiment performed by Heinrich (1992). The experiment includes freely slide down rectangular wedge (0.5 m × 0.5 m) on the plane which is inclined 45° on the horizontal. The water depth in the tank is 1 m, and top of the wedge is initially 1 cm bellow the water surface. The flat length of tank in computational domain is 3 m, and particles size is 0.02 m. The position of the wedge in each time step is estimated by computing vertical velocity of wedge by (Grilli & Watts 1999):

$$\begin{cases} u(t) = c_1 \tan h(c_2 t) & t \leq 0.4s \\ u(t) = 0.6 & t > 0.4s \end{cases} \quad (15)$$

Where c_1 and c_2 are constant values that in our computations are 86 and 0.0175, respectively. In Figure 4, particle configuration due to sliding of rigid wedge at $t = 1$s and $t = 0.5$s is presented, and results of corrected SPH and experimental wave profile is compared.

Figure 5 contrasts vorticity contour in corrected and uncorrected algorithm. As shown in Figure 5a, vorticity in corrected algorithm is more localize rather than vorticity in uncorrected SPH, which is shown in Figure 5b. Furthermore, the tail of vortex above the wedge in Figure 5a represents preservation of the angular momentum and shedding of them. Although differences between water surface profile of corrected

Figure 4. Comparison between experimental and the simulating wave profile for under water rigid landslide at $t = 0.5$s (a) and $t = 1$s (b).

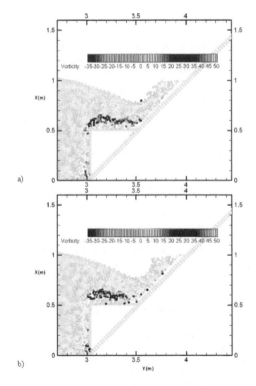

Figure 5. Vorticity contour in under water rigid landslide computed by corrected (a) and uncorrected (b) SPH at $t = 1$s.

and uncorrected SPH are not significant, the results of corrected SPH are closer to the experimental measurements in comparison with uncorrected SPH.

6 CONCLUSION

In this paper, corrected compressible smoothed particles hydrodynamics (SPH) was used for numerical simulation of impulsive waves. Laminar viscosity and Sub-Particle Scale turbulence was employed to simulate viscosity in impulsive wave problems. Kernel gradient of viscous term in momentum equation was corrected in order preservation of angular momentum and improving precision of compressible SPH method. Corrected method was used to simulate Scott Russell wave, under water rigid wedge sliding along an inclined surface. The computational C-SPH results were in good agreement with the experimental data, which are showing the ability of the mesh-less methods to successfully simulate such kind of complex problems. In simulation of impulsive waves, the corrective term changed vorticity patterns and caused smoother water surface; however, this correction technique had not significant role to change water surface elevation.

REFERENCES

Ataie-Ashtiani, B. & Najafi-Jilani, A. 2006. Prediction of submerged landslide generated waves in dam reservoirs: an applied approach. *Dam Engineering* XVII(3): 135–155.

Ataie-Ashtiani, B. & Malek-Mohammadi, S. 2007. Near field amplitude of sub-aerial landslide generated waves in dam reservoirs. *Dam Engineering* XVII(4): 197–222.

Ataie-Ashtiani, B. & Jalali-Farahani, R. 2007. Improvement and Application of I-SPH Method in the Simulation of Impulsive Waves. *Proceedings of the International Conference on Violent Flows*: 116–121, Fukuoka, Japan.

Ataie-Ashtiani, B. & Najafi-Jilani, A. 2007. A higher-order Boussinesq-type model with moving bottom boundary: applications to submarine landslide tsunami waves. *International Journal for Numerical Methods in Fluids* 53(6): 1019–1048.

Ataie-Ashtiani, B. & Najafi-Jilani, A. 2008. Laboratory investigations on impulsive waves caused by underwater landslide. *Coastal Engineering* 55(12): 989–1004.

Ataie-Ashtiani, B. & Nik-khah, A. 2008. Impulsive Waves Caused by Subaerial Landslides. *Environmental Fluid Mechanic* 8(3): 263–280.

Ataie-Ashtiani, B. & Shobeiry, G. 2008. Numerical simulation of landslide impulsive waves by Modified Smooth Particle Hydrodynamics. *International Journal for Numerical Methods in Fluids* 56(2): 209–232.

Ataie-Ashtiani, B. et al. 2008. Modified Incompressible SPH method for simulating free surface problems. *Fluid Dynamics Research* 40(9): 637–661.

Bonet, J. & Lok, T.S. 1999. Variational and momentum preservation aspects of smooth particle hydrodynamic formulation. *Comput. Methods Appl. Mech. Eng* 180: 97–115.

Dalrymple, R.A. & Rogers, B.D. 2006. Numerical modeling of water waves with the SPH method. *Coastal Engineering* 53: 141–147.

Gómez-Gesteira, M. & Dalrymple, R. 2004. Using a 3D SPH method for wave impact on a tall structure. *Journal of Waterway, Port, Coastal and Ocean Engineering* 130(2): 63–69.

Grilli, S.T. & Watts, P. 1999. Modeling of waves generated by a moving submerged body. Applications to underwater landslides. *Journal of Engineering Analysis with Boundary Elements* 23: 645–656.

Gotoh, H. & Shibihara, T. & Sakai, T. 2001. Sub-particle-scale model for the MPS method—Lagrangian flow model for hydraulic engineering. *Computational Fluid Dynamics Journal* 9 (4): 339–347.

Heinrich, P. 1992. Nonlinear water waves generated by submarine and aerial landslides. *Journal of Waterways, Port, Coastal, and Ocean Engineering* 118(3): 249–266.

Khayyer, A. et al. 2008. Corrected Incompressible SPH method for accurate water-surface tracking in breaking waves. *Coastal Engineering* 55(3): 236–250.

Lucy, L.B. 1977. A numerical approach to the testing of the fission hypothesis. *Astron. J* 82: 1013–1024.

Lo, E. & Shao, S. 2002. Simulation of near-shore solitary wave mechanics by an incompressible SPH method. *Applied Ocean Research* 24: 275–286.

Monaghan, J.J. 1992. Smoothed particle hydrodynamics. *Annu. Rev. Astron. Astrophys* 30: 543–574.

Monaghan, J.J. 1994. Simulating free surface flows with SPH. *Journal Computational Physics* 110: 399–406.

Morris, J.P. et al. 1997. Modeling lower Reynolds number incompressible flows using SPH. *Journal Computational Physics* 136: 214–226.

Monaghan, J.J. & Kos, A. 1999. Solitary waves on a Cretan beach. *Journal of Waterway, Port, Coastal and Ocean Engineering* 125(3):145–154.

Monaghan, J.J. & Kos, A. 2000. Scott Russell's wave generator. Physics of Fluids 12: 622–630.

Najafi-Jilani, A. & Ataie-Ashtinai, B. 2008. Estimation of near field characteristics of tsunami generation by submarine landslide. *Ocean Engineering* 35(5–6): 545–557.

Oger, G. et al. 2006. Two-dimensional SPH simulations of wedge water entries. *Journal of Computational Physics* 213(2): 803–822.

SPHysics user guide, version 1.0.002, 2007.

Prediction and Simulation Methods for Geohazard Mitigation – Oka, Murakami & Kimoto (eds)
© 2009 Taylor & Francis Group, London, ISBN 978-0-415-80482-0

Viscosity effect on consolidation of poroelastic soil due to groundwater table depression

T.L. Tsai

National Chiayi University, Chiayi City, Taiwan

ABSTRACT: In this study, the viscosity effect on consolidation of poroelastic soil due to groundwater table depression is examined. A viscoelastic consolidation numerical model is developed to conduct this examination. By nondimensionalizing the governing equations the viscosity number that depends on hydraulic conductivity, viscous moduli, and thickness of soil is obtained to represent the viscosity effect on consolidation of poroelastic soil. The case of clay stratum sandwiched between sandy strata subjected to sudden and gradual groundwater table depressions is used to investigate the importance of viscosity effect to poroelastic consolidation. The results show that the displacement and pore water pressure of clay stratum are strongly related to the viscosity effect. The overestimation of soil displacement will occur if the viscosity effect is neglected. Hence, the viscosity effect needs to be considered in modeling consolidation of poroelastic soil under groundwater table depression.

1 INTRODUCTION

Groundwater is an important water resource, especially for arid or semiarid regions where surface water is highly variable. Due to the increase of water demand and the lack of proper management, land subsidence arising from groundwater overpumping had become a serious problem in many places around the world. From the view point of hydrogeology a soil stratum could be considered as the composition of alternating layers of highly porous sand (aquifers) and highly impervious clay (aquitards), which is usually called a multiaquifer system. Hence, how to achieve the control of land subsidence and the sustainable use of groundwater resource needs to analyze and predict soil consolidation due to groundwater withdrawal in a multiaquifer system. The three-dimensional poroelatic model in which the force balance equation and the continuity equation of fluid and solid are coupled to solve in sand and clay (Lewis & Schrefler 1978, Ng & Mei 1995) could be used for the analysis of soil deformation caused by groundwater withdrawal in a multiaquifer system. With considering the significant differences in permeability and compressibility between sand and clay, the two-step concept (Gambolati & Freeze 1973, Helm 1975, Gambolati et al. 1991, Onta & Gupta 1995, Larson et al. 2001, Tsai et al. 2006, Tseng et al. 2008) was proposed to simulate soil consolidation in a multiaquifer system. In the two-step concept, the variations of groundwater tables in aquifers need to be first obtained from in-situ measurement data or by simulating the groundwater flow in a multiaquifer system. With the known groundwater table variations in aquifers as boundary

conditions of the aquitard, one-dimensional Terzaghi consolidation theory (Terzaghi 1954) is then applied to compute vertical deformation of clay. Due to simplicity and efficiency, the two-step concept was widely used to analyze the soil consolidation due to groundwater withdrawal in a multiaquifer system, especially for practical application to regional land subsidence modeling.

Many researches indicated that the deformation behavior of porous media is related to the viscosity effect (Ehlers & Markert 2000, Xie et al. 2004, Hsu & Lu 2006). Guo (2000) investigated the influence of soil viscosity on pile installation. Hsieh (2006) developed a viscoelastic model for the analysis of dynamic response of soils under periodical surface water disturbance. Hsieh and Hsieh (2007) conducted the study of viscosity effect on the dynamic response of multilayered soils subjected to water wave and flow. However, due to the use of traditional Terzaghi consolidation theory the viscosity effect was not taken into account in modeling soil consolidation in the two-step concept. This could lead to inaccurate assessment of land subsidence. The purpose of this study is to examine the viscosity effect on consolidation of poroelastic soil due to groundwater table variation. A viscoelastic consolidation numerical model is developed herein to conduct this examination. In the following sections, the governing equations for consolidation of poroelastic soil with the consideration of viscosity effect are described first. The governing equations are then nondimensionalized and discretized. Finally, the viscosity effect on consolidation of poroelastic soil under groundwater table depression is investigated.

2 GOVERNING EQUATIONS

The flow equation in saturated porous media without the compressibility of fluid (Bear & Corapcioglu 1981) can be expressed as

$$\nabla \cdot (\overline{\overline{K}} \cdot \nabla P^e) = \rho_w g \frac{\partial}{\partial t} \nabla \cdot u \qquad (1)$$

where $\overline{\overline{K}}$ is hydraulic conductivity tensor. P^e denotes pore water pressure in terms of consolidation-producing incremental values. g is the gravitational acceleration. ρ_w represents ensities of fluid. u is solid displacement.

The equilibrium of forces in incremental state for linear viscoelastic material using Vigot model (Bardet 1992) can be written as

$$\frac{\partial}{\partial x_i}\left[(G+\lambda)\frac{\partial u_j}{\partial x_j}\right] + \frac{\partial}{\partial x_j}\left[G\frac{\partial u_i}{\partial x_j}\right]$$
$$+\frac{\partial}{\partial x_i}\left[(G+\lambda)\frac{\partial^2 u_j}{\partial t \partial x_j}\right] + \frac{\partial}{\partial x_j}\left[G\frac{\partial^2 u_i}{\partial t \partial x_j}\right] \qquad (2)$$
$$= \frac{\partial P^e}{\partial x_i} \quad i,j,k = x,y,z$$

where G and λ are the well-known Lame's constants. G' and λ' are viscous moduli. It must be noticed that the last two terms in left side of Eq. (2) denote the viscosity effect on consolidation of poroelastic media.

A stratum of clay with the thickness of B sandwiched between sandy strata which are highly permeable and much stiffer than the clay shown in Fig. 1 is studied herein. In Fig. 1, the groundwater table depressions h_1 and h_2 respectively occur in sandy strata above and below the clay. Because sand and clay have significant differences in permeability and compressibility, excessive pore water pressure only exists in the clay as consolidation proceeds. All of consolidation is nearly taken place due to the volume change within the clay, while the sandy strata may be considered as rigid media in comparison with the clay. In addition, the horizontal dimension is much larger than the thickness of clay. Hence, one-dimensional consolidation is well assumed herein. This assumption leads to the fact that the flow and strain of clay only occur in the vertical direction. Neuman and Witherspoon (1969) indicated that as compared with three-dimensional flow the error introduced by the assumption of vertical flow in clay stratum of a multiaquifer system is less than five percents when the permeability contrast between neighboring sandy strata and the clay stratum exceeds two orders of magnitude.

Because one-dimensional consolidation is considered, the flow equation shown in Eq. (1) for heterogeneous clay can be simplified as

$$\frac{\partial}{\partial z}\left(K\frac{\partial P^e}{\partial z}\right) = \rho_w g \frac{\partial^2 u_z}{\partial t \partial z} \qquad (3)$$

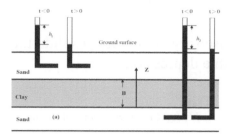

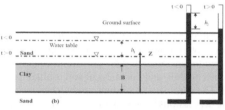

Figure 1. Sketches of soil consolidation due to groundwater table depression: (a) confined case, (b) unconfined case.

The equation for equilibrium of forces given by Eq. (2) becomes

$$\frac{\partial}{\partial z}\left[(2G+\lambda)\frac{\partial u_z}{\partial z}\right] + \frac{\partial}{\partial z}\left[(2G'+\lambda')\frac{\partial^2 u_z}{\partial t \partial z}\right] = \frac{\partial P^e}{\partial z} \qquad (4)$$

Equation 4 indicates that the constitutive relationship for one-dimensional consolidation of viscoelastic soil is modeled by a spring and dashpot in parallel so that they both experience the same deformation or strain and the soil effective stress is the sum of the stresses in spring and dashpot as shown in Fig. 2. In addition, due to the assumption of invariance of total stress, i.e., the neglect of the variation of body force, the decrease of pore water pressure is identical to the increase of soil effective stress. It must be noticed that the traditional Terzaghi consolidation equation can be obtained by substituting Eq. 4 with the neglect of viscosity effect into Eq. 2. The proper initial conditions and boundary conditions for pore water pressure and soil displacement are needed to solve Eqs. (3) and (4). Because only incremental state is taken into consideration, the initial conditions for pore water pressure and soil displacement can be respectively expressed as

$$P^e(z, t = 0) = 0 \qquad (5a)$$

and

$$u_z(z, t = 0) = 0 \qquad (5b)$$

The groundwater table depressions h_1 and h_2 occur in sandy strata above and below the clay. Applying the continuity of pore water pressure, the pore water

208

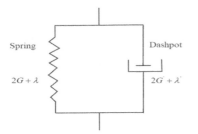

Figure 2. Constitutive relationship for one-dimensional viscoelastic consolidation.

pressure at bottom and top boundaries of clay can be respectively expressed as

$$P^e(z = 0, t) = -\rho_w g h_2(t) \tag{6a}$$

and

$$P^e(z = B, t) = -\rho_w g h_1(t) \tag{6b}$$

The bottom boundary of clay is connected to the nearly rigid sandy stratum. The displacement of clay at $z = 0$ can be expressed as

$$u_z(z = 0, t) = 0 \tag{7}$$

If the overlaying sandy stratum shown in Fig. 1(a) is confined, the top boundary of clay $z = B$ is subjected to an incremental effective stress $-\rho_w g h_1$ due to groundwater table depression, i.e.,

$$(2G + \lambda)\frac{\partial u_z}{\partial z}\bigg|_{z=B} + (2G + \lambda)\frac{\partial^2 u_z}{\partial z \partial t}\bigg|_{z=B} = -\rho_w g h_1(t) \tag{8}$$

However, for the unconfined overlaying stratum (i.e., the existence of free water surface) shown in Fig. 1(b), with considering the decrease in weight by releasing pore water (Corapcioglu & Bear 1983), the incremental effective stress at top boundary of clay becomes

$$(2G + \lambda)\frac{\partial u_z}{\partial z}\bigg|_{z=B} + (2G + \lambda)\frac{\partial^2 u_z}{\partial z \partial t}\bigg|_{z=B}$$
$$= -\rho_w g h_1^{**}(t) \tag{9}$$

3 NONDIMENSIONALIZED AN DISCRETIZED GOVERNING EQUATIONS

With a reference length B, a reference hydraulic conductivity K_f, reference Lame's constants G_f and λ_f, a reference soil displacement $\rho_w g B^2 / (2G_f + \lambda_f)$, and

a reference time $\rho_w g B^2 / (2G_f + \lambda_f)K_f$, Eqs. (3) and (4) can be nondimensionalized as

$$\frac{\partial}{\partial z^*}\left[K^*\frac{\partial P^{e*}}{\partial z^*}\right] = \frac{\partial^2 u_z^*}{\partial t^* \partial z^*} \tag{10}$$

and

$$\frac{\partial}{\partial z^*}\left[(2G^* + \lambda^*)\frac{\partial u_z^*}{\partial z^*}\right] + \frac{\partial}{\partial z^*}\left[N\frac{\partial^2 u_z^*}{\partial t^* \partial z^*}\right] = \frac{\partial P^{e*}}{\partial z^*} \tag{11}$$

where the nondimensionalized pore water pressure $P^{e*} = P^e / \rho_w g B$, the nondimensionalized soil displacement $u_z^* = (2G_f + \lambda_f)u_z / \rho_w g B^2$, the nondimensionalized time $t^* = K_f(2G_f + \lambda_f)t / \rho_w g B^2$, the nondimensionalized coordinate $z^* = z/B$, $N = (2G + \lambda)K_f / \rho_w g B^2$, $K^* = K/K_f$, $G^* = G/(2G_f + \lambda_f)$, and $\lambda^* = \lambda/(2G_f + \lambda_f)$.

It must be pointed out that N shown in Eq. (11) can be called the viscosity number which represents the viscosity effect on soil consolidation. For a homogeneous soil with $B = 10$ m, $\rho_w g = 9.81 \times 10^3$ N/m^2, $K_f = 1.0 \times 10^{-7}$ m/sec, and $(2G + \lambda) = 1.5 \times 10^{12}$ N/m^2/sec yields $N = 0.153$.

The finite difference method is used to solve the nondimensionalized governing equations shown in Eqs. (10) and (11). By applying the central differencing for the spatial derivative terms and backward differencing for the temporal derivative terms, Eqs. (10) and (11) can be respectively discretized. It must note that the with the assumption of invariance of total stress, the viscoelastic consolidation numerical model developed herein can be directly applied to compute deformation of clay layers in a general multiaquifer system including clay strata sandwiched between sand layers.

4 EXAMINATION

The example of identical constant groundwater table depression respectively occurring in sandy strata above and below the clay stratum is first applied to examine the viscosity effect on consolidation of poroelastic soil. For a homogeneous soil (i.e., $K^* = G^* = \lambda^* = 1$) with $h_1/B = h_2/B = 0.01$, the simulated results of soil displacements and pore water pressures with respect to time for different viscosity numbers are shown in Figs. 3 and 4. From Figs. 3 and 4, one can find that the steady-state results are independent of the viscosity number, whereas the transient results are strongly related to the viscosity number. The differences of soil displacements and pore water pressures between with and without considering the viscosity effect become more significant with the increase of viscosity number. When the viscosity number is 0.01, the soil displacements and pore water pressures with and without the viscosity effect have

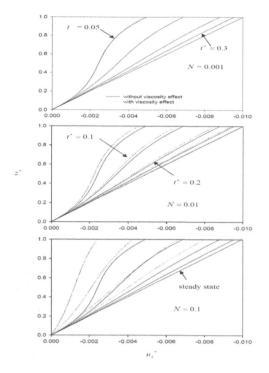

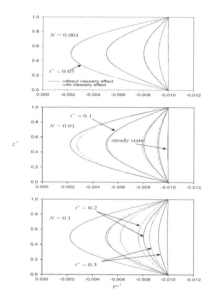

Figure 4. The simulated results of pore water pressures for different viscosity numbers.

Figure 3. The simulated results of soil displacements for different viscosity numbers.

some degree of differences. However, if the viscosity number increases to 0.1 the soil displacements and pore water pressures with the viscosity effect are significantly different from those without the viscosity effect.

Figure 3 shows that at the transient state the soil displacement with the viscosity effect is always less than that without the viscosity effect. This reveals that as compared with considering the soil viscosity the neglect of viscosity effect takes less time to reach the steady-state consolidation as shown in Fig. 5 in which the soil displacements at $z^* = 0.25$ and 0.5 with respect to time for different viscosity numbers are displayed. It can be seen from Fig. 5 that the time to reach the steady state increases with the increase of the viscosity number. Figure 6 depicts that at the beginning of consolidation the pore water pressure depression (i.e., the increase in effective stress of soil) with the viscosity effect is greater than that without the viscosity effect, and the soil with greater viscosity number causes more decrease in pore water pressure. However, when the time is larger than 0.02 the decrease in pore water pressure without the viscosity effect becomes more significant as compared with considering the viscosity effect. This outcome can be explained by analyzing the response of the dashpot and spring stresses with respect to time as shown in Figure 6. It can be

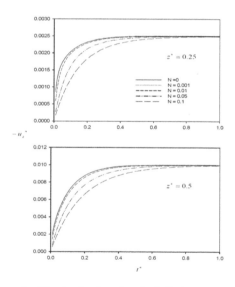

Figure 5. The simulated results of soil displacements with respect to time at $z^* = 0.25$, 0.5 for different viscosity numbers.

seen from Fig. 6 that due to the sudden groundwater table depression the dashpot stress induced by the viscosity effect is maximum at the initial state of consolidation, and then decreases with time to reach zero at the steady state. The spring stresses with and without the viscosity effect, in which the former is always less than the latter at the transient state, are less than the dashpot stress at the early stage of consolidation,

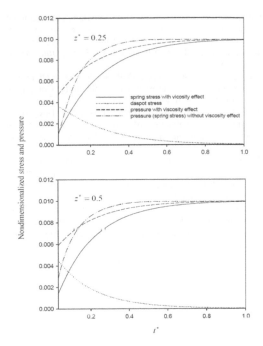

Figure 6. The simulated results of pore water pressures and stresses with $N = 0.1$ at $z^* = 0.25$ and 0.5.

but they gradually increase with time to be greater than the dashpot stress and reach maximum values at the steady state. It can be concluded from mentioned above that the consolidation of poroelastic soil under groundwater table depression is strongly related to the viscosity effect. In addition, it could be concluded from Figs. 3–6 that the soil viscosity seems to need to be taken into account when the viscosity number is greater than 0.001.

5 CONCLUSIONS

Owing to simplicity and efficiency, the analysis of soil consolidation due to groundwater withdrawal in a multiaquifer system was often conducted using the two-step concept in which the traditional one-dimensional Terzaghi consolidation theory is applied to compute vertical deformation of clay with the known groundwater table variations in aquifers that need to be first obtained from in-situ measurement data or by simulating the groundwater flow in a multi-aquifer system. However, with the use of conventional Terzaghi consolidation theory the effect of soil viscosity was not taken into account for the analysis of soil consolidation in the two-step concept. This could result in inaccurate assessment of land subsidence. The goal of this study is to examine the viscosity effect on consolidation of poroelastic soil

under groundwater table depression. The case of clay stratum sandwiched sandy strata subjected to sudden and gradual groundwater table depression is used to investigate the importance of viscosity effect to the consolidation of poroelastic soil. The governing equations are derived from the three-dimensional poroelastic consolidation theory with considering the viscosity effect and the significant differences in permeability and compressibility between sand and clay. The nondimensionalized governing equations indicate that the viscosity effect can be represented by the viscosity number which is related to hydraulic conductivity, viscous moduli, and thickness of soil stratum. The simulated results show that the soil displacement and pore water pressure are strongly affected by the viscosity number. The differences in soil displacements and pore water pressures between with and without the viscosity number are more significant while the viscosity number becomes greater. The neglect of viscosity effect will overestimate soil displacement. The poroelastic soil without viscosity effect takes less time to reach steady-state consolidation as compared with the consideration of viscosity effect.

REFERENCES

Bardet, J.P. 1992. Viscoelastic model for the dynamic behavior of saturated poroelastic soils. *Journal of Applied Mechanics* 59(1): 128–135.

Bear, J. & Corapcioglu, M.Y. 1981. Mathematical model for regional land subsidence due to pumping, 2. Integrated aquifer subsidence equations for vertical and horizontal displacements. *Water Resour Res* 17(3): 947–958.

Corapcioglu, M.Y. & Bear, J. 1983. A mathematical model for regional land subsidence due to pumping, 3. Integrated equations for a phreatic aquifer. *Water Resour Res* 19(4): 895–908.

Ehlers, W. & Markert, B. 2000. On the viscoelastic behavior of fluid-saturated porous materials. *Granular Matter* 2: 153–161.

Gambolati, G. & Freeze, R.A. 1973. Mathematical simulation of the subsidence of venice 1 theory. *Water Resour Res* 9(3): 721–732.

Gambolati, G., Ricceri, G., Bertoni, W., Brighenti G., & Vuillermin, E. 1991. Mathematical simulation of the subsidence of Ravenna. *Water Resour Res* 27(9): 2899–2918.

Guo, W.D. 2000. Visco-elastic consolidation subsequent to pile installation. *Computers and Geotechnics* 26: 113–144.

Gutierrez, M.S. & Lewis, R.W. 2002 Coupling fluid flow and deformation in underground formations. *Journal of Engineering Mechanics* 128(5): 779–787.

Helm, D.C. 1975. One-dimensional simulation of aquifer system compaction near Pixley, California 1. Constant Parameters. *Water Resour Res* 11(3): 465–477.

Helm, D.C. 1987. Three-dimensional consolidation theory in terms of the velocity of soil. *Geotechnique* 37(2): 369–392.

Hsieh, P.C. 2006. A viscoelastic model for the dynamic response of soils to periodical surface water disturbance.

International Journal for Numerical and Analytical Methods in Geomechanics 30: 1201–1212.

Hsieh, P.C. & Hsieh, W.P. 2007. Dynamic analysis of multilayered soils to water waves and flow. *Journal of Engineering Mechanics* 133(3): 357–366.

Hsu, T.W. & Lu, S.C. 2006. Behavior of one-dimensional consolidation under time-dependent loading. *Journal of Engineering Mechanics* 132(4): 457–462.

Larson, K.J., Basagaoglu, H. & Marino, M.A. 2001. Prediction of optimal safe ground water yield and land subsidence in Los Banos-Kettlenman city area, California, Using a calibrated numerical simulation model. *Journal of Hydrology* 242(1): 79–102.

Lewis, R.W. & Schrefler, B. 1978. A fully coupled consolidation model of the subsidence of Venice. *Water Resour Res* 14(2): 223–229.

Neuman, S.P. & Witherspoon, P.A. 1969. Theory of flow in a confined two-aquifer system. *Water Resour Res* 5:803–816.

Ng, C.O. & Mei, C.C. 1995. Ground subsidence of finite amplitude due to pumping and surface Loading. *Water Resour Res* 31(8): 1953–1968.

Onta, P.R. & Gupta, A.D. 1995. Regional management modeling of a complex groundwater system for land subsidence control. *Water Resour Management* 9(1): 1–25.

Terzaghi, K. 1954. *Theoretical Soil Mechanics*, John Wiley, New York.

Thomas, L.H. 1949. *Elliptic Problems in Linear Difference Equations over a Network*, Waston Scientific Computing Laboratory, Columbia, New York.

Tsai, T.L., Chang, K.C. & Huang, L.H. 2006. Body force effect on consolidation of porous elastic media due to pumping. *Journal of the Chinese Institute of Engineers* 29(1): 75–82.

Tseng, C.M., Tsai, T.L., & Huang, L.H. 2008. Effects of body force on transient poroelastic consolidation due to groundwater pumping. *Environmental Geology* (in press, on-line available).

Xie, K.H., Liu, G.B. & Shi, Z.Y. 2004. Dynamics response of partially sealed circular tunnel in viscoelastic saturated soil. *Soil Dynamics and Earthquake Engineering* 24: 1003–1011.

Prediction and Simulation Methods for Geohazard Mitigation – Oka, Murakami & Kimoto (eds)
© 2009 Taylor & Francis Group, London, ISBN 978-0-415-80482-0

Develop of a fully nonlinear and highly dispersive water wave equation set; analysis of wave interacting with varying bathymetry

A. Najafi-Jilani
Islamic Azad University, Islamshahr, Iran

B. Ataie-Ashtiani
Sharif University of Technology, Tehran, Iran

ABSTRACT: Extended Boussinesq-type water wave equations are derived in two horizontal dimensions to capture the nonlinearity effects and frequency dispersion of wave in a high accuracy order. A multi-parameter perturbation analysis is applied in several steps to extend the previous second order Boussinesq-type equations in to 6th order for frequency dispersion and consequential order for nonlinearity terms. The presented high-order Boussinesq-type equation is applied in a numerical model to simulate the wave field transformation due to physical processes such as shoaling, refraction and diffraction. The models results are compared with available experimental data which obtained in a laboratory wave flume with varying bottom in Delft Hydraulic Institute and an excellent agreement is obtained.

1 INTRODUCTION

The Boussinesq-type equations have been applied for water surface wave modeling to increase the order of accuracy for wave frequency dispersion and consequentially for nonlinearity effects. The Boussinesq-type models have been derived using polynomial approximation in the vertical profile of the horizontal velocities. The classic Boussinesq equations rewritten in depth integrated form (Peregrine, 1967) assuming the second order variation of velocity in vertical direction ((0,2) Padé approximant). In these equations, the nonlinearity and frequency dispersion of wave is simulated in first and second order, respectively. As an improvement in (0,2) padé approximation of Boussinesq equations, some researchers rearranged the dispersive terms (Madsen and Schaffer 1998, Chen et al. 2000) or introduced a significant water depth, $Z\alpha$, as a characteristic water depth in which, the horizontal velocity domain is defined. In this (2,2) Padé approximant, the value of $Z\alpha$ optimized using a least-square procedure aimed at minimizing errors in approximated waves phase speed (Beji & Nadaoka 1996). The extension of Boussinesq models to higher accuracy continued by Gobbi and his coworkers (Gobbi et al., 2000). They presented a (4,4) Padé approximant accurate to $O(\mu 4)$ for retaining terms in dispersion, and to all consequential orders in nonlinearity. The work on extending the range of Boussinesq models to higher accuracy continued by Lynett and Liu multi-layer approach (Lynett & Liu 2002). The higher order Boussinesq type models with moving bottom

boundary is developed and applied to study the underwater landslide generated waves (Ataie-Ashtiani & Najafi-Jilani 2006). The main objective of this work is developing of a higher order Boussinesq-type wave equations to capture the frequency dispersion and nonlinearity effects of wave accurately. The higher order Boussinesq-type equation is derived in a depth-integrated form and is investigated in several cases.

2 MATHEMATICAL FORMULATION

The perturbation analysis is used based on the expansion of velocity components to derive the sixth order Boussinesq-type wave model in two horizontal dimensions. A schematic of the computational domain and the main geometric parameters are shown in Figure 1. The dimensionless form of governing equations and boundary conditions in three-dimensional domain can be described as following (Lynett & Liu 2002). In the following equations, x and y are the horizontal coordinates scaled by l_0 which is the horizontal length scale, z is the vertical coordinate scaled by h_0 which is the characteristic water depth, t is time and scaled by $l_0/(gh_0)1/2$, ζ is the water surface displacement scaled by a_0 which is the wave amplitude, h is the total depth based on still water considering the arbitrary bottom boundary $(h(x, y, t))$ and scaled by h_0, u is the vector of horizontal velocity components (u, v) scaled by $\varepsilon \cdot gh_0)^{1/2}$, w is the velocity in vertical direction scaled by $(\varepsilon/\mu) \cdot (gh_0)^{1/2}$, p is the water pressure scaled by

$\gamma \cdot a_0$, and $\nabla = (\partial/\partial x , \partial/\partial y)$ is the horizontal gradient vector.

$$\mu^2 \nabla \cdot \mathbf{u} + w_z = 0 \tag{1}$$

$$\mathbf{u}_t + \varepsilon \mathbf{u} \nabla \cdot \mathbf{u} + \frac{\varepsilon}{\mu^2} w \mathbf{u}_z = -\nabla p \tag{2}$$

$$\varepsilon w_t + \varepsilon^2 \mathbf{u} \cdot \nabla w + \frac{\varepsilon^2}{\mu^2} w w_z = -\varepsilon P_z - 1 \tag{3}$$

We will assume the flow is irrotational. The irrotationality condition can be defined as:

$$(w_y - v_z)\mathbf{i} + (w_x - u_z)\mathbf{j} + (v_x - u_y)\mathbf{k} = \mathbf{0} \tag{4}$$

Using the perturbation analysis in and considering the first three terms of expanded form of velocity component, the sixth-order Boussinesq-type wave equation set is derived.

3 STEEP SLOPE BED EFFECTS

The presented sixth order Boussinesq type wave equations are used to simulate the varying bathymetry effects on wave characteristics. The test is related to the passing of a regular wave over an arbitrary bottom in a long flume. The examination of presented model is made by comparing the results with available experimental data.

The experiments were carried out in Delft Hydraulics Institute (Gobbi et al. 2000). The incident wave amplitude is 2.05 cm, its period is 1.01 s and the water depth is 40 cm. All of the geometrical conditions of submerged sill and the location of numerical wave gauges as well as the result comparison are shown in Figure 1. The dimensions of numerical wave flume are assumed 1.0 (width) × 25.0 (length) meters. The grid size is set to $\Delta x = 20$ cm (along the length) and $\Delta y = 10$ cm (along the width) and the time step is set to $\Delta t = 0.004$ s. As it is shown in Figure 1, the presented model results are in a good agreement with laboratory measurements. The shoaling effects over submerged sill are accurately captured in presented formulation. The effects of wave propagation in various distances from lateral wave maker can be obtained comparing the water surface time series in several wave gauges.

4 MILD BED VARIATION EFFECTS

In the next laboratory case, experiments have been carried out in a 21.0 × 0.45 × 0.6 meter wave tank at the State Key Laboratory of Coastal and Offshore Engineering, Dalian University of Technology (Willis & Wang 2001). A sketch of experimental set-up is shown in Figure 2. The wave tank is set-up using the wave paddle (on left) as a reference point. The model seabed in case A (see Fig. 2) is initially horizontal for 2 m, has a 1:20 slope for the next 3 m, then is again horizontal resulting in a pre-sandbar water depth of 30 cm.

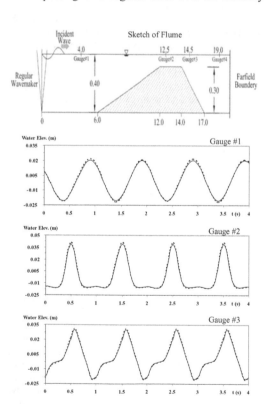

Figure 1. Comparison of presented model (---) with experimental measurements (—).

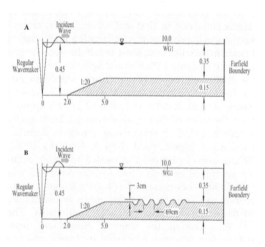

Figure 2. Experimental set-up for wave evolution over varying bathymetry.

214

The model sandbar field in case B (see Fig. 2) contains five sinusoidal undulations, with wavelengths of 0.69 m and amplitudes of 0.03 m. Following the finite sandbar field, the bottom is again horizontal. The end of the wave tank in both cases has an energy-dissipating device.

The grid size of numerical model are set as $\Delta x = 0.1$ m and $\Delta y = 0.05$ m. The comparison of numerical results and experimental measurements are shown in Figure 3. The incidents wave's characteristics are shown in Fig. 3a and the water surface time series are compared in a wave gauge located at 10 m far from wave maker lateral boundary in both cases "A" (horizontal bed) and "B" (sinusoidal bed). The agreement is excellent between numerical results and laboratory measurements and the bottom effects on the wave characteristics are accurately simulated in presented formulation. For further evolution of presented formulation, the numerical simulation is made using two various difference schemes consist of explicit and implicit methods.

5 HIGH-FREQUENCY WAVE PROPAGATION

The above mentioned mathematical formulation is applied to simulate wave propagation in a horizontal bed wave tank to analyze the accuracy of method in high-frequency waves. The incident wave characteristics generate by a lateral wave maker boundary. The experimental data obtained by Shemer et al. (2001). The experiments were performed in a laboratory wave tank, which was 18 m long and 1.2 m wide, with transparent sidewalls and windows at the bottom.

The tank was filled to a mean water depth of 60 cm. Incidents waves were generated by a sea wave simulation RSW 30–60 wave maker. The numerical simulation is performed using both explicit and implicit schemes and the results are compared with the available experimental measurements. The water surface time series are compared in three wave gauges located at 3.45 m, 6.34 m, and 9.23 m from lateral boundary. For better recognition, the envelope curve of measured data is shown in the Figure 4. A good agreement can

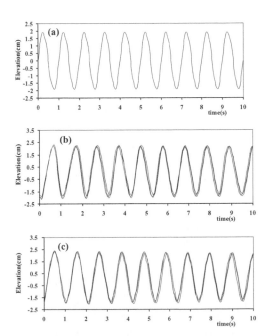

Figure 3. Comparison of numerical results obtained from presented higher-order formulation numerical method using implicit scheme (– – –) and explicit scheme (---) with experimental measurements (—).

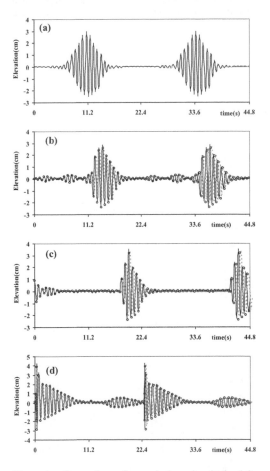

Figure 4. Comparison of numerical results obtained from presented implicit (– – –) and explicit scheme (---) with experimental measurements(°). The incidents wave's characteristics are shown in (a) and the water surface time series are compared at 3.45 m (b), 6.34 m (c) and 9.23 m (d).

be seen generally between numerical and experimental results.

6 WAVE EVALUATION OVER A STEP

To evaluate the presented model in the cases with rapidly varying bathymetry, it has been applied in an experimental case in which, the incident regular wave

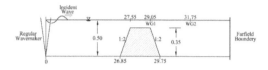

Figure 5. Experimental set up for verification of presented model, the regular incident waves pass over the submerged steep-slope sill.

enters to a flume from left lateral boundary and pass over a submerged step with relatively steep walls [7]. A schematic of experiment set up is presented in Figure 5. The incident wave amplitude in all cases is the same as 2.5 cm but for considering the incident wave period effects on the results, the period has been changed as 1.341 s, 2.012 s, and 2.683 s for cases (1), (2) and (3), respectively. The dimensions of numerical wave flume are assumed 1.0 (width) × 40.0 (length) meters. The grid size is set to $\Delta x = 5$ cm (along the length) and $\Delta y = 5$ cm (along the width). The range of time step is $0.01 \leq \Delta t \leq 0.08$s for case (1), $0.01 \leq \Delta t \leq 0.13$s for case (2), and $0.01 \leq \Delta t \leq 0.22$s for case (3).

The results of numerical modeling (for $\Delta t = 0.01$) are compared with experimental measurements in Figures 6 and 7. The agreement is good for both numerical schemes. Similar to the previous cases,

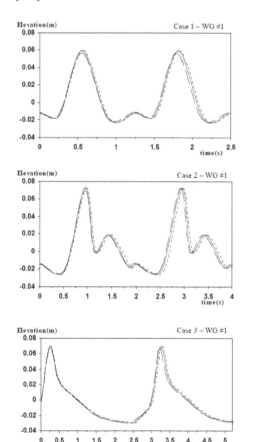

Figure 6. Comparison of numerical results and experimental measurements (——) [7], cases 1, 2, and 3, wave gauge #1.

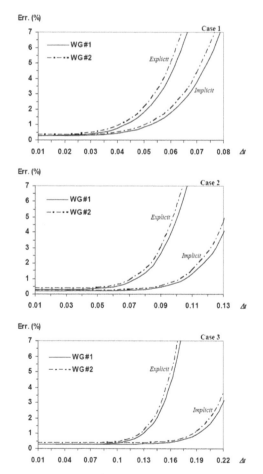

Figure 7. Comparison of numerical results and experimental measurements (——) [7], cases 1, 2, and 3, wave gauge #1.

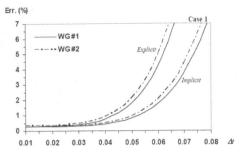

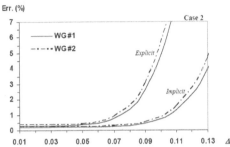

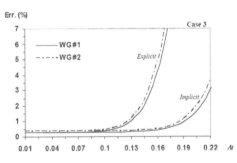

Figure 8. Error analysis of numerical model in various schemes and time steps for passing of wave over submerged steep-slope sill.

numerical error analysis is shown in Figure 8. The effect of rapidly varying bottom boundary as well as incident wave period on the numerical error can be investigated using this figure. As it can be seen, the explicit scheme provides generally more accurate results in a rapidly varying bottom channel. As the wave period decreases, the accuracy of implicit scheme has been increased. For high-frequency incident waves, the accuracy of both schemes is generally close together.

7 CONCLUSION

A two-dimensional depth-integrated fully nonlinear and highly dispersive Boussinesq-type wave equation set is developed. The equations retain terms to $O(\mu^4)$, $\mu = h_0/l_0$ in dispersion and to $O(\varepsilon^5)$, $\varepsilon = a_0/h_0$ in nonlinearity where a_0 is characterizes amplitude, l_0 is

the characterizes horizontal length-scale, and h_0 is the characterizes water depth.

The presented two-horizontal dimension (2HD) equations are applied in a high-order numerical model using Finite Difference Method in high-order discretisation schemes for time and apace. The model is used to analyze the varying bathymetry effects on the passing wave characteristics. The shoaling effects and wave frequency dispersion due to wave propagation in long distances are investigated. It is concluded that the presented Boussinesq-type formulation can accurately simulate the wave characteristics in gradually or rapidly varying bathymetry.

REFERENCES

Ataie-Ashtiani, B. & Najafi-Jilani, A. 2007. A higher-order Boussinesq-type model with moving bottom boundary: applications to submarine landslide tsunami waves. *International Journal for Numerical Methods in Fluids* 53(6), 1019–1048.

Ataie-Ashtiani, B. & Najafi-Jilani, A. 2008. Laboratory Investigations on Impulsive Waves Caused by Underwater Landslide. Coastal Engineering, (Accepted: 8.March.2008)

Ataie-Ashtiani, B. & Malek-Mohammadi, S., February/March 2007. Near field Amplitude of Sub-aerial Landslide Generated Waves in Dam Reservoirs. *Journal of Dam Engineering*.

Ataie-Ashtiani, B. & Najafi-Jilani, A. 2006. Prediction of Submerged Landslide Generated Waves in Dam Reservoirs: An Applied Approach. *Journal of Dam Engineering*, (in Press).

Beji, S. & Nadaoka, K. 1996. A formal derivation and numerical modeling of the improved Boussinesq equations for varying depth. *Journal of Ocean Engineering* 23: 691–704.

Chen, Q., Kirby, J.T., Darlymple, R.A., Kennedy, A.B. & Cawla, A. 2000. Boussinesq modeling of wave transformation, breaking, and runup. II: 2D. *Journal of Waterway, Port, Coast and Ocean Engineering*, (January/February): 48–56.

Enet F., Grilli, S.T. & Watts, P. May 2003. Laboratory Experiments for Tsunami Generated by Underwater Landslides: Comparison with Numerical Modeling. In Proceeding of 13th International Conference on Offshore and Polar Engineering, Honolulu, Hawaii, USA: 372–379.

Fritz, H.M., Hager, W.H. & Minor, H.E. 2004. Near field characteristics of landslide generated impulse waves. *Journal of Waterway, Port, Coastal Ocean Engineering*, 130: 287–302.

Gobbi, M.F., Kirby, J.T. & Wei, G. 2000. A fully nonlinear Boussinesq model for surface waves. II. Extension to O(kh4). *Journal of Fluid Mechanics*, 405: 181–210.

Grilli, S.T. & Watts, P., November/December 2005. Tsunami Generation by Submarine Mass Failure. I: Modeling, Experimental Validation, and Sensitivity Analyses. *Journal of Waterway, Port, Coastal and Ocean Engineering*: 283–297.

Grilli, S.T., Vogelmann, S. & Watts, P. 2002. Development of a 3D numerical wave tank for modeling tsunami generation by under water landslides. *Journal of Engineering Analysis with Boundary Elements*, 26: 301–313.

Grilli, S.T. & Watts, P. 2003. Under Water Landslide Shape, Motion, Deformation and Tsunami Generation. *Journal of International Society of Offshore and Polar Engineers*, ISBN 1-880653-5: 364–371.

Grilli, S.T., Watts, P., Kirby, J.T., Fryer, G.F. & Tappin, D.R. 1999. Landslide Tsunami Case Studies Using a Boussinesq Model and a Fully Nonlinear Tsunami Generation Model. *Journal of Natural Hazards and Earth System Sciences* 3: 391–402.

Lynett, P. & Liu, P.L. 2002. A numerical study of submarine-landslide-generated waves and run-up. Proc. of Royal Society, London, U.K., A458: 2885–2910.

Madsen, P.A. & Schaffer, H.A. 1998. Higher-order Boussinesq-type equations for surface gravity waves: derivation and analysis. *Philosophical Trans.* Royal society, London, U.K., A356: 3123–3184.

Peregrine, D.H. 1967. Long waves on a beach. *Journal of Fluid Mechanics* 27: 815–827.

Shemer, L., Jiao, H., Kit, E. & Agnon, Y. 2001. Evolution of a nonlinear wave field along a tank: experiments and numerical simulations based on the spatial Zakharov equation. *Journal of Fluids Mechanics* 427: 107–129.

Willis, G. & Wang, Y. 2001. Experimental and Numerical Modeling of Wave Propagation and Reflection Over Sinusoidally-Varying Sandbar Field. Research Report, Dalian University of Technology, dalian, China: 1–13. (http://www.clarkson.edu/projects/reushen/reu_china)

An MPM-FDM coupled simulation method for analyzing partially-saturated soil responses

Y. Higo, S. Kimoto, F. Oka, Y. Morinaka & Y. Goto
Kyoto University, Kyoto, Japan

Z. Chen
University of Missouri, Columbia, MO, USA

ABSTRACT: The Material Point Method (MPM), as proposed by Sulsky et al. (1994), has been used to simulate large deformations of materials. A continuum body is divided into a finite number of subregions represented by Lagrangian material points, while the governing equations are formulated and solved with the Eulerian grid. Since this grid can be chosen arbitrarily, mesh tangling does not appear in the MPM. In the present study, the MPM is coupled with the finite difference method (FDM) for simulating unsaturated soil responses based on the simplified three-phase method. The soil skeleton and the pore fluid are discretized by the MPM and FDM, respectively. Seepage and deformation coupled analyses for unsaturated soils are performed, and the potential of the proposed method is demonstrated via example problems.

1 INTRODUCTION

It is well known that large deformation analysis with the finite element method (FEM) may lead to the numerical difficulty due to mesh tangling even when the updated Lagrangian scheme is adopted. Mesh-free methods have been proposed by the research community to circumvent the difficulty by replacing the conventional connectivity matrix with certain innovative treatments in spatial discretization. In the present paper we develop a spatial discretization procedure for the challenging geotechnical problem involving partially fluid-saturated porous media, based on the Material Point Method (MPM) which could be classified as a particle method for coupled computational fluid dynamics (CFD) and computational solid dynamics (CSD) simulation. Within the original framework of the MPM as proposed by Sulsky, Chen & Schreyer (1994), a continuum body is divided into a finite number of sub-regions consisting of Lagrangian material points, and constitutive equations are formulated and solved at the material points while the equation of motion is solved with the use of the background computational grid. Since this computational grid can be chosen arbitrarily, the mesh tangling problem does not appear in the MPM.

Several researchers have applied the MPM to geomechanics problems (e.g., Coetzee et al. 2005, Beuth et al. 2007, Abe et al. 2007), but the published papers with the MPM are limited to single phase soils. The aim of the present work is therefore to develop

the MPM for coupled CFD and CSD model-based simulation of multi-phase materials with application to geotechnical problems. In this paper, we propose that a coupled MPM-Finite Difference Method (FDM) simulation scheme be developed for analyzing partially fluid-saturated porous soil responses. The proposed analysis method is based on the simplified three-phase method (Ehlers et al. 2004, Oka et al. 2007, 2008) in which the compressibility of air is assumed to be very high. Within each time increment, the solid phase is solved by the MPM and pore fluid phase is solved by the FDM via the background mesh of the MPM. As a numerical example, deformation analysis of unsaturated soil such as river embankment under seepage flow has been performed. The applicability of the method to geotechnical problems and future tasks will be discussed.

2 SIMPLIFIED THREE-PHASE METHOD

2.1 Partial stresses for the mixture

When modeling the mechanical behavior of an unsaturated soil, it is necessary to choose appropriate stress variables since they control its mechanical behavior. In this study, the skeleton stress is used for the stress variable in the constitutive relation for the soil skeleton (Kimoto et al. 2007, Oka et al. 2008). The total stress tensor is assumed to be composed of three partial stresses for each phase.

$$\sigma_{ij} = \sigma_{ij}^s + \sigma_{ij}^f + \sigma_{ij}^a \tag{1}$$

where σ_{ij} is the total stress tensor, σ_{ij}^s, σ_{ij}^f, and σ_{ij}^a are the partial stress tensors for solid, liquid, and gas, respectively.

The partial stress tensors for unsaturated soil can be given as follows.

$$\sigma_{ij}^f = -nS_r p^f \delta_{ij} \tag{2}$$

$$\sigma_{ij}^a = -n(1 - S_r)p^a \delta_{ij} \tag{3}$$

$$\sigma_{ij}^s = \sigma_{ij}' - (1 - n)S_r p^f \delta_{ij} - (1 - n)(1 - S_r)p^a \delta_{ij} \tag{4}$$

where σ_{ij}' is the skeleton stress, p^f and p^a are pore water pressure and the pore gas pressure, respectively, n is the porosity and S_r is the degree of saturation.

The skeleton stress is used as the basic stress variable in the model for unsaturated soil. Definitions for the skeleton stress and average fluid pressure are given as follows:

$$\sigma_{ij}' = \sigma_{ij} + P^F \delta_{ij}, \qquad P^F = S_r p^f + (1 - S_r)p^a \tag{5}$$

where P^F is the average pore pressure.

2.2 Mass conservation law

The mass conservation law for the three phases is given by

$$\frac{\partial \bar{\rho}^J}{\partial t} + \frac{\partial (\bar{\rho}^J \dot{u}_i^J)}{\partial x_i} = 0 \tag{6}$$

where $\bar{\rho}^J$ is the average density for the J phase, and $\dot{u}_i^J$ is the velocity vector of the J phase.

$$\bar{\rho}^s = (1 - n)\rho^s \tag{7}$$

$$\bar{\rho}^f = nS_r \rho^f \tag{8}$$

$$\bar{\rho}^a = n(1 - S_r)\rho^a \tag{9}$$

in which J = s, f, and g ; super indices s, f, and g indicate the solid, the liquid, and the gas phases, respectively. ρ^J denotes the mass bulk density of the solid, liquid, and gas, respectively.

2.3 Conservation laws of linear momentum for the three phases

The conservation laws of linear momentum for the three phases are given by

$$\bar{\rho}^s \ddot{u}_i^s - Q_i - R_i = \frac{\partial \sigma_{ij}^s}{\partial x_j} + \bar{\rho}^s b_i \tag{10}$$

$$\bar{\rho}^f \ddot{u}_i^f + R_i - S_i = \frac{\partial \sigma_{ij}^f}{\partial x_j} + \bar{\rho}^f b_i \tag{11}$$

$$\bar{\rho}^a \ddot{u}_i^a + S_i + Q_i = \frac{\partial \sigma_{ij}^a}{\partial x_j} + \bar{\rho}^a b_i \tag{12}$$

where b_i is a body force, Q_i denotes the interaction between solid and gas phases, and R_i denotes the interaction between solid and liquid phases, and S_i denotes the interaction between liquid and gas phases.

These interaction terms Q_i and R_i can be described as

$$R_i = nS_r \frac{\gamma_w}{k^f} \dot{w}_i^f \tag{13}$$

$$Q_i = n(1 - S_r)\frac{\rho^a g}{k^a} \dot{w}_i^a \tag{14}$$

where k^f is the water permeability coefficient, k^a is the air permeability, $\dot{w}_i^f$ is the average relative velocity vector of water with respect to the solid skeleton, and $\dot{w}_i^a$ is the average relative velocity vector of air to the solid skeleton. The relative velocity vectors are defined by

$$\dot{w}_i^f = nS_r(\dot{u}_i^f - \dot{u}_i^s) \tag{15}$$

$$\dot{w}_i^a = n(1 - S_r)(\dot{u}_i^a - \dot{u}_i^s) \tag{16}$$

If the interaction between liquid and gas phases is assumed to be very small, S_i can be described as

$$S_i \cong 0 \tag{17}$$

2.4 Equations of motion for the whole mixture

Based on the above fundamental conservation laws, we can derive equations of motion for the whole mixture.

From the following assumptions,

$$\ddot{u}_i^s \gg (\ddot{u}_i^f - \ddot{u}_i^s), \ddot{u}_i^s \gg (\ddot{u}_i^a - \ddot{u}_i^s) \tag{18}$$

the equations of motion for the whole mixture are obtained as

$$\rho \ddot{u}_i^s = \frac{\partial \sigma_{ij}}{\partial x_j} + \rho b_i \tag{19}$$

where ρ is the density of the whole mixture, i.e., $\rho = \bar{\rho}^s + \bar{\rho}^f + \bar{\rho}^a$.

2.5 Continuity equations for the fluid phase

For the air phase, we assume the compressibility is very large. When the air compressibility is very large, i.e., $K^a \cong 0$, we can set $\dot{p}^a \cong 0$. Hence, we can assume that $p^a = 0$ if the initial air pressure $p_{ini}^a = 0$. This assumption indicates that the continuity equation for air phase is always satisfied (Oka et al. 2008, Kato et al. 2009, to appear).

Since saturation is a function of the suction, the negative pressure head, the time rate for saturation is given by

$$n\dot{S}_r = \frac{C}{\gamma_w}\dot{p}^f \tag{20}$$

where $\theta = V_w/V$ is the volumetric water content, $\psi = p^f/\gamma_w$ is the pressure head, $C = d\theta/d\psi$ is the specific water content.

Using Equation (20), the continuity for water phase is given as:

$$-\frac{\partial}{\partial x_i}\left[\frac{k^f}{\gamma_w}\left(\rho^f \ddot{u}_i^s + \frac{\partial p^f}{\partial x_i} - \rho^f b_i\right)\right] + S_r \dot{\varepsilon}_{ii}^s$$

$$+\left(\frac{nS_r}{K^f} + \frac{C}{\gamma_w}\right) = 0 \qquad (21)$$

$$\frac{1}{\bar{K}^f} = \frac{S_r}{K^f} + \frac{C}{n\gamma_w} \qquad (22)$$

where $\bar{K}^f$ is an apparent volumetric elastic coefficient of pore water.

2.6 Unsaturated seepage characteristics

The soil-water characteristic model proposed by van Genuchten (1980) is used to describe the unsaturated seepage characteristics as:

$$S_e = \left(1 + (\alpha\psi)^{n'}\right)^{-m} \qquad (23)$$

where S_e the effective saturation as:

$$S_e = \frac{\theta - \theta_r}{\theta_s - \theta_r} = \frac{nS_r - \theta_r}{\theta_s - \theta_r} \qquad (24)$$

in which ψ is the negative pressure head (suction), θ_s is the volumetric water content at the saturated state which is equal to porosity n, and θ_r is the residual volumetric water content retained by the soil at the large value of suction head which is disconnected pendular water meniscus.

In Equation (24), α is a scaling parameter which has the dimensions of the inverse of ψ, and n' and m determine the shape of the soil-water characteristics curve. The relation between n' and m leads to an S-shaped type of soil-water characteristics curve, namely, $m = 1 - 1/n'$. Specific permeability coefficient k_r, which is a ratio of the permeability for unsaturated soil to that for the saturated one, is defined as (Mualem 1976)

$$k_r = S_e^{1/2}\left\{1 - (1 - S_e^{1/m})^m\right\}^2 \qquad (25)$$

3 MPM-FDM COUPLED FORMULATION FOR THE SIMPLIFIED THREE-PHASE METHOD

In this section, the MPM, as originally proposed by Sulsky, Chen & Schreyer (1994), is applied to the simplified three-phase method. The equations of motion for the whole mixture are discretized by the MPM, while the continuity equation for the fluid phase is discretized by the FDM. In the computational model, a $u-p$ formulation is adopted in which the displacement of the soil skeleton, u, and the pore fluid pressure, p, are used as the unknown variables.

3.1 MPM formulation of the equations of motion

The assumption in the section 2.5, i.e., $p^a \cong 0$, leads to the following relation between total stress and skeleton stress:

$$\sigma_{ij} = \sigma'_{ij} + S_r p^f \delta_{ij} \qquad (26)$$

Specific stress is adopted for the skeleton stress tensor, namely,

$$\sigma_{ij} = \sigma'_{ij} + S_r p^f \delta_{ij} = \rho\,\sigma'^s_{ij} + S_r p^f \delta_{ij} \qquad (27)$$

where ρ is the density.

The weak form of the equation of motion for the whole mixture is given by

$$\int_\Omega \rho w_i \cdot a_i d\Omega = \int_\Omega \frac{\partial \sigma_{ij}}{\partial x_j} \cdot w_i d\Omega + \int_\Omega \rho w_i \cdot b_i \cdot b_i d\Omega \qquad (28)$$

where w_i is a test function and Ω is the whole domain.

Using Gauss theorem and the specific effective stress Equation (28) yields

$$\int_\Omega \rho w_i \cdot a_i d\Omega = -\int_\Omega \rho\sigma'^s_{ij} : w_{i,j} d\Omega - \int_\Omega p^f w_{i,j} d\Omega$$

$$+ \int_{d\Omega} w_i \cdot \tau_i dS + \int_\Omega \rho w_1 \cdot b_i d\Omega \qquad (29)$$

in which σ'^s_{ij} is specific skeleton stress, and τ_i is traction vector.

In the MPM formulation, a continuum body is divided into a finite number of subregions consisting of Lagrangian material points while the equation of motion is solved using the computational grid (Fig. 1). Thus, the density ρ is written as a sum of point masses as

$$\rho(x) = \sum_{p=1}^{N_p} M_p \bar{\delta}(x - X_p),\ \bar{\delta}(x - X_p) = \delta(x - X_p)/V_p \qquad (30)$$

where δ is Dirac's delta function, N_p is the number of material point, M_p is the mass of each material point, and V_p is the volume occupied by each material point.

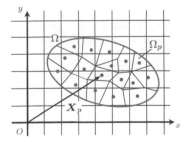

Figure 1. Schematic figure of material points and computational grid.

Substituting Equation (30) into Eq. (29) gives discretized equations of motion in a direct notation as

$$\sum_{p=1}^{N_p} M_p w(X_p) \cdot a(X_p) = -\sum_{p=1}^{N_p} M_p \sigma'^s(X_p) : \nabla w(x)_x = x_p$$

$$-\int_\Omega S_r p^f w \cdot \nabla d\Omega$$

$$+\int_{d\Omega} w.\tau dS$$

$$+\sum_{p=1}^{N_p} M_p w(X_p) \cdot b(X_p) \cdot b(X_p) \tag{31}$$

Let us consider the basis functions for the grid. Acceleration $a(x)$ and test function $w(x)$ at the spatial points x can be represented by the nodal basis functions with spatial points of x as

$$a(x) = \sum_I a_I N_I(x) \tag{32}$$

$$w(x) = \sum_I w_I N_I(x) \tag{33}$$

in which $N_I(x)$ is the basis functions and a_I and w_I are vectors of acceleration and test function at the nodes of grid, respectively. In the present analysis, the basis function is the same as the shape functions for four node is oparametric FEM element.

Substituting these equations into Equation (31) leads to

$$\sum_E \left(\sum_I w_I \sum_J \sum_{p=1}^{N_p} M_p N_I(X_p) \cdot N_j(X_p) a_J \right)$$

$$= \sum_E \left(-\sum_I w_I \sum_{p=1}^{N_p} \sigma'^s(X_p) G_{Ip} - \sum_I w_I S_r p_E^f \{K_v\}_I \right.$$

$$\left. +\sum_I w_I \cdot \tau_I + \sum_I w_I \sum_{p=1}^{N_p} M_p N_I(X_p) b(X_p) \right) \tag{34}$$

where subscript E denotes each grid, p_E^F is average pressure of the fluids defined at the center of the grid.

$$\{K_v\}_I = \int_\Omega \{B_v\}_I d\Omega \tag{35}$$

$$\tau_I = \int_{d\Omega} \tau_I N_I(x) dS \tag{36}$$

$$G_{Ip} = \nabla N_I(x)\big|_{x=X_p^k} \tag{37}$$

in which $\{B_v\}_I$ is the nodal displacement-volumetric strain vector.

Since the test function w_I is arbitrary, Equation (34) for each grid yields

$$\sum_J \sum_{p=1}^{N_p} M_p N_I(X_p) \cdot N_J(X_p) a_J = -\sum_{p=1}^{N_p} M_p \sigma'^s(X_p) G_{IP}$$

$$-S_r p_E^f \{K_v\}_I + \tau_I + \sum_{p=1}^{N_p} M_p N_I(X_p) b(X_p) \tag{38}$$

Although the consistent mass is derived from Equation (38), we use the diagonal mass matrix, i.e., lumped mass for simplicity. The lumped mass at nodes m_I can be given by

$$m_I = \sum_{N_p}^{N_p} M_p N_I(X_p) \tag{39}$$

Using Equation (39) for Equation (38), we obtain the final form of the equation of motion, namely,

$$m_I a_I + S_r \{K_v\}_I p_E^f = -\sum_{p=1}^{N_p} M_p \sigma'^s(X_p) G_{IP} + \tau_I$$

$$+\sum_{p=1}^{N_p} M_p N_I(X_p) b(X_p) \tag{40}$$

3.2 FDM formulation for continuity equation

For the discretization of the continuity equation for fluid phase, we adopt the finite difference method. The finite volume method was used in the liquefaction analysis method by one of the authors (e.g., Oka et al. 2004). In the present analysis, we use the background grid in the MPM for the finite difference method.

3.3 Algorithm of MPM-FDM coupled analysis

The algorithm of MPM-FDM coupled analysis consists of three steps, an initialization phase, a Lagrangian phase, and a convective phase. Schematic figure of the algorithm is illustrated in Figure 2. $k, L, k+1$ indicate the current step, Lagrangian phase, and the next step, respectively. Δt is time increment.

In the initialization phase, information is transferred from the material points to a grid. The mass of the grid nodes m_I^k is determined using the mass of material point M_p by Equation (39), then the velocities at the grid nodes v_I^k is obtained, i.e.,

$$m_I^k v_I^k = \sum_{p=1}^{N_p} M_p V_p N_I(X_p)^k \tag{41}$$

222

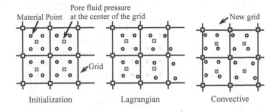

Figure 2. Schematic figure of the algorithm of MPM-FDM coupled analysis (From step k to step $k + 1$).

As for the pore fluid, the pore water pressure at the center of the grid p_E^k is calculated as an average of the pore fluid pressure of each material point in the grid p_p^k as

$$p_E^k = \left(\sum_{p=1}^{N_{pE}} p_p^k \right) / N_{pE} \qquad (42)$$

where N_{pE} is number of material points in the grid.

During the Lagrangian phase, the governing equations are formulated and solved on the grid in which the accelerations at the grid nodes a_I^k and pore fluid pressures at the center of grids p_E^k are calculated. Using the nodal acceleration, the velocities at the grid nodes v_I^L can be updated with an explicit time integration,

$$v_I^L = v_I^k + \Delta t a_I^k \qquad (43)$$

The velocities and the positions of material points V_p^k and X_p^k are updated using a_I^k and v_I^L, respectively,

$$V_p^{k+1} = V_p^k + \Delta t \sum_{I=1}^{N_n} a_I^k N_I (X_p)^k \qquad (44)$$

$$X_p^{k+1} = X_p^k + \Delta t \sum_{I=1}^{N_n} v_I^L N_I (X_p)^k \qquad (45)$$

The strain increments $\Delta \varepsilon_p^{k+1}$ are obtained from gradients of the nodal velocities evaluated at the material points and are used to update the strain for the material points.

$$\Delta \varepsilon_p^{k+1} = \frac{\Delta t}{2} \sum_{I=1}^{N_n} \{ G_{ip} v_I^L + (G_{ip} v_I^L)^T \} \qquad (46)$$

$$\varepsilon_p^{k+1} = \varepsilon_p^k + \Delta \varepsilon_p^{k+1} \qquad (47)$$

Then the stress increment $\Delta \sigma_p^{k+1}$ at the material point is obtained using constitutive equation and the stress is updated as $\sigma_p^{k+1} = \sigma_p^k + \Delta \sigma_p^{k+1}$. As for the pore fluid pressure, since the pressure at the center of grid is the representative of the gird, the pore fluid pressure of the material points in the interior of the grid are updated same as the pore fluid pressure at the center of the grid, namely,

$$p_p^{k+1} = p_E^L \qquad (48)$$

After the Lagrangian phase, the material points are held fixed and the grid is redefined in the convective phase. This motion of the grid relative to the material points models convection.

4 NUMERICAL ANALYSIS OF SEEPAGE-DEFORMATION COUPLED BEHAVIOR BY MPM-FDM COUPLED METHOD

4.1 One-dimensional column

Figure 3 illustrates the boundary conditions for the seepage problem in the one-dimensional unsaturated linear elastic column. The initial saturation is 60%, Young's modulus E is 52 MN/m^2, Poisson's ratio v is 0.3, initial void ratio e_0 is 0.786, saturated permeability coefficient is 1.0×10^{-3} m/s, van Genuchten's parameter are $\alpha = 2.0$ and $n = 4.0$. The height of the soil column is 6 m. The water pressure head of 1.5 m is applied on the top boundary with the rate of 0.3 m/min. The time increment employed in this analysis is 0.001 second.

The deformation and the distribution of the pore water pressure are shown in Figure 4. The initial pore water pressure is -5.52 kPa which corresponds to the initial saturation of 60%. The pore water pressure of saturated soil becomes positive with the advances of infiltration of the water. After the whole soil has been saturated, the distribution of the pore water pressure becomes equal to that of hydrostatic pressure.

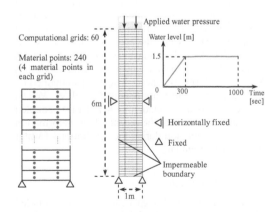

Figure 3. Boundary conditions for the seepage and deformation analysis in one-dimensional unsaturated soil column.

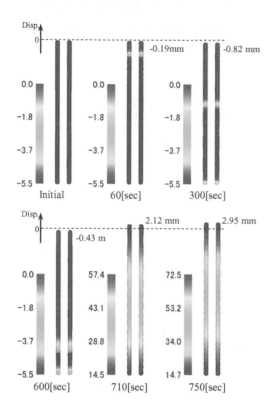

Figure 4. Deformation and distribution of water pressure (Displacement ×100, Unit of pressure: kPa).

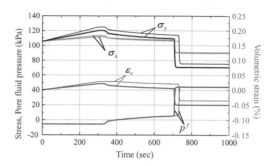

Figure 5. Time history of stress, pore fluid pressure and volumetric strain (thick lines: MPM, thin lines: FEM).

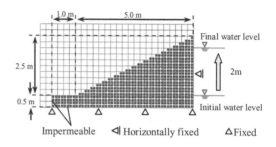

Figure 6. Model of the river embankment and initial configuration of material points and grid.

Table 1. Material parameters.

Initial void ratio	e_0	0.923
Compression index	λ	0.02
Swelling index	κ	0.003
Initial shear modulus ratio	G/σ'_{m0}	225
Saturated permeability coefficient (m/s)	k	1.0×10^{-5}
Failure stress ratio	M_f^*	1.12
Phase transformation stress ratio	M_m^*	0.914
Hardening function parameter	B_0^*	2000
Hardening function parameter	B_1^*	20
Referential strain parameter	γ_{ref}^p	0.008
Referential strain parameter	γ_{ref}^E	0.08
Dilatancy coefficient parameter	D_0	1.5
Dilatancy coefficient parameter	n	4
Bulk modulus of water (kN/m^2)	K^f	2.0×10^6
Wet density (g/cm^3)	ρ	1.91

The time history of vertical and horizontal stresses, pore water pressure, and volumetric strain at the material point located at 3.0 m from the bottom are shown in Figure 5. Before the material point is saturated (330 sec), the traction force due to the water head of 1.5 m and the increase in the unit weight of the soil associated with the saturation induces compressive volumetric strain and the increase in the vertical and horizontal stresses. After saturation, the expansive volumetric strain starts to generate in the soil and rapidly increases when the bottom of the soil is saturated (700 sec). The thin lines indicate the results of finite element analysis under the same conditions (Kato et al. 2009, to appear). The results obtained by MPM are similar to those obtained by FEM. We can say the numerical simulation by MPM-FDM coupled method can reproduce the seepage flow-deformation coupled phenomenon during the infiltration of water from the top boundary.

4.2 River embankment

The two-dimensional seepage-deformation coupled behavior of a river embankment has been simulated (Fig. 6). Water level of the right boundary rises from 0 m to 2 m with the rate of 1 m/min. The river embankment is an unsaturated body modeled by an elasto-plastic constitutive model (Oka et al. 1999). The material parameters used in this analysis are listed in Table 1. The initial saturation is 60% and the initial suction is 5.52 kPa. The time increment used in this analysis is 0.002 second. It is seen in Figure 7 that the proposed method can simulate the

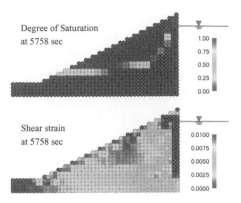

Figure 7.   Distribution of degree of saturation and shear strain.

seepage-deformation behavior during the infiltration of the river water.

5   CONCLUSIONS

We have proposed the coupled MPM-FDM simulation method for analyzing seepage-deformation of unsaturated soil based on the mixture theory. As can be found from the numerical demonstration, the proposed method could well reproduce the seepage-deformation response of unsaturated soil. Using the proposed method, we will simulate the very large deformation, including the post-failure behavior, induced by seepage and/or overflow in order to evaluate the safety of river embankment during flood in the future.

ACKNOWLEDGEMENT

Financial support has been provided by Foundation of River & Watershed Environment Management, River Fund for Survey, Tests and Research, No. 20-1151-003 (2007–2008).

REFERENCES

Abe, K., Johansson, J. & Konagai, K. 2007. A new method for the run-out analysis and motion prediction of rapid and long-traveling landslides with MPM, *Journal of Geotechnical Engineering, JSCE* 63(1): 93–109.

Beuth L., Benz T., Vermeer P.A., Coetzee C.J., Bonnier P. & van den Berg P. 2007. Formulation and validation of a quasi-static Material Point Method, *Proc. NUMOG X* Pande G.N. & Pietruszczak S. eds., Balkema.: 189–195.

Coetzee C.J., Vermeer P.A., & Basson A.H. 2005. The modelling of anchors using the material point method, *Int. J. Numer. Anal. Meth. Geomech.* 29: 879–895.

Ehlers, W., Graf, T. & Amman, M. 2004. Deformation and localization analysis of partially saturated soil, *Computer Methods in Applied Mechanics and Engineering* 193: 2885–2910.

Kato R., Oka F., Kimoto S., Kodaka T. & Sunami S. (to appear), A method of seepage-deformation coupled analysis of unsaturated ground and its application to river embankment, *Journal of Geotechnical Engineering, JSCE*, (In Japanese).

Kimoto, S., Oka, F., Fushita, T. & Fujiwaki, M. 2007. A chemo-thermo-mechanically coupled numerical simulation of the subsurface ground deformations due to methane hydrate dissociation, *Computers and Geotechnics* 34(4): 216–228.

Mualem, Y. 1976, A new model for predicting the hydraulic conductivity of unsaturated porous media, *Water Resources Research* 12: 513–522.

Oka, F., Kodaka, T. & Kim Y.-S. 2004. A cyclic viscoelastic-viscoplastic constitutive model for clay and liquefaction analysis of multi-layered ground, *Int. J. Numerical and Analytical Methods in Geomechanics* 28(2): 131–179.

Oka, F., Kodaka, T., Kimoto, S., Kato, R. & Sunami, S. 2007. Hydro-Mechanical Coupled Analysis of Unsaturated River Embankment due to Seepage Flow, *Key Engineering Materials* 340–341: 1223–1230.

Oka, F., Kimoto, S., Kato, R., Sunami, S. & Kodaka, T. 2008. A soil-water coupled analysis of the deformation of an unsaturated river embankment due to seepage flow and overflow, *Proc. the 12th IACMAG:* 2029–2041.

Oka, F., Feng, H., Kimoto, S., Kodaka, T. & Suzuki, H. 2008. A numerical simulation of triaxial tests of unsaturated soil at constant water and air content by using an elasto-viscoplastic model, Toll, D. and Wheeler, S. (eds), *Proc. of 1st European conference on unsaturated soil:* 735–741.

Oka, F., Yashima A., Tateishi A., Taguchi T. & Yamashita S. 1999, A cyclic elasto-plastic constitutive model for sand considering a plastic-strain dependence of the shear modulus, *Géotechnique* 49(5): 661–680.

Sulsky, D., Chen, Z. & Schreyer H.L. 1994. A particle method for history-dependent materials, *Computer Methods in Applied Mechanics and Engineering* 118: 79–196.

van Genuchten, M.T. 1980, A closed-form equation for predicting the hydraulic conductivity of unsaturated soils, *Soil Science Society of America Journal* 44: 892–898.

Prediction and Simulation Methods for Geohazard Mitigation – Oka, Murakami & Kimoto (eds)
© *2009 Taylor & Francis Group, London, ISBN 978-0-415-80482-0*

Application of SPH method for large deformation analyses of geomaterials

H. Nonoyama, A. Yashima & K. Sawada
Gifu University, Gifu, Japan

S. Moriguchi
Tokyo Institute of Technology, Tokyo, Japan

ABSTRACT: Various types of behaviors of different soils have been predicted by using the FEM with comprehensive constitutive models developed in geomechanics. There are, however, still some problems for the large deformation analysis in the framework of FEM. Numerical instabilities arise due to the distortion of the FE mesh. SPH method is an effective method to solve large deformation problems by a mesh-less Lagrangian scheme. In this study, large deformation analyses of geomaterials using SPH method are carried out. Geomaterials are assumed to be a single-phase Bingham fluid with shear strength of soils. Based on the comparison with the simulated results using SPH method, the theoretical solutions and the existing simulated results, the applicability of SPH method for large deformation analysis of geomaterials is discussed.

1 INTRODUCTION

In a numerical analysis of geotechnical engineering, the behavior of geomaterials can be predicted by introducing suitable constitutive models of soils with consideration of the effect of pore water pressure by using the Finite Element Method (FEM). There are, however, still some problems for the large deformation analyses in the framework of FEM. Numerical instabilities arise due to the distortion of FE mesh.

On the other hand, some numerical methods have been proposed to solve large deformation problems without FE mesh. Eulerian method is one of the solutions for large deformation problems because it is not necessary to take deformation of mesh into consideration. Simulations of large deformation problems of geomaterials, for example, the lateral flow of liquefied ground (Uzuoka 2000, Hadush 2002), the large deformation of slope failure (Moriguchi 2005) and the penetration of rigid body into the ground (Moriguchi 2005), have been carried out based on the Eulerian method. Numerical results obtained in the previous studies were in good agreement with theoretical solutions and experimental results. In the previous studies, the deformation behavior of geomaterials is expressed under the assumption that the geomaterials are a single-phase Bingham fluid with shear strength of soils.

Smoothed Particle Hydrodynamics (SPH) method (Lucy 1977, Gingold 1977), a kind of particle methods, is also an effective method to solve large deformation problems because the method does not require the structured mesh system. Recently, SPH method has been widely used in a variety of fields such as

fluid dynamics (Liu & Liu 2003) or solid mechanics (Libersky et al. 1993). The method has also applied to geotechnical engineering (Maeda et al. 2006). In this study, large deformation analyses of geomaterials using SPH method are carried out. The SMAC algorithm (Amsdam & Harlow 1970) is used to treat geomaterials as incompressible material. In this paper, geomaterials are assumed to be an incompressible single-phase Bingham fluid with shear strength of soils. In order to prevent numerical instabilities due to a large value of the Bingham viscosity for a quasi-rigid material, an implicit calculation procedure is applied to the viscosity term of the equation of motion. Based on the comparison with the simulated results using SPH method, the theoretical solutions and the existing simulated results, the applicability of SPH method for large deformation analyses of geomaterials is discussed.

2 CONSTITUTIVE MODEL

In this study, the geomaterials are modeled as the single-phase Bingham fluid with shear strength of soils proposed by the previous studies (Moriguchi 2005).

$$\tau = \eta_0 \dot{\gamma} + c + p \tan \phi \qquad (1)$$

where τ is the shear stress, η_0 is the viscosity after yield, $\dot{\gamma}$ is the shear strain rate, c is the cohesion of soil, p is the hydraulic pressure and ϕ is the internal friction angle of soil. Because Equation (1) can not be directly calculated, an equivalent viscosity η' obtained from Equation (1) is used. The equivalent viscosity is used for the Newtonian fluid as follows.

$$\eta' = \frac{\tau}{\dot{\gamma}} = \eta_0 + \frac{c + p \tan \phi}{\dot{\gamma}} \quad (2)$$

Figure 1 shows the equivalent viscosity of the Bingham fluid model. The value of the equivalent viscosity dependently changes by the shear strain rate. The constitutive model used in this study can be obtained by introducing the equivalent viscosity into the constitutive model of the Newtonian fluid.

3 NUMERICAL METHOD

3.1 Basic theory of SPH

The foundation of SPH method is an interpolation theory and is divided into two key steps. The first step is a kernel approximation of field functions. The kernel approximations use neighbor particles β located at point x^β within the influence domain of a smoothing function W for a reference particle α located at point x^α. The second step is a particle approximation.

In the first step of interpolation, we define a smoothed physical quantity $< f(x^\alpha) >$ for a physical quantity $f(x^\alpha)$ at reference particle α as bellow.

$$\langle f(x^\alpha) \rangle = \int_\Omega f(x^\beta) W(r, h) dx^\beta \quad (3)$$

where $r = |x^\alpha - x^\beta|$, h is a radius of the influence domain and Ω is the volume of the integral that contains x^α and x^β.

In the second step of interpolation, for the discrete distribution of particles, Equation (3) is approximated by the summation.

$$\langle f(x^\alpha) \rangle = \sum_\beta^N \frac{m^\beta}{\rho^\beta} f(x^\beta) W^{\alpha\beta} \quad (4)$$

where m^β is a mass of neighbor particle, ρ^β is a density of neighbor particle and N is a numbers of particle in influence domain.

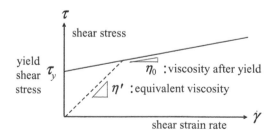

Figure 1. Equivalent viscosity of the Bingham model.

3.2 Incompressibility in the SMAC-SPH method

In general, the equation of continuity and the equation of motion are described as follows.

$$\frac{D\rho}{Dt} = 0 \quad (5)$$

$$\frac{Du_i}{Dt} = -\frac{1}{\rho} \frac{\partial p}{\partial x_i} + \frac{\eta}{\rho} \frac{\partial}{\partial x_j} \left(\frac{\partial u_i}{\partial x_j} \right) + f_i \quad (6)$$

where ρ is the density of fluid, u is the velocity, p is the hydraulic pressure, η is viscosity coefficient and f is external force.

In this study, we used SMAC-SPH method based on fluid dynamics proposed by the previous studies (Sakai et al. 2004). The SMAC algorithm (Amsdam & Harlow 1970) is used to treat geomaterials as incompressible materials. Furthermore, in order to prevent the numerical instabilities due to the large value of the Bingham viscosity for the quasi-rigid materials, an implicit calculation procedure is applied to the viscosity term of the equation of motion. Using the equivalent viscosity η' in place of the viscosity coefficient η is the equation of motion considering spatial gradient of the viscosity term is given,

$$\frac{Du_i}{Dt} = -\frac{1}{\rho} \frac{\partial p}{\partial x_i} + \frac{1}{\rho} \frac{\partial}{\partial x_j} \left[\eta' \left(\frac{\partial u_i}{\partial x_j} + \frac{\partial u_j}{\partial x_i} \right) \right] + f_i \quad (7)$$

In case of the Newtonian fluid, the viscosity coefficient is treated as a constant value and its spatial derivatives are not considered. However, as we can see from Equation (7), the equivalent viscosity has distribution in space. Therefore, the effect of the spatial derivatives is taken into account in Equation (7). Moreover the Equation (7) is separated as follows,

$$\frac{u_i^{**} - u_i^k}{\Delta t} = f_i^k \quad (8)$$

$$\frac{u_i^* - u_i^{**}}{\Delta t} = \frac{1}{\rho} \left[\frac{\partial}{\partial x_j} \left(\eta' \frac{\partial u_i^*}{\partial x_j} \right) + \frac{\partial}{\partial x_j} \left(\eta' \frac{\partial u_j^*}{\partial x_i} \right) \right] \quad (9)$$

$$\frac{u_i^{k+1} - u_i^*}{\Delta t} = -\frac{1}{\rho} \frac{\partial p^{k+1}}{\partial x_i} \quad (10)$$

where Δt is the time increment, subscript k and $k + 1$ indicate the quantities at each calculation time step and * indicate the temporal qualities. Equations (8), (9) and (10) are the external force term, the viscous term and the pressure term of the equation of motion, respectively.

The following is an algorithm used in this study. A temporal value of the velocity u_i^{**} is obtained explicitly using the gravity f_i^k and the velocity u_i^k at previous time from the Equation (8).

$$u_i^{**} = u_i^k + \Delta t f_i^k \quad (11)$$

A temporal value of the velocity u_i^* is obtained using the temporal value u_i^{**} and the temporal value u_i^* of the spatial derivative.

$$u_i^* = u_i^{**} + \frac{\Delta t}{\rho}\left[\frac{\partial}{\partial x_j}\left(\eta'\frac{\partial u_i^*}{\partial x_j}\right) + \frac{\partial}{\partial x_j}\left(\eta'\frac{\partial u_j^*}{\partial x_i}\right)\right] \quad (12)$$

According to the previous studies (Cleary & Monaghan 1999), the right hand side of the temporal value u_i^* of the spatial derivative in Equation (12) can be discretized as follows.

$$\frac{\partial}{\partial x_j}\left(\eta'\frac{\partial u_i^*}{\partial x_j}\right)$$
$$= \sum_{\beta}^{N}\frac{4m^\beta}{\rho^\beta}\frac{\eta'\alpha\eta'\beta}{\eta'\alpha+\eta'\beta}\frac{\left(u_i^{\alpha,*}-u_i^{\beta,*}\right)\left(x_i^\alpha-x_i^\beta\right)\cdot\frac{\partial W^{\alpha\beta}}{\partial x_i}}{\left|x_i^{\alpha\beta}\right|^2} \quad (13)$$

The temporal value of the position x_i^* is obtained by the temporal value u_i^* obtained from Equation (12).

$$x_i^* = x_i^k + \Delta t u_i^* \quad (14)$$

The continuity equation requires that the density of fluid should be constant. This is equivalent to the particle number density being constant, n^0. When the temporal value of a particle number density n^* is not n^0, it is corrected to n^0.

$$n^* + n' = n^0 \quad (15)$$

where n' is a correction value of the particle number density. A correction value of the velocity u_i' occurs in association with the pressure gradient term as follows.

$$u_i' = \Delta t\frac{1}{\rho}\frac{\partial p^{k+1}}{\partial x_i} \quad (16)$$

There is a relationship between the correction value of the velocity u_i' and the correction value of the particle number density n' from the equation of continuity.

$$\frac{1}{n^0}\frac{n'}{\Delta t} + \frac{\partial u_i'}{\partial x_i} = 0 \quad (17)$$

The Poisson equation is obtained from Equations (15), (16) and (17).

$$\frac{\partial^2 p^{k+1}}{\partial x_i^2} = -\frac{\rho}{\Delta t^2}\frac{n^* - n^0}{n^0} \quad (18)$$

According to the previous studies (Cleary & Monaghan 1999), the left hand side of Equation (18) can be discretized as follows.

$$\frac{\partial^2 p^{k+1}}{\partial x_i^2}$$
$$= \sum_{\beta}^{N}\frac{4m^\beta}{\rho^\beta}\left(\frac{1}{\rho^\alpha+\rho^\beta}\right)\frac{(p^\alpha-p^\beta)(x_i^\alpha-x_i^\beta)\cdot\frac{\partial W^{\alpha\beta}}{\partial x_i}}{\left|x_i^{\alpha\beta}\right|^2} \quad (19)$$

By solving Equation (18), a pressure p^{k+1} at present time $k+1$ is obtained. In order to prevent the numerical instabilities due to a negative pressure, the negative value of the pressure p^{k+1} set to zero. Using the pressure p^{k+1} from Equation (16), the correction value

of the velocity u_i' is obtained. Moreover using the correction value of velocity u_i', a velocity u_i^{k+1} and a position x_i^{k+1} at present time $k+1$ are obtained.

$$u_i^{k+1} = u_i^* + u_i' \quad (20)$$

$$x_i^{k+1} = x_i^* + \Delta t u_i' \quad (21)$$

4 NUMERICAL RESULTS

4.1 Two-phase flow problem

Two-phase flow analysis is carried out using two Newtonian fluids in the different density of fluid. The deformation behaviors in the different density of fluid are confirmed. Two cases are considered: Case 1: fluid II is lighter than fluid I and Case 2: fluid II is heavier than fluid I. Figure 2 shows the numerical model in this simulation. The number of particles including the particles comprising the wall is 4,408. The initial interparticle distance is 1.0 cm and the radius of influence domain is 2.0 cm. The acceleration of gravity is 9.81 m/s^2. Tables 1 and 2 summarize the material parameters of fluid I and II, respectively.

Figures 3 (a) and (b) show the simulated time histories of the surface configurations at the different time step for Cases 1 and 2, respectively. In Case 1, it is confirmed that the fluid II float on the surface of fluid I

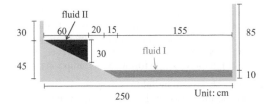

Figure 2. Numerical model for two-phase flow problem.

Table 1. Material parameters of fluid I.

Type of fluid I	Newtonian fluid
$\rho[\text{kg/m}^3]$	1000
$\eta[\text{Pa} \cdot \text{s}]$	0.002

Table 2. Material parameters of fluid II.

Case	1	2
Type of fluid II	Newtonian fluid	
$\rho[\text{kg/m}^3]$	500	1500
$\eta[\text{Pa} \cdot \text{s}]$	0.002	

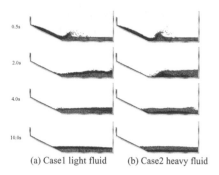

(a) Case1 light fluid   (b) Case2 heavy fluid

Figure 3. Time histories of surface configurations.

after fluid II break down from the slope for the difference between the density of fluid I and II. On the other hand, in Case 2, it is confirmed that the fluid II sink under fluid I. In both cases, it is found that the boundary between the fluid I and II is possible to obviously express.

4.2 Bearing capacity analysis of cohesive ground

Bearing capacity analysis of cohesive ground is carried out. The simulated results are compared with the theoretical solution obtained from Prandtl's theory (Prandtl 1920) as follows.

$$q_u = (2 + \pi)c \cong 5.14c \qquad (22)$$

where q_u is the ultimate bearing capacity, π is the circle ratio, c is the cohesion of soil. It was confirmed that Equation (22) is theoretical solution by some numerical method (Tamura et al. 1994). However, Equation (22) involves some following assumptions.

- Pure cohesive material ($c > 0$, $\phi = 0$)
- No friction between rigid body and ground
- Weightless material

Figure 4 shows the numerical model used in this simulation. A footing is placed on the ground, and the footing is assumed to be a rigid body in this simulation. The footing moves downward at a constant velocity. The constant value of 1.0×10^{-5} m/s is set in this simulation. In order to prevent the penetration, the constant velocity is set to the small value. The number of particles including the particles comprising the wall is 5,146. The initial interparticle distance is 0.01 m and the radius of influence domain is 0.02 m. The density of geomaterial is 1,000 kg/m³. Table 3 summarizes the material parameters. Three different values of cohesion are used in this simulation. In addition, in order to ignore the influence of gravity, the gravity is not acted as the external force. Moreover the boundary conditions between the bottom of the footing and the surface ground are non-slip boundary. The vertical stresses in the soils below the footing are

calculated and the bearing capacities are determined with the average value of the vertical stresses in the particles below the footing.

Figure 5 shows the relationship between cohesion and bearing capacity at the different cohesion. From Figure 5, the simulated results are confirmed that the bearing capacity is changed by the value of the cohesion of soil and is obtained with a certain level of the accuracy.

4.3 Simulation of penetration of rigid body

Simulations of penetration of a rigid body into different materials are performed under an identical condition with the existing simulated results (Moriguchi 2005). Figure 6 shows the numerical model used in this simulation. The number of particles including the particles comprising the wall is 1,250. The initial interparticle distance is 0.005 m and the radius of

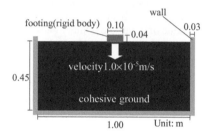

Figure 4. Numerical model for bearing capacity problem.

Table 3. Material parameters.

Case	1	2	3
η_0[Pa · s]		1.0	
c[Pa]	1.0	2.0	3.0
ϕ[deg]		0.0	

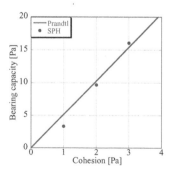

Figure 5. Relationship between cohesion and bearing capacity.

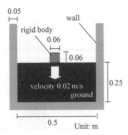

Figure 6. Numerical model for penetration problem.

Table 4. Material parameters.

Case	1	2	3
Type of geomaterials	Newtonian fluid	granular material	cohesive material
η_0[Pa · s]	0.002		1.0
c[Pa]	0.0	0.0	1000.0
ϕ[deg]	0.0	30.0	0.0

influence domain is 0.01 m. The density of geomaterial is 1,000 kg/m³. The acceleration of gravity is 9.81 m/s². Three cases of the geomaterials penetrated into the ground are considered: Case 1: Newtonian fluid and Case 2: granular material and Case 3: cohesive material. Table 4 summarizes the material parameters, respectively. The rigid body moves downwards at a constant velocity at 0.02 m/s and the geomaterials around the rigid body move due to motion of the rigid body.

Figures 7 (a), (b) and (c) shows the simulated time histories of the surface configurations at the different time step with the existing simulated results (Moriguchi 2005) for three cases, respectively. In all cases, it can be seen that the ground deformed due to the motion of the rigid body. In Case 1, the surface configuration become almost flat after the rigid body is completely submerged into the water along with the existing simulated results. On the other hand, in Cases 2 and 3, the surface configurations are not flat after the rigid body penetrated into the ground along with the existing simulated results. The granular material moved to the top of the rigid body after the rigid body penetrated down into the ground. As to the cohesive material in the existing simulated results, it is found that the material stood on by itself, and the shape is kept even if the rigid body penetrated deeply into the ground. However, the material slightly moved after the rigid body penetrated down into the ground in this simulation.

4.4 Flow failure of geomaterial with protecting

Simulation of a flow failure of geomaterials with protecting walls after slope failure is carried out. Two

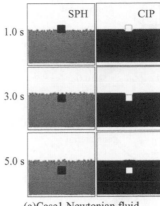

(a)Case1 Newtonian fluid

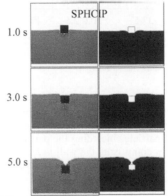

(b)Case2 granular material with $\phi = 30$deg

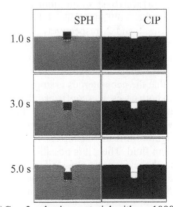

(c)Case3 cohesive material with c =1000Pa

Figure 7. Time histories of rigid material penetration.

cases are considered: Case 1: the slope with lower protective wall and Case 2: the slope with higher protective wall. Figure 8 shows the numerical model in this simulation, respectively. In both cases, black section in Figure 8 will break down. Other sections in Figure 8 are

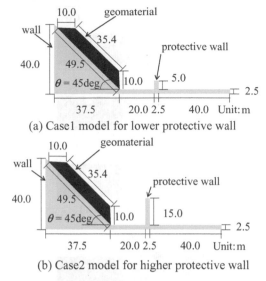

(a) Case1 model for lower protective wall

(b) Case2 model for higher protective wall

Figure 8. Numerical model.

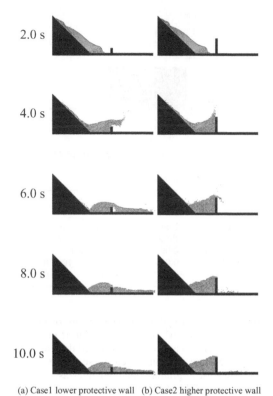

(a) Case1 lower protective wall (b) Case2 higher protective wall

Figure 9. Time histories of surface configurations.

assumed as rigid bodies. The flowed geomaterial set granular material with internal angle ϕ of 30 deg. The number of particles including the particles including the particles comprising the wall is 2,280 and 2,380, respectively. The initial interparticle distance is 0.5 m and the radius of influence domain is 1.0 m. The density of geomaterials is 2,000 kg/m^3. The acceleration of gravity is 9.81 m/s^2.

Figures 9 (a) and (b) show the simulated time histories of the surface configurations at the different time step for Cases 1 and 2, respectively. In Case 1, it is found that the geomaterials flow along the slope and overflow above the protecting wall after the geomaterials hit the protecting wall. On the other hand, in Case 2, it is found that the geomaterials is stopped by protecting wall. In both cases, it is confirmed that the surface configuration after the geomaterials hit the protecting wall shows the characteristics of granular materials. This configuration is the inexpressible configuration in Newtonian fluid. Thus, it is possible to express the deformation behavior of the flow failure of the geomaterials and the difference of the deformation behavior with the different height of the protecting walls.

5 CONCLUSIONS

In this study, the applicability of SPH method based on fluid dynamics for large deformation analyses of geomaterials is discussed. First, two-phase flow analysis is carried out using two Newtonian fluids in the different density of fluid. It is confirmed that the deformation behavior of difference of density of fluid is obviously expressed. Second, bearing capacity analysis of cohesive ground is carried out. The simulated results are confirmed that the bearing capacity is changed by the value of the cohesion of soil and is obtained with a certain level of the accuracy. Third, simulations of penetration of a rigid body into different materials are performed under an identical condition. Based on the comparison between the simulated results and the existing simulated results, it is confirmed that the different deformation behavior by characteristics of geomaterials and the large deformation behavior of geomaterials is expressed. Finally, simulation of the flow failure of geomaterials with protecting walls after slope failure is carried out. It is possible to express the deformation behavior of flow failure of the geomaterials and the difference of the deformation behavior with the different height of the protecting walls. From the numerical results, it is showed that SPH method is possible to predict the large deformation problem of geomaterials.

As the issue on future, we should modify the calculational procedure of pressure and improve the accuracy of analysis in the vicinity of free surface. Moreover we want to work on the expansion for three dimensional analyses and to reproduce real phenomenon of the slope failure.

REFERENCES

Amsdam, A.A. & Harlow, F.H. 1970. The SMAC Method: A Numerical Technique for Calculating Incompressible Fluid Flow. LA-4370. Los Alamos Scientic Laboratory.

Cleary, W. & Monaghan, J.J. 1999. Conduction modeling using smoothed particle hydrodynamics. *J. Comput. Phys.* 148: 227–264.

Gingold, R.A. & Monaghan, J.J. 1977. Smoothed particle hydrodynamics: theory and application to non-spherical stars. *Monthly Notices of the Royal Astronomical Society* 181: 375–389.

Hadush, S. 2002. Fluid dynamics based large deformation analysis in geomechanics with emphasis in liquefaction induced lateral spread. Ph.D. Dissertation. Gifu University.

Libersky, L.D., Petschek, A.G., Carney T.C., Hipp, J.M. & Allahdadi, F.A. 1993. High Strain Lagrangian Hydrodynamics. *J. Comput. Phys.* 109: 67–75.

Liu, G.R. & Liu, M.B. 2003. Smoothed Particle Hydrodynamics: A Meshfree Particle Method. *World Scientific* 449 p.

Lucy, L.B. 1977. A numerical approach to the testing of the fission hypothesis. *Astron. J.* 82: 1023–1024.

Maeda, K., Sakai, H. & Sakai, M. 2006. Development of seepage failure analysis method of ground with smoothed particle hydrodynamics. *Journal of structural and earthquake engineering, JSCE* 23(2): 307–319.

Moriguchi, S. 2005. CIP-based numerical analysis for large deformation of geomaterials. Ph.D. Dissertation. Gifu University.

Prandtl, L. 1920. Über die Härte plastischer Körper: Nachrichten von der Königlichen Gesellschaft der Wissenschaften zu Göttingen, *Math. Phys. KI.* 12: 74–85.

Sakai, Y., Yang, Z.Y. & Jung, Y.G. 2004. Incompressible viscous flow analysis by SPH. *Journal of the Japan Society of Mechanical Engineers Series* B. 70. 696: 1949–1956. (In Japanese).

Tamura, T., Kobayashi, H. & Sumi, T. 1984. Limit analysis of soil structure by rigid plastic finite element method. *Soils and Foundations* 24(1): 34–42.

Uzuoka, R. 2000. Analytical study on the mechanical behavior and prediction of soil liquefaction and flow. Ph.D. Dissertation. Gifu University. (In Japanese).

Prediction and Simulation Methods for Geohazard Mitigation – Oka, Murakami & Kimoto (eds)
© *2009 Taylor & Francis Group, London, ISBN 978-0-415-80482-0*

Effects of initial soil fabric and mode of shearing on quasi-steady state line for monotonic undrained behavior

S. Yimsiri
Burapha University, Chonburi, Thailand

K. Soga
University of Cambridge, Cambridge, UK

ABSTRACT: The effects of initial soil fabric and mode of shearing on quasi-steady state line in void ratio-stress space are studied by employing the Distinct Element Method numerical analysis. The results show that the initial soil fabric and the mode of shearing have a profound effect on the location of the quasi-steady state line. The evolution of the soil fabric during the course of undrained shearing shows that the specimens with different initial soil fabrics reach quasi-steady state at various soil fabric conditions. At quasi-steady state, the soil fabric has a significant adjustment to change its behavior from contractive to dilative. As the stress state approaches the steady state, the soil fabrics of different initial conditions become similar. The numerical analysis results are compared qualitatively with the published experimental data and the effects of specimen reconstitution methods and mode of shearing found in the experimental studies can be systematically explained by the numerical analysis.

1   MONOTONIC UNDRAINED BEHAVIOR OF SAND

The undrained response of loose sand shows that, despite the large difference in the stress-strain behavior at an early stage of loading, the samples tend to exhibit an almost identical behavior at a later stage where the developed axial strain becomes very large. The state of sand deforming continually, keeping the volume constant, under a constant shear stress and confining stress, is called the steady state (Castro 1975). When the sand is loose and is subjected to a large confining stress, it tends to deform fairly largely at the beginning exhibiting contractive behavior with temporary drop in the shear stress, and then starts to exhibit dilative behavior reaching the steady state at the end (flow with limited deformation behavior). The state where a temporary drop in shear stress takes place over a limited range of shear strains is termed as the quasi-steady state (Ishihara 1993).

Ishihara (1993) showed that the characteristic lines sand are different for different methods of specimen reconstitution (see Fig. 1). His results show that the Isotropic Consolidation Line (ICL) and the quasi-steady state line (QSSL) are distinctly different between the two types of specimen reconstitution methods; however, both methods yield a unique steady state line (SSL). Moreover, the QSSL and SSL discussed in most literatures are based mainly on the results obtained from triaxial compression tests. Vaid et al. (1990), Ishihara (1996), and Riemer &

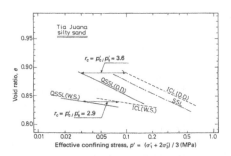

Figure 1.   Characteristic lines from different specimen reconstitution methods (DD-dry deposition and WS-water sedimentation) (Ishihara 1993).

Seed (1997) showed that the mode of shearing also influences the location of QSSL as shown in Fig. 2.

2   DISTINCT ELEMENT METHOD NUMERICAL ANALYSIS

The DEM numerical analysis is performed by PFC^{3D} (Itasca 1999). PFC^{3D} models the movement and interaction of the stressed assemblies of rigid spherical particles. The linear elastic-perfectly plastic contact model is employed in this research. The contact forces and relative contact displacements are linearly related by constant contact stiffnesses. During the analysis,

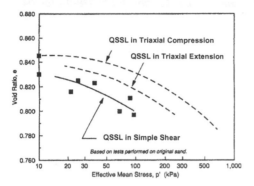

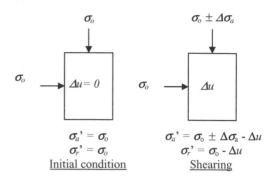

Figure 2. Quasi-steady state lines from simple shear, undrained triaxial compression and extension tests (after Reimer & Seed 1997).

Figure 3. Calculation of Δu from constant volume shearing test in DEM analysis.

several microscopic quantities can be obtained through built-in functions. These quantities include coordination number, porosity, sliding fraction, stress, and strain rate. Additional parameters are introduced in this study and the analysis is done by user-defined functions.

Strains are calculated from the user-defined definition which is derived from the movements of the six rigid walls used as the boundaries of a simulated cubical triaxial specimen. The compressive strain is taken to be positive.

In constant volume shearing with constant total radial stress condition ($\Delta\sigma_r = 0$), the excess pore water pressure Δu can be calculated as shown in Fig. 3 under an assumption of fully saturated condition. This is due to the fact that the axial stress σ_a' and the radial stress σ_r' derived from DEM analysis are in terms of effective stress, as they are fully transmitted through the inter-particle contacts.

Fabric tensor is introduced as an additional index in order to describe the packing structure of spherical particles. According to Oda (1982), the second-order fabric tensor F_{ij} for the assembly of uniform-size spheres is defined as;

$$F_{ij} = \int_0^{2\pi} \int_0^{\pi} n_i n_j E(\gamma, \beta) \sin \gamma \, d\gamma \, d\beta \tag{1}$$

where n_i is the contact normal in i-direction, $E(\gamma,\beta)$ is the contact normal distribution function (spatial probability density function of n), and γ, and β are defined in Fig. 4.

In the case of a cross-anisotropic fabric with its symmetry along the 3-axis in Fig. 4, $E(\gamma,\beta)$ can be expressed by the following Fourier series (Chang et al. 1989).

$$E(\gamma, \beta) = \frac{3(1 + a\cos 2\gamma)}{4\pi(3 - a)} \tag{2}$$

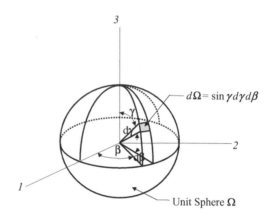

Figure 4. Elementary solid angle.

where a is the degree of fabric anisotropy ($-1 < a < 1$). Eq. (2) has the symmetry $E(\pi + \gamma) = E(\gamma)$ and is independent of β.

The contact normal distribution function according to Eq. (2) for different a values are shown in Fig. 5. When $a > 0$, the contact normals of the particles in the assembly tend to concentrate in the vertical direction (3-direction), whereas, when $a < 0$, they tend to concentrate in the horizontal direction (1- and 2-directions).

By substituting Eq. (2) into Eq. (1) and completing the integration, the fabric tensor F_{ij} is obtained as shown in Eq. (3). It can be seen that the fabric tensor in the cross-anisotropic fabric condition is represented by a single parameter a.

$$F_{ij} = \begin{bmatrix} \frac{3a-5}{5(a-3)} & & \\ & \frac{3a-5}{5(a-3)} & \\ & & \frac{-(5+a)}{5(a-3)} \end{bmatrix} \tag{3}$$

236

Table 1. Numerical analysis program.

Test No.	Initial void ratio	Isotropic confining stress, p'_o(MPa)	Drainage condition	Mode of shearing	Initial degree of fabric anisotropy, a_o
LUC-1	0.76 (loose)	1	undrained	compression	0.09 ($a_o > 0$)
LUC-2	0.76 (loose)	1	undrained	compression	0.01 ($a_o \approx 0$)
LUC-3	0.76 (loose)	1	undrained	compression	$-0.11(a_o < 0)$
LUE-1	0.76 (loose)	1	undrained	extension	$+0.09$ ($a_o > 0$)
LUE-2	0.76 (loose)	1	undrained	extension	$+0.01(a_o \approx 0)$
LUE-3	0.76 (loose)	1	undrained	extension	$-0.11(a_o < 0)$
LDC-1	0.76 (loose)	1	drained	compression	$a_o \approx 0$
LDE-1	0.76 (loose)	1	drained	extension	$a_o \approx 0$
DUC-1	0.61 (dense)	1	undrained	compression	$+0.15$ ($a_o > 0$)
DUC-2	0.61 (dense)	1	undrained	compression	-0.02 ($a_o \approx 0$)
DUC-3	0.61 (dense)	1	undrained	compression	-0.20 ($a_o < 0$)
DUE-1	0.61 (dense)	1	undrained	extension	$+0.15$ ($a_o > 0$)
DUE-2	0.61 (dense)	1	undrained	extension	-0.02 ($a_o \approx 0$)
DUE-3	0.61 (dense)	1	undrained	extension	0.20 ($a_o < 0$)
DDC-1	0.54 (dense)	10	drained	compression	$a_o \approx 0$
DDE-1	0.54 (dense)	10	drained	extension	$a_o \approx 0$

During DEM simulation, the data of contact normal distribution can also be directly obtained and the fabric tensor can be analyzed according to Eq. (4) which is a limited form of Eq. (1).

$$F_{ij} = \frac{1}{N_c} \sum_{N_c} n_i n_j \qquad (4)$$

where N_c is the number of contacts between particles only.

By comparing the fabric tensor obtained from Eq. (4) with the analytical one shown in Eq. (3), the degree of fabric anisotropy a can finally be computed from the DEM analysis.

3 NUMERICAL ANALYSIS PROGRAM AND PROCEDURES

The numerical analysis program is designed to investigate the effects of initial anisotropic soil fabric and mode of shearing on characteristic lines in void ratio—stress space and also to monitor the change of soil fabric during the course of shearing. The specimens are divided into two groups (loose and dense) according to their void ratios. All specimens are prepared under similar isotropic confining stress of 1 MPa, the magnitude of which is arbitrary and does not intend to be quantitatively compared to that of a real soil. The specimens in each group are prepared to have almost identical void ratio but different initial soil fabric prior to shearing. After that, the specimens are sheared in undrained condition with different modes of shearing (compression/extension). Some specimens are subjected to further isotropic confining stress of 10 MPa before drained shearing. The results of these

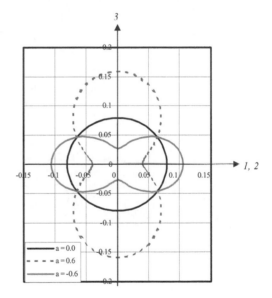

Figure 5. Contact normals distribution function.

drained tests are to help defining SSL. The details of numerical analysis program are listed in Table 1.

Cubical assemblies of uniform-sized spheres contained within six rigid boundary walls are used as simulated soil specimens (see Fig. 6). The input properties are shown in Table 2. Triaxial tests are simulated by moving the six rigid walls in a specified strain-rate as the strain-controlled condition or to a specified stress using servo-controlled as the stress-controlled condition. The undrained condition is achieved by keeping the volume constant during shearing. The specimen

237

Table 2. DEM simulation properties.

Number of particles	Loose specimen: approx. 3500 Dense specimen: approx. 3800
Radius of particle	4 cm
Initial specimen size	$1.2 \times 1.2 \times 1.2$ m^3
Initial void ratio (at $p'_o = 1$MPa)	Loose specimen: 0.76 Dense specimen: 0.61
Particle density	2700 kg/m^3
Inter-particle friction angle	45°
Particle-wall friction angle	0° (smooth wall)
Contact model	Linear elastic—perfectly plastic
Normal contact stiffness	1×10^8 N/m
Tangential contact stiffness	1×10^8 N/m
Wall normal contact stiffness	1×10^8 N/m
Wall tangential contact stiffness	0 (no wall tangential stiffness)

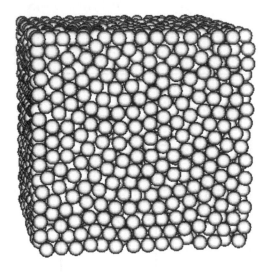

Figure 6. Assemblies of particles after specimen set-up (walls are omitted).

size is chosen to be relatively large compared with the particle size to accommodate around 3500–3800 particles. The inter-particle friction angles are set to be $\phi_\mu = 45°$ which is relatively high compared with that of glass bead (ϕ_μ of glass bead = 15°) because it gives more realistic and stable results. This is due to the spherical nature of the simulated particles which renders excessive rolling in numerical simulation.

4 MONOTONIC UNDRAINED BEHAVIOR

The effects of the initial degree of fabric anisotropy a_o on the undrained behavior are presented in Fig. 7 from the data of dense sand specimen. It can be seen that the

undrained behaviors of the specimens with different initial soil fabrics are very different even though they have almost identical void ratio. In the undrained compression shearing (positive q), the specimen that has contact normals concentrating in vertical direction ($a_o > 0$) has higher stiffness and is more dilative, whereas the specimen that has contact normals concentrating in horizontal direction ($a_o < 0$) has lower stiffness and is more contractive (see Figs. 7b, c). The behavior is opposite in the undrained extension shearing (negative q). The specimen with isotropic fabric ($a_o \approx 0$) behaves as an isotropic elastic material by having the undrained stress path in the vertical direction as shown in Fig. 7a. It is noted that its behavior does not fit between the two extreme of a_o. Higher strength and stiffness of the isotropic-fabric specimen may be due to the higher initial coordination number which may have resulted from some forms of more stable structure due to the packing of uniform-sized particles. Further discussion on the effects of initial degree of fabric anisotropy on the undrained behavior of sand by DEM analysis is presented by Yimsiri & Soga (2001).

5 CHARACTERISTIC LINES IN VOID RATIO—STRESS SPACE

5.1 Effects of initial soil fabric

The DEM results of undrained triaxial tests are presented in the e–log p' space as shown in Fig. 8. The QSSLs of the specimens with different initial degrees of fabric anisotropy a_o are shown together with the SSL obtained from the drained tests. Although the determination of the locations of QSSLs may be quite subjective, it can be seen that the initial degree of fabric anisotropy has a profound effect on the locations of the QSSLs. In compression shearing, the QSSL moves to the right (closer to SSL) as the initial degree of fabric anisotropy a_o increases because the sand specimen becomes more dilative in compression shearing. On the other hand, in extension shearing, the QSSL moves to the left (away from SSL) as the initial degree of fabric anisotropy a_o increases because the sand specimen becomes more contractive in extension shearing.

The change of the degree of fabric anisotropy a with log p' during the course of undrained shearing is shown in Fig. 9. At QSS, the degree of fabric anisotropy a considerably changes without significant change in the effective confining stress. After QSS, the specimens with different initial degrees of fabric anisotropy have similar degree of fabric anisotropy as their stress states approaching SS. The specimens reach QSS with different degrees of fabric, which reflects the dependence of QSSL on the initial degree of fabric anisotropy. At QSS, however, the soil fabric has a significant

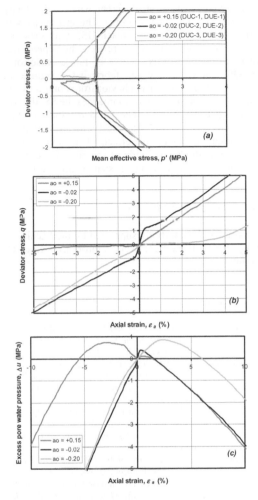

Figure 7. Effects of initial soil fabric on undrained behavior: dense specimen.

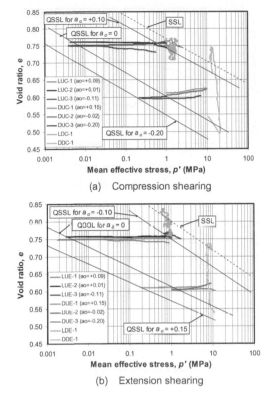

(a)   Compression shearing

(b)   Extension shearing

Figure 8. Effects of initial soil fabric on position of quasi-steady state lines.

adjustment to change its behavior from contractive to dilative. As the stress state approaches the SS, the degree of fabric anisotropy becomes similar.

The DEM results presented in Fig. 8 are qualitatively consistent with the published experimental data discussed earlier. The evolution of soil fabric in Fig. 9 also implies that the quasi-steady state is profoundly influenced by the fabrics formed by different modes of sand deposition, whereas the ultimate steady state is established uniquely for a given sand independently of the mode of sand deposition. The non-uniqueness of the quasi-steady state can be understood with good reasons, if one is reminded of the state of the sand being deformed in the intermediate-strain range still preserving its inherent fabric structure formed during the process of deposition. When the sand is deformed largely, the remnants of initially formed fabric structure are completely destroyed and consequently the

ultimate steady state becomes unaffected by the mode of sand deposition.

5.2   *Effects of mode of shearing*

The DEM results in Figs. 8 and 9 also show the effects of mode of shearing (triaxial compression and extension) on the location of QSSL. The results also indicate that the difference in the location of QSSL in triaxial compression and extension depends on the initial soil fabric. As discussed earlier, the specimens with contact normals concentrating in the vertical direction ($a_o > 0$) are more dilative in compression shearing but more contractive in extension shearing; therefore, the QSSL of the specimens with $a_o > 0$ from triaxial compression should lie to the right of the QSSL from triaxial extension. On the other hand, the specimens with contact normals concentrating in the horizontal direction ($a_o < 0$) are more contractive in compression shearing but more dilative in extension shearing; therefore, the QSSL from triaxial compression should lie to the left of the QSSL from triaxial extension. The DEM results presented in this study are qualitatively consistent with the published experimental data

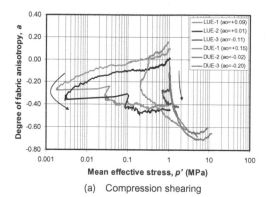

(a)   Compression shearing

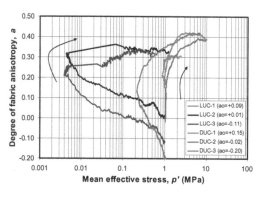

(b)   Extension shearing

Figure 9.   Evolution of degree of fabric anisotropy against mean effective stress.

discussed earlier. It is worth noting here that this conclusion is based only on the published data of triaxial compression tests. It was not possible to find any data from the extension case. However, the DEM results shown in this research indicate that the behavior in the extension mode should be opposite of the compression mode. Ladd (1977) also suggested that a preferred fabric orientation, which results in a higher static strength in compression, will not necessarily result in a higher static strength in extension.

6   CONCLUSIONS

In this research, the DEM analysis is utilized to simulate the undrained triaxial compression/extension tests of specimen with different initial soil fabric and the results are examined to discuss about the effects of initial soil fabric and mode of shearing on the quasi-steady state line in the e–log p' space. It can be seen that the initial soil fabric has a profound effect on the locations of the QSSLs. Moreover, it is found that the quasi-steady state lines from triaxial compression

and triaxial extension are different, whereas the steady state lines are reasonably similar. The specimens with different initial soil fabric reach QSS with different soil fabric conditions, which reflects the dependence of QSSL on the initial degree of fabric anisotropy. At QSS, however, the soil fabric has a significant adjustment to change its behavior from contractive to dilative. As the stress state approaches the SS, the soil fabrics become similar. The DEM results from this study compare qualitatively well with the published experimental data and the effects of specimen reconstitution methods and mode of shearing found in the experimental studies can be systematically explained.

ACKNOWLEDGEMENTS

This research is partially supported by the Faculty of Engineering, Burapha University through its Research Grant #68/2551.

REFERENCES

Castro, G. 1975. Liquefaction and Cyclic Mobility of Saturated Sands. *Journal of the Geotechnical Engineering Division ASCE* 101(GT6): 551–569.

Chang, C.S., Sundaram, S.S., and Misra, A. 1989. Initial moduli of particulated mass with frictional contacts. *International Journal for Numerical and Analytical Methods in Geomechanics* 13: 629–644.

Ishihara, K. 1993. Liquefaction and flow failure during earthquakes. *Geotechnique* 43(3): 351–415.

Ishihara, K. 1996. *Soil Behaviour in Earthquake Geotechnics*. Clarendon Press: Oxford.

Itasca Consulting Group, Inc. 1999. *PFC^{3D}: Particle Flow Code in 3 Dimensions, Version 2.0.* Vol. 1, 2, 3.

Ladd, R.S. 1977. Specimen Preparation and Cyclic Stability of Sands. *Journal of the Geotechnical Engineering Division ASCE* 103(GT6): 535–547.

Oda, M. 1982. Fabric tensor for discontinuous geological materials. *Soils and Foundations* 22(4): 96–108.

Riemer, M.F. and Seed, R.B. 1997. Factors Affecting Apparent Position of Steady-State Line. *Journal of Geotechnical and Geoenvironmental Engineering ASCE* 123(3): 281–288.

Vaid, Y.P., Chung, E.K.F. and Kuerbis, R.H. 1990. Stress path and steady state. *Canadian Geotechnical Journal* 27: 1–7.

Yimsiri, S. and Soga, K. 2001. Effects of soil fabric on undrained behaviour of sands. *Fourth International Conference on Recent Advances on Geotechnical Earthquake Engineering and Soil Dynamics, San Diego, USA.*

Prediction and Simulation Methods for Geohazard Mitigation – Oka, Murakami & Kimoto (eds)
© 2009 Taylor & Francis Group, London, ISBN 978-0-415-80482-0

Study of the interactions between rivers dynamic and slope stability for geohazard prediction: A case in Val Trebbia (Northern Italy)

L. Scesi & P. Gattinoni
Politecnico di Milano, Milano, Italy

ABSTRACT: The paper deals with a methodology to assess the landslide risk along rivers on the basis of geological and hydraulic conditions, recognized as critical (especially as regards the hydrometric level variation and the shear stress for erosion). As an example, the methodology was applied in the middle Trebbia Valley (Northern Italy) where slopes are characterize by several landslides and the Trebbia River, often braided, curves weak rocks. The study allowed to point out the influence of the River Trebbia erosion on the slope evolution providing a methodological approach to introduce hydrologic aspects into landslides hazard assessment.

1 INTRODUCTION

This paper presents the results of a research concerning the hydraulic and hydrogeological aspects involved in the landslide processes. The aim is to quantify the interaction between slope instability and hydraulic and hydrologic parameters (especially as regards the hydrometric level variation and the shear stress linked to the erosion) in connection with the different lithological—structural conditions.

As an example, the study was applied in the middle Trebbia Valley (Northern Italy). In particular, the interaction between the Trebbia River (in between the Aveto stream confluence and the bridge situated to the south of Rivalta) and the slopes nearby the same River was identified.

The paper is divided into three sections:

- the first concerns the geological and morphological study of the area, with particular reference to the evolution of the river-bed and to the location of those areas subjected to landslides;
- the second concerns the hydrologic and hydraulic study of the River, especially in terms of flood contours and shear stress for erosion. So it has been possible to define the critical zones, as regards the hydraulic point of view, and to compare them with the landslides map;
- the third concerns the study of a specific area called "Barberino gorge", where a wide landslide in progress is present and where the shear stress for erosion is particularly high.

2 LANDSLIDE HAZARD ALONG RIVERS

Generally, landslide hazard depends on a number of concomitant factors; among them, water is often very important; therefore, the landslide hazard assessment has to consider both common predisposing factors (slope gradient, lithology, land cover, morphology, drainage area) and the groundwater setting that brings about a hydrogeological susceptibility (Gattinoni 2008).

Besides, to identify the landslide hazard areas along streams it is necessary to define the characteristic of the stream, meant as a complex system constituted by water and alluvial material subject to erosion, transportation and sedimentation.

Usually, the hydrological susceptibility to landslides is aimed to point out the pluviometrical thresholds, that can be assessed either through a deterministic approach (based on the numerical modeling of the infiltration process and the slope stability) or through a statistic approach (based on the time series analysis of landslides and rainfalls). A mix of the two approaches was recently proposed by Benedetti et al. (2006).

In addition to these aspects, linked to the infiltration and the saturation of the superficial deposits, other factors linked to the fluvial dynamic are essential in order to evaluate the slope stability: the erosion degree wielded by the river current at the slope foot and the saturation degree change linked to flood events. Indeed, where the river erodes, the dip of the banks increases and, consequently, a potential slope instability, also large, can occur.

Besides these erosive processes, the forces (e.g. water pressure and material strength) change, in consequence of meteoric water infiltration. This happens because the whole water volume seeped into the slope can gather near the channel so that the water table and the percentage of saturated soil (if thickness is equal) are higher. Where the river erodes and the banks dip increase, the water table can rise to the surface bringing about following consequences:

- the saturation of soil can be reached easier,

- the concentration of the groundwater downflow can trigger underground erosion phenomena (Hagerty 1991),
- the surface erosion phenomena linked to the saturation of the soils having high dip increase the instability conditions.

Therefore, the knowledge of the river dynamic is very important to study the effects of the water flow on landslide susceptibility. The research was aimed at creating a tool for the susceptibility assessment as a function of the river dynamic.

3 GEOLOGICAL, GEOMORPHOLOGIC AND HYDROGRAPHICAL SETTING OF TREBBIA VALLEY

3.1 Geological setting

The studied area is located in the Northern Apennine. The following tectonic Units can be recognized:

Figure 1. Trebbia River meanders.

- "Liguri Unit": characterized by ophiolites, or green magmatic rocks (Jurassic basalts and gabbros), coming from oceanic basins. These Ophiolites are connected to the Cretaceous pelagic sediments (deepwater) and to the Turbidite (Paleocene – Eocene aged).
- "Epiligure Sequence": a thick sedimentary sequence constituted by clastic rocks like conglomerates and sandstones, originated from the Ophiolites and from the sedimentary cover erosion; this sequence settled on the "Unità Liguri";
- "Subliguri Units": constituted by a sequence of Tertiarysedimentary rocks (mainly Oligocene-Miocene);
- "Toscana Unit": mainly constituted by the Bobbio Formation sandstones (San Salvatore Sandstones) Low Miocene aged. These sandstones widely outcrop along the meanders of the Trebbia River.

3.2 Geomorphologic and hydrographic setting

The Trebbia River, the most important stream of the area having the same name of the valley, has its spring in the Ligurian Apennine (Prelà mountain, 1406 m a.s.l.) and flows into the Po River to the west of Piacenza, after flowing a distance of about 115 km. The Trebbia River can be divided in three stretches on the basis of geomorphologic features and hydraulic behaviour:

- a mountain stretch (from the source to Bobbio): this stretch is very embanked with irregular meanders, having a marked bending and a slow evolution, carved in the bedrock (Fig. 1); this portion of Trebbia Valley is characterized by high relief energy (frequent slope breaking), and youthful erosion forms, consequently the river can mobilize

amounts of sediments and transport them downstream rather then deposit them;
- a middle stretch, where slopes and river-bed have a low gradient. Here the bedrock is constituted by weak rocks and the river becomes braided. Flood plains are larger, many fans are present, where tributaries join the Trebbia River, and slopes are characterize by several landslides;
- a floodplain stretch: the river-bed is typically braided, it flows in a wide fan with many flood beds and thick alluvial sediments till the confluence with the Po River; the deposition of sediments are favored by a low flow rate.

In particular, this study regards the middle stretch of the Trebbia River where many landslides occur.

4 LANDSLIDE HAZARD MAP

In the middle-lower Trebbia Valley, the sedimentary turbidites shape not very steep slopes. The presence of those rocks is the cause of wide landslides (active and dormant) located near the Trebbia River and its most important tributaries. Consequently, it is possible to observe many morphologic changes of the Trebbia river-bed, like the extinction of some lateral channels, a narrowing or the local erosion of banks. In the upper Trebbia Valley the number of landslides decreases because of the god quality of the outcropping ophiolites and sandstones. Falls, topples, translational and/or wedge slides take the place of flows. These movements are unlikely to reach the river.

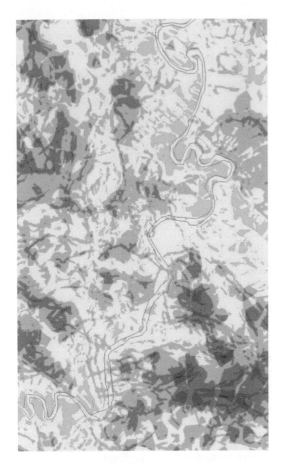

Figure 2. Landslide hazard map. The grey intensity represent the increasing in relative hazard degree.

To map the landslide hazard of the area (Fig. 2) the Zermos methodology was used (Humbert 1977), based on following parameters: landslides location (Regione Emilia Romagna 2002), elevation, slope gradient and exposure, lithology, landslide activity, closeness to landslides, streams, roads and railways, folds and faults, use of soil.

5 FLUVIAL DYNAMIC ANALYSIS

5.1 Morphologic evolution

The in situ surveys allowed to localize the stretches of the river characterized by particular evolutionary trends and, on the basis of existing studies (Bezoari et al. 1984), it was possible to observe that:

– the Trebbia River has a high solid load; for this reason, in the past it was interested by an intensive quarrying activity (Table 1);

Table 1. Characteristics and solid load of Trebbia River, according to the PAI study (Hydrogeological Setting Plain).

Area	920 km^2	Average level	822 m a.s.l.
Average annual rainfall	1398 mm	Solid load	247.2 · 10^3 m^3/year
Specific erosion	0.27 mm/year	Bed solid load	27 · 10^3 m^3/year
Suspended solid load	190.6 · 10^3 m^3/year	Total solid load	217.6 · 10^3 m^3/year

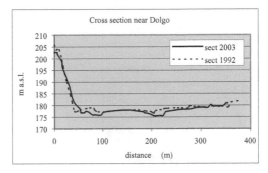

Figure 3. Examples of cross-section.

– in the last few years this capacity decreased, also because of the improvement in watersheds management along many tributaries, especially in the mountain areas of the basin. The quarrying activity together with the basin watersheds management caused changes in the dynamic equilibrium of the River.

Comparing the morphologic cross-sections, referred to different periods, these erosion phenomena can be observed. In particular, cross-sections made in between 1815 and 2000 and the results of a recent morphologic study (Sogni 2003) were examined. The most important phenomena (banks erosion, overflow, etc.) are present in the middle-high part of Trebbia Valley, with:

– a narrowing trend of the river-bed flooding,
– a deep erosion of the river-bed, especially where there is no bedrock, while upstream the lowest altitudes of the river-bed decrease and the medium altitudes increase,
– a deep erosion especially near the river banks (Fig. 3); in the long period, this will trigger instability phenomena.

5.2 Hydrologic and hydraulic study

The hydrologic study was carried out just next to 74 cross-sections, realized along the Trebbia River by AIPO (Interregional Agency for Po River 2001). The peak discharges for assigned return periods, the corresponding water surface elevations and the shear stress for erosion were calculated.

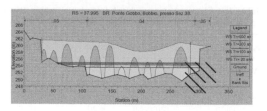

Figure 4. Flood profiles for assigned return periods.

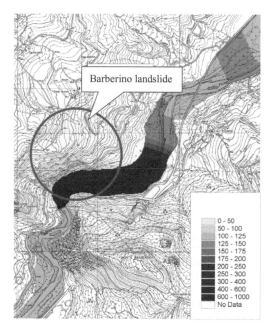

Figure 5. An example of shear stress for erosion distribution along the Trebbia River, corresponding to a 20-year return period. The dark colors means higher shear stress values in Pa.

Figure 6. The grey intensity corresponds to increasing instability index. Two large landslides (in the circles) are situated where the instability index is very high.

landslides are present (for example near the Barberino bridge and to the North of Perino village). In these areas, also the shear stress for erosion are higher as well as downstream the bridges (Fig. 5) and the erosive capacity of the stream, in case of flood, is considerable.

6 INTERACTION BETWEEN RIVER DYNAMIC AND SLOPE STABILITY

The study allowed to identify the areas characterized by geological and hydrologic critical conditions.

As regards the hydrologic critical conditions of the Trebbia River, the analysis of the water surface variations and the shear stress for erosion distribution enabled to identify following elements:

– the areas interested by flood risk and the slope subject to saturation—desaturation cycles,
– the stretches of the river characterised by erosion/sedimentation,
– the areas subject to the highest hydraulic stress along the slopes foot.

The geological critical conditions were identified by thematic maps and by in situ surveys. A correlation between the fluvial dynamic and the slope stability was identified, comparing the hydrologic and geologic critical conditions.

The results of these analyses were validated through the survey of some geomorphologic instability indicators (Masetti 1998, Simon & Downs 1995). These morphologic indicators were surveyed along the right and the left riverbanks and allowed to define a correlation between the fluvial dynamic and the slope stability (Fig. 6) by mean of an Instability Index. This Index is obtained adding up the values assigned to each morphologic indicators.

By layering the landslide hazard map with the instability index distribution (Fig. 7) it is possible to observe that where the instability indexes are higher landslide hazard is not always high. On the contrary, the in situ survey confirms the presence of landslides (Fig. 8).

The direct methods were adopted to estimate the peak discharges, where a sufficient discharge data set was available, combined with indirect methods (using the hydrologic similarity between the examined catchment basin and the next ones or the inflow-outflow transformation mathematical models). The hydraulic profiles (Fig. 4) were determined with the HEC-RAS code (Hydrologic Engineering Center 2002) considering a steady-state flow.

The water surface elevations allowed to define the areas subject to flooding risk, to calculate the shear stress for erosion on the wetted perimeter and, consequently, to identify the areas in which the risk of bank erosion is higher. For these processing, the d_{50} size bed material was considered, that was obtained by the particle-size distribution.

Results showed that the areas characterized by a very high water surface elevation are located where the river-bed shape is narrow and embanked and where the

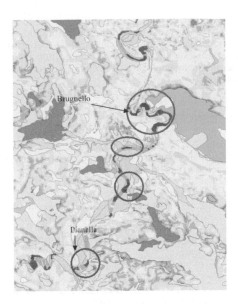

Figure 7. Layering the landslides map and the distribution of the instability index.

Figure 8. A large landslide is present near Brugnello.

Figure 9. Landslide near the Barberino gorge (November 2003).

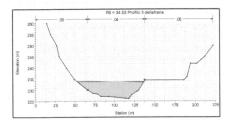

Figure 10. The water surface just next to Barberino landslide.

Figure 11. Erosion produced by the flood on the right bank of the Barberino gorge.

A particularly significant example concerning the interaction between river dynamic and slope instability is represented by the Barberino landslide (Fig. 9), that is located along the left river bank. This stretch of river is very critical from the hydraulic point of view. The movement is "complex" and along the slope it is possible to identify:

- fall, topple, translational and wedge slide movements: these one involved the ophiolites, characterized by a lot of discontinuity sets, tectonites and mylonites, so that they can be considered weak rocks;
- debris flow (occurred during the flood of the year 2001): the crown is placed between 370 and 350 m a.s.l.

where soils (silt, sand with marly limestones blocks medium-sized -0.2 m^3) and a counter slope layer of monogenic and polygenic breccias is present. The debris flow covered the ophiolites, outcropping in the medium-lower portion of the slope (at 290 m a.s.l.) and the municipality road (closed in 1990).

To prove the link existing between slope instability and river dynamic, the flood of the year 2001 was considered. On this occasion the value of the maximum discharge during the flood was about 2350–2400 m^3/s, that corresponds to a return period of about 90 years and to a water surface elevation, near the landslide cross section, of 7.4 m (corresponding to 233.6 m a.s.l.). Therefore, the camping located on the opposite side of the landslides (at an altitude of 234.4 m) was not flooded, but it was damaged by erosive phenomena (Figs. 10–11). Indeed, knowing that:, $d_{50} = 0.028$ m and $\gamma_s = 25.51$ kN/m^3, it is possible to calculate the $u_{critical}$ equal to 0.163 m/s and the $\tau_{critical}$ equal to 26.4 Pa. Comparing the latter value with the shear stress for erosion brought about by river flow during the flood event of the year 2001 (changing from 100 to 700 Pa), it is possible to understand that many damages occurred downstream the Barberino bridge, caused by localized erosion.

7 CONCLUSIONS

The interactions between slope and fluvial dynamic are very complex. On the one hand, slope instability, also if unrelated to the river activity, can play a determinant role in the river-bed evolution; on the other hand, river activities can trigger instability phenomena. Therefore, it is very important to quantify the effects of hydrologic (e.g. heavy rain) and hydraulic factors (e.g. flood plain from water surface elevation data, shear stress, etc.) on the stability conditions and, consequently, to evaluate the geological risk for choosing the most effective mitigation systems in a catchment basin.

This approach is correct if the data quantity and quality (especially the hydraulic cross sections) are sufficient to assure an accurate reconstruction of the river. Otherwise, it is better to carry out a qualitative analysis of the river morphology and to use only the instability index.

The example showed above pointed out that the flood events can produce important shear stress for erosion on the wetted perimeter, these one are higher (over one order of magnitude) than the strength stress of the river-bed sediments.

This phenomenon accelerates the erosion at the slope foot; therefore the landslide progresses from the bottom to the top. Indeed, in the illustrated example:

– the rocks constituting the slopes along the Trebbia River are of the same lithological kind, but it is possible to observe the landslides only where the Trebbia River flow direction is almost orthogonal the left slope;
– the stream erosion at the slope foot determines the development of tension cracks parallel to the river and to the slope;
– these tension cracks cause translational slides of rocks portion (1–10 dm^3), despite the counter slope orientation of the Formations.

In the future, the research will try to quantify the effects of the hydrologic factors (like infiltration, filtration, saturated conditions changes, slide foot erosion) on the slope stability, improving the integrate approach (hydrologic, hydraulic and geotechnical) described previously.

REFERENCES

Benedetti, A.I., Dapporto, S., Casagli, N. & Brugioni, M. 2006. Sviluppo di un modello di previsione di frana per il bacino del fiume Arno. *Giornale di Geologia Applicata* 3 (2006): 181–188.

Bezoari, G., Braga, G., Gervasoni, S., Larcan, E. & Paoletti, A. 1984. Effetto dell'estrazione di inerti sull'evoluzione dell'alveo del fiume Trebbia, In Ministero dei Lavori Pubblici & Magistrato del Po (eds), *Secondo Convegno di Idraulica Padana*, Parma 15–16 giugno 1984.

Gattinoni P. 2008. Landslide hydrogeological susceptibility in the Crati Valley (Italy). In Z. Chen, J. Zhang, Z. Li, F. Wu & K. Ho (eds), *Landslides and Engineered Slopes*, London: Taylor & Francis Group.

Hagerty, D.J. 1991. Piping—sapping erosion I: basic considerations. *J. of Hydr. Eng.* 117: 991–1008.

Hydrologic Engineering Centre 2002. Hec-Ras: *User's Guide and Utility Programs Manual*, U.S. Army Corps of Engineers (eds), Davis CA.

Humbert, M. 1977. La Cartographie ZERMOS. Modalité d'établissement des cartes des zones exposées à des risques liés aux mouvements du sol et du sous-sol. *B.R.G.M.* Bull. Ser. II, sect III(1/2): 5–8.

Interregional Agency for Po River (ed) 2001. *Piano stralcio per l'Assetto Idrogeologico (PAI). Interventi sulla rete idrografica e sui versanti*, Parma.

Masetti, M. 1999. Metodi di analisi dei rapporti tra la dinamica fluviale e dinamica dei versanti. In V. Francani (ed), *Rischio Geologico nella Provincia di Lecco*, *Lecco:* Cattaneo Ed.

Regione Emilia Romagna 2002. *Carta dei dissesti*, Servizio Cartografico e Geologico della Regione Emilia-Romagna (ed), Bologna.

Simon, A. & Downs, P.W. 1995. An interdisciplinary approach to evaluation of potential instability in alluvial channel. *Geomorphology* 12: 215–232.

Sogni, D. 2003. *Evoluzione morfologica recente dell'alveo del tratto inferiore del fiume Trebbia*, Thesis, Università degli Studi di Milano–Bicocca.

Prediction and Simulation Methods for Geohazard Mitigation – Oka, Murakami & Kimoto (eds)
© 2009 Taylor & Francis Group, London, ISBN 978-0-415-80482-0

Numerical simulations of submarine-landslide-induced and sea-floor-collapse-induced tsunami along coastline of South China

K.T. Chau
The Hong Kong Polytechnic University, Hong Kong, China

ABSTRACT: Submarine landslides have recently been identified as one of the major causes for several destructive tsunami, including the 1998 Papua New Guinea tsunami killing over 2,000. According to the published literatures (Mak & Chan, 2007; Zhou & Adams, 1988) and the official record of the State Oceanic Administration of China (SOAC), the number of tsunamis occurred in China is about 15 over the last 2000 years. However, Chau (2008) recently re-reviewed all tsunami data compiled by Lu (1984) and discovered that there are 220 tsunami of unknown origin (no storms, earthquake, or rains were reported during these events), and six of these killing over 10000. It is postulated that some of these devastating historical tsunami events might have been caused by submarine landslides. This paper summarizes the evidence of submarine-induced-tsunami in South China, reports some of huge ground collapses in the form of tiankeng in China and their implications on sea floor collapses, and finally presents some numerical simulation on tsunami on China coastline using TUNAMI-N2.

1 INTRODUCTION

The 2004 South Asian Tsunami is the most deadly tsunami events in history, killing over 225,000 and affecting eleven countries in South Asia including Thailand, Indonesia, Sri Lanka, and India. It was triggered by the second largest earthquake ever recorded by seismograph, with a moment magnitude of 9.3 and an epicentre off the west coast of Sumatra, Indonesia. The rupture fault zone is believed to extend for over 800 km, from Sumatra to Nicobar Islands near Burma in the north. Figure 1 shows a numerical simulation of the tsunami.

A month after the 2004 South Asian Tsunami occurred on December 26, 2004, the author and his colleagues from the Hong Kong Polytechnic University (PolyU) visited Phuket, Phi-Phi Island and

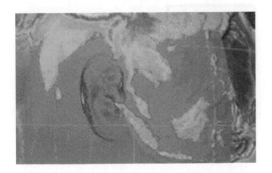

Figure 1. Numerical simulation showing the affected areas in the Indian Ocean (by Vasily V. Titov).

Khao-Lak in Thailand (Figs. 2–3). One thing that we learnt from Thailand trip is that tsunami hazard should not be neglected along any coastline, including South China. After our return from Thailand, a number of related studies have been conducted on the tsunami hazard in South China Sea. In particular, Wai et al. (2005) conducted numerical simulations on the tsunami waves in South China Sea generated by scenario earthquakes, Wong et al. (2005) conducted hydraulic physical tests on the edge wave phenomenon that we observed at Phi-Phi Island, and Chau et al. (2005, 2006) and Chau (2006) considered the field evidences and conducted numerical simulations for the case of Phi-Phi island and compared the results to our physical modeling. It seems that edge wave phenomenon associated with tsunami is not known to most of the local people (general public and scientists alike), so that they mistakenly believed that Victoria Harbor is safe from tsunami attack as seemingly it is sheltered by the Hong Kong Island. History repeatedly shows that such concept is incorrect. They are numerous tsunami events disproved this, including the run-up at Babi Island during the 1992 Flores tsunami, run-up at Okushiri Island during the 1983 Hokkaido tsunami, run-up at Hilo Harbor during the 1960 Chile tsunami. Series of surges and continuous rise of water in tsunami is capable of pushing surges around coastline propagating to so-called sheltered areas and causing severe damages.

More recently, Chau (2008) discovered that the number of historical tsunami in China is far more than that announced by the State Oceanic Administration of China (SOAC). Since there are no earthquakes

Figure 2. Field visit to Phi-Phi Island (Thailand) after the 2004 South Asian tsunami.

Figure 3. Field visit to Khao-Lak (Thailand) after the 2004 South Asian tsunami.

associated with these tsunami (noted that we use "tsunami" for both plural and singular forms following the terminology of Bryant, 2001), the present study investigates the possibility of tsunami induced by submarine landslides and sea floor collapses in South China Sea.

1.1 Tsunami hazard studies for China

The 2004 South Asian Tsunami alerted us that tsunami hazard cannot be ignored along any coastline. For unknown reasons, tsunami hazard in China have largely been underestimated. According to Zhou & Adams (1988), Mak & Chan (2007) and, the State Oceanic Administration of China (SOAC), the number of tsunami occurred in China is about 15 in the last 2000 years. Liu et al. (2007) investigated the tsunami for South China Sea using numerical simulations. More recently, Chau (2008) showed that many tsunami events reported in the book "Historical tsunamis and related events in China" by Lu (1984) were not associated with earthquake and have been ignored by previous studies. About 220 tsunami events are of unknown

origin in the last 2000 years. No storms, earthquake, or rains were reported during these events. Clearly, a lot of historical records reported in this valuable book have been ignored by previous authors and by SOAC, as they restricted themselves to earthquake-induced tsunami.

The economic and infrastructural development in China in recent years has progressed at an unprecedented pace since the "open China" policy implemented about 30 years ago. Formerly un-resided coastlines have been developed into big coastal cities, with much denser population and built with important infrastructures, such as cross-harbor tunnels and subways. These recent developments are highly vulnerable to tsunami. If an event similar to 2004 South Asian Tsunami happens along the coastline of China, the outcome would be much more devastating than any historical tsunami occurred in China. Before the 2004 Boxing Day Tsunami, not many residents (if there is any) in South Asian countries, such as Thailand and Sri Lanka, would believe that tsunami is a real threat to their lives. Therefore, a more thorough study on tsunami hazard (not restricted to earthquake-induced ones) in China is of utmost importance.

Figure 4 shows the evidence of huge historical tsunami along the China Sea—a boulder field bought up by historical tsunami at the coastline of Miyako Island close to Taiwan and the Zhejiang coastline.

1.2 Historical tsunami in China

In this section, the finding by Chau (2008) is summarized briefly. According to the historical records compiled by Lu (1984), Chau (2008) discovered that there are over 220 tsunami of unknown causes occurred along China's coastline since 48 BC, and at least 6 of these tsunami events that killed more than 10,000 people (in the years of 1045, 1329, 1458, 1536, 1776 and 1782). However, the number of events reported by Mak & Chan (2007) and SOAC are far less than the actual number. They ignored tsunami

Figure 4. Boulder field at Miyako Island, which is close to the China Sea, location shown by red dot.

events (even with devastation and high fatality) that cannot be linked directly to a felt earthquake. According to *Webster's Ninth New Collegiate Dictionary*, a tsunami is "a great sea wave produced by submarine earth movement or volcanic eruption". Bryant (2001) also concurred that a tsunami can be caused by earthquake, submarine landslides, volcanic eruption, or meteor impact. Therefore, the tsunami hazard in China is highly underrated. Figure 5 shows the locations of neglected tsunami events in South China compiled by Chau (2008). It is postulated that some of these historical tsunami events in China are caused by submarine landslides.

2 SUBMARINE-LANDSLIDE-INDUCED TSUNAMI

2.1 *Submarine landslide hazard in the world*

Submarine landslides have been identified as one of the probable sources for several destructive tsunami, including the 1998 Papau New Guinea Tsunami, (Lynett et al., 2003). In other parts of the world, the tsunami hazard from submarine landslides has been investigated seriously, such as the COSTA project financed by the European Commission from 2000–2004 (Mienert, 2004). Many submarine mass movements have been studied under some international and national projects, such as ADFEX (Artic Delta Failure Experiment, 1989–1992), GLORIA (a side-scan sonar survey of the US Exclusive Economic Zone, 1984–1991), STEAM (Sediment Transport on European Atlantic Margins, 1993–1996), ENAM II (European North Atlantic Margin, 1993–1999), STRATAFORM (1995–2001), and Seabed Slope Process in Deep Water Continental Margin (northwest Gulf of Mexico, 1996–2004) (Locat & Lee, 2002). McMurtry et al. (2004) had discussed the evidence of historical giant submarine landslides and its mega-tsunami in Hawaiian Islands. Gusiakov (2003) have discussed how to identify slide-generated tsunami in the historical catalogue. The far-field wave patterns generated from earthquake and submarine

landslide are of fundamental difference (Okal & Synolakis, 2003).

2.2 *Submarine landslide hazard in China*

According to Liu (1992), 5 sea floor debris fans have been found at the continental self in the South China Sea, ranging from 38–98 km in length and 11–39 km in width. Near the Pearl River Delta areas, there are 4 debris fans ranging 54–66 km in length and 14–50 km in width (Fig. 6). Based on 3.5 Hz echograms conducted by the Lamont-Doherty Geological Survey ships from 1965–1980 in South China Sea, the largest single debris slump near the edge of the continental self is up to 4,500 km^2 (Damuth, 1980). These submarine landslides might induce some historical tsunami of unknown sources.

Figure 7 further shows the cross-sections of some slumps and scars from submarine landslides in the South China Sea reported by Damuth (1980).

In addition to Figures 6–7, there is also an unexplained tsunami that may relate to submarine landslide

Figure 6. Topographical map of South China Sea, showing huge debris fans at the Pearl River Delta and the edge of continental shelf (shown in red circles) (after Liu, 1992).

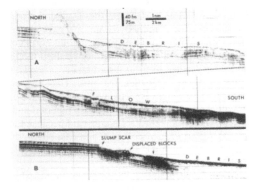

Figure 7. Sonar cross-sections of some locations in South China Sea, showing evidences of submarine landslides (after Damuth, 1980).

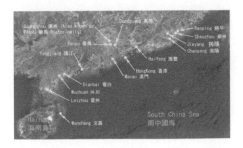

Figure 5. Locations of historical tsunami in South China.

close to Hong Kong waters. On June 24, 1988, a moderate earthquake of magnitude 5.4 occurred at north Luzon Island, generating a tsunami in Hong Kong. The tide gauge records at Quarry Bay Station and Tai Po Kau Station are 0.65 m and 1.03 m respectively. It is normally believed that only earthquakes of magnitude larger than 7.3 are likely to generate tsunami. In addition, the epicenter of the earthquake is 18.606°N and 121.013°E, which is about than 820 km from the stations in Hong Kong. Tsunami was only recorded in Hong Kong, but not in locations closer to the epicenter, like Philippines and Taiwan. Therefore, it is likely that a submarine landslide near Hong Kong was triggered by the earthquake and generating a local tsunami.

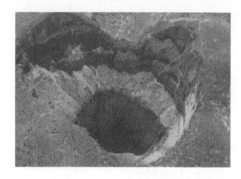

Figure 8.   Xiaozhai Tiankeng in Sichuan (size = 650 m × 535 m and depth = 662 mandvolume = 119.3 Mm^3).

3   SEA FLOOR COLLAPSE IN CHINA

Before we look at the possibility of sea floor collapse and its associated tsunami, the occurrence of huge tiankeng or huge sinkhole (resulting from ground collapse) in China will be summarized in next section.

3.1   Tiankengs in China

The Chinese term "tiankeng" is for "huge collapse doline". In China, tiankengs that were first widely known are in Sichuan Province. In 1994, a number of tiankengs were found unexpectedly in China in caving expeditions in Guangxi and Sichuan Provinces. There are 50 known tiankengs in China accounting for 64% of 78 found in the world (Zhu & Chen, 2006). Three of these are giant tiankengs of more than 500 m deep and 500 m in entrance diameter. They are the Xiaozhai Tiankeng in Fengjie of Sichuan, Dashiwei Tiankeng in Leye of Guangxi and Haolong Tiankeng in Bama of Guangxi. There are also five large tiankengs more than 300 m in depth and in entrance diameter. Most of these tiankengs were formed in soluble rocks (such as limestone and marble) in which massive amounts of rock material were dissolved and eroded away at depth by a highly dynamic cave river system. If the local geological layers were mainly horizontal with vertical joints, the collapsed roof will form very impressive vertical cliffs at the entrance, as shown in Figures 8–11 for Xiaozhai, Dashiwa, Haolong, and Dacaokou Tiankengs respectively. The size, maximum and volume of these tiankengs are also reported in the figures. These tiankengs are significant because they were related to ground collapse, and data shows that majority of these huge ground collapse occurred in China.

If such huge doline collapse occurs at sea floor, tsunami of considerable size may be generated. Since about 2/3 of the tiankengs of the world are found in China, the possibility of sea floor along China coastline is of particular interest.

Figure 9.   Dashiwei Tiankeng in Guangxi (size = 600 m × 420 m and depth = 613 m and volume = 75 Mm^3).

Figure 10.   Haolong Tiankeng in Guangxi (size = 800 m × 600 m and depth = 509 m and volume = 110 Mm^3).

Figure 11.   Dacaokou Tiankeng in Guizhou (size = 920 m × 240 m and depth = 220 m and volume = 25 Mm^3).

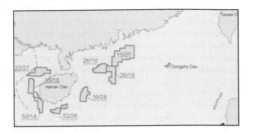

Figure 12. Locations of natural gas fields in South China Sea.

3.2 Natural gas extraction in China

Whether the huge tiankengs summarized in the last section and shown in Figures 8–11 existed at sea bottom is not known. But even they do, they would be covered by sediments at sea bottom and would not be so visible, unless the local sea current is strong enough to carry away all sediments.

Since the discovery of natural gas in the South China Sea in the eighties, offshore platforms were started to deploy to extract natural gas from sea bottom. Figur 12 shows the locations of the natural gas fields in South China Sea. For example, the Panyu 30-1 gas field is within 100 km of Hong Kong and is being mined by the China National Offshore Oil Corporation (CNOOC). Gas extraction from sea bottom may induce intensive flow in the rock stratum below the sea bottom. In the long run, the possibility of sea floor collapse increases, which in turn may lead to local tsunami events. As remarked by Guo (1991), submarine cave-in is not uncommon at where marine deposits are found in abundant. Bursting of submarine shallow-seated gas can also led to catastrophic tsunami.

In view of the increasing activities of gas and petroleum extraction from the South China Sea, sea-floor-collapse-induced tsunami should be considered seriously in the future.

4 PRELIMINARY RESULTS OF NUMERICAL SIMULATIONS

As a preliminary study, a number of scenarios of were conducted using a shallow water theory. In particular, the computer program TUNAMI-N2 (Tohoku University Numerical Analysis Model for Investigation of Near-field tsunamis version 2) was used to generate the tsunami run-up along the South China coastline.

First, bathymetry data was downloaded from the National Geophysical Data Centre (NGDC) of the National Oceanic and Atmospheric Administration (NOAA) with a grid spacing of 2 minutes of latitude by 2 minutes of longitude (3.6 km by 3.6 km). Numerical simulations for three different scenarios are considered: they are circular source (sea-floor collapse),

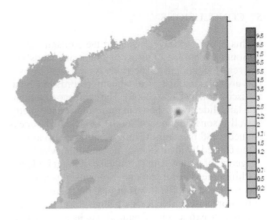

Figure 13. Maximum water level for circular source (sea-floor collapse).

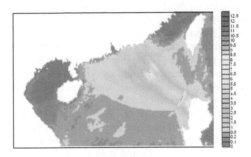

Figure 14. Maximum water level for line source (submarine landslide).

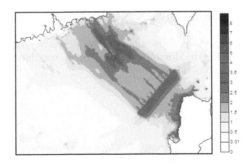

Figure 15. Maximum water level for line source (earthquake).

line source (submarine landslide), and Okada (1985) solution for reverse fault (earthquake). The results are shown in Figures 13–15 respectively.

The source locations for these scenarios are chosen along the Manila trench, which is along the tectonic plate boundary between the Eurasian plate and the Philippine plate and is believed capable of generating huge earthquakes and tsunami for South China Sea.

5 CONCLUSIONS

In this paper, we have presented the tsunami hazard of South China Sea in terms of submarine landslide, and sea-floor collapse. The study is motivated by over 220 unexplained historical tsunami records in China, as discovered by Chau (2008). The evidence of submarine landslides were shown in terms of the bathymetry as well as sonar sections of South China Sea. The unexplained 1988 tsunami in Hong Kong was also argued being caused by local submarine landslide, which was triggered by a moderate earthquake of magnitude 5.4 in Philippine. The large number of tiankengs in China was reviewed briefly. It was argued that China seems to be more susceptible for such huge ground collapse. The recent extraction of natural gas in South China Sea is interpreted as a catalyst for the occurrence of sea floor collapse, similar to the tiankeng formation. Our preliminary numerical simulations show that all submarine landslide, sea floor collapse and earthquake can induce tsunami that affect Hong Kong and the South China coastline. The present numerical analysis is preliminary, and more refined analyses are needed in prescribing the precise initial sea surface deformation due to submarine landslides of various volumes, shapes and sliding velocities, and to sea floor collapse of various shapes and sizes.

Nevertheless, this study is the first submarine-landslide-induced tsunami for South China Sea, and the first study investigating the sea-floor-collapse-induced-tsunami. It should path the way for future research work along the same direction.

ACKNOWLEDGEMENTS

The paper was fully supported by Research Grants Council of the Hong Kong SAR Government through GRF No. PolyU 5002/08P.

REFERENCES

Bryant, E. 2001. *Tsunami: The Underrated Hazard*. Cambridge University Press.

Chau, K.T. 2006. Edge wave at Phi-Phi Island during the December 26, 2004 South Asian Tsunami. *Third Tsunami Symposium*, May 22–28, 2006, University of Hawaii, Hawaii, USA.

Chau, K.T. 2008. Tsunami hazard along coastlines of China: a re-examination of historical data. *The 14th World Conference on Earthquake Engineering, 14 WCEE*, October 12–17, 2008 (full paper 015-0041 in CD-ROM).

Chau, K.T., Wong, R.H.C., Wai, O.W.H., Lin, H.Y., Jhan, J.M. & Dong, Z. 2006. Tsunami hazard estimation for South China Sea. *The second Guangdong-Hong Kong-Macau conference on Earthquake Technology*, March 1–2, 2006, Macau, China (full paper in CD-ROM).

Chau, K.T., Wai, O.W.H., Li, C.W. & Wong, R.H.C. 2005. Field evidences of edge waves caused by the December 26, 2004 tsunami at Phi-Phi island Thailand. *2005 ANCER Annual Meeting*, November 10–13, Jeju, Korea, 2005.

Damuth, J.E. 1980. Quaternary sedimentation processes in the South China Basin as Revealed by echo-character mapping and piston-core studies. In D.E. Hayes (ed), *The Tectonic and Geologic Evolution of Southeast Asian Sea and Islands*. Geophysical Monograph 23: 105–125. American Geophysical Union: Washington DC.

Guo, X. (ed) 1991. *Geological Hazards of China and Their Prevention and Control. Ministry of Geology & Mineral Resources*, State Science & Technology Commission, State Planning Commission, Geological Publishing House: Beijing.

Gusiakov, V.K. 2003. Identification of slide-generated tsunamis in the historical catalogues. In Yalcmer and Pelinovsky, Okal & Synolakis (eds). *Submarine Landslide and Tsunamis*: 17–24. Kluwer Academic Publishers: Netherlands.

Liu, Y.C., Santo, S.A., Wang, S.M., Shi, Y.L., Liu, H.L. & Yuen, D.A. 2007. Tsunami hazards along Chinese coast from potential earthquakes in South China Sea. *Physics of the Earth and Planetary Interiors* 163(1–4): 233–244.

Liu, G. 1992. *Geologic-Geophysic Features of China Seas and Adjacent Regions*. Science Press Beijing (in Chinese).

Locat, J. & Lee, H.J. 2002. Submarine landslides: advances and challenges. *Can. Geotech. J.* 39: 193–212.

Lu, R. (ed) 1984. *Historical Tsunamis and Related Events in China*. Oceanaaic Press: Beijing (in Chinese). (in Chinese 陸人驥編, 1984. 中國歷代災害性海潮史料. 北京: 海洋出版社).

Lynett, P.J., Borrero, J.C., Liu, P.L.-F. & Synolakis, C.E. 2003. Field survey and numerical simulations: A review of the 1998 Papua New Guinea tsunami. *Pure Appl. Geophys* 160: 2119–2146.

Mak, S. & Chan, L.S. 2007. Historical tsunamis in South China. *Natural Hazards* 43: 147–164.

McMurtry, G.M., Watts, P., Fryer, G.J., Smith, J.R. & Imamura, F. 2004. Giant landslides, mega-tsunamis and paleo-sea level in the Hawaiian Islands. *Marine Geology* 203: 219–233.

Mienert, J. 2004. COSTA-continental slope stability: major aims and topics. *Marine Geology* 213: 1–7.

Okada, Y. 1985. Surface deformation due to shear and tensile faults in a half-space. *Bull. Seism. Soc. Am.* 75: 1135–1154.

Okal, E.A. & Synolakis, C.E. 2003. A theoretical comparison of tsunamis from dislocations and landslides. *Pure and Applied Geophysics* 160: 2177–2188.

Wai, O.W.H., Dong, Z., Zhan, J.M., Li, C.W., Chau, K.T. & Wong, R.H.C. 2005. Numerical prediction of tsunami generated wave patterns in Hong Kong waters. *2005 ANCER Annual Meeting*, November 10–13, Jeju, Korea.

Wong, R.H.C., Lin, H.Y., Chau, K.T., Wai, O.W.H. & Li, C.W. 2005. Experimental simulations on edge wave induced by tsunami at PP island (Thailand) on Dec 26, 2004. *2005 ANCER Annual Meeting*, November 10–13, Jeju, Korea.

Zhou, Q. & Adams, W.M. 1988. Tsunamigenic earthquakes in China: 1831 BC to 1980 AD. *Science of Tsunami Hazards* 4(3): 131–148.

Zhu, X. & Chen, W. 2006. Tiankengs in the karst of China. *Speleogenesis and Evolution of Karst Aquifers* 4 (1): 1–18.

Advanced constitutive modeling
of geomaterials and laboratory and field testing

Prediction and Simulation Methods for Geohazard Mitigation – Oka, Murakami & Kimoto (eds)
© 2009 Taylor & Francis Group, London, ISBN 978-0-415-80482-0

A thermo-elasto-viscoplastic model for soft sedimentary rock

S. Zhang & F. Zhang
Nagoya Institute of Technology, Nagoya, Japan

ABSTRACT: In this paper, a new thermo-elasto-viscoplastic model is proposed, which can characterize thermodynamic behaviors of soft sedimentary rocks. Firstly, as is the same as Cam-clay model, plastic volumetric stain which consists of two parts, stress-induced part and thermodynamic part, is used as hardening parameter. Both parts of the plastic volumetric stain can be derived from an extended e-ln*p* relation in which the thermodynamic part is deduced based on a concept of 'equivalent stress'. Secondly, regarding soft sedimentary rocks as a heavily overconsolidated soil in the same way as the model proposed by Zhang et al. (2005), an extended subloading yield surface (Hashiguchi and Ueno, 1977) and an extended void ratio difference are proposed based on the concept of equivalent stress. Furthermore, time-dependent evolution equation for the extended void ratio difference is formularized, which consider both the influences of temperature and stresses. Finally, it is proved that the proposed model satisfies the thermodynamic theorems in the framework of non-equilibrium thermodynamics, which makes the proposed model reasonable, sophisticated but comprehensive.

1 INTRODUCTION

There always exist some crucial problems in treating nuclear waste disposal. Deep burying of the nuclear waste in relative intact sedimentary rock ground is considered to be a practical way. The heat emission due to the radioactivity of the nuclear waste disposal, however, will increase the temperature of surrounding ground and endanger the surrounding. It is known that for some geomaterials, e.g., soft sedimentary rock, temperature and its change may affect the mechanical behaviors of the rock, especially the long-term stability. Therefore, it is necessary to evaluate long-term stabilities of the ground subjected to heating process due to radioactivity of the nuclear waste disposal. In the analysis, the first but a key important factor is that a simple and reasonable thermo-elasto-viscoplastic constitutive model should be established to characterize the thermo-dynamic behaviors of soft rocks. Some experimental results can be found in literatures, e.g., Okada (2005 & 2006); Fujinuma et al. (2003). Temperature and its change affect the stress-strain relation of soft rocks in the way that as the temperature decreases, the peak value of stress difference increases; meanwhile, the stress-strain relation changes from ductility to brittle. The temperature and its change affect the creep behavior of soft rocks greatly in the way that as temperature increases about 40 degrees, the creep failure time may decreases with 2–4 orders (Okada 2005 & 2006).

In order to simulate thermodynamic behavior of geomaterials, some thermo-elasto-viscoplastic models have been proposed, most of which are based on thermodynamic theorems at first, using the second thermodynamic theorem to establish a series of restricted relations for the variables involved in the models, e.g. stress, strain and entropy etc. and then deduced the model with common concepts such as flow rule, yielding function, plastic potential, normality rule and etc. In proposing a thermodynamic model, the most important but very difficult step is to formulize the thermodynamic functions, which satisfies above-mentioned restricted relations for the variables, which always makes the model too complicated and difficult to understand.

A reversed research approach has been used in this paper. Unlike the models proposed in the past researches in which the thermodynamic theorems are the tools to formulize constitutive model themselves as restricted conditions, the thermodynamic theorems in the newly proposed model are only used as the tool to verify the logicality of the model in the meaning of thermodynamics after the model is established. Therefore, in establishing the model, the flow rule, yielding function, plastic potential, evolution equation for subloading surface are chosen without considering the restrictions controlled by the thermodynamic theorems at the very beginning. It is therefore possible for us to choose reasonable formulations based on very simple physical meanings and practical equations in soil mechanics such as the e-ln*p* relation in consolidation tests, so that the establishment of a new model become much easier.

In this paper, based on the works by Zhang et al. (2005), a new thermo-elasto-viscoplastic model is proposed to describe the thermodynamic behaviors of soft sedimentary rocks, in which, not only the thermodynamic characteristics, but also the normal mechanical behaviors of soft rocks that have already been clarified in experiments and been modeled in the previous researches, can be described properly.

2 CONCEPT OF EQUIVALENT STRSS AND THERMO-ELASTOPLASTIC MODEL FOR NORMALLY CONSOLIDATED SOILS

One of the important characteristics of the Cam-clay model is that the plastic volumetric strain is explored as a hardening parameter for the elastoplastic model of soils. Furthermore, it is also shown in the work by Zhang et al. (2005) that the plastic volumetric strain can also be used as a hardening parameter in a constitutive model for soft rocks. Therefore, the plastic volumetric strain is also assumed as the hardening parameter in the newly proposed model. Change of temperature may generate both elastic and plastic volumetric strains, It is reasonable to assume that the plastic volumetric strain of geomaterials is made up from two independent parts: thermodynamic and stress-induced, which can be expressed as:

$$\varepsilon_v^p = \varepsilon_v^{p\sigma} + \varepsilon_v^{p\theta} \quad \text{or} \quad d\varepsilon_v^p = d\varepsilon_v^{p\sigma} + d\varepsilon_v^{p\theta} \tag{1}$$

where, ε_v^p is the total plastic volumetric strain. $\varepsilon_v^{p\sigma}$ is the stress-induced plastic volumetric strain and $\varepsilon_v^{p\theta}$ is the thermodynamic plastic volumetric strain. The second part of Equation 1 expresses the incremental relation of the plastic volumetric strains.

Firstly, it is assumed that the relation between the stress and the stress-induced plastic volumetric strain still can be expressed by the Cam-clay model as:

$$f_1(\sigma, \varepsilon_v^{p\sigma}) = \ln\frac{\sigma_m}{\sigma_{m0}} + \frac{\sqrt{3}\sqrt{J_2}}{M\sigma_m} - \frac{1}{C_p}\varepsilon_v^{p\sigma} = 0 \tag{2}$$

where, σ_{m0} is a reference pressure and equals to 98 kPa, which is the standard atmospheric pressure. M is the ratio of shearing stress at critical state. $C_p = E_p/(1+e_0)$, where, e_0 is the reference void ratio at σ_{m0}; E_p is a plastic modulus and physically it equals to the value of $\lambda - \kappa$, in where, λ is compression index and κ is swelling index.

In order to consider the effect of temperature and its change, it is necessary to formulize the relation between the temperature and the thermodynamic plastic volumetric strain, based on the concept of equivalent stress. It is known that under the condition of constant-stress state, change of temperature may also generate thermodynamic volumetric strain, including elastic volumetric strain $\varepsilon_v^{e\theta}$ and plastic volumetric strain $\varepsilon_v^{p\theta}$, as were the material subjected to a real mean stresses. Accordingly, it is assumed that thermodynamic volumetric strain is induced by an imaginary stress $\tilde{\sigma}_m$, namely, the equivalent stress. In evaluating the stress-induced volumetric strain, it is necessary to define a reference pressure σ_{m0} at a

reference temperature θ_0, which represents the global average temperature and is assumed here to be 288 K. Considering the limitation of the variation range for temperature that θ should be larger or equal to 273 K, a linear relation between the change of temperature $\theta - \theta_0$ and the thermodynamic elastic volumetric strain $\varepsilon_v^{e\theta}$ is assumed as:

$$\varepsilon_v^{e\theta} = 3\alpha(\theta - \theta_0) \tag{3}$$

where, α is linear thermo-expansion coefficient, and takes a negative value because a compressive volumetric strain is assumed as positive in geomechanics.

Based on the concept of equivalent stress, the relation between equivalent stress and thermo- dynamic elastic volumetric strain can then be evaluated with Hooke's law. Considering Equation 3, this relation can be expressed as:

$$\tilde{\sigma}_m = \sigma_{m0} + d\tilde{\sigma}_m = \sigma_{m0} + K\varepsilon_v^{e\theta} = \sigma_{m0} + 3K\alpha(\theta - \theta_0) \tag{4}$$

where, K is volume elastic modulus, and is equal to $E/3/(1-2\nu)$, in which E is the Young's modulus and ν is the Poisson's ratio.

On the other side, the relation between the equivalent stress and the thermodynamic plastic volumetric strain $\varepsilon_v^{p\theta}$ is evaluated by e-ln$p$ relations in both compression and swelling processes based on the equivalent stress. Detailed process to evaluate the relation between the plastic thermodynamic void ratio $\Delta e^{p\theta}$ and the equivalent stress under the condition of constant-stress state is shown in Figure 1. Considering Equation 4, this relation can be expressed as:

$$\begin{aligned}\varepsilon_v^{p\theta} &= C_p \ln\frac{\tilde{\sigma}_m}{\sigma_{m0} + 3K\alpha(\theta_0 - \theta_0)}\\ &= C_p \ln\frac{\sigma_{m0} + 3K\alpha(\theta - \theta_0)}{\sigma_{m0}}\end{aligned} \tag{5}$$

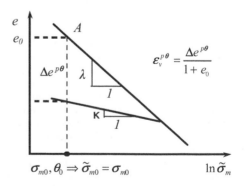

Figure 1. Illustration of the relation between equivalent stress and void ratio difference.

Similar to the Cam-clay model, the plastic potential function related to temperature is assumed in the following way based on Equation 5:

$$f_2(\theta, \varepsilon_v^{p\theta}) = \ln \frac{\sigma_{m0} + 3K\alpha(\theta - \theta_0)}{\sigma_{m0}} - \frac{1}{C_p}\varepsilon_v^{p\theta} = 0 \quad (6)$$

Considering both the effects of temperature and stress, the total plastic volumetric strain is made up from thermodynamic and stress-induced, and substituting Equations 5 and 6 into Equation 1, a new thermoplastic potential function can then be obtained as follow:

$$f(\sigma, \varepsilon_v^{p\sigma}, \theta, \varepsilon_v^{p\theta}) = f_1(\sigma, \varepsilon_v^{p\sigma}) + f_2(\theta, \varepsilon_v^{p\theta})$$

$$= \ln \frac{\sigma_m}{\sigma_{m0}} + \frac{\sqrt{3}\sqrt{J_2}}{M\sigma_m} + \ln \frac{\sigma_{m0} + 3K\alpha(\theta - \theta_0)}{\sigma_{m0}}$$

$$- \frac{1}{C_p}\varepsilon_v^p = 0 \quad (7)$$

Because associated flow rule is adopted, the proposed thermoplastic potential is also the yield function. The following equations can then be obtained as:

$$df = \frac{\partial f}{\partial \sigma_{ij}} d\sigma_{ij} + \frac{\partial f}{\partial \theta} d\theta + \frac{\partial f}{\partial \varepsilon_v^p}(d\varepsilon_v^{p\theta} + d\varepsilon_v^{p\sigma}) = 0 \quad (8)$$

$$d\varepsilon_{ij}^{p\sigma} = \Lambda \frac{\partial f}{\partial \sigma_{ij}} \quad \text{and} \quad d\varepsilon_v^{p\sigma} = \Lambda \frac{\partial f}{\partial \sigma_{ii}} \quad (9)$$

Considering the factor that the change of temperature can only generate volumetric strain, the stress increment can be calculated by the Hooke's law as:

$$d\sigma_{ij} = E_{ijkl}\left(d\varepsilon_{kl} - d\varepsilon_{kl}^{p\sigma} - \frac{d\varepsilon_v^{p\theta}\delta_{kl}}{3} - \frac{d\varepsilon_v^{e\theta}\delta_{kl}}{3}\right) \quad (10)$$

where, $\varepsilon_{kl}^{e\theta}$ is the thermoelastic strain tensor. It can be rewritten as:

$$d\sigma_{ij} = E_{ijkl}d\varepsilon_{kl} - E_{ijkl}\frac{\partial f}{\partial \sigma_{kl}}\Lambda - K\delta_{ij}d\varepsilon_v^{p\theta} - 3K\alpha\delta_{ij}d\theta \quad (11)$$

where, δ_{ij} is the Kronecker tensor. From Equation 11, positive valuable Λ can be easily calculated and detailed incremental expression for the incremental stress tensor can be obtained:

$$d\sigma_{ij} = \left(E_{ijkl} - E_{ijkl}^{p\sigma}\right)d\varepsilon_{kl} - \left(E_{ijkl} - E_{ijkl}^{p\theta}\right)\frac{d\varepsilon_v^{e\theta}}{3}\delta_{kl} \quad (12)$$

$$h_p = \frac{1}{C_p}\frac{\partial f}{\partial \sigma_{mm}} + \frac{\partial f}{\partial \sigma_{mn}}E_{mnpq}\frac{\partial f}{\partial \sigma_{pq}} \quad (13)$$

$$h_\theta = \left[\left(E_{ijmn}\frac{\partial f}{\partial \sigma_{mn}}\frac{\partial f}{\partial \sigma_{ij}}\right)/h_p - 1\right]\frac{C_p \cdot K}{\sigma_{m0} + 3K\alpha(\theta - \theta_0)}$$

$$+ \left(E_{ijmn}\frac{\partial f}{\partial \sigma_{mn}}\frac{\partial f}{\partial \sigma_{ij}}\right)/h_p \quad (14)$$

$$E_{ijkl}^{p\theta} = h_\theta E_{ijkl}, \quad E_{ijkl}^{p\sigma} = \left(E_{mnkl}\frac{\partial f}{\partial \sigma_{pq}}\frac{\partial f}{\partial \sigma_{mn}}E_{ijpq}\right)/h_p \quad (15)$$

In considering overconsolidated materials, The concept of subloading yield surface (Hashiguchi and Ueno, 1977) is also introduced in this paper. Void ratio difference ρ^σ due to stress and overconsolidated ratio OCR^σ can be expressed as:

$$\rho^\sigma = C_p(1+e_0) \cdot \ln \frac{\sigma_{N1e}}{\sigma_{N1}} = C_p(1+e_0) \cdot \ln OCR^\sigma \quad (16)$$

Using the concept of equivalent stress, it is easy to define an equivalent void ratio difference ρ^θ and an equivalent overconsolidated ratio OCR^θ, taking after those induced by real stresses shown in Equation 16. Both ρ^θ and OCR^θ are caused by the change of equivalent stress due to change of temperature and can be expressed as:

$$\rho^\theta = C_p(1 + e_0) \cdot \ln \frac{\tilde{\sigma}_{N1e}}{\tilde{\sigma}_{N1}}$$

$$= C_p(1 + e_0) \cdot \ln \frac{\sigma_{m0} + 3K\alpha(\theta_{N1e} - \theta_0)}{\sigma_{m0} + 3K\alpha(\theta_{N1} - \theta_0)} \quad (17)$$

Accordingly, both the extended void ratio difference ρ and the extended overconsolidated ratio OCR include two parts: the thermodynamic and the stress-induced, and can be expressed as:

$$\rho = \rho^\sigma + \rho^\theta = C_p(1 + e_0) \cdot \ln \frac{\sigma_{N1e}}{\sigma_{N1}}$$

$$= C_p(1 + e_0) \ln OCR \quad (18)$$

In fact, the equivalent stress is a stress state that includes the influence of temperature. Therefore, in representing a real present stress state in stress space $(\sigma_m, \sqrt{J_2})$, both present stress state and present temperature state are considered simultaneously. The expression for the extended subloading yield surface can then be given in the following way:

$$f_s = C_p \ln \frac{\sigma_m}{\sigma_{N1}} + C_p \ln \frac{\sigma_{m0} + 3K\alpha(\theta - \theta_0)}{\sigma_{m0} + 3K\alpha(\theta_{N1} - \theta_0)}$$

$$+ C_p \frac{\sqrt{3}\sqrt{J_2}}{M\sigma_m} = 0 \quad (19)$$

Stress-induced plastic volumetric strain due to the change of stress from σ_{m0} to σ_{N1e} can be evaluated as:

$$\varepsilon_v^{p\sigma} = C_p \ln \frac{\sigma_{N1e}}{\sigma_{m0}} \quad (20)$$

Meanwhile, the thermodynamic plastic volumetric strain due to the change of temperature from θ_0 to θ_{N1e} can be evaluated as:

$$\varepsilon_v^{p\theta} = C_p \ln \frac{\sigma_{m0} + 3K\alpha(\theta_{N1e} - \theta_0)}{\sigma_{m0}} \quad (21)$$

257

The extended subloading yield surface can be written as:

$$f(\sigma, \theta, \varepsilon_v^p) = \ln \frac{\sigma_m}{\sigma_{m0}} + \frac{\sqrt{3}\sqrt{J_2}}{M\sigma_m}$$
$$+ \ln \frac{\sigma_{m0} + 3K\alpha(\theta - \theta_0)}{\sigma_{m0}} - \frac{1}{C_p}\left(\varepsilon_v^p - \frac{\rho}{1+e_0}\right) = 0 \quad (22)$$

The consistency equation is then expressed as:

$$df = \frac{\partial f}{\partial \sigma_{ij}} d\sigma_{ij} + \frac{\partial f}{\partial \theta} d\theta - \frac{1}{C_p}\left(d\varepsilon_v^p - \frac{d\rho}{1+e_0}\right) = 0 \quad (23)$$

An evolution equation for the extended void ratio difference ρ is expressed here by the sum of actual stress σ_m and the equivalent stress increment $(\tilde{\sigma}_m - \sigma_{m0})$ in the following way:

$$-\frac{1}{1+e_0} d\rho = \frac{a\rho^2}{\sigma_m + (\tilde{\sigma}_m - \sigma_{m0})} \cdot \Lambda$$
$$= \frac{a\rho^2}{\sigma_m + 3K\alpha(\theta - \theta_0)} \cdot \Lambda \quad (24)$$

By substituting this evolution equation into the Equation 23, it is easy to obtain the relation:

$$\Lambda = \frac{\partial f}{\partial \sigma_{ij}} d\sigma_{ij} \Big/ \left(\frac{h_p\text{sub}}{C_p}\right), \quad h_p\text{sub} = \frac{\partial f}{\partial \sigma_{ii}}$$
$$+ \frac{a\rho^2}{\sigma_m + 3K\alpha(\theta - \theta_0)} \quad (25)$$

Similar to the work by Zhang et al. (2005), the time dependent evolution equation for the extended void ratio difference ρ, which includes the thermodynamic part and stress-induced part, can be given as:

$$\frac{\dot{\rho}}{1+e_0} = -\Lambda \cdot \frac{G(\rho, t)}{\sigma_m + 3K\alpha(\theta - \theta_0)} + h(t) \quad (26)$$

where,

$$\begin{cases} h(t) = \dot{\varepsilon}_v^0 [1 + t/t_1]^{-\bar{\alpha}} = (\dot{\varepsilon}_{v0}^\theta + \dot{\varepsilon}_{v0}^\sigma)[1 + t/t_1]^{-\bar{\alpha}} \\ G(\rho, t) = a\rho^{1+C_n \ln(1+t/t_1)} = a(\rho^\theta + \rho^\sigma)^{1+C_n \ln(1+t/t_1)} \end{cases} \quad (27)$$

Total plastic volumetric strain rate is expressed as:

$$\dot{\varepsilon}_v^p = \dot{\varepsilon}_v^{p\sigma} + \dot{\varepsilon}_v^{p\theta} \quad (28)$$

Associated flow rule is adopted and therefore the viscoplastic potential is expressed as:

$$\dot{\varepsilon}_{ij}^{p\sigma} = \Lambda \frac{\partial f}{\partial \sigma_{ij}} \quad \text{and} \quad \dot{\varepsilon}_v^{p\sigma} = \Lambda \frac{\partial f}{\partial \sigma_{kk}} \quad (29)$$

The consistency equation can be written as:

$$\dot{f} = \frac{\partial f}{\partial \sigma_{ij}} \dot{\sigma}_{ij} + \frac{\partial f}{\partial \theta} \dot{\theta} - \frac{1}{C_p}\left(\dot{\varepsilon}_v^p - \frac{\dot{\rho}}{1+e_0}\right) = 0 \quad (30)$$

By which the positive variable Λ can be obtained as:

$$\Lambda = \left(\dot{f}_\sigma + \frac{h(t)}{C_p}\right) \Big/ \frac{h_{sub}^p}{C_p} \quad (31)$$

Compared with the model proposed by Zhang et al. (2005), the newly proposed model only adds one parameter, the linear thermo-expansion coefficient α, which can be definitely determined with clear physical meanings. All other parameters are determined in the same way as in the previous model. Detailed discussion on this issue can be referred to the corresponding reference (Zhang et al., 2005).

3 THERMODYNAMIC BEHAVIOR OF PROPOSED MODEL

Being different from most of existed thermo-elasto-viscoplastic models in which the thermodynamic theorems are used to establish a series of restricted relations for the variables involved in the models, the thermodynamic theorems are not discussed in formulating the newly proposed model at the beginning. Therefore, it is necessary to verify if the new model satisfy the thermodynamic theorems, especially the 1st and 2nd thermodynamic theorems, so that the rationality of new model can be assured.

When field equations of thermodynamics for a body are considered, state variables such as stresses, strains and temperature are in general inhomogeneous and change constantly. It is necessary to describe properly energy exchange happened between an arbitrary element and its 'external system'. It is, however, very difficult, sometime even impossible to define the external system. For instance, when we consider the heating effect of nuclear radiation, it is related to mass energy conversion of the element itself which is not taken into consideration within the framework of this research. It is necessary to use non-equilibrium thermodynamics (Charles Kittel, 1980 and David Jou et al., 2008) to describe the thermodynamic behaviors of any arbitrary element in the body.

In the 1st thermodynamic theorem, it is stated that total energy flowing in/out the element is equal to the energy store/lost of the element, namely, internal energy E. The total energy flow in/out the element is made up from two parts: the work W done by external forces and the heat Q. The heat includes external heat and internal heat. The former one is the heat flux h_i that generates from external heat source and flows in/out through the surfaces surrounding the element; while the latter one is generated from the internal heat source such as radioactivity of nuclear waste disposal within the element. Because of the conservation of energy in the element, the following equation can be obtained:

$$E = W + Q \quad (32)$$

It also can be expressed in rate form as:

$$\dot{E} = \dot{W} + \dot{Q} \Leftrightarrow D \cdot \dot{e} = D \cdot \dot{w} + D \cdot \dot{q} \quad (33)$$

where, D is the density of element, e is the internal energy per unit mass, w is the work per unit mass and q is the heat per unit mass. The relation between θ and h_i obeys the Fourier's law, that is:

$$h_i = -k \cdot \partial\theta/\partial x_i \qquad (34)$$

where, k is heat conductivity coefficient. Then, $\dot{W}$ and $\dot{Q}$ can be expressed as:

$$\dot{W} = D \cdot \dot{w} = \sigma_{ij} \cdot \dot{\varepsilon}_{ij} = \sigma_{ij} \cdot \left(\dot{\varepsilon}^e_{ij} + \dot{\varepsilon}^p_{ij}\right)$$
$$= \sigma_{ij} \cdot \left(\dot{\varepsilon}^{e\sigma}_{ij} + \dot{\varepsilon}^{e\theta}_{ij}\right) + \sigma_{ij} \cdot \left(\dot{\varepsilon}^{p\sigma}_{ij} + \dot{\varepsilon}^{p\theta}_{ij}\right) \qquad (35)$$

$$\dot{Q} = D \cdot \dot{q} = -\partial h_i/\partial x_i + r \cdot D$$
$$= \partial(k\ \partial\theta/\partial x_i)/\partial x_i$$
$$+ r \cdot D = k \cdot (\partial^2\theta/\partial x_i^2) + r \cdot D \qquad (36)$$

where, r is the internal heat supply per unit time per unit mass.

The changing rate of internal energy $\dot{E}$ includes reversible work rate $\dot{W}^\circ$ and the changing rate of thermal energy $\dot{Q}^\circ$. In the element, $\dot{W}^\circ$ is stored in the form of elastic potential energy:

$$\dot{W}^\circ = \sigma_{ij} \cdot \dot{\varepsilon}^e_{ij} = \sigma_{ij} \cdot \left(\dot{\varepsilon}^{e\sigma}_{ij} + \dot{\varepsilon}^{e\theta}_{ij}\right) \qquad (37)$$

The changing rate of thermal energy $\dot{Q}^\circ$ can be obtained as:

$$\dot{Q}^\circ + \dot{W}^\circ = \dot{W} + \dot{Q} \Rightarrow$$
$$\dot{Q}^\circ = \sigma_{ij} \cdot (\dot{\varepsilon}^e_{ij} + \dot{\varepsilon}^p_{ij}) + (k \cdot \frac{\partial^2\theta}{\partial x_i^2} + r \cdot D) - \sigma_{ij} \cdot \dot{\varepsilon}^e_{ij}$$
$$= \sigma_{ij} \cdot (\dot{\varepsilon}^{p\sigma}_{ij} + \dot{\varepsilon}^{p\theta}_{ij}) + (k \cdot \frac{\partial^2\theta}{\partial x_i^2} + r \cdot D) \qquad (38)$$

The changing rate of thermal energy per unit mass $\dot{q}^\circ$ can then be expressed as:

$$\dot{q}^\circ = \frac{\sigma_{ij} \cdot (\dot{\varepsilon}^{p\sigma}_{ij} + \dot{\varepsilon}^{p\theta}_{ij})}{D} + (\frac{k}{D} \cdot \frac{\partial^2\theta}{\partial x_i^2} + r) \qquad (39)$$

Based on the definition of entropy, the material time derivative of the entropy $\dot{\eta}$ can be calculated:

$$\dot{\eta} = \frac{\dot{q}^\circ}{\theta} = \left(\frac{\sigma_{ij} \cdot (\dot{\varepsilon}^{p\sigma}_{ij} + \dot{\varepsilon}^{p\theta}_{ij})}{D} + \frac{k}{D} \cdot \frac{\partial^2\theta}{\partial x_i^2} + r\right) \cdot \frac{1}{\theta} \qquad (40)$$

In non-equilibrium thermodynamics, the changing rate of entropy density is made up from three parts: the first part comes from the irreversible course in the element, the second part comes from the heat flux and the third part comes from inner heat source such as radioactivity of nuclear waste disposal within the element. The relation among them can be shown as:

$$D\dot{\eta} = \dot{\gamma} + \left[-\left(\frac{h_i}{\theta}\right)_{,i} + D\frac{r}{\theta}\right] \quad \text{or}$$

$$\dot{\gamma} = D\dot{\eta} - \left[-\left(\frac{h_i}{\theta}\right)_{,i} + D\frac{r}{\theta}\right] \qquad (41)$$

where, $\dot{\gamma}$ is called as entropy production in the irreversible course of the element and must be greater than or equal to zero (Clausius-Duhem inequality). Substituting the Equations 34 and 40 into Equation 41, $\dot{\gamma}$ can be expressed as:

$$\dot{\gamma} = D \cdot \left(\frac{\sigma_{ij} \cdot (\dot{\varepsilon}^{p\sigma}_{ij} + \dot{\varepsilon}^{p\theta}_{ij})}{D} + \frac{k}{D} \cdot \frac{\partial^2\theta}{\partial x_i^2} + r\right) \cdot \frac{1}{\theta} \qquad (42)$$

$$-k\frac{1}{\theta}\frac{\partial^2\theta}{\partial x_i^2} + k\frac{1}{\theta^2}\left(\frac{\partial\theta}{\partial x_i}\right)^2 - D\frac{r}{\theta} = \frac{1}{\theta}.$$

$$\sigma_{ij} \cdot \left(\dot{\varepsilon}^{p\sigma}_{ij} + \dot{\varepsilon}^{p\theta}_{ij}\right) + k\frac{1}{\theta^2}\left(\frac{\partial\theta}{\partial x_i}\right)^2$$

$$\text{or} \quad \dot{\gamma} = \frac{1}{\theta} \cdot \dot{W}^p + k\frac{1}{\theta^2}(\frac{\partial\theta}{\partial x_i})^2 \qquad (43)$$

where, $\dot{W}^p$ is the rate of plastic work that can be proved to be positive. Therefore the inequality $\dot{\gamma} \geq 0$ is valid.

4 PERFORMANCE OF THE NEWLY PROPOSED MODEL

In order to check the performance of the proposed model, drained conventional triaxial compression tests of soft rock under constant shear strain, is simulated. The physical properties of the soft rock and the material parameters involved in the model are listed in Table 1. Because viscosity is not considered here, time dependent parameters C_n and $\bar{\alpha}$ take the values of zero.

Figure 2 shows show the stress-strain relations of the soft rocks at difference constant temperatures at which the shearing is carried. It is seen from the figure that the peak value of stress difference increases as

Table 1. Physical properties of soft rock and material parameters involved in thermo-elasto-viscoplastic model.

α(1/K)	8.0×10^{-6}	E (MPa)	900.0
β	1.50	E_p $(\lambda - \kappa)$	0.040
C_n	$-(0.025)$	$\bar{\alpha}$	$-(0.70)$
v	0.0864	a	500.0
R_f	11.0		
OCR	150.0	Void ratio at reference state e_0	0.72
σ'_{30} (MPa)	0.098	Initial yielding stress of consolidation p'_c (MPa)	15.0

259

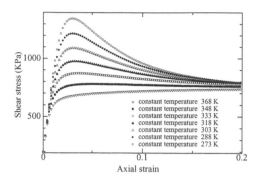

Figure 2. Stress-strain relations at the difference constant temperatures.

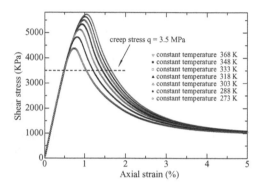

Figure 3. Simulated stress-strain relations at difference constant temperatures.

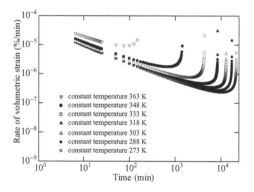

Figure 4. Time histories of creep rates at difference constant temperatures.

temperature decreases in drained conventional triaxial compression tests at different constant temperatures. It is also know that the stress-strain relation changes from ductility to brittle as temperature decreases, which is coincident with the experimental results the thermodynamic behaviors of soft rocks obtained from the tests.

In order to check the viscoplastic behaviors of the proposed model, drained conventional triaxial compression tests controlled with constant strain rate and consequent drained triaxial pure creep tests, are simulated with the newly proposed model under the condition that temperatures are kept in constant with different values during shearing while the shear rate is $\dot{\varepsilon} = 0.10\%$/min. Physical properties of soft rock and material parameters involved in thermo-elasto-viscoplastic model are also listed in Table 1, in which the time dependent parameters C_n and $\bar{\alpha}$ are not to be zero and take the values in parenthesis.

Figure 3 shows the relations between the stress difference and strain of soft rocks in drained conventional triaxial compression tests under constant strain rate ($\dot{\varepsilon} = 0.10\%$/min) at different constant temperatures. The calculated results also can describe the thermodynamic characteristics of soft rocks, which are observed in the experiments by Okada (2005 & 2006) and Fujinuma et al. (2003). In simulating drained creep tests, a shear stress of $q = 3.5$ MPa is chosen as the creep stress which is not applied to the specimen abruptly but loaded with drained shearing at the same strain rate ($\dot{\varepsilon} = 0.10\%$/min) as is used in the simulation of the drained compression tests shown in Figure 3.

Figure 4 shows the simulate time histories of creep rates of the soft rocks at difference constant temperatures. It is seen from this figure that the general characteristics of creep behavior, such as initial creep rate, steady creep and creep rupture, can be simulated properly. Moreover, the calculated results can describe properly the factors that the creep failure time is largely dependent on temperature, and that the higher the temperature is, the faster the creep rupture will be.

5 CONCLUSIONS

In this paper, a thermo-elasto-viscoplastic model for soft sedimentary rocks is proposed based on a concept of equivalent stress. It can describe the thermodynamic behaviors of soft rocks not only in drained conventional triaxial compression tests but also in drained triaxial creep tests. In order to verify if the new model satisfy the thermodynamic theorems, especially the 1st and 2nd thermodynamic theorems, Non-equilibrium thermodynamics is used. It is illustrated that the first thermodynamic theorem, relating to the conservation of energy in arbitrary element, can be satisfied by the model. It is proved that entropy production of the element is always greater or equal to zero, that is, the second thermodynamic theorem is satisfied. The model is developed based on simple physical meaning and then the requirement for satisfying thermodynamic theorems is verified after the model is established, which makes it possible to propose the model in a reasonable, sophisticated but comprehensive way.

REFERENCES

Charles Kittel. 1980. *Thermal Physics*. W. H. Freeman & Co.

David Jou, José Casas-Vázquez, & Georgy Lebon. 2008. *Understanding Non-equilibrium Thermodynamics*. Springer.

Fujinuma, S., Okada, T., Hibino, S. & Yokokura, T. 2003. Consolidated undrained triaxial compression test of sedimentary soft rock at a high temperature. *Proceedings of the 32nd Symposium on Rock Mechanics*: 125–130. Japan (in Japanese).

Hashiguchi, K., & Ueno, M. 1977. Elastoplastic constitutive laws of granular material. Spec. Ses. 9, Murayama, S. and Schofield, A.N. (eds.), *Constitutive Equations of Soils, Pro. 9th Int. Conf. Soil mech. Found. Engrg.*: 73–82. Tokyo.

Nakai, T. & Hinoko, M. 2004. A simple elastoplastic model for normally and overconsolidated soils with unified materials parameters. *Soils and Foundations* 44(2): 53–70.

Okada, T. 2005. Mechanical properties of sedimentary soft rock at high temperatures (Part1)-Evaluation of temperature dependency based on triaxial compression test. *Civil Engineering Research Laboratory Rep. No. N04026* (in Japanese).

Okada, T. 2006. Mechanical properties of sedimentary soft rock at high temperatures (Part2)-Evaluation of temperature dependency of creep behavior based on unconfined compression test. *Civil Engineering Research Laboratory Rep. No. N05057* (in Japanese).

Zhang, F., Yashima, A., Nakai, T., Ye, G.L. & Aung, H. 2005. An elasto-viscoplastic model for soft sedimentary rock based on tij concept and subloading yield surface. *Soils and Foundations* 45(1): 65–73.

Prediction and Simulation Methods for Geohazard Mitigation – Oka, Murakami & Kimoto (eds)
© 2009 Taylor & Francis Group, London, ISBN 978-0-415-80482-0

Particle crushing and deformation behaviour

D. Muir Wood
University of Bristol, Bristol, UK

M. Kikumoto
Nagoya Institute of Technology, Nagoya, Japan

A. Russell
University of New South Wales, Sydney, Australia

ABSTRACT: Particle crushing occurs in granular materials in various engineering applications. The effect of particle crushing is to broaden the grading of particle sizes and lower the critical state line and other characteristics of the volumetric response in the compression plane. An existing constitutive model, Severn-Trent sand, in which the critical state line plays a central role, has been extended to include the effects of particle crushing. Severn-Trent sand is a distortional hardening Mohr-Coulomb model described within a kinematic hardening, bounding surface framework. Strength is a variable quantity, dependent on the current value of state parameter which varies with shearing-induced dilatancy and changes in stress level. Particle crushing lowers the critical state line and hence tends to produce increase in state parameter and hence a looser response even without any change in volumetric packing. Dense material having crushed may liquefy and lose strength—thus leading to increased run-out of debris flows.

1 INTRODUCTION

Particle crushing occurs in granular materials in various engineering applications such as the driving of piles in carbonate sands (where the strength of the particles is low) and debris flows (where the energy levels are high). The occurrence of particle crushing implies that the material being tested or being subjected to geotechnical loads is actually changing. The changes are usually irreversible so that the material that exists at the end of a pile driving exercise is quite different from the material that was there to begin with and the properties will also in general be altered.

We will concentrate here on the modelling of particle crushing in a sand-like material. An existing constitutive model, Severn-Trent sand, which provides satisfactory simulations for the distortional behaviour of sand at low stresses, will be modified in order to accommodate the effects of particle crushing. It is suggested that the primary effect of particle crushing is the broadening of the grading leading to the possibility of more efficient packings and hence higher limiting densities or lower limiting void ratios or porosities. The critical state line which forms a central component of the Severn-Trent sand model is a locus of limiting void ratios. Linking its position with the current grading gives a simple route for the incorporation of effects of particle crushing into this model.

2 SEVERN-TRENT SAND

Severn-Trent sand (Gajo & Muir Wood 1999a,b) is a distortional hardening elastic-plastic extended Mohr-Coulomb model which brings together four simple relationships in order to describe the response of sands of different densities and at different stress levels with a rather modest number of soil parameters. The elements of the model are shown in Figure 1. First, it is proposed that there exists a locus of limiting states—critical states—defined in terms of density or specific volume and mean effective stress (Figure 1 (a)). Having defined such a locus, it is then possible to define a state parameter ψ_g as the volumetric distance of the current state of the soil from the critical state line at the current mean stress. This state parameter is the driver for all aspects of the soil response: it seems intuitively reasonable that the soil should have a feeling for the limiting packings at its current stress level and adjust its response accordingly.

Second, it is clear that strength of soils has to be seen as a dependent rather than an independent quantity. A typical relationship between state parameter and strength is shown in Figure 1 (b). Since state parameter will in general vary during a test so also will the current strength vary.

Having introduced a variable strength, the third element is the monotonic hardening law (Figure 1 (c))

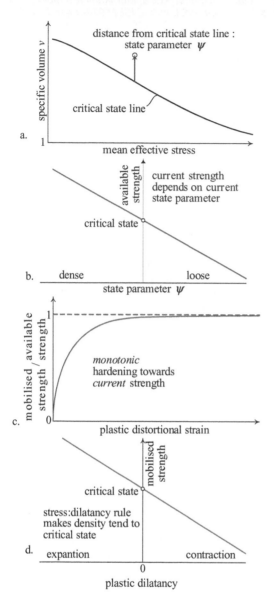

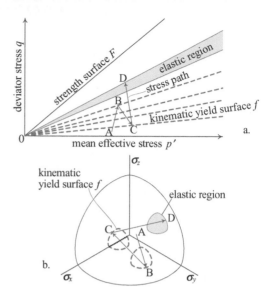

Figure 2. Severn-Trent sand: (a) yielding driven by mobilized friction—incorporation of kinematic hardening; (b) deviatoric view of principal stress space showing kinematic distortional yield surface.

forces volume change towards the critical state whenever distortional straining occurs. There may be dilation or contraction depending on the current value of stress ratio relative to the critical state stress ratio.

Severn-Trent sand is a distortional hardening model which has a purely frictional description of yielding (Figure 2). As the soil shears plastically it is hardening towards the current strength. As the soil shears plastically, volume changes occur of a sign and magnitude dependent on the mobilised friction and the current value of state parameter. The effect of this volume change is always to encourage the soil to approach its critical state. The nonlinear link between dilatancy and intermediate densities allows the model to predict non-monotonic stress-strain response, with a clear peak strength followed by strain softening even though the hardening law is monotonic.

Figure 1. (a) Compression plane, critical state line and definition of state parameter ψ; (b) link between current strength and current state parameter; (c) monotonic hardening law; (d) stress-dilatancy relationship.

which describes the monotonic approach to the current strength of the soil with increasing plastic distortional strain. But of course the strength that is sought is not constant because the density or the stress level of the soil will be changing and with it the current value of state parameter and hence strength (Figure 1 (b)). The primary mechanism for volume change is plastic dilatancy and the fourth element of the model is a stress-dilatancy relationship (Figure 1(d)) which

3 EVOLUTION OF PARTICLE CRUSHING

It is evident from Figure 2 that yielding is tied to mobilised friction alone: mere increase in stress level will not necessarily produce plastic strains. This may be a satisfactory assumption for sands with unbreakable particles or at low stress levels. For more general application a separate, volumetric, mechanism is required. It is proposed that this volumetric mechanism links plastic volumetric strains with effects of particle crushing—so that in the absence of particle crushing there will be no plastic volumetric

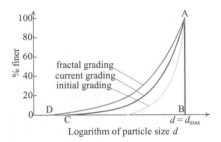

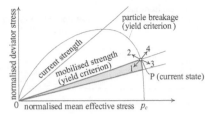

Figure 3. Definition of grading state index I_G as ratio of areas ABC and ABD.

Figure 4. Strength locus, distortional yield locus and crushing yield locus in normalised stress space.

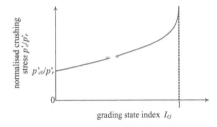

Figure 5. Variation of crushing stress with I_G.

strains other than those resulting from plastic dilatancy (Figure 1 (d)).

In order to incorporate effects of particle crushing in the model it is necessary to have some means of describing the varying grading of the soil: a grading state index, I_G. Figure 3 suggests a definition linked to the location of the current grading between two limiting gradings: the single sized distribution at one extreme and a fractal self-similar distribution at the other extreme. We define I_G as the ratio of areas ABC and ABD where area ABD is the area under a fractal limiting particle size distribution in which the structure (arrangement of particles) looks the same at all scales (McDowell et al. 1996, Muir Wood 2007). It might be suggested that this limiting grading is difficult to attain and difficult to evaluate. However, since it only appears in the definition of I_G as a quantity to scale the area under the current grading it can be seen as another soil parameter.

We need some volumetric mechanism which describes the evolution of crushing and the resulting deformations. Hardin (1985) introduces a breakage potential as a way of monitoring the process of particle crushing. He proposes a function of current stresses as a breakage criterion and, while we do not use the idea of breakage potential, the breakage criterion can be introduced as a second yield mechanism as shown in Figure 4. Because the strength of the soil is a variable dependent on the current value of state parameter ψ, it is convenient to formulate much of the theoretical development of the model in a normalised stress space. Diagrams here will show simply the triaxial section to illustrate the general principles, using mean effective stress p' and deviator stress q in the usual way. If strength is dependent on state parameter according to a linear relationship: $\eta_p = M(1-k\psi)$ where M is the critical state strength and k a soil parameter, then we can normalise deviator stresses by dividing by $(1-k\psi)$ and correspondingly introduce a normalised stress ratio $\bar{\eta} = \bar{q}/p' = q/[p'(1-k\psi)]$.

$$p'\left[1 + \frac{1}{2}(\bar{\eta}/M)^3\right] - p_c = 0 \qquad (1)$$

and this is introduced as the second yield mechanism in Figure 4. It is assumed for convenience that plastic strains for this mechanism obey an associated flow rule. The yield loci are scaled in such a way that the normal is vertical (implying purely distortional response) where they cross the strength line, as shown in Figure 4. The plastic strains for the distortional mechanism follow a non-associated flow rule as hinted in Figure 1(d).

The size p_c of the crushing surface is linked with the current grading expressed through the grading state index I_G:

$$I_G = 1 - \exp\left[-\left(\frac{p_c - p_{c0}}{p_r}\right)^{k_2}\right] \qquad (2)$$

where p_{c0} is the size of the crushing surface, describing the initiation of particle crushing, for $I_G = 0$; k_2 is a soil constant and p_r is a reference stress (Figure 5). Equations (1) and (2) are empirical equations: a theoretical link between particle grading and the stress conditions causing particle crushing is needed.

For a given stress state P which is sitting on both the distortional yield locus and the particle crushing yield locus there are four possibilities which need to be tested in numerical implementation (Figure 4): 1: the stress is unloading for both mechanisms and purely elastic response is predicted; 2: only the distortional mechanism is engaged and the response is the same as that emerging from Severn-Trent sand; 3: only the particle crushing mechanism is engaged; 4: both mechanisms are engaged.

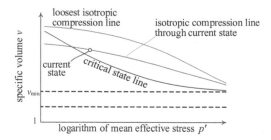

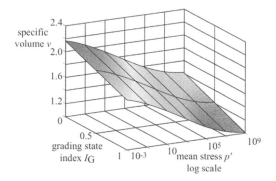

Figure 6. Critical state and normal compression lines.

Figure 8. Critical state surface: $p' : v : I_G$.

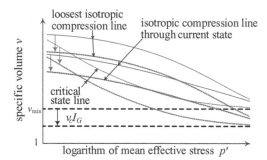

Figure 7. Critical state line falls as I_G rises.

(2005)). We expect that any critical packings including the critical state packing will become more dense with increase in I_G. The values of v_{max} and v_{min} in (3) are thus not constant: as I_G *increases* from 0 to 1 the values of both v_{max} and v_{min} *decrease* by the same amount v_c.

We add an assertion about the volume changes occurring in the sand as the particles break. It is axiomatic that we cannot have specific volumes greater than v_{max} (or less than v_{min}) whatever we may do to our soil. We can find a compression line of the form:

$$v = (v_{min} - v_c I_G) + \left[v_{max}^* - (v_{min} - v_c I_G) \right]$$
$$\times \exp[-\left(p'/p_{cs}\right)^{k_1}] \qquad (4)$$

where p_{nc} is a reference stress which controls the shape of the compression line. This will fit our current state (specific volume v and mean effective stress p') through the single volumetric variable v_{max}^*. In order to ensure that only acceptable values of specific volume can be attained, we shift the compression line by a scaled amount dependent on v_{max}^*:

$$\delta v = -K_c v_c \delta I_G = -\frac{v_{max}^* - v_{min}}{v_{max} - v_{min}} v_c \delta I_G \qquad (5)$$

which then implies that the volumetric strain linked with the crushing yield mechanism is:

$$\delta \varepsilon_p^{pc} = \frac{K_c v_c \delta I_G}{v} \qquad (6)$$

We now need to define a hardening criterion for the particle crushing mechanism. Traditionally the critical state line has always been chosen as linear in a semi logarithmic compression plane, plotting specific volume v and mean effective stress $\ln p'$ (or possibly in a doubly logarithmic plane of $\ln v$ and $\ln p'$). The logarithmic form is unsatisfactory because it does not recognise physically reasonable limits. We know that there must be a lower bound to the specific volume $v = 1$ for very large stresses and we expect that there will be a limiting maximum specific volume at very low stresses, beyond which there is no possibility of the granular material actually transmitting stresses. Let us introduce a modified form of critical state line following Gudehus (1997) (Figure 6):

$$v = v_{min} + (v_{max} - v_{min}) \exp[-(p'/P_{cs})^{k_1}] \qquad (3)$$

where p_{cs} is a reference stress controlling the shape of the critical state line. This equation ensures that the critical state line exists for the range of specific volume $v_{max} > v > v_{min}$.

The effect of particle crushing is to broaden the grading making the packings of the particles more efficient. Fine particles occupy the spaces between the coarser particles: the ideal limiting fractal grading would be able to pack so efficiently that the porosity was equal to zero and the specific volume equal to 1. In another idealised case, the minimum specific volume for a packing of single sized spheres ($I_G = 0$) is 1.35 (the recently proved Kepler conjecture, Hales

Because crushing occurs as the stress level increases, and because crushing causes a change in grading and hence a change in the position of the critical state line, the shape of the critical state line that would be observed as the stress level of triaxial testing (say) was progressively increased becomes a critical state *surface* (Figure 8) in terms of specific volume v, mean effective stress p', and grading index I_G, because the limiting values of specific volume v_{max} and v_{min} in eq. (3) fall as the grading index increases. We cannot ignore the change in grading. The critical state line seen in the $v : \ln p'$ compression plane in a series

266

of tests with steadily increasing stress level is thus a projection onto this plane of a section through this critical state surface.

4 SIMULATIONS

Two sets of simulations are shown here. First, a series of drained triaxial compression tests with constant mean effective stress p' have been performed with varying amounts of initial isotropic precompression to produce different changes in crushing index I_G (Figure 9). Figure 9 (d) shows the histories imposed in the compression plane $\ln p' : v$. Sample AF was sheared at constant mean stress from A to F without any precompression; sample ABAF was precompressed from A to B and back again and then sheared—this is assumed to be a precompression excursion within the current crushing surface; samples ACAF and ADAF were correspondingly precompressed from A to C and D respectively and back again to A and then sheared—for both these samples crushing occurs and the grading state index increases (Figure 9 (c)). (Point F denotes a failure state which will be different for each sample.) The effect of the precompression is to crush some of the particles and hence to increase I_G and hence to lower the critical state line. The initial state parameter at A increases as the extent of the precompression increases and the subsequent stress:strain responses (Figure 9 (a)) and volumetric responses (Figure 9 (b)) change gradually from those expected of a dense material to those expected of a loose material: the peak disappears from the stress:strain response and the amount of volumetric expansion accompanying shear reduces. Without particle crushing the response of all four samples would be identical to that of sample A.

Second, a series of undrained triaxial compression tests have been performed on samples with different initial densities but the same initial mean effective stress. Figures 10 (a) and (c) show the paths in the compression plane and the effective stress paths, respectively. For loose samples (with initially positive values of state parameter) undrained constant volume shearing has to end up on the critical state line at a low stress level. Such a test hardly engages with the crushing criterion so that the effect of including particle crushing is negligible. On the other hand tests on initially dense samples (with initially negative values of state parameter) show typically some contractive initial behaviour, with the mean effective stress falling, followed by some volumetric dilation which translates into mean effective stress increase in these samples. Increase in mean effective stress in the sample with initial specific volume of 1.8 is within the current crushing surface. However, for the simulations shown in which the initial specific volume is less than 1.7, the effective mean stress increases, some particle

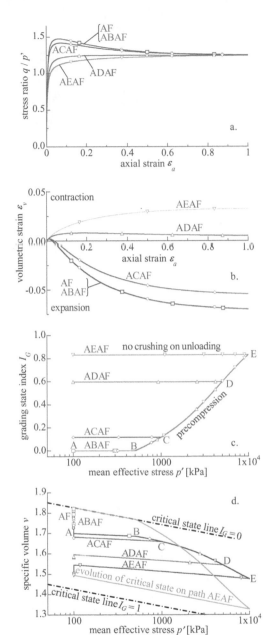

Figure 9. Shearing at constant mean effective stress p' of samples with different precompression histories: (a) stress:strain response; (b) volumetric response, (c) evolution of grading state index I_G; (d) compression plane $\ln p' : v$.

crushing occurs, the grading state index increases, and the critical state line falls (Figure 10 (b)). The importance of the downward evolution of the critical state line in Figure 10 (a) is obvious: without this down-turn the effective stress paths of the denser samples would in principle meet the unchanging critical

267

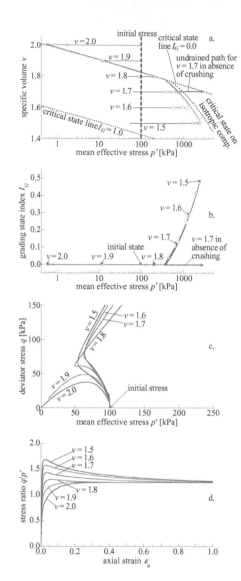

Figure 10. Undrained compression with different initial densities: (a) compression plane lnp : v, (b) evolution of grading state index I_G, (c) effective stress paths, (d) stress strain response.

state line at extremely high effective stresses, as indicated by the arrow for $v = 1$ in Figure 10 (a). (In practice, cavitation of the pore fluid might limit the development of negative pore pressure and the accompanying increase in mean effective stress.)

5 CONCLUSIONS

A model has been developed which incorporates necessary descriptions of aspects of constitutive behaviour of granular materials. In particular these include (a) a critical state line which exists for all possible specific volumes that are present in the soil samples; (b) a dependency of current strength on current value of state parameter which in general changes as the soil dilates as it is sheared; (c) the evolution of limiting states with particle crushing described through a grading state index. The effect of crushing is to shift the critical state line downwards in the compression plane because broader gradings are able to pack more efficiently. As a consequence the crushing soil feels looser and dense samples as they crush show aspects of drained and undrained behaviour which are more reminiscent of loose materials.

ACKNOWLEDGEMENTS

This research would not have been possible without the generous support of the Japan Society for the Promotion of Science enabling collaboration visits of Mamoru Kikumoto to Bristol and David Muir Wood to Nagoya.

REFERENCES

Gajo, A. & Muir Wood, D. 1999a. A kinematic hardening constitutive model for sands: the multiaxial formulation. *Int. J. for Numerical and Analytical Methods in Geomechanics* 23(5): 925–965.

Gajo, A. & Muir Wood, D. 1999b. Severn-Trent sand: a kinematic hardening constitutive model for sands: the q–p formulation. *Géotechnique* 49(5): 595–614.

Gudehus, G. 1997. Attractors, percolation thresholds and phase limits of granular soils. In: R.P. Behringer & J.T. Jenkins (eds). *Powders and Grains '97*: 169–183. Rotterdam: Balkema.

Hales, T.C. 2005. A proof of the Kepler conjecture. *Annals of Mathematics* 162: 1065–1185.

Hardin, B.O. 1985. Crushing of soil particles. *J Geotech Eng,* ASCE 111(10): 1177–1192.

McDowell, G.R., Bolton, M.D. & Robertson, D. 1996. The fractal crushing of granular materials. *J Mech Phys Solids* 44(12): 2079–2102.

Muir Wood, D. 2007. The magic of sands: 20th Bjerrum Lecture presented in Oslo 25 November 2005 *Canadian Geotechnical Journal* 44(11): 1329–1350.

Prediction and Simulation Methods for Geohazard Mitigation – Oka, Murakami & Kimoto (eds)
© 2009 Taylor & Francis Group, London, ISBN 978-0-415-80482-0

A study of the compression behaviour of structured clays

S. Horpibulsuk, J. Suebsuk & A. Chinkulkijniwat
Suranaree University of Technology, Nakhon Ratchasima, Thailand

M.D. Liu
The University of New South Wale, Sydney, Australia

ABSTRACT: The compression behaviour of soils has always been a topic of investigation in geotechnical engineering. A study of the compression behaviour of soils with natural structures and with artificially treated structures is made in this paper. The voids ratio of any structured soil is the summation of the voids ratio of the same soil in a reconstituted state and the additional voids ratio sustained by the soil structure. This finding provides a useful framework for development of a compression model quantifying the influence of soil structures. Based on this model, a comparative study is performed on the influence of different types of soil structures on the compression behaviour. Therefore, the compressibility is investigated and quantified of soils in different states, such as reconstituted state, natural state, and chemical treated states and cemented state.

1 INTRODUCTION

The compression behaviour of soils has always been a topic of investigation in geotechnical engineering because many field problems are analyzed solely on the basis of soil properties obtained from compression tests. For example, the great majority of settlement calculations for geotechnical structures is made based on the coefficient of compressibility of the soil layers beneath the foundations (Butterfield & Baligh 1996). Modelling the compression behaviour of soils is, in reality, very complicated because of the wide ranges involved in two major variables of the materials found or used in the engineering practice. One is the variation in material composition, i.e., grain size and mineralogy. The other is the variation in the structure of the soils, i.e., the arrangement and bonding of the constituents. Structure is defined here in a very broad sense in that it encompasses all features of a soil that are different from those of the material with the same mineralogy at a selected reference state. Therefore, the structure of soil and its influence, as defined in this paper, are relative quantities and depend on the selected reference state. The removal of geomaterial structure is commonly referred to as destructuring. The destructuring due to stress excursions is usually an irrecoverable and progressive process. A study of the compression behaviour of clays with natural structures and with artificially treated structures is made in this paper, and the behaviour of soils in reconstituted states is adopted as the reference state behaviour.

2 COMPRESSION BEHAVIOUR OF STRUCTURED CLAYS

A comparison of the compression behaviour of clay in reconstituted states, naturally structured states and cemented states is shown in Figure 1 (A_w stands for cement content by weight. Test data from Lorenzo & Bergado 2004). Similar features of the compression behaviour of clays with various naturally and artificially structures have widely observed (e.g., Terzaghi 1953, Cotecchia & Chandler 2000, Horpibulsuk et al. 2004). The following features on the mechanical properties of structured clays are observed.

1. Generally speaking, the voids ratio for a structured soil at a given stress level is higher than that of the reconstituted soil of the same mineralogy. When soil undergoes destructuring, the additional voids ratio sustained by soil structure decreases. The difference in the compression behaviour between a natural clay and the soil with same mineralogy with artificially cementation is essentially quantitative in this aspect. The additional voids ratio sustained by cementation structure is much higher than that sustained by natural soil structure; the rate of destructuring of induced cemented clays is generally lower than that of naturally structured clays.

2. As σ_v' increases, the compression curves corresponding to the structured soils approaches to the curve for the reconstituted soil. For most natural soft clays, their compression curves appear to be

asymptotic to the curve for the reconstituted soil, i.e., the influence of soil structure tends to diminish as σ'_v increases. However, for artificially cemented clays the compression behaviour of cemented clay is usually not asymptotic to that of the parent material in a reconstituted state. It is seen that part of the cementation structure does not disappear. The type of the induced cementation structure may be classified as meta-stable (Cotecchia & Chandler 2000).

Consequently, a new material idealisation of the compression behaviour of soils with both naturally and artificially formed structures is proposed and shown in Fig. 2. A modification of the work by Liu & Carter (1999) is made so that the compression behaviour of a structured clay and that of a reconstituted clay are not necessarily asymptotic. The voids ratio for a structured soil, e, can be expressed in terms of the

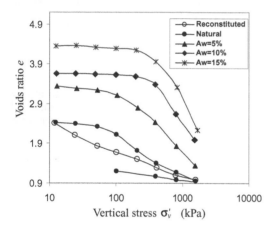

Figure 1. One dimensional compression behaviour of soft Bangkok clay with various structures (Lorenzo & Bergado 2004).

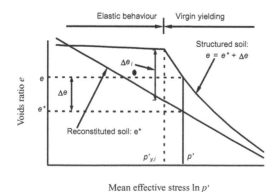

Figure 2. Compression behaviour of structured clays.

corresponding voids ratio for the reconstituted soil, e^*, and the component due to the structure, Δe, i.e.,

$$e = e^* + \Delta e \qquad (1)$$

The properties of a reconstituted soil are called the intrinsic properties, and are denoted by the symbol * attached to the relevant symbols. Hence, under all conditions the influence of soil structure can be measured by comparing its behaviour with the intrinsic behaviour. The behaviour of reconstituted soil is regarded as the reference behaviour and the associated properties are regarded as the reference properties in this study. The difference, Δe, identifies the effect of soil structure. In order to model the variation in the destructuring rate for a variety of structured soils during virgin compression, a modified virgin compression equation is obtained as follows,

$$\Delta e = (\Delta e_i - c)\left(\frac{p'_{y,i}}{p'}\right)^b + c \qquad \text{for } p' \geq p'_{y,i} \quad (2)$$

Δe_i is the additional voids ratio at $p' = p'_{y,i}$, where virgin yielding begins (Figure 2). b is a parameter quantifying the rate of destructuring, referred to as the *destructuring index*. c is a soil parameter, describing the part of the additional voids ratio that can not be removed by increasing the compression stress. Parameter c is defined by the following equation,

$$c = \lim_{p' \to \infty} \Delta e \qquad (3)$$

It is usually convenient to substitute the vertical effective stress σ'_v for the mean effective stress p' when describing one-dimensional compression behaviour. Consequently, for one-dimensional compression behaviour, Equation (2) can be written as

$$\Delta e = (\Delta e_i - c)\left(\frac{\sigma'_{vy,i}}{\sigma'_v}\right)^b \qquad \text{for } \sigma'_v \geq \sigma'_{vy,i} \quad (4)$$

where $\sigma'_{vy,i}$ is the initial vertical yield stress. For a given type of soil structure it is assumed that the compression destructuring index b will take the same value in Equations (2) and (4). It may be noticed that this assumption implies the following constraint on the stress state in the soil during a compression test,

$$\frac{p'}{\sigma'_v} = \frac{p'_{y,i}}{\sigma'_{vy,i}} \qquad (5)$$

$p'_{y,i}$ and $\sigma'_{vy,i}$ are two stress quantities that refer to the same yielding stress state. Condition (5) is the definition of pure compression loading, and may only be satisfied approximately for one-dimensional compression tests.

Table 1. Soil parameters for the three soils.

Soil and Figures	Tests	Intrinsic soil parameters			Structural soil parameters			
		e^*_{IC}	λ^*	κ^*	Δe_i	$p'_{yi}(\sigma'_{v,yi})$	b	c
Soft Bangkok clay, Fig. 3	Reconstituted	3.0	0.28	0.05				
	Natural				0.53	80 kPa	1.0	0.0
	$A_w = 5\%$				1.60	180 kPa	0.7	0.1
	$A_w = 10\%$				2.15	370 kPa	0.7	0.4
	$A_w = 15\%$				2.86	400 kPa	0.7	0.6
Stiff Vallericca clay, Fig. 4	Reconstituted	1.58	0.28	0.06				
	Natural				0.085	2950 kPa	0.45	0.0
Soft Louiseville clay treated with lime, Fig. 5	Reconstituted	3.9	0.7	0.1				
	2% lime				9.5	0.37 kPa	0.3	0.0
	5% lime				9.5	4.2 kPa	0.3	0.0
	10% lime				8.8	16 kPa	0.3	0.0

3 ANALYSIS OF EXPERIMENTAL DATA

Based on the introduced framework for quantifying the influence of soil structures on compression behaviour, in this section the compressibility is investigated of soils in different states, such as reconstituted state, natural state, and chemical treated states. For simplicity, it is assumed that soil behaviour for $p' \leq p'_{yi}$ is elastic and elastic properties are independent of soil structures. Therefore, both reconstituted and structured soils behave linearly in the e–$\ln p'$ with the same gradient.

The results of eleven tests on three soils are now considered. They are one dimensional compression tests on soft Bangkok clay in reconstituted states, natural states, and artificially cemented states (test data from Lorenzo & Bergado 2004), isotropic compression tests on stiff Vallericca clay (test data from D'Onofrio et al. 1998), and one dimensional compression behaviour of soft Louiseville clay treated with lime (test data after Locat et al. 1996). All values of the parameters determined are listed in Table 1. Comparisons between the theoretical equations and the experimental data are shown in Figures 3 to 5.

Considering that the simulations cover a very wide range of stress levels and material structures and loading variations, it is seen that the compression model provides a very satisfactory description of the behaviour of structured soils. It is also seen that the reduction in the additional voids ratio resulting from the destructuring in virgin compression has been simulated with good accuracy. It is also seen that the assumption that elastic deformation is independent of the material structure appears to be an acceptable approximation.

The structures investigated in this section include natural soft structure, natural strong structure, artificial soft structure and artificial strong structure. It is seen that the compression behaviour of all these

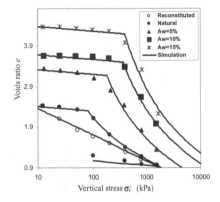

Figure 3. Simulated one dimensional compression behaviour of soft Bangkok clay with various structures.

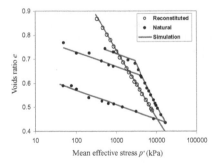

Figure 4. Simulated isotropic compression behaviour of stiff Vallericca clay (data from D'Onofrio et al. 1998).

structured soils can be described successfully with the behaviour of the same soil in reconstituted soils as a base. Consequently, the focus of structured soil behaviour will be the reduction of the additional voids

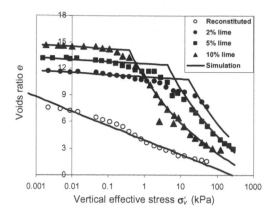

Figure 5. Simulated one dimensional compression behaviour of lime treated Louiseville clay (data from Locat et al. 1996).

ratio associated with destructuring, which is modeled with structure-dependent soil parameters.

From the tests simulated, it is seen that part of the voids ratio associated with strong cementation structure doe not diminish with increasing stress level and the rate of destructuring for soils with strong or stable structure is generally lower than that of soils with soft structure. This is in consistent with experimental observation reported, i.e., Huang & Airey (1998), (Cotecchia & Chandler 2000), and Horpibulsuk et al. (2001).

4 CONCLUSIONS

The compression behaviour of soils with various natural and artificial structures has been investigated. The compression model proposed by Liu & Carter (1999) is employed as a base for this study; the special feature of artificially cemented soils has been considered. A comparative study is performed on the influence of different types of soil structures on the compression behaviour. The ability of this compression model to simulate the response of the following structured soils during virgin yielding has been demonstrated: soft and stiff natural clays, and soft clay treated by lime and soft clay with cementation.

REFERENCES

Butterfield, R. & Baligh, F. 1996. A new evaluation of loading cycles in an oedometer, *Géotechnique* 46: 547–553.

Cotecchia, F. & Chandler, R.J. 2000. A general framework for the mechanical behaviour of clays. *Geotechnique* 50(4): 431–447.

A.D'Onofrio, F.S. de Magistris & Olivares, L. 1998. Influence of soil structure on the behaviour of two natural stiff clays in the pre-failure range. In Evangelista and Picarelli (eds), *The Geotechnics of Hard Soils—Soft Rocks*: 497–505.

Horpibulsuk, S., Bergado, D.T. & Lorenzo, G.A. 2004. Compressibility of cement admixed clays at high water content. *Geotechnique* 54(2): 151–154.

Huang, J.T. & Airey, D.W. 1998. Properties of an artificially cemented carbonate sand, *Journal of Geotechnical and Geoenvironmental Engineering*, ASCE 124(6): 492–499.

Liu M.D. & Carter J.P. 1999. Virgin compression of structured soils. *Géotechnique* 49(1): 43–57.

Liu M.D. & Carter J.P. 2000. Modelling the destructuring of soils during virgin compression. *Géotechnique* 50(4): 479–483.

Locat J., Tremblay H. & Leroueil S. 1996. Mechanical and hydraulic behaviour of a soft inorganic clay treated with lime. *Canadian Geotechnical* Journal 33(3): 654–669.

Lorenzo G.A. & Bergado D.T. 2004. Fundamental parameters of cement-admixed clay: new approach, *Journal of Geotechnical Engineering Division*, ASCE 130(10): 1042–1050.

Terzaghi, K. 1953. Fifty years of subsoil exploration. *Proc. 3rd Int. Conf. Soil Mechanics and Foundation Engineering*: 227–238.

Prediction and Simulation Methods for Geohazard Mitigation – Oka, Murakami & Kimoto (eds)
© 2009 Taylor & Francis Group, London, ISBN 978-0-415-80482-0

Mechanistic picture of time effects in granular materials

P.V. Lade

The Catholic University of America, Washington, DC, USA

ABSTRACT: Based on observations from creep, stress relaxation and strain rate experiments on sand, a mechanistic picture of time effects in granular materials emerges. The mechanisms of time effects in granular materials are unique and appear to depend on interparticle friction, grain crushing and grain rearrangement. This mechanistic picture is based on measured behavior in drained triaxial compression tests in which strain rate effects are observed as small to negligible, and creep and relaxation are caused by the same phenomenon, but it appears that one cannot be obtained from the other

1 INTRODUCTION

Time dependent behavior of granular materials is quite different from the viscous behavior observed in clays. The effects of strain-rate, creep and stress relaxation of clays follow a classic pattern of viscous behavior observed for most materials. For such materials the stiffness and strength increase with increasing strain rate and phenomena such as creep, relaxation and strain rate effects are governed by the same basic mechanism. This behavior is denoted as "isotach" behavior, i.e. there is a unique stress-strain-strain rate relation for a given clay. For such materials, the creep properties may be obtained from, say, a triaxial compression test and used for prediction of stress relaxation in another experiment.

For granular materials the effects of strain rate or loading rate are observed to be small to negligible, and while creep and stress relaxation are caused by the same phenomenon, namely grain crushing followed by grain rearrangement, the prediction of one phenomenon can apparently not be accomplished on the basis of the other. Such behavior is referred to as "nonisotach" behavior. The role of grain crushing is explained, and the transfer of forces through the grain structure is different in creep and relaxation tests. While each phenomenon follows similar patterns, the strains produced after one day of creep does not produce a result that correlates with the changes in stress due to one day of stress relaxation.

To throw further light on the effects of time on the behavior of sand, presented here is a study of strain rate effects, creep, and stress relaxation in crushed coral sand.

2 PREVIOUS STUDIES

Comprehensive reviews of time dependent behavior of soils and models for characterization of this behavior have recently been presented in the literature (Augustesen et al., 2004, Liingaard et al., 2004). The essence of these reviews is that clay and sand behave differently with respect to time. They show that strain rate has important influence on the stress-strain behavior of clay, while widely different strain rates produce essentially the same stress-strain relation for sand, as seen in experiments presented by e.g. Tatsuoka et al. (2002), Kuwano and Jardine (2002), and Kiyota and Tatsuoka (2006). Changes in strain rate have permanent effects in clay, where switches from one to another stress-strain curve occur in response to changes in strain rate. Only temporary changes occur in the stress-strain relations for sand, as also observed by the authors listed above

Observed behavior shows (Augustesen et al., 2004) that the phenomena of creep and stress relaxation are also different in clay and sand. For clay, creep and relaxation properties can be obtained from constant rate of strain tests, and vice versa, as shown by e.g. Leroueil and Margues (1996). The fact that creep, relaxation and strain rate effects can be modeled by the same basic viscous mechanism is referred to as "isotach" behavior.

An investigation of strain rate effects in dense Cambria sand under drained and undrained conditions at high pressures performed by Yamamuro and Lade (1993) showed no significant rate effects on the stress-strain relations. The Hostun and Toyoura sands tested by Matsushita et al. (1999) exhibited noticeable amounts of creep and relaxation but no strain rate effects. This led to one of the main conclusions: The phenomena of creep and relaxation cannot be predicted based on results of constant rate of strain tests. This is because the changes in stress-strain relations due to changes in strain rate are temporary. This behavior of sand is labeled "non-isotach" behavior.

To provide a more comprehensive background for development of an appropriate constitutive model for time effects in sand, which shows non-isotach behavior, experiments have been performed to study time effects in sand.

3 EXPERIMENTAL STUDY OF SAND

The time-dependent behavior of sand was studied in a conventional triaxial apparatus. Modifications to this equipment were made to improve its capability to carry out long-term tests with steady stresses and accurate measurements at a constant temperature. Mechanical equipment with negligible drift in applied pressures and loads and measurement systems without zero drift or devices in which the zero position could be verified during experiments were employed for all testing. The triaxial equipment, the loading systems, the deformation measurement systems, and the temperature control were explained by Lade and Liu (1998).

The sand tested was crushed coral sand. The gradation consisted of grain sizes between the No. 30 and No. 140 U.S. sieves (0.60 to 0.106 mm) with a nearly straight line gradation between these two sizes. The maximum and minimum void ratios were 1.22 and 0.70. The specific gravity of sand grains was 2.88. The tests on crushed coral sand were performed on specimens with a relative density of 60% corresponding to a void ratio of 0.91.

4 STRAIN RATE EFFECTS

Triaxial compression tests were performed on crushed coral sand with an effective confining pressure of 200 kPa and with five different, constant axial strain rates varying from 0.00665%/min to 1.70%/min, corresponding to a 256-fold increase in strain rate. The results of these tests are shown in Figs. 1(a) and 1(b). They indicate that the influence of strain rate on the characteristics of the stress-strain and volume change curves is negligible. Thus, the slopes of the curves as well as the strengths are very little affected by the strain rate. Similar results for sand have been found by Yamamuro and Lade (1993) and by Matsushita et al. (1999). This departure from classic time-dependent behavior, according to which the stiffness and the strength increase with increasing strain rate, is significant, because it indicates that it may not be possible to employ conventional viscous type models to capture the time-dependent behavior of sands. Such models

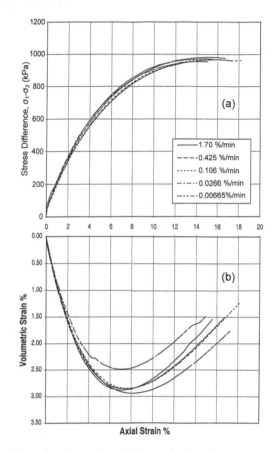

Figure 1. Comparison of (a) stress-strain and (b) volume change relations for drained triaxial compression tests on crushed coral sand performed with five different strain rates.

have been successfully used to characterize a number of other materials, including soils such as clays (see Liingaard et al. (2004) for comprehensive review of time effect models).

5 CREEP EXPERIMENTS

Conventional creep experiments were performed after the specimen had been loaded corresponding to the average strain rate of 0.106%/min. Once the desired deviator stresses of 500, 700, and 900 kPa had been reached, the specimen was allowed to creep for approximately one day (= 1440 min). After the creep stage, the deviator stress was again increased sufficiently to join the virgin or primary stress-strain curve before another creep test was initiated.

Fig. 2(a) shows the stress-strain and volume change curves, and superimposed on these diagrams are the results of corresponding load controlled experiment. As creep proceeds at a given stress, the plastic yield

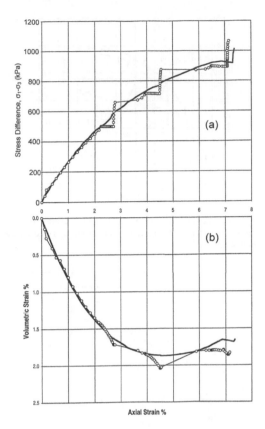

900 kPa had been reached, the axial deformation was held constant to observe the stress difference relax. Relaxation periods of 1000 min (i.e. a little less than one day) were employed in all but a few experiments in which longer relaxation periods were used. After the stress relaxation stage, the stress difference was again increased sufficiently to join the virgin or primary stress-strain curve before another relaxation test was initiated.

Fig. 3(a) shows the stress-strain relations obtained from the basic experiment performed with stress relaxation at the three desired stress differences and at a stress point beyond peak failure. Reloading after stress relaxation exhibits structuration effects similar to those observed after periods of creep, i.e. a temporary increase in the deviator strength beyond that corresponding to the primary loading curve. In all cases, the stress-strain curve appears to unite with or become the primary stress-strain relation well before the next stress relaxation point is reached.

Fig. 3(b) shows that volumetric contraction of the specimen is associated with relaxation of the stress difference, whether or not the specimen is contracting or dilating during primary loading. These periods of volumetric contraction produce offsets in the volume change curve and like the stress-strain relation, the volume changes also seem to recover to the basic volume change curve upon reloading.

Figure 2. (a) Stress-strain and (b) volume change relations from load controlled drained triaxial compression test on crushed coral sand with creep stages of 1440 min. at stress differences of 500, 700, and 900 kPa. Comparison with results from load controlled test without creep stages.

surface moves out to higher stresses. This may be seen from the fact that further loading first produces what appears to be elastic reloading.

The volume change curves corresponding to creep, shown in Fig. 2(b), do not follow the reference curve, unlike the previous experiments on Antelope Valley sand presented by Lade and Liu (1998). This means that the potential for inelastic creep strains cannot be taken to be the same as the potential for plastic strains, as was the case for the Antelope Valley sand. The potential for inelastic strains for crushed coral sand must be inclined such that the creep volumetric strains are more contractive than those obtained from the plastic potential at the same stress point.

6 STRESS RELAXATION EXPERIMENTS

Stress relaxation experiments were performed after primary loading with a strain rate of 0.106 %/min. Once the desired stress differences of 500, 700, and

7 COMPARISON OF CREEP AND STRESS RELAXATION

The stress relaxation may be compared with the creep observations by plotting the points of initiation and the end points after a certain amount of time on the same diagram. To overcome the small differences in the primary stress-strain curves from the two experiments in Figs. 2 and 3, the stress-strain curve shown in Fig. 3(a) is used as the base curve from which creep and stress relaxation are initiated.

The comparison of stress relaxation and creep after one day is shown in Fig. 4 in which the points of initiation of creep have been located on this base curve, while the end points obtained after approximately 1 day of creep are shown relative to the initiation points. The data from both types of tests are very consistent, and they show how much the axial strains change due to creep and how much the axial stresses change due to relaxation, respectively. It is clear that the amount of creep and the amount of relaxation resulting after 1 day define curves that are located at quite different positions.

Fig. 5 shows a comparison of creep and stress relaxation effects after one day plotted from a common stress-strain curve for comparable experiments performed to study time effect on Antelope Valley sand (after Lade 2007). As for the crushed coral sand, the

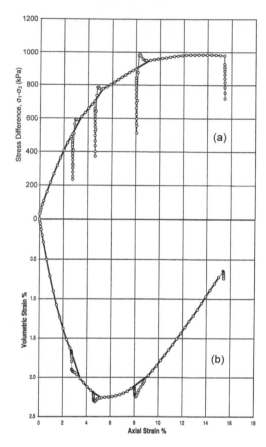

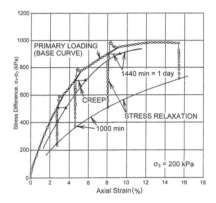

Figure 4. Comparison of creep and stress relaxation experiments performed in two triaxial compression tests on crushed coral sand.

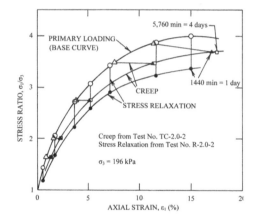

Figure 5. Comparison of creep and stress relaxation experiments performed in two triaxial compression tests on Antelope Valley sand.

Figure 3. (a) Stress-strain and (b) volume change relations from deformation controlled drained triaxial compression test on crushed coral sand with stress relaxation stages of 1000 min. at stress differences of 500, 700, and 900 kPa. Comparison with results from deformation controlled test without stress relaxation stages.

experiments on Antelope Valley sand showed that strain rate effects are negligible (Lade 2007), and Fig. 5 indicates that the observed stress relaxation behavior does not correspond with the measured creep behavior. It is concluded that neither the crushed coral sand, nor the Antelope Valley sand exhibits classic viscous effects. Note also that the amounts of disagreement between the stress-strain relations after one day of creep or stress relaxation are quite different for the two sands.

The fact that the same basic mechanism can account for creep, stress relaxation, and rate dependency and can serve as basis for prediction of one from the other, as is the case for clays, indicates that the material complies with the "correspondence principle" according to Sheahan and Kaliakin (1999). The experiments presented here showed noticeable amounts of creep and relaxation but no strain rate effects. Further, the stress

relaxation and the creep responses do not appear to follow the correspondence principle, i.e. two different stress-strain relations are obtained after 1 day, as indicated for two different sands in Figs. 4 and 5. Thus, it appears that the phenomena of creep, stress relaxation and strain rate effects in sand cannot be predicted from the same type of test using a viscous type model. The type of behavior observed for sands is referred to as "nonisotach" behavior.

8 PROPOSED MECHANISTIC PICTURE OF TIME EFFECTS IN SANDS

Particle breakage has often been observed to be associated with time effects in granular materials, and a mechanistic picture of time effects may be constructed on the basis of this phenomenon. Particle

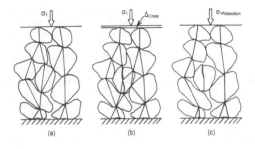

Figure 6. (a) Initial force chains in particle structure and effects of grain crushing in (b) creep test and in (c) stress relaxation test.

breakage may not occur and time effects are negligible in granular materials at very low stresses. Time effects become significant with increasing confining pressure and increasing stress difference, as has been observed in several studies (see Yamamuro and Lade 1993). These studies also noted the association between particle crushing and its occurrence with time.

Fig. 6(a) shows an assembly of grains that have been loaded up to a given stress difference and either creep or stress relaxation occurs from this point. The diagram shows the force chains down through the assembly. The grain in the middle fractures in the beginning of either of these two types of time effects.

The responses of the grain assembly are quite different for the two phenomena. Fig. 6(b) shows what happens during creep in which the vertical stress is held constant. The assembly adjusts its structure to carry the vertical stress. This requires adjustment of the grains and it results in some vertical deformation and new force chains are created to match the externally applied stress. The redundancy in the grain structure allows new force chains to be created and engage other grains that may break. But slowly the amount of breakage will reduce and the creep will slow down with time, just as observed in the experiments.

Fig. 6(c) shows what happens in the stress relaxation experiment. After the grain has broken, the grain structure is not able to carry the vertical stress, but since the assembly is prevented from vertical deformation, the stress relaxes. New force chains are created around the broken grain which does not carry any load. It is the small amount of grain movement in the creep tests that allows new contact to be created and forces to be carried through the grain skeleton. Without this adjustment and consequent deformation to achieve the adjustment, the grain structure is able to transmit only a reduced load and stress relaxation is the consequence. It can also be seen that if a lower limit to the relaxed stress exists, then it depends on the grain strength rather than its frictional properties. Thus, the amount of difference between the creep and stress relaxation for the two sands shown in Figs. 4

and 5 may be explained in terms of different grain strengths.

It may be seen that a relation between creep and stress relaxation does not exist, because the two explanations do not allow a transition from one to the other phenomenon. Thus, Figs. 4 and 5 show that the amounts of creep and the amounts of stress relaxation after one day do not converge towards the same curves. In fact, the two sets of curves are quite different and one cannot be obtained from the other. This nonisotach behavior is quite different from that of the isotach, viscous behavior exhibited by clays.

With only friction (and slippage when the frictional resistance is overcome) and particle breakage (when the strengths of the particles are overcome) as basic behavior constituents, how can the observed time effects be explained for granular materials? Experiments on rock specimens have clearly shown that their strengths are strongly dependent on time. Current research on single sand particles has already indicated that they behave similar to rock specimens in the sense that their crushing strengths are time-dependent. This will be explained elsewhere.

9 CONCLUSION

Observations from experiments show that strain rate effects are negligible for crushed coral sand, unlike for clays in which strain rate effects are significant. Further, the observed stress relaxation behavior was not in "correspondence" with the measured creep behavior. Therefore, the amount of stress relaxation predicted on the basis of model parameters determined from creep experiments will be too small. It is concluded that sands do not exhibit classic viscous effects, and their behavior is indicated as "nonisotach", while the typical viscous behavior of clay is termed "isotach". Thus, there are significant differences in the time-dependent behavior patterns of sands and clays.

A mechanistic picture of time effects in sands is proposed in which interparticle friction, grain crushing, and grain rearrangement play the key roles.

Grain crushing is a time dependent phenomenon and this accounts for the time dependency observed in granular materials. Additional experimental research is required to understand the behavior of sand and to develop a more correct constitutive framework for the time-dependent behavior of sand.

REFERENCES

Augustesen, A., Liingaard, M. & Lade, P.V. 2004. Evaluation of Time Dependent Behavior of Soils. *Int. J. Geomech.* ASCE 4(3): 137–156.
Kiyota, T. & Tatsuoka, F. 2006. Viscous Property of Loose Sand in Triaxial Compression, Extension and Cyclic Loading. *Soils and Found.* 46(5): 665–684.

Kuwano, R. & Jardine, R. 2002. On measuring creep behaviour in granular materials through triaxial testing. *Can. Geotech. J.* 39(5): 1061–1074.

Lade, P.V. 2007. Experimental Study and Analysis of Creep and Stress Relaxation in Granular materials. *Proc. Geo-Denver,* Denver, Colorado, February (CD-rom).

Lade, P.V. & Liu, C.-T. 1998. Experimental Study of Drained Creep Behavior of Sand. *J. Engr. Mech. ASCE* 124(8): 912–920.

Leroueil, S. & Marques, M.E.S. 1996. State of the Art: Importance of Strain Rate and Temperature Effects in Geotechnical Engineering. *Measuring and Modeling Time Dependent Soil Behavior*, Geotech. Spec. Publ. No. 61, Sheahan, T.C., and Kaliakin, V.N., eds, ASCE, Reston, VA, 1–60.

Liingaard, M., Augustesen, A. & Lade, P.V. 2004. Characterization of Models for Time-Dependent Behavior of Soils. *Int. J. Geomech.* ASCE, 4(3), 157–177.

Matsushita, M., Tatsuoka, F. Koseki, J., Cazacliu, B., Benedetto, H. & Yasin, S.J.M. 1999. Time effects on the pre-peak deformation properties of sands. *Pre-failure Deformation Characteristics of Geomaterials*, Jamiolkowski, Lancelotta, and Lo Presti, eds., Balkema, Rotterdam, 681–689.

Sheahan, T.C. & Kaliakin, V.N. 1999. Microstructural Considerations and Validity of the Correspondence Principle for Cohesive Soils. *Engineering Mechanics, Proceedings of the 13th Conference*, N. Jones and R. Ghanem, eds., ASCE publ., Baltimore, MD, USA.

Tatsuoka, F., Shihara, M., Di Benedetto, H. & Kuwano, R. 2002. Time-dependent shear deformation characteristics of geomaterials and their simulation. *Soils and Found.* 42(2): 103–129.

Yamamuro, J.A. & Lade, P.V. 1993. Effects of Strain Rate on Instability of Granular Soils. *Geotech. Test. J.* 16(3): 304–313.

Prediction and Simulation Methods for Geohazard Mitigation – Oka, Murakami & Kimoto (eds)
© 2009 Taylor & Francis Group, London, ISBN 978-0-415-80482-0

Role of Coulomb's and Mohr-Coulomb's failure criteria in shear band

T. Tokue & S. Shigemura
Nihon University, Tokyo, Japan

ABSTRACT: The roles of Coulomb's and Mohr-Coulomb's failure criteria in a shear band, i.e. a usual slip line, were examined by conducting tilting failure tests of a model slope and several kinds of shear tests with the following main results: 1) Coulomb's failure criterion determined by a simple shear test should be used in the limit equilibrium analysis only when the boundary planes of a shear band are not actually-formed failure surfaces; 2) Mohr-Coulomb's failure criterion should be used in the limit equilibrium analysis when the boundary planes of a shear band become the actual failure surfaces because of weak planes, e.g. clayey seams or joints; and 3) Mohr-Coulomb's failure criterion plays substantial roles in the formation of a shear band and in the occurrence of a large deformation in a shear band.

1 INTRODUCTION

1.1 Coulomb's and Mohr-Coulomb's failure criteria in the limit equilibrium analysis

The result of limit equilibrium analysis depends primarily on the shape and position of the slip surface and the failure criterion for soil adopted in the analysis. However, in many cases, the analysis assumes a priori a slip surface, i.e. a failure surface, and uses both Mohr-Coulomb's and Coulomb's failure criteria without a clear mechanical basis for either choice. The usual definition of Coulomb's failure criterion is the linear relationship between the shear and normal stresses at failure on the slip surface, and a typical method for determining this failure criterion has been a box shear test, which simply considers a shear band formed between the upper and lower boxes as the failure surface. On the other hand, the usual definition of Mohr-Coulomb's failure criterion is the linear envelope line of a group of Mohr's stress circles at failure, where the point of contact between the envelope line and the Mohr's stress circle shows the stresses on the failure surface. The typical method to determine this failure criterion has been a triaxial compression test. Though the confining and deforming conditions of the specimens in both tests are very different, some have argued that these two criteria should agree with each other, since they are the failure criteria on common failure surfaces. Accordingly, it is important to clarify the relationship between these two criteria.

1.2 Shear band and objectives of this paper

Many previous shear band studies relate to a thin shear band with the thickness of several tens of a soil-grain diameter at most, which is formed in an almost homogeneous soil specimen under uniform stress conditions in a shear test (e.g. Yoshida & Tatsuoka 1997, Sasda et al. 1999, Oda et al. 2004). On the contrary, we have studied shear bands from the standpoint that the usual slip surface is not a surface, but rather a shear band with some thickness in the ground under non-uniform stress conditions (Tokue & Shigemura 2003, Shigemura & Tokue 2005). From the viewpoint of deformation, a shear band can be thought of as a chain of soil elements in simple shear.

Our previous study clarified the geometrical and mechanical characteristics of a shear band and the relationship between Coulomb's and Mohr-Coulomb's failure criteria, and offered a structural model of a shear band (Fig. 1) (Tokue 1999, Tokue & Shigemura 2003, 2006, 2007, Shigemura & Tokue 2005, 2006). Moreover, we strongly suggested that the boundary planes of a shear band in the ground coincide with the maximum shear stress planes considering the correspondence between the boundary planes of a shear band and the upper and lower planes of a simple shear specimen (Tokue & Shigemura 2003, Shigemura & Tokue 2005).

Failure according to Coulomb's failure criterion
(Failure on the boundary planes of a shear band)

Failure according to Mohr-Coulomb's failure criterion
(Failure on the maximum mobilized plane)

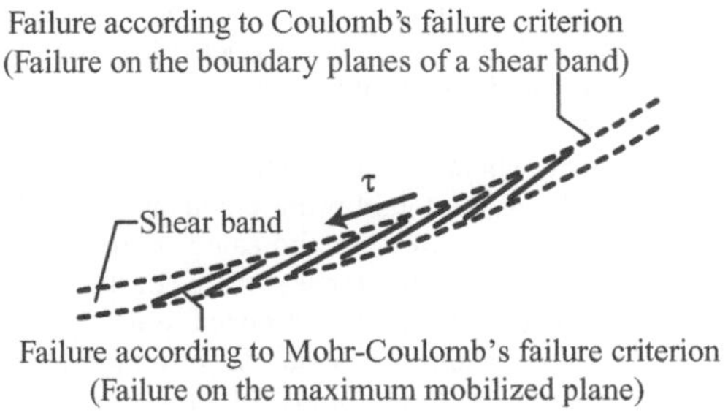

Figure 1. Structural model of a shear band.

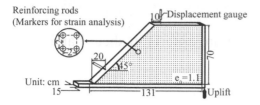

Figure 2. Tilting failure test of a model slope.

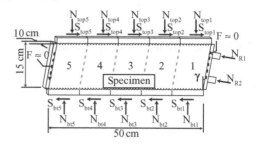

Figure 3. Simple shear test using a large specimen with five coupled elements.

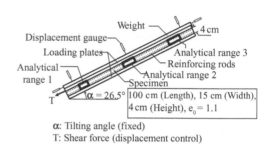

α: Tilting angle (fixed)
T: Shear force (displacement control)

Figure 4. Tilted simple shear test using a long specimen.

The objectives of this paper are to establish the validity of this proposition and to clarify the roles of Coulomb's and Mohr-Coulomb's failure criteria in a shear band by conducting tilting failure tests of a model slope and several kinds of shear tests.

2 OUTLINE OF TESTS

2.1 Tilting failure test of a model slope

A sandy soil was prepared by mixing Gifu sand, specific gravity (G_s) 2.64 and maximum and minimum void ratios (e_{max} and e_{min}) of 1.17 and 0.73, with bentonite and machine oil. Sandy soil model slopes of plane-strain type, 70 cm high, 15 cm wide, and an elevation angle of 45°, were constructed in a very loose state with an initial void ratio e_0 of 1.1 (Fig. 2). A number of reinforcing rods, which had the ends attached with thin plastic discs and were also used as markers for deformation analysis of a shear band, were set placed at intervals of 2 cm within the slope (Fig. 2). This enabled the slope to stand on its own bottom when the sideboards of a test box were removed just before tilting. The test box was tilted one degree at a time; after each movement the displacement gauges were allowed to equilibrate.

2.2 Simple shear test using a large specimen with five-coupled elements

The test materials were the same sandy soil as used in the tilting failure test, a dry mixture of sand and gravel, and Oh-iso gravel. A large specimen ($15 \times 50 \times 10$ cm) consisting of five coupled elements (Fig. 3) was prepared. The tests were conducted by displacement control under constant overburden pressures of 25, 49, and 75 kN/m². The average stress state of each element, which was inevitably influenced by the interaction between neighboring elements, could be determined by measurement and calculation because load cells were set on the three surfaces of the specimen (Fig. 3), and friction was reduced to nearly zero on the surfaces of the specimen except on the upper and lower surfaces. Accordingly, Mohr's stress circle at failure of each element can be drawn. A transparent acrylic plate for the front wall of the specimen enabled

us to observe the deformation and failure within the specimen.

2.3 Tilted simple shear test using a long specimen

The test material was the same sandy soil as mentioned above. In order to obtain in detail the inner configuration of a shear band according to shear deformation, we conducted a tilted simple shear test (Fig. 4) using a one-meter long specimen under conditions very close to those of the shear band in a model slope. After tilting by 26.5° considering the tilting failure test of a model slope, the specimen was sheared by a large shear strain of 100% by pulling the upper loading plate by displacement control.

3 SHEAR BAND AND MAXIMUM SHEAR STRESS PLANE

3.1 The maximum shear strain vector

We have proposed that the horizontal plane of a simple shear specimen almost agrees with the maximum shear stress plane, τ_{max}-plane, at the time of failure (Tokue & Shigemura 2003). Considering that there might be some relation between the maximum shear stress, τ_{max}, and strain, γ_{max}, we introduce a new index, "the maximum shear strain vector", γ_{max}-vector, and examine its mechanical meaning.

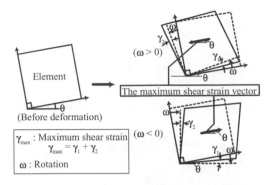

Figure 5. Definition of the maximum shear strain vector.

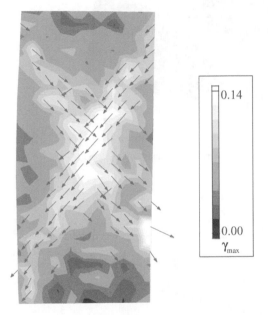

Figure 6. Distribution of γ_{max}-vectors at the peak in the biaxial compression test (dense sand, $e_0 = 0.7$).

The direction of the γ_{max}-vector in each element consisting of four markers is that of the coordinates before deformation which show the maximum shear strain, γ_{max}, after deformation (Fig. 5), and the magnitude of γ_{max}-vector is in proportion to the value of γ_{max}. To test this concept experimentally, the markers were set beforehand in soil specimens and a model slope. Figures 6 and 7 show the distributions of the γ_{max}-vectors in the concentrated zones of γ_{max}, i.e. the failure surfaces due to Mohr-Coulomb's failure criterion (Tokue & Shigemura 2007), at the time of the peak in the biaxial compression test and the simple shear test, respectively. The directions of the γ_{max}-vectors nearly agree with those of the τ_{max}-planes because the average angle of the γ_{max}-vectors from the horizontal

plane was almost ±45° in the biaxial compression and 0° in the simple shear. These results show that the γ_{max}-vector was parallel to the τ_{max}-plane.

3.2 Boundary planes of a shear band

Figure 8 shows the distribution of γ_{max}-vectors in the concentrated zone of γ_{max}, i.e. the shear band, which was formed just before the collapse of the model slope. It can be seen that the directions of γ_{max}-vectors approximately agreed with the direction of the shear band, and accordingly the boundary planes of the shear band were the planes of maximum shear stress as proposed in this work. In other words, a shear band was formed in a linking direction of the maximum shear stress planes in the ground.

3.3 Verification of the shear band model

Figure 9 shows a photograph of the shear band formed in a sandy soil model slope. Detailed observation of Figure 9 finds a number of actual failure surfaces (cracks) in the shear band. It is clear that the shear band is the envelope band of a number of actual failure surfaces. It was confirmed that these actual failure surfaces were formed according to Mohr-Coulomb's failure criterion in the simple shear specimens (Tokue & Shigemura 2007). In this way, Mohr-Coulomb's failure criterion plays a substantial role in the formation of a shear band. The results shown in Figure 9 demonstrate that the proposed model of Figure 1 is valid for the present case of a modeled sandy soil slope.

3.4 Role of Coulomb's failure criterion

Considering that the usual slip surface is a shear band with some thickness and the usual definition of Coulomb's failure criterion is the failure criterion on a slip surface, Coulomb's failure criterion should be applied to a shear band. Taking account of the thickness of a shear band, the failure criterion determined by a simple shear test is actually valid as Coulomb's failure criterion. In fact, when Bishop's simplified method and the conventional slice method were applied to the shear bands of the model slope, the safety factor, F_s, became almost unity using Coulomb's failure criterion, but exceeded 1.4 using Mohr-Coulomb's failure criterion (Tokue & Shigemura 2003).

As indicated in paragraph 3.3, a shear band is the envelope band of a group of actual failure surfaces formed by Mohr-Coulomb's failure criterion (Tokue & Shigemura 2007). Moreover, it seems generally true that the boundary planes of a shear band are not the actual failure surfaces in real soils, as seen at the boundary planes of the shear band of Figure 9 and the upper and lower surfaces of Figure 10-(1).

All the above indicators suggest that Coulomb's failure criterion is the failure criterion to be applied

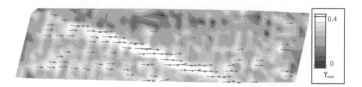

Figure 7. Distribution of γ_{max}-vectors at the peak in the simple shear test with a large specimen (loose sandy soil, $e_0 = 1.1$).

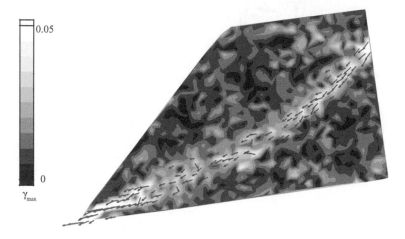

Figure 8. Distribution of γ_{max}-vectors in the model slope just before collapse (loose sandy soil, $e_0 = 1.1$).

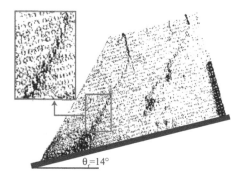

Figure 9. Shear band formed in a sandy soil model slope just before collapse.

to the boundary planes of a shear band in the limit equilibrium analysis only when the boundary planes are not actual failure surfaces.

4 ROLE OF MOHR-COULOMB'S FAILURE CRITERION

4.1 *Failure criterion on an actual failure surface*

Figure 10 shows the actual failure surfaces formed at the peak in the simple shear specimens and Mohr's

stress circles at the peak in the center element of the simple shear specimen (Fig. 3). It seems that the actual failure surface is tilted in Figure 10-(1) because of the near complete homogeneity of the sandy soil specimen, and is almost horizontal in Figure 10-(2) because of the horizontally layered specimen formed by separation of a dry mixture of sand and gravel at the time of compaction. Moreover, the actual failure surface is also almost horizontal in Figure 10-(3) because of likely slips between the notched loading and pedestal plates and the surfaces of the Oh-iso gravel specimen, consisting of rounded grains with high strength

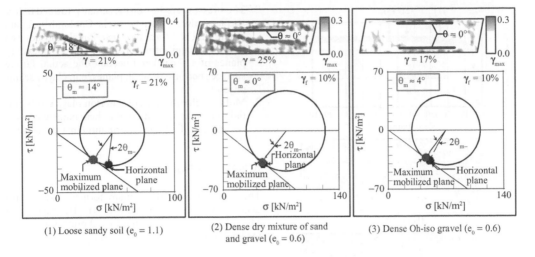

Figure 10. Actual failure surfaces and Mohr's stress circle at the peak in simple shear.

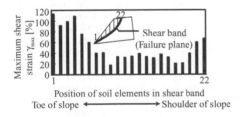

Figure 11. Distribution of γ_{max} within the shear band in a sandy soil model slope just before collapse.

in a very dense state. In all these cases, the actual failure surface nearly coincides with the maximum mobilized plane on Mohr's stress circle, i.e. $(\tau/\sigma)_{max}$-plane, because $\theta_{m-} \approx \theta$. The maximum mobilized plane almost agrees with the failure plane due to Mohr-Coulomb's failure criterion because the cohesion, c, is very small in these three materials. Accordingly, the actual failure surfaces in these specimens are formed by Mohr-Coulomb's failure criterion.

All the above indicators suggest that the stresses at failure on an actual failure surface always satisfy Mohr-Coulomb's failure criterion regardless of the direction of the surface. Accordingly this criterion should be used in the limit equilibrium analysis in the special case of the boundary planes of a shear band becoming the actual failure surfaces because of weak planes, e.g. clayey seams or joints.

4.2 *Relationship between Mohr-Coulomb's and Coulomb's failure criteria*

As shown in paragraph 3.3, a shear band is an enve-lope band of a number of actual failure surfaces formed according to Mohr-Coulomb's failure criterion. That is, the stresses at failure on the actual failure surfaces always satisfy Mohr-Coulomb's failure criterion. On the contrary, Coulomb's failure criterion should be used only when the boundary planes of a shear band are not the actually-formed failure surfaces. In this way, the usual slip surface, i.e. a shear band, is not always an actual failure surface. Taking these results into account, the confused recognition relating to these two failure criteria, i.e. whether or not these two crite-ria should agree with each other, seems to be caused by the vagueness in the relationship between "a slip sur-face" and "a failure surface". However, the proposed shear band model of Figure 1 gives a clear and well-grounded relationship between these two surfaces, and accordingly a clear relationship between Mohr-Coulomb's and Coulomb's failure criteria as indicated above.

4.3 *Mechanism of a large deformation in a shear band*

Figure 11 gives the distribution of γ_{max} within the shear band in a sandy soil-model slope just before the col-lapse of the slope. The distribution of γ_{max} changes significantly from over 100% near the toe of the slope to 20~30% near the middle of the slope even in a small model slope with a height of 70 cm. Figure 12 shows the inner configuration of a one-meter long specimen according to shear deformation in a tilted-simple shear test. The conditions of the specimen, such as shape and size of the cross section, lateral confinement by rein-forcing rods, void ratios, stress level, etc., were made very close to those of the shear band in a model slope. We can see, although not always clearly by eye, a group

283

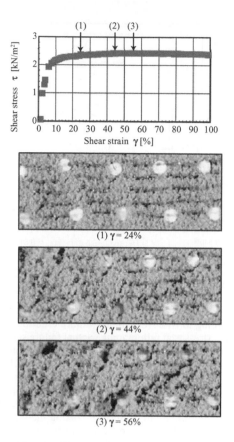

(1) γ = 24%

(2) γ = 44%

(3) γ = 56%

Figure 12. Development of actual failure suface (cracks) in a long specimen according to shear deformation in simple shear (loose sandy soil, $e_0 = 1.1$).

of actual failure surfaces (cracks) formed by Mohr-Coulomb's failure criterion in Figure 12-(1) where the shear stress has nearly reached a peak. At higher strain levels, the actual failure surfaces, which hardly change their number with additional shear deformation, rotate and open gradually with the development of shear deformation as shown in Figures 12-(2) to 12-(3). In this way, a group of actual failure surfaces formed by Mohr-Coulomb's failure criterion plays an important role in a large deformation occurring in a shear band.

5 CONCLUSIONS

1. A shear band, i.e. a usual slip line, is formed in a linking direction of the maximum shear stress planes in the ground when the boundary planes of a shear band are not actually-formed failure surfaces. Coulomb's failure criterion determined by a simple shear test should be used in the limit equilibrium analysis in this case.

2. The stresses at failure on the actually-formed failure surfaces always satisfy Mohr-Coulomb's failure criterion. Therefore, this criterion should be used in the limit equilibrium analysis when the boundary planes of a shear band become the actual failure surfaces because of weak planes, e.g. clayey seams or joints.
3. Mohr-Coulomb's failure criterion plays a substantial role in the formation of a shear band and in the occurrence of a large deformation in a shear band.

ACKNOWLEDGMENTS

The authors wish to thank many former graduate students, especially Mr. T. Furuya, Mr. H. Tanaka and Mr. J. Fuwa, for performing the laboratory tests and the analysis.

REFERENCES

Oda, M., Takemura, T. and Takahasgi, M. 2004. Microstructure in shear band observed by microfo-cus X-ray computed tomography, *Geotechnique* 54(8): 539–542.

Saada, A.S., Liang, L., Figueroa, J. and Cope, C.T. 1999. Bifurcation and shear band propagation on sands, *Geotechnique* 49(3): 367–385.

Shigemura, S. and Tokue, T. 2005. Failure Mechanism and Characteristics of Soil Subjected to Interaction between Soil Elements in Simple Shear, *Proceedings of the 16th International Conference on Soil Mechanics and Geotechnical Engineering* (4): 2575–2578. ISSMGE.

Shigemura, S. and Tokue, T. 2006. Limit equilibrium analysis considering deformation of shear band in model slope, *Advances in earth structures, Geotechnical Special Publication No.151., (Proceedings of the GeoShanghai Conference)*: 83–89. ASCE.

Tokue, T. 1999. A New Concept of Solving Ground Failure Problems and Failure Propagation Phenomena, *Proceedings of the 44th Symposium on Geotechnical Engineering*: 161–166 (in Japanease). JGS.

Tokue, T. and Shigemura, S. 2003. Progressive Failure and Failure Criteria for Soil in Limit Equilibrium Analysis For Slope Stability, *Proceedings of The International Workshop on Prediction and Simulation Methods in Geomechanics (IWS-Athens 2003)*: 85–88. ISSMGE.

Tokue, T. and Shigemura, S. 2006. Types of progressive failures by external force condition and their failure mechanism, *Advances in earth structures, Geotechnical Special Publication No.151., (Proceedings of the GeoShanghai Conference)*: 98–104. ASCE.

Tokue, T. and Shigemura, S. 2007. Mechanical characteristics and failure criteria of a shear band, *Proceedings of the 13th Asian Regional Conference on Soil Mechanics and Geotechnical Engineering 1(1)*: 244–247. ISSMGE.

Yoshida, T. and Tatsuoka, F. 1997. Deformation property of shear band in sand subjected to plane strain compression and its relation to particle characteristics, *Proceedings of the 14th International Conference on Soil Mechanics and Foundation Engineering (1)1*: 237–240. ISSMFE.

Prediction and Simulation Methods for Geohazard Mitigation – Oka, Murakami & Kimoto (eds)
© 2009 Taylor & Francis Group, London, ISBN 978-0-415-80482-0

Collapse of soils: Experimental and numerical investigations

A. Daouadji, M. Jrad & A. Lejeune
Université Paul Verlaine, Metz, France

F. Darve
Institut National Polytechnique, Grenoble, France

ABSTRACT: The aim of this study is to detect the collapse point of soil when submitted to given loading program. An extensive series of tests is performed on loose sand under conventional drained and undrained triaxial compression tests and also under constant shear drained tests. It is found that a non-associated elastoplastic multimechanism model can accurately reproduce not only the global stress–strain behaviour of the loose sand but also the collapse point. This analysis is based on the second order work criterion.

1 INTRODUCTION

Failures of masses of soils are of major interest for engineers and geologists. The social and financial impacts impose a fully understanding of the mechanisms which triggered some failures occurring well before the ultimate state. An essential issue is to predict them. Even if the mechanisms explaining the localization of deformations are still explored, it is now possible to predict their occurrence by using criteria such as the vanishing of the determinant of the so-called acoustic tensor which is derived from the tangential stiffness tensor (Rice 1976). However, a non-negligible amount of catastrophic failures, or collapses, can not be predicted using these criteria or classical elasto-plastic theory. Hence, Darve (1987) and Darve et al. (2004) have proposed a criterion based on the sufficient condition of stability of Hill (1958). On another hand, Nova (1994) and Imposimato and Nova (1998) have shown that an arbitrary applied loading program can lead (or not) to collapses for proper control variables.

2 EXPERIMENTAL ANALYSIS

2.1 Test procedures

The aim of our experimental study is to detect the occurrence of collapses which occur under a diffuse mode of failure. The localisation of the deformation has thus to be prevented. So the density of the samples has to be the lowest as possible, i.e. the void ratio $e = V_v/V_s$ where V_s the volume of the solid and V_v the volume of the voids ($= V_w$, the volume of the water phase as the sample is fully saturated), has to be chosen as high as possible otherwise it will favour strain localisation. For that purpose, the grain size distribution of the granular material has to be as uniform as possible to achieve a high value of the void ratio (the uniformity coefficient C_u is equal to 2). Finally, a slenderness ratio ($= H/D$ where H is the height and D the diameter of the cylindrical specimen) of 1 combined with enlarged frictionless plates and free ends were used at the bottom and at the top to ensure homogeneous deformation.

The chosen granular material is Hostun sand which is a sand containing 99.2% of silica and is widely used in France. The principal characteristics of the sand are summarised in Table 1. The moist tamping method was used to prepare the specimen. It is well known that a strong anisotropy of physical properties is generated when using this method (Vaid et al., 1999). However, data obtained in our study are compared with those obtained by other researchers on the same material (for approximately the same void ratio) so the same preparation method was used (Chu and Leong 2002, Matiotti et al. 1995, di Prisco and Imposimato 1997). Hence, as far as the method used for preparing specimens and especially filling the mould is the same the initial fabric does not influence the comparison.

A latex membrane of 0.2 mm was used to be filled by the sand. The sand with a moisture content of about 3% was gently compacted in five layers inside the mould as described by Chu and Leong (2002) in order to reach the maximum void ratio $e_{max} \approx 1$ before isotropic consolidation. Samples were then flushed with carbon dioxide during 15 minutes and then with de-aired water. Samples were assumed fully saturated if B-Skempton coefficient is greater than 98%.

2.2 Testing program

All test specimens first underwent isotropic compressions to the desired initial confining pressure and then drained specimens were sheared under constant

Table 1. Principal properties of Hostun sand (S28).

Type	d_{50} (mm)	Cu	Specific density	e_{max}	e_{min}	ϕ (°)
Quartz	0.33	2	2.65	1	0.656	32

confining pressure. Axial displacements and volume changes were measured using Linear Variable Differential Transformer (LVDT) and Pressure/Volume Controllers (PVC) respectively.

2.2.1 Conventional displacement controlled drained tests

Isotropically consolidated drained (CD) triaxial tests on loose Hostun sand samples were carried out to determine the failure state. The three displacement-controlled CD tests (CD1, CD2 and CD3) were realized under effective cell pressures of 100, 300 and 750 kPa respectively (see Table 2). The stress-strain curves of these three CD tests are given in Figure 1. It can be seen from $q–p'$ plane that a critical state is reached (in tests CD1 and CD2) or nearly reached (in test CD3) when both the deviator stress and the volumetric strain tend to constant asymptotic values. The critical state line (CSL), which is also the failure line for the loose granular materials, can thus be established. This CSL is plotted together with the stress paths of the three CD tests on a $q–p'$ diagram.

For a small axial strain, the slope of the curves in the $q–\varepsilon_1$ plane increases with the increase of the initial effective mean normal pressure whereas the slope of the curves in the $\varepsilon_v –\varepsilon_1$ plane remains approximately constant. This highlights that the Young modulus is non-linear and depends on the effective mean pressure. In this plane, a very slight dilation can be observed for the lower value of the effective mean pressure. This behavior is conformed to previous experimental observations.

The computed friction angle at failure is about 32° (see Table 1) which is a classical value for a loose sample made of quartz sand. Moreover, the line joining the peaks in the $q–p'$ plane passes through the origin.

2.2.2 Isotropically consolidated undrained compression tests

The results obtained for load-controlled tests are compared to those obtained by Matiotti et al. (1995), Doanh et al. (1997) and described by Servant et al. (2005) and Khoa et al. (2006) for displacement-controlled tests on same sand with a very close density. The experimental device described by Servant et al. (2005) and Khoa et al. (2006) is a displacement-controlled device which was modified so that a constant stiffness could be applied at peak shear stress. The constant stiffness results from a spring attached to the displacement-controlled device. At the peak of

stress, it is noted that $\eta = q/p'$ is equal to 0.6, which corresponds to a mobilised angle of friction of approximately 16°. Good agreement is shown for all these experimental results in Figure 2.

2.2.3 Constant shear drained test

Constant shear drained tests were carried out by Sasitharan et al. (1993) on saturated loose Ottawa sand by using dead loads applied *via* a plate. Mattioti et al. (1995), Di Prisco and Imposimato (1997), Nova and Imposimato (1997) and Gajo (2004) have performed similar t As done in previous studies, the method we employed consists in increasing the pore pressure whilst the total stresses σ_1 and σ_3 were kept constant tests on Hostun sand using dead loads and also a pneumatic chamber for the later.

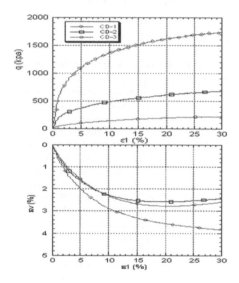

Figure 1. Experimental results on loose Hostun sand during drained compression triaxial tests.

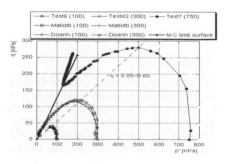

Figure 2. Comparison of our results with those of Matiotti et al. (1995), Doanh and al. (1997) and Servant et al. (2005) for p'_o = 100, 300, 750 kPa.

The void ratios of their specimen were approximately same void ratio as in experiments we have carried on. This type of test consists in shearing the specimen to a prescribed stress ratio along a drained compression triaxial path and then to unload the specimen by increasing slowly the pore water pressure to simulate the loading of soils within a slope or an embankment when a low increase in pore pressure occurs. As done in previous studies, the method we employed consists in increasing the pore pressure whilst the total stresses σ_1 and σ_3 were kept constant.

Typical results of a load-controlled test are presented in Figures 3 and 4. Control variables are presented in Figure 3. After an isotropic consolidation stage to a desired initial effective mean pressure p'_o of 300 kPa, a drained triaxial compression test was applied to the sample to a prescribed value of the deviatoric stress of 119 kPa (point A in Figure 4). The corresponding axial and volumetric strains are 0.42% and +0.32% respectively. As described above, the total radial (σ_3) and axial (σ_1) stresses where $\sigma_1 = \sigma_3 + q$ and $q = F/S$ (F is the axial force and S the section of the sample) are maintained constant whereas the pore water pressure u increases. The imposed rate of increase of the pore water pressure is equal to 0.5 kPa per second and is applied by mean of an injection of water inside the sample at a rate of 0.5 mm^3 per second. This increase of water inside the sample corresponds to a volumetric strain rate of $-6.55.10^{-4}$% per second (the negative sign correspond to a dilation of the sample). This stage is labeled AB in Figure 3. At point B, the deviatoric stress can no more be kept constant and the axial strain rate highly increases. The test is no more controllable (as defined by Nova, 1994) in the sense that the loading program can not be maintained. This collapse is also noticeable in these Figures as jumps in axial strain are observed. Indeed, all experimental points are presented so a bigger interval between to consecutive points corresponds to a

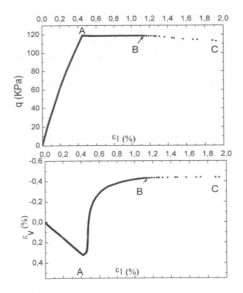

Figure 4. Evolution of the deviatoric stress versus axial strain during CSD test on Hostun sand ($p'_o = 300$ kPa). Collapse occurs at point B. Jumps in axial strain are observed.

jump. Such behavior can only be observed with a load-controlled test.

After the shearing stage, the volumetric strain is increased as it is clearly shown in Figure 2 (point A). The axial strain is almost constant (0.45%) and then begins to increase. In the ε_v–ε_1 plane (see Figure 4), this stage corresponds more or less to a vertical line, i.e. a dilation for a constant axial strain. Then, the axial strain increases more rapidly to reach a value of 1.1% corresponding to a sudden failure. It is remarkable to notice that collapse happens at the maximum of the volumetric strain as it was formally established by Darve et al. (2000) without any localization. During BC, the strain rate increases and becomes higher than 1 per second so the behavior changes from a quasi static regime to a dynamic regime.

3 NUMERICAL ANALYSIS

3.1 *Hujeux elastoplastic multimechanism model*

The non-associated constitutive model used for the simulation of the mechanical behavior of an element of soil is an elastoplastic model which combines four mechanisms: three deviatoric mechanisms of plane strain and an isotropic mechanism. Isotropic and kinematic hardenings are incorporated inside the model to accurately reproduce the mechanical behavior of sands, gravels and clays.

A brief description is given hereafter in the case of monotonic loading. A more detailed description can be

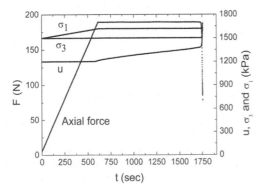

Figure 3. Evolution of the axial load, the total axial and radial stresses and the pore pressure versus time for Hostun sand ($p'_o = 300$ kPa).

found in Aubry et al. (1982), Hujeux (1985), Daouadji et al. (2001) or Hamadi et al. (2008).

Classical additive decomposition of the strain rate tensor $(d\varepsilon)$ is assumed which can thus be decomposed into an elastic strain rate tensor $(d\varepsilon^e)$ and a plastic strain rate tensor $(d\varepsilon^p)$. Thus, it can be written as

$$d\varepsilon = d\varepsilon^e + d\varepsilon^p \tag{1}$$

The hypo-elastic part of the constitutive relation is isotropic, non-linear and depends on the effective mean normal pressure:

$$K(p') = K_{ref}\left(\frac{p'}{p'_{ref}}\right)^n, \quad G(p') = G_{ref}\left(\frac{p'}{p'_{ref}}\right)^n \tag{2}$$

3.1.1 Elastic part

The elastic part of the strain rate can be decomposed as a volumetric and a deviatoric elastic parts.

$$d\varepsilon^e = d\varepsilon_v^e + d\varepsilon_d^e \tag{3}$$

where K and G are respectively the bulk and shear moduli, K_{ref} and G_{ref} are the bulk and shear moduli measured at the effective mean reference pressure p'_{ref} and n is a material constant $(0 \leq n \leq 1)$.

3.1.2 Plastic part

The plastic part of the deformation is assumed to be governed by three deviatoric plane strain mechanisms in three orthogonal planes and a purely volumetric mechanism. Only one elementary deviatoric mechanism is presented as it has the main influence on the macroscopic mechanical behavior of the material.

a) *Yield function and flow rule for the deviatoric mechanism*

For the deviatoric mechanism k, in the ij-plane, plane strain and strain rates are produced. For each k-plane $(k = 1, 2, 3)$, the yield surface is given by:

$$f_k\left(s_k, p'_k, \varepsilon_{vk}^p, r_k\right) = \|\tilde{s}_k\| - r_k \tag{4}$$

where

$$\begin{cases} \tilde{s}_{k1} = \frac{s_{k_1}}{F_k(p'_k, p_c)} \\ \tilde{s}_{k2} = \frac{s_{k_2}}{F_k(p'_k, p_c)} \end{cases} \tag{5}$$

The factor of normalisation F_k, is given by:

$$F_k\left(p'_k, p_c\right) = \sin\phi \, p'_k \left(1 - b \, \text{Log} \, \frac{p'_k}{p_c}\right) \tag{6}$$

r_k is an internal variable allowing a progressive mobilisation of the deviatoric mechanism k and pc is the critical pressure related to the plastic volumetric strain ε_v^p by the following equation:

$$p_c = p_{co} \, \exp(\beta \, \varepsilon_v^p) \tag{7}$$

where $1/\beta$ is the slope of the critical state line in the ε_{vp}–Log p' plane. In the k-plane, the plastic deviatoric

strain rate tensor de_k^p and the plastic volumetric strain rate $(d\varepsilon_v^p)_k$ are defined as:

$$\begin{cases} de_{k_1}^p = \left(d\varepsilon_i^p\right)_k - \left(d\varepsilon_j^p\right)_k = d\lambda_k^p \, \Psi_{k_1}^d \\ de_{k_2}^p = 2\left(d\varepsilon_{ij}^p\right)_k = d\lambda_k^p \, \Psi_{k_2}^d \end{cases} \tag{8}$$

$$\left(d\varepsilon_v^p\right)_k = \left(d\varepsilon_i^p\right)_k + \left(d\varepsilon_j^p\right)_k = d\lambda_k^p \, \Psi_k^v \tag{9}$$

where $d\lambda_k^p$ is the plastic multiplier and Ψ_k gives the direction of plastic strains. It is assuming that this tensor Ψ_k can be decomposed into a deviatoric part and a volumetric part. In the normalised deviatoric plane, Ψ_k^d is defined as:

$$\Psi_k^d = \frac{\partial f_k}{\partial s_k} = \frac{s_k}{\|s_k\|} \tag{10}$$

The plastic potential relationship allowing contraction and dilation is given by:

$$\Psi_k^v = \frac{d\varepsilon_{vk}^p}{d\varepsilon_{dk}^p} \, I = \alpha_k \left(\sin\phi - \frac{s_k : \Psi_k^d}{p'_k}\right) I \tag{11}$$

b) *Yield function and flow rule for the isotropic mechanism*

Experimental results demonstrate the existence of plastic strains during an isotropic compression. Because they are not activated under this stress path, the deviatoric mechanisms cannot generate purely volumetric plastic strains. An isotropic mechanism is thus introduced

$$f_4(p', p_c, r_4) = |\tilde{p}| - r_4 \tag{12}$$

where

$$\tilde{p} = \frac{p'}{F_4(p_c)} \tag{13}$$

and F_4 (pc) is a factor of normalization given by

$$F_4(p_c) = d \, p_c \tag{14}$$

r_4 is the degree of mobilization of this mechanism. Its value increases continuously from r_{el4} (elastic) to 1 and its evolution law is a hyperbolic form given by

$$\dot{r}_4 = \dot{\lambda}_p^4 \frac{(1 - r_4)^2}{c} \frac{P'_{ref}}{P_c} \tag{15}$$

The mechanism generates plastic volumetric strains:

$$\left(\dot{\varepsilon}_{ij}^p\right)_4 = \dot{\lambda}_4^p \Psi_4^v \frac{\delta_{ij}}{3} \tag{16}$$

c and d are constant parameters of the model.

During loading, the mechanisms can be activated and are coupled by the density hardening pc as

$$\dot{\varepsilon}_v^p = \sum_{m=1}^{4} \left(\dot{\varepsilon}_v^p\right)_m \tag{17}$$

288

For each of these mechanisms, the equation of compatibility gives the plastic multipliers by:

$$\dot{f_k}\left(q_k, p'_k, \varepsilon^p_{v_k}, r_k\right) = \frac{\partial f_k}{\partial q_k} dq_k + \frac{\partial f_k}{\partial p'_k} dp'_k$$

$$+ \frac{\partial f_k}{\partial \varepsilon^p_{v_k}} d\varepsilon^p_{v_k} + \frac{\partial f_k}{\partial r_k} dr_k = 0 \quad (18)$$

3.2 Numerical results

Numerical simulations of collapses are provided using the elastoplastic constitutive model described in the previous section. For the sake of readability, only the simulations of the experimental data obtained for CSD test under an initial effective mean pressure of 300 kPa are presented in Figures 5.

The CSD test presented in section 2.2 is simulated and comparison between experimental data and numerical results are given in Figure 5. The deviatoric stress versus the axial strain is given together with the volumetric strain versus the axial strain. The former is accurately reproduced while there is a dilatancy given by the model is higher than the observed one. This is mainly due to the fact that the model calibration is done using compression tests and not using extension or loading—unloading tests. However, the second order work is nil for an axial strain or 0.93% which is close to the experimental collapse point (1.1%). One should notice that there exists a plateau in the ε_v–ε_1 plane and that the points are not easily defined.

The second order work is computed and negative or nil values are represented by dots in Figure 4. It is clearly shown that the first nil value of the second order work appears at the peak of the volumetric strain as clearly illustrated by the curve in the ε_v–ε_1 plane in these Figures. Loss of stability can occur after this point if using the relevant control variables during the experiments. The second order work is nil at the beginning of the CSD stage meaning that the material is inside a bifurcation domain and that the stress increment direction belongs to a cone (Darve et al. 2007).

4 CONCLUSION

Compression triaxial tests and constant shear drained tests on Hostun sand under loose conditions have been carried out. During constant shear drained tests, it is found that, for a given void ratio and an initial effective mean pressure, the volumetric behavior of the specimen is dependent on the value of the shear stress (of the stress ratio precisely). Indeed, the tendency of the material to dilate observed for lower values of the shear stress during the constant shear stress step vanishes for higher values. At the peak of the volumetric

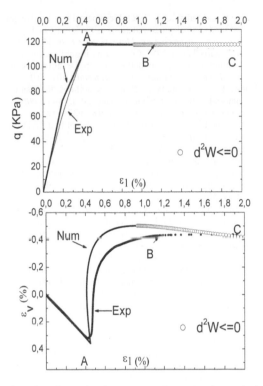

Figure 5. Comparison between experimental and numerical results of constant shear drained tests on loose Hostun sand.

strain, it is no more possible to keep the deviatoric stress constant which corresponds to a loss of controllability as defined by Nova (1994). At that point, the axial strain rate increases to change from a quasi static regime to a dynamic regime without any strain localization.

In order to simulate the behaviour of the tested material, an elastoplastic multimechanism model is used. The numerical results obtained under the same stress path qualitatively reproduce the macroscopic behaviour of the specimens. The second order work is computed and it is found that the first nil value coincide with the peak of the volumetric strain. Thus, when the relevant control variables are used, the loss of stability corresponds to the loss of controllability.

REFERENCES

Aubry D., Hujeux J.-C., Lassoudiere F. and Meimon Y. 1982. A double memory model with multiple mechanisms for cyclic soil behaviour. Int. Symp. Num. Mod. Geomech., Balkema, Zurich: 3–13.
Chu J. and Leong W.K. 2002 Effect of fines on instability behaviour of loose sand. Géotechnique 52(10): 751–755.
Daouadji A., Hicher P.-Y. and Rahma A. 2001. An elastoplastic model for granular materials taking into account grain breakage. Eur. J. Mech. A/Solids 20: 113–137.

Darve F. and Labanieh S. 1982. Incremental constitutive law for sands and clays. Simulations of monotonic and cyclic tests. *Int. J. Numer. Anal. Meth. Geomech.* 6: 243–275.

Darve F. and Chau B. 1987. Constitutive instabilities in incrementally non-linear modeling. In C.S. Desai (ed.), *Constitutive Laws for Engineering Materials*: 301–310.

Darve F. and Laouafa F. 2000. Instabilities in granular materials and application to landslides. *Mechanics of Cohesive—Frictional Materials* 5(8): 627–652.

Darve F. and Sibille L., Daouadji A. and Nicot F. 2007. Bifurcations in granular media: macro-and micro-mechanics approaches. *C.R. Mecanique* 335: 496–515.

Darve F. and Servant G., Laouafa F. and Khoa H.D.V. 2004. Failure in geomaterials, continuous and discrete analyses. *Comp. Meth. In Appl. Mech. and Eng.* 193(27–29): 3057–3085.

Doanh T., Ibraim E., and Matiotti R. 1997. Undrained instability of very loose Hostun sand in triaxial compression and extension. Part 1: experimental observations. *Mech. Cohes. Frict. Mater.* 2: 47–70.

Di Prisco C. and Imposimato S. 1997. Experimental analysis and theoretical interpretation of triaxial load controlled loose sand specimen collapses. *Mechanics of Cohesive—Frictional Materials* 2: 93–120.

Gajo A. 2004. The influence of system compliance on collapse of triaxial sand samples. *Canadian Geotechnical Journal* 41: 257–273.

Hamadi, A., Modaressi-Farahmand Razavi, A. and Darve, F. 2008. Bifurcation and instability modelling by a multimechanism elasto-plastic model. *Int. J. Numer. Anal. Meth. Geomech.* 32(5): 461–492.

Hill R. 1958. A general theory of uniqueness and stability in elastic-plastic solids. *J. Mech. Phys. Solids* 6: 239–249.

Hujeux J.-C. 1985. Une loi de comportement pour le chargement cyclique des sols, Génie parasismique, Editions V. Davidovici, Presses ENPC : 287–302.

Imposimato S, Nova R. 1998. An investigation on the uniqueness of the incremental response of elastoplastic models for virgin sand. *Mechanics of Cohesive—Frictional Materials* 3: 65–87.

Khoa H.D.V., Georgopoulos I.O., Darve F. and Laouafa F. 2006. Diffuse failure in geomaterials: Experiments and modelling. *Computers and Geotechnics* 33:1–14.

Matiotti R., Di Prisco C., and Nova R. 1995. Experimental observations on static liquefaction of loose sand. In Ishihara (ed.) *Earthquake Geotechnical Engineering*: Balkema. 817–822.

Nova R. 1994. Controllability of the incremental response of soil specimens subjected to arbitrary loading programmes. *J. Mech. behav. Mater.* 5(2): 193–201.

Nova R. and Imposimato S. 1997. Non-uniqueness of the incremental response of soil specimens under true-triaxial stress paths. In Pande and Pietruszczak (eds.). *NUMOG VI*: 193–197: Balkema.

Rice, J. 1976. The localization of plastic deformation. In Koiter (ed.): (1) 207–220. Delft: IUTAM Proceedings.

Sasitharan S., Robertson P.K., Sego D.C., and Morgenstern N.R. 1993. Collapse behavior of sand. *Canadian Geotechnical Journal* 30: 569–577.

Skopek P., Morgenstern N.R., Robertson P.K. and Sego D.C. 1994. Collapse of dry sand. *Canadian Geotechnical Journal* 31: 1008–1014.

Vaid Y.P., Sivathayalan S., and Stedman D. 1999. Influence of specimen-reconstituting method on the undrained response of sand. *Geotechnical Testing Journal* 22(3): 187–195.

Prediction and Simulation Methods for Geohazard Mitigation – Oka, Murakami & Kimoto (eds)
© 2009 Taylor & Francis Group, London, ISBN 978-0-415-80482-0

Water-submergence effects on geotechnical properties of crushed mudstone

M. Aziz, I. Towhata, S. Yamada & M.U. Qureshi
The University of Tokyo, Tokyo, Japan

ABSTRACT: Slope geometry and topographical studies possibly are not the perfect tools for landslide risk assessment in widespread tertiary sedimentary soft rocks in Japan and Pakistan because these rocks are not yet fully lithified and are highly susceptible to weathering and rainfall effects. This paper discusses the strength and deformation response of crushed mudstone to water-submergence because of its extremely high vulnerability to water induced deterioration. Drained torsional shear tests on saturated soils revealed the decay of grains and change in particle shape during consolidation and shearing stages with decrease in particle disintegration at higher confining pressures. Peak shear strengths and friction angles are considerably higher with governing dilatant behaviour for dry specimens as compared to the saturated ones. This study is a caution to conventional soil mechanics in which decay of grains and loss of strength with time are often uncared.

1 INTRODUCTION

While dealing with rapidly degradable geomaterials like soft rocks, more emphasis should be given to the internal causes of natural slope failures which result in a decrease of the shearing resistance (e.g. progressive failure, weathering-induce deterioration, seepage erosion). With reference to Figure 1, this paper focuses on the time-dependent decay of material, i.e. negative ageing due to weathering and water-induced deterioration. It is an essential cause of decrease in shearing resistance since topographical risk assessment is not adequate enough due to continuous changes in lithology, tectonic actions and infrastructure developments in hilly areas.

With extensive developments in hilly areas in recent decades, the research focus has got some inclinations towards macro and micro-scale geotechnical engineering design parameters instead of only dealing with mega-scale geological studies. Geotechnical design parameters are now becoming an important part of any design process in such areas because of the development of more sophisticated and robust laboratory and in-situ testing techniques and powerful microscopes for microfabric studies as well.

2 OBJECTIVES OF THE RESEARCH

The conditions under which soft rocks are degraded due to physical or chemical weathering had been well established (Hawkins et al. 1988). However, very few efforts have been made so far towards constitutive modelling of degradable geomaterials e.g. Feda (2002), Castellanza & Nova (2004) and Pinyol et al. (2007). The motivation behind this study was from some shear tests conducted on conventional non-degradable geomaterials (Toyoura silica sand) which showed no change in stress-strain response in dry and saturated test conditions. Such results are also verified by Fioravante & Capoferri (1997). The main objective of this research is to formulate the process of loss of strength and stiffness of degradable geomaterials for their better stress-strain response modeling, especially sedimentary soft rocks due to their rapid water-induced deterioration. As shown in Figure 2, a piece of mudstone rapidly disintegrates and transforms from rock to soil only after a few minutes of submergence in water.

The alterations that take place in soft rocks due to water-submergence are of unusual importance since such a progression is sometimes rapid and resulting

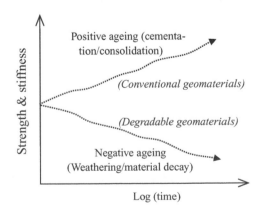

Figure 1. Time effects on geomaterials.

Figure 2. Water-induced deterioration of weathered mudstone.

Figure 3. Mega torque HCTS apparatus.

consequences due to the strength reduction can be disastrous. Apart from the soil mechanics of conventional geomaterials, this work is an attempt to facilitate the design and construction engineers to better understand the idea of including decay of geotechnical properties with time for better risk assessment and hazard mapping of natural slopes as well as to foresee the long-term behavior of construction material borrowed from such geology.

3 TESTING PROGRAM

3.1 *Apparatus*

The results presented and discussed in this paper are from a hollow cylindrical torsional shear apparatus, hereafter referred to as HCTS. The apparatus shown in Figure 3 is equipped with mega-torque motor system to allow strain controlled monotonic/cyclic torsional shear to a sufficiently large strain level.

Some studies in the past have been done on shear behaviour of degradable geomaterials using direct shear and triaxial test apparatuses, but in this particular study HCTS device has been used because of its many advantages over conventional shear tests.

The advantages and limitations of HCTS devices are presented in a state-of-the-art paper by Saada (1988).

3.2 *Geology of the material*

Rock pieces of weathered mudstone were collected from two different geological formations in Yokosuka. The names are Zushi formation of late Miocene and Morito formation of early to middle Miocene age according to the geological map of the area issued by Geological Survey of Japan. Materials collected from Zushi formation have sand intrusion in mudstone whereas Morito formation mostly consists of massive mudstone, hereafter referred to as S01-MS and S02-MM respectively. These rock pieces were later crushed and reconstituted in the laboratory for various tests.

3.3 *Preparation of tested materials*

After oven drying of the collected material, it was crushed using a ball-mill and a hammer. The particles between 2 mm and 0.075 mm sizes were recovered from the crushed material. The fines were observed to be non-plastic (NP). Crushed rock material has been used in this research for two main reasons; firstly, it is almost impossible to retrieve undisturbed intact cores of weathered soft rocks, especially hollow cores for HCTS and secondly the objective of the research being lies in the stress-strain behavior of degradable geomaterials. It requires more surface area of the tested material to excel the rate of degradation in the laboratory. The grain size distribution (GSD) curves of the finally test-ready crushed materials are given in Figure 4.

Because of uncontrolled manual crushing, GSD of the two materials could not be exactly matched with each other. To compare our test material with conventional geomaterial, GSD and index properties of Toyoura sand are also given. Physical properties of the prepared materials determined according to the standards of JGS are summarized in Table 1.

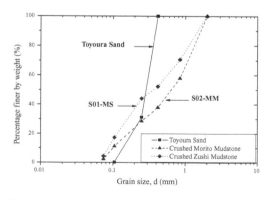

Figure 4. Grain size distribution curves of the test materials.

Figure 5 obtained with the help of a digital microscope, gives a general idea about the particle shape which is angular to sub-angular before the test. Here, the interesting feature to observe is that even after crushing and mechanical remoulding, the intrinsic bedding planes of sedimentary rocks are still preserved in relatively large grains.

3.4 Sample preparation and test procedure

Oven-dried crushed material was thoroughly mixed before sample preparation to make certain that the physical properties of each specimen are representative of the above mentioned values. Hollow specimens ($20 \times 10 \times 6$ cm) were prepared at desired initial relative densities by air-pluviation and gentle rod tamping. A typical HCTS test sample before and after the test is shown in Figure 6.

Table 1. Physical properties of the test materials.

Soil type	G_s	e_{max}	e_{min}	D_{50} (mm)	C_u
Toyoura Sand	2.65	0.98	0.61	0.22	1.3
S01-MS	2.52	1.724	1.174	0.37	6.8
S02-MM	2.39	1.669	1.029	0.65	8.2

Figure 5. Microscopic observations of decay of grains before and after test (Zoom: 75X).

Figure 6. (a) Free standing specimen before test under 20 kPa suction (b) Sheared specimen after test.

Table 2. Summary of test results on initial relative density, eff. confining stress, peak shear strength and peak friction angle.

Sr No	Test sample	Test conditions	Dr_i (%)	P' (kPa)	τ_{peak} (kPa)	ϕ_{peak} (deg)
1	MS02	Saturated	90.8	50	29.8	37.6
2	MS03	Saturated	88.6	100	61.7	38.7
3	MS04	Saturated	89.1	150	89.8	37.3
4	MS09	Dry	93.0	50	41.6	60.6
5	MS10	Dry	90.1	75	65	60.1
6	MS06	Dry	90.2	100	83.7	59.7
7	MM01	Saturated	81.1	50	32.4	40.2
8	MM07	Saturated	74.0	75	48.5	40.9
9	MM04	Saturated	77.4	100	64.2	40.0
10	MM10	Dry	77.8	50	41.9	51.7
11	MM08	Dry	77.4	75	57.4	49.5
12	MM03	Dry	77.5	100	77.3	50.0

All the specimens were normally consolidated at the desired effective stress, P' and thereafter, drained monotonic torsional shear under constant mean effective principal stress was performed at constant strain rate of 0.018%/min. GSD and particle shape were determined for each dry and saturated tests before and after the shearing. The summary of tests under dry and saturated conditions conducted on S01-MS and S02-MM is given in Table 2.

4 TEST RESULTS AND DISCUSSIONS

The test results presented in this paper in the following sections are focused on comparison of strength and deformation characteristics of dry and saturated degradable crushed/weathered rocks. Therefore, instead of conducting a vast parametric study, only the effects of confining pressure (i.e. equivalent to the depth of weathered zone) were investigated keeping the initial relative density (D_{Ri}) constant both for the dry and saturated specimens. Capacity of the drained volume measurement system did not allow making initially loose specimens because of large amount of drained volume in such a case, so denser packing was made during specimen preparation. Instead of using relative density after consolidation (D_{Rc}) as a parameter of comparison, D_{Ri} has been used because saturated specimens have shown change in particle shape and size due to water-submergence.

4.1 Strength and deformation characteristics

4.1.1 Deformation during isotropic consolidation
Deformation behavior of the materials was studied during isotropic consolidation and shearing of the specimens. The consolidation and shearing time were kept constants for all the tests for comparison. Vertical strains are plotted against mean effective stress

in Figures 7–8. A large difference of vertical strains amongst dry and saturated specimens at the same effective confining pressure during isotropic consolidation can be observed.

Although the rate of consolidation (increase in effective confining pressure) was kept constant for all the tests i.e. 3 kPa/min. but saturated tests have shown much faster increase in vertical strains as compared to dry tests. This supports the idea of decay of grains under saturated test conditions. For this reason, the respective volumetric compression is also observed to be significantly higher in case of saturated tests during consolidation stage. Hence, from engineering point of view, this phenomenon can be of great concern in progressive failure of slopes, settlement of rockfill dams and foundations.

4.1.2 Torsional shear test results

The results obtained from drained monotonic torsional shear tests on dry and saturated specimens at constant effective confining stress and shear strain rate are presented in this section. To limit the scope of work to surface weathered soft rocks, the confining pressures have been kept below 150 kPa. The volumetric

changes have been investigated in terms of vertical strains, ε_z, where $+\varepsilon_z$ denotes volumetric compression ($-ve$ dilatancy) and $-\varepsilon_z$ refers to volumetric expansion ($+ve$ dilatancy). The shear stress-shear strain and vertical strain-shear strain relationships for both test materials have been shown in Figures 9–12. Irrespective of the test conditions (saturated or dry), it is observed that the stress-strain curve for both materials becomes stiffer and the peak shear strength becomes higher with increase in effective confining pressure level. But, due to disintegration of grains during saturated tests, shear resistance decreases drastically and an increase in the compressibility is observed. Average drop in peak shear stress values for S01MS and S02MM are 38% and 23% respectively between dry and saturated test conditions. This process, if extrapolated in terms of scale and time, gives the idea about the process of strength loss during transformation of rock to soil due to many physiochemical weathering actions, water being the most severe one.

Although the specimens were tested at relatively higher densities, but they have shown low peak shear strengths and the stress-strain behaviors are similar to those of loose soils. This is because of very high maximum and minimum void ratios of the

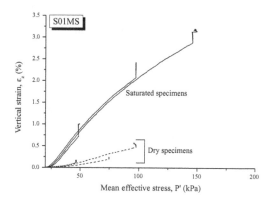

Figure 7.   Isotropic consolidation of S01MS specimens.

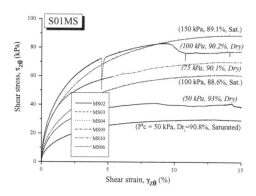

Figure 9.   Stress-strain behavior of S01MS specimens.

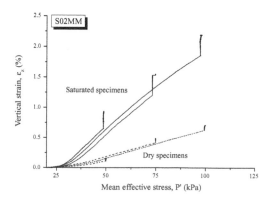

Figure 8.   Isotropic consolidation of S02MM specimens.

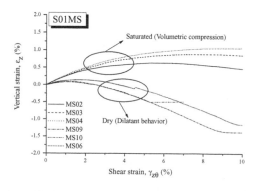

Figure 10.   Volume change of S01MS specimens.

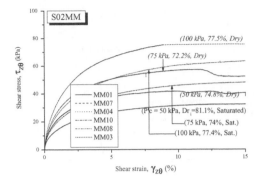

Figure 11. Stress-strain behavior of S02MM specimens.

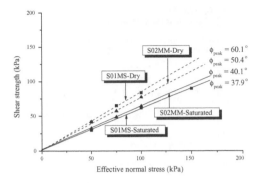

Figure 13. Strength parameters of S01MS and S02MM.

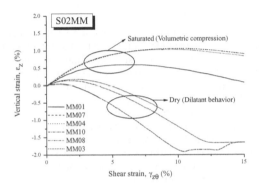

Figure 12. Volume change of S02MM specimens.

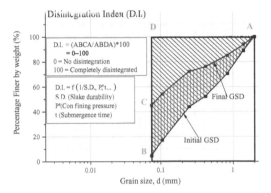

Figure 14. Particle disintegration due to water-submergence (Particle crushing due to shear by Hardin, 1985).

crushed materials which bring in the possibility of high volume changes during the increase in effective confining pressure and shearing. Moreover, the volume change during progressive increase in shear strain is predominantly compressive for saturated specimens and dilatant behavior is observed for dry specimens after slight initial compression. An increase in compression and decrease in dilatancy can be noted as the effective stress is increased.

Mohr stress circles were drawn to obtain the strength parameters and failure envelope. Soils being predominantly non-cohesive have shown high values of friction angle. Due to disintegration of the grains during water-submergence, the mobilized angle of shear resistance is also found to change significantly between saturated and dry tests as shown in Figure 13. ϕ_{peak} drops from 60.1° to 37.9° (reduction of 59%) and from 50.4° to 40.1° (reduction of 26%) for S01MS and S02MM specimens respectively.

The data also reveals that even for the crushed material the geology remains the governing parameter for strength and deformation characteristics because it can be seen that the failure envelope of material from late Miocene Zushi mudstone has a greater strength loss

upon submergence as compared to early to middle Miocene Morito formation.

4.2 Particle disintegration during saturated tests

Soft rock deposits as well as the weathered slope surfaces are comprised of granular geomaterials which are strongly susceptible to further disintegration upon cyclic wetting and drying. Because engineering properties of granular materials (e.g. stress-strain behavior, volume change, pore water pressure development, etc) are strongly dependent on the GSD in working conditions, therefore for long term behavior and failure risk assessment, it is necessary to identify and quantify the effects of grains disintegration due to water-induced deterioration. As mentioned earlier in Sec. 3.4, the GSD were determined both before and after the shear tests. On the basis of area under initial and final GSD curves, a disintegration index (D.I.) has been defined (Fig. 14) to quantify the amount of disintegration due to water-submergence in terms of change in shape of the initial GSD curve.

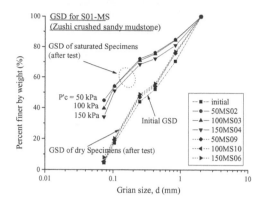

Figure 15.   GSD curves of S01MS before and after the tests.

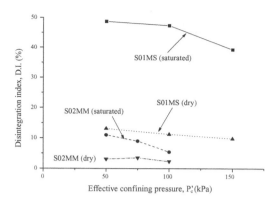

Figure 17.   Disintegration index for S01MS & S02MM.

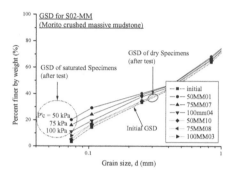

Figure 16.   GSD curves of S02MM before and after the tests.

This kind of index has been perceived from previous work of various researchers like grading index (I_G) by Muir et al (2007) and relative breakage index (Br) by Hardin (1985) who have used such indices for quantifying particle crushing at very high shear strains in ring shear tests. The GSD curves of S01-MS and S02-MM after torsional shear tests at different confining pressures are given in Figure 15–16. Change in shape of the GSD curve and the quantity of grains less than 0.075 mm at the end of torsional shear test indicates the extent of disintegration of the material which is also presented in form of D.I. (Fig. 17).

Only saturated tests have shown enormous change in GSD curve (i.e. increase in fines due to disintegration of particles), whereas no/little changes are observed in dry tests. Therefore, it confirms the idea that disintegration of particles is largely taking place due to decay of grains caused by water-submergence (i.e. slaking of the particles) rather than crushing of the grains due to shearing because grain crushing at a shear strain level of 15% is expected to be of negligible amount. It is depicted from the given GSD curves that material S01-MS have shown very high degree of degradation relative to S02-MM with an

average increase in fines of 35.2% and 12.4% respectively. It is interesting to note that rate of disintegration has decreased with increase in confining pressure in both materials. This is quite rational because naturally extent of weathering decreases at greater depths (i.e. higher confining pressures).

Furthermore, it has been observed that due to water induced disintegration the particle shape has also been affected (Fig. 5). Microscopic observations have shown that larger particles are sub-rounded to rounded after saturated tests while particles remain angular to sub-angular after dry tests.

5   CONCLUSIONS

In all the models currently used for prediction and simulation of geohazard mitigation, there is still a strong need to bridge the gap between geology and geotechnical engineering for better visualization of the response of geomaterial over the life span of the geotechnical structures. Apart from conventional soil mechanics in which the strength loss due to decay of grains with time is often ignored, geotechnical properties of geomaterials which are highly vulnerable to water-induced deterioration have been presented. The main conclusions are:

– Zushi sandy mudstone have shown very high degree of disintegration due to water-submergence relative to massive mudstone of Morito formation with an average increase in fines of 35.2% and 12.4% respectively after the tests.
– The amount of disintegration of grains lessens at higher effective confining stress. It reveals that water-induced deterioration near the slope surfaces is more severe.
– Microscopic observations have shown that larger particles are more or less sub-rounded to rounded

after saturated tests whereas particles remain angular to sub-angular after dry tests. This phenomenon can be attributed to one of the reasons of decrease in shear resistance upon water-submergence for weathered materials and is likely to initiate shallow slope failures as well.
− For non-conventional degradable geomaterials, shear strengths are significantly affected by the test conditions (i.e. saturated or dry) which should be taken into account at design stage of any geotechnical project for such geologies. For example, ϕ_{peak} drops from 60.1° to 37.9° (reduction of 59%) and from 50.4° to 40.1° (reduction of 26%) upon water-submergence for Zushi and Morito formations respectively.

REFERENCES

Castellanza, R. & Nova, R. 2004. Oedometric tests on artificially weathered carbonatic soft rocks. *Journal of Geotechnical and Geoenvironmental Engineering.* 130(7): 728–739.

Feda, J. 2002. Notes on the effect of grain crushing on the granular soil behaviour. Engineering Geology 63: 93–98.
Hardin, B.O. 1985. Crushing of soil particles. *Journal of Geotechnical Engineering,* ASCE 111(10): 1177–1192.
Hawkins, A.B., Lawrence, M.S. & Privett, K.D. 1988. Implications of weathering on the engineering properties of the Fuller's Earth formation. Geotechnique 38(4): 517–532.
McDowell, G.R. & Bolton, M.D. 1998. On the micromechanics of crushable aggregates. *Geotechnique* 48(5): 667–679.
Saada, A.S. 1988. Hollow cylinder torsional devices: Their advantages and limitations. ASTM special technical publication 977:766–795.

Mitigating embankment failure due to heavy rainfall using L-shaped geosynthetic drain (LGD)

S. Shibuya, M. Saito & N. Torii
Kobe University, Kobe, Japan

K. Hara
Taiyokogyo Corporation, Hirakata, Japan

ABSTRACT: In recent years, geotechnical engineers in Japan confront frequently with the embankment failure induced by heavy rainfall. For the purpose of mitigating such embankment failure, Shibuya et al. (2008) has recently proposed a remedial measure to protect embankment from the invasion of seepage water flow by using L-shaped geosynthetic drain (LGD). In this paper, the effectiveness of LGD system in preventing the seepage flow into the embankment is demonstrated by performing a series of numerical analysis. The behavior of a large-scale test embankment (3.9 m long, 2 m wide and 2.5 m high) subjected to seepage flow from the back of the fill was also examined in the laboratory. The significance of the LGD was manifested by performing a couple of comparative tests with and without the LGD. The seepage flow characteristics of the test fill was interpreted by comparing the test result with the result of numerical analysis. It is suggested that the LGD is effective in mitigating the embankment failure in the event of heavy rainfall.

1 INTRODUCTION

A failure of Terr Armée wall reported recently by Shibuya et al. (2007) was induced by seepage water flow in the event of heavy rainfall, which in turn weakened the initially unsaturated fill material and also pushed the wall by the accumulated water pressure behind the wall. Based on this experience, the use of an L-shaped geosynthetic drain (LGD) comprising a set of vertical and horizontal planar-geosynthetics has been proposed by Shibuya et al. (2008). Saito et al. (2008) has performed numerical simulation in which the efficiency of the LGD system in preventing seepage water flow into the fill was successfully demonstrated.

In this paper, the effectiveness of LGD in preventing the seepage flow into the embankment is demonstrated in a series of numerical analysis, in which the result with LGD was compared to the case comprising horizontally layered geosynthetic drain system. A full-scale seepage flow test using an embankment (3.9 m in length, 2.0 m in width and 2.5 m in height) made of a well-graded decomposed granite soil was carried out. The behavior of the test embankment undertaking seepage flow from the back was carefully examined by monitoring vertical settlement, lateral deformation, pore pressures and moisture content at several points in the test embankment.

2 NUMERICAL SIMULATION

2.1 Governing equation and boundary conditions

In the seepage analysis, the following Richards' equation (Richards 1931) is employed as the governing equation;

$$(C + \beta S_s) \frac{\partial \psi}{\partial t} = \nabla \cdot [K \cdot (\nabla \psi + \nabla Z)] \qquad (1)$$

where C is the specific water capacity ($= \phi dS_w/d\psi$) (n.b., ϕ: porosity, and S_w: the degree of saturation ($0 \leq S_w \leq 1$)), S_s is the specific storage coefficient, K is the hydraulic conductivity tensor, ψ is the pressure head, and Z is the elevation head. Note that β is equal to unity in the saturated zone involved with $S_w = 1$, and to zero in the unsaturated zone with $S_w \neq 1$. The K can be expressed in terms of the relative permeability k_r and the saturated hydraulic conductivity Ks, i.e.,

$$K = k_r \cdot Ks \qquad (2)$$

The boundary condition on Γ_l, where the pore pressure head is defined, is given by

$$\psi = \psi_1 \quad \text{on} \quad \Gamma_1 \qquad (3)$$

On the boundary Γ_2, the flux q is defined in the following form;

$$q = q_2 = -n \cdot \mathbf{K} \cdot (\nabla \psi + \nabla Z) \quad \text{on} \quad \Gamma_2 \qquad (4)$$

where **n** denotes the outwardly directed unit normal vector.

The relative permeability k_r is an essential parameter in the seepage analysis. This property is considered as a function of S_w, whereas the S_w is considered as a function of capillary pressure ψ_c ($\equiv -\psi$). Among many mathematical models previously proposed to describe the water retention curve, the van Genuchten equation (van Genuchten 1980) (i.e., VG model) is employed in the present study; i.e.,

$$S_e = \frac{S_w - S_r}{S_f - S_r} = \left\{1 + (\alpha \psi_c)^n\right\}^{-m} \tag{5}$$

where S_e is the effective saturation, S_r is the residual saturation, S_f is the saturation at $\psi_c = 0$, and α, n and m are parameters. The parameters, n and m, are both dimensionless, whereas α has the dimension that can be defined as the reciprocal of the pressure head. The parameters n and m are not independent to each other, and they are related by

$$m = 1 - 1/n \tag{6}$$

The relative permeability and effective saturation are interrelated as shown in the following form (Maulem 1976) ;

$$k_r = S_e^{\varepsilon}\left\{1 - (1 - S_e^{1/m})^m\right\}^2 \tag{7}$$

where ε is a parameter regarding the degree of interconnection among voids. Generally, a value of 0.5 is used for ε. The water retention curve and relative permeability can be calculated when the parameters α, n, S_r and S_f are all given. More details regarding the numerical techniques are given by Saito et al. (2008).

2.2 Effectiveness of LGD

Figure 1 shows three cases for which the seepage flow was simulated. The idealized flow domain was assumed in a manner that impervious boundary was considered at the bottom, the flux $q = 0$ at the seepage surface boundary at the right-hand vertical wall when $\psi < 0$, and $\psi = 0$ when $q < 0$. The left-hand vertical plane is the boundary to generate the water pressure varying with time. It was postulated that the water pressure rose linearly with time to reach the maximum value corresponding to $H = 0.9$ m. The S_w prior to seepage flow was set 0.46. In the first case, no geo-drain was employed in the model ground having the $k_s = 2.3 \times 10^{-3}$ cm/s, the length of 2 m and the height of 1.0 m (see Fig. 1a). In the second case, a series of geosynthetic drain was employed to form horizontal layers (see Fig. 1b). The permeability of the geosynthetic drain was assumed $k_d = 1.0$ cm/s and $k_d = 10$ cm/s. The LGD was employed in the third case (see Fig. 1c). Figure 2 shows the distribution of degree of saturation at steady flow state. It is obvious in that the LGD is effective in preventing seepage water

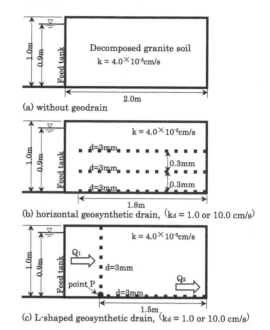

(a) without geodrain

(b) horizontal geosynthetic drain, (k_d = 1.0 or 10.0 cm/s)

(c) L-shaped geosynthetic drain, (k_d = 1.0 or 10.0 cm/s)

Figure 1. Cases of numerical simulation performed.

flow into the protected region. The effect is more significant when the permeability of geosynthetic drain is large. Note also that the horizontally layered geodrain system is less efficient compared to the LGD system in terms of reducing the area of saturation.

3 EXPERIMENT

3.1 A large-scale embankment

The test embankment was made by using approximately 20 m³ of decomposed granite soil having the mean particle diameter of 1.15 mm, and the maximum dry density of 1.938 g/cm³ involved with the optimum water content of 11.6% (for details, see Saito et al. 2008). The embankment was constructed at stages, each in which the prescribed amount of the air-dried soil having the initial water content of 8% on average was compacted to form a 0.25 m thick layer.

A couple of comparative tests with and without the LGD, i.e., case 1 and case 2, respectively, were performed. Figures 3 and 4 show the configuration of these two embankments. The dry density was about 85% and nearly 90% of the maximum dry density for case 1 and case 2, respectively. It should be mentioned that in case 2 embankment with LGD, the compaction of the upstream portion facing the water supply pipes was not easy owing to the limited working space so that the degree of compaction was about 86%, which was lower than the downstream portion with the compaction of nearly 90%.

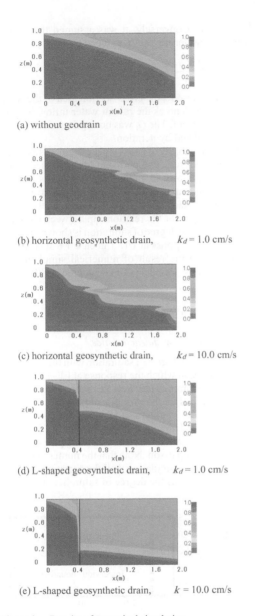

(a) without geodrain

(b) horizontal geosynthetic drain, $k_d = 1.0$ cm/s

(c) horizontal geosynthetic drain, $k_d = 10.0$ cm/s

(d) L-shaped geosynthetic drain, $k_d = 1.0$ cm/s

(e) L-shaped geosynthetic drain, $k = 10.0$ cm/s

Figure 2. Results of numerical simulation.

In both the cases, the seepage flow was initiated by raising the water level in the water supply pipes installed at one end of the embankment. The other end was reinforced by using five rectangular-shaped EPS blocks, each wrapped with a sheet of geogrid. It should be mentioned that the EPS blocks were light in weight, and not connected to each other. Accordingly, the wall facing was rather considered flexible. In case 2 embankment, a set of geosynthetic sheets comprising a 11 mm thick core material covered with non-woven geotextiles jacket was installed vertically and horizontally to form the LGD system.

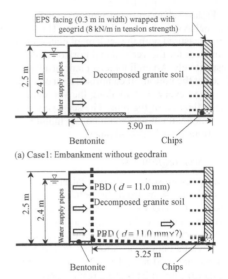

(a) Case1: Embankment without geodrain

(b) Case2: Embankment with L-shaped geosynthetic drain

Figure 3. Configuration of test embankments.

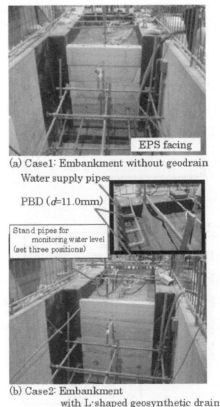

(a) Case1: Embankment without geodrain

(b) Case2: Embankment with L-shaped geosynthetic drain

Figure 4. Photos of test embankments.

Figure 5 shows instrumentations used for monitoring the behavior of the embankment. The deformation was measured with the settlement plates and the inclinometers for the vertical and horizontal deformation, respectively. An inclinometer made of a PVC pipe was fixed at the bottom. It comprised a set of strain gauges mounted every 0.25 m along the vertical to measure the deflection of the pipe. The seepage water level in the embankment was continuously monitored at three positions by using stand pipes.

3.2 Seepage flow test

The upstream water level was raised quickly to reach a level of 2.4 m from the bottom (see Fig. 3). In case 2 with LGD, the filling and de-watering cycle was repeated. Figure 6 shows the idealized flow domain in case 2 test. Following the (x, z) co-ordinate in Fig. 6, impervious boundaries involved with $q = 0$ are considered at $z = 0$ m and $z = 2.5$ m at the bottom and surface of the embankment, respectively. As for seepage surface at $x = 3.6$ m, the condition with $q = 0$ is satisfied when the pressure head $\psi < 0$, and $\psi = 0$ when $q < 0$.

The upstream vertical plane at $x = 0$ m is the boundary to generate the water pressure varying with time. It was postulated that the water pressure rose linearly with time to reach the maximum value of

$H = 2.4$ m. Conversely, it was assumed that in the event of dewatering, the H reduced by satisfying;

$$\frac{dH}{dt} = -\frac{Q_i(t, H)}{A} \tag{8}$$

where A represents the cross-section area of the water tank, and Q_i denotes the rate of water inflow into the test embankment. The Q_i was not a priori given so that it was calculated by iteration.

The permeability of geosynthetic drain was assumed 8.5×10^0 cm/sec. The permeability soil used in the analysis are shown in Table 1. Porosity ϕ as fixed at the measured value of 0.35, implying that no volume change took place throughout the seepage test.

An example for the relationship between capillary pressure and the degree of saturation is shown in Fig. 7, in which the response at UU-1 in case 1 test is shown, together with the result of numerical simulation by means of the VG model (van Genuchten 1980).

4 DISCUSSIONS

4.1 Seepage flow characteristics

The variation of degree of saturation with time is examined in Fig. 8, in which the response at LL-1 and LL-4 in case 1 is examined (see Fig. 5). Figure 9 shows similar observations for the variation of pore pressure head with time. The contours of ψ since the commencement of water filling are shown in Fig. 10, in which the measured water level in the standing pipes are also plotted for comparison. Note that the numerical analysis is capable of simulating well the variations of pore pressure as well as the degree of saturation with time.

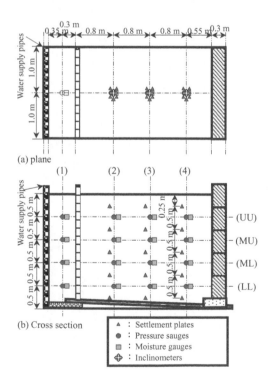

(a) plane

(b) Cross section

▲ : Settlement plates
● : Pressure sauges
■ : Moisture gauges
⊕ : Inclinometers

Figure 5.   Instrumentations employed.

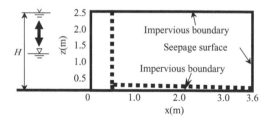

Figure 6.   Flow domain assumed in the numerical analysis.

Table 1.   Permeability of decomposed granite soil.

Item	Permeability
Measured	5.3×10^{-4} cm/sec
Simulation	
case 1	8.5×10^{-3} cm/sec
case 2	
– upstream	2.3×10^{-3} cm/sec
– downstream	6.0×10^{-4} cm/sec

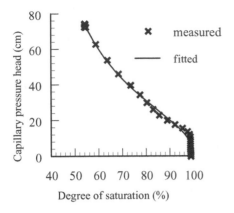

Figure 7. Water retention curves (observation vs simulation).

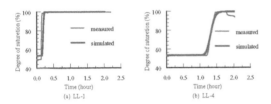

Figure 8. Variation of S_w with time (case 1).

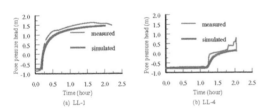

Figure 9. Variation of ψ with time (case 1).

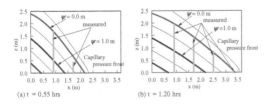

Figure 10. Contours of ψ (case 1).

Comparisons between the observation and the numerical simulation for case 2 test are made in Figs. 11 and 12, in which the variation of S_w with time at UU1 and LL-3 (see Fig. 5) as well as the rate of discharge are examined, respectively. Similar to case 1, the numerical analysis was good enough for simulating not only the variations of pore pressure and

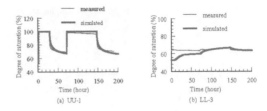

Figure 11. Variation of S_w with time (case 2 with LGD).

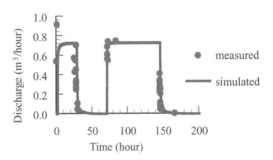

Figure 12. Discharge with time (case 2).

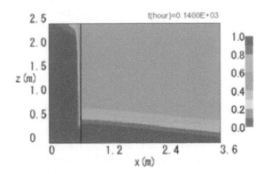

Figure 13. Simulation of S_w at t=140 hrs (case 2).

the degree of saturation with time but also the rate of discharge.

Figure 13 shows the contours of S_w at the elapsed time t = 80 hrs in case 2 test, which corresponds to a steady flow at the second stage of water filling (see Fig. 12). It is obvious in this Figure that the LGD is highly effective in reducing the downstream water level behind the vertical drain.

4.2 Deformation of test embankment

In case 1 embankment without the geosynthetic drain, the embankment failed after about 80 mins since the start of water filling. The failure occurred as soon as the seepage front reached to the chips mounted underneath the EPS block wall. As seen in Fig. 14, a tension crack developed parallel to the wall at a distance of

303

(a) First collapse

(b) Subsequent collapse

Figure 14. Collapse of case 1 embankment.

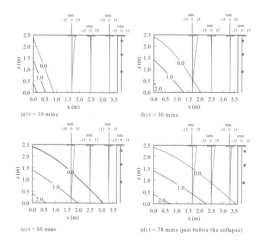

(a) t = 10 mins

(b) t = 30 mins

(c) t = 60 mins

(d) t = 78 mins (just before the collapse)

Figure 15. Horizontal deformation with contours of pore pressure head (case 1).

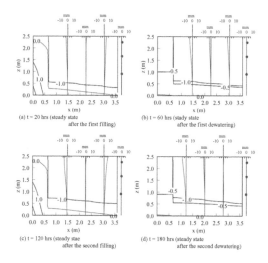

(a) t = 20 hrs (steady state after the first filling)

(b) t = 60 hrs (stady state after the first dewatering)

(c) t = 120 hrs (steady stae after the second filling)

(d) t = 180 hrs (steady state after the second dewatering)

Figure 16. Horizontal deformation with contours of pore pressure head (case 2).

about 50 cm away from the edge of the EPS wall. The failure may well be triggered by the compression failure of soil adjacent to the EPS block facing. Note also that the EPS wall reinforced with short geogrid was still effective in preventing collapse of the wall facing.

The variation of horizontal deformation with time in case 1 embankment, together with the contours of pore pressure head is shown in Fig. 15. On water filling, the upstream soil at the surface deformed by about 13 mm towards the wall, whereas the other part of the embankment exhibited no deformation at all (refer to the instant at t = 10 min). As the seepage flow gradually progressed towards the wall, the deformation also propagated towards the downstream possibly due to the increase in water pressure behind the embankment at $x = 0$ m. At 78 mins after the filling (i.e., a few minutes before the collapse), the wall was pushed away by about 12 mm at the middle and upper measuring points. The amount of displacement of the wall virtually coincided with those of three inclinometers at the corresponding level, which in turn suggests no resistance of the wall against the earth pressure. Conversely, the need for a rigid facing is strongly suggested in order to resist against the increase in earth pressure induced by the water pressure at the other end of the embankment.

In case 2 embankment with LGD, the embankment showed no sign of large deformation as examined over a period of one week. As stated earlier, the LGD was effective in reducing the downstream water level. As it can be seen in Fig. 12, the water discharged as much as $0.7 \, m^3$ per hour from the tip of the bottom geosynthetic drain.

The response of horizontal deformation in case 2 embankment, together with the contours of pore pressure head is shown in Fig. 16. A trend was obvious for the horizontal deformation that the embankment deformed towards the wall and away from the wall in a synchronized manner with the filling and dewatering, respectively. However, the amount of the overall deformation was insignificant as compared to the case 1. The efficiency of the LGD in reducing significantly the deformation of the protected region of the embankment is well demonstrated in these tests.

5 CONCLUSIONS

The LGD is effective in preventing seepage water flow into the enbankment. The effect is more significant when the permeability of geosynthetic drain is large. The horizontally layered geo-drain system is less efficient compared to the LGD system in terms of reducing the area of saturation in the embankment. It was successfully demonstrated in the seepage flow test that the LGD was effective in reducing significantly the water level as well as the soil deformation of the protected zone against the attack of seepage flow from the back. It was also manifested that the numerical analysis is capable of simulating many aspects of the seepage characteristics such as the variations of soil suction and capillary pressure with time. When the L-shaped geosynthetic drain LGD was not employed, the test embankment failed as soon as the seepage flow reached to the wall. Therefore, the need for a rigid facing, together with the LGD is strongly suggested.

ACKNOWLEDGEMENT

The research was supported by research grant-in-aid No. 19360214 from the Ministry of Education and Science in Japan.

REFERENCES

Luckner, L., van Genuchtn, M.Th. & Nielsen, D.R. 1989. A Consistent Set of Parametric Models for the Subsurface, *Water Resources Research* 25: 2187–2193.

Maulem, Y. 1976. A new model for predicting the hydraulic conductivity of unsaturated porous media, *Water Resources Research* 12: 513–522.

Richards, L.A. 1931. Capillary Conduction of Liquids through Porous Mediums, *Physics* 1: 318–333.

Saito, M., Shibuya, S., Mitsui, J. & Hara, K. 2008. L-shaped geodrain in embankment -model test and numerical simulation-, *Proceedings of the 4th Asian Regional Conference on Geosynthetics, Shanhai*: 428–433.

Scott, P.S., Farquhar, G.J. & Kouwen, N. 1983. Hysteretic Effects on Net Infiltration, *In Advances in Infiltration, Am. Soc. Agric. Eng., St. Joseph, MI*: 163–170.

Shibuya, S., Kawaguchi, T. & Chae, J.G. 2007. Failure of Reinforced Earth as Attacked by Typhoon No.23 in 2004, *Soils and Foundations*, 47–1: 153–160.

Shibuya, S., Saito, M., Hara, K. & Masuo, T. 2008. Watertight embankment using L-shaped geosynthetic drain -Model test and numerical simulation-, *Geosynthetics Engineering Journal 23, Japan Chapter of International Geosynthetics Society*: 139–146.

van Genuchten, M.T. 1980. A closed-form equation for predicting the hydraulic conductivity of unsaturated soils, *Soil Science Society American Journal*: 44: 892–898.

Prediction and Simulation Methods for Geohazard Mitigation – Oka, Murakami & Kimoto (eds)
© 2009 Taylor & Francis Group, London, ISBN 978-0-415-80482-0

Shear strength behavior of unsaturated compacted sandy soils

Md. A. Alim & M. Nishigaki
Okayama University, Okayama, Japan

ABSTRACT: Embankments which carry high-way or rail-way need firm foundations, for that reason they are generally constructed with compacted sandy soils. Generally the saturation conditions of soils of embankment during raining season are more than 70%. However, during heavy rains, the soils become more saturated thereby reducing shear strength and consequently failure occurs. The purpose of this study is to observe the shear strength variation of unsaturated sandy soils with degree of saturation of more than 70%. A series of triaxial shear strength tests conducted on 2 types of sandy soils (taken from Okayama and Hiroshima areas of Japan). In both, 9 specimens of each type of soils were considered for undrained triaxial tests with pore-water pressure measurement. The specimens were prepared by static compaction with different initial degree of saturation ranging from 70% to 100% but with the same void ratio for each soil type (0.61 and 0.65). The chosen void ratio gives around 90% proctor compaction. Experimental results show that, the shear strength decreases linearly with increase in degree of saturation.

1   INTRODUCTION

Geographically, unsaturated soils are widely distributed in semi-arid areas of the world (Fredlund & Rahardjo 1993). The upper surface of groundwater is called phreatic surface or groundwater table (WT). Failure of earth structures occurs due to change in WT and saturation condition of soils.

The WT inside the earthen embankment fluctuates all the time due to environmental conditions, however, in raining season it fluctuates frequently, because this time, much water infiltrates into the embankment due to heavy rain and increased river stage. The potential slip surface may be formed at which unsaturated soil with negative pore-water pressures exist above groundwater table. A prolonged period of rainfall or increased river stage may lead to the change in pore-water pressures and then may result in not only the change in the location of a potential slip surface, but also the local or global instability of embankment due to shear strength reduction.

Many different types of soils may be suitable for use in the construction of an embankment, ranging from granular to more finely sized soils; however, granular soils are highly desirable. Normally, the coarser fill materials are placed at or near the bottom or base of the embankment in order to provide a firm foundation for the embankment and also to facilitate drainage and prevent saturation.

Embankments which carry high-way or rail-way need firm foundations, for that reason they are generally constructed with compacted sandy soils for which shear strength plays a very important role in stability analysis. However, there is limited available information in existing literature on shear strength behavior of unsaturated sandy soils. Considerable researches have been carried out on the behavior of unsaturated fine-grained soils as reported in many literatures (Fredlund et al. 1995, Vanapalli et al. 1996, Miao et al. 2002, Cunningham et al. 2003, Rahardjo et al. 2004, Zhan & Ng 2006, Thu et al. 2007, Ng et al. 2007, Sun et al. 2007 etc.). On the other hand, a few researches have been conducted on sandy soils (Donald 1956, Rohm & Vilar 1995, Drumright & Nelson 1995, Gan & Fredlund 1996, Oberg & Sallfours 1997, Lu & Wu 2005).

Generally the saturation conditions of soils of embankment during raining season are more than 70%. However, during heavy rains, the soils become more saturated thereby reducing shear strength and consequently failure occurs.

The purpose of this study is to observe the shear strength variation of unsaturated sandy soils with degree of saturation more than 70% and to develop a numerical model for predicting shear strength with respect to degree of saturation. A series of triaxial shear strength tests conducted on 2 types of sandy soils with the same void ratio but varying degree of saturation. In both, 9 specimens of each type of soils were considered for undrained triaxial tests with pore-water pressure measurement using modified triaxial apparatus. The undrained triaxial test was chosen to keep the constant water content into the specimens.

2 SHEAR STRENGTH EQUATION OF UNSATURATED SOIL

The shear strength of unsaturated soils is based on Mohr–Coulomb criterion. Bishop (1959) proposed shear strength equation for unsaturated soils by extending Terzaghi's principle of effective stress for saturated soils. Bishop's original equation can be arranged as shown.

$$\tau_f = c' + (\sigma - u_a)\tan\phi' + \chi(u_a - u_w)\tan\phi' \quad (1)$$

Where τ_f = shear strength of unsaturated soil; c'= effective cohesion for saturated soil; ϕ' = angle of frictional resistance; $(\sigma - u_a)$ = net normal stress; $(u_a - u_w)$ = matric suction; and χ = a parameter dependent on the degree of saturation.

The last part of Equation (1) is the shear strength contribution due to unsaturation of soil called suction strength(τ_{us}). Therefore, rearranging Equation (1), it becomes

$$\tau_f = c' + (\sigma - u_a)\tan\phi' + \tau_{us} \quad (2)$$

Where,

$$\tau_{us} = \chi(u_a - u_w)\tan\phi' \quad (3)$$

Now from Equation (2), the equation of total cohesion (C) can be written as

$$C = c' + \tau_{us} \quad (4)$$

Then the shear strength Equation (2) becomes,

$$\tau_f = C + (\sigma - u_a)\tan\phi' \quad (5)$$

Equation (4) represents the shear strength contribution due to unsaturation of soils. For stability analysis of any unsaturated slope, the suction strength (τ_{us}) variation of unsaturated soils with respect to matric suction or degree of saturation has been using.

3 MATERIALS

The 2 types of sandy soils used in this study are called Sample-1 and Sample-2. The grain size distributions of the soils are shown in Figure 1. The basic properties of the 2 samples are presented in Table 1.

4 DETERMINATION OF MATRIC SUCTION

Matric suctions were determined using a Tempe pressure cell which operates on the same principle as the conventional pressure plate apparatus. The schematic diagram of Tempe pressure cell is shown in Figure 2.

The Tempe pressure cell was placed on a support and the water level maintained at the bottom of the soil specimen. When the air pressure was set to desired matric suction, water started to drain from the soil

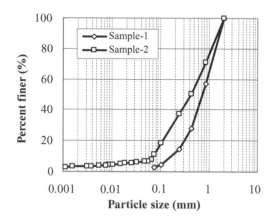

Figure 1. Particles size distribution curves of samples.

Table 1. Basic properties of soils.

Properties	Sample-1	Sample-2
Specific gravity, G_s	2.63	2.64
Maximum dry density, (g/cm^3)	1.85	1.79
Optimum water content, (%)	14.7	15.5
Grain size analysis		
$\quad$ D$_{60}$ (mm)	0.90	0.58
$\quad$ D$_{30}$ (mm)	0.44	0.17
$\quad$ D$_{10}$ (mm)	0.17	0.08
Coefficient of uniformity, C_u	5.29	7.25
Coefficient of curvature, C_c	1.26	0.62
Soil properties used in SWCC and $\quad$ triaxial tests		
$\quad$ Dry density, (g/cm^3)	1.63	1.60
$\quad$ Void ratio, e	0.61	0.65
Saturated permeability, K_s (cm/s)	2.2E-3	9.4E-4

Table 2. Estimated matric suction corresponding to degree of saturation (S_r).

	Sample-1	Sample-2
Matrci suction/ Degree of saturation	$(u_a - u_w)$ in kPa	$(u_a - u_w)$ in kPa
$S_r = 0.7$	2.5	5
$S_r = 0.8$	1.5	3.3
$S_r = 1$	0	0

specimen through the membrane filter until equilibrium was attained. The outflow of water from the soil specimen was measured by an electronic balance. The procedure was repeated at higher applied air pressures until the degree of saturation less than 0.70 was obtained. The plot of degree of saturation against corresponding matric suctions gave the soil-water characteristic curve (SWCC) and from the SWCCs the matric suctions were estimated as shown in Table 2.

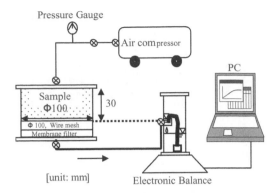

Pressure Gauge
Air compressor
PC
Sample
Φ100
30
Φ 100, Wire mesh
Membrane filter
[unit: mm]
Electronic Balance

Figure 2. The schematic diagram of Tempe pressure cell.

5 TRIAXIAL TEST PROGRAM AND PROCEDURE

A series of 18 triaxial tests were carried out on 2 soil samples refereed to as Sample-1 and Sample-2 (9 tests for each sample). The compacted unsaturated specimen's specifications are presented in Table 3. The specific amount of soil was mixed with desired quantity of water and then compacted statically in a stainless mold 100 mm height and 50 mm diameter in 3 layers. To get saturated specimen, the unsaturated specimen with initial degree of saturation 0.95 was saturated by flowing water through the bottom valve of pedestal and keeping the water head higher than the specimen height and assumed it is field saturation condition. A double-walled strain-controlled modified triaxial apparatus, which allowed measurement of pore-water pressure and volume change, was used in this study for undrained triaxial tests. Each test was commenced by saturating the high air entry disc (2 bars) with de-aired distilled water. Prior to the tests, the specimen was mounted on the pedestal of the triaxial cell and rubber membrane was placed around the specimen. O-rings were placed over the membrane on the bottom pedestal and upper cap. The specimens were then brought to confining pressures of 100 kPa, 150 kPa and 200 kPa and started shearing under undrained triaxial tests with pore-water pressure measurement, while the strain-state was 0.01 mm/min. During shearing process, the axial deformation, axial load, total volume change and pore-water pressure were measured. All the specimens have been sheared up to 13% strain and peak deviator stresses were found within this range.

6 PRESENTATION OF TEST RESULTS AND DISCUSSION

All the specimens were sheared under undrained triaxial shear with pore-water pressure measurement.

Table 3. Specifications of tested specimens.

Sample No.	Specimen No.	S_r*	Void ratio*	σ_3 (kPa)	$(u_a - u_w)$ (kPa)
Sample-1	Specimen-1			100	
	Specimen-2	1	0.61	150	0.0
	Specimen-3			200	
	Specimen-1			100	
	Specimen-2	0.8	0.61	150	1.5
	Specimen-3			200	
	Specimen-1			100	
	Specimen-2	0.7	0.61	150	2.5
	Specimen-3			200	
Sample-2	Specimen-1			100	
	Specimen-2	1	0.65	150	0.0
	Specimen-3			200	
	Specimen-1			100	
	Specimen-2	0.8	0.65	150	3.3
	Specimen-3			200	
	Specimen-1			100	
	Specimen-2	0.7	0.65	150	5.0
	Specimen-3			200	

*After specimen preparation

Table 4. Shear strength parameters of tested soils.

	Sample-1			Sample-2		
S_r	0.7	0.8	1	0.7	0.8	1
C (kPa)	14	11	8.5	19.5	15	9.8
τ_{us} (kPa)	5.5	2.5	0	10.2	5.2	0
Φ^0	38.9	39.1	39.5	36.3	36.8	37.8

The suction before shearing start is defined as initial suction and it is assumed that there is no change in soil suction during shearing. Degree of saturation (S_r), confining pressure (σ_3) and other specifications for the tested specimens are tabulated in Table 3.

Deviator stress versus axial strain at different confining pressure but at the same degree of saturation are presented in Figures 3 and 4 of soil Sample-1 and Sample-2 respectively. All the experimental results show that deviator stress increases with increase in confining pressure. Figures 5 and 6 represent the comparison of test results at different degree of saturation but at the same confining pressure (200 kPa) for soil Sample-1 and Sample-2 respectively. All figures show that at the same confining pressure, deviator stress decreases as degree of saturation increases. It means that unsaturation or matric suction is contributing to the shear strength of unsaturated soils.

From the experimental results, shear strength parameters are calculated for unsaturated soil using pore air pressure while for saturated soil using pore water pressure in both cases using Equation (4) and Equation (5) and presented in Table 4.

309

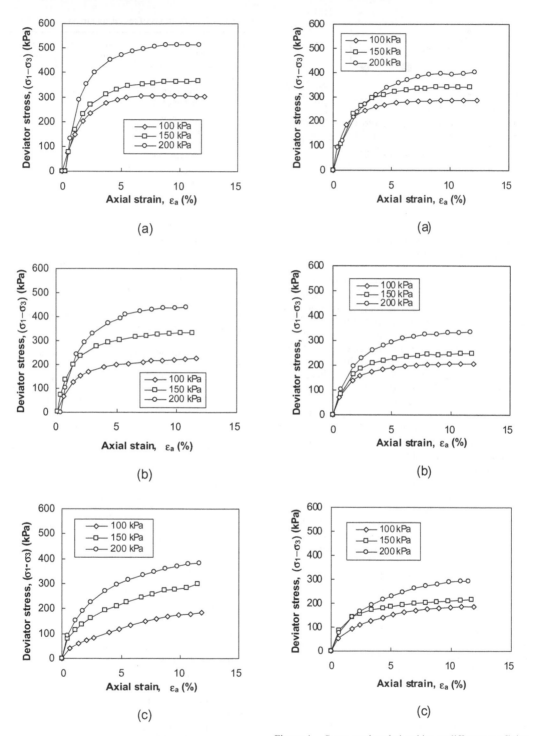

Figure 3. Stress-strain relationships at different confining pressure (100 kPa, 150 kPa and 200 kPa) for sample-1 (a) specimen with $S_r = 0.7$ (b) specimen with $S_r = 0.8$ (c) saturated specimen.

Figure 4. Stress-strain relationships at different confining pressure (100 kPa, 150 kPa and 200 kPa) for sample-2 (a) specimen with $S_r = 0.7$ (b) specimen with $S_r = 0.8$ (c) saturated specimen.

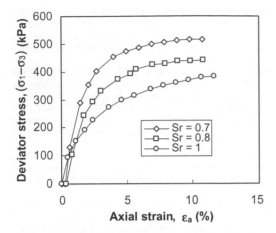

Figure 5. Comparison of stress-strain relationship for different degree of saturation at the same confining pressure of 200 kPa (for sample-1).

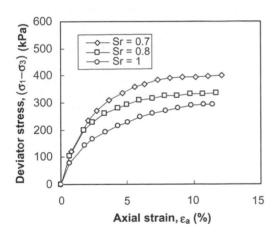

Figure 6. Comparison of stress-strain relationship for different degree of saturation at the same confining pressure of 200 kPa (for sample-2).

Calculated total cohesion and frictional angle are plotted against degree of saturation in Figure 7, which represent the total cohesion versus degree of saturation. Figure 8 shows the relationship between frictional angle and degree of saturation.

From the experimental results it was observed that, the total cohesion increases with decrease in degree of saturation linearly. On the other hand, variation of frictional angle with degree of saturation is almost the same. However, slight variation of frictional angle indicates that it maybe has a little role to suction strength contribution of unsaturated sandy soils.

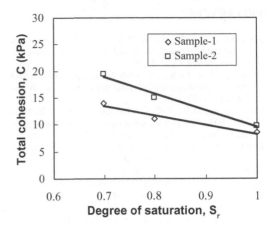

Figure 7. Relationship between total cohesion and degree of saturation.

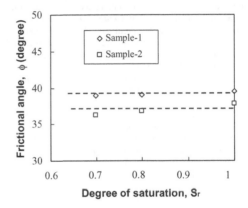

Figure 8. Relationship between frictional angle and degree of saturation.

7 CONCLUSIONS AND RECOMMENDATIONS

The following conclusions can be drawn from this study:

1. It is found that the degree of saturation affects the shear strength of sandy soils, and the cohesion increases with decrease in degree of saturation linearly within the range of degree of saturation from 70% to 100%.
2. Friction angle has a very little role to suction strength contribution of unsaturated sandy soils.
3. A simple parameter, degree of saturation could be used for estimating the suction strength of unsaturated sandy soils within the range of degree of saturation from 70% to 100%.
4. Further studies are however required to examine the aforesaid behaviors on others sandy soils.

REFERENCES

Alim, M.A. & Nishigaki, M. 2007. A simplified procedure to determine hyperbolic model parameters. *Australian Geomechanics* 42(4): 51–56.

Bishop, A.W. & Blight, G.E. 1963. Some aspects of effective stress in saturated and partly saturated soils. *Ge'otechnique* 13(3):177–197.

Cunningham, M.R., Ridley, A.M., Dinnen, K., & Burland, J.B. 2003. The mechanical behavior of a reconstituted unsaturated silty clay. *Geotechnique* 53(2): 183–194.

Donald, I.B. 1956. Shear strength measurements in unsaturated non-cohesive soils with negative pore pressures. *Proceedings, 2nd Australia and New Zealand Conference on Soil Mechanics and Foundation Engineering, Christchurch, New Zealand*: 200–205.

Drumright, E.E. & Nelson, J.D.1995. The shear strength of unsaturated tailings sand. *Proceedings of the first international conference on unsaturated soils/unsat'95/Paris/France/6–8 September*: 45–50.

Fredlund, D.G. & Rahardjo, H. 1993. *Soil Mechanics for Unsaturated Soils*. New York: John Wiley & Sons, Inc.

Fredlund, D.G. 1979. Appropriate concepts and technology for unsaturated soils. *Can. Geotech. J.* 16:121–39.

Fredlund, D.G., Xing, A., Fredlund, M.D. & Barbour, S.L. 1995. The relationship of the unsaturated soil shear strength to the soil-water characteristic curve. *Can. Geotech. J.* 32: 440–448.

Gan, J.K.-M. & Fredlund D.G. 1996. Shear strength characteristics of two saprolitic soils. *Can. Geotech. J.* 33: 595–609.

Lu, N. & Likos, W.J. 2004. *Unsaturated Soil Mechanics*, Hoboken, New Jersey: John Wiley & Sons, Inc.

Lu. N. & Wu, B. 2005. Unsaturated shear strength behavior of a fine sand, *Geomechanics-II, American Society of Civil Engineering, 1801 Alexander Bell Drive, Reston, VA 20191 USA*: 488–499.

Mashhour, M.M., Ibrahim, M.I. & El-Emam, M.M. 1995. Variation of unsaturated soil shear strength parameters with suction, Proceedings of the first international conference on unsaturated soils/unsat'95/Paris/France/6–8 September: 1487–1493.

Miao, L., Liu, S. & Lai, Y. 2002. Research of soil-water characteristics and shear strength features on Nanyang expansive soil. *Engineering Geology* 65(4): 261–267.

Ng, C.W.W., Cui, Y., Chen, R. & Delage, P. 2007. The axis-translation and osmotic techniques in shear testing of unsaturated soils: a comparison. *Soils and Foundations* 47(2): 675–684.

Oberg, A. & Sallfors, G. 1997. Determination of Shear Strength Parameters of Unsaturated Silts and Sands Based on the Water Retention Curve. *Geotechnical Testing Journal* 20(1): 40–48.

Rahardjo, H., Heng, O.B. & Choon, L.E. 2004. Shear strength of a compacted residual soil from consolidated drain and constant water content triaxial tests. *Can. Geotech. J.* 41: 421–436.

Rohm, S.A. & Vilar, O.M. 1995. Shear strength of an unsaturated sandy soil, Proceedings of the first international conference on unsaturated soils/unsat'95/Paris/France/ 6–8 September: 189–193.

Sun, D., Sheng, D. & Xu, Y. 2007. Collapse behavior of unsaturated compacted soil with different initial densities. *Can. Geotech. J.* 44: 673–686.

Thu, T.M., Rahardjo, H. & Leong, E.C. 2007. Critical state behavior of a compacted silt specimen. *Soils and Foundations* 47 (4): 749–755.

Vanapalli, S.K., Fredlund, D.G., Pufahl, D.E. & Clifton, A.W. 1996. Model for the prediction of shear strength with respect to soil suction. *Can. Geotech. J.* 33: 379–392.

Zhan, T. L.T. & Ng, C.W.W. 2006. Shear strength characteristics of unsaturated expansive clay. *Can. Geotech. J.* 43: 751–763.

Prediction and Simulation Methods for Geohazard Mitigation – Oka, Murakami & Kimoto (eds)
© *2009 Taylor & Francis Group, London, ISBN 978-0-415-80482-0*

Modeling the behavior of artificially structured clays by the Modified Structured Cam Clay model

J. Suebsuk, S. Horpibulsuk & A. Chinkulkijniwat
Suranaree University of Technology, Nakhon-Ratchasima, Thailand

M.D. Liu
University of Wollongong, Australia

ABSTRACT: A practical constitutive model for artificially structured clays is formulated within the Structured Cam Clay (SCC) model framework, referred to as Modified Structured Cam Clay (MSCC) model. Based on experimental observation, it is considered that the effect of cementation is as the increase in mean effective stress. The strain softening behavior after the peak strength state caused by dilation and crushing of cementation structure for both normally and over-consolidated states can be simulated by the reduction in effective stress and the shrinkage of yield surface. The model parameters can be divided into those describing destructured properties and those for cementation structure, which can be simply determined from the conventional triaxial tests on artificially structured samples. The capability of the model is illustrated by the comparison of the simulated and measured undrained and drained shear responses of artificially structured Ariake clay.

1 INTRODUCTION

For the improvement of the soft ground by the chemical admixture such as in-situ deep mixing technique, the natural clay is disturbed by mixing wings and mixed with cement or lime. The natural structure is destroyed and taken over by the cementation structure. The cement or lime admixed clay is thus designated as "Artificially structured clay".

In recent years, the rapid advances in computer hardware and the associated reduction in cost has resulted in a marked increase in the use of numerical methods to analyze geotechnical problems. The ability for such methods to provide realistic predictions is dependent upon the accuracy of the constitutive model used for representing the mechanical behavior of the soil. There has been great progress in constitutive modeling of the behavior of soil with natural structure, such as those proposed by Gens & Nova (1993), Whittle (1993), Wheeler (1997), Rouainia & Wood (2000) and Kavvadas & Amorosi (2000), Chai et al. (2004). However, most of these constitutive models are generally complicated and their model parameters are difficult for practical identification.

Recently, Liu & Carter (2002) and Carter & Liu (2005) have introduced a simple predictive model, the Structured Cam Clay (SCC) model, for natural structured clay. It has been formulated simply by introducing the influence of natural structure into the Modified Cam Clay (MCC) model. The SCC model is suitable for naturally structured clay with negligible cohesion

intercept and without softening behavior when stress states on virgin yielding state. However, for artificially structured clay, the cohesion intercept is generally very significant and the softening behavior is realized for both inside and on the yield surface (Wissa et al. 1965, Clough et al. 1981, Kasama et al. 2000, Miura et al. 2001, Horpibulsuk et al. 2004, & Horpibulsuk et al. 2005). Horpibulsuk et al. (2009) have introduced the influence of cementation on the cohesion intercept into SCC model.

The present paper attempts to extend the work of Horpibulsuk et al. (2009) for artificially structured clays. It focuses on the development of a new flow rule to account for the influence of the crushing of cementation structure on strain softening behavior. The proposed model is designated as the Modified Structured Cam Clay (MSCC) model. A theoretical study of the behavior of artificially structured clay is then made in the paper. The verification of the model is finally done by simulating the undrained and drained shear behavior of structured clays in artificial states. The artificially structured clays are cemented Araike clay (Horpibulsuk et al. 2004 & Horpibulsuk 2001).

2 CONCEPTUAL FRAMEWORK OF MSCC MODEL

2.1 *Material idealization*

In the proposed MSCC model, structured clay is idealized as an isotropic material with elastic and virgin yielding behavior. The yield surface varies isotropically with plastic volumetric deformation. Soil behavior is assumed to be elastic for any stress excursion inside the current yield surface. Virgin yielding occurs for a stress variation originating on the yield surface and causing it to change. During virgin yielding, the current stress of a soil stays on the yield surface.

Based on extensive compression test results, a material idealization for the compression behavior of structured clays whose void ratios are in meta-stable state is introduced in Figure 1a. The compression behavior is inverse S-shaped curve (Nagaraj et al., 1990; Liu & Carter, 1999 and 2000; and Horpibulsuk et al., 2007). The compression strain is negligible up to the yield stress, $p'_{y,i}$, due to the contribution of cementation. Beyond this yield stress, there is sudden compression of relative high magnitude, indicated by the steep slope and caused by the break-up of the cementation. On further loading, the difference in void ratio between structured and destructured states (Δe) with logarithm of stress decreases and finally reaches nearly a constant value at a high effective stress. This constant value is designated as the residual additional void ratio sustained by structure, c. Hence, the virgin compression behavior of a structured soil can be expressed by the following equation,

$$e = e^* + \Delta e \qquad (1)$$

The destructured line (e^* vs log p') is used as a reference line for estimating compression curves of structured clay. It is theoretically obtained from the compression test on the completely remolded sample. In practice, it is possible to assume that critical state line of destructured and structured clays is identical (Nagaraj & Miura, 2001). As such, the destructured line is simply estimated from the critical state line of the structured clay (generally obtained from triaxial test) by assuming that the behavior of destructured clay follows MCC model in which the slope of critical state and isotropic compression line is the same.

$$e^*_{IC} = e_{cs} + (\lambda^* - \kappa) \ln 2 \qquad (2)$$

where e^*_{IC} and e_{cs} is the void ratio at mean effective stress of 1 kPa of isotropic compression and critical state lines, respectively, λ^* is the compression index of critical state line, which is the same as that of destructured line, and κ is the swelling index of structured clay.

It is found that the compression equation for the additional void ratio of naturally structured clay proposed by Liu & Carter (1999, and 2000) is also applicable for the artificially structured clays. Considering the consistency with the introduction of modified mean

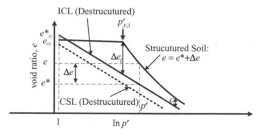

(a) Compression behavior of structured soil

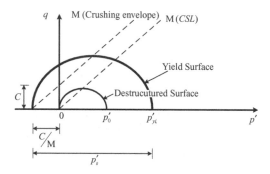

(b) Yield and Destructured Surfaces

Figure 1. Material idealization for Modified Structured Cam clay model.

effective stress parameter $\bar{p}'$, the following compression equation for structured clays is proposed,

$$e = e^* + (\Delta e_i - c) \left(\frac{p'_{y,i} + C/M}{\bar{p}'} \right)^b + c \qquad (3)$$

where b and c are soil parameters describing the additional void ratio sustained by cementation. Δe_i is the value of the additional void ratio at the start of virgin yielding (Figure 1a). Parameter c is defined by the following equation,

$$c = \lim_{p' \to \infty} \Delta e \qquad (4)$$

Following the tradition of the Modified Cam Clay model, the yield surface of structured soil in $q-p'$ plane is assumed to be elliptical and is described as (Figure 1b),

$$f = q^2 - M\bar{p}' (p'_s - \bar{p}') = 0 \qquad (5)$$

where p'_s is the size of the yield surface, which is effectively shifted to the left along the p' axis by a distance of C/M. The size of the initial yield surface, $p'_{s,i}$ is linked to the initial mean effective yield stress, $p'_{y,i}$ by the following equation (Figure 1b),

$$p'_{s,i} = p'_{y,i} + C/M \qquad (6)$$

314

2.2 Stress states on yield surface

2.2.1 Hardening behavior

For stress states on the yield surface and with $dp'_s > 0$, virgin yielding occurs. The plastic volumetric strain increment for the model is derived from the assumption that the plastic volumetric strain is dependent upon the change in the size of the yield surface and the magnitude of current shear stress. The plastic volumetric strain increment is thus derived from Eq. (3) as follows

$$d\varepsilon_v^p = \left\{ \begin{array}{l} (\lambda^* - \kappa) \\ + b\langle\Delta e - c\rangle\left[1 - \frac{\gamma\bar{\eta}}{M}\right] \end{array} \right\} \frac{dp'_s}{(1+e)p'_s} \quad (7)$$

where

$$\langle\Delta e - c =\rangle = \begin{cases} \Delta e - c & \text{if } \{|\Delta e| - |c|\} \geq 0 \\ 0 & \text{if } \{|\Delta e| - |c|\} < 0 \end{cases} \quad (8)$$

$|x|$ represents the absolute value of the quantity x.

The term $\frac{\bar{\eta}}{M}$ is introduced to take the effect of current shear stress into account as has been similarly done by Liu & Carter (2002). Parameter γ is a soil parameter reflecting the degree of cementation and anisotropic behavior. It is equal to zero for isotropic response. For a given structured clay, γ can be determined from anisotropic consolidation tests at different η.

Based on elastoplastic theory, the volumetric strain is determined in the following form:

$$d\varepsilon_v = d\varepsilon_v^e + \left\{ \begin{array}{l} (\lambda^* - \kappa) \\ + b\langle\Delta e - c\rangle\left[1 - \frac{\gamma\bar{\eta}}{M}\right] \end{array} \right\} \times \frac{dp'_s}{(1+e)p'_s} \quad (9)$$

In Eq. (9), the additional void ratio will diminish as the soil reaches the critical state and then the cementation structure of the soil is completely removed.

Even though MSCC model employs the yield function of Modified Cam Clay model as a yield surface, its flow rule is not taken for determination of shear strain. This is because the flow rule of Modified Cam Clay model generally produces too much shear strain and therefore overprediction of at rest earth pressure (McDowell, 2000; and McDowell & Hau, 2003). A new flow rule is introduced based on that proposed by McDowell & Hau (2003); and Liu & Carter (2002) and (2003) as follows

$$\frac{d\varepsilon_s^p}{d\varepsilon_v^p} = \frac{k\bar{\eta}}{|M^2 - \bar{\eta}^2| + \left|1 - \sqrt{\frac{p'_d}{p'_s}}\right|} \quad (10)$$

where k is the model parameter depending upon the degree of cementation and p'_d is the size of the yield surface of destructured clay. Based on the destructured isotropic compression equation and considering the influence of the additional void ratio c, the following equation for p'_d is found,

$$p'_d = \frac{\exp\left(\frac{e^*_{IC} - e + c}{\lambda^* - \kappa}\right)}{p'\left(\frac{\kappa}{\lambda^* - \kappa}\right)} \quad (11)$$

It is noted from Eq. (10) that the lower the k value, the lower the shear strain at failure (the higher the stiffness). The term $\left|1 - \sqrt{p'_d/p'_s}\right|$ is introduced so that $\frac{d\varepsilon_s^p}{d\varepsilon_v^p} \neq \infty$ when the stress states reach the crushing envelope ($M = \bar{\eta}$). p'_s is always higher than p'_d at precritical state due to the cementation effect ($C > 0$). At critical state where no cementation effect, $p'_s = p'_d$. Thus, the asymptotic feature does not occur at crushing state and hence the strain softening behavior can be simulated. For destructured clay whose $C = 0$ and $p'_s = p'_d$, when $k = 2.0$, the flow rule becomes that of the Modified Cam Clay (MCC) model. From Eqs. (7 and 10), the shear strain can be determined in the form:

$$d\varepsilon_d = d\varepsilon_d^e + \frac{k\bar{\eta}}{|M^2 - \eta^{-2}| + \left|1 - \sqrt{p'_o/p'_s}\right|}$$

$$\times \left\{ (\lambda^* - \kappa) + b\langle\Delta e - c\rangle\left[1 - \frac{\gamma\bar{\eta}}{M}\right] \right\}$$

$$\times \frac{dp'_s}{(1+e)p'_s} \quad (12)$$

2.2.2 Softening behavior

The softening behavior after the peak strength state is caused by two processes: the dilation and the crushing of cementation structure. The crushing of the cementation structure leads to the removal of cementation structure in the transition to a critical state. Based on experimental observation (Huang, 1994; Horpibulsuk, 2001; and Horpibulsuk et al., 2004), the dilation mainly occurs for the stress state on the dry side of crushing envelope (heavily over-consolidated range). At this state, the failure criterion is the bulging failure and eventually the samples split associated with shear failure (Horpibulsuk et al. 2004). The softening behavior due to dilation is captured in the same way as done by MCC model in which $d\bar{p}' < 0$ and yield surface shrinks ($d\bar{p}'_s < 0$) but without reduction in cohesion.

For the stress state on the wet side of the crushing envelope (lightly over-consolidated and normally consolidated ranges), the dilation is insignificant. The failure criterion of the artificially structured clay in this condition is the shear failure with a distinct failure plane (Horpibulsuk et al., 2004). When stress states reach the crushing envelope and $d\bar{p}' < 0$ for both on the dry and the wet sides of crushing envelope, the $\bar{\eta}$

decreases due to the reduction in cohesion, C, caused by the removal of the cementation structure. Consequently, the yield surface shrinks. It is assumed that stress state is always on the yield surface and $\bar{\eta} = M$ during softening. Based on extensive test results of naturally and artificially structured clays (at various cement contents), it is found that the reduction in cohesion can be expressed in the form:

$$dC = -\varpi \left(\frac{C}{C_{in}} \right) \frac{|d\bar{p'}|}{\sqrt{\left(\frac{q}{p'} - M \right)}} \qquad (13)$$

where C_{in} is the value of the initial cementation strength and ϖ is a model parameter reflecting softening behavior. The higher the ϖ, the faster the softening in $q - \varepsilon_s$ relationship.

During the softening process, the plastic volumetric deformation is described by the same equation as that for virgin yielding, i.e., Eq. (12). However, a modification to the shear strain increment is made to ensure that the shear deformation is always positive, i.e.,

$$d\varepsilon_d^p = -\frac{k\bar{\eta}}{|M^2 - \bar{\eta}^2| + \left| 1 - \sqrt{p_o'/p_s'} \right|}$$
$$\times \left\{ (\lambda * -\kappa) + b \langle \Delta e - c \rangle \left[1 - \frac{\gamma \bar{\eta}}{M} \right] \right\} \frac{dp_s'}{(1+e)p_s'} \qquad (14)$$

3 APPLICATION AND VERIFICATION OF MSCC MODEL

In this section, the performance of the MSSC model is examined by simulating the shear behavior of artificially structured clays. Cemented Ariake clay (Horpibulsuk et al. 2004 & Horpibulsuk 2001) is employed for this simulation. The capability of the proposed model for describing the shear behavior of structured clay

Table 1. Values of MSCC model parameters for cemented clay.

Model Parameters	Ariake clay	
	$A_w = 6\%$	$A_w = 18\%$
λ^*	0.6	0.49
κ	0.078	0.03
e_{cs}	6.19	6.56
b	0.5	0.5
Δe_i	0.15	0.48
c	0.05	0.05
M	1.4	1.4
C (kPa)	73	600
p_y' (kPa)	43	1800
k	1	0.4
γ	0.5	0.5
G	5000	35000
ϖ	0.5	0.5

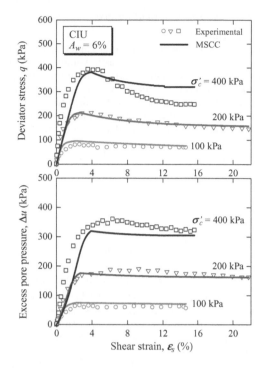

Figure 2. Comparison of experimental and simulated undrained shear behavior for $A_w = 6\%$.

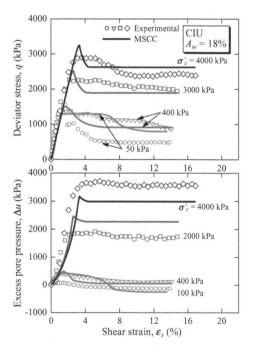

Figure 3. Comparison of experimental and simulated undrained shear behavior for $A_w = 18\%$.

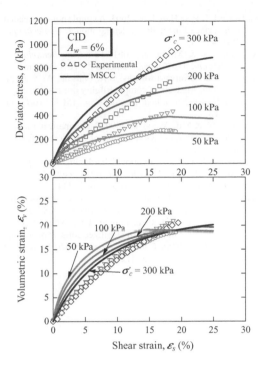

Figure 4. Comparison of experimental and simulated drained shear behavior for $A_w = 6\%$.

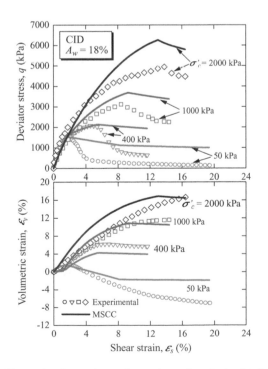

Figure 5. Comparison of experimental and simulated drained shear behavior for $A_w = 18\%$.

is evaluated based on the comparison between model simulation and experimental data. Values of model parameters identified are listed in Table 1 for both 6% and 18% cement content (A_w).

Comparisons of undrained and drained shear responses under low and high effective confining stresses (σ_3') are shown in Figures 2 through 5 for both cement contents. Overall speaking, the shear behavior has been simulated highly satisfactorily considering that all model parameters were essentially identified according to their definitions.

4 CONCLUSIONS

In this paper, a study of modeling the behavior of artificially structured clay is made. Modifications of the Structured Cam Clay model are proposed so that the influence of cementation on soil behavior can be incorporated. It has been demonstrated that this simple predictive MSCC model has captured the main features of the behavior of structured clays reasonably well. Because the parameters can be determined from conventional tests, the MSCC model is a valuable tool for geotechnical practitioners.

ACKNOWLEDGMENTS

The financial support provided from the Commission on Higher Education (CHE) and the Thailand Research Fund (TRF) under contract DIG5180008 is appreciated. The first author acknowledges the financial support from the Commission on Higher Education (CHE) for his Ph.D. study. The authors wish to express their gratitude to Professor J.P. Carter for his very valuable work which leaded to this paper.

REFERENCES

Carter, J.P. & Liu, M.D. 2005. Review of the Structured Cam Clay model. *Soil constitutive models: evaluation, selection, and calibration, ASCE, Geotechnical special publication* 128: 99–132.

Chai, J.C., Miura, N. & Zhu H.H. 2004. Compression and consolidation characteristics of structured natural clays. *Canadian Geotechnical Journal* 41(6): 1250–1258.

Clough, G.W., Sitar, N., Bachus, R.C. & Rad, N.S. 1981. Cemented sands under static loading. *Journal of Geotechnical Engineering Division, ASCE* 107(GT6): 799–817.

Gens, A. & Nova, R. 1993. Conceptual bases for constitutive model for bonded soil and weak rocks. *Geotechnical Engineering of Hard Soil-Soft Rocks*. Balkema.

Horpibulsuk, S. 2001. Analysis and Assessment of Engineering Behavior of Cement Stabilized Clays. *Ph.D. dissertation, Saga University*. Saga, Japan.

Horpibulsuk, S., Miura, N. & Bergado, D.T. 2004. Undrained shear behaviour of cement admixed clay at high water content. *Journal of Geotechnical and Geoenvironmental Engineering, ASCE* 130(10): 1096–1105.

Horpibulsuk, S., Miura, N. & Nagaraj, T.S. 2005. Clay-water/cement ratio identity of cement admixed soft clay. *Journal of Geotechnical and Geoenvironmental Engineering, ASCE* 131(2): 187–192.

Horpibulsuk, S., Liu, M.D., Liyanapathirana, D.S. & Suebsuk, J. 2009. Behavior of cemented clay simulated via the theoretical framework of the SCC model. *Computer and Geotechnics*. (Under review).

Horpibulsuk, S., Shibuya, S., Fuenkajorn, K. & Katkan, W. 2007. Assessment of engineering properties of Bangkok clay. *Canadian Geotechnical Journal* 44(2) 173–187.

Huang, J.T. 1994. The Effects of Density and Cementation on Cemented Sands. *Ph.D. Thesis*. Sydney University.

Kasama, K., Ochiai, H. & Yasufuku, N. 2000. On the stress-strain behaviour of lightly cemented clay based on an extended critical state concept. *Soils and Foundations* 40(5): 37–47.

Kavvadas, M. & Amorosi, A. 2000. A constitutive model for structured soils. *Géotechnique* 50(3): 263–273.

Liu, M.D. & Carter, J.P. 1999. Virgin compression of structured soils. *Géotechnique* 49(1): 43–57.

Liu, M.D. & Carter, J.P. 2000. Modelling the destructuring of soils during virgin compression. *Géotechnique* 50(4): 479–483.

Liu, M.D. & Carter, J.P. 2002. Structured cam clay model. *Canadian Geotechnical Journal* 39(6): 1313–1332.

Liu, M.D., & Carter, J.P. 2003. The volumetric deformation of natural clays. *International Journal of Geomechanics, ASCE* 3(3/4): 236–252.

McDowell, G.R. 2000. A family of yield loci based on micro mechanics. *Soils and Foundations* 40(6): 133–137.

McDowell, G.R. & Hau, K.W. 2003. A simple non-associated three surface kinematic hardening model. *Geotechnique* 53(4): 433–437.

Nagaraj, T.S. & Miura, N. 2001. *Soft Clay Behaviour—Analysis and Assessment*. Netherlands: A.A.Balkema.

Nagaraj, T.S., Srinivasa Murthy, B.R., Vatsala, A. & Joshi, R.C. 1990. Analysis of compressibility of sensitive clays. *Journal of Geotechnical Engineering, ASCE* 116(GT1): 105–118.

Rouainia, M. & Muir Wood, D. 2000. A kinematic hardening model for natural clays with loss of structure. *Géotechnique* 50(2): 153–164.

Wheeler, S.J. 1997. A rotational hardening elasto-plastic model for clays. *Proc. 14th Int. Conference Soil Mechanics and Foundation Engineering* 1: 431–434.

Whittle, A.J. 1993. Evaluation of a constitutive model for overconsolidated clays. *Géotechnique* 43(2): 289–314.

Wissa, A.E.Z., Ladd, C.C. & Lambe, T.W. 1965. Effective stress strength parameters of stabilized soils. *Proceedings of 6th International Conference on Soil Mechanics and Foundation Engineering*: 412–416.

Prediction and Simulation Methods for Geohazard Mitigation – Oka, Murakami & Kimoto (eds)
© 2009 Taylor & Francis Group, London, ISBN 978-0-415-80482-0

Behavior of sands in constant deviatoric stress loading

A. Azizi, R. Imam, A. Soroush & R. Zandian
Amirkabir University of Technology, Tehran, Iran

ABSTRACT: Many instances of flow-type slope failures are believed to be caused by changes in soil stresses resulting from rises in pore water pressure; among them are submarine flowslides and mining waste dump failures. In a soil mass, a rise in pore water pressure leads to a decrease in mean effective stresses, while the vertical gravity load remains unchanged. Similar loading may be applied to a soil sample in the triaxial apparatus by keeping the deviatoric stress constant, while the confining pressure is decreased. Previous research has shown that under such loading, loose sands initially dilate slightly, and then start to contract substantially as failure is approached. These contractions can lead to the increase in pore pressure and, consequently, failure of the soil mass. The current study shows that these volume contractions are affected by factors such as void ratio, confining pressure, level of deviatoric stress, and anisotropic consolidation.

1 INTRODUCTION

1.1 The constant deviatoric stress loading

The rise in pore water pressure is believed to be responsible for many cases of flow-type slope failures (see e.g. Morgenstern 1994, Eckersley 1990, Dawson et al. 1992, Castro et al. 1992, Lade 1993, and Anderson & Sitar 1995). These failures have been observed to occur following pore water pressure rises resulting from the infiltration of rainwater, snowmelt, cyclic loading, etc. and are often of the types referred to as flow slides or debris flows.

A rise in pore water pressure within the soil mass leads to a reduction in the soil confining stresses. This may occur while the vertical loads due to soil weight remain unchanged, or increase as a result of soil wetting. Such stresses applied to the soil may be represented in the triaxial apparatus by applying a constant load to the top of the soil sample, while reducing the confining stresses. Since in such tests the deviatoric stress applied to the soil sample remains approximately constant, the loading may be referred to as a "Constant Deviatoric Stress" or CDS loading. CDS loading is applied under drained conditions since preventing drainage will casue the soil sample to follow a stress path with an increasing or decreasing shear strength depending on soil density and, therefore, a constant deviatoric stress cannot be maintained.

Previous studies have shown that fine-grained and medium to dense coarse-grained soils generally dilate under CDS loading, while very loose and loose sands initially experience small or no volume changes, and then start to contract substantially as failure is approached (see e.g. Sasitharan et al. 1994, Skopek

1994, and Anderson & Riemer 1995). In loose saturated sands, such contractions can lead to the generation of substantial pore water pressure and loss of strength. The same can happen in the field under poor drainage conditions, and the loss of soil strength may result in the static liquefaction and sudden catastrophic failure of slopes.

Conventional limit equilibrium methods of slope stability analysis can not predict such failures since the volume contractions and loss of strength usually occur well before the mobilization of failure shear strength. It is important, therefore, that the conditions leading to the volume contractions, and the factors that affect them are determined such that the potential for such failures can be determined.

This study examines the factors that affect the aforementioned volume contractions by applying CDS loading to loose and very loose samples of a local sand called the Firoozkooh sand. The relationship between the contractive behavior of dry samples and the collapse of saturated samples in CDS loading is also examined by conducting CDS tests on saturated samples of the same sand.

1.2 Undrained behavior of loose sand

Figure 1(a) illustrates a typical stress path for a sample of very loose liquefiable sand in monotonic triaxial compression loading. The shear strength initially increases to a peak at P, and then starts decreasing (strain softening) until it reaches the critical state condition at F, where it continues to experience shear strain under constant shear stress (q), mean effective normal stress (p'), and void ratio (e) as described by Casagrande (1936).

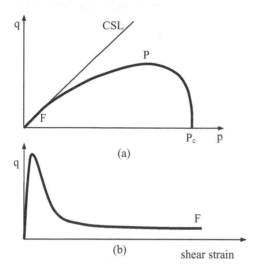

Figure 1. Typical results of undrained triaxial compression test on loose sand (a) Stress paths (b) stress-strain relationship.

Figure 1(b) shows that before the maximum q is reached at P, the rate of development of shear strain is small, but it increases substantially afterwards. At F, the soil is left with small shear strength and almost "liquefies." Such liquefied soil may behave similar to a liquid and experience large shear strains, or "flow" even under small loads.

2 DESCRIPTION OF THE TESTS

2.1 *The sand samples*

Tests were carried out on a local uniform, subangular, quartzic sand called the Firoozkooh sand. Its specific gravity is $G_s = 2.65$, mean grain size $D_{50} = 0.25$ millimeters, and maximum and minimum void ratios are $e_{max} = 0.87$ and $e_{min} = 0.55$ respectively. All samples were prepaed using the moist tamping procedure, in which the 71 mm diameter, 160 mm high samples were prepared by pouring 8 layers of sand with 2.5 percent moisture. After pouring each layer, it was lightly tamped using a metal tamper with a diameter equal to that of the mould. The initial layers were tamped once and more gently, and the last layers were tamped twice and using a somewhat higher energy such that the sample will have as uniform a density as possible.

2.2 *Loading procedure*

All tests were carried out using a GCTS triaxial testing apparatus loacted in the soil mechanics laboratory of Amirkabir University of Technology.

Most tests were carried out on dry samples, but a few were done on saturated samples. Most dry samples were first consolidated isotropically and then subjected to a dead axial load that remained constant throughout the test. The dead load was applied by placing specified weights on a round plate supported by a shaft screwed to the upper platen of the triaxial sample. A few samples were also consolidated anisotropically before applying the dead load.

In the tests on dry samples, following application of the dead load, the cell pressure was gradually reduced until the critical state condition was reached. In the tests on saturated samples, the soil was first subjected to a conventional strain-controlled undrained triaxial compression loading until the specified deviatoric stress was reached. Shearing was then stopped and the deviatoric load applied to the top of the sample was locked at the current state. A round metal plate was then screwed to the top of the metal shaft connected to the upper triaxial platen, and sufficient dead weight was placed on the round plate such that the same deviatoric stress that was applied by the loading frame prior to stopping the strain-controlled test was applied to the top of the sample.

The drainage port was then opened and a back pressure equal to the current pore pressure was applied such that volume of the sample remained unchanged. A drained testing was then initiated by gradually increasing the back pressure while the cell pressure was kept constant. This resulted in the application of an almost constant deviatoric stress to the top of the sample, while reducing the mean effective normal stress under drained conditions.

3 SAND BEHAVIOR IN CDS LOADING

3.1 *CDS test on dry sand*

Figure 2 shows a typical result of tests on dry sand. The sample is consolidated isotropically to 250 kPa (point A), and subjected to a deviatoric stress (point B). The mean normal stress is then gradually reduced while the deviatoric stress remains constant. It may be seen that during the initial stage of the CDS stress path, the sample experiences slight dilation, but as it reaches a certain state prior to reaching point C, it begins experiencing small volumetric contraction. Upon further reduction of p, the rate of these contractions increases substantially (at point C) until the CSL is reached at D. Changes in sample height also take place similarly, as they initially occur at a smaller rate and then the rate increases substantially (see Figure 2(c))

It is noted that the location of the CSL both in the e–p plane and in the p–q plane was determined using results of undrained triaxial compression tests similar to that shown in Figure 1.

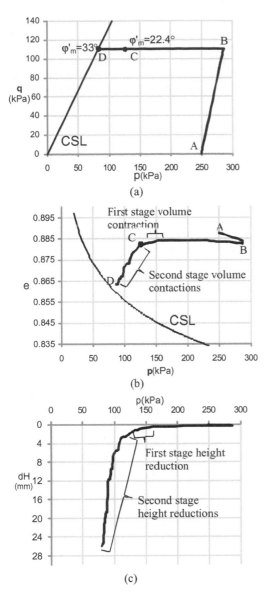

(a)

(b)

(c)

Figure 2. Typical behavior of dry sand in CDS loading: (a) Stress path (b) Void ratio v.s. mean effective normal stress (c) Changes in sample height v.s. mean effective normal stress.

Imam et al. (2002) showed that after application of the deviatoric stress (at B), the sample will have a yield surface which passes through points B and C, and while the stress path moves between these two stress states, the sample experiences unloading, which results in volume expansion as observed in the test results. Imam et al. (2005) used similarly derived yield

surfaces to model the behavior of sand under various loading conditions.

3.2 CDS test on saturated sand

Figure 3 shows results of a CDS test on a saturated sample of the same sand. The sample is consolidated isotropically to 250 kPa and void ratio of 0.901. It is then subjected to undrained loading until the deviatoric stress reaches 73 kPa at an axial strain of 0.38 percent. A dead load resulting in the same deviatoric stress of 73 kPa is then applied to the top of the sample, the back pressure is set equal to the current pore pressure, and the back pressure is gradually decreased under drained condition. The mean effective stress is decreased from 157 kPa to 86 kPa, where the sample suddenly collapses. The collapse occurs at such high speed that no data could be recorded during this stage. At collapse, the axial strain is 1.5 percent, the void ratio is 0.909 and the mobilized friction angle is 22 degrees, which is substantially smaller than the mobilized friction angle at critical state, which was measured to be 33 degrees.

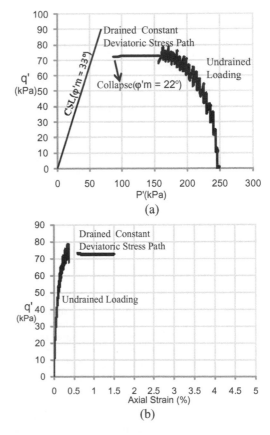

(a)

(b)

Figure 3. Behavior of saturated sand in CDS loading: (a) Stress path (b) Deviatoric stress v.s. axial strain.

It is of interest to correlate the behavior of the dry sample with that of the saturated sample under CDS loading, in order to examine the cause of failure of the latter sample. As shown in Figure 2, the dry sample experiences contraction as it reaches a certain state prior to failure, and the same is expected to occur with the saturated sample which is subjected to drained CDS loading, provided full drainage can occur to allow all the volume contractions which would take place in a dry sample to occur. However, due to limitations in the drainage capacity of the triaxial sample, some of the volume changes can not occur sufficiently rapidly and, as a result, some pore pressure develops which results in the saturated sample to follow a stress path similar to that of an undrained sample, leading to a reduction in shear strength as in Figure 1(a). This results in the inability of the saturated sample to sustain the constant deviatoric load and leads to its sudden collapse well before its reaching the failure envelope.

4 FACTORS AFFECTING BEHAVIOR IN CDS LOADING

4.1 Void ratio

Figure 4 shows results of CDS tests on samples consolidated to 250 kPa and subjected to a constant deviatoric stress of 110 kPa. Void ratios at initial sample preparations (e_p) for all tests are shown and void ratios after consolidation may be determined from the curves shown. It may be noticed that as the confining pressure is decreased, samples with smaller void ratios experience less volume contractions such that the sample with $e_p = 0.846$ exhibits no volume contraction, but tends to dilate as it approaches the CSL. The onset of volume contraction also occurs at smaller mean effective stress in samples with lower void ratios, indicating that instability occurs at a higher mobilized friction angle. This is consistent with past research indicating that static liquefaction occurs only in loose sands (see e.g. Castro 1969).

4.2 Confining pressure

In order to examine the effect of confining pressure, samples were prepared with the same void ratio of $e_p = 0.972$ and then consolidated to mean normal stresses of 150, 250, and 350 kPa. It is noted that due to difference in consolidation pressures, void ratios of the samples after consolidation were not the same. All samples were then subjected to the same CDS loading of 110 kPa.

Results of these tests shown in Figure 5 indicate that the onset of volume contraction in all the samples occurs at about the same mean normal stress, and therefore, the same mobilized friction angle. The amounts of volume contractions in these samples are also similar. Although in the samples with higher confining pressures, the void ratios during the CDS loading are smaller, the higher confining pressure appears to compensate for the decrease in void ratio, resulting in about the same volume contraction and mobilized friction angle at onset of contraction. These results are consistent with the findings of Anderson and Riemer (1995), who indicated that the potential for collapse (volume contraction) is related to the state parameter of the sample. The state parameter is defined as the difference between the current void ratio and the void ratio at critical state at the same mean effective stress (ie. the vertical distance between the current state and the critical state at the same p in an e–p plot). It may be noticed from Figure 5 that all samples have similar state parameters at consolidation, regardless of their confining pressure.

It may also be noticed from Figure 5 that the volume expansion during the initial stage of the CDS loading is higher in the tests with the higher confining pressures. This is also consistent with the notion that as

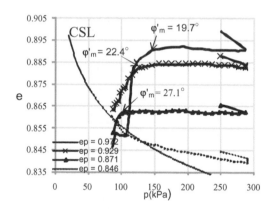

Figure 4. Effet of void ratio on the potential for volume contraction in CDS loading.

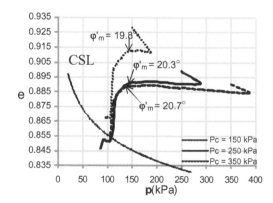

Figure 5. Effet of confining pressure on sand behavior in the CDS loading.

the confining pressure increases, the size of the elastic region that develops after consolidation is larger, resulting in more unloading during the initial part of the CDS loading (Imam et al. 2002).

4.3 Deviatoric stress

Samples prepared at similar void ratios and consolidated to the same confining pressure of 250 kPa were subjected to deviatoric stresses of 94, 110, 153, and 200 kPa during the CDS loading test. Results of these tests shown in Figure 6 indicate that in general, for samples subjected to higher deviatoric stress, as the confining pressure is decreased, both the volume contractions are higher, and they start earlier.

4.4 Anisotropic consolidation

A number of CDS tests were carried out on anisotropically consolidated samples of Firoozkooh sand. Compared to isotropically consolidated samples, such samples are closer to those which occur in the field, and studying their behavior is of more interest.

Figure 7 shows results of CDS tests on isotropically consolidated and anisotropically consolidated samples of the same sand. The anisotropically consolidated sample was subjected to increments of stresses with a constant ratio of major to minor principal stresses of 2.5. The increments were applied in steps unil the minor principal stress reached 250 kPa. The deviatoric stress at the end of consolidation was 375 kPa. The other sample was consolidated isotropically to 250 kPa and then subjected to CDS loading.

As shown in Figure 7, the stress path of the anisotropically consolidated sample is closer to the CSL compared to that of the isotropically consolidated sample. As a result, after a small decrease in mean effective stress, volume contraction is initiated. This behavior indicates that actual soils in the field

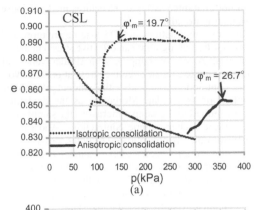

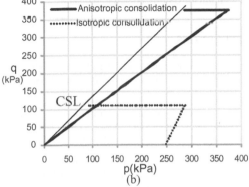

Figure 7. Effect of anisotropic consolidation on sand behavior in CDS loading: (a) Stress path (b) Variation of void ratio with mean normal stress.

are more subject to loss of strengh compared to the soil samples prepared in the laboratory by isotropic consolidation.

On the other hand, due to the higher shear stresses applied during consolidation, the anisotropically consolidated sample reaches a smaller void ratio following consoldation compared to the isotropically consolidated sample, and will therefore experience smaller volume contractions.

5 CONCLUSIONS

Results of constant deviatoric stress (CDS) tests carried out on sample of a uniform, fine grained, subangular sand called the Firoozkooh sand indicated that volume contractions of dry sands subjected to CDS loading initiate at mobilized friction angles substantially smaller than the friction angle at failure. Since such volume contractions can lead to loss of shear strength and instability of slopes, these results indicate that conventional limit equilibrium stability analyses

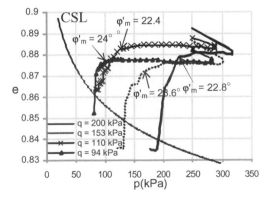

Figure 6. Effet of deviatoric stress on sand behavior in the CDS loading.

can not guarantee saftey of structures made of loose granular materails.

Volume contractions in CDS loading are more and occur earlier (at higher mean effective stress) in looser samples. Also, for samples with similar values of state parameter, mean effective stress at onset of volume contractions, and also the amount of volume contraction is similar. Moreover, as the deviatoric stress applied to the sample increases, the sample will experinece higher volume contractions. And finally, in anisotropically consolidated samples, onset of volume contractions occurs after a smaller decrease in mean normal stress. Such samples, which represent more closely stress state of soils in the field, may be on the verge of volume contraction and collapse at their current anisotropiclly consolidated state.

REFERENCES

Anderson, S.A. & Sitar, N. 1995. Analysis of rainfall-induced debris flows. *Journal of Geotechnical Engineering*, ASCE, 121(7): 544–552.

Anderson, S.A. & Riemer, M.F. 1995. Collapse of Saturated Soil Due to Reduction in Confinement. *Journal of Geotechnical Engineering*, ASCE, 121(2): 216–219.

Casagrande, A. 1936. Characteristics of Cohesionless Soils Affecting the Stability of Slopes and Earth Fills. *Journal of Boston Society of Civil Engineers*, Jan.: 13–32.

Castro, G., Seed, R.B., Keller, T.O. & Seed, H.B. 1992. "Steady State strength analysis of the Lower San Fernando Dam Slide. *Proc. of the 53rd Canadian Geotechnical Conference*, Montreal: 169–176.

Castro, G., 1969. Liquefaction of sand, Harvard Soil Mechanics series, No. 81, Harvard university, Cambridge, MA.

Dawson, R.F., Morgenstern, N.R. & Gu, W.H. 1992. Instability Mechanics initiating Flow Failure in Mountainous Mine Waste Dumps. Contract Report, Phase 1, to Energy, Mines and Resources Canada.

Eckersley, J.D. 1990. Instrumented Laboratory Flowslides. *Geotechnique* 40(3): 489–502.

Imam, S.M.R., Morgenstern, N.R., Robertson, P.K. & Chan, D.H. 2005. A Critical State Constitutive Model for Liquefiable Sand, *Canadian Geotechnical Journal* 42: 830–855.

Imam, S.M.R., Morgenstern, N.R., Robertson, P.K. & Chan, D.H. 2002. Yielding and flow liquefaction of loose sand. *Soils and Foundations* 42(3): 19–31.

Lade, P.V. 1993. Initiation of Static Instability in the Nerlerk berm. *Canadian Geotechnical Journal* 30(6): 895–904.

Morgenstern, N.R. 1994. Observations on the collapse of granular materials. "The Kersten Lecture" 42nd Annual Geotechnical Engineering Conference, Minneapolis, Minnesota.

Sasitharan, S., Robertson, P.K. Sego, D.C. & Morgenstern, N.R. 1994. Collapse behavior of sand, *Canadian Geotechnical Journal* 30(4): 569–577.

Skopek, P. 1994. Collapse behavior of very loose dry sand. Ph.D.thesis, The University of Alberta.

Sladen, J.A., D'Hollander, R.D. & Krahn, J. 1985a. Back analysis of the Nerlerk berm liquefaction slides, *Canadian Geotechnical Journal* 22: 564–578.

Prediction and Simulation Methods for Geohazard Mitigation – Oka, Murakami & Kimoto (eds)
© *2009 Taylor & Francis Group, London, ISBN 978-0-415-80482-0*

Instability and failure in soils subjected to internal erosion

P.-Y. Hicher
Ecole Centrale Nantes, Nantes, France

C.S. Chang
University of Massachusetts, Amherst, USA

ABSTRACT: On a worldwide scale, the mechanisms of soil erosion are the main cause of failure of major hydraulic works. To study these mechanisms, a constitutive modeling method for granular materials based on particle level interactions has been developed using a homogenization technique. The model requires a limited number of parameters, which can easily be determined from conventional laboratory testing. Comparisons between numerical simulations and experimental results have shown that the model can accurately reproduce the overall mechanical behavior of loose and dense granular assemblies. The previous model has been modified so that it can incorporate the physical changes due to internal erosion. Different numerical simulations have been performed in order to study the mechanical properties of eroded soils. The results show that a progressive deformation of the soil volume takes place during erosion. Depending on the loading condition, large strains can be developed up to failure. Special attention is given to constant deviatoric stress paths, which are representative of a change in hydraulic conditions within the slope of an embankment. Numerical results show that an instability condition can be reached along such stress paths when the soil undergoes internal erosion.

1 INTRODUCTION

On a worldwide scale, the mechanisms of soil erosion are the main cause of failure of major hydraulic works (CFGB, 1997). Internal erosion occurs when particles are pulled off by seepage forces and transported downstream. Two fundamental types of internal erosion can be distinguished: piping and suffusion. This paper focuses on the latter which corresponds to the transport of fine particles through the pores of the coarse matrix. The aim of this work is to develop a numerical method capable of predicting the mechanical behavior of soils subjected to suffusion. The approach is based on the use of a homogenization technique in order to derive the stress-strain relationship of the granular assembly from forces and displacements at the particle level. Our basic idea is to view the packing as represented by a set of micro systems which correspond to the contact planes. The overall stress-strain relationship of the packing is obtained from averaging the contact plane behaviors. Along these lines, we recently developed a new stress-strain model which considers inter- particle forces and displacements along a set of contact planes (Chang & Hicher 2005). In this paper, we extend the capability of the model in order to incorporate the structural changes due to internal erosion.

2 STRESS-STRAIN MODEL

The microstructural model developed by Chang & Hicher (2005) is briefly described here. In this model, a soil is viewed as a collection of non-cohesive particles. The deformation of a representative volume of the material is generated by mobilizing particle contacts in various orientations. On each contact plane, an auxiliary local coordinate can be established by means of three orthogonal unit vectors n, s, t. The vector n is outward normal to the contact plane. Vectors s and t are on the contact plane.

2.1 Inter-particle behavior

Elastic Stiffness: The contact stiffness of a contact plane includes normal stiffness, k_n^α, and shear stiffness, k_r^α. The elastic stiffness tensor is defined by

$$f_i^\alpha = k_{ij}^{\alpha e}\delta_j^{\alpha e} \tag{1}$$

which can be related to the contact normal and shear stiffness

$$k_{ij}^{\alpha e} = k_n^\alpha n_i^\alpha n_j^\alpha + k_r^\alpha(s_i^\alpha s_j^\alpha + t_i^\alpha t_j^\alpha) \tag{2}$$

The value of the stiffness for two elastic spheres can be estimated from Hertz-Mindlin's formulation.

For sand grains, a revised form was adopted (Chang et al. 1989), given by

$$k_n = k_{n0} \left(\frac{f_n}{G_g l^2} \right)^n ; \quad k_t = k_{t0} \left(\frac{f_n}{G_g l^2} \right)^n \qquad (3)$$

where G_g is the elastic modulus for the grains, f_n is the contact force in normal direction, l is the branch length between two particles, k_{no}, k_{ro} and n are material constants.

Plastic Yield Function: The yield function is assumed to be of the Mohr-Coulomb type, defined in a contact-force space (e.g. f_n, f_s, f_t)

$$F(f_i, \kappa) = T - f_n \kappa(\Delta^P) = 0 \qquad (4)$$

where $\kappa(\Delta^P)$ is a hardening/softening parameter. The shear force T and the rate of plastic sliding Δ^P are defined as

$$T = \sqrt{f_s^2 + f_t^2} \quad \text{and} \quad \Delta^P = \sqrt{\left(\delta_s^p\right)^2 + \left(\delta_t^p\right)^2} \qquad (5)$$

The hardening function is defined by a hyperbolic curve in $\kappa - \Delta^P$ plane, which involves two material constants: ϕ_p and k_{p0}.

$$\kappa = \frac{k_{p0} \tan \phi_p \, \Delta^P}{|f_n| \tan \phi_p + k_{p0} \Delta^P} \qquad (6)$$

Plastic Flow Rule: plastic sliding often occurs along the tangential direction of the contact plane with an upward or downward movement; thus, shear-induced dilation/contraction takes place. The dilatancy effect can be described by

$$\frac{d\delta_n^p}{d\Delta^P} = \frac{T}{f_n} - \tan \phi_0 \qquad (7)$$

where the material constant ϕ_0 can be considered in most cases to be equal to the inter-particle friction angle ϕ_μ. On the yield surface, under a loading condition, the shear plastic flow is determined by a normality rule applied to the yield function. However, the plastic flow in the direction normal to the contact plane is governed by the stress-dilatancy equation in Eq. (7). Thus, the flow rule is non-associated.

Elasto-plastic Relationship: With the elements discussed above, the incremental force-displacement relationship of the inter-particle contact can be obtained. Including both elastic and plastic behavior, this relationship is given by

$$\dot{f}_i^\alpha = k_{ij}^{\alpha p} \dot{\delta}_j^\alpha \qquad (8)$$

Detailed expression of the elasto-plastic stiffness tensor can be derived in a straightforward manner from yield function and flow rule, and, therefore, is not given here.

2.2 Influence of void ratio on the mobilized friction angle

Resistance against sliding on a contact plane depends on the degree of interlocking by neighboring particles. The resistance can be related to the packing void ratio e by

$$\tan \phi_p = \left(\frac{e_c}{e} \right)^m \tan \phi_\mu \qquad (9)$$

where m is a material constant.

e_c corresponds to the critical void ratio for a given state of stress. For dense packing, e_c/e is greater than 1 and, therefore, the apparent inter-particle friction angle ϕ_p is greater than the internal friction angle ϕ_μ. When the packing structure dilates, the degree of interlocking and the apparent frictional angle is reduced, which results in a strain-softening phenomenon. For loose packing, the apparent frictional angle ϕ_p is smaller than the internal friction angle ϕ_μ and increases during the material contraction.

The critical void ratio e_c is a function of the mean stress applied to the overall assembly and can be written as follows

$$e_c = \Gamma - \lambda \log \left(p' \right) \quad \text{or} \quad e_c = e_{\text{ref}} - \lambda \log \left(\frac{p'}{p_{\text{ref}}} \right) \qquad (10)$$

where Γ and λ are two material constants, p' is the mean effective stress of the packing, and $(e_{\text{ref}}, p_{\text{ref}})$ is a reference point on the critical state line.

2.3 Micro-macro relationship

The stress-strain relationship for an assembly can be determined by integrating the behavior of inter-particle contacts in all orientations. In the integration process, a micro-macro relationship is required. Following the Love-Weber approach, the stress increment can be obtained by the contact forces and branch vectors for all contacts (Christofferson et al. 1981, Rothenburg, L. & Selvadurai, A. P. S. 1981), as follows

$$\dot{\sigma}_{ij} = \frac{1}{V} \sum_{\alpha=1}^{N} \dot{f}_j^\alpha l_i^\alpha \qquad (11)$$

The mean force on the contact plane of each orientation is

$$\dot{f}_j^\alpha = \dot{\sigma}_{ij} A_{ik}^{-1} l_k^\alpha V \qquad (12)$$

where the branch vector l_k^α is defined as the vector joining the centres of two particles, and the fabric tensor is defined as

$$A_{ik} = \sum_{\alpha=1}^{N} l_i^\alpha l_k^\alpha \qquad (13)$$

Using the principle of energy balance, which states that the work done in a representative volume element

326

is equal to the work done on all inter-cluster planes within the element,

$$\sigma_{ij}\dot{u}_{j,i} = \frac{1}{V}\sum_{\alpha=1}^{N}f_j^{\alpha}\dot{\delta}_j^{\alpha} \qquad (14)$$

From Eqs 12 and 14, we obtain the relation between the global strain and inter-particle displacement

$$\dot{u}_{j,i} = A_{ik}^{-1}\sum_{\alpha=1}^{N}\dot{\delta}_j^{\alpha}l_k^{\alpha} \qquad (15)$$

2.4 Stress-strain relationship

From Eqs.15, 18, and 20, the following relationship between stress increment and strain increment can be obtained:

$$\dot{u}_{i,j} = C_{ijmp}\dot{\sigma}_{mp} \qquad (16)$$

$$C_{ijmp} = A_{ik}^{-1}A_{mn}^{-1}V\sum_{\alpha=1}^{N}(k_{jp}^{ep})^{-1}l_k^{\alpha}l_n^{\alpha} \qquad (17)$$

When the contact number N is sufficiently large in an isotropic packing, the summation of the compliance tensor in Eq. 17 and the summation of the fabric tensor in Eq. 13 can be written in integral form, given by

$$C_{ijmp} = A_{ik}^{-1}A_{mn}^{-1}\frac{NV}{2\pi}\int_0^{\pi/2}\int_0^{2\pi}k_{jp}^{ep}(\gamma,\beta)^{-1}$$
$$\times\, l_k(\gamma,\beta)\,l_n(\gamma,\beta)\sin\gamma\,d\gamma\,d\beta$$

$$A_{ik} = \frac{N}{2\pi}\int_0^{\pi/2}\int_0^{2\pi}l_i(\gamma,\beta)\,l_k(\gamma,\beta)\sin\gamma\,d\gamma\,d\beta \qquad (18)$$

The integration of Eq. 18 in a spherical coordinate can be carried out numerically by using Gaussian integration points over the surface of the sphere. We found that the results were more accurate by using a set of fully symmetrical integration points. Observing the performance of different numbers of orientations, we found 37 points to be adequate.

2.5 Model parameters

One can summarize the material parameters as:

- Normalized contact number per unit volume: Nl^3/V
- mean particle size, $2R$
- Inter-particle elastic constants: k_{n0}, k_{t0} and n;
- Inter-particle friction angle: ϕ_μ and m;
- Inter-particle hardening rule: k_{p0} and ϕ_0;
- Critical state for packing: λ and Γ or e_{ref} and p_{ref}

Except for critical state parameters, all the parameters are inter-particles. The inter-particle elastic constant k_{n0} is assumed to be equal to 61000 N/mm. The value

of k_{t0}/k_{n0} is commonly about 0.4, corresponding to a Poisson's ratio of the soil $\nu = 0.2$ and the exponent $n = 0.5$. (Hicher 1996; Hicher & Chang 2006).

Standard values for k_{p0} and ϕ_0 are the following: $k_{p0} = k_{n0}$ and $\phi_0 = \phi_\mu$ (Chang & Hicher 2005). Therefore, for dry or saturated samples, only five parameters have to be obtained from experimental results and these can all be determined from the stress-strain curves obtained from triaxial tests.

The number of contacts per unit volume Nl^3/V changes during the deformation. Using the experimental data by Oda (1977) for three mixtures of spheres, we can approximate the total number of contact per unit volume related to the void ratio by the following expression

$$\frac{N}{V} = \left(\frac{N}{V}\right)_0\frac{(1+e_0)\,e_0}{(1+e)\,e} \qquad (19)$$

where e_0 is the initial void ratio of the granular assembly. This equation is used to describe the evolution of the contact number per unit volume. The initial contact number per unit volume can be obtained by matching the predicted and experimentally measured elastic modulus for specimens with different void ratios (Hicher & Chang 2005, 2006).

3 NUMERICAL SIMULATIONS

3.1 Soil characteristics

Internal erosion in soil masses occurs when the smaller particles are pulled off by seepage forces and transported downstream. This phenomenon happens in well graded materials having a significant percentage of fine particles. In this study, we selected a soil made of non-cohesive particles, typically a silt-sand-gravel mixture, with a grain size distribution corresponding to $c_u = d_{60}/d_{10} = 10$. Then, the material parameters have to be assumed, for which we used the correlations proposed by Biarez & Hicher (1994) relating the physical properties of a granular assembly to its mechanical properties. The maximum and minimum void ratios of a widely graded granular material are typically $e_{max} = 0.6$ and $e_{min} = 0.25$. We can, then, derive the values of the two parameters corresponding to the position of the critical state in the $e-p'$ plane: $\lambda = 0.05$ and $p_{ref} = 0.01$ MPa for $e_{ref} = e_{max} = 0.6$. The friction angle at critical state ϕ_μ is considered equal to $30°$. In Eq. 16, the value of m is taken equal to 1. The

Table 1. Model parameters for the selected soil.

e_{ref}	p_{ref}(Mpa)	λ	$\phi_\mu(°)$	$\phi_0(°)$	m
0.6	0.01	0.05	30	30	1

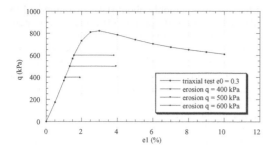

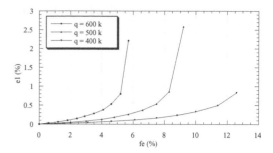

Figure 2. Evolution of axial strain during erosion tests at constant deviatoric stress ratios.

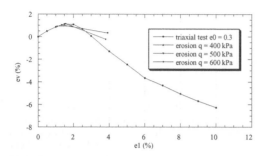

Figure 1. Monotonic triaxial test on the selected soil and straining during erosion process.

set of parameters for the selected soil is presented in Table 1.

We considered for this study a well compacted material with an initial void ratio $e_0 = 0.3$, corresponding to a relative density $Dr = 85\%$. Figure 1 shows the results of the simulation of a triaxial test on a specimen subjected to an initial isotropic confining stress equal to 200 kPa.

3.2 Triaxial behavior of eroded soil

The soil selected above is now subjected to internal suffusion. For this purpose we define what we call the *eroded fraction fe* as follows

$$fe = Wf/Ws \tag{20}$$

where Wf is the weight of the eroded particles and Ws is the initial total solid weight per unit volume. When suffusion takes place, the eroded fraction *fe* increases progressively, creating a change in the void ratio. This change is taken into account in the model by using Eq. 9. If the material is subjected to a constant state of external stresses, the evolution of the sliding resistance introduced by Eq. 9 creates a disequilibrium at each contact point leading to local sliding. All the local displacements are then integrated to produce the macroscopic deformation of the soil specimen. Several numerical tests have been conducted. They consisted in shearing the specimen up to a given value of the deviatoric stress and then progressively eroding the

specimen while the external stresses are kept constant. Figure 1 shows the evolution of the axial deformation during this process. One can see in Figure 2 that the erosion induced deformations develop more rapidly when the stress ratio is higher. At a given state of erosion, corresponding to a given value of the eroded fraction *fe*, the axial strain starts to increase very fast leading the specimen to failure. The evolution of the volumetric strain depends also on the stress level. For high stress levels, the volume increases, corresponding to a dilatant behavior of the eroded soil, while for lower stress levels, a small contractancy is obtained (Figure 1).

3.3 Instability and diffuse failure

Instability in granular materials has become during these last years an important topic of study. Several authors (see for example Darve & Chau 1987, Darve et al. 1995, Nova 1994) have demonstrated that granular materials could exhibit domains of instability inside the plastic limit. This type of failure mode, called diffuse failure, is usually found in loose and medium dense sand. It is far different from strain softening in dense sand which corresponds to strain localization in shear bands. This mode of failure can lead to catastrophic collapse of earth slopes. We will study here the possibility of an unstable regime appearing along a specific stress path: a constant-q test in triaxial condition. This stress path can simulate the loading condition of a soil element within a slope when a progressive increase in pore pressure occurs. The occurrence of instability can be linked to Hill's sufficient condition of stability (Hill 1956), which states that a material, progressing from one stress state to another, is stable if the second-order work is strictly positive

$$d^2W = d\sigma_{ij}\,d\varepsilon_{ij} > 0 \tag{21}$$

Thus, according to Hill's condition, whether a material is stable or not depends not only on the current stress state but also on the direction of the stress increment. For constant-q tests ($dq = 0$), according to Eq. 1, the second-order work is reduced to $d^2W = dp\,d\varepsilon_v$. Since the mean stress is progressively decreased (i.e., $dp < 0$), the second-order work becomes negative if and only if $d\varepsilon_v \leq 0$ (i.e., the volume contracts). We will examine the possibility of occurrence of this condition of instability for specimens of intact and eroded soils.

3.4 Failure condition in intact soil

Figure 3 shows the stress paths followed by the soil specimens during constant-q testing. The specimen is first isotropically consolidated up to 200 kPa. Then, a deviatoric stress q is applied up to a given value. During the third and last stage, the deviatoric stress q is maintained constant while the mean effective stress p' is progressively decreased. Several values of q have been applied. The results are shown in Figure 4. One can see that the decrease of the mean effective stress is accompanied by a decrease of the volumetric strain, which corresponds, according to Eq. 21, to a positive value of the second-order work. This condition persists all along the constant-q stress path. At the end of the test, the axial strain starts to increase rapidly, leading to the failure of the soil specimen. This failure condition is obtained when the stress state reaches the plastic limit, as one can see in Figure 3. We obtain in this case a classic mode of failure in dense sand, above the critical state line.

3.5 Failure condition in eroded soil

The specimens have now been eroded in the same condition as previously. The process is stopped before reaching the failure during the erosion process. Then, the specimens were subjected to the same constant-q loading as shown in Figure 5. The results are presented in Figure 6. For the test with $q = 200$ kPa, the initial part of the $p-\varepsilon_v$ curve shows that, as the mean stress p decreases, the volume increases. This trend continues up to a certain point where the volume starts to decrease. According to Eq. 21, this point corresponds to the onset of instability, which is obtained

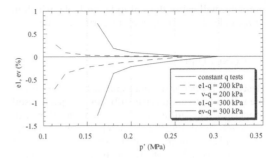

Figure 4. Axial strain and volumetric change during constant-q tests on intact dense specimens.

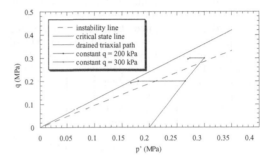

Figure 5. Stress paths during constant-q tests on eroded specimens.

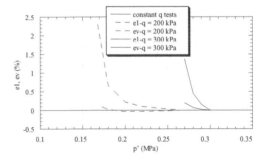

Figure 6. Axial strain and volumetric change during constant-q tests on eroded specimens.

at a stress level much lower than the critical state failure line. When the instability condition is reached, the axial strain starts to increase rapidly, indicating that the material is leading to a diffuse failure state. The test corresponding to the higher deviatoric stress level represents an interesting example, since the erosion process causes it to be in a domain where it is unstable as soon as the mean effective stress starts to decrease. Therefore, any perturbation in the hydraulic conditions, leading to the slightest increase in pore

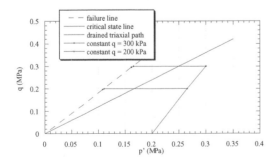

Figure 3. Stress paths during constant-q tests on intact dense specimens.

pressure, will cause the soil to be unstable, which in turn can lead to a collapse of the earth structure.

4 CONCLUSION

Based on a microstructural approach, we examined the influence of an internal suffusion on specific soil properties. The suffusion process is modeled by progressively increasing the granular assembly void ratio. This leads to a decrease of the sliding resistance of each inter-particle contact, which creates disequilibrium between the external applied loading and the internal contact forces. As a consequence, local sliding occurs which leads to macroscopic deformations of the soil specimen. The amplitude of the suffusion induced deformations depends on the amount of eroded mass, as well as on the applied stress level. At elevated stress levels, large deformations can develop and lead to soil failure. Another type of failure called diffuse failure can occur in eroded soil masses. Numerical simulations of constant-q loading, corresponding to a pore pressure increase within the soil mass, show that the condition of instability can be reached when the eroded fraction becomes high enough. These numerical results appear to agree with observations made on embankment dams which suffered internal erosion, and in particular with the description of their modes of failure.

REFERENCES

Biarez, J. & Hicher, P.-Y. 1994. Elementary Mechanics of Soil Behaviour, Balkema, p. 208.

CFGB 1997. Internal erosion: typology, detection, repair. *Barrages et Réservoirs*, n°6. Comité français des grands barrages.

Chang, C.S., Sundaram, S.S. & Misra, A. 1989. Initial Moduli of Particulate Mass with Frictional Contacts. *Int. J. for Numerical & Analytical Methods in Geomechanics*, John Wieley & Sons 13(6): 626–641.

Chang, C.S. & Hicher, P.-Y. 2005. An elastoplastic model for granular materials with microstructural consideration. *Int. J. of Solids and Structures* 42(14): 4258–4277.

Christofferson, J., Mehrabadi, M.M. & Nemat-Nassar, S. 1981. A micromechanical description on granular material behavior. ASME *Journal of Applied Mechanics* 48: 339–344.

Darve, F. & Chau, B. 1987. Constitutive instabilities in incrementally non linear modeling. in *Constitutive Laws for Engineering Materials*, C.S. Desai ed.: 301–310.

Darve, F. Flavigny, E. & Meghachou, M. 1995. Constitutive modelling and instabilities of soil behaviour. *Comp. and Geotech.*, 17(2): 203–224.

Hicher, P.-Y. 1996. Elastic properties of soils. *Journal of Geotechnical Engineering*, ASCE 122(8): 641–648.

Hicher, P.-Y. & Chang, C. 2006. An anisotropic non linear elastic model for particulate materials. *J. Geotechnical and Environmental Engineering*, ASCE 132, n°8.

Hill, R. 1956. A general theory of uniqueness and stability in elasto-plastic solids. *J. Mechanics and Physics of Solids* 6: 236–249.

Nova R. 1994. Controllability of the incremental response of soil specimens subjected to arbitrary loading programs. *J. Mechanical Behaviour of Materials* 5(2): 193–201.

Oda, M. 1977. Co-ordination Number and Its Relation to Shear Strength of Granular Material. *Soils and Foundations* 17(2): 29–42.

Rothenburg, L. & Selvadurai, A.P.S. 1981. Micromechanical definitions of the Cauchy stress tensor for particular media," *Mechanics of structured media*, (Selvadurai, A.P.S., eds.): 469–486. Amsterdam: Elsevier.

Prediction and Simulation Methods for Geohazard Mitigation – Oka, Murakami & Kimoto (eds)
© 2009 Taylor & Francis Group, London, ISBN 978-0-415-80482-0

Shear failure behavior of compacted bentonite

T. Kodaka & Y. Teramoto
Meijo University, Nagoya, Japan

ABSTRACT: Compacted bentonite is planned for use as a buffer material in the geological disposal of high-level radioactive waste. In this research, a series of constant volume direct shear tests was performed in order to study the failure behavior of compacted bentonite. At the same time, shear bands were observed during direct shearing using a CCD microscope and a PIV image analysis. In addition, the insides of specimens were observed with a micro-focus X-ray CT scanner. The over-consolidated unsaturated compacted bentonite specimen showed brittle failure behavior from an early stage of shearing, and a shear band with a number of cracks developed during shearing. On the other hand, according to the results of the X-ray CT observation, no shear bands or cracks were generated in the saturated specimen.

1 INTRODUCTION

The geological disposal of high-level radioactive waste (HLW) is considered to be a global standard disposal method. Japanese government policy for such geological disposal states that HLW is to be buried very deep in the ground, i.e., more than 300 m, using a multi-barrier system. Compacted bentonite is one of the promising buffer materials for the artificial barrier, which is expected to be able to hold the overpack of HLW and to prevent the leakage of nuclide from the overpack over a period of ten to one hundred thousand years.

Figure 1 illustrates two possible and typical conditions for the multi-barrier system over a long period from the closing of the HLW disposal site, in which the conditions refer to Komine's arguments (e.g., Komine and Ogata 1999). Since the compacted bentonite used for the buffer material is produced from unsaturated powder bentonite, the buffer material is thought to remain in an unsaturated condition until the reflooding of the disposal site, *see* Figure 1(a). It will take tens of years for the buffer material to gradually become saturated by the reflooding of the HLW disposal site. Bentonite has very high swelling and expanding properties; therefore, a high level of swelling pressure will develop in the buffer material during the saturation process of the compacted bentonite. After saturation, the buffer material will be acted upon by the supposed external forces caused by the creep deformation of the surrounding rock mass and the corrosion expansion of the overpack over the long period, *see* Figure 1(b).

It is important, therefore, to study not only the permeability, but also the deformation and the failure characteristics of compacted bentonite under both unsaturated and saturated conditions. Up to now, many series of triaxial compression tests have been performed by the Japan Nuclear Cycle Development

Institute (JNC) (e.g., Namikawa and Sugano 1997; Takaji and Suzuki 1999) in order to investigate the deformation and strength characteristics of saturated compacted bentonite. JNC was reorganized to the Japan Atomic Energy Agency (JAEA) in 2005. On the other hand, almost no triaxial compression tests have been conducted on unsaturated compacted bentonite because of the difficulty involved in these tests. Therefore, the deformation and strength characteristics of unsaturated compacted bentonite have not yet been clarified. An unsaturated condition will be maintained at the HLW disposal site for tens of years. Consequently, it is important to grasp the shear failure behavior of unsaturated compacted bentonite from the viewpoint of the site design.

In the present study, a series of direct shear tests on both unsaturated and saturated compacted bentonite are performed. Direct shear tests under constant volume conditions constitute a powerful tool for grasping the shear properties of unsaturated geomaterials, because constant volume tests can produce similar effective stress paths even for unsaturated geomaterials. In addition, direct shear tests can easily show shear bands during shearing. Shear band formation and development in buffer materials is assumed to be caused by the faulting of earthquakes. It seems that the performance of the buffer material will deteriorate due to the shear band development; therefore, the shear band properties in the compacted bentonite should be investigated in detail. In this study, a highly confining pressure direct shear testing apparatus is newly developed. Shear bands were observed during direct shearing using the image processing system, which consisted of a microscope and a PIV digital image analysis. Furthermore, the insides of the sheared specimens were observed with a micro-focus X-ray CT scanner.

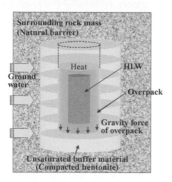

(a) Period from the closing of the disposal site until the reflooding

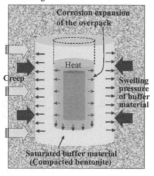

(b) Period after the reflooding of the disposal site

Figure 1.   Supposed conditions for the multi-barrier system at the HLW disposal site over a long period.

2   TESTING PROCEDURE

2.1   *Testing apparatus*

Photo 1 shows a highly confining pressure direct shear apparatus, which has been newly developed for this research. Mega-torque motors are adopted for vertical and lateral loads with capacities of 10 and 8 MPa, respectively. The displacement rate can be controlled in the range of 0.001 to 1 mm/min. The mega-torque motors can be stopped at any position, so that ideal constant volume tests can be carried out. A load cell for vertical stress is placed in the bottom of the shear box.

2.2   *Image analysis system for the observation of local shear behavior*

One of the advantages of direct shear tests is that the local shear behavior can be easily observed by generating a shear band by force. In this study, rectangular specimens are adopted. The transverse sections are 5 cm squares, 4 cm in height. A clear acrylic plate, 10 mm in thickness, is placed in front of each specimen to observe the shear behavior, as shown in

Photo 2. The image processing system, developed for the observation, consists of a CCD microscope, *see* Photo 3, and a PIV (Particle Image Velocimetry) image analysis method. An example of the digital image is illustrated in Photo 4; it corresponds to the red box in Photo 2. Since the prescribed silica sand is mixed with bentonite, as will be discussed later, there is a suitable pattern for the PIV image analysis in the digital image.

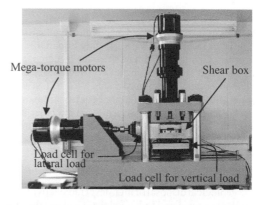

photo 1.   High pressure direct shear test apparatus.

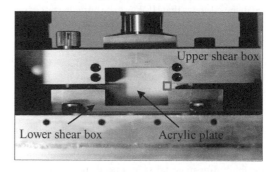

photo 2.   Shear box for observing the shear bands which developed in the rectangular specimens.

photo 3.   Observation of the shear bands using the CCD microscope.

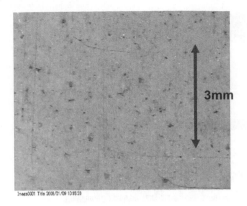

photo 4. An example of a digital image by the CCD microscope.

2.3 Testing specimens for the unsaturated compacted bentonite

The specimens used in this study were prepared by the following method: Powder bentonite (70% wt) and silica sand (30% wt) were mixed well. The mixed sample was put in the direct shear box and then compressed by vertical compression. The mass of the sample was determined from the dry density of each compressed complete specimen.

Two types of specimens were used in this study. One is a normally consolidated specimen and the other is an over-consolidated specimen. The over-consolidated specimen was unloaded to 0 MPa after the vertical compression during the specimen preparation. The normally consolidated specimen was slightly unloaded by means of spacing between the upper the and lower shear boxes just before shearing.

3 RESULTS OF DIRECT SHEAR TESTS

3.1 Normally consolidated specimen

Dry densities of 1.4, 1.5, and 1.6 Mg/m3 were adopted for the normally consolidated specimen for the tests in this section. The shearing was conducted up to a lateral displacement of 6 mm at a rate of 0.4 mm/min under constant volume conditions.

Figure 2(a) shows the relations between shear stress and lateral displacement. Figure 2(b) shows the stress paths. The shear stress rapidly increases just after initiating the shear for all specimens. The larger density specimen has the higher shear stress. A decrease in shear stress appears after the peak strength in these cases. The stress paths obtained from the constant volume tests, shown in Figure 2(a), are similar to the effective stress paths of the saturated geomaterials. In the case of the normally consolidated specimen,

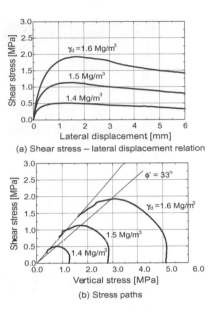

Figure 2. Shear test results for the normally consolidated (NC) specimen of unsaturated bentonite with various dry densities.

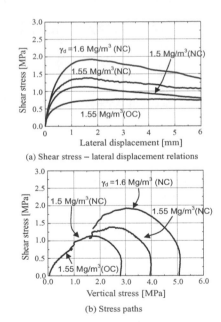

Figure 3. Shear test results for the over-consolidated (OC) and the normally consolidated (NC) specimens.

a decrease in vertical stress appears from an early stage of the shearing, in which this tendency is similar to that for normally consolidated clay. After the peak shear stress, apparent strain softening can be observed. The

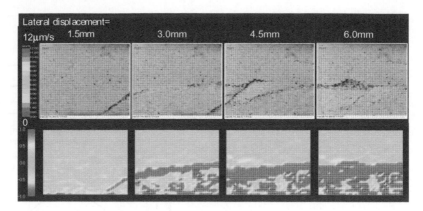

Figure 4. Observation of shear bands in the normally consolidated specimen during shearing; displacement velocity distribution (upper) and shear strain distribution (lower).

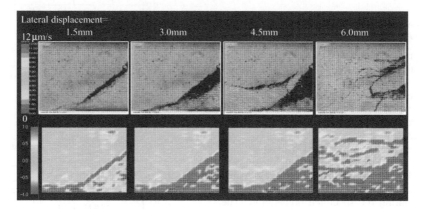

Figure 5. Observation of the shear band in the over-consolidated specimen during shearing; displacement velocity distribution (upper) and shear strain distribution (lower).

softening occurs with a plastic volumetric compression, which suggests that the normally consolidated unsaturated specimen consists of highly structural soil. The angle of shear resistance, evaluated from the peak strength, is 33 degrees. This is about two times the 17 degrees obtained from the triaxial compression tests for the saturated specimen.

3.2 Over-consolidated specimen

A dry density of 1.55 Mg/m^3 was adopted for the over-consolidated specimen for the tests in this section. Figure 3(a) shows the relations between shear stress and lateral displacement. Figure 3(b) shows the stress paths. For comparison, the results for the normally consolidated specimen of 1.5, 1.55, and 1.6 Mg/m^3 are also shown in these figures. The shear stress monotonically increases and the vertical stress increases from zero. In the case of the over-consolidated specimen,

the shear stress gradually increases with an increase in the vertical stress, which is caused by the constraint of the volume expansion due to the dilatancy during shearing.

Before the reflooding of the HLW disposal site, the buffer material will be under unsaturated and over-consolidated conditions. Therefore, it should be noted that the buffer material does not have such a large shear resistance and that the shear behavior is accompanied by plastic expansion.

4 OBSERVATION OF SHEAR BANDS

Figures 4 and 5 show the shear bands observed in the cases of 1.55 Mg/m^3 for the normally consolidated and the over-consolidated specimens, respectively. Displacement velocity vectors and shear strain distributions are shown in upper and lower of each figure, respectively. The shear strain distributions are calculated from the displacement increments at fixed points obtained by the PIV analysis. Consequently, the shear strain does not correspond to the real total strain.

An inclined shear band can be seen for the displacement of 1.5 mm in each specimen. In the case of the normally consolidated specimen, the shear band gradually develops horizontally and is accompanied by small cracks. Figure 4 indicates that a concentration of horizontal thin shear strain appears in this case. In the case of the over-consolidated specimen, on the other hand, the shear band was generated from an early stage of shearing, and then shear failure developed over a relatively wide region with the collapse of massive fragments.

5 OBSERVATION OF UNSATURATED COMPACTED BENTONITE BY X-RAY CT

From the results of the image analysis shown in the previous chapter, it is suggested that there is volumetric expansion even in the shear band in the normally consolidated specimen. The most important task of the buffer material is to prevent the leakage of nuclide from the overpack, since the generation of cracks and expansion seems to cause fatal damage. In this chapter, the insides of the specimens were observed using a micro focus X-ray CT. The apparatus is KYOTO-GEO μXCT (TOSHIBA, TOSCANER-32250μHDK) at Kyoto University. The specimens were vacuum-packed just after the shear tests to maintain a constant water content and to be confined by negative pressure. Then, they were taken to Kyoto University so the density distributions in the specimens could be observed by a micro-focus X-ray CT scanner.

Figure 6 shows the CT image for the normally consolidated specimen. It can be seen that several shear bands developed from the edge of the shear box. With an increase in displacement, stress is released around

the edge. Consequently, shear bands with small cracks were generated around the edge, as observed by the microscope.

6 SHEAR BEHAVIOR OF SATURATED COMPACTED BENTONITE

For a comparison with the unsaturated specimens, a series of direct shear tests using the saturated specimen was performed. The compacted bentonite was saturated maintaining constant volume conditions for two months in a rigid mold. The large expansive pressure

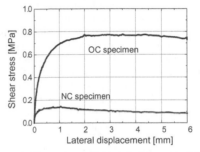

(a) Shear stress – lateral displacement relations

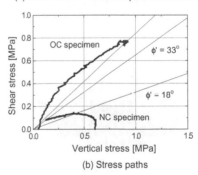

(b) Stress paths

Figure 7. Shear test results for the normally consolidated (NC) saturated specimen and the over-consolidated (OC) unsaturated specimen.

Tube voltage: 120 kV, Tube current: 90 μA, Number of view: 2400, Number of total sheets: 20, Size of unit boxel: 36.7 μm

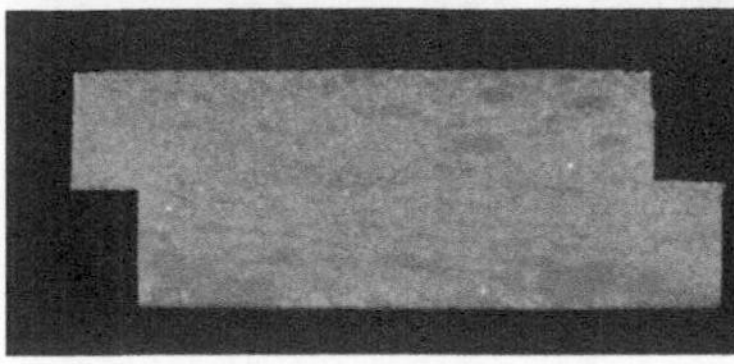

Figure 8. X-ray CT image of the normally consolidated specimen of saturated bentonite.

Tube voltage: 12 kV, Tube current: 90 μA, Number of view: 1200, Number of total sheets: 10, Size of unit boxel: 36.7 μm

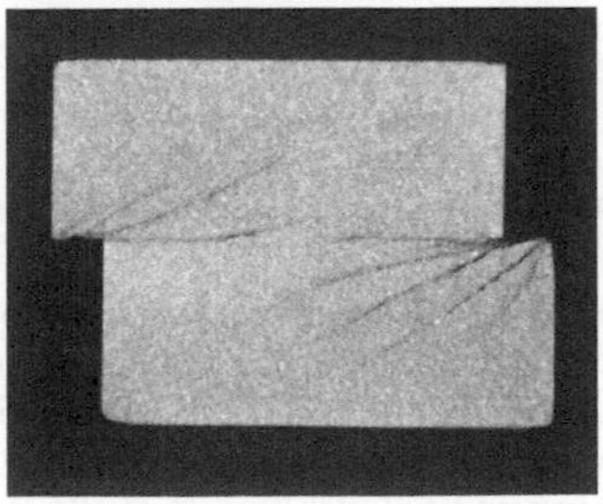

Figure 6. X-ray CT image of the normally consolidated specimen of unsaturated bentonite.

was measured during the saturation process. For the direct shear tests using the saturated specimen, the measured expansive pressure was adopted as the initial vertical stress for the normally consolidated specimen.

Figure 7 shows the results of the direct shear tests for the saturated specimen. The shear stress of the saturated specimen rapidly decreased with the saturation. The angle of shear resistance is evaluated to be 18 degrees from the normally consolidated saturated specimen. This angle is almost the same as that seen in the triaxial tests of previous research works.

The X-ray CT image of the normally consolidated saturated specimen after the shear tests is shown in Figure 8. No shear bands or cracks are seen inside the specimen even though a 6 mm gap exists in the center of the specimen. Then, a higher resolution X-ray CT scan was done by cutting a small cylindrical specimen, as shown ins Figures 9 and 10. The CT image of the transversal section of the cylindrical specimen is shown in Figure 11. The vertical lines seen in the upper and the lower parts are just marks left by the cutter knife. The mixed silica sand, whose maximum diameter is about 0.4 mm, can be seen as a lot of white dots. Even though a high resolution observation was carried out, no change in density can be seen. This result suggests that the properties of the buffer material will not deteriorate due to the development of a shear band even if the buffer material is saturated by the reflooding of the disposal site.

Figure 9. Cutting a cylindrical specimen for the high resolution X-ray CT.

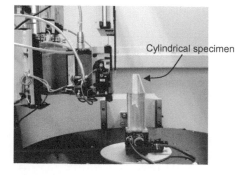

Figure 10. Setting the cylindrical specimen on the micro-focus X-ray CT apparatus.

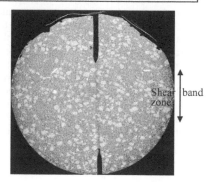

Figure 11. X-ray CT image of the saturated cylindrical specimen.

7 CONCLUSIONS

In order to investigate the deformation and strength characteristics of compacted bentonite, direct shear tests and the observation of shear bands were carried out. It was found that unsaturated compacted bentonite should be evaluated before the reflooding of a HLW disposal site in order to grasp the expected performance. In particular, since it is possible for shear bands to develop in the buffer material due to faulting, the properties of the shear bands in unsaturated compacted bentonite should be clarified.

The results of direct shear tests showed that the shear strength of the unsaturated specimens is relatively large. However, the shear strength rapidly decreases with saturation. Using a microscope and an image analysis method, the shear band generation and development could be directly observed. The shear bands in the unsaturated specimens consist of an expansive region and a large number of small cracks. From the evaluation by the X-ray CT image of the unsaturated specimens after shearing, several low-density bands were clearly observed. On the other hand, no shear bands or cracks were seen in the saturated specimen.

ACKNOWLEDGEMENTS

The authors wish to express their sincere appreciation to Professor Ohnishi of Kyoto University and Professor Komine of Ibaragi University for their variable pieces of advice, and to Professor Oka, Dr. Higo, and Mr. Sanagawa of Kyoto University for their great support regarding the experimental work with the microfocus X-ray CT. Financial support for this study has been provided by the Radioactive Waste Management Funding and Research Center, Japan.

REFERENCES

Komine, K. & Ogata, N, 1999. Experimental study on swelling characteristics of sand-bentonite mixture for nuclear waste disposal, *Soils and Foundations*, 39(2), 83–97.

Namikawa, T. & Sugano, T. 1997. Shear properties of buffer material 1, Technical Report of Power Reacter and Nuclear Fuel Development Corporation, PNC TN8410 97-074 (in Japanese).

Takaji, K. & Suzuki, H. 1999. Static mechanical properties of buffer material, Technical Report of Japan Nuclear Cycle Development Institute, JNC TN8400 99-041 (in Japanese).

Prediction and Simulation Methods for Geohazard Mitigation – Oka, Murakami & Kimoto (eds)
© *2009 Taylor & Francis Group, London, ISBN 978-0-415-80482-0*

Stress-strain and water retention characteristics of micro-porous ceramic particles made with burning sludge

K. Kawai, A. Iizuka & S. Kanazawa
Kobe University, Kobe, Japan

A. Fukuda
Kankyo-souken Co. Ltd., Tokyo, Japan

S. Tachibana
Saitama University, Saitama, Japan

S. Ohno
Kajima Co. Ltd., Tokyo, Japan

ABSTRACT: When troublesome sludge discharged from industries and purification plants are burnt by 1300 degree, they change to innocuous micro-porous ceramic materials. The authors are seeking for practical use of such micro-porous ceramic materials in the geotechnical engineering field, particularly for geohazard mitigation. Then, in order to grasp fundamental mechanical properties of the micro-porous ceramic particles, a series of laboratory tests, that is, triaxial compressive shear tests, permeability tests and water retention tests, were performed. It is found that the stress-strain characteristics and permeability of the micro-porous ceramic particles are very similar to those of standard sandy materials but they have very high water retention capability. Next, applicability of the unsaturated elasto-plastic constitutive model proposed by Ohno et al. (2007) to the mechanical behavior of micro-porous ceramic particles was examined in this paper.

1   INTRODUCTION

Paper sludge (PS) discharged from paper factories and mud deposits from purification plants are very troublesome materials from the environmental view point. They settle around the mouth of rivers and coastal areas and it is said that they cause serious damage for ecological system there. These sludge materials with a small amount of clay are burned by 1300 degree and then they change to stable and innocuous micro-porous ceramic particles. Then, the authors are seeking for recycled use of such micro-porous ceramic particles in the geotechnical engineering field, particularly for geohazard mitigation. In this paper, for such a purpose in the authors' mind, fundamental mechanical properties of the micro-porous ceramic particles are examined through a series of laboratory tests, that is, triaxial compressive shear tests, permeability test and water retention test are performed.

2   MATERIALS

Materials used in this study are burnt mixture of PS and clay and burnt sludge that settled in the bottom of clean water reservoirs in the purification plant, see Photo 1. Herein, we shall call 'white' sample and 'red' sample, respectively. According to SEM observation, see Photo 2, they have innumerable voids in the diameter of 5 to 50 μm . And the specific gravity of particle is 2.58 of 'white' sample and 2.61 of 'red' sample. The maximum/minimum dry density is 1.45/1.14 g/cm³ of 'white' sample and 0.98/0.75 g/cm³ of 'red' sample, respectively. Figure 1 indicates the particle size distribution. It is found that both samples consist of particles of which size widely distributes, that is, according

(a)   (b)

Photo 1.   Materials used in this study, (a) burnt mixture of PS and clay, 'white' sample, (b) burnt sludge from purification plant, 'red' sample.

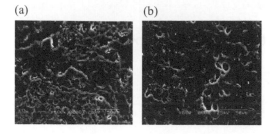

Photo 2. SEM pictures, (a) 'white' sample, (b) 'red' sample.

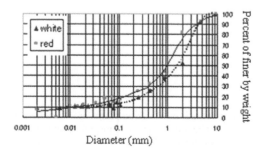

Figure 1. Particle size distribution curves.

to the engineering particle size classification, clay of 10%, silt of 9%, sand of 57% and gravel of 24% for the 'white' sample and clay of 11%, silt of 3%, sand of 44% and gravel of 42% for the 'red' sample.

3 STRESS-STRAIN CHARACTERISTCS AND PERMEABILITY

A series of triaxial shear tests (CD test) is performed to examine the stress-strain characteristics of the materials. Figure 2 shows the test apparatus used in experiment. Dense (Dr = 80%) and loose (Dr = 30%) specimens are prepared for both 'white' and 'red' samples. They were set in the triaxial apparatus and were permeated by distilled water for 24 hours to raise the degree of saturation. But, B value resulted in staying around 0.7 to 0.8, because of existence of many micro voids in the particles which is thought to prevent saturating. And, the specimens were sheared under the drain condition with various confining pressures, 100, 200 and 300 kPa, at the rate of 0.1 mm/min. Figures 3 and 4 show the principal stress difference (shear stress) and the shear strain relations and shear stress and volumetric strain relations for the 'white' sample. Likewise, Figures 5 and 6 are for the 'red' sample. Figures 7 to 10 indicate plots of the principal stress difference, q, and the effective mean stress, p', at the peaks of principal stress difference and the volumetric strain for

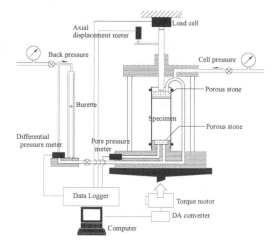

Figure 2. Triaxial test apparatus.

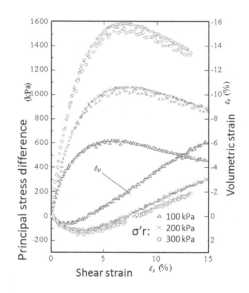

Figure 3. Stress and strain relations (Dr = 80%, 'white' sample).

the 'white' and 'red' samples, respectively. According to the figures, the critical state theory would be applicable to the samples used in experiment.

Since particles of specimens have many micro voids, they are easily crushable. Therefore, it is thought that particle-crushing occurs during shearing.

Before shear tests, the permeability test is carried out for each specimen in the triaxial test apparatus. The permeability coefficients were found to be 2.7×10^{-4} cm/sec for the dense 'white' sample (Dr = 80%), 6.9×10^{-4} cm/sec for the loose 'white' sample (Dr = 30%), 9.1×10^{-4} cm/sec for the dense 'red' sample (Dr = 80%) and $1.2 \times$

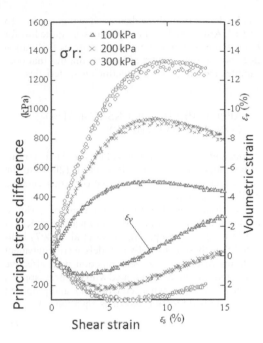

Figure 4.   Stress and strain relations (Dr = 30%, 'white' sample).

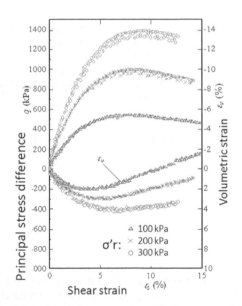

Figure 5.   Stress and strain relations (Dr = 80%, 'red'sample).

10^{-3} cm/sec for the loose 'red' sample (Dr = 30%), respectively. They have relatively high permeability, although it is very difficult to completely saturate them.

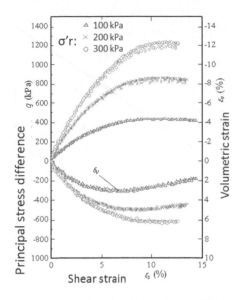

Figure 6.   Stress and strain relations (Dr = 30%, 'red' sample).

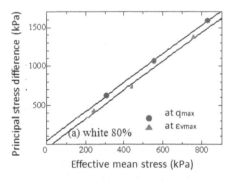

Figure 7.   Stress state at failure (Dr = 80%, 'white' sample).

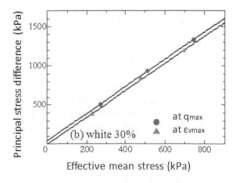

Figure 8.   Stress state at failure (Dr = 30%, 'white' sample).

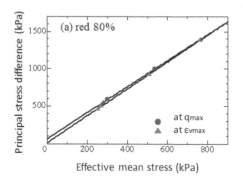

Figure 9. Stress state at failure (Dr = 80%, 'red' sample).

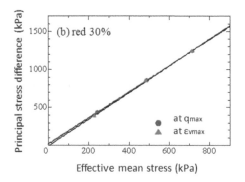

Figure 10. Stress state at failure (Dr = 30%, 'red' sample).

4 WATER RETENTION CHARACTERISTCS

In order to examine the water retention characteristics, a model test was performed. Figure 11 shows the model test equipment to measure the water retention characteristics of the materials. The material was spread over in a mould and compacted to the same dry density with the dense samples for the triaxial shear test in the mould, of which inner diameter and height are 20 cm respectively, and then such four moulds were piled up as shown in Figure 11. ADR censor and tensiometer (water pressure gauge) are equipped in each mould to measure the water content and the pore water pressure. In order to saturate the specimen, distilled water is supplied from the bottom up to the top surface of the specimen. After the water level is kept at the top surface of the sample for several hours, the water is slowly drained with measuring the water content and the pore water pressure and finally the water level is lowered from −3 m from the bottom of the equipment.

Obtained test results are summarized in Figures 12 to 15. The change of volume water content with time and the change of pore water pressure with time are shown in these figures. The water retention curves

can be drawn from these data as shown in Figures 16 and 17. Although the air entry values are quite low for both 'white' and 'red' samples, the degree of saturation does not go down under 50%. This implies that both samples have high water retention characteristics.

5 APPLICABILITY OF UNSATURATED CONSTITUTIVE THEORY

Applicability of the unsaturated elasto-plastic constitutive model proposed by Ohno et al. (2007) to the stress and strain behavior of micro-porous ceramic particles is examined.

Constitutive models for unsaturated soil (Alonso et al. 1990, Kohgo et al. 1993, and Karube & Kawai 2001) have been proposed. Ohno et al. (2007) indicated that these constitutive models were equivalent in terms of expressing yield function with effective stress and parameters associated with stiffness, applied effective stress, shown as equation (1), and proposed a general yield function shown as equation (4).

$$\sigma' = \sigma^{net} + p_s 1 \qquad (1)$$

$$\sigma^{net} = \sigma - p_a 1, \quad p_s = S_{re} s \qquad (2)$$

$$s = p_a - p_w, \quad S_{re} = \frac{S_r - S_{rc}}{1 - S_{rc}} \qquad (3)$$

In the above equations, σ' is the effective stress tensor; σ^{net} is the net stress tensor; 1 is the second rank unit tensor; σ: is the total stress tensor; s is suction, p_s is the suction stress; p_a is pore-air pressure; p_w is the pore-water stress, S_r is the degree of saturation; S_{re} is the effective degree of saturation; and S_{rc} is the degree of saturation at $s \rightarrow \infty$.

$$f\left(\sigma', \zeta, \varepsilon_v^p\right) = MD \ln \frac{p'}{\zeta p'_{sat}} + D\frac{q}{p'} - \varepsilon_v^p = 0 \qquad (4)$$

$$p' = \frac{1}{3}\sigma' : 1, \quad q = \sqrt{\frac{3}{2}\mathbf{s} : \mathbf{s}}, \quad \mathbf{s} = \sigma' - p'1 = \mathbf{A} : \sigma',$$

$$\mathbf{A} = \mathbf{I} - \frac{1}{3}1 \otimes 1 \qquad (5)$$

where M is q/p' at the critical state; D is the dilatancy coefficient; $\mathbf{I}$ is the four rank unit tensor; and : and $\otimes$ are the inner and outer product operators, respectively. Increase of yield stress due to desaturation is expressed as the product of yield stress in the saturated state, p'_{sat}, and a parameter contributing to hardening, ζ. Ohno et al. (2007) introduced this parameter, ζ, shown as

$$\zeta = \exp\left[(1 - S_{re})^n \ln a\right] \qquad (6)$$

Here, equation (4) corresponds to the Cam-Clay Model for the saturated state ($S_{re} = 1$ and $\zeta = 1$).

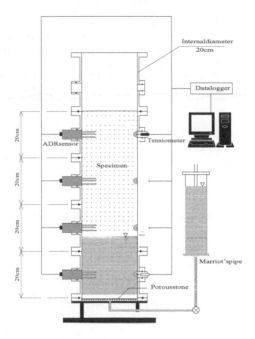

Figure 11. Water retention model test apparatus.

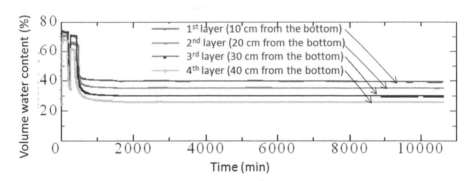

Figure 12. Volume water content change with water being drained off for dense 'white' sample.

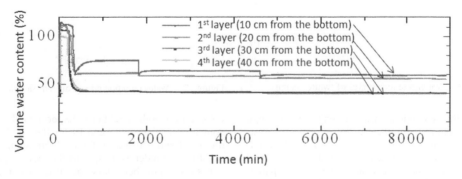

Figure 13. Volume water content change with water being drained off for dense 'red' sample.

343

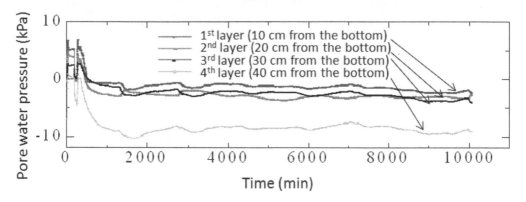

Figure 14. Pore water pressure change with water being drained off for dense 'white' sample.

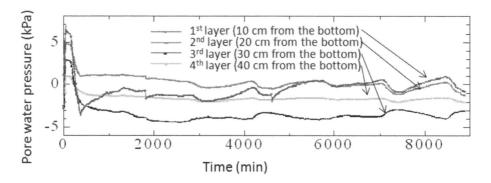

Figure 15. Pore water pressure change with water being drained off for dense 'red' sample.

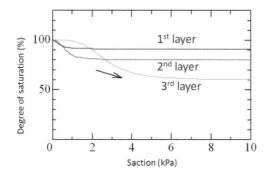

Figure 16. Water retention curves of 'white' sample.

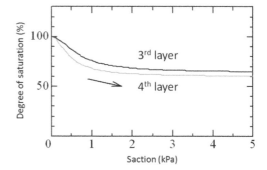

Figure 17. Water retention curves of 'red' sample.

For the specimens used in experiment, input parameters needed in the constitutive model are estimated as follows: $\lambda = 0.04$, $\kappa = 0.012$, $M = 1.73$ for 'white' specimens and $\lambda = 0.05$, $\kappa = 0.012$, $M = 1.76$ for 'red' specimens, in which other parameters, $\nu = 0.33$, $m = 1.2$, $n = 1.0$, $a = 150$ are common for both specimens. Experimental data are compared in Figures 18 and 19 for the 'white' sample and Figure 20 and 21 for the 'red' sample. The constitutive model well explains the monitored stress and shear strain behaviors, but it does not explain volumetric change with shearing.

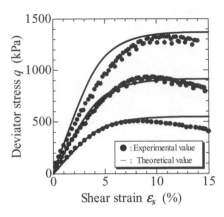

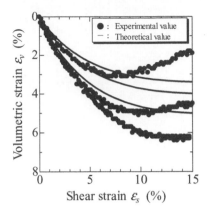

Figure 18. Comparison of stress and strain relation for 'white' specimen.

Figure 21. Comparison of volumetric strain change for 'red' sample.

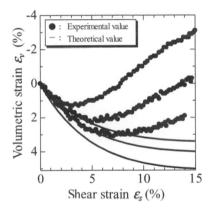

Figure 19. Comparison of volumetric strain change for 'white' sample.

REFERENCES

Alonso, E.E., Gens, A. & Josa, A.1990. A constitutive model for partially saturated soils, *Géotechnique* 40(3): 405–430.

Karube, D. and Kawai, K. 2001. The role of pore water in the mechanical behavior of unsaturated soils, *Geotechnical and Geological Engineering* 19(3): 211–241.

Kawai, K., Iizuka, A. & Tachibana, S. 2007. The influences of evapo-transpiration on the ground, *Proc.3rd Asian Conf. on Unsaturated Soils:* 359–364.

Kohgo, Y., Nakano, M. & Miyazaki, T. 1993. Verification of the generalized elasto-plastic model for unsaturated soils, *Soils and Foundations* 33(4): 64–73.

Ohno, S., Kawai, K. & Tachibana, S. 2007. Elasto-plastic constitutive model for unsaturated soil applied effective degree of saturation as a parameter expressing stiffness, *Journal of JSCE* 63(4): 1132–1141 (in Japanese).

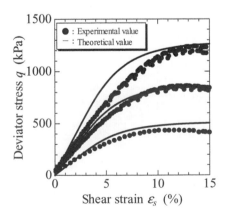

Figure 20. Comparison of stress and strain relation for 'red' specimen.

Prediction and Simulation Methods for Geohazard Mitigation – Oka, Murakami & Kimoto (eds)
© *2009 Taylor & Francis Group, London, ISBN 978-0-415-80482-0*

Rate effect on residual state strength of clay related with fast landslide

M. Suzuki & T. Yamamoto
Yamaguchi University, Ube, Japan

Y. Kai
International Development Consultants, Chuo, Tokyo, Japan

ABSTRACT: In order to clarify residual state strength characteristic of reconstituted clay subjected to a fast shearing, a series of ring shear tests was carried out by changing different shear displacement rates with several specimens. The shear displacement rate was set within the range of 0.02 and 10.0 mm/min. In addition, another ring shear test was carried out by changing step-by-step the shear displacement rate in the residual state which a specimen had once reached. The positive rate effects of the residual strength were observed in both conditions. Such the rate dependency can be changed by soil physical property.

1 INTRODUCTION

In 2004, a great earthquake occurred in the middle part of the Niigata Prefecture. The magnitude of the earthquake was 6.8, and the occurrence of strong aftershocks continued for one month. Fast landslide occurred at many sites in this area. These events have shown the importance of developing methods for estimating slope stability during or after an earthquake.

The residual state strength of clay, which is one of the most important strength parameters in estimating the stability of a reactivated landslide, is conventionally determined by either a ring shear test or a reversal box shear test. According to previous researches, the residual strength of clay can be more or less affected by shearing speed. Lemos et al. (1985) pointed out that the residual strength of soils with a high clay content showing "sliding shear" increased as the shear displacement rate increased, whereas the residual strength of soils with low clay content showing "turbulent shear" exhibited a reversed rate effect. They also stated that the rate effects are quite small but may play a major part in preventing the rapid movement of a landslide. Therefore, it is very important to clarify the influence of the shear displacement rate on the residual strength. However, the residual strength characteristics of soil subjected to fast shearing has not been investigated in great detail.

In this study, a series of ring shear tests was carried out using different shear displacement rates with several different clay specimens. The shear displacement rate was set within the range of 0.02 and 10.0 mm/min. Reconstituted samples tested were kaolin clay, and five natural clays causing a reactivated landslide. Another test was carried out by changing, step-by-step, the shear displacement rate in the residual state reached by a specimen. This paper describes the influence of the shear displacement rate on the residual strength of reconstituted clay based on the ring shear test results.

2 RING SHEAR TEST

2.1 *Preparation of soil samples*

Soil samples used in this study are kaolin and five kinds of natural clays that were collected in a disturbed state at two landslide sites in Yamaguchi and Okinawa Prefectures. The physical properties of these samples are listed in Table 1. Several samples of these clays were passed through a 0.425-mm sieve. The reconstituted sample was prepared by mixing a powdered sample with pure water at the water content equal to two times its liquid limit, and then pre-consolidating under the required consolidation pressure using a large tank. A specimen was cut from the pre-consolidated sample and then formed.

Table 1. Physical properties of soil samples used.

Soil sample	Site	ρ_s g/cm³	w_L %	I_P	F_{clay} %
Kaolin B*	Okayama	2.618	62.0	21.8	35.3
Shimajiri mudstone	Okinawa	2.586	91.4	61.4	55.5
Yuya A	Yamaguchi	2.606	89.0	53.1	87.0
Yuya B	ditto	2.592	43.3	19.9	32.0
Yuya C	ditto	2.519	49.0	28.6	34.5
Yuya D	ditto	2.554	44.7	26.9	47.3

* Some of these data were published in Suzuki et al. (2007).

2.2 Test apparatus and procedure

The residual strength of clay has conventionally been determined by the reversal direct box shear test (e.g. Nakamori et al. 1996) and the ring shear test (e.g. Bishop et al. 1970). Table 2 summarizes the essential features of the two apparatuses. The ring shear test apparatus has the main advantage of enabling the continuous application of a large deformation to a specimen. This continuity is very important when investigating the influence of the shear displacement rate. Figure 1 shows the ring shear apparatus employed in this study. This apparatus is similar to the Bishop-type apparatus (Bishop et al. 1971). The ring-shaped specimen has an inner diameter of 6 cm, outer diameter of 10 cm, and wall thickness of 2 cm. The specimen is sheared at a level 1 cm above the base plate. During the test, shear stress, normal stress, frictional force and normal displacement are monitored and recorded. The apparatus used can measure the frictional force between the specimen and rings using the load cell attached to the loading frame via a linking yoke connected to the upper ring (Suzuki et al. 1997). Frictional force along the rings is induced by the displacement of the specimen relative to the rings. In this apparatus, the load-receiving plate does not rotate but is supported using a ball bearing to prevent deviation of normal forces.

In order to investigate the influence of the shear displacement rate, a series of tests was carried out by setting different shear displacement rates using multiple specimens. The shear displacement rate was changed within the range of 0.02 and 10 mm/min. The normal stress, σ_N, was fixed to 196 kPa. The influence of the method of changing the shear displacement rate was investigated using a single specimen. The shear displacement rate was continuously increased and decreased in the residual state. The normal stress was set within the range of 98 and 392 kPa.

Figure 1. Front view of ring shear test apparatus used and a typical specimen.

First, each specimen was normally consolidated under a constant consolidation stress until the end of primary consolidation was confirmed by judgment based on the $3t$ method. To avoid swelling of the specimen due to submergence, pure water was poured into a water bath immediately after applying the consolidation stress. Subsequently, the specimen was sheared under a constant total normal stress, until the shear displacement, δ, became a sufficient value to reach the residual state. Here δ is defined as an intermediate circular arc between the inner and outer rings. To minimize the friction between the upper and lower rings and the outflow of the sample from the shear surface, the gap between the rings was fixed at 0.20 mm. The frictional force along the perimeter of the specimen acted upward in the case of a negative dilatancy and downward in the case of a positive dilatancy. In this test, net normal stress acting on the shear surface was calculated based on the frictional force measured by the load cell. When determining the gap between the upper and lower rings before the start of the shear test, changes in frictional force along the rings were measured for modifying normal stress accordingly.

2.3 Determination of residual strength

It is necessary to consider the shear displacement in determining the residual strength without an error based on the test results. In this paper, an improved method was proposed for objectively determining the residual strength regardless of the shear displacement by applying hyperbolic approximation to the measurement of the post-peak shear behavior (Suzuki et al. 1997). The hyperbolic approximation was applied to the measured relationship between the stress ratio, τ/σ_N, and the shear rotation angle, θ. Here θ is defined as an angle rotating on a vertical axis. An asymptotic value was defined as the stress ratio at the residual state, $(\tau/\sigma_N)_r$. Hyperbolic approximation parameters, a and b, are given by the segment and gradient of the line fitted to the measurement of the relationship between $\theta/(\tau/\sigma_N)$ and θ by the least square method. If the approximated hyperbola is in good agreement

Table 2. Comparison in features of reversal direct box shear and ring shear tests (after Suzuki et al., 2007).

Features / Type of test		Reversal direct box shear test (RDBST)	Ring shear test (RST)
Shear mode		σ_N	σ_N
Soil sample	Quantity	Little	Much
	Condition*	Undisturbed	Remolded / Reconstituted
Increment of shear displacement		Intermittent	Continuous
Cross sectional area of specimen		Changed	Unchanged
Leakage of soil sample		Little	Much
Reorientation of platy clay particles parallel to the direction of shearing		Straight	Circumferential

* The condition of sample that facilitates testing. Both RDBST and RST are, however, applicable to undisturbed samples, remolded samples and reconstituted samples.

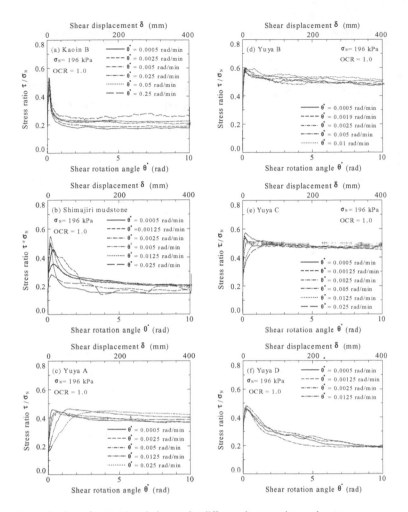

Figure 2. Ring shear behaviors of reconstituted clays under different shear rotation angle rates.

with the measurement, $(\tau/\sigma_N)_r$ is given as the inverse of b. The validity of the data fitting can be assessed using the correlation coefficient, r. The lowest value of r among all test cases was 0.9. The applicability of the method was supposed based on the test results of kaolin and natural clays under various test conditions. Therefore, the residual strength of a soil sample can be determined by this method.

3 RESULTS AND DISCUSSIONS

3.1 Ring shear behavior of reconstituted clay under different shear rates

Skempton (1985) reported that variation in the residual strength within the usual range of the slow laboratory test (0.002–0.01 mm/min) was negligible. Since then,

the rate effect on the residual strength of various soils has been examined using a ring shear test apparatus (Lemos et al. 1985; Yatabe et al. 1991, Tika et al. 1996, Suzuki et al. 2000, 2001). Among these studies, Lemos et al. (1985) pointed out that the residual strength of soils with a high clay content showing sliding shear increased as the shear displacement rate increased, whereas the residual strength of soils with a low clay content showing turbulent shear exhibited the reverse rate effect. These modes were first recognized by Lupini et al. (1981). Suzuki et al. (2000) reported that the residual strength could either increase, decrease or remain constant as the shear displacement rate increased: according to the ring shear test results, the tendency depended on the clay fraction and the kind of dominant clay mineral. Suzuki et al. (2007) also reported that the residual strength of overconsolidated and cemented clays was dependant on

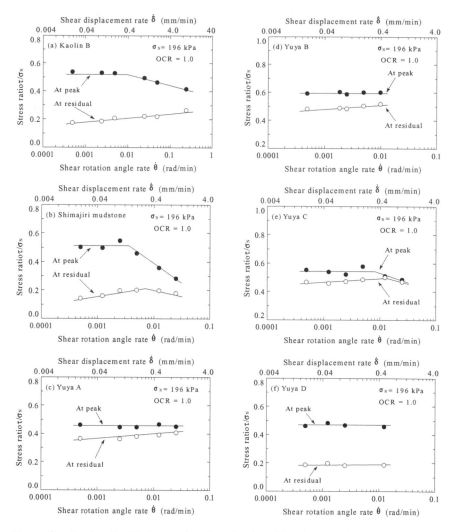

Figure 3. Stress ratios at peak and residual states plotted as a function of shearing speed for reconstituted clays.

the shear displacement rate based on the results of the reversal direct box shear test. It should be emphasized, however, that this rate dependency of the residual strength has not been sufficiently clarified in terms of the physical properties of the sample. Therefore, it is important to examine the rate dependency in terms of physical properties based on ring shear test results.

Figure 2 shows the relationships between the stress ratio and shear rotation angle, θ, for these samples. Here, the stress ratio is defined as the value of the shear stress normalized by the normal stress. $\theta = 10$ rad is equivalent to $\delta = 400$ mm. As the shearing proceeded, the stress ratio rose rapidly or moderately, and then began to decline, accompanied by a fluctuation of the stress ratio. Finally the specimen reached

an almost constant value. Several specimens of these samples exhibited remarkable strain softening. It can be seen from this figure that the stress ratio and shear rotation angle curves are more or less affected by the shear rotation angle rate. In the case of Yuya D shown in Fig. 2 (f), however, the stress ratio became almost same, irrespective of the shear rotation angle rate.

3.2 Relationship between stress ratio and shear displacement rate of reconstituted clay

Figure 3 shows the relationships between the stress ratio at peak and residual states and the shear displacement rate on a log scale for these reconstituted

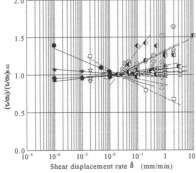

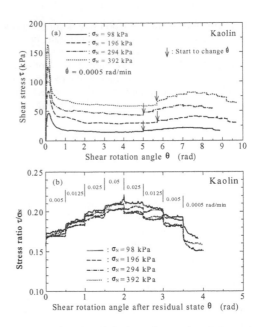

Figure 4. Rate effect of residual strength observed in various soils (after Nakamura and Shimizu 1978, Scheffler & Ullrich 1981, Lemos et al. 1985, Okada & So 1988, Yatabe et al. 1991, Suzuki et al. 2000).

Figure 5. Ring shear behaviors of reconstituted clays subjected to change of shear displacement rate in the residual state.

clays. The data plotted in this figure correspond to those in Fig. 2. In the case of Kaolin B shown in Fig. 3(a), the peak value decreased with an increase in the shear displacement rate above 0.01 rad/min. This tendency may be induced by the generation of excess pore water pressure at the shear surface. The effective normal stress for that was thought to be decreased. The stress ratio at the peak state becomes almost constant within 0.01 rad/min. In the range which the peak stress ratio was unchanged, the excess pore water pressure was considered to be fully dissipated, so that a great change in effective normal stress was not caused on the shear surface. In the case of Shimajiri mudstone shown in Fig. 3(b), the threshold value showing a dramatic change of peak value was around 0.004 rad/min. In the other cases, the peak value remains constant in the whole range. The difference may be due to the permeability of these samples. Therefore, it should be allowed to achieve a fully drained condition in this range.

On the other hand, it can be seen from this figure that the stress ratio at the residual state increases linearly with an increase in the shear displacement rate. This finding is in good agreement with those obtained by the ring shear test (Suzuki et al. 2000) and reversal direct box shear test (Suzuki et al. 2007). Therefore the positive rate effect was not essentially related to any change of effective normal stress. It is considered to be attributable to either a decrease in the void ratio or the viscosity of high plastic clay.

Figure 4 shows the normalized stress ratio at the residual state plotted against the shear displacement rate. Here $(\tau_r/\sigma_N)_{0.02}$ is defined as a value of the residual stress ratio at 0.02 mm/min. The data from previous studies (Nakamura and Shimizu 1978, Scheffler & Ullrich 1981, Lemos et al. 1985, Okada and So 1988, Yatabe et al. 1991, Suzuki et al. 2000) are also indicated in this figure. The increase, decrease or constancy of the residual strength with increasing shear displacement rate depended on the physical properties of the soil sample. As above mentioned, cohesive soils tend to exhibit a positive rate effect, probably because of the viscosity of the material. On the other hand, we have data for non-cohesive soils showing a negative rate effect.

3.3 Effect of acceleration and deceleration in residual state

In the ring shear test, the shear displacement rate was changed step-by-step once a specimen reached the residual state. The results of Kaolin B are shown in Figure 5. Fig. 5(a) shows the relationships between shear stress and shear rotation angle under different normal stresses. The shearing speed was accelerated and reduced within the range of 0.005 rad/min to 0.05 rad/min in the residual state. Fig. 5(b) shows a change in the stress ratio after the residual state. Here the rotation angle at which the speed began to be changed was set at zero. As the shear rotation rate is continuously increased and decreased in the residual state, $(\tau/\sigma_N)_r$ is evidently increased and decreased. Although the results of Yuya A and D were omitted for lack of space, similar changes in the stress ratio at the residual state were observed in both samples. This rate effect is not related to the method for changing the shearing speed.

4 CONCLUSIONS

The following conclusions can be drawn from our results.

1. For six kinds of reconstituted clay, the stress ratio in the residual state increases linearly with an increase in the shear displacement rate on a log scale. This tendency was in agreement with those observed in cemented clays obtained by the reversal direct box shear test.
2. The change of the stress ratio in the residual state was linked to the acceleration and deceleration in the residual state. Therefore, the rate dependency of the residual strength was observed in both tests.
3. According to our and others' data, the residual strength of soils with more plasticity increased with an increased shear displacement rate, whereas the residual stress ratio of soils with less plasticity exhibited either a reversed rate effect or no clear change. Thus, the tendency shown by Lemos et al. (1985) is fundamentally supported by our data, and is dependent on the physical properties of the sample.

REFERENCES

Bishop, A.W., Green, G.E., Garga, V.K., Andresen, A. & Brown, J.D. 1971. A new ring shear apparatus and its application to the measurement of residual strength, *Géotechnique* 21(4): 273–328.

Lemos, L.J.L., Skempton, A.W. & Vaughan, P.R. 1985. Earthquake loading of shear surfaces in slopes, *Proc. of the 11th I.C.S.M.F.E.* (4): 1955–1958.

Lupini, J.F., Skinner, A.E. & Vaughan, P.R. 1981. The drained residual strength of cohesive soils, *Géotechnique* 31(2): 181–213.

Nakamori, K., Yang, P. & Sokobiki, H. 1996. Strength characteristics of undisturbed landslide clays in tertiary mudstone, *Soils and Foundations* 36(3): 75–83.

Nakamura, H. & Shimizu, K. 1978. Soil tests for the determination of shear strength along the sliding surface, *Journal of the Japan Landslide Society* 15(2): 25–32 (in Japanese).

Okada, F. & So, E. 1988. Relation between residual strength and microstructure, *Proc. of 23rd Japan National Conference on Geotechnical Engineering*, JGS: 227–228 (in Japanese).

Scheffler, H. & Ullrich, W. 1981. Determination of drained shear strength of cohesive soils, *Proc. of 10th I.C.S.M.F.E* 10: 775–778.

Skempton, A. W. 1964. Long-term stability of clay slopes, *Géotechnique* 14(2): 77–102.

Skempton, A. W. 1985. Residual strength of clays in landslides, folded strata and the laboratory, *Géotechnique* 35(1): 3–18.

Suzuki, M., Umezaki, T. & Kawakami, H. 1997. Relation between residual strength and shear displacement of clay in ring shear test, *Journal of Geotechnical Engineering*, JSCE, 575(III-40): 141–158 (in Japanese).

Suzuki, M., Umezaki, T., Kawakami, H. & Yamamoto, T. 2000. Residual strength of soil by direct shear test, *Journal of Geotechnical Engineering*, JSCE, 645(III-50): 37–50 (in Japanese).

Suzuki, M., Yamamoto, T., Tanikawa, K., Fukuda, J. & Hisanaga, K. 2001. Variation in residual strength of clay with shearing speed, *Memoirs of the Faculty of Engineering, Yamaguchi University*, 52(1): 45–49.

Suzuki, M., Kobayashi, K., Yamamoto, T., Matsubara, T. & Fukuda, J. 2004. Influence of shear rate on residual strength of clay in ring shear test, *Memoirs of the Faculty of Engineering, Yamaguchi University*, 55(2): 49–62.

Suzuki, M., Tsuzuki, S. & Yamamoto, T. 2007. Residual strength characteristics of naturally and artificially cemented clays in reversal direct box shear test, *Soils and Foundations* 47(6): 1029–1044.

Tika, T.E., Vaughan, P.R. & Lemos, L.J.L. 1995. Fast shearing of pre-existing shear zones in soil, *Géotechnique* 46(2): 197–233.

Yatabe, R., Yagi, N. & Enoki, M. 1991. Ring shear characteristics of clays in fractured-zone-landslide, *Journal of Geotechnical Engineering*, JSCE, 436(III-16): 93–101 (in Japanese).

Prediction and Simulation Methods for Geohazard Mitigation – Oka, Murakami & Kimoto (eds)
© 2009 Taylor & Francis Group, London, ISBN 978-0-415-80482-0

Constitutive model and its application of flow deformation induced by liquefied sand

Y.M. Chen, H.L. Liu, G.J. Shao & X.B. Sha
Hohai University, Nanjing, Jiangsu, China
Geotechnical Institute of Hohai University, Nanjing, China

ABSTRACT: Based on the torsional shear tests of post-liquefied deformation of the saturated sand, the definition of strain rate and the apparent viscosity were introduced to analyze the flow characteristics of the post-liquefied sand. The relationship between shear stress with strain rate, and the apparent viscosity properties, in static and dynamic torsional processes were analyzed. The results showed that, in the zero effective stress state, the apparent viscosity decreased with the shear strain rate increasing, and the liquefied sand belonged to the shear-thinning Non-Newtonian fluid. The power equation could be used to fit the relationship between the shear stress with the shear strain rate in zero effective stress state. The flow constitutive model in zero effective stress state was implemented in FLAC3D and the Liquefy model which can describe the fluid deformation of the liquefied sand was obtained. Based on FLAC3D, the simplified analyzing method of the large flow deformation was proposed. The numerical simulations of shaking table tests were performed. The calculating results showed that the large deformation induced by liquefied sand was related with many factors, such as the boundary conditions, geometry parameters, mechanical properties, soil layer profiles and so on.

1 INTRODUCTION

The term liquefaction, which is defined by the Committee on Soil Dynamics of the Geotechnical Engineering Division, ASCE (1978), is the act or process of transforming any substance into a liquid. In cohesionless soils, the transformation is from a solid state to a liquefied state as a consequence of increased pore pressure and reduced effective stress. Based on the above definition, two suppositions can be obtained. Firstly, the liquefied state may be described as a kind of fluid. Secondly, the process of liquefaction involves its occurrence, continuation and finish. Many natural disasters due to soil liquefaction are observed during strong earthquakes. A new International Standard, ISO23469, is being developed, which adopts a performance based approach in order to provide a general and flexible framework for designing geotechnical works (Liu et al. 2002). And the proper calculating method of the large deformation induced by the post-liquefaction is one of the important issues (Chen & Zhou 2007).

The idea of viewing the liquefied sand as a kind of fluid is a new method for the soil liquefaction research (Sasaki et al. 1992). Some laboratory tests and numerical analysis were performed based on fluid mechanics. (Orense et al. 1998, Uzuoka et al. 1998, Timate & Towhata 1999)

The laboratory tests of liquefied sand, determined using hollow torsion-shear tests, are analyzed from the view of fluid mechanics (Chen et al. 2006). The main flow characteristics, including the apparent viscosity and the relation between the stress and strain rate, are summarized. The results show that, for the liquefied sand, the relation between the stress and the strain rate of the liquefied sand can be fitted well by the constitutive model of the shear thinning non-Newtonian fluid (pseudo-plastic fluid), which is a formulation of the power law. According to the results of the laboratory tests, an experimental model was proposed to describe the flow deformation of the liquefied sand.

In this paper, the experimental model is introduced. And then the model is extended to three dimensional equations, and formulated into a finite-difference algorithm (FLAC^{3D}) to obtain the new model named Liquefy Model. The Liquefy Model gives an accurate description of the liquefied phenomena. The simplified method of large flow deformation induced by liquefied sand is proposed based on Liquefy model in FLAC^{3D} framework.

For the simplified method, the process of liquefaction and the strength recovery of liquefied sand were not considered. In some boundary conditions, the flow deformation processes of sand in the complete liquefaction state were analyzed. There were two steps in the analysis: (1) to view the material of soil as an elastic continua, and set initial stress, calculate the initial stress distribution in the pre-liquefaction state; (2) to view the liquefied layer soil as Liquefy model, and perform liquefied solution for a certain time to get the result of deformation in the liquefied state.

2 FLOW CONSITITUTIVE MODEL OF LIQUEFIED SAND

The fluid constitutive model for the liquefied sand adopted in the present paper is developed by Chen et al. 2007. The model introduces the view of fluid mechanics into the liquefaction research. The laboratory fundamentals of the constitutive model were the hollow dynamic torsional shear tests performed on the multifunctional triaxial test equipment and the pulling ball tests performed on the small shaking table. The experimental model in this paper was concentrated on the zero effective stress state in the soil liquefaction process. The apparent viscosity of the liquefied sand decreases with the shear strain rate increasing. The sand in the zero effective stress is a shear thinning non-Newtonian fluid. The power formulation can be used to fit the relationship between the shear stress with the shear strain rate as Equation 1.

$$\tau = k_0 \left(\dot{\gamma} \right)^{n_0} \tag{1}$$

where, τ = shear stress, $\dot{\gamma}$ = shear strain rate, k_0 and n_0 = two laboratory parameters in zero effective stress state.

For the convenience of programming implementation, Equation 2 and 3 should be generalized to three-dimensional condition. The generalized shear stress q is used to be replaced the shear stress τ,

$$q = \sqrt{\frac{3}{2} S_{ij} S_{ij}} \tag{2}$$

where S_{ij} = the deviatoric part of the stress tensor σ_{ij}.

The shear strain rate tensor $\dot{\varepsilon}_{ij}$ can be decomposed as mean shear strain rate component $\dot{\varepsilon}_{ij}^p$ and deviatoric shear strain rate component $\dot{\varepsilon}_{ij}^d$.

$$\dot{\varepsilon}_{ij} = \dot{\varepsilon}_{ij}^p + \dot{\varepsilon}_{ij}^d \tag{3}$$

Then the deviatoric shear strain rate tensor can be decomposed into two components: elastic component $\dot{\varepsilon}_{ij}^{de}$ and fluid component $\dot{\varepsilon}_{ij}^{df}$.

$$\dot{\varepsilon}_{ij}^d = \dot{\varepsilon}_{ij}^{de} + \dot{\varepsilon}_{ij}^{df} \tag{4}$$

The fluid shear strain rate $\dot{\varepsilon}$ is used to replace the shear strain rate $\dot{\gamma}$ in Equation 1. The fluid shear strain rate is coaxial with the deviatoric stress tensor (normalized by the generalized shear stress), and is given by

$$\dot{\varepsilon} = \frac{2}{3} \dot{\varepsilon}_{ij}^{df} \left(\frac{q}{S_{ij}} \right) \tag{5}$$

The fluid constitutive model in three-dimensional condition can be formulated by

$$q_0 = k_0 \left(\dot{\varepsilon}_0 \right)^{n_0} \tag{6}$$

where, q_0 = generalized stress in zero effective stress state.

The fluid constitutive model is implemented in FLAC3D with VC++ environment. For the details of the basic procedures and the programming essentials of constitutive model redefining the reader is referred to Chen & Liu 2007 and Chen & Xu 2009.

3 DEVELOPMENT OF LIQUEFY MODEL

The simplified method is proposed on the FLAC3D code. The FLAC3D code is an explicit finite difference scheme for the continua and it is suitable for the solution of large deformation problem. The real time for the liquefaction process can be simulated by FLAC3D. The interfaces can be modeled as the interaction between the two different materials, for example the interaction between the liquefied layer and non-liquefied layer.

The open framework of the user defined model is adopted in FLAC3D The defined model is written in C++, and compiled as a DLL file (dynamic link library) that can be loaded whenever it is needed.

3.1 Iterative procedure

Rewrite Equation 1 as:

$$\dot{\varepsilon} = \left(\frac{q}{k} \right)^{1/n} \tag{7}$$

In the process, the strain rate tensor $\dot{\varepsilon}_{ij}$ is referred, the relation between partial strain rate tensor and total strain rate tensor is:

$$\dot{\varepsilon}_{ij}^d = \dot{\varepsilon}_{ij} - \frac{\dot{\varepsilon}_{kk} \delta_{ij}}{3} \tag{8}$$

The relation between the flexibility section of deviatoric strain rate tensor and deviatoric stress rate tensor is:

$$\dot{\varepsilon}_{ij}^{de} = \frac{\dot{\sigma}_{ij}^d}{2G} \tag{9}$$

In the equation, G is elastic shear modulus, $\dot{\sigma}_{ij}^d$ is deviatoric strain rate tensor:

$$\dot{\sigma}_{ij}^d = \dot{\sigma}_{ij} - \frac{\dot{\sigma}_{kk} \delta_{ij}}{3} \tag{10}$$

Flow deformation of deviatoric strain rate tensor and deviatoric stress are coaxial, and the relation between both can be described as Equation 11.

$$\dot{\varepsilon}_{ij}^{df} = \frac{3}{2} \left(\frac{\sigma_{ij}^d}{q} \right) \dot{\varepsilon} \tag{11}$$

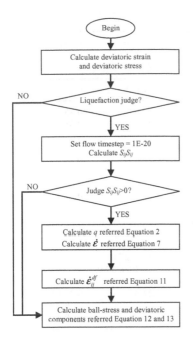

Figure 1. Flow diagram of simplified method model.

Unit: cm

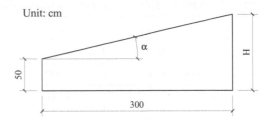

Figure 2. Calculation sketch of shaking table test.

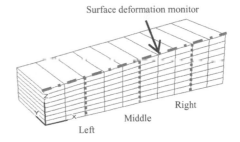

Figure 3. Calculation grid and monitoring of shaking table test.

The iterative procedure can be described as:

If the stress tensor is $\sigma_{ij}^{(t)}$ in time t, stress tensor at the time of $t + \Delta t$ can be calculated by the following approach:

Ball-stress components:

$$\sigma_{kk}^{(t+\Delta t)} = \sigma_{kk}^{(t)} + 3K\dot{\varepsilon}_{kk}\Delta t \tag{12}$$

Deviatoric stress components:

$$\sigma_{ij}^{d(t+\Delta t)} = \sigma_{ij}^{d(t)} + 2G\left(\dot{\varepsilon}_{ij} - \dot{\varepsilon}_{ij}^{df}\right)\Delta t \tag{13}$$

3.2 Programming diagram

The methodology of writing a constitutive model in C++ for operation in FLAC3D includes descriptions of the base class, member functions, registration of models, information passed between the model and FLAC3D, and the model state indicators. Because the user defined model is shared with the same base class of in-built model, there is a same implementation efficiency level between user defined model and in-built model. The implementation of a DLL model is described and illustrated in Figure 1.

4 APPLICATION OF FLOW MODEL

The idea of treating the liquefied sandy soil as a liquid derived from the result of shaking table tests. So, the analyses of large deformation in the shaking table tests were performed using the proposed flow constitutive model.

4.1 Analysis model

The shaking table model considered in the calculation has length 300 cm and the width 100 cm. The sand soil is layered is in the shaking table box. The height of the sand soil in the left side is 50 cm, and in the right is different according to the surface slope angel. In the calculation, the initial stress is set to get a balance state. The stress distribution in the pre-liquefaction state is obtained. Then the sandy soil material is changed to Liquefy model proposed in the paper, the flow deformation calculation is performed for a certain time.

The sketch of the calculated shaking table test is showed in Figure 2 and the difference grid in FLAC3D is showed in Figure 3. The number of the zone is 80. The velocities in Y and Z directions of the model are rigid and that of X direction is free. The velocities in the normal directions of the four model surfaces are rigid. The surface deformation is monitored in the calculation. The horizontal deformations in three different cross-sections (showed in Figure 3) are also monitored.

Table 1. Calculation plan on simulation of shaking table test.

Factor	Height of right side H (cm)	Flow time T (s)	Consistency coefficient k ($\times 10^3$)	Flow Index n (−)	Elastic modulus E (MPa)	Poisson ratio μ (−)
T	80	1~10	3.1054	0.3225	27.0	0.35
H	60,80, 100, 120,140	1~10	3.1054	0.3225	27.0	0.35
k	80	1~10	1.0, 2.0, 3.1054, 4.0, 5.0	0.3225	27.0	0.35
n	80	1~10	3.1054	0.1, 0.2, 0.3225, 0.4, 0.5, 0.6	27.0	0.35
E	80	1~10	3.1054	0.3225	0.01, 0.1, 1.0, 10.0, 20.0, 27.0, 40.0, 50.0, 60.0	0.35
μ	80	1~10	3.1054	0.3225	27.0	0.2, 0.25, 0.3, 0.35, 0.4, 0.45

4.2 Analysis plan

The analysis plan in the calculation is showed in Table 1. The calculated time of the liquefaction is from 1s to 10s. The parameters of the flow model comes from the laboratory result of the static shear experiments, the consistency coefficient k = 3.1054E3 and the flow index is n = 0.3225.

There are six factors considered in the calculation plan. The gradient of the liquefied layer soil, the parameter of the flow model (consistency coefficient k and flow index n) and elastic parameter (the modulus of elasticity E and the Poisson ratio μ) into consideration.

4.3 Analysis results

4.3.1 Influence of flow time
The relation between the horizontal displacement in the three cross-sections with the liquefaction time is showed in Figure 4. It can be easily found that the horizontal displacement develops quickly at the beginning of the liquefaction (1~3s), due to the large generalized shearing stress. The displacement forms in the three cross sections are different. The maximum displacement in the left side occurs closely to the model surface, the deform curve is similar to the sine function. As the middle cross section, the horizontal shift occurs under the 1/2 width and in the 1/2 depth. The value of horizontal deformation above the 1/2 depth almost maintains const. The deformation curve is not corresponding to the sine function. For the right cross section, horizontal displacement maximum occurs under the surface, and on the top side of the model, the deformation value is lessened compared to the maximum.

4.3.2 Influence of surface slope
The horizontal displacements of the standard cross section are monitored in the calculation. Figure 5 shows the deformed curves when the time equals to 10s. It shows that the horizontal displacement of the soil enhanced with the gradient increasing. Except for the cross section curve in the left of the model is similar to the sine function, the appearance of curves in the other positions are quite different from the sine function curve. The main change is that the middle position and the node at the bottom on the right side of the model have a large deformation. On the left side of the model, with the depth increases, the horizontal displacement reduces gradually, and the largest displacement occurs at the top of the model. In the middle cross-section, the maximum horizontal departure occurs in 1/2 depth and on the right side, with the liquefied layer slope increases, the relative position of the maximum horizontal departure moves down gradually. In the case with H 100 cm and 120 cm, the maximum horizontal displacement occurs at the bottom of the model, and with the depth increases, the horizontal departure increases gradually.

4.3.3 Influence of consistency coefficient k
Parameter k is used to describe the consistency coefficient of the non-Newtonian fluid. With the increasing k, it means the apparent viscosity of the fluid is

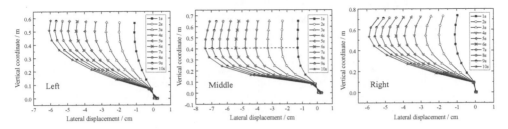

Figure 4. Relationship between the horizontal displacements of the model with liquefaction time.

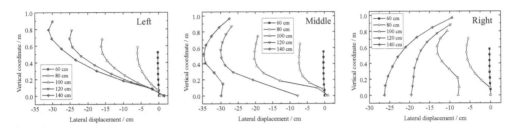

Figure 5. Horizontal displacement curve in different surface slope at liquefaction time is 10s.

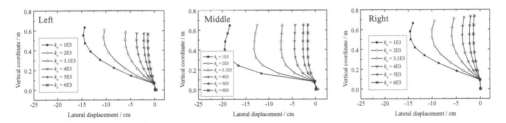

Figure 6. Influence of parameter k on horizontal displacement curve.

higher and the ability to resist deformation is stronger. Figure 6 shows the displacement curve of three typical cross-sections which the time of liquefaction 10s. It can be founded that with the consistency coefficient k increase sing, the horizontal displacement decreases rapidly. At the same time, in regard to the different consistency coefficient k, the horizontal displacement curves on each section are similar and the positions for the maximum horizontal departure occurs are almost same. It can be concluded that the consistency coefficient just affects the size of the liquefied deformation, but it will not affect the pattern of the liquefied deformation.

4.3.4 Influence of flow index n

Parameter n, which called the flowing index, describes the level of non-linear of non-Newtonian fluid. Because the liquefied sandy soil belongs to a shear thinning non-Newtonian fluid, its parameter n value range is between 0 and 1.0. In regard to the shear thinning non-Newtonian fluid, when n value is larger, it means

the non-linear of the fluid is stronger and the ability to resist deformation is lower. Figure 7 shows the horizontal displacement curve of three typical cross-sections with the time of calculation 10s. It can be founded that with the parameter n increasing, the level deformations of each section increase gradually.

As similar to the consistency coefficient k, the change of flow index n will only cause the numerical change of the liquefied deformation, and it will not affect the deformation model. The horizontal displacement curves of each section are similar, and the position where the maximum horizontal departure occurs will not change with the flow index n changing.

4.3.5 Influence of elastic parameters E and μ

The change situation of maximum horizontal departure curve in a typical cross-section with the deformation modulus E and the Poisson ratio μ shows in the Figure 8. It can be founded that with the deformation

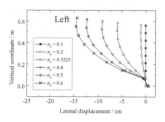 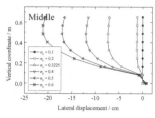 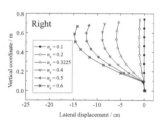

Figure 7.   Influence of parameter n on horizontal displacement curve.

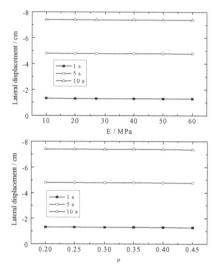

Figure 8.   Influence of elastic parameters on maximum horizontal displacement.

modulus and the Poisson ratio changing, the deformation situation will not change in the each section. Therefore, in the liquefied flow distortion analysis, the deformation modulus E and the Poisson ratio μ have no affection on the result.

5   CONCLUSIONS

The idea of viewing the liquefied sand as a kind of fluid is a relatively new research method. In this paper, the flow characteristics of liquefied sand are analyzed and an experimental constitutive model is proposed based on Non-Newtonian fluid mechanics. The flow model is developed in FLAC3D code and a simplified method of large flow deformation under the liquefied state is proposed. Some conclusions are obtained as followings:

1. The numerical simulations of the post-liquefied deformation on shaking table experiment are performed. According to the calculation results, the displacement of liquefied soil increases gradually with the time of liquefaction elapsed. And the speed of deformation reduces gradually.
2. The flow parameters, including the consistency coefficient k and flow index n have the important influence on the size of the liquefaction deformation. The deformation increases with the consistency coefficient k reducing and the flow index n increasing. The changes of flow parameters have no effect on the change the pattern of liquefied flow deformation.
3. The elastic parameters, including the elastic modulus E and the Poisson ratio μ, have no affection about the size and pattern of the liquefied deformation.
4. The proposed simplified method describes the process of the flowing deformation, and it can be used to the original characteristic of FLAC3D program, carrying out complex boundary conditions, complex contact conditions and taking the liquefied deformation analysis between liquefied soil and structure. This will be the direction of our works for next step.

ACKNOWLEDGEMENT

This paper was financially supported by the National Science Fund for Distinguished Young Scholars (Grant No. 50825901); the Doctorial Fund of Jiangsu Province (Grant No. 0801002C); the China Postdoctoral Science Foundation (Grant No. 20080441020) and the Open Laboratory Fund of Institute of Engineering Mechanics, China Earthquake Administration (Grant No.: 2007A03).

REFERENCES

Chen, Y.M. & Zhou, Y.D. 2007. Advance in sand postliquefaction research based on fluid mechanics method. *Journal of Hohai University* 35(4): 418–421.

Chen, Y.M., Liu, H.L., Zhou, Y.D. 2006. Analysis on flow characteristics of liquefied and post-liquefied sand. *Chinese Journal of Geotechnical Engineering* 28(9): 1139–1143.

Chen, Y.M. & Liu, H.L. 2007. Development and implementation of Duncan-Chang constitutive model in FLAC3D. *Rock and Mechanics* 28(10): 2123–2126.

Chen, Y.M. & Xu, D.P. Fundamentals and applications of FLAC/FLAC3D. Beijing: China Water Power Press.

Liu, H.L., Zhou, Y.D. & Gao, Y.F. 2002. Study on the behavior of large ground displacement of sand due to seismic liquefaction. *Chinese Journal of Geotechnical Engineering* 24(2): 142–146.

Orense, R.P. & Towhata, I. 1998. Three dimensional analysis on lateral displacement of liquefied subsoil. *Soils and Foundations* 38(4): 1–15.

Sasaki, Y., Towhata, I., Tokida, K.I., Yamada, K., Matsumoto, H. & Tamari, Y. 1992. Mechanism of permanent displacement of ground caused by seismic liquefaction. *Soils and Foundations* 32(3): 79–96.

Tmate, S. & Towhata, I. 1999. Numerical simulation of ground flow caused by seismic liquefaction. *Soil Dynamics and Earthquake Engineering* 18: 473–485.

The Committee of Soil Dynamics of Geotechnical Engineering Division. 1978. Definition of terms related to liquefaction. *Journal of Geotechnical Engineering. ASCE,* 104(GT9): 1197–1120.

Uzuoka, R., Yashima, A., Kawakami, T. & Konrad, J.M. 1998. Fluid dynamics based prediction of liquefaction induced lateral spreading. *Computers and Geotechnics* 22(3/4): 243–282.

Evaluation of shear strain in sand under triaxial compression using CT data

Y. Watanabe & J. Otani
Kumamoto University, Kumamoto, Japan

N. Lenoir
Laboratoire 3S-R, CNRS, Grenoble, France

T. Nakai
Nagoya Institute of Technology, Nagaya, Japan

ABSTRACT: The purpose of this paper is to evaluate three dimensional displacement and strain properties of sand in three dimensions under triaxial compression using X-ray CT. Complete 3D images of the specimens are recorded at several stages throughout the test. Here in this paper, the movements of real soil particles are traced on the CT images, so that more precise displacement property of soils can be discussed. The test which has been done is triaxial compression test and the material used in the test is sandy soil called "Yamazuna sand". Based on the results of its displacement property, the strain field in the soil specimen are calculated using finite element method, in which the results of displacement fields from X-ray CT is the input data for the analysis. Finally, more quantified shear strain of sand is obtained in three dimensions.

1   INTRODUCTION

In geotechnical engineering, there are two approaches to solve mechanical property of soils. One is the use of continuum mechanics and the other is that of theory of granular materials. In case of numerical analysis, finite element method (FEM) is in the former category while distinct element method (DEM) is the latter ones. Even if either of those approaches is chosen, it can be said that the precise discussion on the strain field in the soil is very important in order to characterize the deformation or the failure of soils. Especially, for the discussion on the failure of soils, the strain localization in the soil is one of key issues. There are some research activities, which tried to investigate strain field including displacement property. As for the experimental studies, element test such as triaxial compression or plane strain tests and model loading test have been conducted and the movements of the surface of the soils were traced (Yamamoto et al. 2001 and Nielsen et al. 2003). However, the strain field obtained is restricted to in two dimensions.

In the mean time, X-ray CT (Computed Tomography) scanner which is one of nondestructive testing methods has been applied to engineering fields. Using this apparatus, the inside three dimensional behaviors of the materials could be investigated without any destructions. In particular, the X-ray CT has much higher resolution because of high power of the x-rays. So far, authors have conducted a series of studies on the application of industrial X-ray CT scanner to

geotechnical engineering (Otani et al. 2003). There are also some researches on the use of X-ray CT in order to characterize the displacement field in three dimensions (Takano et al. 2006). However, the behavior obtained from this study itself seems not to be the behavior of the soil because of the use of artificial makers which were traced on CT images.

The objective of this paper is to visualize the strain field in soils using X-ray CT. Here, the movement of real soil particles are traced on the CT images, so that more precise displacement property of soils can be discussed. The test which has been done is triaxial compression test and the material used in the test is sandy soil called "Yamazuna sand". Based on the results of its displacement property, the strain field in the soil specimen is calculated using finite element method, in which the results of displacement fields from X-ray CT is the input data for the analysis. Finally, more quantified shear strain of sand is obtained in three dimensions.

2   SUMMARY OF TEST PROCEDURE

A triaxial compression test under drained condition was performed and at the same time, a series of CT scanning of the soil specimen were conducted during the process of the compression. Dry Yamazuna sand which has a wide grain size distribution was used in the test. Figure 1 shows the grain size distribution of this sand in which the minimum particle size was

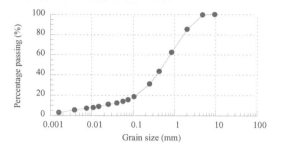

Figure 1. Grain size distribution curve.

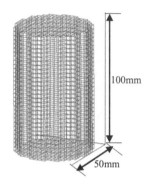

Figure 2. 3-D finite element mesh.

Table 1. Material parameteres for t_{ij} model.

λ	0.0084
κ	0.0060
R_{cs}	4.7
β	−0.60
α	0.85

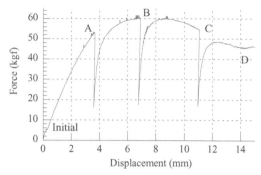

Figure 3. Force-displacement relationship.

0.001 mm and its maximum size was 10 mm. The size of soil specimen was 50 mm in diameter and 100 mm in height with a relative density of 90%. The loading speed was 0.3%/min. and the confined pressure of 50 kPa was applied. For the condition of CT scanning, the voltage used was 150 kV and the electric current was 4 mA. In fact, a spatial resolution was the size of one voxel, $0.073 \times 0.073 \times 0.3$ mm^3, and the specimen was scanned with every 0.3 mm thickness from the bottom to the top of the specimen with the total number of 330 slices.

3 FINITE ELEMENT ANALYSIS

Based on the resulted displacement field from CT scanning in the test with image analyses, a strain field of soil specimen was calculated using 3-D finite element method. Here, the computer code developed by Nakai was used. In the analysis, elasto-plastic constitutive law called "t_{ij} model" was used (Nakai 2007). The obtained displacement field from CT scanning was applied at the locations of nodes in the finite element mesh. Figure 2 shows 3-D finite element mesh which is the total number of 7936 cubic elements with eight nodes in each element. The soil parameters for "t_{ij} model" were listed in Table 1. After calculating

the displacements in the specimen, the strains were calculated, automatically.

4 RESULTS AND DISCUSSION

4.1 Test results

Figure 3 shows force-displacement relationship from triaxial compression test. As shown in this figure, strain levels of Initial, A, B, C and D are the steps of when the CT scanning were conducted. Although there is a stress relaxation due to stopping the loading during CT scanning, the quality of the test has been checked with that of monotonic loading for the same triaxial compression test (Otani et al. 2002). In order to investigate the initiation of shear band in the soil, the CT images at the levels of B, C and D with a different height (h = 40, 55 and 70 mm) from the bottom of the specimen are shown in Figure 4. It is noted that CT image is made by the spatial distribution of so called "CT value" which corresponds to the material density and is total of 256 black and white color gradations in which black color means the low density while white color shows the high density. Low density areas appear around the upper part of the specimen at B, while the strain localization appears from the upper to the bottom of the specimen at the level C. Also, at the level D, it can be realized that the total shear band is formed in the soil.

Since a large number of the cross sectional images were obtained for all the strain levels, any cross sectional images including vertical cross sectional ones can be obtained. Figure 5 shows vertical images for

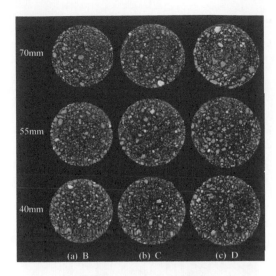

Figure 4. Horizontal cross sectional images.

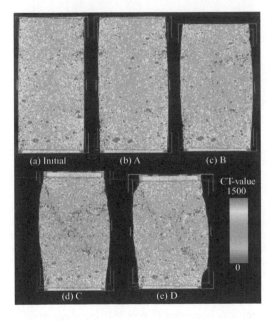

Figure 5. Vertical reconstruction images.

all the levels. These images were made by the density change using seven different colors. The distribution of the density in the soil is clearly evaluated with the existence of soil particles. For the case of strain level B, some banded low density areas appear in the middle of the specimen and these areas become one linear band from left top to the right bottom at the level C. And this banded area is extended with the volume change of the specimen at the level D. According to those images, it is concluded that the shape of the shear band inside

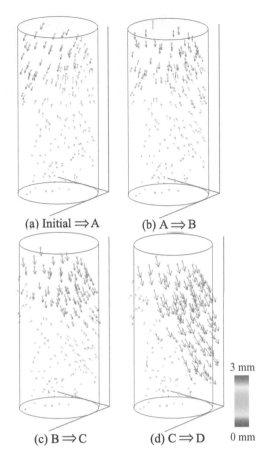

Figure 6. Displacement vectors in three dimensions.

specimen could be well investigated using X-ray CT as a density change region.

4.2 Results of image processing analysis

A new method of image processing analysis has been developed for the purpose of tracing soil particles in the CT images by the authors (Watanabe et al. 2007) and this method was applied to the results of triaxial compression test in this paper. And as a result of tracing all the represented soil particles, the displacements of the particles in three dimensions are denoted by vector notations. Figure 6 shows 3-D displacement vectors for each of the strain levels ((a) Initial–A, (b) A–B, (c) B–C, and (d) C–D). The movements of the soil particles around the upper part of the specimen are more obvious than those of lower part. And it can be said that the displacements before the strain level B do not have any specific direction but after the level B, all the vectors move to the direction of right below from the left top in the specimen. It is confirmed that this

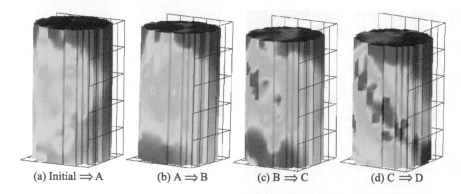

(a) Initial $\Rightarrow$ A   (b) A $\Rightarrow$ B   (c) B $\Rightarrow$ C   (d) C $\Rightarrow$ D

Figure 7.   Distribution of shear strain in three dimensions.

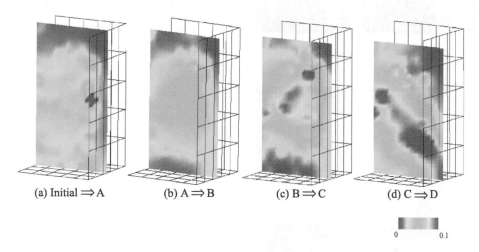

(a) Initial $\Rightarrow$ A   (b) A $\Rightarrow$ B   (c) B $\Rightarrow$ C   (d) C $\Rightarrow$ D

0   0.1

Figure 8.   Distribution of shear strain at one vertical cross section.

behavior seems to be well correlated with the results of density change by CT images which was shown in Figure 5. Finally, it can be said that the movements of the particles was visualized in three dimensions with fairly good accuracy.

4.3   *Results of numerical analysis*

Figure 7 shows 3-D distribution of shear strain in the soil for all the loading steps. And it is confirmed that the strain localization is progressive as the load increases. More precisely, it is observed that the shear strain appears around the center of the specimen at level A and is spread towards whole of the specimen at level B. And at level C, more localized strain area around the center of the specimen is observed and finally, the total shear band is shaped around the liner area from left top to the right bottom of the specimen at level D. Figure 8 shows another shear strain distribution at one of the vertical cross section of the specimen.

It is observed that the shear strain appears around the center of the specimen at level A and at level B. And at level C, more localized strain area around the center of the specimen is observed and shear band is shaped toward right top in the specimen. And finally, the total shear band is shaped around the liner area from left top to the right bottom of the specimen at level D. Therefore, it is important to note that the shear strain is initiated around the center of the specimen and it can be said that the process of shear banding is realized from inside of the specimen towards the other areas. Finally, it is concluded that the strain field in the soil under triaxial compression was evaluated in detail.

5   CONCLUSIONS

In this paper, the displacements in the soil were visualized under triaxial compression and the strain field in the soil specimen was calculated using finite element method based on the quantified CT data in three dimensions.

The conclusions drawn from this paper are listed as follows:

1. The displacement field of sand was visualized properly based on the analysis of the movement of soil particles on CT images;
2. The strain field of sand was calculated based on the displacement field from the CT images by finite element analysis; and
3. The strain localization in sand was visualized in three dimensions.

However, only one test case was analyzed in this paper and it is expected that more different soil conditions such as different confined pressures have to be analyzed. And finally, it is confirmed that the strain field in the soil could be evaluated using the method here in this paper and the investigation of stress field will be the next target for this research.

REFERENCES

Nakai, T. 2007. Modeling of soil behavior based on t_{ij} concept, 13th Asian Regional Conference on Soil Mechanics and Geotechnical Engineering.

Nielsen, S.F. Poulsen, H.F., Beckmann, F., Thorning, C. & Wert, J.A. 2003. Measurements of plastic displacement gradient components in three dimensions using marker particles and synchrotron X-ray absorption microtomography, Acta Materiala 51: 2407–2415.

Otani, J., Mukunoki, T. & Obara, Y. 2002. Characterization of failure in sand under triaxial compression using an industrial X-ray CT scanner, International Journal of Physical Modeling in Geotechnics: 15–22.

Otani, J. & Obara, Y. 2003. X-ray CT for Geomaterials, BALKEMA.

Takano, D., Khoa Dang Pham & Otani, J. 2006. 3-D visualization of soil behavior due to laterally pile using X-ray CT, Journal of Applied Mechanics JSCE 9: 513–520.

Watanabe, Y., Otani, J., Lenoir, N. & Takano, D. 2007. Displacement property in sand under triaxial compression using X-ray CT, Journal of Applied Mechanics JSCE 10: 505–512.

Yamamoto, K. & Otani, J. 2001. Microscopic observation on progressive failure on reinforced foundations, Soils and Foundations 41(1): 25–37.

Elastoplastic modeling of geomaterials considering the influence of density and bonding

T. Nakai, H. Kyokawa, M. Kikumoto, H.M. Shahin & F. Zhang

Nagoya Institute Technology, Nagoya, Japan

ABSTRACT: To understand easily the method for modeling the behavior of structured soils, one-dimensional description of elastoplastic behavior is firstly shown, and then it is extended to a constitutive model in general three-dimensional stress conditions. To describe the stress-strain behavior of structured soils, attention is focused on the density and the bonding as the main factors that affect a structured soil, because it can be considered that the soil skeleton structure in a state, which is looser than that of a normally consolidated soil, is formed by bonding effects including interlocking between soil particles and others. The extension from one-dimensional model to three-dimensional model can be done only by defining the yield function using the invariants of modified stress 't_{ij}' instead of one-dimensional stress 'σ' and assuming the flow rule in modified stress space t_{ij}.

1 INTRODUCTION

Cam-clay model (Schofield and Wroth 1968) is certainly very simple, i.e., the number of material parameters is few, and the meaning of each parameter is clear. However, Cam-clay model has problems to describe the following features: (1) Influence of intermediate principal stress on deformation and strength of soils, (2) Influence of density and/or confining pressure on the deformation and strength of soils, (3) Behavior of structured soils and aged soils.

The authors developed an elastoplastic model which is called subloading t_{ij} model (Nakai and Hinokio, 2004). This model can describe the above futures (1) and (2) properly, only adding one parameter to the same as those of Cam clay model. The feature (1) is considered using t_{ij} concept (Nakai & Mihara, 1984), and the feature (2) is considered referring to Hashiguchi's subloading surface concept (Hashiguchi 1980) and revising it. However, this model cannot describe the typical behavior of structured soils (feature (3)) which is observed in natural deposited clays.

It is known that under one-dimensional consolidation (or isotropic consolidation), remolded normally consolidated clay shows typical strain-hardening elastoplastic behavior, so that clay is assumed to be non-linear elastic in the region where the current stress is smaller than the yield stress (over consolidation region). However, real clay shows elastoplastic behavior even in over consolidation region. Furthermore, natural clay behaves intricately compared with remolded clay which is used in laboratory tests, because natural clay develops a complex structure in its deposition process. Such structured clay can exist in a region where its void ratio is greater than that of

non-structured normally consolidated clay under the same stress condition. Such type of structured clay shows more brittle and more compressive behavior than non-structured clay.

In the present paper, by introducing a state variable to represent the influence of density, a simple method to describe the elastoplastic consolidation behavior of over consolidated soil is firstly presented. To describe the consolidation behavior of structured soil, attention is focused in the real density and the bonding as the main factors that affect a structured soil, because it can be considered that the soil skeleton structure in a looser state than that of a normally consolidated soil is formed by bonding effects. A simple method to describe consolidation behavior of structured soil is then shown. Finally, these one-dimensional models are extended to ones which can describe three-dimensional stress-strain behavior of over consolidated soil and structured soil in general stress conditions.

2 ONE-DIMENSIONAL MODEL FOR OVER CONSOLIDATED SOIL

Figure 1 shows the $e - \ln \sigma$ relation in over consolidated soil schematically. Even in over consolidation region, there occurs elastoplastic deformation, the void ratio of soil approaches to the normally consolidation line (NCL) gradually with increasing stress. Figure 2 shows the change of void ratio when the stress condition moves from the initial state I ($\sigma = \sigma_0$) to the current state P ($\sigma = \sigma$). Here, e_0 and e are the initial and current void ratios of the over consolidated soil, and e_{N0} and e_N are the

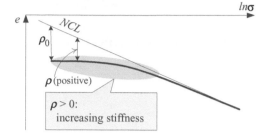

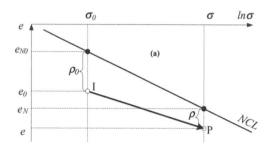

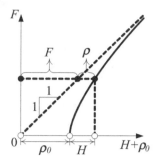

Figure 1. Void ratio (e): ln σ relation in OC clay.

Figure 2. Change of void ratio in OC clay.

Figure 3. Explanation of F and H in OC clay.

corresponding void ratio on the normally consolidation line. The difference of void ratio between normally and over consolidated soils, then, is expressed as the change from $\rho_0 (= e_{N0} - e_0)$ to $\rho = (e_N - e)$. Here, it can be assumed that the recoverable change of void ratio Δe^e (elastic component) for over consolidated soils is the same as that for normally consolidated soils and given by the following expression using the swelling index κ:

$$(-\Delta e)^e = \kappa \ln \frac{\sigma}{\sigma_0} \qquad (1)$$

So, the plastic change of void ratio Δe^p of over consolidated soil is obtained on referring to Fig. 2.

$$
\begin{aligned}
(-\Delta e)^p &= (-\Delta e) - (-\Delta e)^e \\
&= \{(e_{N_0} - e_N) - (\rho_0 - \rho) - (-\Delta e)^e\} \\
&= \lambda \ln \frac{\sigma}{\sigma_0} - (\rho_0 - \rho) - \kappa \ln \frac{\sigma}{\sigma_0} \quad (2)
\end{aligned}
$$

Here, λ is the compression index. It can be considered that the difference ρ is the state variable which represents the influence of density. By defining the term of the stress and the term of void ratio as

$$F = (\lambda - \kappa) \ln \frac{\sigma}{\sigma_0}, \quad H = (-\Delta e)^p \qquad (3)$$

Eq. (2) can be written as follows:

$$F + \rho = H + \rho_0 \quad \text{or} \quad f = F - \{H + (\rho_0 - \rho)\} = 0 \quad (4)$$

The solid line in Fig. 3 shows the relation as Eq. (4) represented in term of the relation between F and $(H + \rho_0)$. This line is approaching the broken line $(F = H)$ for normally consolidated soil, with the development of plastic deformation. The value of state variable ρ, which represents the difference of void ratio between over consolidated soil and normally consolidated soil at the same stress condition, decreases monotonously from ρ_0 to zero with development of plastic deformation. The tangential slope dF/dH of the solid line gives an idea of the stiffness against the plastic change of void ratio for over consolidated soil. This can be compared with the stiffness of a normally consolidated soil given by the slope dF/dH of the broken line, which is always unity in this diagram.

From the consistency condition ($df = 0$) at the occurrence of plastic deformation with satisfying Eq. (4), the following equation is obtained:

$$
\begin{aligned}
df &= dF - \{dH - d\rho\} \\
&= (\lambda - \kappa) \frac{d\sigma}{\sigma} - \{d(-\Delta e)^p - d\rho\} = 0 \quad (5)
\end{aligned}
$$

From this equation, the increment of the plastic change of void ratio is expressed as

$$d(-\Delta e)^p = (\lambda - \kappa) \frac{d\sigma}{\sigma} + d\rho \qquad (6)$$

As shown in Figs. 1 and 3, it can be assumed that the state variable ρ representing the density decreases ($d\rho < 0$) with development of plastic deformation (volume contraction) and finally becomes zero (normally consolidated state). Further, it can be considered that the larger the value of ρ is, the faster the rate of degradation of ρ is. Then, the evolution rule of ρ can be given in the following form:

$$d\rho = -G(\rho) \cdot d(-\Delta e)^p \qquad (7)$$

Here, $G(\rho)$ should be an increasing function of ρ with satisfying $G(0) = 0$, from the condition above mentioned.

From Eqs. (1), (6) and (7), the increment of total change of void ratio is given by the summation of the plastic component and the elastic component.

$$d(-\Delta e) = d(-\Delta e)^p + d(-\Delta e)^e$$

$$= \left\{ \frac{\lambda - \kappa}{1 + G(\rho)} + \kappa \right\} \frac{d\sigma}{\sigma} \qquad (8)$$

Equation (8) means that $G(\rho)$ increases the stiffness of soil, and its effect becomes large with the increase of the value of ρ. The method to consider the influence of density presented here corresponds to one-dimensional interpretation of the subloading surface concept by Hashiguchi (1980).

3 ONE-DIMENSIONAL MODEL FOR STRUCTURED SOIL

Solid curve in Fig. 4 shows a typical $e - n\,\sigma$ relation of natural clay schematically. Natural clay behaves intricately compared with remolded clay which is used in laboratory tests, because natural clay develops a complex structure in its deposition process. Such structured soil can exist in a region where its void ratio is greater than that of non-structured normally consolidated soil under the same stress condition. Such type of structured soil shows more brittle and more compressive behavior than non-structured soil. Asaoka et al. (2002) and Asaoka (2005) developed a model to describe such structured soils, introducing subloading surface and superloading surface concepts to the Cam-clay model. In their modeling, a factor related to the over consolidation ratio (corresponding to imaginary density) has been introduced to increase the stiffness, and a factor related to the soil skeleton structure has been introduced to decrease the stiffness. By controlling the evolution rules of these factors, they described various features of consolidations and shear behaviors of structured soils.

In the present paper, attention is focused in the real density and the bonding as the main factors that affect the behavior of structured soil, because it can be considered that the soil skeleton structure in a looser state than that of a normally consolidated soil is formed by bonding effects. Figure 5 shows the change of void ratio when the stress condition moves from the initial state I ($\sigma = \sigma_0$) to the current state P ($\sigma = \sigma$) in the same way as that in Fig. 2. Here, e_0 and e are the initial and current void ratios of structured soil, and e_{N0} and e_N are the corresponding void ratio on the normally consolidation line. The arrow with broken line denotes the same change of void ratio as that in Fig. 2 for over consolidated soil. Now, it can be understood that the structured soil is stiffer than over consolidated soil, even if their densities (ρ) are the same. The change of void ratio indicated by the arrow

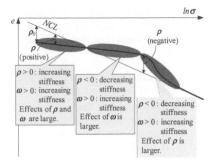

Figure 4. Void ratio (e): In σ relation in structured clay.

with solid line is smaller than that for over consolidated soil (arrow with broken line). Such increase in stiffness will be expressed by introducing the imaginary density ω which represents the effect of the bonding, in addition to the real density ρ. Considering the factor of the bonding as well as the factor of density and referring to Eq. (4), we can assume the following relation between F and H:

$$F + \rho + \omega = H + \rho_0 + \omega_0$$

$$\text{or } f = F - \{H + (\rho_0 - \rho) + (\omega_0 - \omega)\} = 0 \qquad (9)$$

A typical relation between F and $(H + \rho_0 + \omega_0)$ is shown in Fig. 6. The horizontal distance between the solid line and the dash-dotted line indicates the magnitude of the bonding effect ω, starts from the initial value ω_0, and decreases with the development of plastic deformation. Furthermore, the horizontal distance between the dash-dotted line and the broken line implies the influence of the current density ρ.

From the consistency condition ($df = 0$) at the occurrence of plastic deformation with satisfying Eq. (9), the following equation is obtained:

$$df = dF - \{dH - d\rho - d\omega\}$$

$$= (\lambda - \kappa)\frac{d\sigma}{\sigma} - \{d(-\Delta e)^p - d\rho - d\omega\} = 0 \qquad (10)$$

The increment of the plastic change of void ratio for structured soil is then expressed as

$$d(-\Delta e)^p = (\lambda - \kappa)\frac{d\sigma}{\sigma} + d\rho + d\omega \qquad (11)$$

Since it can be assumed that the value of ω decreases with the development of plastic deformation and finally becomes zero in the same way as the evolution of ρ, the evolution rule of ω can be expressed as

$$d\omega = -Q(\omega) \cdot d(-\Delta e)^p \qquad (12)$$

369

Here, $Q(\omega)$ is a monotonously increasing function of ω satisfying $Q(0) = 0$, such as $Q(\omega) = b\omega$ (b: material parameter). Substituting Eqs. (7) and (12) into Eq. (11) gives

$$d(-\Delta e)^p = \frac{\lambda - \kappa}{1 + G(\rho) + Q(\omega)} \cdot \frac{d\sigma}{\sigma} \qquad (13)$$

As shown in this equation, the increase of the stiffness of structured soil from remolded normally consolidated soil is governed by the evolution functions of $G(\rho)$ and $Q(\omega)$. Finally the increment of total change of void ratio is given by the summation of the elastic component in Eq. (1) and the plastic component in Eq. (13)

$$d(-\Delta e) = d(-\Delta e)^p + d(-\Delta e)^e$$
$$= \left\{ \frac{\lambda - \kappa}{1 + G(\rho) + Q(\omega)} + \kappa \right\} \frac{d\sigma}{\sigma} \qquad (14)$$

If $G(\rho) = 0$ and $Q(\omega) = 0$, Eq. (14) expresses the relation for remolded normally consolidated soil (abbreviated as NC soil). It is also apparent that positive values of $G(\rho)$ and $Q(\omega)$ have the effect to increase the stiffness of soil.

Now, Eq. (11) implies that the increment of the plastic change of void ratio for structured soil is given by $d\rho$ and $d\omega$ in addition to the same component as that of the normally consolidated soil (first term of right-hand side of the equation). Therefore, the increment of real density ($\Delta \rho$) for structured soil is given not only by the effect of current density ρ but also by the effect of imaginary density ω due to bonding as follows:

$$\Delta \rho = d\rho + d\omega = - \{G(\rho) + Q(\omega)\} \cdot (-\Delta e)^p \quad (15)$$

The value of the state valuable ρ can also be updated as the difference between the void ratio corresponding to the normally consolidation line (NCL) at the current stress and the current void ratio.

Here, we will briefly discuss about the consolidation behavior of structured soil in Fig. 4. Assume the initial state with positive ρ_0 and positive ω_0. At the first stage ($\rho > 0$ and $\omega > 0$), the stiffness of the soil is much larger than that of NC soil, because of the positive values of $G(\rho)$ and $Q(\omega)$. When the current void ratio becomes the same as that on NCL ($\rho = 0$), the stiffness of the soil is still larger than that of NC soil ($\omega > 0$), so it is possible to have the state of structured soil which is looser than that on NCL. In this stage ($\rho < 0$ and $\omega > 0$), the effect to increase the stiffness by positive value of ω is larger than the effect to decrease the stiffness by negative value of ρ. After this stage the effect of ω becomes small with development of plastic deformation. On the other hand, the effect of ρ to decrease the stiffness becomes prominent because of the negative value of ρ. Finally the void ratio approaches to that on NCL, because ρ and ω converge to zero.

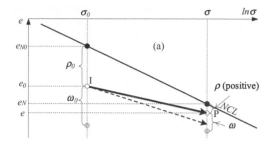

Figure 5. Change of void ratio in structured clay.

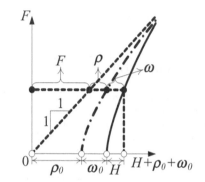

Figure 6. Explanation of F and H in structured clay.

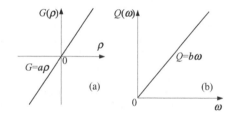

Figure 7. $G(\rho)$ and $Q(\omega)$ given by linear functions.

The one-dimensional loading condition of soil is presented from the condition that there never occurs the reduction of plastic change of void ratio as follows:

$$\begin{cases} d(-\Delta e)^p \neq 0 & \text{if } d(-\Delta e)^p > 0 \\ d(-\Delta e)^p = 0 & \text{if } d(-\Delta e)^p \leq 0 \end{cases} \qquad (16)$$

In order to check the validity of the present model, numerical simulations of one-dimensional compression tests are carried out. The functions of $G(\rho)$ and $Q(\omega)$ in Eqs. (7) and (12) are given by simple linear functions of ρ and ω as shown in Fig. 7. Here, though $Q(\omega)$ is defined in positive side alone, $G(\rho)$ is defined both in positive and negative side. This is because it is possible that ρ becomes negative, as mentioned above. Assuming Fujinomori clay which is used in the previous experimental verification of

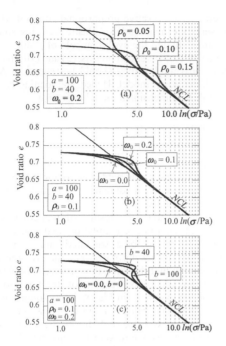

Figure 8. Calculated results of clay with different ρ_0, ω_0 and b.

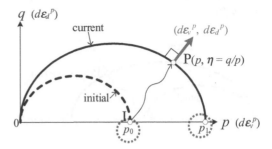

Figure 9. Yield surface on (p, q) plane and direction of plastic strain increment.

constitutive models (e.g., Nakai and Hinokio, 2004; Nakai, 2007), following material parameters are employed in the numerical simulations: compression index $\lambda = 0.104$, swelling index $\kappa = 0.010$ and void ratio on NCL at $\sigma = Pa = 98$ kPa (atmospheric pressure) N = 0.83. Figure 8(a) shows the calculated $e - \ln \sigma$ relation of the one-dimensional compression using initial bonding parameter ($\omega_0 = 0.2$) and different initial void ratio. Figure 8(b) shows the calculated results using the same initial void ratio ($\rho_0 = 0.1$) and different bonding parameter. In these figures, the parameters (a and b) which represent the degradation rate of ρ and ω are fixed. It can be seen from these Figure that it is possible to describe the deformation of structured clay only considering the effect of density and bonding and their evolution rules. Further, Figure 8(c) shows the results in which the initial void ratio and the initial bonding are the same but the parameter b is different. We can see that the result with large value of b (=100) describe void ratio-stress relation with strain softening.

4 EXTENSION TO THREE-DIMENSIONAL MODEL

In most of three-dimensional models such as Cam clay model, their yield functions are formulated using the stress invariants (mean stress p and deviatoric stress q)

or (p and $\eta = q/p$), and flow rule is assumed in ordinary σ_{ij} space as shown in Fig. 9. Here, p and q correspond to the normal and parallel components of σ_{ij} to the octahedral plane. However, it is known that constitutive models formulated in such a way cannot consider the influence of intermediate principal stress on the deformation and strength of soils properly.

For considering the influence of intermediate principal stress automatically, the concept of modified stress t_{ij} is proposed (Nakai and Mihara, 1984). In this concept, the yield function is formulated using the stress invariants based on t_{ij} concept (t_N and t_S) or (t_N and $X = t_S/t_N$) instead of (p and q) or (p and $\eta = q/p$), and the following flow rule is assumed in the modified stress t_{ij} space instead of the ordinal σ_{ij} space, as shown in Fig. 10:

$$d\varepsilon_{ij}^p = \Lambda \frac{\partial f}{\partial t_{ij}} = \Lambda \left(\frac{\partial f}{\partial t_N} \frac{\partial t_N}{\partial t_{ij}} + \frac{\partial f}{\partial X} \frac{\partial X}{\partial t_{ij}} \right) \quad (17)$$

Here, (t_N and t_S) correspond to the normal and parallel components to the spatially mobilized plane (Matsuoka and Nakai, 1974). The detail of t_{ij} concept has been described in the previous papers (e.g., Nakai and Mihara, 1984; Nakai and Hinokio, 2004), and the three-dimensional model for structured soil based on t_{ij} concept has also been described in another paper (Nakai, 2007; Nakai et al. 2008). So, we will present here only the points different from the one-dimensional model.

In three-dimensional model, the yield functions is fundamentally the same as Eq. (4) and (9). Here, F and H in these equations are given by Eq. (3) as the functions of stress σ and the plastic change of void ratio $(-\Delta e)^p$ in the one-dimensional model, On the other hand, in three-dimensional model based on t_{ij} concept, F is given as follows by replacing σ and σ_0 with t_{N1} and t_{N10}, respectively:

$$F = (\lambda - \kappa) \ln \frac{t_{N1}}{t_{N0}} = (\lambda - \kappa) \left\{ \ln \frac{t_N}{t_{N0}} + \varsigma(X) \right\} \quad (18)$$

where, t_{N1} and t_{N0} are the values of the current and initial yield loci at t_N axis, and $\xi(X)$ is a monotonically increasing function of $X = t_S/t_N$. Also,

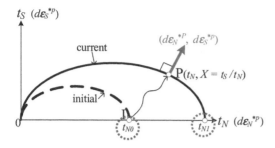

Figure 10. Yield surface on (t_N, t_S) plane and direction of plastic strain increment.

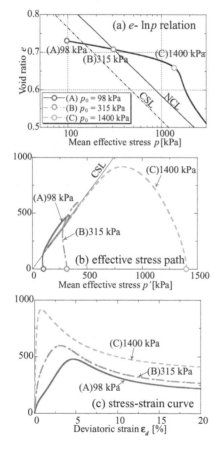

(a) e- $\ln p$ relation

○—(A) $p_0 = 98$ kPa
○- -(B) $p_0 = 315$ kPa
○—(C) $p_0 = 1400$ kPa

(b) effective stress path

(c) stress-strain curve

Figure 11. Calculated results of isotropic compression and undrained shear tests on structured clay.

H is described in the same way as that in one-dimensional model. Then, H and its increment are expressed as follows referring to Eq. (17):

$$H = (-\Delta e)^p = (1 + e_0) \cdot \varepsilon_v^p, \quad dH = (1 + e_0)\Lambda \frac{\partial f}{\partial t_{ii}} \quad (19)$$

Here, Λ is a positive proportional constant which represents the magnitude of plastic stain increment. Therefore, though the evolution rule of ρ and ω are related to $(-\Delta e)^p$ as shown in Eqs. (5) and (10) in one-dimensional model, they should be related to Λ in three-dimensional model. Hence, the loading condition in three-dimensional model is given as

$$\begin{cases} d\varepsilon_{ij}^p \neq 0 & \text{if } \Lambda > 0 \\ d\varepsilon_{ij}^p = 0 & \text{if } \Lambda \leq 0 \end{cases} \quad (20)$$

There is nothing new but the above mentioned procedures to extend one-dimensional model to three-dimensional model.

Figure 11 shows the results of numerical simulations of isotropic compression and succeeding undrained shear tests on a structured Fujinomori clay. Diagram (a) is the results of isotropic compression test. It can be seen that the three-dimensional model simulates well compression behavior of structured soil in the same way as those in Fig. 8. in one-dimensional compression. Diagrams (b) and (c) show the result of effective stress paths and stress- strain curves in undrained triaxial compression tests on clays which are sheared from stress conditions (A), (B) and (C) in diagram (a). The model simulates well typical undrained shear behavior of structured soil. i.e., the differences of stress paths and stress- strain curves depending on the magnitude of confining pressure, rewinding of stress path after increasing of mean stress and deviatoric stress and others.

5 CONCLUSIONS

As a one-dimensional model, a method to describe the behavior of over consolidated soil is presented by using the state valuable of density and its monotonous evolution rule. Next, introducing the state valuable which represent the effect of bonding as well as the state valuable of density, a model to describe the behavior of structured soil is developed. Furthermore, it is shown that by introducing the t_{ij} concept proposed before, these models can easily be extended to three-dimensional ones. The validity of these models is checked by the simulations of one-dimensional compression and undrained shear tests on structured soils with different densities and bonding effects.

REFERENCES

Asaoka, A. 2005. Consolidation of clay and compaction of sand—an elastoplastic description, *Proc of 12th Asian Regional Conf. on Soil Mech. and Geotechnical Eng.*, Keynote Paper, Singapore, 2: 1157–1195.

Asaoka, A., Noda, T., Yamada, E., Kaneda, K. & Nakano, M. 2002. An elasto-plastic description of two distinct volume change mechanisms of soils, *Soils and Foundations*. 42(3): 47–57.

Hashiguchi, K. 1980. Constitutive equation of elastoplastic materials with elasto-plastic transition. *Jour. of Appli. Mech., ASME*, 102(2): 266–272.

Matsuoka, H. & Nakai, T. 1974. Stress-deformation and strength characteristics of soil under three different principal stresses. *Proc. of JSCE*, 232: 59–70.

Nakai, T. 2007. Modeling of soils behavior based on t_{ij} concept, *Proc. of 13th Asian Regional Conf. on Soil Mech. and Geotechnical Eng.*, Keynote Paper, Kolkata, Preprint 1–25.

Nakai, T. & Hinokio, T. 2004. A simple elastoplastic model for normally and over consolidated soils with unified material parameters. *Soils and Foundations*, 44(2): 53–70.

Nakai, T. & Mihara, Y. 1984. A new mechanical quantity for soils and its application to elastoplastic constitutive models. *Soils and Foundations*, 24(2): 82–94.

Nakai, T., Zhang, F., Kyokawa, H., Kikumoto, M. & Shahin, H.M. 2008. Modeling the influence of density and bonding on geomaterials, *Proc. of 2nd International Symposium on Geotechnics of Soft Soils*, Keynote Paper, Glasgow, 65–76.

Schofield, A.N. & Wroth, C.P. 1968. *Critical State Soil Mechanics*, McGrow-Hill, London.

Prediction and Simulation Methods for Geohazard Mitigation – Oka, Murakami & Kimoto (eds)
© 2009 Taylor & Francis Group, London, ISBN 978-0-415-80482-0

Cyclic strength of sand with various kinds of fines

M. Hyodo, UkGie Kim & T. Kaneko
Yamaguchi University, Ube, Japan

ABSTRACT: In most design codes, soils are classified as either sand or clay, and appropriate design equations are used to represent their behaviour. For example, the behaviour of sandy soils is expressed in terms of the soil's relative density, whereas consistency limits are often used for clays. However, sand-clay mixtures, which are typically referred to as intermediate soils, cannot be easily categorized as either sand or clay and therefore a unified interpretation of how the soil will behave at the transition point, i.e., from sandy behaviour when fines are few to clay behaviour at high fines content, is necessary. In this paper, the cyclic shear behaviour of sand and fines mixtures with various plasticities was investigated by considering variations in fines content and compaction energy, while paying attention to the void ratio expressed in terms of sand structure. Then, by using the concept of equivalent granular void ratio, the contribution of fines on the cyclic shear strength of the soil was estimated.

1 INTRODUCTION

Since the Anchorage (Alaska) and Niigata (Japan) earthquakes, great strides have been made in understanding the mechanism behind liquefaction and the conditions that make soils susceptible to it. However, most of the studies have concentrated on clean sands or sands with relatively small amounts of fines, with the assumption that the behaviour of sandy soils is similar to that of clean sands. Moreover, many studies have focused on the behaviour of sand and clay separately to simplify things.

Such assumption is not valid in actual grounds because of the presence of soils with various characteristics. Most soils in natural state are generally composed of combinations of sand, clay, silt, etc. It is hard to classify such soils into either sand or clay, because they possess both properties of sand and fines. These kinds of soil are called *intermediate soils*. The properties of intermediate soils change in various ways due to the effect of density or fines content. Therefore, difficulties arise in understanding their dynamic characteristics and liquefaction potential. Numerous laboratory studies on sand-fine mixtures have been performed, and have produced what appear to be conflicting results. Some studies have reported that increasing the fine content in sand will either increase or decrease the liquefaction resistance of the sand; others indicated that the liquefaction resistance will decrease until some limiting fine content is reached, after which it will increase again. Therefore, a single interpretation appears to be difficult. Such differences in effects have been attributed to the presence or absence of activity in finer grains, i.e.,

inactive fines decreases the liquefaction strength of sands, while the opposite tendency is observed for active fines (Matsumoto et al. 1999). Additionally, several studies have shown that the liquefaction resistance of such mixture is more closely related to its sand skeletal structure than to its fine content. However, a unified framework has not been established yet on how the interaction between coarse-grained and fine-grained particles affects the overall behavior of the mixture.

In this paper, silica sand and three kinds of fines were mixed together at various proportions, and a wide range of soil structures, ranging from one with sand dominating the soil structure to one with fines controlling the behaviour, were prepared. Then, using the concept of granular void ratio, undrained cyclic shear tests were performed on the mixtures.

2 CHARACTERISTICS OF SOILS USED

2.1 Test materials

In the experiments, Silica sand with adjusted grain size and fine-grained materials, consisting of Iwakuni clay (natural clay), Tottori silt (non-plastic silt) and kaolin clay, were mixed separately at various proportions in order to form wide range of soil types, i.e., from sand to clay/silt. Initially, No. V5, R5.5, V6 and V3 Mikawa silica sands were mixed at 1:2:2:5 ratios by dry weight in order to modify the sand's grain size distribution. The silica sand with adjusted grain size distribution has $e_{max} = 0.850$ and $e_{min} = 0.524$.

Table 1. Physical properties of test materials.

Sample	Fines content (%)	Clay content (%)	G_s	I_p	d_{50} (mm)	U_c
Silica sand	0	0.0	2.652	NP	0.861	4.04
Iwakuni clay	98	38.8	2.610	47.54	0.006	–
Tottori silt	98	6.0	2.665	NP	0.019	2.85
Kaolin clay	100	72.3	2.618	21.80	0.002	–

Table 2. Relation between mixed proportions and fines content.

Silica sand	Iwakuni clay		Tottori silt		Kaolin clay	
Mixture rate by weight (%)	Mixture rate by weight (%)	Fines content (%)	Mixture rate by weight (%)	Fines content (%)	Mixture rate by weight (%)	Fines content (%)
100	0	0.0	0	0.0	01	0.0
90	10	9.8	10	9.8	10	10.0
85	15	14.7	15	14.7	15	15.0
83	17	16.7	–	–	–	–
80	20	19.6	20	19.6	20	20.0
70	30	29.4	30	29.4	30	30.0
50	50	49.0	50	49.0	–	–
0	100	98.0	100	98.0	100	100.0

Complete index properties for the component soils are presented in Table 1, while the grain size distribution curves are illustrated in Figure 1. Fines content of the soil mixtures were varied from 0 to 100% based on dry weights, with clay contents varying from 0 to 72.3%. The relation between mixed proportions and fine contents of the samples used in the experiment are shown in Table 2. Since Iwakuni clay and Tottori silt have an original sand content of 2%, the fines content of each sample of sand-fines mixture is smaller than the mixture rate by weight. For Iwakuni clay mixture with $F_c > 20\%$ and Kaolin clay mixture with $F_c > 30\%$, the soil mixture has activity, while Tottori silt mixtures are classified as non-plastic (ML).

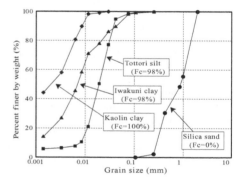

Figure 1. Grain size distribution curves of samples.

2.2 Skeletal structure of sand-clay mixture

When fines content is low, sand dominates the structure of the mixture and fines occupy the void spaces between the coarse grains. Even for constant fines content, the void ratio of the sand structure can change freely, with the strength dependent on the void ratio of the sand structure. When F_c is high, on the other hand, the fines comprise the soil matrix with sand particles scattered in between, and the strength of the mixture is governed chiefly by the characteristics of the fine-grained soils. Thus, sand structure has a large effect on the strength of sand-clay mixtures and therefore, it is

more appropriate to pay attention on the void ratio of the sand structure rather than on fines content (Hyodo et al. 2006). A fully saturated sand-clay mixture has a three-phase composition, namely the coarse-grained particles, fine-grained particles and pore water, as shown in Figure 2. In this paper, the state of the sand structure formed within the soil mixture is defined by considering the fines as voids and using the concept of granular void ratio, e_g (Georgiannou et al. 1990, Itoh et al. 2001), i.e.:

$$e_g = \frac{V_w + V_{sf}}{V_{ss}} \qquad (1)$$

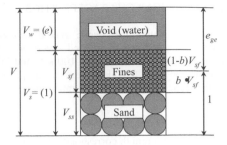

Figure 2. Phase diagram.

where V_w is volume of water, V_{ss} is volume of the coarse grains and V_{sf} is the volume of the fine grains.

2.3 Specimen preparation

For soil mixture with active portions, each sample was mixed at water content about twice its liquid limit, after which it was placed in a pre-consolidation cell and subjected to a vertical pressure of 50 kPa. After consolidation, a specimen measuring 5 cm in diameter and 10 cm in height was formed. On the other hand, to prepare soil specimen for samples with non-plastic portion, the soil mixtures were placed in a mold in 5 layers with each layer compacted at a prescribed number of blows using a steel rammer. The compaction energy, E_c, was calculated as follows (Adachi et al. 2000):

$$E_c = \frac{W_R \cdot H \cdot N_L \cdot N_B}{V} \tag{2}$$

In the above equation, W_R is the rammer weight ($= 0.00116$ kN), H is the drop height (m), N_L is the number of layers ($= 5$), N_B is the number of blows per layer, and V is the volume of mold (m^3). In the experiment, various compaction energies, E_c, were obtained by changing H and N_B. Note that each soil mixture was thoroughly mixed at an initial water content of $w = 11\%$.

Figure 3(a) shows the relation between void ratio and fines content for Iwakuni clay mixtures after consolidation at an effective confining pressure of $\sigma_c' = 100$ kPa. For mixtures with $F_c \leq 20\%$, there is only one relation for the prescribed consolidation pressure. Depending on the normal consolidation condition, the void ratio of the pre-consolidated sample is unique for a given effective confining pressure. On the other hand, samples with $F_c < 20\%$ were formed at constant compaction energies, corresponding to $E_c = 22$–504 kJ/m^3. It is possible to form specimens with different void ratios by changing the compaction energy applied. It can be observed that for soil samples prepared under low compaction energies ($E_c = 22$ and 51 kJ/m^3), the void ratio becomes smaller when F_c increases. However, for compaction energies higher than $E_c = 113$ kJ/m^3, the void ratio decreases with

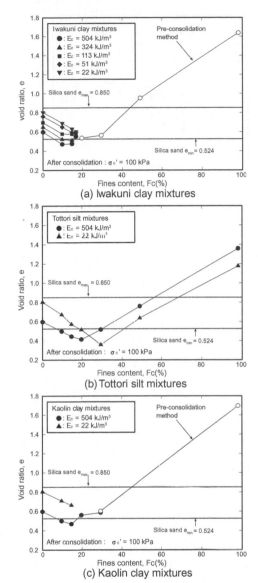

(a) Iwakuni clay mixtures

(b) Tottori silt mixtures

(c) Kaolin clay mixtures

Figure 3. Relation between void ratio and fine content.

increase in F_c and attains its minimum value when the fines content is about $F_c = 10$–15%; afterward, the relation between fines content and void ratio bends upward. Note that for all compaction energies, the void ratios of the specimens seem to converge at a single point.

The relations between void ratio and fines content for Tottori silt mixtures are shown in Figure 3(b). Since these mixtures are non-plastic, sample preparation by pre-consolidation method was not possible; instead,

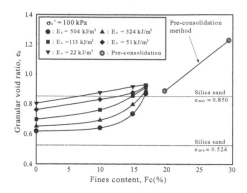

Figure 4. Relation between void ratio and fines content for Iwakuni clay mixtures ($Fc \leq 30\%$).

moist compaction method was employed on all samples ($F_c = 0$–100%). After consolidation, the void ratio of specimen tend to change at $F_c > 30\%$. This is because during the saturation process, remarkable volumetric compression occurred on specimens with high silt content or with loose initial density. The void ratio of both specimens after consolidation is lowest at $F_c = 20$–30%.

Figure 3(c) shows the same relations for Kaolin clay mixtures. In this case, mixtures with $F_c \leq 20\%$ were non-plastic, while when $F_c = 30, 100\%$, the mixture is considered to have activity. Therefore, each specimen for soil mixture with $F_c \leq 30\%$ was prepared by pre-consolidation method. On the other hand, soil samples with $F_c = 0$–30% were formed at constant compaction energies, corresponding to $E_c = 22, 504$ kJ/m³. It can be observed that under low compaction energy ($E_c = 22$ kJ/m³), the void ratio becomes smaller when the fines content increases. However, under high compaction energy ($E_c = 504$ kJ/m³), the void ratio decreases with increase in fines content and attains its minimum value when the fines content is about $F_c = 15\%$. Note that for soil samples with $F_c = 30\%$ prepared by compaction (at $E_c = 504$ kJ/m³) and by pre-consolidation, the void ratios of the specimens indicate almost similar values.

Figure 4 shows the relation between granular void ratio, e_g, and fines content, F_c, for Iwakuni clay mixtures for $F_c < 30\%$. For each compaction energy, the granular void ratio increases at different rates with increasing fines content. It is observed that for soil samples with $F_c = 0$–17%, it is possible to form specimens with different granular void ratios by changing the compaction energy applied. For soil samples with granular void ratios less than the e_{max} ($= 0.85$) of Silica sand, the coarse-grained soils comprise the soil mixture. On the other hand, for those specimens with granular void ratios exceeding e_{max} of silica sand, it can be considered that the soil matrix of these soils is dominated by the clay particles.

3 CYCLIC SHEAR CHARACTERISTICS

Several series of undrained cyclic triaxial tests were conducted on the above-mentioned specimens with effective confining pressure $\sigma'_c = 100$ kPa and loading frequency $f = 0.02$ Hz using an air pressure controlled-type cyclic triaxial test apparatus.

3.1 Results of cyclic shear tests

The cyclic shear strength corresponding to 20 cycles (hereinafter referred to as cyclic shear strength ratio, $R_{L(N=20)}$) is plotted against the fines content, as shown in Figure 5. The test results for Iwakuni clay mixtures with soil types ranging from sand to clay are summarized in Figure 5(a). Firstly, when $F_c > 20\%$, the cyclic shear strength increases rapidly, and when $F_c > 30\%$, the $R_{L(N=20)}$ asymptotically approaches the strength of clay particles. This is because the fines dominate the structure of the soil matrix as the fines content increases. Secondly, for specimens with $F_c = 0$–20%, subjected to constant compaction energy, it is observed that as the fines content increases, there is a corresponding decrease in $R_{L(N=20)}$ for dense samples (compacted under high energies), while an increase is noted for loose samples (compacted under low energies). Such response can be interpreted as follows: for dense condition, the skeletal structure of the sand is gradually lost as the fines content increases, resulting in decrease in strength. For loose samples, the increase in the amount of fines leads to an increase in density of the soil, which results in an increase in strength.

Figure 5(b) shows the same relations for Tottori silt mixtures. For specimens prepared under low compaction energy, any increase in fines content results in increase in liquefaction strength. In contrast, for specimens prepared under high compaction energy, liquefaction strength shows complex change as the fines content increases.

The results for Kaolin clay mixtures are illustrated in Figure 5(c). For a given compaction energy, it can be seen that as the fines content increases, there is a decrease in cyclic strength. In this case, the granular void ratio of the soil mixture increases as the fines content increases, resulting in decrease in cyclic strength. For pre-consolidated samples with $F_c \leq 30\%$, the value of $R_{L(N=20)}$ is almost constant. In this case, the fines form the main matrix of the soil and sand has almost no influence on the strength.

3.2 Relation between skeletal structure and cyclic shear strength

Based on the results of cyclic shear tests discussed above, it can be summarised that the fines content and the resulting skeletal structure have significant effects on the cyclic strength of sand-clay mixtures.

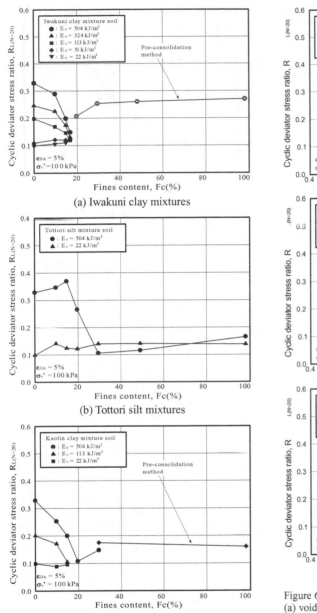

(a) Iwakuni clay mixtures

(b) Tottori silt mixtures

(c) Kaolin clay mixtures

Figure 5.   Relation between cyclic shear strength ratio and fines contents.

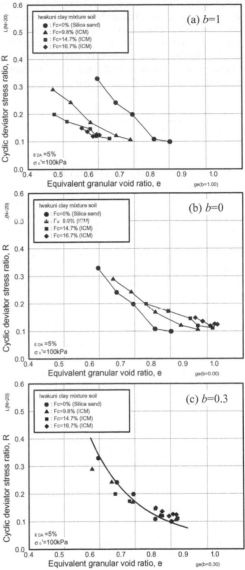

Figure 6.   Relation between cyclic shear strength ratio and (a) void ratio ($b = 1$); and (b) granular void ration ($b = 0$); and (c) equivalent granular void ratio ($b = 0.3$).

Figure 6 illustrate the relationship of the cyclic shear strength for Iwakuni clay mixtures and the void ratio or granular void ratio, for samples with $F_c < 20\%$. From Figure 6(a), it is observed that there is a remarkable decrease in $R_{L(N=20)}$ with increase in void ratio and fines content. Figure 6(b) shows that for a given cyclic strength, the granular void ratio increases when F_c increases. Moreover, an increase in fines content results in an increase in strength for a particular e_g. When the granular void ratio exceeds $e_{max} = 0.85$, the clay particles dominate the soil matrix and a significant strength increase is not observed. In other words, after a threshold value of fines content is reached, contacts between the sand grains totally disappear and they start to "float" within the fine-grain matrix. As can be seen in both figures, there is no unique relationship between e, e_g and cyclic shear strength for sand-clay mixture. In order to provide a coherent way of characterizing the strength of soil mixtures, the way how fines affect the strength of the host sand needs to be incorporated into the definition of granular void ratio. Thevanayagam et al. (2002) introduced the following equation:

$$e_{ge} = \frac{e + (1 - b)f_c}{1 - (1 - b)f_c}; \quad f_c = \frac{V_{sf}}{V_{sf} + V_{ss}} \quad (3)$$

where e_{ge} is the equivalent granular void ratio, f_c is the fines content (in terms of volume) and b denotes the portion of the fines that contributes to the active inter-grain contacts. Basically, $b = 0$ means that none of the fine grains actively participates in supporting the coarse-grain skeleton (i.e. the fines act exactly like voids); and $b = 1$ implies that all fines actively participate in supporting the coarse/grain skeleton (i.e. the fines are indistinguishable from the host sand particles). The magnitude of b depends on grain size disparity and grain characteristics.

By assigning different values of b, the void ratios in Figure 6(a) and (b) are re-calculated. From the results, a b value of 0.3 indicates that almost all the data fall into a narrow band that surrounds the data for clean sand ($F_c = 0\%$), as illustrated in Figure 6(c). This implies that 30% of the fines of Iwakuni clay mixture contribute to the cyclic strength of the soil.

Figure 7 shows the variation in cyclic shear strength, $R_{L(N=20)}$ for the three kinds of fines with respect to equivalent granular void ratio, e_{ge}. As a result of changing the parameter b, good correlation between $R_{L(N=20)}$ and e_{ge} was obtained at around $b = 0.3$ for Iwakuni clay, 0.43 for Tottori silt and 0.14 for Kaolin clay.

In Figure 8, the relation between $R_{L(N=20)}$ and equivalent granular relative density is illustrated for

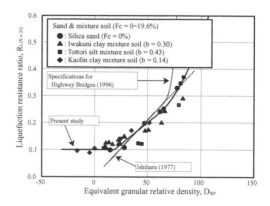

Figure 8. Relation between liquefaction resistance ratio and equivalent granular relative density.

all the soil mixtures used. The equivalent granular relative density. D_{rge}, is defined as:

$$D_{rge} = \frac{e_{max, HS} - e_{ge}}{e_{max, HS} - e_{min, HS}} \times 100(\%) \quad (4)$$

where, $e_{max, HS}$ and $e_{min, HS}$ are the maximum and minimum void ratios of the host sand (without any fines), respectively, and e_{ge} is the equivalent granular void ratio. If the equivalent void ratio in Fig. 7 is re-calculated by assigning a D_{rge} for each mixture, the results indicated that all the data fall within a small band, including those for clean sand ($F_c = 0\%$).

4 RELATION BETWEEN CONTRIBUTION FACTOR AND PHYSICAL PARAMETERS

The above results mean that the b parameter is related to the ratio of the void size distribution of the host sand and the particle size distribution of the fines. A simpler way is to look at the mean value of fines size and mean value of voids in the host sand. The mean value of fines size can be represented by the d_{50} of the fines, while d_{10} of the host sand may be used as a rough index of the mean void size. Ni et al. (2004) proposed the grain size ratio, χ, as:

$$\chi = \frac{d_{10, Hostsand}}{d_{50, Fines}} \quad (5)$$

The expectation is that the b value will decrease with increasing χ. The contribution factors from the studies by Ni (2004) and those obtained in this experiment are presented in Table 3, while the relation between contribution factor and grain size is illustrated in Figure 9. In the figure, it is observed that as the grain size ratio increases, there is a corresponding decrease in contribution factor, regardless of the plasticity index of the soil mixture.

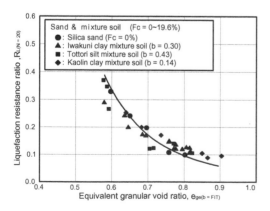

Figure 7. Relation between liquefaction resistance ratio and equivalent granular void ratio.

Table 3. Contribution factor and grain size ratio.

Soil	Reference	b	e_g	x
Old Alluvium	Ni, Q., et al (2004)	0.70	0.62~0.75	4.4
Toyoura sand	Zlatovic & Ishihara (1995)	0.25	0.85~1.15	11
with silt	Thevanayagam & Mohan (2002)	0.25	0.460~0.947	16
Silica-iwakuni clay	Present study	0.30	0.597~1.227	24.7
Silica-tottori silt	Present study	0.43	0.597~1.167	14.3
Silica-kaolin clay	present study	0.14	0.597~1.228	34

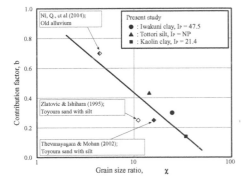

Figure 9. Relation between contribution factor and grain size ratio.

5 CONCLUSIONS

Undrained cyclic shear tests were performed in order to investigate the effects of fines content on the liquefaction strength of sand-fines soils. Based on the test results, the following conclusions were reached.

1. For soil samples with F_c beyond a certain threshold value, the strength was governed by the fines, which control the soil matrix.
2. Increasing fines content resulted in decrease in cyclic strength for dense soils, while an increase in strength was generally observed for loose soils.
3. When sand particles dominated the soil matrix, a unified formula representing the cyclic shear strength and equivalent granular void ratio for the sand-fines mixtures with various sand structures and fine contents were obtained. For Iwakuni clay, Tottori silt, and Kaolin clay mixed with sand, the contributions of the fines to the active inter-grain contact were 30, 43, and 14%, respectively.

4. The contribution factor is dependent on the grain size ratio of the soil mixture.

REFERENCES

Adachi, M., Yasuhara, K. and Shimabukuro, A. 2000. Influences of sample preparation method on the behavior of non-plastic silts in undrained monotonic and cyclic triaxial tests. *Tsuchi-to-Kiso* 48(11): 24–27 (in Japanese).

Georgiannou, V.N., Burland, J.B. and Hight, D.W. 1990. The undrained behaviour of clayey sands in triaxial compression and extension. *Geotechnique* 40(3): 431–449.

Hyodo, M., Orense, R., Ishikawa, S., Yamada, S., Kim, U.G. and Kim, J.G. 2006. Effects of fines content on cyclic shear characteristics of sand-clay mixtures. *Proc., Earthquake Geotechnical Engineering Workshop-Canterbury*: 81–89.

Itoh, S., Hyodo, M., Fujii, T., Yamamoto, Y. and Taniguchi, S. 2001. Undrained monotonic and cyclic shear characteristics of sand, clay and intermediate soils. *JSCE* 680/III-55: 233–243 (in Japanese).

Matsumoto, K., Hyodo, M. and Yoshimoto, N. 1999. Cyclic shear properties of intermediate soil subjected to initial shear. *Proc., 34th Japan National Conference on Geotechnical Engineering*: 637–638 (in Japanese).

Ni, Q., Tan, T.S., Dasari, G.R. & Hight, D.W. 2004. Contribution of fines to the compressive strength of mixed soils. *Géotechnique* 54(9): 561–569.

Thevanayagam, S., Shenthan, T., Mohan, S. and Liang, J. 2002. Undrained fragility of clean sands, silty sands and sandy silt. *J. Geotechnical & Geoenvironmental Engineering* 128(10): 849–859.

Rheological study of nanometric colloidal silica used for grouting

A. Guefrech, N. Saiyouri & P.Y. Hicher
Ecole Centrale de Nantes, Nantes, France

ABSTRACT: Colloidal silica is a grout having controllable gel time. Thus, it can be injected in low permeable soils. This study is aimed to characterize silica sols and to understand how formulation parameters affect the gel time. The kinetics of silica gel depends on concentration and nature of salt, silica concentration and particle size. The gel time was measured by visual observation and rheological investigation.

1 INTRODUCTION

Micro cement grout is a commonly used material for reinforcing and waterproofing the soils. The penetrability of this grout is very limited especially in low permeable soils. Therefore, it is necessary to find a new generation of grout with more fluidity and controllable setting time, easy to inject and durable (Perssof et al.1988, 1996).

"Colloidal silica" is attractive as a potential grouting material because of its controllable gel time and excellent durability. In addition, it is also chemically and biologically inert and nontoxic.

The colloidal silica is a stable suspension of discrete, dense particles of amorphous silica of uniform particle size from 3 to 500 nm. The density and viscosity of this material are approaching to those of the water in dilute solutions (Iler 1979, Bergna 2006).

When silica sol is used as injection grout in the presence of saline solution, the silica particles coagulate and form a solid continuous skeleton with a controllable setting time. Sodium chloride NaCl and calcium dichloride $CaCl_2$ are the most frequently used ones for gelling.

The main characteristics of silica sols are: the particle size, specific surface area, the concentration of SiO_2 in silica sol (% by weight), pH and the particle charge. The colloidal silica can be used for many purposes (Iler 1979): paper production, coating, polishing electronic products, concrete ...

2 MATERIALS

In our experimental study, we tested eight silica sols supplied by different manufacturers. The nomenclature of silica sol used in our experimental study is summarized in Table 1 and each supplier is designated by a capital letter. The stability of silica suspension is very sensitive to the presence of salts. The positive counter ions balancing the negative surface charge are diffusely oriented around the particle. Thus the repelling forces between particles extend for some distance out from the particle surface. As salt is added, the counter ions move much closer to the particle surface, which reduces the distance through which the stability forces act. This causes a reduction in sol stability by increasing the probability of interparticle collision, the flocculation and gelling (Allen et al. 1969, 1970, Depasse & Watillon 1970) of silica particles.

Negative Silica sols can be coagulated by other flocculating agents as cationic organic polymers, cationic surfactants which form micelles, nonionic polymer, ...

In our experiments, three salts were used to coagulate silica sols: sodium chloride NaCl, potassium chloride KCl and calcium dichloride $CaCl_2$. The mean difference between this salts are the chemical nature of cation and the valency. These salts are purchased from Fischer Scientific Bioblock.

3 TESTS AND METHODS

3.1 *Particle size distribution*

In order to determine the particle size distribution, the technique of dynamic light scattering was used. The Nicomp 380 instrument supplied by CAD Instrument has a large size range from 1 to 500 nm. Before measurement, the concentrated silica sols must be diluted by purified water.

Table 1. Typical proprieties of silica sols.

Products		
Silica A1	Silica B1	Silica C1
Silica A2	Silica B2	Silica C2
	Silica B3	Silica C3

3.2 Gel time measurement

To use silica sol as injectable grout in soil, the particles have to aggregate and form a gel with controllable sitting time. The grout is made by combining silica sol and salt solution.

The purpose of this series of experiments is to measure the gel time and to study how formulation parameters affect the behavior of sol-gel reaction.

To measure gel time, we used two techniques: the visual observation and rheological investigation (Agren & Rosenholm1998).

3.3 Visual time gel determination

The gelling time was determined by visual observation. The sample was put in test tube and the gel time was noted when the sample stopped flowing. The sample was remained unagitated in order to avoid breaking of gel and links between silica particles.

3.4 Gel time determination by rheological tests

By rheological measurement it is possible to have information on structural transition sol-gel. With this technique, we followed the evolution of the elastic (storage) G′ and viscous (loss) G″ module with time. The sample was considered as visco-elastic material and the transition from viscous liquid to elastic solid was referred to as the cross point between the two modules. The scientific best founded definition of gel time is to determine the point of gelation as the point of frequency independent loss factor tan (δ) as function of time. However, in our study, the equality of G′ and G″ is used to determine the gel time for these measurements.

Typical result was presented by the Figure 1 and the transition to the gel state is defined as the intersection between curves.

The sample was subjected to time sweep; the frequency (1 Hz) and the strain were kept constant during the oscillatory measurement. Just a small deformation

(0.01%) was imposed so that gel structure may not be damaged.

The dynamic oscillatory measurements were performed by controlled stress rheometer (TA Instruments) using two geometries: concentric coaxial cylinder geometry and parallel plats (ø = 40 mm) using a constant gap of 1 mm for all tests.

The temperature was kept constant at 20°C during measurements. To prevent evaporation during the measurement, a plastic lid was placed on the top of the geometry.

The gelling time was also determined by visual observation. The sample was put in test tube and the gel time was noted when the sample stopped flowing. The sample was remained unagitated in order to avoid breaking of gel and links between silica particles. All samples were prepared by combining silica sol and salt solution with constant volume ratio and only the concentration and the nature of salt were changed.

4 RESULTS AND DISCUSSION

4.1 Particle size distribution of silica sols

An example of visual aspect of the silica sols is shown in Figure 2.

The appearance of concentrated silica sol is white and milky and if the solid content is reduced the suspension is opalescent. The sol is as clear as water for small solid content (10 to 15%).

The distribution of Silica A1 obtained after more than one hour of measurement running is shown in Figure 3. The distribution of all silica sols is analyzed by multimodal approach.

The mean diameter calculated from this analysis is 90 nm. Figure 4 represent the particle size distribution of Silica B1 having 50% solid content of SiO_2 and it has 117 nm average diameter.

It is known that the specific surface area is inversely proportional to the particle size. According to the technical data sheet, we point out that Silica A1 has

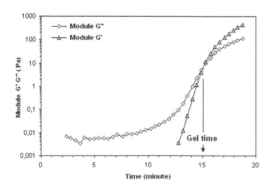

Figure 1. Determined rheological gel time by evolution of the elastic and viscous modulus.

Figure 2. The visual aspect of silica suspension B1, B2 and B3.

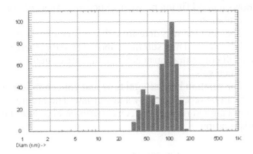

Figure 3. The intensity weighted distribution analysis result for Silica A1.

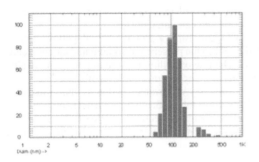

Figure 4. The intensity weighted distribution analysis result for Silica B1.

a specific surface greater than Silica B1. This is in agreement with measurement of the average diameter done by dynamic light scattering.

4.2 Gel time of colloidal silica

In this section, we study the evolution of gelling time of colloidal silica combined with saline solution. For this reason, two techniques were used to estimate the transition between liquid and solid phase.

4.2.1 Effect of salt concentration: visual observation and rheological investigation

The proportion between silica and salt solution was kept constant for all samples, while the concentration and nature of salt were changed. Tables 2 and 3 show the visual gelling time of silica sol in presence of sodium chloride and potassium chloride with different concentrations.

In Figures 5 and 6 we can see the dependence of gel time as a function of concentration of salt. The gelling time was reduced with concentration for all samples and with the two salts. As the concentration of salt was increased, the gel time became short.

The reduction of gel time was explained by the number of cations adsorbed on the surface of silica particle and may be the bridging factor of coagulation.

Table 2. Visual gelling time (minutes) of Silica A2, B2 and B3 with Sodium Chloride NaCl.

	Concentration of NaCl (g/L water)				
	120	100	90	80	60
Silica A2	20	50	30	80	510
Silica B2	1	5	5	25	250
Silica B3	7	27	45	90	425

Table 3. Visual gelling time (minutes) of Silica C1, C2 and C3 with Potassium Chloride KCl.

	Concentration of KCl (g/L water)				
	120	100	90	80	60
Silica C1	6	15	36	74	580
Silica C2	2	6	11	20	107
Silica C3	2	5	11	22	124

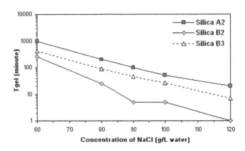

Figure 5. The visual gelling time of silica suspension A2, B2 and in presence of sodium chloride.

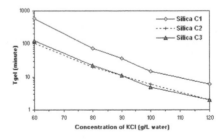

Figure 6. The visual gelling time of silica suspension C1, C2 and C3 with potassium chloride.

The positive charge of ion is attracted by the negative surface of particle which permits the flocculation of particles and forming a link between surfaces.

The gelling time was also determined by rheological methods. Figure 7 shows an example of evolution of elastic G' and viscous G'' module with different concentrations of sodium chloride for Silica C2. The gel time corresponds to the cross point between the two modulus. We note the same tendency as the visual observation, which means that the gelling becomes fast if the concentration of salt is increased.

In order to compare those two techniques, we plot in Figure 8 the gelling time for Silica C1, C2 and C3 obtained by the two means versus concentrations of sodium chloride. We can see that the curves are similar and the gelling time did not vary much if we use the visual observation or rheological measurement.

This means that we can easily use the visually determined gelling times as quick guide for rheological measurements of the transition behavior at the gel point. For low concentration of salt for which the gel is long, we noted some difference between measurements obtained by those two techniques because it is difficult to evaluate visually and exactly the transition and the structural changes in the sol gel reaction.

The gel time was determined by visual observation and rheological measurement and we find that increasing salt concentration accelerates the rate of gelling for all silica sols.

The relation between gel time and concentration of salt can be represented by a power law. The Figure 9

shows a good fitting between experimental points and this relation.

The gel time was determined by visual observation and rheological measurement and we find that increasing salt concentration accelerate the rate of gelling for all silica sols.

The relation between gel time and concentration of salt can be represented by a power law. The Figure 10 shows a good fitting between experimental points and this relation.

4.2.2 Effect of the cation charge

In this section, we focus on the effect of valency of cation used for coagulation on the gel time of colloidal silica.

The salt used to coagulate silica sols are sodium chloride and calcium chloride

In Figure 10, we plot the visual gelling time of Silica B4 in the presence of sodium and calcium chloride. The rate of gelling is more important in divalent cation Ca^{2+} for all concentration of salts.

A property of divalent cation is that when it is adsorbed on the surface silica particle, only a single negative charge is neutralized (Iler 1975, 1979). So the divalent cation represents a positive charge site on

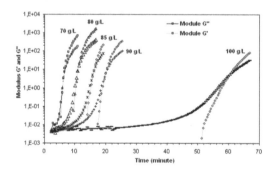

Figure 7. The evolution of storage G′ and loss G″ module of versus time for Silica C2 with different sodium chloride concentration.

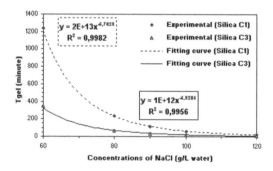

Figure 9. Power law: gelling time versus concentration of salt.

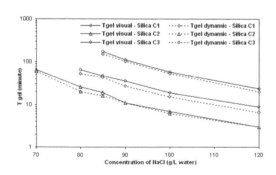

Figure 8. Comparaison between visual and rheological gel time with sodium chloride for Silica C1, C2 and C3.

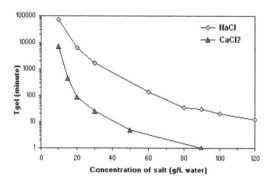

Figure 10. The visual gelling time of silica B4 evolution versus sodium and calcium chloride concentration.

the surface wich ca attract a negative nearby particles. This acts as a bridge by reacting with two particles at their points of contact.

5 CONCLUSIONS AND DISCUSSIONS

By visual observation and rheological measurement, we tried to understand the evolution of the gelling time of silica sol with concentration and charge of salt. Three salts and eight silica sols were used for this study. We observed that the gelling time of all silica suspension is short if the concentration of salt is increased.

The charge of ion influences the kinetics of the sol gel transition. The polyvalent cations (Ca^{2+}) are more reactive than monovalent one (Na^+, K^+). Also, the gelling time can be affected by the chemical nature of salt.

There are many parameters, which modify the kinetics of gelling and the reaction of coagulation.

The evolution of the viscosity in time is another rheological parameter which must be studied in order to control the injectability of the material in the soil.

REFERENCES

Agren, P. & Rosenholm, J.B. 1998. Phase behavior and structural changes in tetraetylorthosilicate derived gels in the presence of polyethylene glycol, studied by rheological techniques and visual observations, *Journal of colloid and interface science* 98 (204): 45–52.

Allen, L.H. & Matijevic, E. 1969. Stability of colloidal silica I effect of simple electrolytes, *Journal of colloid and interface science* 69 (31): 287–296.

Allen, L.H. & Matijevic, E. 1970. Stability of colloidal silica II Ion exchange, *Journal of colloid and interface science* 70 (33): 420–429.

Bergna, H.E. 2006. Colloid chemistry of silica : an overview, Surfactant science series 131, Taylor & Francis.

Depasse, J. & Watillon, A. 1970. The stability of amorphous colloidal silica, *Journal of colloid and interface science* 70 (33): 430–438.

Iler, R.K. 1975. Coagulation of colloidal silica by calcium ions, mechanism, and effect of particle size, *Journal of colloid and interface science* 75 (53): 476–488.

Iler, R.K. 1979. The chemistry of silica, Wiley-Interscience.

Persoff, P. Moridis, G.J. Apps, J.A. & Pruess, K. 1988. Evaluation tests for colloidal silica for use in grouting, the American Society for Testing and Materials 98 (21): 264–269.

Persoff, P. Apps, J.A. Moridis, G.J. & Whang, J.M. 1996. Effect of dilution and contaminants on strength and hydraulic conductivity of sand grouted with colloidal silica gel, International Containment Technology Conference and Exbibition St Petersburg, 9–12 February, 1997.

Direct observation of patchy fluid distribution: Laboratory study

M. Lebedev & B. Gurevich
Curtin University of Technology, Perth, Australia

B. Clennell, M. Pervukhina, V. Shulakova & T. Müller
CSIRO Petroleum, Perth, Australia

ABSTRACT: The geohazards predictions during oil and gas extraction are becoming extremely important recently. Reservoir development changes fluid distribution inside reservoir and in some cases facilitates major geohazards. Exploration geophysics methods and especially seismic method are usually applied for investigation of underground fluid distribution and thus prevention of possible hazards. Interpretation of exploration seismograms requires understanding of the relationship between distribution of the fluids patches and acoustic properties of rocks. The sizes of patch as well as its distribution play a significant role in seismic response. In this paper we reported the results of laboratory experiments of simultaneous observation of fluid saturation by X-ray computer tomography and ultrasonic velocity measurements. Results are compared with numerical simulations.

1 INSTRUCTIONS

Maximising the recovery of known hydrocarbon reserves is one of the biggest challenges facing the petroleum industry today. Optimal production strategies require accurate monitoring of production-induced changes of reservoir saturation and pressure over the life of the field. Time-lapse seismic technology is increasingly used to map these changes in space and time. However until now, interpretation of time-lapse seismic data has been mostly qualitative. In order to allow accurate estimation of the saturation, it is necessary to know the quantitative relationship between fluid saturation and seismic characteristics (elastic moduli, velocity dispersion and attenuation). The problem of calculating acoustic properties of rocks saturated with a mixture of two fluids has attracted considerable interest (Gist 1994, Mavko & Holen-Hoeksema 1994, Knight et al. 1998); for a comprehensive review of theoretical and experimental studies of the patchy saturation problem see Toms et al. (2006).

For a porous rock whose matrix is elastically homogeneous, and inhomogeneities are caused only by spatial variations in fluid properties, two theoretical bounds for the P-velocity are known (Mavko & Mukerji 1998, Mavko et al. 1998). In the static (or low-frequency) limit, saturation can be considered as homogeneous, and hence the rock may be looked at as saturated with a homogeneous mixture of the fluids. In this case the bulk modulus of the rock is defined by the Gassmann equation with the fluid bulk modulus given by Wood's formula i.e., the saturation-weighted harmonic average of the bulk moduli of fluids. The Gassmann-Wood bound is valid when the characteristic patch size is small compared to the fluid diffusion length. The diffusion length is primarily controlled by rock permeability, fluid viscosity and wave frequency. In the opposite case, when the patch size is much larger than the diffusion length, there is no pressure communication between fluid pockets and consequently no fluid flow occurs. In this no-flow (or high-frequency) limit, the overall rock behaves like an elastic composite consisting of homogeneous patches whose elastic moduli are given by Gassmann's theory. Since all these patches have the same shear modulus, the effective P-wave modulus can be obtained using Hill's equation, i.e., the saturation-weighted harmonic average of the P-wave moduli. Whereas the Gassmann-Wood and Gassmann-Hill bounds apply in the low- and high-frequency limits, respectively, for intermediate frequencies, uneven deformation of fluid patches by the passing wave results in local pressure gradients and hence in wave-induced fluid flow. Such wave-induced flow cases wave attenuation and thus velocity dispersion. The effects of regularly distributed fluid patches of simple geometry (spheres, flat slabs) were first studied by White (1975), White et al. (1975), and Dutta & Ode (1979). More recently, the effect of regularly distributed patches of more general shape was modelled by Johnson (2001) and Pride et al. (2004). For randomly distributed fluid patches, Müller & Gurevich (2004) and Müller et al. (2008) showed how the effect of wave-induced flow controls the transition from the Gassmann-Wood to the Gassmann-Hill bounds.

While theoretical poroelastic models can predict the acoustic response for a given spatial distribution of fluid patches, the factors controlling the formation of the patches are less well understood. These factors can be studied using fluid-injection experiments in the laboratory. Previously reported laboratory observations demonstrate a qualitative link between fluid patch distribution and acoustic velocities (Cadoret et al. 1995, 1998; Monsen & Johnstad 2005). In order to get a deeper insight into the factors influencing the patch distribution and the associated wave response, we perform simultaneous measurements of P-wave velocities and rock sample X-ray computer tomography (CT) imaging. The CT imaging allows us to infer the fluid distribution inside the rock sample during saturation (water imbibition). We then show that the experimental results are consistent with theoretical predictions and numerical simulations.

2 EXPERIMENTAL SET UP

Experiments are performed on a cylindrical sample (38 mm in diameter and 60 mm long) cut from a

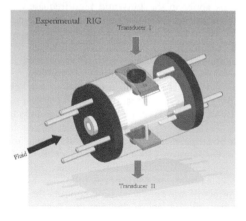

Figure 1.   Core sample jacketed inside experimental rig: 1 – X-ray transparent jacket; 2 – injection pipe; 3 – ultrasonic transdusers.

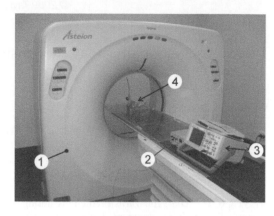

Figure 2.   Simultaneous acquisitions of acoustic properties and fluid saturation using ultrasonic transducers and CT scan: 1 – Computer tomograph; 2 – ultrasonic pulse-receiver; 3 – oscilloscope; 4 – jacketed sample.

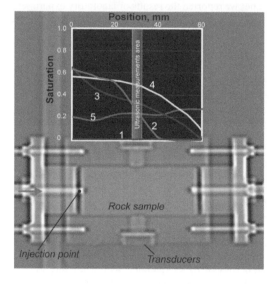

Figure 3.   X-ray image of the whole core sample. The embedded plot shows the profiles of water saturation along the 60 mm long sample at different times (increasing from 1 to 5) during the saturation experiment.

Casino sandstone (Otway Basin, Australia). The sample is dried at 100°C under reduced pressure for 24 h. The petrophysical properties are measured using a Coretest AP-608 automatic permeameter/porosimeter (Table 1). Then, the sample is sealed with a thin epoxy layer in order to prevent fluid leakage through the surface. Longitudinal (Vp) and shear wave (Vs) velocities at a frequency of 1 MHz are measured in the direction across to the core axis (perpendicular to the fluid flow) using broadband ultrasonic transducers. Intermediate aluminum "guide-pins" are placed between the sample and transducers to secure

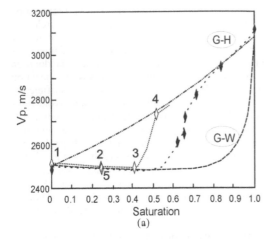

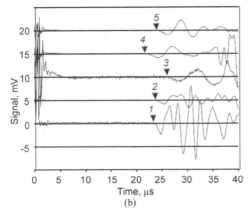

Figure 4. a) Velocity versus saturation for the Casino Otway Basin sandstone from the quasi-static saturation experiment (black diamonds) and dynamic saturation experiment (white diamonds). The numbers from 1 to 5 indicate acoustic measurements corresponding to the CT images shown in Figure 5, and saturation profiles shown in Figure 3. Theoretical Gassmann-Wood (G–W) and Gassmann-Hill (G–H) bounds are shown by dashed and dash-dotted lines, respectively. b) Output signals corresponding to five stages during the dynamic fluid injection experiment.

sufficient and constant coupling, as well as to provide transparency for X-ray radiation.

Two different saturation methods—referred to as dynamic and quasi-static saturation—are used in this study. In the dynamic saturation experiments, the samples are jacketed in the experimental cell made out of X-ray transparent material PMMA (Figure 1).

Distilled water is injected into the sample from one side. The injection rate is 10 mL per day. In quasi-static saturation experiments the samples are saturated during a long period of time (up to 2 weeks) under reduced pressure in order to achieve near-uniform fluid distribution for a given saturation level.

For quasi-static experiments the saturation level is determined by measurement of the volume (weight) of water fraction divided by total volume of pores. The fluid distribution in the dynamic saturation experiments (both spatial and time dependence) is characterized using X-ray computer tomography. (Figure 2).

The resolution is $0.2 \times 0.2 \times 1$ mm^3 (voxel size). This resolution is not sufficient to image the exact fluid patch geometry; however, it shows overall character of the fluid distribution. The fluid saturation is estimated as a difference in average CT-number (a value related to material density) between the saturated and dried sample divided by the volume of porous fraction.

The temporal evolution of the saturation profile along the rock sample is shown in Figure 3. We see that the overall saturation profile becomes progressively more uniform over time. All experiments are performed at laboratory environment at a temperature of 25°C. Velocity is obtained by first break picking on output signal. Figure 4 shows the P-wave velocity as a function of saturation for the Casino Otway sandstone from the dynamic saturation experiment (empty symbols) and quasi-static saturation experiment (solid symbols). A transition from the Gassmann-Wood to Gassmann Hill bound is clearly observed for water saturations of 40–50% for dyn-amic saturation and 60–70% for quasi-static saturation. Figure 5 shows CT images recorded during the dynamic saturation experiments. Scans 1–5 correspond to points 1–5 on the velocity-saturation curve (Figure 4): increasing saturation from the dried sample (1) up to 50% (point 4) and then decreasing saturation down to 26% (point 5). At low saturations, the velocity can be well described by the Gassmann Wood relation; however, a sharp increase in velocity is observed when saturation exceeds 40% (points 3 and 4 in the Figure 4). Then, the velocity-saturation approaches the Gassmann-Hill bound.

We also observe that for relatively large fluid injection rates (dynamic experiment) the transition from homogeneous to patchy saturation (from the Gassmann-Wood to Gassmann-Hill) occurs at smaller degrees of saturation as compared with low injection rates (quasi-static experiment). This observation suggests that the process of fluid patch formation is controlled by the injection rate.

3 NUMERICAL SIMULATIONS

In order to validate our interpretation of the velocity-saturation relation in terms of wave-induced flow we also perform numerical simulations of wave propagation in a 2-D poroelastic solid. The simulations are performed using 2D finite-difference solver of Biot's equations of dynamic poroelasticity (Krzikalla & Müller 2007).

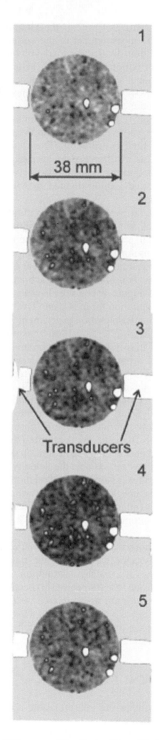

We numerically simulate plane wave propagation through 2D water saturated media with circle gas inclusion (Figure 6b). Wave propagation is initiated by applying a normal stress at the top of the domain. As a loading function we use $\sin(3x)$, where the duration of source signal (1 s) and time of wave propagation (5 s with $2 \cdot 10^{-4}$ s time step) are chosen sufficiently long to model quasistatic conditions and avoid resonance. The shape of outer water-saturated medium is a circle for analytical solution and a square for computational model. The size of model is 40×40 grid points. Periodical boundary conditions (PBC) are applied in lateral direction. Receiver configuration consists of 20 receiver lines with 20 receivers in each line. Distance between lines and receivers—2 grid points. Material properties for the numerical model are shown in the Table 1.

As a result of numerical simulations we obtain seismograms of velocity divergence $v_{xx} + v_{yy}$, velocity difference $v_{yy} - v_{xx}$, and stresses $-s_{xx} - s_{yy}$, $s_{yy} - s_{xx}$. In order to obtain frequency-dependent velocity and attenuation curves we used the method suggested by Masson (2007). At first, integration of the bulk velocity divergence and velocity difference yields mean and differential strains. Then, stresses and strains are averaged within the model and transformed to the frequency domain by Fourier transform. Finally, effective bulk and shear moduli, fast P-wave velocity and

Table 1. Model parameters fo 2D medium.

	Rock frame	Gas	Fluid
Porosity	0.2		
Permeability (m^2)	$1 \cdot 10^{-12}$		
Bulk modulus			
$\quad K_{dry}$ (Pa)	$5 \cdot 10^9$	$0.1 \cdot 10^9$	$2.25 \cdot 10^9$
$\quad K_{grain}$ (Pa)	$35 \cdot 10^9$		
Shear modulus			
$\quad G_{dry}$ (Pa)	$11 \cdot 10^9$		
Density (kg/m^3)	2650	100	990
Viscosity (Pa $\cdot$ s)		$0.1 \cdot 10^{-6}$	$1 \cdot 10^{-3}$

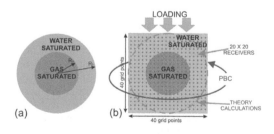

(a) (b)

Figure 5. CT images of the Casino Otway Basin sandstone at different times after start of injection at the same position: 1 – dried sample; 2–1 h; 3–24 h; 4–72 h. Contour of guide-pins are visible at right- and left-hand side of the sample.

Figure 6. Schematic plot of patchy-saturated model: a 2D periodic geometry with inner circle filled with gas and outer medium (a – circle for analytical solution and b – square for computer model) filled with water.

attenuation are calculated from these averaged fields according to Wenzlau (in press)

Velocity dispersion and attenuation are numerically simulated for a wide range of frequencies from $10^{-2} - 10^3$ Hz. The numerical results for water saturation of 0.75 are shown by black dots in Figure 7 in comparison with the analytical solutions for three slightly different media.

The analytical results for 3 different media are shown because the analytical and numerical results are obtained for similar but still slightly different models (Figure 6a, b) where the model at Figure 6a does not fill a plane.

Dispersion and attenuation curves (dots) for four frequencies (0.4, 0.6, 1, 2 Hz) versus water saturation are shown in the Figure 8 together with analytical solution calculated with the same S/V ratio (the ratio of the boundary area between two phases to the sample volume). Variations in water saturation (0.19, 0.44, 0.64, 0.75, 0.84, 0.94) are obtained by variation of the inner circle radius. Velocities are bounded by Biot-Gassmann-Wood (BGW) curve for low frequencies and by Biot-Gassmann-Hill (BGH) curve for high frequencies Johnson (2001).

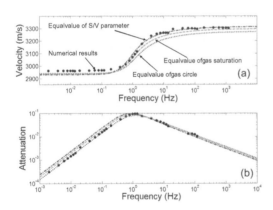

Figure 7. Three variants of theoretical prediction (dashed lines) and numerical calculations (dots) of P-wave velocity (a) and attenuation (b) vs. frequency.

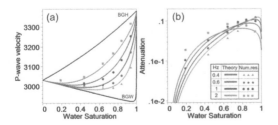

Figure 8. Theoretical prediction (solid lines) and numerical calculations (dots) of P-wave velocity (a) and attenuation (b) vs. water saturation for different frequencies.

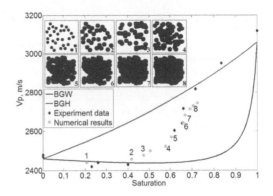

Figure 9. Velocity-saturation relation determined from numerical simulations of wave propagation in a poroelastic solid with randomly distributed patches that cluster for larger saturation values (see inset). Experimentally determined velocities for the quasi-static injection experiment are also shown.

Figures 7 and 8 show that our numerical results are in a reasonably good agreement with the theory. The slight existing discrepancies between the numerical and analytical results are most probably caused by different geometries rather than having a physical significance.

Figure 9 shows the modeled velocity-saturation relation obtained for material constants and wave frequency used in laboratory experiment described above. The degree of saturation is varied by increasing the size of the randomly distributed fluid patches. It can be seen that for large saturations, the fluid patches begin to cluster resembling the clustering observed in the CT scans. Despite the simplified numerical setup (2D modeling only and patch distribution not derived directly from the CT scans) the simulation results reproduce the overall behavior of the measured velocity-saturation relation.

This confirms that the transition observed in the ultrasonic measurements can be attributed to the mechanism of wave-induced flow. We plan to extend our numerical simulations to 3D so that the 3D patch distributions reconstructed from CT images can be directly used as input model in our simulations.

4 CONCLUSIONS

The experimental results obtained on low-permeability samples show that at low saturation values the velocity-saturation dependence can be described by the Gassmann-Wood relationship. At intermediate saturations there is a transition behavior that is controlled by the fluid patch arrangement and fluid patch size. Also, the fluid patch size is controlled by the injection rate. In particular, we show that for relatively large fluid injection rate this transition occurs at smaller degrees of saturation as compared with high injection rate.

The results illustrate the non-unique relationships between saturation and velocity in sandstones dependent on texture and fluid displacement history: fuller understanding of these phenomena are needed for accurate assessment of time lapse seismic measurements, be they for oil and gas recovery or for CO_2 disposal purposes.

REFERENCES

Cadoret, T., Marion D. & Zinszner B. 1995. Influence of frequency and fluid distribution on elastic wave velocities in partially saturated limestones. *Journal of Geophysical Research* 100: 9789–9803.

Cadoret, T., Mavko, G. & Zinszner, B. 1998. Fluid distribution effect on sonic attenuation in partially saturated limestones. *Geophysics* 63: 154–160.

Dutta, N.C. & Odé, H. 1979. Attenuation and dispersion of compressional waves in fluid-filled porous rocks with partial gas saturation (White model)—Part I: Biot theory. *Geophysics* 44: 1777–1788.

Gassman, F. 1951. Elastic waves through a packing of spheres. *Geophysics* 16: 673–685.

Gist, A.G. 1994. Interpreting laboratory velocity measurements in partially gas-saturated rocks. *Geophysics* 54: 1100–1108.

Johnson, D.L. 2001. Theory of frequency dependent acoustics in patchy-saturated porous media. *Journal of the Acoustical Society America* 110: 682–694.

Knight, R., Dvorkin, J. & Nur, A. 1998. Acoustic signatures of partial saturation. *Geophysics* 63: 132–138.

Krzikalla, F. & Müller, T.M. 2007. High-contrast finite-differences modeling in heterogeneous poroelastic media. *77th SEG Annual Meeting, San Antonio, Extended Abstracts*: 2030–2034.

Masson, Y.J. & Pride, S.R. 2007. Poroelastic finite-difference modeling of seismic attenuation and dispersion due to mesoscopic-scale heterogeneity. *Journal of Geophysical Research—Solid Earth* 112: B03204.

Mavko, G. & Nolen-Hoeksema, R. 1994. Estimating seismic velocities at ultrasonic frequencies in partially saturated rocks. *Geophysics* 59: 252–258.

Mavko, G. & Mukerji, bT. 1998. Bounds on low frequency seismic velocities in partially saturated rocks. *Geophysics* 63: 918–924.

Mavko, G., Mukerji, T. & Dvorkin, J. 1998. *The Rock Physics Handbook: Tools for seismic analysis in porous media.* Cambridge: Cambridge University Press.

Monsen, K. & Johnstad, S.E. 2005. Improved understanding of velocity-saturation relationships using 4D computer-tomography acoustic measurements. *Geophysical Prospecting* 53: 173–181.

Müller, T.M. & Gurevich, B. 2004. One-dimensional random patchy saturation model for velocity and attenuation in porous rocks. *Geophysics* 69: 1166–1172.

Müller, T.M., Toms-Stewart, J. & Wenzlau, F. 2008. Velocity-saturation relation for partially saturated rocks with fractal pore fluid distribution. *Geophysical Research Letters* 35: L09306.

Pride, S.R., Berryman, J.G. & Harris, J.M. 2004. Seismic attenuation due to wave- induced flow. *Journal of Geophysical Research* 109: B01201.

Toms, J., Müller, T.M., Ciz, R. & Gurevich, B. 2006. Comparative review of theoretical models for elastic wave attenuation and dispersion in partially saturated rocks. *Soil Dynamics and Earthquake Engineering.* 26: 548–565.

Wenzlau, F., & Müller, T.M. (in press). Finite-difference modeling of wave propagation and diffusion in poroelastic media. *Geophysics.*

White, J.E. 1975. Computed seismic speeds and attenuation in rocks with partial gas saturation. *Geophysics* 40: 224–232.

White, J.E., Mikhaylova, N.G. & Lyakhovitskiy, F. M. 1976. Low-frequency seismic waves in fluid-saturated layered rocks. *Physics of the Solid Earth Transactions. Izvestiya* 11: 654–659.

Visualization of leakage behavior through a hole in geomembrane using X-ray CT scanner

T. Mukunoki, K. Nagata, M. Shigetoku & J. Otani
Kumamoto University, Japan

ABSTRACT: The objective of this study is to visualize the leakage behavior of leachate in the ground underneath geomembrane with defects and to evaluate its behavior quantitatively. In this study, the model tests to simulate the leakage flow of leachate through the defects in geomembrane were performed using potassium and iodine liquid with specific gravity of 1.25. X-ray CT image of leakage flow through geomembrane defect was discussed and the 3-D finite difference analysis with respect to advective and diffusive transport and the X-ray CT image analysis quantitatively.

1 INTRODUCTION

Geomembranes used at the bottom of landfill are aged and damaged with the traffic loads of dump tracks during dumping waste and thermal stress with sum beam during the process of installing geomembranes (Rowe 1998). The damage factors would cause holes and crack in geomembrane and finally, leachate in the landfill starts to leak into the ground. Leakage volume of leachate would be affected with defect configuration and ground properties underneath geomembrane sheet. In order to predict the leakage volume of leachate in the ground, many researchers proposed the two dimensional (2-D) numerical model in the several boundary conditions (Giroud et al. 1989 and Touze-Foltz et al. 2001). In fact, it is difficult to observe the inner behavior of soil materials without any destruction. The industrial X-ray CT scanner is a powerful tool to distinguish the density distribution in the engineering materials such as soils, rocks and concrete (Otani et al. 2000, Otani and Obara 2003 and Desrues et al. 2006). X-EARTH Center newly established in Kumamoto University since 2008 possesses the industrial one and has kept contributing to visualize the inner condition of soil and rock materials due to loading and permeating fluid.

The objective of this study is to visualize the leakage behavior of leachate in the ground underneath geomembrane with defects and to evaluate its behavior quantitatively. In this study, the model tests to simulate the leakage flow of leachate through the defects in geomembrane are performed using potassium and iodine liquid. The X-ray CT image of leakage flow through geomembrane defect would be discussed and the 3-D finite element analysis with respect to advective and diffusive transport is conducted to evaluate the leakage behavior obtained from the X-ray CT image analysis.

2 METHODS

2.1 X-ray CT scanner

The output of X-ray CT analysis is an x-ray attenuation coefficient (μ) at each different special location in the specimen and it is well-known that the x-ray attenuation coefficient (μ) is a linier relation to the density of the material. Lenoir et al. (2007) described the issue of x-ray attenuation coefficient in detail. In the CT system installed at Kumamoto University, the CT-value is defined by the following equation:

$$\text{CT-value} = (\mu_t - \mu_w) K / \mu_w \qquad (1)$$

where μ_t is coefficient of absorption at scanning point: μ_w is coefficient of absorption for water; and K: material constant. It is noted that the coefficient of absorption for air is zero for the condition and then, the CT-value of the air is -1000.

CT-value also has linear relationship to the density of bulk density regulated in this study. The CT-value for each location in the specimen is given to the voxel and the aggregation of voxel is the X-ray CT image. The CT-values can be shown on each voxel as black and white in 256 grey level colors. Noted that the display level can be changed by a CT users valuably; hence the CT users should check the range of CT-value obtained to show the image accurately and to do quantitative discussion.

2.2 Model leakage test

Cylindrical Soil box made of acrylic material was prepared in this test as shown in Fig. 1. Inner diameter and height of model ground 200 mm each. At the bottom of soil box, the gravel was placed as drainage layer and the vertical flow was regulated. In order to fix the defected membrane sheet at the top of the model ground, this soil box was a flange type. In the above

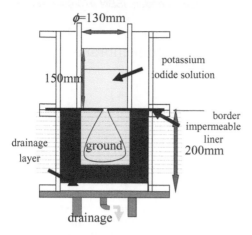

Figure 1. Schematic of leakage test apparatus.

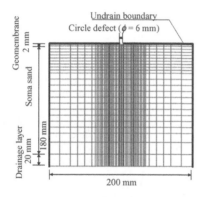

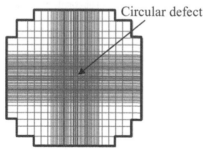

Figure 2. Cross section of cylindrical mesh for 3-D modeling.

on defect part of the membrane sheet, another cylinder was placed and contaminant solutions were installed. At first, the defect part was plugged to keep initial condition as the condition of no contaminated ground.

The soil box described in the above was fixed on the specimen table in the X-ray CT room. After scanning the model ground at initial condition, the plug on the defect part was opened and the leakage test

was started. After 30 seconds, the defected part was plugged again and then, the model ground was scanned again at the same height as the initial condition. Likewise, the model ground was scanned eventually, at the initial, 30 seconds, 1 min, 2 min, 4 min and 6 min after.

2.3 Materials

Souma sands were used in this test. Mean diameter of soil particle was 106 μm and the hydraulic conductivity of water was 2.0–3.0 $\times 10^{-6}$ m/s as the porosity of 0.43. Souma sands were installed with compaction method and saturation degree of model ground was 95–100%. Specific gravity was 2.65 t/cm³ and the dry density of model ground was 1.51 t/m³. 30% dilute solution of Iodine and potassium solution (as referred KI solution herein) was used as contaminant. The density of the solution was 1.25 greater than water but the viscosity was same as water.

2.4 Test cases

Three different defect models were prepared in this study, namely; a circular hole, two circular holes and rectangular hole. The defect areas were all same and hence, it was possible to discuss the effect of defect configuration.

2.5 Numerical analysis

In order to model the phenomena of model test with 3 dimensional advection-dispersion, the open code called 3-Dtransu was used. The governing equation should be simply given in the following for the model test.

$$\frac{\partial c}{\partial t} = \frac{\partial}{\partial x_i}\left(D_{ij}\frac{\partial c}{\partial x_j}\right) - V_i\frac{\partial c}{\partial x_i} \quad (2)$$

$$V_i = -k_{ij}\left\{\frac{\partial h}{\partial x_i} + \left(\frac{\rho}{\rho_f} - 1\right)n_j\right\} \quad (i, j = x, y, z) \quad (3)$$

$$D_{ij} = \alpha_T \|V\| \delta_{ij} + (\alpha_L - \alpha_T)\frac{V_i V_j}{\|V\|} + D_e\delta \quad (4)$$

where c is the concentration, D_{ij} is the hydrodynamic dispersion coefficient, V_i is the true flow velocity, k_{ij} is the hydraulic conductivity, h is the total head, ρ is

Table 1. Parameters for advection-dispersion modelling.

	Soma sand	Gravel
k of KI (m/s)	6.67×10^{-6}	1.0×10^{-3}
n	0.43	0.40
a_L(m)	0.4	0.4
α_T (m)	0.04	0.04

the fluid density, ρ_f is the density of water, and n_j is the unit vector for vertical direction. The hydraulic conductivity of KI solution was obtained from the hydraulic conductivity test with constant head. Table 1 summarized all parameters used in this study. Total number of nodes was 77,496 and the total number of elements was 71,668.

3 RESULTS AND DISCUSSION

3.1 3-D image analysis after leakage

Fig. 3 shows 3-D images of leakage volume reconstructed from the 2-D X-ray CT images at each depth from the bottom of membrane sheet for Cases 1, 2 and 3 after 360 seconds. Mukunoki et al. (2008) reported that the leakage area for Case 1was the least in other two cases and the values of area were 2954, 3863 and 4209 mm² for each case, respectively. The discussion point of this issue was that the area of defect part was same for all. Despite of the condition, leakage volume with rectangular defect was the greatest in all cases.

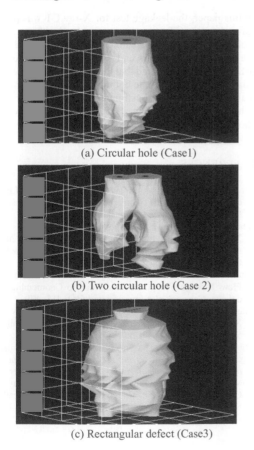

(a) Circular hole (Case1)

(b) Two circular hole (Case 2)

(c) Rectangular defect (Case3)

Figure 3. 3-D image of leakage volume.

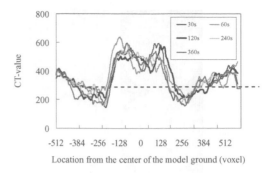

Figure 4. Profile of CT value for each elapsed time.

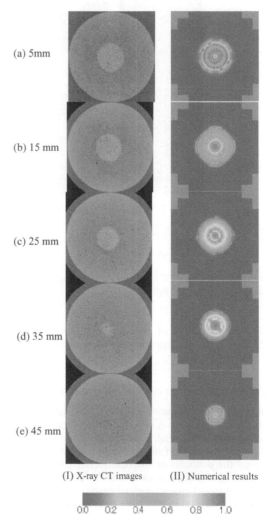

(a) 5mm

(b) 15 mm

(c) 25 mm

(d) 35 mm

(e) 45 mm

(I) X-ray CT images   (II) Numerical results

Figure 5. (I) X-ray CT images and (II) numerical results of model ground for leakage test.

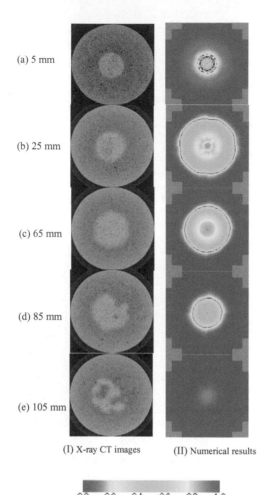

(a) 5 mm

(b) 25 mm

(c) 65 mm

(d) 85 mm

(e) 105 mm

(I) X-ray CT images   (II) Numerical results

0.0   0.2   0.4   0.6   0.8   1.0

Figure 6. (I) X-ray CT images and (II) numerical results of model ground for leakage test.

The outline length of each defect is 18.8, 26.3 and 58 mm so the leakage amount flow would be related to the outline length.

3.2 Image analysis and numerical results

In order to quantitatively evaluate the leakage area in the CT image, it is effective to investigate the CT value distribution. Fig. 4 shows CT-value distribution for the 5 mm depth from the bottom of geomembrane along the diameter of soil box obtained from X-ray CT images at each elapsed time for Case1. The CT value of model ground at initial condition was approximately from 200 to 300. After leakage, the CT values

distribute from 450 to 600 between the voxel number of −265 and 192 for each elapsed time.

Figs. 5 and 6 (I), (II) show X-ray CT image and the normalized concentration contour by initial concentration obtained from numerical analysis of model ground with leaking KI solution after 30 seconds and 360 seconds for Case 1. For discussion to the vertical direction, the numerical results slightly over estimate X-ray CT image after 30 seconds; however, X-ray CT images indicated that KI solution migrated into the ground more than numerical results after 6 minutes. On contrary, the leakage area of KI solution at the depth of 25 mm as shown Fig. 6 (b) is greater than that of X-ray CT image.

CT-value can evaluate the density change but not low concentration so it is important to quantitatively assess the density change area using suitable numerical model. In this study, more discussion should be needed about numerical model used, in particular dispersivity of soil tested.

4 CONCLUSIONS

In this paper, the leakage test for X-ray CT was performed and the CT images were compared with numerical results. X-ray CT scanner could well visualize the leakage area in soils under a geomembrane hole. 3-D CT images indicated that the leakage volume was related to not area of hole but the outline. The numerical model slightly over estimated the migration of KI solution in the beginning. Definition of dispersivity should be discussed more detail.

REFERENCES

Giroud, J.P. & Bonaparte, R. 1989. Leakage through Liners Constructed with Geomembranes-partII. Composite Liners. Geotextiles and Geomembranes 8: 71–111.

Desrues, J. Viggiani, G. & Besuelle, P. 2006. Advances in X-ray Tomography for Geomaterials, Proc. of the 2nd international workshop on X-ray CT for Geomaterials, GeoX 2006.

Touze-Foltz, N., Rowe, R.K., & Duquennoi, C. 1999. Liquid Flow Through Composite Liners due to Geomembrane defects. Geosynthetics International 6(6): 455–479.

Otani, J., Mukunoki T. & Obara. Y. 2000. Application of X-ray CT method for characterization of failure in soils. Soils & Foundations 40(2): 113–120.

Otani J. & Obara, Y. 2003. X-ray CT for Geomaterials soils, concrete, rocks. Proc. of the 1st international workshop on X-ray CT for Geomaterials, GeoX 2003.

Rowe, R.K. 1998. Geosynthetics and the Minimization of Contaminant Migration through Barrier Systems Beneath Solid Waste. Keynote Lecture for the 6th Int. Conf. on Geosynthetics: 64–77.

Thermo-hydro-mechanical instabilities

A seepage-deformation coupled analysis method for unsaturated river embankments

R. Kato & S. Sunami
Nikken Sekkei Civil Engineering Ltd., Osaka, Japan

F. Oka & S. Kimoto
Kyoto University, Kyoto, Japan

T. Kodaka
Meijo University, Nagoya, Japan

ABSTRACT: In the present paper, a water-soil coupled elasto-plastic finite element analysis method is proposed for unsaturated river embankments by incorporating unsaturated soil-water characteristics. The simplified three-phase method is developed based on the multi-phase porous theory by assuming that the compressibility of pore air is very high. Using the proposed method, the behavior of a river embankment, consisting of sandy soil in different conditions, is numerically analyzed while the river water level is high. From the numerical results, the seepage-deformation characteristics of the river embankment and the effects of permeability and the initial saturation are clarified.

1 INTRODUCTION

Recently, frequent floods have occurred due to heavy rains and typhoons that bring about a rise in the water level of rivers that exceed the design high water level. In Japan, it is expected that the tendency for localized torrential rains will continue to increase. Therefore, it is necessary to reconsider the evaluation method for river dike embankments during floods for the safety of the embankments. However, in the design method for embankments, the seepage analysis and the stability analysis (circular arc method) are performed separately, and the deformation is not directly considered (Japan Institute of Construction Engineering 2002).

In the present paper, a water-soil coupled elasto-plastic finite element analysis method is proposed for unsaturated river embankments by incorporating unsaturated soil-water characteristics. A simplified three-phase method is developed based on the multi-phase porous theory by assuming that the compressibility of pore air is very high (Ehlers et al. 2004, Oka et al. 2007). Using the proposed method, we analyze the behavior of a river embankment, consisting of sandy soil in different conditions, while the river water level is high.

2 SEEPAGE-DEFORMATION COUPLED ANALYSIS OF UNSATURATED SOIL

2.1 *Partial stress for the mixture*

The total stress tensor is assumed to be composed of three partial stress values for each phase.

$$\sigma_{ij} = \sigma_{ij}^s + \sigma_{ij}^f + \sigma_{ij}^a \tag{1}$$

where σ_{ij} is the total stress tensor, and σ_{ij}^s, σ_{ij}^f, and σ_{ij}^a are the partial stress tensors for solid, liquid, and gas, respectively.

The partial stress tensors for unsaturated soil can be given as follows:

$$\sigma_{ij}^f = -nS_r p^f \delta_{ij} \tag{2}$$

$$\sigma_{ij}^a = -n(1 - S_r)p^a \delta_{ij} \tag{3}$$

$$\sigma_{ij}^s = \sigma_{ij}' - (1 - n)S_r p^f \delta_{ij} - (1 - n)(1 - S_r)p^a \delta_{ij} \tag{4}$$

where σ_{ij}' is the skeleton stress (Jommi 2000), p^f and p^a are the partial stress levels for the pore water pressure and the pore gas pressure, respectively, n is the porosity, and S_r is the degree of saturation.

The skeleton stress is used as the basic stress variable in the model for unsaturated soil. Definitions for the skeleton stress and the average fluid pressure are given as follows:

$$\sigma'_{ij} = \sigma_{ij} + P^F \delta_{ij}, \qquad P^F = S_r p^f + (1 - S_r)p^a \qquad (5)$$

where P^F is the average pore pressure.

Adopting the skeleton stress provides a natural application of the mixture theory to unsaturated soil. The definition in Equation (5) is similar to Bishop's definition for the effective stress of unsaturated soil. In addition to Equation (5), the effect of suction on the constitutive model should always be taken into account. This assumption leads to a reasonable consideration of the collapse behavior of unsaturated soil, which has been known as a behavior that cannot be described by Bishop's definition for the effective stress of unsaturated soil. Introducing suction $(p^c = -(p^a - p^f))$ into the model, however, makes it possible to formulate a model for unsaturated soil, starting from a model for saturated soil, by using the skeleton stress instead of the effective stress. An elasto-plastic model (Oka et al. 1999) should be extended to unsaturated soil using the skeleton stress in order to consider the effect of suction. For the simple model in this study, however, the effect of suction has not been considered.

2.2 Mass conservation law

The mass conservation law for the three phases is given by

$$\frac{\partial \bar{\rho}^J}{\partial t} + \frac{\partial (\bar{\rho}^J \dot{u}_i^J)}{\partial x_i} = 0 \qquad (6)$$

where $\bar{\rho}^J$ is the average density for the J phase and $\dot{u}_i^J$ is the velocity vector for the J phase.

$$\bar{\rho}^s = (1 - n)\rho^s \qquad (7)$$

$$\bar{\rho}^f = nS_r \rho^f \qquad (8)$$

$$\bar{\rho}^a = n(1 - S_r)\rho^a \qquad (9)$$

where $J = s$, f, and g in which super indices s, f, and g indicate the solid, the liquid, and the gas phases, respectively. ρ^J is the mass bulk density of the solid, the liquid, and the gas.

2.3 Conservation laws of linear momentum for the three phases

The conservation law of linear momentum for the three phases is given by

$$\bar{\rho}^s \ddot{u}_i^s - Q_i - R_i = \frac{\partial \sigma_{ij}^s}{\partial x_j} + \bar{\rho}^s b_i \qquad (10)$$

$$\bar{\rho}^f \ddot{u}_i^f + R_i - S_i = \frac{\partial \sigma_{ij}^f}{\partial x_j} + \bar{\rho}^f b_i \qquad (11)$$

$$\bar{\rho}^a \ddot{u}_i^a + S_i + Q_i = \frac{\partial \sigma_{ij}^a}{\partial x_j} + \bar{\rho}^a b_i \qquad (12)$$

where b_i is a body force, Q_i denotes the interaction between solid and gas phases, R_i denotes the interaction between solid and liquid phases, and S_i denotes the interaction between liquid and gas phases.

These interaction terms Q_i and R_i can be described as

$$R_i = nS_r \frac{\gamma_w}{k^f} \dot{w}_i^f \qquad (13)$$

$$Q_i = n(1 - S_r)\frac{\rho^a g}{k^a} \dot{w}_i^a \qquad (14)$$

When we assume that the interaction between liquid and gas phases is too small, S_i can be described as

$$S_i \cong 0 \qquad (15)$$

where k^f is the water permeability coefficient, k^a is the air permeability, $\dot{w}_i^f$ is the average relative velocity vector of water with respect to the solid skeleton, and $\dot{w}_i^a$ is the average relative velocity vector of air to the solid skeleton. The relative velocity vectors are defined by

$$\dot{w}_i^f = nS_r(\dot{u}_i^f - \dot{u}_i^s) \qquad (16)$$

$$\dot{w}_i^a = n(1 - S_r)\left(\dot{u}_i^a - \dot{u}_i^s\right) \qquad (17)$$

Using Equation (16), Equation (11) becomes

$$\bar{\rho}^f \left(\ddot{u}_i^s + \frac{1}{nS_r}\ddot{w}_i^f\right) + R_i = \frac{\partial \sigma_{ij}^f}{\partial x_j} + \bar{\rho}^f b_i \qquad (18)$$

When we assume that $\ddot{w}_i^f \cong 0$ and use Equations (3), (6), and (16), Equation (18) becomes

$$nS_r \rho^f \ddot{u}_i^s + nS_r \frac{\gamma_w}{k^f} \dot{w}_i^f = -nS_r \frac{\partial p}{\partial x_i} + nS_r \rho^f b_i \qquad (19)$$

After manipulation, the average relative velocity vector of water to the solid skeleton and the average relative velocity vector of air to the solid skeleton are shown as

$$\dot{w}_i^f = -\frac{k^f}{\gamma_w}\left(\frac{\partial p}{\partial x_i} + \rho^f \ddot{u}_i^s - \rho^f b_i\right) \qquad (20)$$

$$\dot{w}_i^a = -\frac{k^a}{\rho^a g}\left(\frac{\partial p}{\partial x_i} + \rho^a \ddot{u}_i^s - \rho^a b_i\right) \qquad (21)$$

2.4 Equation of motion for whole mixture

Based on the above fundamental conservation laws, we can derive equations of motion for the whole mixture. Substituting Equations (7), (8), and (9) into the given equation and adding Equations (10)–(12), we have

$$\rho \ddot{u}_i^s + nS_r \rho^f(\ddot{u}_i^f - \ddot{u}_i^s) + n(1 - S_r)\rho^a(\ddot{u}_i^a - \ddot{u}_i^s)$$

$$= \frac{\partial \sigma_{ij}}{\partial x_j} + \rho b_i \qquad (22)$$

where ρ is the mass density of the mixture as $\rho = \bar{\rho}^s + \bar{\rho}^f + \bar{\rho}^a$, and $\ddot{u}_i^s$ is the acceleration vector of the solid phase.

From the following assumptions,

$$\ddot{u}_i^s \gg (\ddot{u}_i^f - \ddot{u}_i^s), \ddot{u}_i^s \gg (\ddot{u}_i^a - \ddot{u}_i^s) \qquad (23)$$

the equations of motion for the whole mixture are defined as

$$\bar{\rho}\ddot{u}_i^s = \frac{\partial \sigma_{ij}}{\partial x_j} + \bar{\rho} b_i \qquad (24)$$

2.5 Continuity equations for the fluid phase

Using the mass conservation law for the solid and the liquid phases, given in Equation (6), and assuming the incompressibility of soil particles, we obtain

$$\frac{\partial \{nS_r(\ddot{u}_i^f - \ddot{u}_i^s)\}}{\partial x_i} + S_r \dot{\varepsilon}_{ii}^s + nS_r \frac{\dot{\rho}^f}{\rho^f} + n\dot{S}_r = 0 \qquad (25)$$

Incorporating Equation (20) and $p^f = -K^f \varepsilon_{ii}^f$ (K^f: volumetric elastic coefficient of liquid, ε_{ii}^f: volumetric strain of liquid) into the above equation leads to the following continuity equation for the liquid phases:

$$-\frac{\partial}{\partial x_i}\left[\frac{k^f}{\gamma_w}\left(\rho^f \ddot{u}_i^s + \frac{\partial p^f}{\partial x_i} - \rho^f b_i\right)\right]$$

$$+ S_r \dot{\varepsilon}_{ii}^s + n\dot{S}_r + nS_r \frac{\dot{p}^f}{K^f} = 0 \qquad (26)$$

Similarly, we can derive the continuity equation with the assumption that the spatial gradients of porosity and saturation are sufficiently small.

$$-\frac{\partial}{\partial x_i}\left[\frac{k^a}{\rho^a g}\left(\rho^a \ddot{u}_i^s + \frac{\partial p^a}{\partial x_i} - \rho^a b_i\right)\right]$$

$$+ (1 - S_r)\dot{\varepsilon}_{ii}^s - n\dot{S}_r + n(1 - S_r)\frac{\dot{p}^a}{K^a} = 0 \qquad (27)$$

2.6 Air pressure

When we assume that the air is elastic and that its constitutive equation is given by $p^a = -K^a \varepsilon_{ii}^a$ (K^a: volumetric elastic coefficient of air, ε_{ii}^a: volumetric

strain of air), Equation (27) becomes

$$K^a \left\{-\frac{\partial}{\partial x_i}\left[\frac{k^a}{\gamma_w}\left(\rho^a \ddot{u}_{ii}^s + \frac{\partial p^a}{\partial x_i} - \rho^a b_i\right)\right]\right.$$

$$\left. + (1 - S_r)\dot{\varepsilon}_{ii}^s - n\dot{S}r\right\} + n(1 - Sr)\dot{p}^a = 0 \qquad (28)$$

When the air compressibility is very large, compared with the other phases, we can set $K^a \cong 0$. In other words, from Equation (28), we obtain $\dot{P}^a \cong 0$. The above discussion means that we can assume that $P^a = 0$ if initial air pressure $P_{ini}^a = 0$. This shows that the continuity in Equation (28) is always satisfied.

Since saturation is a function of the pressure head, the time rate for saturation is given by

$$n\dot{S}_r = n\frac{dS_r}{d\theta}\frac{d\theta}{d\psi}\frac{d\psi}{dp}\dot{p}^c = n\frac{1}{n}C\frac{1}{\gamma_w}\dot{p}^c = \frac{C}{\gamma_w}\dot{p}^c \qquad (29)$$

where $\theta = \frac{V_w}{V}$ is the volumetric water content, $\psi = \frac{p^c}{\gamma_w}$ is the pressure head, and $C = \frac{d\theta}{d\psi}$ is the specific water content.

Using Equation (28), Equation (29) becomes

$$-\frac{\partial}{\partial x_i}\left[\frac{k^f}{\gamma_w}\left(\rho^f \ddot{u}_i^s + \frac{\partial p^f}{\partial x_i} - \rho^f b_i\right)\right]$$

$$+ S_r \dot{\varepsilon}_{ii}^s + \left(\frac{nS_r}{K^f} + \frac{C}{\gamma_w}\right) = 0 \qquad (30)$$

The apparent volumetric elastic coefficient of pore water, $\bar{K}^f$, is defined as

$$\frac{1}{\bar{K}^f} = \frac{S_r}{K^f} + \frac{C}{n\gamma_w} \qquad (31)$$

Then, the continuity equation can be written as

$$-\frac{\partial}{\partial x_i}\left[\frac{k^f}{\gamma_w}\left(\rho^f \ddot{u}_i^s + \frac{\partial p^f}{\partial x_i} - \rho^f b_i\right)\right] + S_r \dot{\varepsilon}_{ii}^s + \frac{n}{\bar{K}^f}\dot{p}^f = 0 \quad (32)$$

3 UNSATURAED SEEPAGE CHARACTERISTICS

The soil-water characteristic model proposed (van Genuchten 1980) is used to describe the unsaturated seepage characteristics for which effective saturation S_e is adopted as

$$S_e = \frac{\theta - \theta_r}{\theta_s - \theta_r} = \frac{nS_r - \theta_r}{\theta_s - \theta_r} \qquad (33)$$

where θ_s is the volumetric water content at the saturated state, which is equal to porosity n, and θ_r is the residual volumetric water content retained by the soil at the large value of suction head which is a disconnected pendular water meniscus.

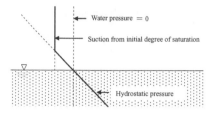

Figure 1. Initial pressure head.

In order to determine the soil-water characteristics, effective saturation S_e can be related to negative pressure head ψ through the following relation:

$$S_e = \left(1 + (\alpha\psi)^{n'}\right)^{-m} \qquad (34)$$

where α is a scaling parameter which has the dimensions of the inverse of ψ, and n' and m determine the shape of the soil-water characteristic curve. The relation between n' and m leads to an S-shaped type of soil-water characteristic curve, namely,

$$m = 1 - \frac{1}{n'} \qquad (35)$$

Specific water content C, used in Equation (29), can be calculated as

$$C = \alpha(n'-1)(\theta_s - \theta_r)S_e^{1/m}(1 - S_e^{1/m})^m \qquad (36)$$

Specific permeability coefficient k_r, which is a ratio of the permeability of unsaturated soil to that of saturated soil, is defined (Mualem 1976) by

$$K_r = S_e^{1/2}\left\{1 - (1 - S_e^{1/m})^m\right\}^2 \qquad (37)$$

Applying the above-mentioned relations, we can describe the unsaturated seepage characteristics.

In the analysis, the unsaturated region is treated in the following manner:

In the embankment, the initial suction, i.e., the initial negative pore water pressure, is assumed to be constant. Below the water level, the pore water pressure is given by the hydrostatic pressure. In the transition region between the water level and the constant suction region, we assume that the pore water pressure is linearly interpolated (Figure 1).

When the pressure head is negative, the increase in soil modulus due to suction is considered.

4 NUMERICAL ANALYSIS OF A RIVER EMBANKMENT

4.1 Analysis condition

Figure 2 shows the model of the river embankment and the finite element mesh used in the analysis. The soil parameters for Toyoura sand are listed in Table 1. The air permeability is not specified because the analysis is simplified such that the air pressure is always zero, as indicated in Equation (28). We follow the Guide for Structural Investigations of River Embankments[1], and set van Genuchten's parameters at $\alpha = 2$ and $n' = 4$. The rising rate of the water level is 1/3 (m/hour) until the water level reaches the crest of the embankment (18 hours), and then it remains constant for 100 hours and/or until the accumulated plastic deviatoric strain ($\gamma^p = \int d\gamma^p$, $d\gamma^p = (de_{ij}^p\, de_{ij}^p)^{1/2}$ reaches 5%.

The analysis cases and the results from this study are listed in Table 2. In Case-2 and Case-3, the initial degrees of saturation of the embankment are different from that in Case-1. In Case-4 and Case-5, the coefficients of permeability for the saturated soil of the embankment are different from that in Case-1. Case-1a and Case-4a are for the embankment and the foundation ground consisting of dense sand with a relative density of 90%.

4.2 Results

4.2.1 Case-1

The distributions of saturation and accumulated plastic deviatoric strain are illustrated in Figures 3 and 4,

Table 1. Material parameters.

Toyoura sand	Dr60%	Dr90%
Initial void ratio e_s	0.756	0.689
Compression index λ	0.012	0.012
Swelling index κ	0.0025	0.0025
Int.shear coefficient ratio G_0/σ'_{m0}	827.3	949.5
Permeability k (m/s)	1.0×10^{-5}	1.0×10^{-5}
Gravity acceleration g (m/s²)	9.8	9.8
Density ρ(t/m³)	1.9	1.9
Stress ratio at PT M_m^*	0.792	0.766
Stress ratio at failure M_f^*	0.987	1.153
Hardening parameter B_0^*	2700	3950
Hardening parameter B_1^*	130	130.5
Hardening parameter C_f	750	600
Elastic modulus of water K_f (kPa)	2.0×10^5	2.0×10^5
Quasi-OCR OCR*	1.2	1.2
Anisotoropy parameter C_d	2000	2000
Dilatancy parameters D_0^*, n	1.1, 1.0	2.0, 1.0
Plastic ref. strain γ_{ref}^{P*}	0.04	0.2
Elastic ref. strain γ_{ref}^{E*}	0.5	1.8

Table 2. Analysis cases and the results.

Case No.	Relative density (%)	Initial degree of saturation (%)	Saturated coefficient of permeability (m/s)	Horizontal local hydraulic gradient (%)	Accumulated deviatric plastic strain (%)
		Embankment		Maximam value (at bottom of slope)	
1	60	60	1.00E-05	0.575 (69 hours)	5.00 (69 hours)
2	60	80	1.00E-05	0.594 (31 hours)	5.00 (31 hours)
3	60	50	1.00E-05	0.572 (94 hours)	5.00 (94 hours)
4	60	60	1.00E-04	0.569 (25 hours)	5.00 (25 hours)
5	60	60	1.00E-06	0.441 (100 hours)	0.18 (100 hours)
1a	90	60	1.00E-05	0.624 (100 hours)	0.72 (100 hours)
4a	90	60	1.00E-04	0.571 (100 hours)	0.23 (100 hours)

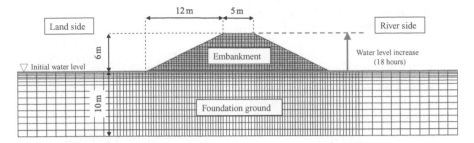

Figure 2. Model of the embankment and the finite element mesh.

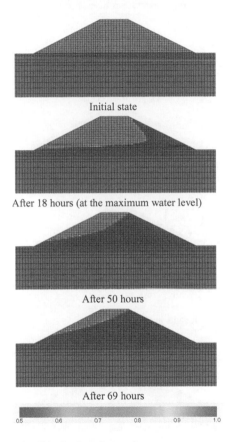

Initial state

After 18 hours (at the maximum water level)

After 50 hours

After 69 hours

Figure 3. Distribution of saturation.

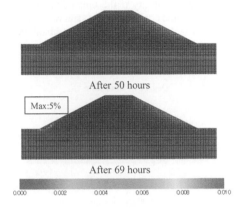

After 50 hours

Max:5%

After 69 hours

Figure 4. Distribution of deviatoric strain.

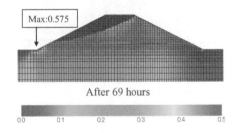

Max:0.575

After 69 hours

Figure 5. Distribution of horizontal hydraulic gradient.

respectively. The water level of the river reached the crest of the embankment after 18 hours. Then, the seepage flow infiltrated from the river side to the land side. The deviatoric strain increased at the toe of the embankment on the land side after the seepage flow reached the toe of the embankment. Then, the deviatoric strain increased rapidly after 69 hours. The distribution of the horizontal local hydraulic gradient after 69 hours is illustrated in Figure 5. The horizontal local hydraulic gradient at the toe of the embankment on

the land side was over 0.5. In the Guide for Structural Investigations of River Embankments, the allowable gradient for Piping is 0.5[1].

In this section, the following numerical results for other cases are compared with that for Case-1. The distributions of saturation and accumulated plastic deviatoric strain are illustrated in Figures 6–11 for different cases. For a higher initial degree of saturation, the deformation of river embankments grows rapidly (Figure 6). On the other hand, for a lower initial degree of saturation, the deformation develops slowly (Figure 7). For the case of a higher coefficient of permeability, the seepage point is higher than that for the

405

lower permeability on the land side of the river, and large strain occurs over a wider area (Figure 8).

4.2.2 *Other cases*

A lower coefficient of permeability drastically improves the stability at the toe of the embankment on the land side (Figure 9). In Case-1a and Case-4a, which consist of dense sand, the seepage flow does not provide a large deformation at the toe of the embankment on the land side (Figures 10 and 11).

The results listed in Table 2 show that the embankment is unstable for higher levels of initial saturation and higher permeability if the embankment and the foundation ground consist of loose sand whose relative density is 60%. On the other hand, the embankment is relatively stable for higher coefficients of permeability

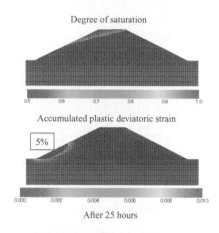

After 25 hours

Figure 8. Results in Case-4.

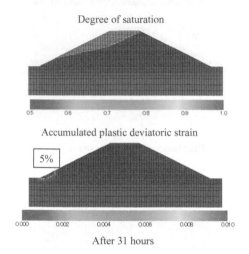

After 31 hours

Figure 6. Results in Case-2.

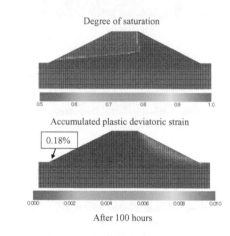

After 100 hours

Figure 9. Results in Case-5.

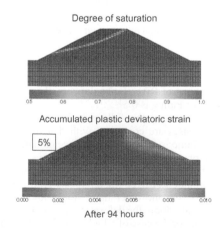

After 94 hours

Figure 7. Results in Case-3.

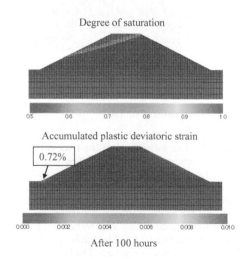

After 100 hours

Figure 10. Results in Case-1a.

Degree of saturation

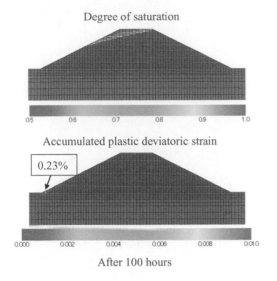

Accumulated plastic deviatoric strain

0.23%

After 100 hours

Figure 11.   Results in Case-4a.

if the embankment and the foundation ground consist of dense sand whose relative density is 90%.

5   CONCLUDING REMARKS

In the present study, the deformation and the stability of a river embankment during a flood have been investigated. From the numerical results, it is seen that stability and deformation depend on the initial degree of saturation and the permeability coefficient. From the present study, it has been found that the proposed seepage-deformation coupled analysis method can be applied to accurately simulate the deformation behavior of river embankments.

REFERENCES

Japan Institute of Construction Engineering, 2002. Guide for Structural Investigations of River Embankments.

Ehlers, W., Graf, T. and Amman, M., 2004. Deformation and localization analysis of partially saturated soil. *Computer methods in applied mechanics and engineering* 193: 2885–2910.

Oka. F., Kodaka, T., Kimoto, S., Kato, R. and Sunami, S. 2007. Hydro-Mechanical Coupled Analysis of Unsaturated River Embankment due to Seepage Flow. *Key Engineering Materials* 340–341: 1223–1230.

Jommi, C. 2000. Remarks on the constitutive modeling of unsaturated soils. In Tarantino, A. & Manvuso, C. (eds), *Experimental Evidence and Theoretical Approaches in Unsaturated Soils*, Balkema: 139–153.

Oka, F., Yashima, A., Tateishi, A, Taguchi, Y. and Yamashita, S. 1999. A cyclic elasto-plastic constitutive model for sand considering a plastic-strain dependence of the shear modulus, *Geotechnique* 49(5): 661–680.

van Genuchten, M.T. 1980. A closed-form equation for predicting the hydraulic conductivity of unsaturated soils, *Soil Science Society of America Journal* 44: 892–898.

Mualem, Y. 1976. A new model for predicting the hydraulic conductivity of unsaturated porous media, *Water Resources Research* 12: 513–522.

Prediction and Simulation Methods for Geohazard Mitigation – Oka, Murakami & Kimoto (eds)
© 2009 Taylor & Francis Group, London, ISBN 978-0-415-80482-0

Coupled thermo-mechanical analysis for slope behavior during freezing and thawing

T. Ishikawa, S. Miura & S. Akagawa
Hokkaido University, Sapporo, Japan

M. Sato
Geoscience Research Laboratory Co. Ltd., Tokyo, Japan

S. Kawamura
Muroran Institute of Technology, Muroran, Japan

ABSTRACT: This paper presents a new analytical procedure with a coupled thermo-mechanical FE analysis to simulate the mechanical behavior of frost heaved soil slope during freezing and thawing. First, a coupled thermo-mechanical FE analysis, which can consider the change in deformation-strength characteristics due to freeze-thaw action, was newly developed. Next, to examine the applicability of the coupled analysis and to analyze the mechanical behavior of frost heaved soil slopes during freezing and thawing, a numerical simulation of a freeze-thawing test for model slope was conducted. As the results, it is revealed that the coupled analysis proposed in this paper is an effective method to simulate the slope behavior during freezing and thawing, and that freeze-thaw action has a profound influence on the slope failure at a subsurface layer in cold regions.

1 INTRODUCTION

In cold regions like Hokkaido, the north area of Japan, a natural disaster such as slope failure at cut slope or landslide at natural slope is often occurred in snow-melting season. The natural disaster in cold regions is deemed to be caused by the increase in degree of saturation arising from snow-melting and/or the change in deformation-strength characteristics resulting from freeze-thaw action. Up to now, a large number of experimental studies have been made on the mechanical behavior of frost-heaving geotechnical materials such as clay and silts (Aoyama et al. 1979, Nishimura et al. 1990, Ono et al. 2003). Also, Ishikawa et al. (2008) has examined the mechanical behavior of crushable volcanic coarse-grained soils exposed to the freeze-thaw action, which lack in frost heave characteristics. These researches indicate that the freeze-thaw action has strong influences on the deformation-strength characteristics of geotechnical materials regardless of frost heave characteristics of soils.

In general, a coupled hydro-mechanical analysis is employed for numerical simulations of rainfall-induced slope failure occurred in relatively warm regions. However, as mentioned above, it is indispensable for the precise prediction of slope failure occurred in cold regions at snow-melting season to take into account the influence of freeze-thaw action of pore fluid on the mechanical behavior of soil ground. Accordingly, it is more desirable for numerical models simulating slope failure induced by water infiltration due to rainfall and/or snowmelt to insert a coupled thermo-mechanical analysis, which can reproduce the freeze-thaw process of soil ground, into a conventional coupled hydro-mechanical analysis. So far, some analytical studies have been made on the application of numerical simulations on the thermo-hydro-mechanical behavior of geotechnical materials. For examples, Neaupane & Yamabe (2001) proposed a nonlinear elasto-plastic analysis by a two-dimensional Finite Element analysis which can simulate the mechanical behavior of rocks during freezing and thawing, and Kimoto et al. (2007) proposed a coupled thermo-hydro-mechanically FE analysis with the thermo-elasto-viscoplastic model which can simulate the thermal consolidation of heating water-saturated clay. However, little attention has been given to the mechanical behavior of frost heaved soil ground during thawing such as slope failure in snow-melting season, and the development of a numerical model that can simulate the frost heave of soil ground has been a subject of study for a long time.

At present, we develop a new analytical procedure with a coupled thermo-hydro-mechanical FE analysis to examine the slope failure occurred in cold regions suffered from freeze-thaw action at snow-melting season. As part of the research, this paper presents a new analytical procedure to simulate the mechanical behavior of frost heaved soil slopes during freezing and thawing. First, a coupled thermo-mechanical FE analysis, which can consider the change in deformation-strength characteristics due to freeze-thaw action, was newly developed. Next, to examine the applicability of the coupled analysis and to grasp the qualitative tendency for the mechanical behavior of frost heaved soil slopes during freezing and thawing, a numerical simulation of a freeze-thawing test for model slope was conducted.

2 THERMO-HYDRO-MECHANICAL MODEL FOR FREEZE-THAWING SOIL

2.1 Preconditions for numerical modeling

Numerical modeling in this paper is based on the following assumptions:

- Water-saturated soil is considered as two-phase material consisting of an interstitial fluid and a solid skeleton, namely pore water and soil skeleton.
- Soil skeleton and pore water, respectively, assume to be a continuum and a homogeneous isotropic elastic body.
- The infinitesimal strain theory holds in the range of the deformation of soil skeleton and pore water intended for this paper.
- At an unfrozen part of soil, the principle of effective stress is assumed to govern the behavior of soil skeleton, and the flow of pore water follows Darcy's law.
- A drop in temperature of soil is assumed to cause in-situ freezing of pore water without seepage flow. In the strict sense of the word, this paper does not deal with "frost heave phenomena" of soils though the volume of frozen soil will dilate.
- When pore water freezes, the bulk modulus of pore water approaches that of soil skeleton, and as the result pore water is considered to be compressible.
- A numerical simulation for slope failure in snow-melting season consists of two analytical procedures, namely freeze-thawing procedure with coupled thermo-mechanical FE analysis and water infiltration procedure with coupled hydromechanical FE analysis.

Note that compression of stress and pore water pressure is given a negative sign in this paper.

2.2 Governing equations

2.2.1 Law of conservation of thermal energy
The heat-conduction equation is given as:

$$C\rho \frac{\partial T}{\partial t} = -\frac{\partial}{\partial x_i}\left(\lambda \frac{\partial T}{\partial x_j}\right) + C_w \rho_w \frac{\partial (v_{wi}T)}{\partial x_i} \quad (1)$$

where C = specific heat at constant volume of soil; ρ = density of soil; λ = thermal conductivity of soil; T = temperature; C_w = specific heat at constant volume of pore water; ρ_w = density of pore water; and v_{wi} = pore water velocity. Here, the right side of Equation 1 stands for the influx of heat added to the infinitesimal soil element from the surroundings per unit time.

2.2.2 Constitutive law
In this study, for the simplicity of numerical analysis, the soil is assumed to be a linear isotropic elastic material. Accordingly, the stress-strain relationships of soil can be written as:

$$\Delta \varepsilon_v = \Delta \varepsilon_{ii} = \frac{\Delta \sigma_{ii}}{3K_d} - \frac{\Delta u_w}{K_d} + \alpha_s \Delta T, \quad (2)$$

$$\Delta \gamma /2 = \Delta \varepsilon_{ij}' = \frac{\Delta \sigma_{ij}'}{2G} = \frac{1}{2G}\left(\Delta \sigma_{ij} - \frac{1}{3}\Delta \sigma_{kk}\delta_{ij}\right)$$

$$\Delta \sigma_{ij} = 2G\Delta \varepsilon_{ij}' + \left(K_d - \frac{2G}{3}\right)\Delta \varepsilon_{kk}\delta_{ij}$$

$$+ \Delta u_w \delta_{ij} - K_d \alpha_s \Delta T \delta_{ij} \quad (3)$$

where $\Delta \varepsilon_v$ = increment in volumetric strain of soil; $\Delta \sigma$ = increment in total stress; K_d = bulk modulus of soil under drainage condition; u_w = pore water pressure; α_s = coefficient of thermal expansion of soil; ΔT = increment in temperature; $\Delta \gamma$ = increment in shear strain; G = shear modulus of soil; and δ_{ij} = Kronecker's delta.

2.3 Mathematical modeling for freeze-thawing

In lowering the temperature of soil from air temperature, pore water in the soil freezes at about 0°C, and as the result various mechanical characteristics of soil change through the phase-transition process from fluid to solid. This paper focuses on mathematical modeling for freeze-thawing of soil in terms of the effect of latent heat, the change in volume and rigidity of soil associated with the phase-transition of pore water.

Figure 1 explains mathematical modeling of latent heat during temperature drop. We assume that a part of the heat quantity $(-\Delta q)$ lost in unit time from a unit volume of soil by cooling was expended to heat of solidification (Q_l) per unit volume of soil, and the rest is contributed to temperature drop of soil. Here, the rate of the expenditure for heat of solidification to the total outflow heat quantity $(-\Delta q)$ is defined as a variable (a). In addition, we assume that the influence of latent heat during temperature drop is negligible at temperatures under the final freezing temperature of soil (T_f). These assumptions are based on the consideration of inhomogeneity of temperature distribution due to a discrepancy in thermal conductivity between soil skeleton and pore water, and the existence of unfrozen water in soil. Accordingly, the soil temperature change

410

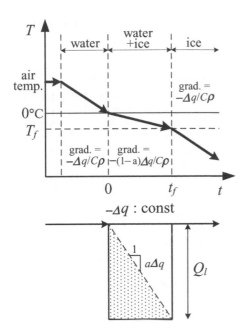

Figure 1. Mathematical modeling of latent heat.

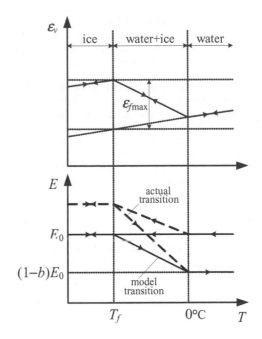

Figure 2. Mechanical properties during freezing and thawing.

over the temperature range of 0°C to T_f is given as:

$$T = -\Delta q \times (1 - a)\, t/C\rho \qquad (4)$$

where t = elapsed time from when T is equal to 0°C. Here, with Equation 4 and Figure 1 in mind, the elapsed time (t_f) until $T = T_f$ can be written as:

$$t_f = \frac{-C\rho T_f}{(1 - a)\,\Delta q} = \frac{Q_l}{a\Delta q} \qquad (5)$$

Solving Equation 5 for a gives:

$$a = \frac{Q_l}{Q_l - C\rho T_f} \qquad (6)$$

Consequently, if the value of T_f is determined by frost heave tests of soils and so on, the variable (a) can be found.

Furthermore, we assume that the frost expansion strain (ε_f), which is generated by frost heave of soils over the temperature range of 0°C to T_f, can be expressed as (Figure 2):

$$\varepsilon_f = \varepsilon_{f\,max} \frac{a\Delta qt}{Q_l} \qquad (7)$$

where $\varepsilon_{f\,max}$ = maximum frost expansion strain of soil. Here, the maximum frost expansion strain ($\varepsilon_{f\,max}$) is closely related to the frost heave ratio obtained from frost heave tests of soils.

On the other hand, in case of temperature rising process, the phenomena opposite to the above-mentioned phenomena during temperature drop will be observed

in soils. Besides, this paper introduces a new algorithm that explains the change in deformation-strength characteristics of soils caused by freeze-thaw action into the analytical procedure with a coupled thermo-mechanical FE analysis (Figure 2). We assume that the deformation modulus (E) for soils exposed to freeze-thaw action drops during thawing, namely over the temperature range of 0°C to T_f, as follows:

$$E = E_0 \left(1 - b\frac{a\Delta qt}{Q_l}\right) \qquad (8)$$

where E_0 = deformation modulus before freeze-thawing; and b = decreasing ratio of rigidity by freeze-thaw action. Since the growth of ice lens within soils may lead to a weakening of the structure of soil skeleton for frost heaved soils such as the formation of local cracks and the density reduction (Qi et al. 2006), the decreasing ratio of rigidity by freeze-thaw action (b) is closely related to the frost heave ratio. Note that the increase in soil rigidity caused by freezing has not been incorporated into mathematical modeling in this paper because it has little influence on slope failure.

3 NUMERICAL EXPERIMENT

3.1 Analytical conditions

A numerical simulation of a freeze-thawing test for model slope is performed with the above-mentioned coupled thermo-mechanical FE analysis. For detail information about real testing methods of the freeze-thawing test conducted separately from this paper, Kawamura et al. (2009) is to be referred.

Figure 3 shows the size, dimension and boundary condition of two-dimensional FE model under plane-strain condition, together with the element mesh which consists of four-node quadrilateral elements. A slope of FE model, which material is a volcanic coarse-grained soil, is the same size as that of the actual model test. For the displacement boundary condition, the base side boundary of FE model is fixed in the vertical direction, and the left-hand side boundary is fixed in the horizontal direction. For the thermal boundary condition, the isothermal boundary is imposed on the surface of both right-hand side and upper side of FE model, and the adiabatic boundary is imposed on the surface of both left-hand side and lower side. Table 1 summarizes the input parameters of FE simulation performed in this paper. Note that, as the first step of the research, the input parameters in this paper were not set to the values obtained from the element tests under experimental conditions similar to model tests, but set to the common values employed in past researches and the like.

A FE simulation of freeze-thawing test for model slope was performed as follows. First, a stability analysis was done by gravity force of 9.80 m/s². Here, the state of FE model after stability analysis is called "the initial loading state." Next, the initial temperature of all elements was set at 22°C equal to the room temperature of laboratory. Subsequently, for representing an installation of dry ice on the slope surface in the model test, the temperature of the isothermal boundary on the slope surface was dropped into −30°C and kept constant for 5 hours. Finally, for simulating the thawing process of the frozen soil slope at the room temperature, the temperature of the isothermal boundary was returned to the room temperature of 22°C, and the FE model was left as it was for 5 hours.

3.2 Results and discussions

Figure 4 shows time histories of ground temperature (T) obtained from FE simulation for two measuring points (T4, T5) as shown in Figure 3, which differ in depth from the slope surface. Figure 5 shows temperature distributions inside the soil slope of FE model at four different elapsed times (t_{fp}) of 1 hour, 5 hours, 6 hours and 7 hours, respectively. Here, the elapsed time (t_{fp}) is defined as the elapsed time from the start of freezing (t_{fp}). From these figures, it is recognized that a frozen soil layer parallel to the slope surface is formed at the subsurface layer of FE model in freezing process, and the frozen soil layer is gradually thickening with the passage of time during temperature drop. Whereas, it can be observed that though the frozen soil layer is gradually thawing from the slope surface in thawing process, after a lapse of about 2 hours into

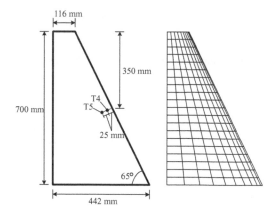

Figure 3. Schematic diagram of two-dimensional FE model.

Table 1. Input parameters of FE simulation.

Parameter name	Value
Density of soil ρ	1.40 g/cm³
Specific heat at constant volume of soil C	1.0 kJ/kgK
Thermal conductivity of soil λ	1.0 W/mK
Coefficient of thermal expansion of soil α_S	5.0×10^{-6}
Deformation modulus before freeze-thawing E_0	100.0 Mpa
Poisson's ratio of soil ν	0.20
Decreasing ratio of rigidity by freeze-thaw action b	0.90
Maximum forst expansion strain of soil $\varepsilon_{f\max}$	0.02
Final freezing temperature T_f	−5.0°C
Heat of solification per unit volume of soil Ql	1.6×10^4 kcal/m³

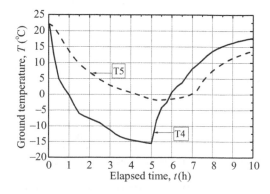

Figure 4. Time histories of ground temperature.

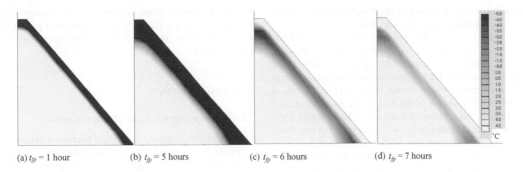

(a) t_{fp} = 1 hour   (b) t_{fp} = 5 hours   (c) t_{fp} = 6 hours   (d) t_{fp} = 7 hours

Figure 5. Temperature distributions inside soil slope during freezing and thawing.

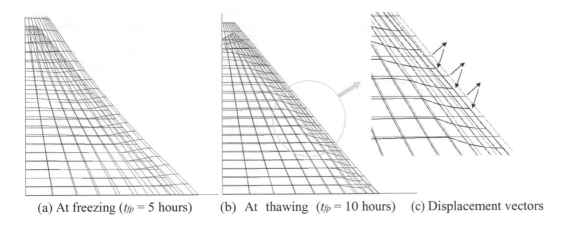

(a) At freezing (t_{fp} = 5 hours)   (b) At thawing (t_{fp} = 10 hours)   (c) Displacement vectors

Figure 6. Deformation behavior of FE model during freezing and thawing.

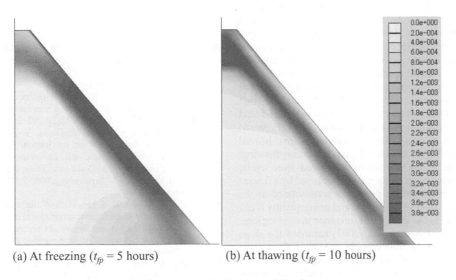

(a) At freezing (t_{fp} = 5 hours)   (b) At thawing (t_{fp} = 10 hours)

Figure 7. Contour maps of increment in shear strain during freezing and thawing.

413

thawing, a thin frozen soil layer parallel to the slope surface remains inside the soil slope.

Figure 6 shows the deformation behavior of FE model at freezing (t_{fp} = 5 hours) and at thawing (t_{fp} = 10 hours), respectively. Figure 6 also shows typical examples for the displacement vectors of some finite elements close to the slope surface. Here, a displacement vector is defined as the movement of a node of a finite element from the initial loading state to the state at freezing (t_{fp} = 5 hours) or the one from the state at freezing (t_{fp} = 5 hours) to the state at thawing (t_{fp} = 10 hours). According to Figure 6, the finite elements located in the frozen area of soil slope near the surface tend to expand in freezing process and contract in thawing process, and as the result the subsurface layer of FE model upheaves in a direction normal to the slope surface during temperature drop and subsides in the nearly vertical direction during temperature rise. This indicates that the freeze-thaw action causes the residual displacement parallel to the slope surface at a subsurface layer where frost heave occurs. The above-mentioned analytical results qualitatively agree well with past experimental researches (Harris & Davies 2000, Kawamura et al. 2009). Moreover, it is worth noting that a conventional coupled thermo-mechanical FE analysis without incorporating a new algorithm expressed by Equation 8 into the analytical procedure could not reproduce the above-mentioned slope behavior during freezing and thawing. Accordingly, it seems reasonable to conclude that the coupled analysis proposed in this paper is an effective method to simulate the slope behavior during freezing and thawing.

Figure 7 shows contour maps of the increment in shear strain from the initial loading state inside the soil slope of FE model at two different elapsed times (t_{fp}) of 5 hours in freezing process and 10 hours in thawing process, respectively. In Figure 7, it is observed that the shear strain is developed at the subsurface layer of FE model by freezing, and the shear strain remains there even if the temperature distribution of FE model approaches the temperature distribution at the initial loading state by temperature rise. This development of shear strain seems to be caused by the tendency that the deformation in the vertical direction by dead load becomes outstanding at the subsurface layer (Figure 6) as a result of decreasing the rigidity of frost-heaved finite elements as shown in Equation 8 during thawing. Under the severe environment suffered from cyclic freeze-thaw action, the shear strain caused by a freeze-thaw action might be gradually accumulated at the subsurface layer of soil slope, and consequently it can be harmful to the stability of soil slope in cold regions. Besides, Kawamura et al. (2007) regarded the slope behavior at shear strain of 4–6% as slope failure, since the deformation of soil slope was rapidly developed over the shear strain of 4–6% in rainfall tests of volcanic soil slope. Accordingly, in case of adhering to

this definition of slope failure, it seems reasonable to conclude that a freeze-thaw action has a profound influence on the slope failure at subsurface layer in cold regions.

4 CONCLUSIONS

The following conclusions can be obtained;

1. This paper proposed a coupled thermo-mechanical FE analysis into which a new algorithm that explains the degradation in deformation modulus of soils caused by freeze-thaw action was incorporated.
2. In FE simulation with the newly proposed coupled analysis, the subsurface layer of FE model upheaves in a direction normal to the slope surface during temperature drop and subsides in the nearly vertical direction during temperature rise.
3. The newly proposed coupled analysis is an effective method to simulate residual deformation caused by freeze-thaw action at subsurface layer, while a conventional analysis can not reproduce the slope behavior during freezing and thawing.
4. Since shear strain caused by a freeze-thaw action might be gradually accumulated at subsurface layer of soil slope, cyclic freeze-thawing has a profound influence on the stability of soil slope in cold regions.

ACKNOWLEDGMENTS

This research was supported in part by Committee on Advanced Road Technology from Ministry of Land, Infrastructure, Transport and Tourism.

REFERENCES

Aoyama, K., Ogawa, S. & Fukuda, M. 1979. Mechanical properties of soils subjected to freeze-thaw action. *Proc. of the 34th annual conference of the Japan society of Civil Engineers* 3: 719–720. (in Japanese)

Harris, C. & Davies, M.C.R. 2000. Gelifluction: Observations from large-scale laboratory simulations. *Arctic, Antarctic, and Alpine Research* 32(2): 202–207.

Ishikawa, T., Miura, S., Ito, K. & Ozaki, Y. 2008. Influence of freeze-thaw action on mechanical behavior of saturated crushable volcanic soil. *Deformation Characteristics of Geomaterials (IS-Atlanta 2008); Proc. intern. symp., Atlanta, 22–24 September 2008.* 2, Rotterdam: Millpress. 557–564.

Kawamura, S., Kohata, K. & Ino, H. 2007. Rainfall-induced slope failure of volcanic coarse-grained soil in Hokkaido. In N. Son (ed), *13th Asian Regional Conference on Soil Mechanics and Geotechnical Engineering; Proc. intern. conf., Kolkata, 10–14 December 2007*: 931–934.

Kawamura, S., Miura, S., Ishikawa, S. & Ino, H. 2009. Failure mechanism of volcanic slope due to rainfall and freeze-thaw action. *Prediction and simulation methods* for geohazard mitigation (IS-Kyoto 2009); Proc. intern. symp., Kyoto, 25–27 May 2009. Rotterdam: Balkema. (in prep.)

Kimoto, S., Oka, F., Kim, Y.S., Takada, N. & Higo, Y. 2007. A finite element analysis of the thermo-hydro-mechanically coupled problem of a cohesive deposit using a thermo-elasto-viscoplastic model. *Key Engineering Materials,* Vols. 340–341, Switzerland: Trans Tech Publications. 1291–1296.

Neaupane, K.M. & Yamabe, T. 2001. A fully coupled thermo-hydro-mechanical nonlinear model for a frozen medium. *Computers and Geotechnics* 28: 613–637.

Nishimura, T., Ogawa, S. & Wada, T. 1990. Influence of freezing and thawing on suction of unsaturated cohesive soils. *Journal of Geotechnical Engineering* 424/III-14: 243–250. (in Japanese).

Ono, T. & Wada, T. 2003. The properties of normally and overconsolidated clay after freezing and thawing history. *Journal of Geotechnical Engineering* 743/III-64: 47–57. (in Japanese).

Qi, J., Vermeer, P.A. & Cheng, G. 2006. A review of the influence of freeze-thaw cycles on soil geotechnical properties. *Permafrost and periglacial processes* 17: 245–252.

Semi 3D simulation of ground water advection and diffusion for alluvial layer in part of Kyoto basin

T. Kitaoka & H. Kusumi
Department of Civil and Environmental Engineering Kansai University, Osaka, Japan

I. Kusaka
Chuo Fukken consultant Co. Ltd., Osaka, Japan

ABSTRACT: The research of my paper on the ground water is to predict the extent of the pollutant source. This study will be constructed the semi 3D model and proceeded the semi 3D seepage flow analysis, and finally, performed the advection-diffusion of pollutant. The area of the simulation model is located the southern part of Kyoto, it has a typical ground water basin shape. That's why the underground water has been used for daily life for a long time in this area. However, if ground water is polluted, most of the people in Joyo city will suffer from illness caused by the contamination. This contamination for this study is voc, mercury, arsenic.

1 INTRODUCTION

In Kyoto, the underground water has been used for daily life for a long time, such as making tofu and dyeing kimono. Kyoto has a typical groundwater basin where there are three main rivers; Uji River, Kizu River and Katsura River. These rivers flow into the area between Tennou-zan and Otoko-yama mountains, then flow out to Osaka. For this reason, Kyoto basin saves enough underground water. Due to the fact that the underground water in Kyoto basin flows out from this point only, it is estimated that abundant underground water is saved.

The quantity of underground water is calculated to 21.1 billion tons and is comparable to the volume of water in Lake Biwa (25 billion tons). This research is focused on Joyo city which is located in the southern part of Kyoto prefecture. In Joyo city 80 percent of water for daily life comes from ground water. So, if ground water is polluted, most of the people in Joyo city will suffer from illness caused by the contamination. To predict the extent of the pollutant source, this study will be constructed the semi 3D model and proceeded the semi 3D seepage flow analysis, and finally, performed the advection-diffusion of pollutant analysis.

2 GEOLOGICAL CONDITIONS IN KYOTO

Kyoto basin is surrounded by mountains made of basement rock such as Paleozoic strata and granite, as is shown in Fig. 1. These mountains penetrate into

the centre of the basin with a gentle gradient. The basement rock has a bowl shape and sedimentary layers made after the tertiary era accumulated on it. The basement rock is an impermeable bed; and, upon it, there are diluvial formations and alluvial formations consist of alternate layers of permeable and impermeable layers. This means that the sediment can save a

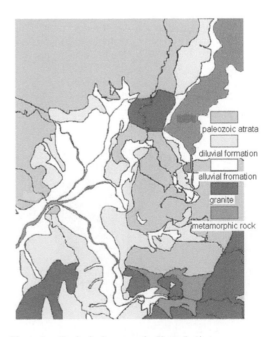

paleozoic atrata

diluvial formation

alluvial fromation

granite

metamorphic rock

Figure 1. Geological map on the Kyoto basin.

lot of underground water. The layer of this simulation model consists of alluvial sand and then dilluvial clay and sand.

3 GROUNDWATER RESOURCES IN JOYO CITY

3.1 Water supply project in Joyo city

The water supply project in Joyo city was authorized in 1962. After that, many water supply projects were established. In recent years underground water has been allocated for the public water supply as the population has grown. Since the area has abundant underground water, the first water treatment plant was established in 1971, the second in 1973, and the third in 1978.

3.2 Private wells

There are 505 wells known in Joyo city. Fig. 2 shows the location of these wells. Fig. 3 shows the total

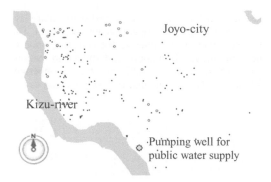

Figure 2. The position of wells in Joyo city.

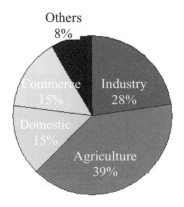

Figure 3. Total amount using underground water.

amount of wells using underground water. The wells are used for agricultural and industrial purposes and for life's daily needs. There are many shallow wells, less than 100 meters in depth and there some industrial wells, some of which are very deep (more than 100 meters). Most wells are used for agriculture, and used in the farming seasons (spring, summer and autumn). The wells are second most used for industry, and are used all year round for this purpose. The remaining wells are for commercial and domestic use. It is also thought that there are other wells that exist but, these wells are not recorded or analyzed this paper. The quantity of total pumped underground water is calculated to be approximately 25,000,000 m^3. This amount of water is equal to over 2.5 times the public water supply in Joyo city.

3.3 Changes to the underground water level of monitoring well

Monitoring wells are used for the measurement of the water level, and not for the pumping of underground water. Surrounding wells and climate conditions influence changes in the underground water level, and the underground water level in an aquifer can be exactly calculated. There are two kinds of monitoring wells, shallower monitoring wells for unconfined underground water and deeper monitoring wells for confined underground water. The relationship between the changes of water level in shallow wells with precipitation, and the relationship between changes of water level in deep wells with the amount of pumped underground water for public water supply will be analyzed.

3.3.1 Changes of underground water in shallow wells

This study focuses on shallow layer. Shallow wells are generally intended for unconfined aquifer, near the ground. The underground water level of the aquifer is influenced by precipitation. In Joyo city, there are two shallow monitoring wells; Yamazaki Well and

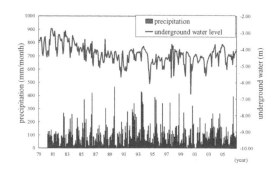

Figure 4. Yamazaki monitoring well and precipitation.

Tsuji Well. Fig. 4 shows the relationship between the underground water level in the Yamazaki Well and precipitation. As Fig. 4 shows, a large drawdown was caused when there were bad draughts in 1994 and 2000. However since then, the underground water level has been generally stable. It is evident; therefore, evident that the underground water level has been influenced by precipitation.

4 THE WATER LEVEL OF JOYO

The location of monitoring wells for the analysis range is shown in Fig. 5. We measured the ground water level for 3 months during agricultural off-season. (See Fig. 6 in which indicates isoline of water level.) The isoline

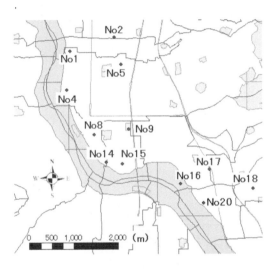

Figure 5. The location of monitoring wells.

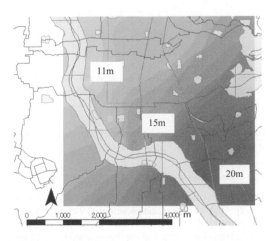

Figure 6. The isoline figure of underground water.

of Fig. 6 is the interpolated data measured simultaneously from 12 monitoring wells. And this figure can predict the direction of ground water flow. We can guess the direction of the water flow go to northwest. And we can guess the groundwater is flowing in from east and west in this area.

5 THE SIMULATION MODEL

The simulation model has been established by examining the material of the stratum taken from the data which is called "KANSAI GROUND INFORMATION DATABASE" in 2007. This study targets at the shallow layer. This model has been established by dividing 2 layers. And especially this study focuses on alluvial layers which is shallow layer. To understand the strata in Joyo we use 130 wellborn data.

5.1 *The area of the analysis*

The analysis range is the inside of the heavy line Fig. 7. The range of east to west is 5900 m, south to north is 5600 m, south to north is 5600 m.

5.2 *Mesh division*

The mesh model is shown Fig. 8. This figure's mesh model stand for 10 times actual mesh in vertical. The mesh size is 100 m. And the model is divided into 1788 node points and 1675 elements.

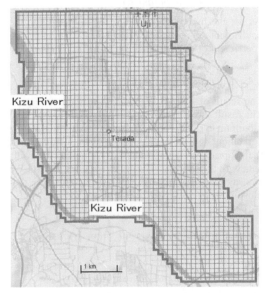

Figure 7. The analysis range.

Figure 8. The mesh model.

6 THE RESULT OF ANALYSIS SEEPAGE

6.1 *The outline of seepage flow analysis*

We performed semi-3D seepage flow analysis by the FEM, and can predict water level and direction of underground water flow.

6.2 *Boundary condition*

The boundary conditions are set as follows. Kizu River is treated as prescribed head boundary in west. South and East are firstly treated as prescribed head boundaries. And we can find the steady discharge. Both of them are given a set discharge boundary condition. As for north side, we set free boundary condition, because of lack of hydrological data. Therefore, it is set by result of a steady analysis with the boundary condition.

6.3 *Hydrological parameter*

As for the hydrological parameter in this study, the coefficient of permeability is 1.3E-03 (cm/s). The coefficient of storage is 1.0E-02 (1/m).

6.4 *The result of analysis seepage flow*

The result of seepage flow analysis is shown Fig. 9. The result shows that underground water flow in from mountain to river. The analysis result is approximately agreement with isoline figure. And then, we examined the propriety by comparing the 12 monitoring wells data with the analysis. The graph is shown Fig. 10. We can say that this model is accurate. It is because that most analysis result's error is within 1 m.

7 THE RESULT OF ADVECTION DIFFUSION ANALYSIS

7.1 *The outline of advection diffusion analysis*

The present study predicts how 1 p.p.m source of the contamination spreads in 50 years in alluvial. This

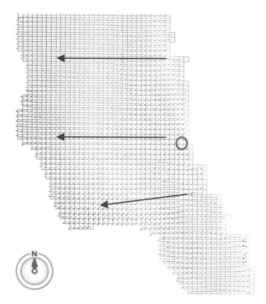

Figure 9. The figure of flow vector.

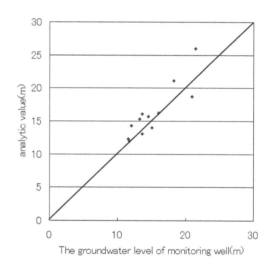

Figure 10. The comparison between the water level of monitoring wells and the water level of analysis.

analysis can predict the behavior of voc, mercury, and, arsenic.

7.2 *The analysis range*

The assumed contaminated locations are set in red circle Fig. 9. And, this model is selected 9 elements, and that is divided 5 m mesh. What's more, this model is divided into 1 m and 0.5 m meshes around the contaminated location to get the detailed analysis of the result.

7.3 *Hydrologic parameter*

Hydrologic parameter is set like Table 1. The vector for the speed of moving ground water could be obtained by seepage analysis. The location of each groundwater flow vector show Fig. 11.

7.4 *The result of advection diffusion analysis*

Voc diffuse over 100 m. But we can predict the extent of VOC behavior for 50 years is estimated at about 300 m, Fig. 12. We can predict the extent of mercury behavior for 50 years is estimated at about 20 m, Fig. 13. We can predict the extent of arsenic behavior for 50 years is estimated at about 5 m, Fig. 14. Lastly Figs. 15, 16, 17 shows the comparison of the concentration and time. The concentration of VOC is steady at each distance within 20 year. Mercury can be predicted that is steady 0.45 p.p.m at 5 m from the source of the pollutant. Arsenic can be predicted that it don't diffuse over 5 m from the source of the pollutant. It is recognized that the concentration of each contaminant can predict steady.

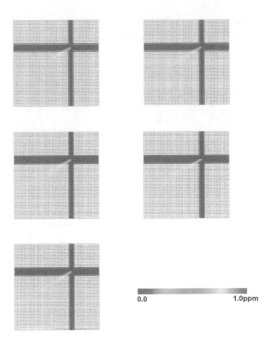

Figure 12. The extent of VOC behavior for 50 year.

Table 1. Hydrologic parameter.

	Vx (m/day)	Vy (m/day)
1	−9.746E-03	−6.299E-03
2	−9.696E-03	−6.587E-03
3	−9.258E-03	−8.226E-03
4	−9.782E-03	−5.645E-03
5	−1.004E-02	−5.733E-03
6	−1.058E-02	−5.757E-03
7	−9.813E-03	−4.984E-03
8	−1.008E-02	−4.968E-03
9	−1.045E-02	−4.855E-03

Figure 13. The extent of mercury behavior for 50 year.

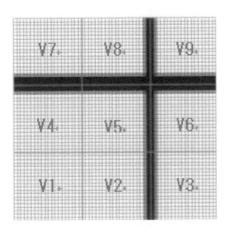

Figure 11. The location of the speed of moving groundwater.

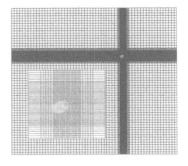

Figure 14. The extent of arsenic behavior for 50 year.

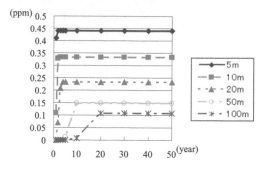

Figure 15. The concentration and time (VOC).

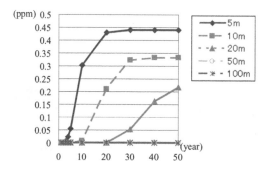

Figure 16. The concentration and time (mercury).

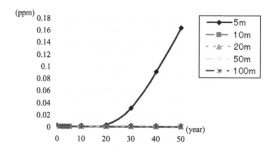

Figure 17. The concentration and time (arsenic).

8 CONCLUSIONS

The present study, the simulation model has been established. Furthermore, we succeeded to predict the ground water flow and the expansion of the each pollutant. This simulation model predicts the extent of VOC behavior for 50 years is estimated 300 m. This simulation model predicts the extent of mercury behavior for 50 years is estimated 20 m. This simulation model predicts the extent of arsenic behavior for 50 years is estimated 5 m.

However, the analysis model should be made more accurately by altering the parameters. The challenges for the future will be improvement of the model and reviewing the hydrological parameters.

REFERENCES

Karlheinz, S. & Joanna, M. 2003. The modeling of Underground water Environment due to Civil Engineers (in Japanese), Gihodo published C: 328–334.

Kansai Ground Information Conference. 2007. Kansai Ground Information Database (in Japanese) CD-ROM.

Prediction and Simulation Methods for Geohazard Mitigation – Oka, Murakami & Kimoto (eds)
© 2009 Taylor & Francis Group, London, ISBN 978-0-415-80482-0

A finite element study of localized and diffuse deformations in sand based on a density-stress-fabric dependent elastoplastic model

R.G. Wan & M. Pinheiro
University of Calgary, Calgary, Canada

ABSTRACT: The occurrence of various bifurcation modes in sand is numerically studied in a boundary value setting via finite element computations. It is found that an elastoplastic constitutive model simply enriched with an accurate description of dilatancy with density, stress and fabric level sensitivities is sufficient to capture bifurcated failure modes in sand. We illustrate some numerical examples of sand samples in biaxial testing conditions where strain localization naturally emerges from a completely homogeneous density and material property distribution field. Interestingly, it is found that the bifurcation point and post-bifurcation path are sensitive to small changes in initial conditions. Turning to axisymmetric stress-strain conditions, we demonstrate that a diffuse type of failure prevails over shear localization close to the plastic limit surface along certain loading directions in drained testing conditions. The spatial distribution of second-order work correlates very well with the zone of diffuse deformations in the sample.

1 INTRODUCTION

In geomechanics, instability manifests itself through various deformation modes. For example, a natural slope may undergo large movements as a result of deformations either localizing into shear bands or developing in a diffuse manner throughout the entire mass. In both cases, the problem underlies a material instability phenomenon that originates in the small scale due to the microstructural features of the geomaterial. At such a microscale, mechanisms of energy dissipation and inter-granular force transmission lead to local instabilities that reflect at the macroscopic scale as shear dilatancy, localization of deformations, and liquefaction, among others.

In the case of localization, one finds distinctive forms of concentrated deformations such as shear bands, compaction bands, and dilation bands (Rudnicki & Rice 1975, Vardoulakis et al. 1978 and Vardoulakis & Sulem 1995). These bands lead to an unstable response that is associated to a bifurcation at the material point level. On the other hand, it has been recently found that other types of instabilities may also appear before plastic localized failure is reached (Darve 1994). This means that other forms of failure may as well occur within the Mohr-Coulomb plastic limit surface. For example, when loose sand is sheared in undrained conditions, it may collapse as a result of a spontaneous loss in strength at stress levels far from the plastic limit surface. In this type of instability, normally coined as diffuse instability, no localization of deformations appears but a rather generalized failure takes place due to sufficiently large losses of inter-granular contact numbers. Such a diffuse instability may precede strain localization which can be seen to be a special case of the former. The implication is that stress states deemed to be safe with respect to a limit condition or localization can be still vulnerable to another type of instability such as of the diffuse type.

One of the main objectives of this study is to establish plausible models that capture numerically both localized and diffuse instabilities and explore the non-uniqueness associated to the underlying field equations in a boundary value problem. As anticipated, non-uniqueness can be triggered by introducing small perturbations in either the constraints of the boundary value problem or the initial conditions. It is shown in plane strain conditions that these small perturbations impact directly on the position of the shear band that is formed, and hence on the overall response of the structure. Next, diffuse type of instability is examined under both axisymmetric boundary and loading conditions. One aspect that is investigated is the role of the initial void ratio distribution in triggering instability. Both homogeneous and heterogeneous void ratio distributions (uniform and Gaussian types) are considered. A density-stress-fabric dependent elasto-plastic model (Wan & Guo 2004) is used to represent the constitutive behaviour of the material. All simulations are carried within the finite element framework.

2 DENSITY-STRESS-FABRIC DEPENDENT ELASTOPLASTIC MODEL

The ability of a constitutive model to describe soil behaviour within a high degree of fidelity is crucial in any failure analysis. For instance, a constitutive model for geomaterials must be able to capture essential behavioural features such as non-linearities, irreversibilities, and dependencies of the mechanical behaviour on stress level, density and fabric. The model herein used has many of the attributes described in the above. It was developed by Wan & Guo (2004) to describe the mechanical behaviour of granular materials, especially sands. For the sake of brevity, only the core framework of this model, referred to as the WG-model, is recalled.

The WG-model is a two-surface elastoplastic model that describes both deviatoric and compaction behaviours of granular materials. Rather than advocating advanced concepts such those found in micropolar and gradient plasticity, the model is founded on Rowe's (1962) stress-dilatancy theory for predicting volumetric changes under deviatoric loading. The original theory has been further enriched to address density, stress level, and fabric dependencies as well as cyclic loading regime conditions (Wan & Guo 2004). Recent enhancements to the model include a new cap yield surface and its extension to three-dimensional conditions (Pinheiro, unpubl.). In the current model, there are two yield surfaces: a shear-yield surface, $f^{(s)}$, and a cap-yield surface, $f^{(c)}$, as shown in Figure 1.

The first yield surface treats deviatoric loading dominated by dilatancy whereas the second surface accounts solely for isotropic loading producing plastic volumetric compressive strains. A non-associated flow rule derived from the enriched stress-dilatancy theory is used to calculate the increment of plastic shear deformations. The updating of both shear-yield and cap-yield surfaces is governed by two different hardening-softening laws. For the shear-yield surface,

Table 1. Main equations in WG-model for loading involving the shear-yield surface.

Yield surface	$f^{(s)}$	$:= q - M_\varphi\, p;\; M_{TC} := \dfrac{6\sin\varphi_m}{3 - \sin\varphi_m}$
	M_φ	$:= \dfrac{2\mu}{(1+\mu) - (1-\mu)t} M_{TC}$
Potential function	$g^{(s)}$	$:= q - M_\psi p;\; M_\psi := \dfrac{\sin\varphi_m - \sin\varphi_f}{1 - \sin\varphi_m \sin\varphi_f}$
	$\sin\varphi_f$	$:= \dfrac{\alpha_F + \gamma^p}{\alpha_0 + \gamma^p}\left(\dfrac{e}{e_{cs}}\right)^{n_f}\sin\varphi_{cs}$
Hardening law	$\sin\varphi_m$	$= \dfrac{\gamma^p}{\alpha_0 + \gamma^p}\left(\dfrac{e}{e_{cs}}\right)^{-n_m}\sin\varphi_{cs}$
Other evolution law	e_{cs}	$:= e_{cs0}\exp\left[-h_{cs}\left(\dfrac{p}{p_0}\right)^{n_{cs}}\right]$

where: p = mean effective stress; q = deviatoric stress; $t = \sin 3\theta$ (θ = Lode's angle); φ_m, φ_f, φ_{cs} = friction angles mobilized, at failure, and at critical state, respectively; e, e_{cs} = void ratio at current and at critical states, respectively; γ^p = deviatoric plastic strain; e_{cs0}, μ, α_F, α_0, n_f, n_m, n_{cs} and h_{cs} are material parameters; $p_0 = 1$ kPa.

the mobilized friction angle acts as plastic hardening variable, whereas for the cap-yield surface, the preconsolidation pressure, p_c, governs plastic hardening. Table 1 summarizes the equations used in the model.

3 LOCALIZED AND DIFFUSE INSTABILITIES

Stability in its simplest form in a physical system generally evokes the simple idea that a small perturbation input or load results into a small-bounded output or response (Lyapunov 1907). In the realm of geomechanics, the well-known phenomenon of plastic failure can be viewed as a problem of instability whereby a small or even null stress input leads to large deformations. Thus, according to Lyapunov's concept, a material would exhibit unstable behaviour whenever its stress state reaches a certain limit state defined by a failure criterion such as the Mohr-Coulomb. However, experimental tests have shown that a material may present an unstable behaviour even before any failure criterion is violated (Darve 1994, Lade 2002, Chu et al. 2003). The next sub-sections define briefly two such types of instability in geomechanics, namely, localized instability and diffuse instability.

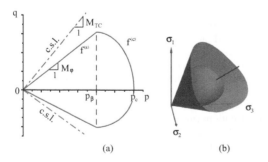

(a)   (b)

Figure 1. Generalized WG-model: (a) Trace of yield surfaces in the meridional plane. (b) Three-dimensional view in principal stress space. The acronym c.s.l. stands for critical state line.

3.1 Localized instability

The localized stress-strain response of geomaterials essentially results from their underlying micro- and macro-mechanical properties. Similarly, boundary conditions, the imposed stress path, relative density and fabric are among the most important factors that would trigger localization. It is often questioned whether the phenomenon is an inherent effect of the constitutive behaviour, or a consequence of the boundary value problem, or an outcome of both.

A great amount of experimental effort has been expended in an attempt to answer to the above questions. A good overview can be found in Desrues & Viggiani (2004). Parallel with experimental laboratory investigations, there have been numerous mathematical and numerical treatments of strain localization; see for example the works of Rudnicki & Rice (1975), Vardoulakis (1980) and Borja (2002). In these works, localized instability is viewed as a bifurcation of the underlying continuum field equations. In other words, an alternate mode of non-homogeneous deformation is possible apart from the homogeneous one due to loss of uniqueness of the governing incremental equilibrium equations. The non-homogeneous mode involves a surface with a kinematic discontinuity upon which interfacial slip occurs. By imposing continuity of tractions across the discontinuity surface and the same constitutive relationship throughout the body, Rudnicki & Rice's (1975) criterion for strain localization emerges as:

$$\det(\mathbf{n} : \mathbf{D} : \mathbf{n}) = 0 \tag{1}$$

where $\mathbf{D}$ is the tangent constitutive matrix in the sense that $d\boldsymbol{\sigma} = \mathbf{D} : d\varepsilon$, and $\mathbf{n}$ is the normal to the discontinuity surface or shear band.

As discussed earlier, the triggering of localization is a function of the material constitutive behaviour in a local sense. Localization presents a material instability which can be formulated within energy principles as will be discussed next.

3.2 Diffuse instability

In contrast to localized deformations, diffuse instability is normally characterized by a runaway deformation mode by which a decrease in stress bearing capacity occurs in the absence of localized strains. Static liquefaction is an example of a diffuse failure.

Diffuse instability can be mathematically described using the energy-based Hill's stability criterion which is defined by the sign of the second-order work (Hill 1958). Whenever the second-order work given by the product of incremental stress, $d\boldsymbol{\sigma}$, and strain, $d\varepsilon$ becomes negative at a material point during a loading increment, there is a potential for material instability. If the loss of positiveness of the second-order work becomes pervasive within the structure, collapse will eventually occur. Locally, the second-order work criterion is given by:

$$W_2 := d\boldsymbol{\sigma} : d\varepsilon \tag{2}$$

This notion of second-order work which in fact underlies a bifurcation problem can be extended to highlight the directional character of the loading and response behaviours of a material and its relationship to instability. The domain of stress-strain states for which at least one loading/response direction exists such that the second-order work becomes zero or negative gives rises to a so-called bifurcation domain. This concept, though, has to be brought hand-in-hand with that of controllability. Controllability, according to Nova (1994), highlights the fact that instability condition is also a function of the type of loading programme followed; therefore it depends on parameters controlled, either force, or displacement, or a mix of both. For example, in stress controlled tests, failure will not occur before the plastic limit condition. Furthermore, it can be shown that localization is a special case of diffuse instability as demonstrated numerically in sub-section 4.2.

4 FEM STUDY OF LOCALIZED AND DIFFUSE INSTABILITY

In the following sub-sections, the occurrence of various localized and diffuse instability modes in sand, as discussed above, is numerically studied in a boundary value setting via finite element computations. The density-stress-fabric dependent elastoplastic model described in Section 2 of this paper was implemented into the commercial finite element code Abaqus (2006) through the user material subroutine facility called UMAT. As such, full advantage of the capabilities of Abaqus can be taken in terms of searching algorithm for locating limit or bifurcation points. However, for ensuring robustness in the plasticity calculations, an implicit stress return and consistent tangent operator algorithm together with a spectral decomposition of the stress tensor were developed and implemented in UMAT.

4.1 Plane strain conditions

The initial boundary value problem (BVP) illustrated in Figure 2 is examined. This BVP represents a displacement controlled biaxial test on a hypothetical $10 \times 22 \times 1$ cm sand specimen isotropically compressed to 100 kPa and submitted to a 3 cm vertical displacement on its top surface. Deformations are only allowed to occur along components y and z. The strain component in the direction x is constrained to zero so as to reproduce plane strain conditions. Material parameters and initial conditions are identical for all elements; therefore the specimen is perfectly homogeneous. A total of 220 three-dimensional C3D8 elements with eight nodes were used in these simulations with full integration.

In view of evaluating the effects of boundary conditions on the loss of homogeneity in deformations under the presence of a perfectly homogeneous field during loading history, the following subtle displacement constraints that block the lateral displacement u_y were examined as shown in Figure 3.

Among all configurations for imposed kinematical constraints shown in Figure 3, the case (g) with both top and bottom middle nodes restrained laterally refers to a perfectly symmetrical configuration. Together with the constant material parameter field, case (g) represents a perfectly homogeneous specimen. Under such circumstance, any numerical results yielding localization would correspond to a genuine loss of homogeneous deformations during the course of loading of the specimen.

Figure 3 also shows the numerical results in terms of deformed configuration and values of plastic deviatoric strains for a homogeneous and initially dense specimen (initial void ratio, $e_0 = 0.55$). Although

homogeneous conditions are imposed, localized deformation in the form of an inclined shear band appears in all cases (a)–(g) of the specimens. As anticipated, the inclination of the shear band in all cases is approximately equal to 56° to the horizontal, which is predictable from the mobilized friction angle of the sand at the instant of bifurcation.

The above numerical results from a theoretical viewpoint clearly confirm that, under the presence of material instability, there is a loss of uniqueness in the solution of the underlying field equations governing the structural response of the biaxial specimen. This loss of uniqueness allows a shear band to emerge with its position being dictated by the boundary constraint imposed. It appears that the loss of uniqueness is not only governed by small subtleties in artificially introduced boundary kinematic constraints, but also by a material instability as evidenced in the perfectly homogeneous case (g). In the latter case strain localization is still triggered in spite of perfect symmetry and initial homogeneous conditions. It would be plausible to attribute this bifurcation to the influence of the constitutive model that has the characteristic of strong dependency of material behaviour on stress, density and fabric levels with the added feature of strength degradation through softening. As a matter of comparison, the case (h) in Figure 3 illustrates the fundamental solution which refers to the homogeneous deformation mode. This mode was obtained by rerunning the whole boundary value problem as a single element problem so that shear band formation would be overcome.

Figure 4a shows the effective stress path followed by elements on the top boundary for all cases (a)–(h). Given the various positions and configurations of the shear band obtained in the finite element computations, the compelling consequence on the load versus displacement curves is that various post peak results (branches) are obtained with seemingly slightly different bifurcation points, as shown in Figure 4b. The curves essentially represent the mean deviatoric stress q for all elements located on the top boundary of the specimen. Note that in the pre-peak regime, specimens respond identically for all cases despite of small subtleties in the boundary kinematic constraint.

It is worth noting that the above numerical results seem to point to the fact that small imperfections gives rise to distinctively different post peak response as soon as localization is implicated. The practical significance of this is that no two samples tested in the lab can give rise to the same response curve as a consequence of initial imperfections and boundary effects. The open question here is whether there exists a 'unique' solution for the post peak behaviour irrespective of the nature of the initial imperfections and the post-peak searching algorithm. Material perturbation in which material properties are assumed to scatter around a specific mean value in space may be a

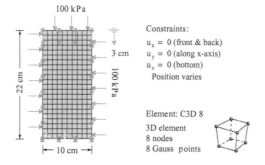

Figure 2. Geometry, mesh, boundary conditions (loads and constraints), and element type used in the finite element simulations.

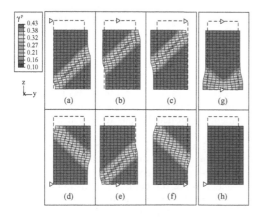

Figure 3. Plastic deviatoric strain: (a)–(g) Deformed meshes showing different localized deformation responses for the distinct initial boundary conditions. (h) Theoretical homogeneous response with no shear band.

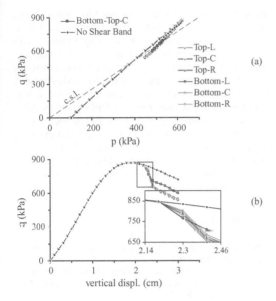

Figure 4. Average response for elements located on the top boundary.

more natural method whereby the bifurcated solution would be more objectively found irrespective of the imperfections in boundary conditions.

4.2 Axisymmetric conditions

In this section we essentially examine the same prototype problem as in sub-section 4.1, except that the geometry and loading conditions are now axisymmetric just like in a so-called 'triaxial' test in soil mechanics. The specimen's size is the same as previously, i.e. 22 cm tall and 10 cm of diameter. The initial imposed constraints are: $u_z = 0$ on all nodes of the bottom surface and $u_x = u_y = 0$ on one node located at the axis of geometrical symmetry of the problem. A total of 550 three-dimensional C3D8 elements with eight nodes and full integration were used in these simulations.

In order to erase the influence of initial imperfections as explored in the previous sub-section, material perturbation is introduced through a small fluctuation in material properties in the form of a random distribution of initial void ratio throughout the specimen. To specify a certain void ratio variation throughout the specimens, both a Uniform and a Gaussian (normal) stochastic void ratio distribution are used as illustrated in Figure 5. In the Uniform distribution case, the void ratio varies between a minimum value of 0.55 and a maximum value of 0.65 with the approximately the same frequency, corresponding to a dense sand. By contrast, in the Gaussian distribution, the void ratio is made to spread about a mean value of 0.6 with a

standard deviation of 0.025. In either case, the void ratio distribution inside the specimen is random and non-spatially correlated.

Figure 6 shows the deformed shapes of the cylindrical sand specimen at the end of the application of 3 cm of vertical displacement for three cases, i.e. (a) initial homogeneous condition ($e_0 = 0.60$), (b) initial Uniform void ratio distribution and (c) initial Gaussian void ratio distribution. As anticipated, the unperturbed case (a) gives the fundamental homogeneous response. However, both the initial Uniform and Gaussian distribution cases give a diffuse mode. Note that depending on the distribution type, either bulging at the top or at the bottom parts of the specimen is obtained. There is no clear strain localization even though the limit plastic state has been attained. If there were at all any localization, it would be in the form of multiple bands obscured by bulging of the top or bottom parts of the specimen.

The post-peak response is depicted in Figure 7 which gives the effective stress paths for two elements located on the top and bottom portions of the specimen. The stress increments in the post peak seem to point along a certain direction that gives a negative second-order work. As discussed previously, diffuse failure in the absence of strain localization is obtained as evidenced by the bulging modes shown in Figure 8.

Figure 8 shows the evolution of the second-order work throughout the middle section of the sample for

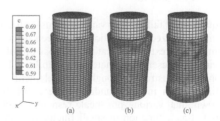

Figure 5. Clustered frequency plots of void ratio for Uniform and Gaussian distributions for dense sand.

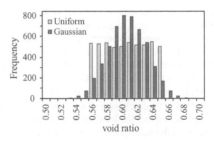

Figure 6. Undeformed-deformed configurations for: (a) Homogeneous, (b) Uniform, and (c) Gaussian void ratio distributions for dense sand.

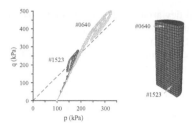

Figure 7. Effective stress paths for two elements inside the cylindrical specimen

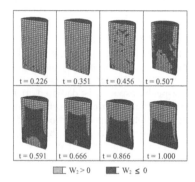

Figure 8. Temporal evolution of second-order work for dense sand specimen with Gaussian void ratio distribution.

the case of Gaussian void ratio distribution. It is noticeable how the zone of negative second-order work coincides with the bulged region of the specimen where diffuse failure manifests more intensely after pseudo time $t = 0.866$. At the plastic limit condition ($t = 0.48$ for the element #0640 and $t = 0.58$ for the element #1523), for any strain localization that would occur, the negative second-order work criterion is violated. This confirms the discussion in sub-section 3.2 where it was mentioned that the localization condition is contained within the bifurcation domain derived from the second-order work criterion.

5 CONCLUSIONS

The question of failure in its various forms such as diffuse and localized deformation have been discussed within the context of bifurcation and loss of uniqueness in the presence of material instability. These notions are illustrated through finite element computations of an initial boundary value problem. It is shown that various post-peak responses can be computed as a function of small fluctuations in the initial boundary conditions. Interestingly, it is possible to capture the bifurcation of a homogeneous deformation field into a non-homogeneous one with

shear localization, without the need to introduce any artificial non-homogeneity in the calculations. This is due to enriched features such as strong dilatancy with pyknotropy, barotropy and anisotropy level sensitivities of the elastoplastic constitutive model that are sufficient to capture bifurcated failure modes in sand. Under axisymmetric stress-strain conditions, it was demonstrated that, according to Hill's stability criterion, diffuse instability prevails over shear localization close to the plastic limit surface along certain loading directions in drained testing conditions. Finally, it was shown that a correlation exists between the spatial distribution of second-order work and the zone of diffuse deformations in the sample.

REFERENCES

Abaqus, 2006. *ABAQUS/CAE: User's Manual.* Version 6.6, ABAQUS Inc, USA.

Borja, R.I. 2002. Bifurcation of elastoplastic solids to shear band mode at finite strain. *Comput. Methods Appl. Mech. Engrg.* 191: 5287–5314.

Chu, J., Leroueil, S. & Leong, W.K. 2003. Unstable behaviour of sand and its implication for slope instability. *Can. Geotech. J.,* 40: 873–885.

Darve, F. 1994. Stability and uniqueness in geomaterials constitutive modeling. In Chambon, Desrues, Vardoulakis (eds), *Localisation and Bifurcation Theory for Soils and Rocks:* 73–88, Rotterdam: Balkema.

Desrues, J. & Viggiani, G. 2004. Strain localization in sand: an overview of the experimental results obtained in Grenoble using stereophotogrammetry. *Int. J. Num. Anal. Methods in Geomechanics* 28(4): 279–321.

Hill, R. 1958. A general theory of uniqueness and stability in elastic-plastic solids. *J. Mech. Phys. Solids,* 6: 236–249.

Lade, P.V. 2002. Instability, shear banding, and failure in granular materials. *Intl. J. Solids Structures* 39: 3337–3357.

Lyapunov, A.M. 1907. *Problème général de la stabilité des mouvements.* Annals of the Faculty of Sciences in Toulouse, 9: 203–274.

Nova R. 1994. Controllability of the incremental response of soil specimens subjected to arbitrary loading programs. *J. Mech. Behavior of Materials,* 5(2): 193–201.

Rowe, P.W. 1962. The stress-dilatancy relation for static equilibrium of an assembly of particles in contact. *Proc. of the Royal Society of London.* Series A, Mathematical and Physical Sciences, 269(1339): 500–527.

Rudnicki, J.W. & Rice, J.R. 1975. Conditions for the localization of deformation in pressure sensitive dilatant material. *J. Mech. Phys. Solids* 23: 371–394.

Vardoulakis, I. 1980. Shear band inclination and shear modulus of sand in biaxial tests. *Int. J. Num. and Anal. Methods in Geomechanics* 4: 103–119.

Vardoulakis, I., Goldscheider, M. & Gudehus, G. 1978. Formation of shear bands in sand bodies as a bifurcation problem. *Intl J. Num. Anal. Methods in Geomech.* 2: 99–128.

Vardoulakis, I. & Sulem, J. 1995. *Bifurcation Analysis in Geomechanics,* Chapman & Hall.

Wan, R.G. & Guo, P.J. 2004. Stress dilatancy and fabric dependencies on sand behavior. *J. Eng. Mech.* 130(6): 635–645.

Prediction and Simulation Methods for Geohazard Mitigation – Oka, Murakami & Kimoto (eds)
© 2009 Taylor & Francis Group, London, ISBN 978-0-415-80482-0

Thermo-poro mechanical analysis of catastrophic landslides

I.Vardoulakis & E. Veveakis
National Technical University of Athens, Athens, Greece

ABSTRACT: Catastrophic landslides are considered to slide dynamically under the presence of some weakening mechanism, like thermal pressurization, which reduces the strength of the slide near zero. In this study, based on energy considerations we model the run-off of a catastrophic landslide to obtain an estimate on the time that the slide will enter the catastrophic regime. Based on first principles of thermodynamics, we show that soon after the onset of accelerating sliding, pressurization sets in and the slide collapses catastrophically.

1 INTRODUCTION

An estimate of the run-out of a potential slide can be computed by a simple energy consideration (Dade & Huppert 1998, Figure 1): One considers the sliding earth masses lumped together as single object that is moving downslope. The total mass of moving earth-materials is M and its dead weight is $W = Mg$, where g is the gravity acceleration. Mobilized by one of a number of possible trigger mechanisms this mass slides under the action of gravity with variable speed $V(t)$. This motion of the earth mass may be described by the energy balance equation, that postulates the balance between changes in kinetic energy, the work done by the externally applied forces (i.e. the dead weight of the landslide) and the energy dissipated due to the frictional contact of the earth mass along its track. Let dz be the change in elevation of the center of gravity of the slide and dQ the energy dissipated for this increment of motion. Energy balance requires that the gain in kinetic energy equals to the work supplied by the dead weight during its fall (i.e. the loss in potential energy) minus the work dissipated:

$$d\left(\frac{1}{2}MV^2\right) = Wdz - dQ \tag{1}$$

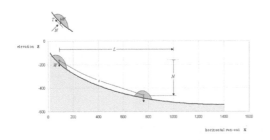

Figure 1. Run-out of a landslide on a variable topography.

Eq. (1) can be formally integrated along the path of the sliding mass. Let the total length of this path be s and the corresponding total fall height be H. At the origin and at the end of the sliding motion the velocity of the sliding mass is equal to zero; i.e. $V(0) = 0$ and $V(s) = 0$ Thus Equation (1) yields

$$M\int_0^s VdV = W\int_0^H dz - \int_0^s dQ \Rightarrow 0 = WH - \int_0^s dQ \tag{2}$$

We assume that the work dissipated during this motion is only due to basal friction,

$$\int_0^s dQ = \int_0^s Tds \tag{3}$$

where T is the friction force that is acted upon the sliding body by the track as a tangential to the track reaction contact force. Eq. (3) means that all energy dissipated within the sliding mass is neglected. The resulting model is then called a sliding (rigid) block model for the landslide. With this assumption eq. (2) becomes,

$$\int_0^s Tds = WH \tag{4}$$

i.e. all the potential energy of the rock mass is dissipated by the shear force that develops at the base of it as frictional reaction. For purely frictional sliding, a simple Coulomb friction law may be assumed,

$$T = N\mu \tag{5}$$

where μ is a friction coefficient, that is assumed here to be constant. Note that the introduction of μ implies no constraints on the details of the physics of the run-out. In eq. (5) with N we denote the force that is acted upon the sliding body by the track as a normal to the

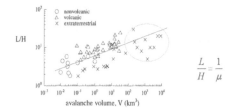

<figure>Figure 2. Relative runout L/H as a function of the rockfal volume V. Figure taken from Dade & Huppert (1998).</figure>

track reaction contact force. In a first approximation we may set that,

$$N \approx W \cos \beta \tag{6}$$

The approximation refers to the neglect of centripetal accelerations (static equilibrium in normal direction). Upon substitution of this closure for T and dQ into eq. (4) one gets,

$$WH = \mu W \int_0^s \cos \beta ds = \mu W \int_0^L dx = \mu WL \tag{7}$$

where L is the horizontal run-out. Thus,

$$\frac{H}{L} = \mu \tag{8}$$

As was first noticed by Heim 1932, catastrophic landslides are characterized by a very low (H/L)-value, thus indicating a very small apparent friction coefficient. Indeed field data suggest a reduction of the (H/L)-value with the volume of the landslide (Figure 2). This observation has triggered intensive research efforts for the disclosure of possible mechanisms that would explain this severe reduction in frictional resistance. The idea that a heat generating mechanism might account for the total loss of strength of large earth slides due to thermal pressurization of the pore fluid inside the failure zone has been discussed in the past by Uriel & Molina 1974, Goguel 1978, Anderson 1980, Voight & Faust 1982 and more recently by Vardoulakis 2000, 2002, Garagash & Rudnicki 2003, Goren & Aharonov 2007 and de Blasio 2008. Here we present a simplified analysis of the thermal pressurization phenomenon in order to narrow down the most important factors that should influence its occurrence.

2 PROBLEM DEFINITION

2.1 General considerations

As demonstrated in Veveakis et al. 2007, thermal creep and run-away lead to the phenomenon of localization of the deformation within the shear zone at the

base of the sliding mass into an ultra thin band of intense shear. Thus for the study of the final pressurization run-away phase of the slide we consider this basal shear-band with thickness $2d_B$, that is embedded between two half-spaces of the clay gouge material (Figure 3). The velocity field inside the shear-band is assumed to be isochoric and anti-symmetric with respect to the shear-band axis

$$v_x = v(y) = -v(-y); \quad v_y = 0 \tag{9}$$

The shear band is considered as a heat and pore-pressure source. Fluid flux q_i and heat flux Q_i are assumed to take place only in the direction normal to the shear-band axis and the corresponding flux vectors are assumed to be also anti-symmetric with respect to the shear-band axis

$$q_x = 0, \quad q_y = q(y) = -q(-y)$$
$$Q_x = 0, \quad Q_y = Q(y) = -Q(-y) \tag{10}$$

Heat production inside the shear-band is driven by the applied shear stress and the velocity gradient

$$D_{\text{heat}} = \beta \tau \frac{\partial v_x}{\partial y} \quad -d_B < y < d_B \tag{11}$$

where $0 < \beta \leq 1$ the Taylor-Quinney coefficient of the shear-band material, τ is the applied shear stress and v_d is the velocity of the particles at the shear-band boundaries.

Inside the shear-band the shear stress obeys the dynamic equation

$$\frac{\partial \tau}{\partial y} = \rho_m \frac{\partial v_x}{\partial t} \tag{12}$$

where ρ_m is the total density of the clay-water mixture. However a parametric study of the momentum equation (12) indicates that inside the shear band the shear stress is practically constant 0,0,

$$\tau = \tau_d(t) \tag{13}$$

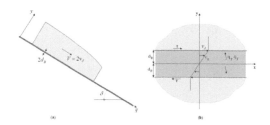

<figure>Figure 3. a) Sliding block along a basal shear band. b) Basal shear-band (rotated to horizontal) as a source of heat and pore-pressure.</figure>

430

By neglecting heat convection, local energy balance considerations require that (Vardoulakis & Sulem 1995)

$$-\frac{\partial Q_y}{\partial y} = j\,(\rho C)_m\,\frac{\partial \theta}{\partial t} - D_{\text{heat}} \tag{14}$$

where $(\rho C)_m$ is the specific heat of the shear-band material and $j = 4.2$ J/cal is the mechanical equivalent of heat.

By neglecting mass convection, fluid mass conservation inside the shear band is expressed by the following equation 0,0,

$$-\frac{\partial q_y}{\partial y} = c\left(\frac{\partial \Delta p_w}{\partial t} - \lambda_m\,\frac{\partial \theta}{\partial t}\right) \tag{15}$$

where Δp_w is the excess pore-water pressure, θ is the temperature, λ_m is the pressurization coefficient Vardoulakis 2002, Rice 2006, Sulem et al. 2007 and c is the compressibility of the shear-band material.

Let $\bar{p}$ and $\bar{\theta}$ are the mean excess pore-water pressure and temperature inside the shear-band

$$
\begin{aligned}
\bar{p}(t) &= \tfrac{1}{d_B}\int_0^{d_B} \Delta p_w(y,t)\,dy \\
\bar{\theta}(t) &= \tfrac{1}{d_B}\int_0^{d_B} \theta(y,t)\,dy
\end{aligned}
\tag{16}
$$

Due to the fact that the shear band thickness is assumed to be constant we get,

$$
\begin{aligned}
\int_0^{d_B} \frac{\partial \Delta p_w}{\partial t}\,dy &= \frac{\partial}{\partial t}\int_0^{d_B} \Delta p_w\,dy \approx d_B\,\frac{d\bar{p}}{dt} \\[2mm]
\int_0^{d_B} \frac{\partial \theta}{\partial t}\,dy &= \frac{\partial}{\partial t}\int_0^{d_B} \theta\,dy \approx d_B\,\frac{d\bar{\theta}}{dt}
\end{aligned}
\tag{17}
$$

With these remarks and notation, the corresponding "height-averaged" form of eqs. (14) and (15) yield the corresponding equations for the fluxes at the shear band boundaries,

$$Q_y\big|_{y=d_B} = -j\,(\rho C)_m\,d_B\,\frac{d\bar{\theta}}{dt} + \beta \tau_d v_d \tag{18}$$

$$q_y\big|_{y=d_B} = c d_B\left(\lambda_m\,\frac{d\bar{\theta}}{dt} - \frac{d\bar{p}}{dt}\right) \tag{19}$$

where we used the fact that that the fluxes are zero at the axis of the band

$$\int_0^{d_B} \frac{\partial Q_y}{\partial y}\,dy = Q_y\big|_{y=d_B}\,;\int_{-d_B}^{0} \frac{\partial Q_y}{\partial y}\,dy = -\,Q_y\big|_{y=-d_B} \tag{20}$$

$$\int_0^{d_B} \frac{\partial q_y}{\partial y}\,dy = q_y\big|_{y=d_B}\,;\int_{-d_B}^{0} \frac{\partial q_y}{\partial y}\,dy = -\,q_y\big|_{y=-d_B} \tag{21}$$

and for simplicity we consider only the upper boundary, assuming that the material properties of the two half-spaces are the same. Note that in the adiabatic limit from eq. (18) we get

$$Q_y\big|_{y=d_B} = 0 \quad\Rightarrow\quad \frac{d\bar{\theta}}{dt} = \frac{\beta}{j\,(\rho C)_m}\,\tau_d\,\frac{v_d}{d_B} \tag{22}$$

Similarly in the undrained limit from eq. (19) we get,

$$q_y\big|_{y=d_B} = 0 \quad\Rightarrow\quad \frac{d\bar{p}}{dt} = \lambda_m\,\frac{d\bar{\theta}}{dt} \tag{23}$$

2.2 Heat diffusion

Temperature and heat flux is continuous at the internal boundaries between the deforming shear-band and the half-spaces above and beneath it. From Fourier's law of heat conduction, applied for the half-space we have

$$Q_y = -k_F\,\frac{\partial \theta}{\partial y} \tag{24}$$

where k_F is the coefficient of thermal conductivity for the clay-water mixture. From eqs. (18) and (24) we can compute the temperature gradient at this internal boundary De Blasio 2008:

$$\frac{\partial \theta}{\partial y}\bigg|_{y=d_B} = -\theta_d'(t) \tag{25}$$

where

$$\theta_d' = \frac{d_B}{\kappa_m}\left(\frac{\beta}{j\,(\rho C)_m}\,\tau_d(t)\,\frac{v_d(t)}{d_B} - \frac{d\bar{\theta}}{dt}\right) \tag{26}$$

where κ_m is the thermal diffusivity of the of the clay material that surrounds the shear-band and for the derivation of eq. (35) we used the formula,

$$\kappa_m = \frac{k_F}{(\rho C)_m} \tag{27}$$

By neglecting convection due to pore-fluid flow, the heat equation in the half-space reads,

$$\frac{\partial \theta}{\partial t} = \kappa_m\,\frac{\partial^2 \theta}{\partial \tilde{y}^2}, \quad \tilde{y} = y - d_B \ge 0 \tag{28}$$

Eq. (28) is solved with the following boundary conditions:

$$(b.c.)_0:\quad \tilde{y}=0:\quad \frac{\partial \theta}{\partial \tilde{y}} = -\theta_d'(t) \tag{29}$$

$$(b.c.)_\infty:\quad \tilde{y}=\infty:\quad \theta = \theta_0 \tag{30}$$

where $\theta_d'(t)$ is given by eq. (26) and θ_0 is the initial temperature that is also found as ambient temperature at a large distance from the shear-band boundary.

The solution of eq. (28) with the above b.c., eqs. (29) and (30), will provide the information for the value of the temperature at the half-space/shear-band interface. We assume that approximately the height-averaged temperature inside the shear-band differs but little from this value; i.e. we assume that,

$$\bar{\theta}(t) \approx \theta(\tilde{y}, t)|_{\tilde{y}=0} \tag{31}$$

2.3 Pore-Pressure diffusion

Pore-pressure and fluid flow is also continuous at the internal boundaries between the deforming shear-band and the half-spaces above and beneath it. From Darcy's law, applied for the upper half-space we have

$$q_y = -\frac{k}{\eta_w} \frac{\partial \Delta p_w}{\partial y} \tag{32}$$

where k is the physical permeability of the clay and η_w is the viscosity of the pore-water. In eq. (32) Δp_w is the excess pore-water pressure

$$\Delta p_w = p_w - p_{w0} \tag{33}$$

where p_{w0} is the initial pore-pressure that is also found as ambient pore-pressure at a large distance from shear-band boundary.

From eqs. (19) and (32) we can compute the excess pore-pressure gradient at this internal boundary:

$$\left.\frac{\partial \Delta p_w}{\partial y}\right|_{y=d_B} = -\Delta p'_d(t) \tag{34}$$

where

$$\Delta p'_d = \frac{d_B}{c_v}\left(\lambda_m \frac{d\bar{\theta}}{dt} - \frac{d\bar{p}}{dt}\right) \tag{35}$$

where c_v is the consolidation coefficient of the clay material that surrounds the shear-band and for the derivation of eq. (35) we used the formula,

$$c_v = \frac{k}{c\eta_w} \tag{36}$$

Note that in general the consolidation coefficient in a clay is found to be an increasing function of temperature,

$$c_v = c_v(\theta) \uparrow \tag{37}$$

The amount of drainage of the shear-band material into the surrounding half-spaces is governed by the corresponding pore-water pressure diffusion equation, which in turn derives from mass balance considerations and Darcy's law Vardoulakis 2002

$$\frac{\partial \Delta p_w}{\partial t} = \frac{\partial}{\partial \tilde{y}}\left(c_v(\theta)\frac{\partial \Delta p_w}{\partial \tilde{y}}\right) \tag{38}$$

Eq. (38) is solved with the following boundary conditions:

$$(b.c.)_0: \quad \tilde{y} = 0: \quad \frac{\partial \Delta p_w}{\partial \tilde{y}} \approx -\Delta p'_d(t) \tag{39}$$

$$(b.c.)_\infty: \quad \tilde{y} = \infty: \quad \Delta p_w = 0 \tag{40}$$

where $\Delta p'_d(t)$ is given by eq. (35) and we assumed that the excess pore-water pressure at a large distance from the shear-band boundary is zero.

The solution of eq. (38) with the above b.c., eqs. (39) and (40), will provide the information for the value of the excess pore-pressure at the half-space / shear-band interface. We assume that approximately the height-averaged excess pore-pressure inside the shear-band differs but little from this value; i.e. we assume that,

$$\bar{p}(t) \approx \Delta p_w(\tilde{y}, t)|_{\tilde{y}=0} \tag{41}$$

2.4 Sliding-Block dynamics

In order to close the problem we must restrict the shear stress $\tau_d(t)$ and the velocity $v_d(t)$ as they appear in the thermal b.c., eq. (26). The shear stress is assumed to obey a friction law,

$$\tau_d = (\sigma'_0 - \bar{p}(t))\,\mu_{res} \tag{42}$$

where σ'_0 the initial effective stress normal to the shear band plane and μ_{res} is the residual friction coefficient inside the shear-band. For the sake of simplicity we consider here the single-block mechanism on single planar surface with constant slope angle δ. In that case the dynamic equation for the sliding block yields Veveakis et al. 2007,

$$\frac{dV}{dt} = \left(1 - \frac{\mu_{res}}{\tan\delta}\frac{\sigma'_0 - \bar{p}(t)}{\sigma_0}\right) g \sin\delta \tag{43}$$

Where σ_0 is the total normal geostatic stress at the failure plane, and

$$V = 2v_d(t) \tag{44}$$

is the velocity of the sliding block. Note that this computation can be easily generalized for variable slope topography.

3 ANALYSIS AND RESULTS

3.1 Dimension analysis

We select d_B for the scaling of the y-coordinate that relates to the diffusion processes

$$y^* = \frac{y}{d_B} \tag{45}$$

This assumption means that heat- and pore-water diffusion affect only a small domain in the neighborhood of the shear band, that scales with the width of the source.

The time scale of the problem is governed essentially by the sliding block dynamics, that is gravity driven. This observation is suggesting to introduce a reference length scale, H_{ref} such that,

$$t^* = \frac{t}{t_{\text{ref}}} \quad ; \quad t_{\text{ref}} = \sqrt{\frac{H_{\text{ref}}}{g \sin \delta}} \tag{46}$$

For the dependent variables we introduce the following scalings,

$$V^* = \frac{V(t)}{v_{\text{ref}}} \quad ; \quad v_{\text{ref}} = \sqrt{H_{\text{ref}} g \sin \delta} \tag{47}$$

$$\tau^* = \frac{\tau_d(t)}{\tau_{\text{ref}}} \quad ; \quad \tau_{\text{ref}} = \tau_d(0) = \sigma_0' \tag{48}$$

$$p^* = \frac{\Delta p_w}{p_{\text{ref}}} \quad ; \quad p_{\text{ref}} = \sigma_0' \tag{49}$$

$$\theta^* = \frac{\theta}{\theta_{\text{ref}}} \quad ; \quad \theta_{\text{ref}} = \theta_0 \tag{50}$$

We may select t_{ref} such that the resulting dimensionless heat diffusivity is equal unity, thus yielding a time scale that is in turn determined by the shear band thickness and the heat diffusivity

$$t_{\text{ref}} = \frac{d_B^2}{\kappa_m} \tag{51}$$

and with that from eqs. (46) and (47) we get,

$$H_{\text{ref}} = \left(\frac{d_B^2}{\kappa_m}\right)^2 g \sin \delta \tag{52}$$

$$v_{\text{ref}} = \frac{d_B^2}{\kappa_m} g \sin \delta \tag{53}$$

Let also

$$\kappa^* = \frac{c_{v0}}{\kappa_m} \tag{54}$$

This choice yields to the following set of equations[1],

1) Heat conduction:

$$\frac{\partial \theta}{\partial t} = \frac{\partial^2 \theta}{\partial y^2}$$

$$(b.c.)_0 : \theta(0, t) = \bar{\theta}(t)$$

$$\frac{d\bar{\theta}}{dt} = n_1 \tau(t) V(t) + \left.\frac{\partial \theta}{\partial y}\right|_{y=0} ;$$

$$n_1 = \frac{1}{2 j} \frac{\beta \tau_{\text{ref}}}{(\rho C)_m \theta_{\text{ref}}} \frac{d_B^3}{\kappa_m^2} g \sin \delta$$

$$(b.c.)_\infty : \theta(\infty, t) = 1$$

---

[1] We suppressed the superimposed asterix, since all quantities appearing hereafter are meant to be dimensionless.

$$(i.c.) : t = 0, \ 0 < y < \infty : \quad \theta = 1 \tag{55}$$

2) Excess pore-water pressure diffusion:

$$\frac{\partial p}{\partial t} = \kappa \frac{\partial}{\partial y}\left(\exp(n_2\theta) \frac{\partial p}{\partial y}\right) ;$$

$$\kappa = \frac{c_{v0}}{\kappa_m}, \quad n_2 = \frac{\theta_{\text{ref}}}{\theta_c}$$

$$(b.c.)_0 : p(0, t) = \bar{p}(t)$$

$$\frac{d\bar{p}}{dt} = n_3 \frac{d\bar{\theta}}{dt} + \kappa \exp(n_2\theta) \left.\frac{\partial p}{\partial y}\right|_{y=0} ;$$

$$n_3 = \lambda_m \frac{\theta_{\text{ref}}}{p_{\text{ref}}}$$

$$(b.c.)_\infty : p(\infty, t) = 0$$

$$(i.c.) : t = 0, \ 0 < y < \infty : \quad p = 0 \tag{56}$$

3) Sliding block dynamics:

$$\tau(t) = \mu_{\text{res}}(1 - \bar{p}(t))$$

$$\frac{dV}{dt} = 1 - n_4\tau(t) \quad ; \quad n_4 = \frac{1}{\tan \delta} \frac{\sigma_0'}{\sigma_0}$$

$$(i.c.) : \quad \bar{p}(0) = 0 \quad ; \quad V(0) = 0 \tag{57}$$

We remark that for the process to start we must have that,

$$\left.\frac{dV}{dt}\right|_{t=0} > 0, \ \tau(0) = 1 \tag{58}$$

$$\Rightarrow \quad n = n_4 \, \mu_{\text{res}} = \frac{\sigma_0'}{\sigma_0} \frac{\tan \phi_{\text{res}}}{\tan \delta} < 1$$

This number n is identified as a "*safety factor*".

3.2 The Undrained-Adiabatic limit

In the undrained-adiabatic limit the set of governing equations is,

$$\frac{d\bar{\theta}}{dt} = n_1 \tau(t) V(t)$$

$$\frac{d\bar{p}}{dt} = n_3 \frac{d\bar{\theta}}{dt}$$

$$\tau(t) = \mu_{\text{res}}(1 - \bar{p}(t))$$

$$\frac{dV}{dt} = 1 - n_4\tau(t) \tag{59}$$

By eliminating $\tau(t)$ and $\theta(t)$ we get,

$$\frac{d\bar{p}}{dt} = m(1 - \bar{p}(t)) V(t) \quad ; \quad m = n_1 n_3 \mu_{\text{res}}$$

433

$$\frac{dV}{dt} = 1 - n\left(1 - \bar{p}(t)\right) \quad ; \quad n = n_4 \mu_{\text{res}} \qquad (60)$$

We may also eliminate $V(t)$, finally yielding a single o.d.e. for the excess pore-water pressure,

$$(1 - \bar{p})\frac{d^2\bar{p}}{dt^2} - \left(\frac{d\bar{p}}{dt}\right)^2 + m\left(1 - \bar{p}\right)^2$$
$$\times \left(1 + n\left(1 - \bar{p}\right)\right) = 0 \qquad (61)$$

with the following initial an final conditions

$$(i.c.): \quad \bar{p}(0) = 0$$
$$(f.c.): \quad \bar{p}(\infty) = 1 \qquad (62)$$

The solution of eq. (61) that fulfills the above conditions is

$$t = \frac{1}{\sqrt{2m}} \int_0^p \frac{dx}{(1-x)\sqrt{(-\ln(1-x) - nx)}} \qquad (63)$$

This equation allows us to get a fair estimate for the duration of the process (Figure 4). We select conventionally that the process has fully developed when the excess pore-water pressure reaches a high percentage of the in situ effective normal stress. Here we select this to be 99%-tile (4):

$$t_{sat} \approx \frac{1}{\sqrt{2m}} T(n) \quad ;$$

$$T(n) = \int_0^{0.99} \frac{dx}{(1-x)\sqrt{(-\ln(1-x) - nx)}},$$

$$0 < n < 1 \qquad (64)$$

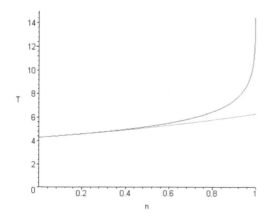

Figure 4. 99% saturation time as function of the safety factor n, eq. (58).

Table 1. Dimensionless parameters.

Parameter	n_1	κ^*	n_2	n_3	μ_{res}
Value	1213.07	0.3966	0.0308	0.35	0.18

with

$$T(n) = 4.292 + 1.307n + 0.709n^2 + O(n^3) \qquad (65)$$

For example, for the values listed in Table 1 we get that the "saturation time" for the undrained and adiabatic process is estimated to 1.25 sec!

4 CONCLUSIONS

In this study we performed a dynamic analysis of the run-off of catastrophic behaviour due to shear heating induced thermal pressurization. We concluded that in short time after the onset of the process, the block slides on a frictionless base. The onset of the sliding is determined by a safety factor $\sigma_0' \tan \varphi_{\text{res}} < \sigma_0 \tan \delta$, which is derived from the governing equations of the problem at hand. This formulation has the advantage that it may easily be generalized to include variable topographies in a dynamic setting.

REFERENCES

Anderson, D.L. 1980. *An earthquake induced heat mechanism to explain the loss of strength of large rock and earth slides. Int.* Conf. on Engineering for Protection from natural disasters, Bangkok.

Dade, W.B. and H.E. Huppert 1998, Long-runout rockfalls, *Geology* 26(9): 803–806.

De Blasio, F.V. 2008. Production of frictional heat and hot vapour in a model of self-lubricating landslides. *Rock Mechanics and Rock Engineering* 41: 219–226.

Garagash, D.I. and Rudnicki, J.W. 2003. Shear heating of a fluid-saturated slip-weakening dilatant fault zone: 1. Limiting regimes. *Journal of Geophysical Research* 108 (B2): 2121, doi: 1029/2001JB001653.

Goguel, J. 1978. Scale-dependent rockslide mechanisms, with emphasis on the role of pore fluid vaporization. Ch. 20, pp. 693–705, in Barry Voight, (ed.) *Developments in Geotechnical Engineering, Vol. 14a, Rockslides and Avalances, 1, Natural Phenomena.* Elsevier.

Goren, L. and Aharonov, E. 2007. Long runout landslides: The role of frictional heating and hydraulic diffusivity. *Geophysical Research Letters* 34(7): L07301.

Heim, A. 1932. *Bergsturz und Menschenleben. Fretz u. Wasmuth,* Zürich.

Rice, J. 2006, Heating and weakening of faults during earthquake slip, *J. Geophys. Res.* 111: B05311, doi:10.1029/2005JB004006.

Sulem, J., Lazar, P. and Vardoulakis I. 2007. Thermo-poromechanical properties of clayey gouge and application to rapid fault shearing. Int. *J. Num. Anal. Meth. in Geomechanics* 31: 523–540.

Vardoulakis, I. 2000. Catastrophic landslides due to frictional heating of the failure plane. *Mech. Coh. Frict. Mat.* 5: 443–467.

Vardoulakis, I. 2002. Dynamic thermo-poro-mechanical Analysis of Catastrophic Landslides. *Géotechnique* 52(3): 157–171.

Vardoulakis, I. and Sulem, 1995. *J. Bifurcation Analysis in Geomechanics*, Chapman & Hall.

Veveakis, E., Vardoulakis, I. and Di Toro G. 2007. Thermo-poro-mechanics of creeping landslides: the 1963 Vaiont slide, Northern Italy. *Journal of Geophysical Research* 112: F03026.

Voight, B. and C. Faust 1982. Frictional heat and strength loss in some rapid landslides. *Géotechnique* 32: 43–54.

Uriel, R.S. and Molina, R. 1974. Kinematic aspects of the Vaiont slide. Proc. 3rd Congress Int. Soc. Rock Mechanics, Denver, Vol. II, Part B, 865–870.

Finite element prediction of the effects of faulting on tunnels

A. Cividini

Politecnico (Technical University) di Milano, Milano, Italy

ABSTRACT: A numerical estimation is presented of the effects induced in an existing tunnel by the development of a fault from the deep bedrock. The spreading of the fault within the deposit hosting the tunnel can be studied in static conditions through a series of elastic-plastic, plane strain finite element analyses. They account for the reduction of the shear strength and stiffness characteristics of the faulting zone with increasing irreversible strains.

1 INTRODUCTION

The problem here considered derives from a previous study (Cividini et al. 2007a), concerning the possible damage to a building deriving from a faulting process. In fact, a trench excavated in the vicinity of a building, hosting a public facility, showed a vertical band of remolded soil underneath it. A geological and geophysical investigation was carried out to identify the origin of the band. It turned out that the band was probably generated during an earthquake about three hundred years ago. That event apparently produced a differential vertical displacement of about 50 cm across the band.

On the basis of the data obtained in the above mentioned investigation, the Authority decided to consider the effects of faulting for lifelines, including shallow and deep tunnels, located in the region. The spreading of the fault within the alluvial deposit hosting a shallow tunnel, and the consequent effects on its permanent liner and at the surface of the soil deposit, are presented elsewhere (Cividini et al. 2007b, Cividini 2008), whilst here some preliminary results are presented of the investigation for a deep tunnel located in a cohesive deposit.

The faulting process is analyzed in plane strain conditions through a series of nonlinear finite element analyses that account for the decrease of the shear strength and stiffness parameters of the geotechnical medium with increasing plastic strains. This permits following the development of the shear band in the soil mass and evaluating its consequences in terms of the internal forces in the tunnel liner.

The main characteristics of the solution procedure and of the finite element model are first outlined in the following. Then the results of some numerical analyses, based on different assumptions for the parameters, are presented.

2 MATERIAL BEHAVIOUR

The material model adopted in this study (Cividini & Gioda 1992) derives from the interpretation of direct shear tests on stiff soil samples. A qualitative representation of these results is shown in Figure 1 through the variation of the average shear stress τ with increasing horizontal displacement δ (under constant normal stress σ) and the consequent change of the failure envelope. Under constant normal stress the average shear stress τ first increases with increasing horizontal displacement until its peak value is reached. In the average normal stress-shear stress plane (Figure 1b), this corresponds to the so-called peak failure envelope characterized by peak cohesion c_p and friction angle ϕ_p. Small increments of displacement produce a reduction of the shear resistance, mainly caused by a substantial reduction of the peak cohesion. This leads to a second (fully softened) failure condition. Finally, further large increments of the horizontal displacement bring the friction angle to its residual value ϕ_r, while the cohesion vanishes.

The material behavior is introduced in the stress analysis through the simple law depicted in Figure 2, that governs the reduction of the friction angle ϕ with increasing square root of the second invariant of the deviatoric plastic strains J_2.

The friction angle remains constant and equal to its peak value ϕ_p until the invariant reaches a first limit $\sqrt{J_{2p}}$. Then a reduction occurs until a second limit $\sqrt{J_{2r}}$ is attained, which corresponds to the ultimate condition.

Similar laws, perhaps with different limits on the plastic strain invariant, can be adopted also for the cohesive component, for the angle of plastic dilation and for the (unloading) modulus of elasticity.

The described scheme of material behavior has been introduced in the finite element program **SoSIA2** for

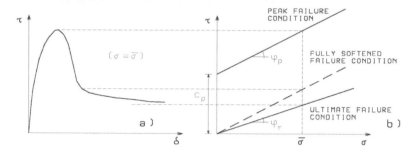

Figure 1. (a) Schematic shear stress τ versus horizontal displacement δ diagram from a direct shear test on stiff soil and (b) relevant failure envelopes.

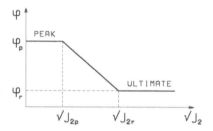

Figure 2. Variation of the friction angle ϕ with increasing plastic deformation J_2.

Soil Structure Interaction Analysis (Cividini & Gioda 1992).

The calculations are carried out by applying the external loads or the imposed boundary displacements in small increments. At each step, first an elasto-perfectly plastic analysis is performed, based on the current values of the shear strength parameters. This leads, in general, to an increment of the plastic strains and to a reduction of the shear strength and elastic coefficients according to the scheme in Figure 2.

The yield criterion is then modified for each Gauss integration point on the basis of the new parameters; the portion of the stress-state exceeding it is converted into equivalent (in the finite element sense) nodal forces and is subtracted from the total stresses. The solution process for the current displacement increment continues with a further elasto-plastic analysis in which the mesh is subjected to the previously evaluated forces and the updated stresses and mechanical parameters are considered for all the elements. The iterations for the current load increment terminate when the variation of the mechanical parameters becomes negligible.

This solution technique is straightforward to implement and is stable even in the presence of sharp variations of the mechanical parameters. A limit is that it cannot follow the overall load-displacement curve if snap-back occurs. In this case, when the

analysis is carried out under imposed displacements, a sudden change of the corresponding reactions takes place and the analysis continues along the post snap-back branch of the curve.

This method of analysis has been already applied to actual field problems, e.g. (Cividini et al. 2004, Cividini 2003), and was used also for the analysis of the faulting process here considered. The calculations were carried out in plane strain regime. The influence of the water table, which is deeply located, is neglected.

The shear modulus at small strain level G_0 of the soil deposit was determined on the basis of the shear wave velocities recorded in the field. Considering the appreciable difference existing between the elastic parameters derived from dynamic and static tests, the peak and residual shear moduli G_p, G_r introduced in the calculations are a fraction of G_0. The peak and residual elastic moduli E_p, E_r were derived from G_p, G_r assuming a Poisson ratio $\upsilon = 0.25$.

As to the shear strength parameters, lacking any experimental information, the same peak ϕ_p and residual ϕ friction angles where assumed for the deposit. The residual cohesion c_r can be reasonably neglected, while the peak cohesion c_p was varied to verify its influence on the faulting process. To avoid an excessive plastic volume increase during shearing, the angle of plastic dilation is 1/10 of the friction angle in both peak and residual conditions.

Two additional parameters of the adopted material scheme (cf. Figure 2) are the limits $\sqrt{J_{2p}}$ and $\sqrt{J_{2r}}$ of the second invariant of the deviatoric plastic strains The values assumed in the analyses here recalled correspond to a brittle behavior of the soil mass as discussed by Cividini et al. (2007b).

3 NUMERICAL MODEL

The analyses are based on a finite element grid that contains 4592 quadrilateral, four node, isoparametric elements, 16 beam elements for modeling the permanent liner and 4713 nodes.

The tunnel has a diameter of 8 m, its center is located 200 m below the ground surface and the grid, having width of 124 m, extends 20 m above and 40 m below its center.

The horizontal displacements are constrained on the two vertical sides of the grid, whilst both displacement components are constrained at its base.

The movement along the deep fault was simulated by imposing a vertical settlement to the right part of the bottom boundary. To limit the mesh dependency of the results (e.g. Pietruszczak & Stolle 1985, 1987), elements of constant size were used in the vicinity of the tunnel where the shear band is likely to develop.

It should be observed that the two-dimensional scheme implies a deep fault of unlimited length parallel to the tunnel axis, whilst in the vast majority of cases a non vanishing angle exist between tunnel and fault. Consequently this study represents only an initial attempt to investigate the problem and should be completed by more meaningful, but cumbersome, three-dimensional calculations.

The plane strain analyses begin evaluating the initial stress state in the geotechnical medium based on its unit weight and on the coefficient of earth pressure at rest. Then the stress state in the elements to be excavated is converted into the so-called excavation nodal forces and the tunnel excavation is simulated in two steps on the basis of the convergence-confinement method (Panet et al. 2001).

The excavated elements are removed from the mesh and the 70% of the nodal forces is applied to the contour of the opening. This first step simulates the excavation prior to the installation of the support and the percentage was chosen on the basis of previous experience in the analysis of tunnels (e.g. Gioda & Locatelli 1999, Contini et al. 2007).

The support is then introduced through beam elements, accounting for the effects of shear on their deformation, the stiffness of which is equivalent to that of a 0.60 m thick reinforced concrete lining. Finally, the remaining part 30% of the excavation forces is applied to the contour of the tunnel.

A thin layer of interface elements (with $c = 0$ and $\phi = \phi_r$) was introduced around the liner to account for the disturbance caused by the tunnel excavation.

The last stage of the analysis consists in applying a differential settlement to the mesh bottom, so that the calculations provide the evolution of the shear band.

4   RESULTS OF CALCULATIONS

The results of two series of finite element analyses are recalled here to verify the influence of some material parameters on the spreading of the shear band.

In the first series of analyses the influence of the elastic parameters is investigated with reference to the case of a cohesive deposit having $\phi_p = 15°$,

$c_p = 500$ kPa and $\phi_r = 12°$ as reported in Corigliano et al. (2007). Considering a layered deposit and a homogeneous one, the spreading of the shear band for four different values of the base settlement are shown in Figures 3 and 4, through the contour lines of the square root of the second invariant of the deviatoric plastic strains.

The shear band propagates almost vertically up to the tunnel in both cases. As expected it can be observed the limited influence of the elastic parameters on the progress of the fault within the cohesive deposit.

The second series of analyses investigates the faulting process in a homogeneous rock mass. From the diagram in Figures 5 it can be noted that the band in analysis C2 ($\phi_p = 40°$, $c_p = 25$ MPa and $\phi_r = 30°$) propagates almost horizontally up to a base settlement of 0.20 m, when a second fault develops quickly upright.

Finally, Figure 6 shows the influence of the rock mass strength on the shear band propagation observed when the settlement at the base reaches 0.50 m. Note that in the analyses C1 and C3 the cohesion value is respectively one half and twice the value assumed in the case labeled C2 and that, for comparison, Figure 6

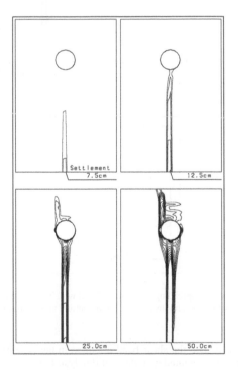

Figure 3.   Spreading of the shear band with increasing base settlement in a layered and cohesive deposit (the value of the minimum contour line of the square root of the second invariant of the deviatoric plastic strains is 1% and the contour line interval has value equal to 0.5%).

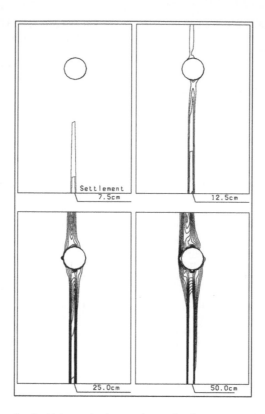

Figure 4.   Spreading of the shear band with increasing base settlement in a homogeneous deposit (the value of the minimum contour line of the square root of the second invariant of the deviatoric plastic strains is 1% and the contour line interval has value equal to 0.5%).

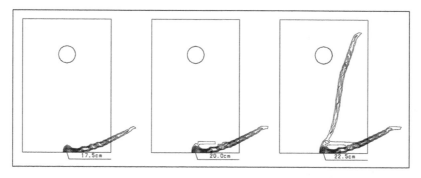

Figure 5.   Spreading of the shear band with increasing base settlement in analysis C2 (minimum contour line of the square root of the second invariant of the deviatoric plastic strains 1%, contour line interval 0.5%).

reports also the shear band obtained in the case of a weakly cemented granular soil (analysis C0).

Quite obviously the spreading of the shear band and the deformation of the soil mass induce variations of the internal forces within the permanent lining and the results of the analyses allow evaluating the variation of the maximum axial force and bending moment taking place in the lining for the entire displacement process.

For sake of briefness the results are not reported here, however they can be used for a quantitative evaluation of the effects of faulting on the stability of the tunnel, as shown for instance in Cividini et al. (2007b).

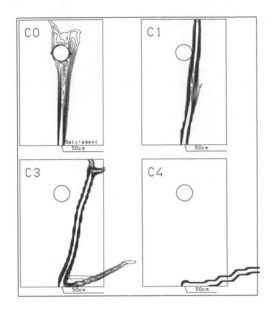

Figure 6. Influence of cohesion on shear band propagation for base settlement equal to 0.50 m (minimum contour line of the square root of the second invariant of the deviatoric plastic strains 1%, contour line interval 0.5%).

It is worthwhile to recall that the above mentioned estimations were dependent on parameters the values of which have been simply guessed due to the difficulties met in the mechanical characterization of the deposit.

It seems therefore advisable to address the problem of the quantitative evaluation of the material constants for those soil deposits that cannot be subjected to standard in situ tests and that cannot be properly sampled for carrying out adequate laboratory tests.

Another relevant drawback is also present, which is inherent to the analysis procedure. In fact some model parameters, i.e. the limits on the plastic deviatoric strains $\sqrt{J_{2r}}$ and $\sqrt{J_{2p}}$, do not represent solely material characteristics, but depend also on the size of the finite elements adopted in the discretization as discussed for instance by Pietruszczak & Stolle (1985, 1987). To overcome these two drawbacks recourse could be made to a back analysis, or parameter identification, procedure (e.g. Gioda & Locatelli 1999, Gioda & Sakurai 1987) for an in situ tests as recalled in Cividini et al. (2007a, 2007b).

5 CONCLUSIONS

Some results of a study have been presented on the propagation of a fault from the deep bedrock throughout a deposit hosting a tunnel. Lacking an adequate mechanical characterization of the soil some plane strain finite element analyses of the faulting process were attempted based on an elastic plastic material model allowing for strain softening effects. The numerical results show that the faulting process has a marked influence on the spreading of the plastic zone within the soil mass, but do not allow reaching a quantitative conclusion on the overall stability of the opening, since for the problem here considered the results of a standard geological and seismic investigation do not provide a proper soil characterization, as required when engineering calculations should account for the reduction of its shear strength and stiffness parameters with increasing plastic strains.

The results of this initial investigation suggest some further steps toward a proper analysis of the problem at hand. First the mechanical parameters of the deposit should be evaluated through the back analysis of a relatively large in situ test that involves a sufficiently large portion of the soil mass.

Having reached a proper calibration of the numerical model, the finite element analyses should be extended towards a three dimensional discretization of the tunnel and of the surrounding soil. In spite of the consequent computational burden this seems in fact a mandatory step considering that quite rarely the tunnel axis is parallel to the faults present in the deep bedrock.

ACKNOWLEDGEMENTS

This study is supported by the Ministry of University and Research of the Italian Government.

REFERENCES

Cividini, A. 2008. Finite element prediction of the effects of faulting on a shallow tunnel. *Proceedings of the 4th International Conference on Structural Engineering and Mechanics, ASEM'08,* Jeju (Korea): 1093–1102, 26–28 May, CD-ROM, ISBN 978-89-89693-21-5-98530.

Cividini, A. & Gioda, G. 1992. A finite element analysis of direct shear tests on stiff clays. *International Journal for Numerical and Analytical Methods in Geomechanics* 16: 869–886.

Cividini, A., Borgonovo, G. & Gioda, G. 2004. Finite element analysis of geotechnical rehabilitation works. *JSCE Rock Mechanics Newsletter No. 5* (http://www.jsce.or.jp/committee/rm/News/news5).

Cividini, A., Gioda, G. & Petrini, L. 2007a. Finite element prediction of the effects of faulting below an existing building", *Proceedings of the International Workshop on Constitutive Modelling —Development, Implementation, Evaluation and Application* (J-H. Yin, X-S. Li, A.T. Yeung and C.S. Desai Eds.), Hong Kong (China), ISSN 978-988-99537-0-6, Advanced Technovation Limited, Hong Kong: 708–717.

Cividini, A., Gioda, G. & Petrini, V. 2007b. Finite element evaluation of the effects of faulting on a shallow tunnel in alluvial soil. *Proceedings of the ECCOMAS Thematic Conference on Computational Methods in Tunnelling (EURO:TUN 2007)* (J. Eberhardsteiner, G. Beer, C. Hellmich, H.A. Mang, G. Meschke and W. Schubert Eds.), Vienna (Austria), August 27–29, CD-ROM, ISBN-10-3-9501554-7-3, ISBN-13-978-3-9501554-7-1,.

Cividini, A. 2003. Application of numerical procedures to slope stability analysis. *Proceedings of the International Workshop on Prediction and Simulation Methods in Geomechanics* (F. Oka, I. Vardoulakis, A. Murakami and T. Kodaka Eds.), Athens (Greece), October 14–15, ISBN: 4-88644-811-9: 101–104.

Contini, A., Gioda, G. & Cividini, A. 2007. Numerical evaluation of the surface settlements due to grouting and tunnel excavation. *ASCE International Journal of Geomechanics,* 7: 217–226.

Corigliano, M., Barla, G. & Lai, C. 2007. Seismic vulnerability of rock tunnels using fragility curves. *Proceedings of the 11th International Congress on Rock Mechanics,* Lisbon (Portugal), July 8–12.

Gioda, G. & Locatelli, L. 1999. Back analysis of the measurements performed during the excavation of a shallow tunnel in sand. *International Journal for Numerical and Analytical Methods in Geomechanics* 23: 1407–1426.

Gioda, G. & Sakurai, S. 1987. Back analysis procedures for the interpretation of field measurements in geomechanics. *International Journal for Numerical and Analytical Methods in Geomechanics* 11: 555–583.

Panet, M. et al. 2001. Recommendations on the convergence-confinement method. AFTES report, Version 1.

Pietruszczak, S. & Stolle, D.F.E. 1985. Deformation of strain softening materials. Part I: objectivity of finite element solution based on conventional strain softening formulation. *Computers and Geotechnics* 1: 99–115.

Pietruszczak, S. & Stolle, D.F.E. 1987. Deformation of strain softening materials. Part II: modeling of strain softening response. *Computers and Geotechnics* 4: 109–123.

Monitoring and non-destructive investigation methods

Prediction and Simulation Methods for Geohazard Mitigation – Oka, Murakami & Kimoto (eds)
© 2009 Taylor & Francis Group, London, ISBN 978-0-415-80482-0

Development of new technology for flood disaster mitigation in Bangladesh

M.Z. Hossain
Mie University, Tsu, Japan

ABSTRACT: The development of new technology for flood disaster mitigation in Bangladesh is highly indispensable as compared to industrialized countries due to its lack of construction materials and insufficient financial support to tackle the flood problems. This paper proposes a new technique called as composite technique with the use of recycled materials and cement-sand mortar for protection of earthen embankments constructed for flood disaster mitigation. The development of the composite technique offers in this paper is crucial due to its synergetic action from three components such as aggregates, cement matrix and mesh. The mesh provides the required tensile strength to the composites, whereas the interfacial friction between the soil and the cement matrix along with aggregates provides adequate frictional capacity enabling an optimum design of the embankments to flood disaster mitigation in Bangladesh.

1 INTRODUCTION

It is well known that Bangladesh is a flood prone country having complex river system (Fig. 1). Flood disaster in Bangladesh is a chronic problem occurring frequently during rainfall, causing serious damages to livelihoods, agricultural and engineering infrastructures (Saifullah 1988). Construction of earthen embankments for flood disaster mitigation has been the history of Bangladesh since time immemorial.

Over the last few decades, more than 13,000 km of earthen embankments have been constructed because of their cheapest form to protect people's health, homes, agricultures and city dwellers from flooding (Islam 1994). It is evident that the earthen embankments in Bangladesh are overwhelmed with multi-facetted problems. These are not only unsuccessful to serve the purpose for which they are constructed but also create many other new problems because the earthen embankments are breached easily due to rainfall splash and wave action (Fig. 2). To minimize the impact of flood disasters, sustainable and cost-effective protection measures of these embankments are now crucial for flood disaster mitigation in Bangladesh.

In accordance with the above, this paper proposes a new technique called as composite embankment with the use of recycled materials to flood disaster mitigation in Bangladesh. The main components of this technique are recycled aggregates, solid wastes and cement composites for embankment protection against rainfall splash and wave action. Embankment protection with various materials still remains an art in its rudimentary level, and ideas are evolving towards assessing the uniqueness of an optimal reinforcement

system thus far (Fukuoka 1998, Jones 1996, Hossain & Sakai 2008, Hossain 2008). Composite reinforcement made of cement mortar reinforced with mesh is gaining much popularity lately for effective application in reinforced earth structures (Koerner 1994, Murray & Irwin 1981, Hossain & Inoue 2003, Hossain 2007). In a composite reinforcement, two or more different types of materials are rationally combined to produce a new composite that derives benefits from each of the components and exhibits a synergetic response (Hossain & Kajisa 2006). Composite reinforcement using single steel wire in cement mortar for reinforced

Figure 1. River system of Bangladesh.

Figure 2. Flood protection embankment failed due to rainfall splash and wave action.

soil application is one of the examples (Sivakumar et al. 2003). Thin cement composite is the matrix acting compositely with an elasto-plastic material made of high tensile mesh encased with sand-cement mortar. If the composite material properly applied in soil reinforcement, it attains its optimal reinforcing capability owing to the synergetic action of mortar with backfill and mortar with mesh. Thin cement composite elements with enough tensile resistance provided by the mesh and enough frictional resistance provided by the interfacial friction between the cement mortar and backfill can be a potential reinforcing material for reinforced earth structures as compared to conventional reinforcements (Hossain & Sakai 2007, Kakao et al. 2001).

This paper deals with the development of various composite reinforcement systems for earth reinforcing material and their comparative study with other conventional reinforcements such as geogrids and geosynthetics. To fully understand the interface shear behavior of reinforcements in a given situation, shear tests of reinforcements embedded in soil are usually performed (Milligan & Palmeria 1987, Yasufuku & Ochiai 2005). In this research, shear tests of composite made with plain surface and rough surfaces along with ordinary reinforcements embedded in soil are carried out. Field application of the composite technology is demonstrated and a comparison between the proposed technique and the conventional one is made in terms of cost, strength, durability and frictional resistances. The paper also shows in what ways the composite embankment to flood disaster mitigation in Bangladesh is better than the existing structures.

2 FIELD VISIT AND DATA COLECTION

Various types of information on embankment failure such as geometry, soil conditions, river position, flood water levels, construction procedure, materials used, hydraulic and hydrologic condition are collected from

different sources for instance, local people of the areas of embankment failure, officials of Bangladesh Water Development Board (BWDB), contractor connected to the design and construction of the embankments. To understand the physical properties of the embankment materials, samples from the failed location are collected and analyzed.

3 MATERIALS AND METHODS

3.1 Properties of soil

The particle size distribution curve of the soil used in this research work (Fig. 3) revealed that nearly 8% of the soil is clay, 41% is silt and 51% is sand. According to the unified classification system, the soil is classified as SF. The other properties of soil are depicted in Table 1.

3.2 Preparation of composite for laboratory tests

The composite for laboratory tests were prepared in wooden moulds. The requisite amounts of sand and cement were dry-mixed in a pan, followed by the gradual addition of requisite quantity of water while the mix was continuously stirred.

Ordinary Portland cement and river sand passing through No. 8 (2.38 mm) sieve, having a fineness modulus of 2.33 were used for casting. Both the cement-sand ratio and water-cement ratio were 0.5 by weight. The square mesh obtained from the market was cut

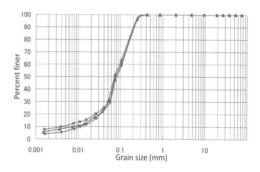

Figure 3. Particle size distribution curve of soil used.

Table 1. Basic properties of soil.

Particle density	2.747 g·cm^{-3}
Liquid limit	25.8%
Plastic limit	NP
Maximum size	425 μm
Sand (75 μm–2 mm)	51%
Silt (5 μm–75 μm)	41%
Clay <5 μm	8%
Soil type	SF

Figure 4. Composite reinforcement made of sand-cement mortar with plain and rough surfaces.

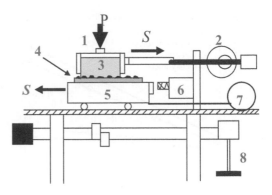

Figure 5. Shear testing apparatus.

to the desired size. The diameter of wire was 1.0 mm with center-to-center opening of 10 mm. The sand-cement mortar layer was spread at the base of the mould on which the first mesh was laid. It was then covered by further application of the mortar. Composite reinforcements with ordinary plain and rough surfaces (thickness of the rough surface was approximately 2–4 mm made of stone). The thickness and size of the composite reinforcement were 10.0 mm and 315 × 380 mm respectively (Fig. 4).

3.3 Test apparatus

The apparatus used in this study is shown in Fig. 5. For convenience of the readers, the important components of the testing equipment are numbered numerically starting from top-left to right-down in ascending manner as the number from [1] to [8], where the number [1] is the normal load application plate for upper box, [2] is the shear stress measuring device, [3] is the upper box filled with soil, [4] is the composite reinforcement, [5] is the lower box, [6] is the electrically operated shear jack, [7] is the displacement measuring dial gauge and [8] is the device taking normal load which acted on the upper box.

3.4 Method of testing

The composite reinforcement panels were made to obtain rectangular pieces of 316 mm by 380 mm in size with 120 mm extended mesh. The specified length of the pieces was selected in order to facilitate clamping with the shear apparatus. The composites were clamped in the box in such a way that the embedded length of the panel was 380 mm in the loading direction and 316 mm in the transverse direction. Water was added gradually to the soil, and thoroughly mixed to obtain desired uniform water content throughout the soil. After embedding the composite on the lower box, the upper box was set on the panel, and then the soil was filled in the upper box. The shear tests were

carried out in the way of pushing out the composite along with the lower box from the soil with constant selected speed of 1.0 mm per minute by means of screw jack under electrically operated constant pressure. The shear forces were measured using a tension load cell with the least count of 5 N. The displacements were measured by means of a mechanical dial gage with least count of 0.001 mm. All the shear tests were conducted according to the standard of the Japanese Geotechnical Society (JGS), T941-199X under four normal stress conditions such as 80, 120, 160 and 200 kPa.

4 RESULTS AND DISCUSSION

4.1 Soil-reinforcement interaction

For the sake of clear perception on the ultimate shear strength of the composite-soil interfaces, the ultimate shear strengths corresponding to the different overburden pressures (normal stresses) of the composite panels with plain and rough surfaces are plotted in Fig. 6. It is evident that the ultimate shear strength is increasing with the increase in the normal stress for both the composite reinforcements. The rate of increase of the ultimate shear strength for the composite reinforcements with rough surface is more than that of the composite reinforcements with plain surface. The composite with stone show the highest increasing rate of the ultimate shear strength with an increase in the normal stress. This may be the effect of greater roughness of the composite reinforcement as well as more frictional resistance due to small stone on the surface. This figure indicates the applied normal stresses as the controlled variables as given in abscissa and ultimate shear strengths as the random variables as given in the ordinate. It is noted here that the R-square or the coefficient of determination of the regression analysis in Fig. 6 shows that the R-square value for all the cases close to 1.0 indicates that the test data are fitted

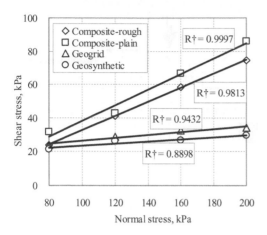

Figure 6.   Normal stress versus shear stress relationships.

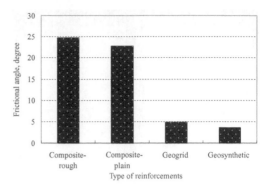

Figure 7.   Friction resistances of different reinforcements.

well for almost all of the variability with the variables specified in this paper.

In calculating the interaction resistances such as internal friction under shear test, generally, a number of identical specimens are tested under different normal stresses. The shear stress required to cause failure is determined for each normal stress. The failure envelope is obtained by plotting the points corresponding to the shear strengths at different normal stresses and joining them by a straight line. The straight lines plotted in Fig. 6 are similar to those of the method of failure envelope for direct shear test. The inclination of the linear lines (failure envelope) to the horizontal gives the angle of the shear resistances (frictional angle). For the sake of clarity, the angle of shear resistance in terms of frictional angles are calculated and plotted in Fig. 7. It is evident from this figure that the angle of shear resistance (frictional angle) is large for composite technology than the ordinary reinforcements. The composite technology with small stone on the surface gives the highest shear resistance among the four cases studies in this paper.

4.2   Embankments protection

The cross-sectional view of the composite structure showing the mortar, mesh and the anchoring pin is given in Fig. 8. The length of the anchored pin ranges from 200 to 400 mm with diameter of 13 to 16 mm depending on the slope conditions and position of the pin. The anchored pin is used to hold the mesh on the surface of the embankment during construction and to facilitate the composite structures to be remained of the slope during on-service. The average thickness of the composite structures used for embankment protection is usually 6 to 7 mm. However, this may be little bit thicker due to the unevenness of the surface of the

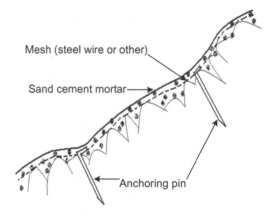

Figure 8.   Cross-sectional view of the composite technology used for embankment protection.

slope. A part of the completed composite structures for embankment surface protection is shown in Fig. 9. This figure also shows a portion of the mesh placed on the slope of the embankment and the application technique of the mortar on the mesh.

4.3   Cost, strength and durability

The cost of composite structures varies widely depending on the type of mesh used. According to the present market price, the cost of composite structure varies from 200 to 500 yen per square meter for steel mesh reinforced mortar and from 400 to 800 yen for high performance carbon or polymeric mesh reinforced mortar. The strength of composite structures was found as up to 35 MPa (300 kN/m) for 10 mm thickness composite structures. On the other hand, the cost of geogrids or geosynthetics varies from 1000 to 2000 yen. Actually, it depends on type of material, thickness, grid size and strength. The strength of geogrids

Figure 9. Application of composite technology made of sandcement mortar for embankment protection.

Figure 10. Unsucessful use of sandbags for embankment protection.

or geosynthetics varies from 30 to 800 kN/m. Corrosive mesh reinforced composite structures may be less durable than the geogrids or geosynthetics but non corrosive mesh reinforced composite structures are expected to be more durable because of their more frictional resistance in the composite-soil interface.

4.4 Composite with sandbags

The sand filled geotextile bags (sandbags) and geo-carpeting are often used to protect the flood embankment from failure in Bangladesh but it was not successful (Fig. 10). Depending on the design, combination of the sandbags and the cement composite may be an effective composite structure for successful protection of flood embankment in Bangladesh.

4.5 Discussion

The recent studies in the year 2007 showed that many flood embankments breached during the rainy seasons (Table 2) and thus, these are not successful to flood disaster mitigation. To date, the main aim of embankments construction is to seal off the water from

Table 2. Embankments breached in 2007.

Name of Embankmen and location	Date of breach	Damages
Jamuna embankment, Bogura	May 2	Over 1,50,000 families have been affected
Baufal Embankment, Patuakhali	May 18	12 villages flooded and 2000 acres land damaged
Cross Dams, Teknaf	June 5	Damages 150 houses, 40 fishermens
Gabtoli, Dhaka	June 11	Connecting road damaged
Khosbari, Sirajgonj	June 13	Nearly 1200 meters breached
Rajbari, Rajshahi	June 16	Nearly 50 km has been damaged

neighboring rivers. However, owing to the wrong construction method followed, these embankments lead to several new problems. First, these embankments could not solve the flood problem effectively and permanently because of their ease of vulnerability to rain splash and wave action of flood water. Thus, failure of embankments not only increases the siltation in the river beds and floodplains but also gives a rise of flood water level which in turn increases the water current. Second, wrong construction creates a risky situation for the inhabitants inside the boundary of the embankments. Finally, it costs huge amount of money every year for reconstruction and repair. Also, sudden breaches of embankments are continuously destroying huge lives, crops, agricultures, poultry, fisheries etc.

In order to cope with the above problems of embankments which are based on wrong construction; the present article suggests a technique for alternative construction of embankments and planning to control and mitigate flood for long term basis. The basic premise of this technique is that the embankments either being fully reinforced or the surface be protected by the composite technique as demonstrated above so that it would be durable against rain splash and water current. The use of recycled materials, cement composites (sand-cement mortar, soil-cement etc.) and their combination can be a significant step in this direction (Vipulanandan & Elton 1998, Hossain et al. 2006). The disposal of solid wastes has become a major problem in Bangladesh. Proper use of these solid wastes as such as recycled aggregates may lead not only to quality embankments at considerable savings but also to solutions for environmental problems (Hossain & Sakai 2008, Kakao, et al. 2001).

The main advantages of this technology are as follows. 1) It uses locally available recycled materials to reinforce the embankments soil that reduces the cost of the construction materials and conserves environment. 2) It prevents damages of embankments by increasing the strength of soil. By proper construction, flood mitigation can be possible though durable embankment which reduces the siltation on riverbeds and floodplains. It also neither creates a risky situation for the inhabitants of floodplains nor brings in any new problem for wastage of huge amount of money.

5 CONCLUSIONS

For all types of soil-structures interaction performed under shear test in this study, the common feature is that there is an increase in shear stress with an increase in the normal stress. The composite-soil interface shear resistances are more than that of the ordinary soil reinforcing materials such as geogrids and geosynthetics. The shear resistance at the soil-structure interface with rough surfaces of composite structure is the highest among the reinforcements tested in this study. Composite structure that has been demonstrated in this study is an example that can be constructed for embankment slope protection in Bangladesh using composite technology. Results obtained have shown that the utility and economy can both be achieved with simple techniques using locally available materials. It is expected that the observations made in this study will bring the new concept in gaining wide acceptance of cement composite for the construction of strong, durable and cost-effective composite structures for protection of embankment to flood disaster mitigation in Bangladesh.

ACKNOWLEDGMENTS

This research is partly supported by the Research Grant No. 18580243 and 19405036 with funds from Grants-in-Aid for Scientific Research given by the Japanese Government. The author grateful acknowledges these supports. Any opinions, findings, and conclusions expressed in this study are those of the authors and do not necessarily reflect the views of the sponsor.

REFERENCES

Fukuoka, M. 1998. Earth Reinforcement-West and East, *Int. Geotech. Symp. on Theory and Practice of Earth Reinforcement*: 33–47. Fukuoka, Japan.

Hossain, M.Z. & Inoue S. 2003. Ferrocement-Soil Interface Shear Behavior. *Journal of Ferrocement* 33(2): 79–90.

Hossain, M.Z. & Sakai, T. 2007. A Study on Pullout Behavior of Reinforcement Due to Variation of Water Content of Soil. *Agricultural Engineering International Journal*. 19(89): 1–15.

Hossain, M.Z. & Sakai, T. 2008. Severity of Flood Embankments in Bangladesh and Its Remedial Approach. *Agricultural Engineering International Journal* 10(37): 1–11.

Hossain, M.Z. & Sakai, T. 2008. The Effectiveness of Nominal Dosage of Ordinary Cement on Strength and Permeability of Clayey Soil. *J. of Soil Physics* (110): 1–11.

Hossain, M.Z. 2007. A Comparative Study on Pullout Behavior of Reinforcements for Effective Design of Reinforced Soil Structures. *International Agricultural Engineering Journal* 16(3): 123–138.

Hossain, M.Z. 2008. A Potential Composite Material for Possible Applications in Earth Reinforcement. *Agricultural Engineering International Journal* 10(36): 1–18.

Hossain, M.Z. 2008. Pullout Response of Ferrocement Members Embedded in Soil. *ACI Material Journal* 105(2): 115–124.

Hossain, M.Z. & Kajisa, T. 2006. Development of Ferrocement Elements for Soil Reinforcement Applications, *Proc. of the Int. Conf. on Ferrocement and Laminated Composites, 6–8 Feb., Bangkok*: 377–388.

Hossain, M.Z., Narioka, H. & Sakai, T. 2006. Effect of Ordinary Portland-Cement on Properties of Clayey Soil in Mie Prefecture. *J. of Soil Physics* (103): 31–38.

Islam, M.Z. 1994. Embankment Failure and Sedimentation Over the Flood Plain in Bangladesh: Field Investigation and Basic Model Experiments. *J. of Natural Disaster Science* 16(1): 27–53.

Jones, C.F.W.B. 1996. *Earth Reinforcement and Soil Structures*. Buterworths Advanced Series in Geotech. Engg. London.

Kakao, B.G., Shimizu, H. & Nishimura, S. 2001. Residual Strength of Colluvium and Stability Analysis of Farmland Slope. *Agricultural Engineering International Journal* 3(3):1–12.

Koerner, R.M. 1994. *Designing With Geosynthetics*. Third Edition: Prentice Hall Inc.

Milligan, G.W.E. & Palmeria, E. 1987. Prediction of Bond Between Soil and Reinforcement; *Proc. of Int. Conf. on Prediction and Performance in Geotech. Eng. Calgary*: 147–153.

Murray, R.T. & Irwin, M.J. 1981. *A Preliminary Study of TRRL Anchored Earth*, Transport and Road Research Laboratory Report SR 674.

Saifullah, A.M.M. 1988. Embankments for Flood Protection: Success and Failure; *Proc. "Floods in Bangladesh", Institution of Engineers Bangladesh, Dhaka:* 10–12.

Sivakumar Babu, G.L., Shridharan, A. and Kishore Babu, K. 2003. Composite Reinforcement for Reinforced Soil Applications. *Soils and Foundation* 43(2): 123–128.

UNDP. 1988. Report of the Mission on 1987 Flood Occurrence. *Analysis and Recommendation Action* 2: 12–13.

Vipulanandan, C. & Elton, D.J. 1998. Recycled Materials in Geotechnical Applications; *Proc. of sessions sponsored by the Geo-Institute of ASCE in Conj. with the ASCE Annual Conv.* 18–21 October, Geo. Special Pub. (79): 1–230.

Yasufuku, N. & Ochiai, H. 2005. Sand-Steel Interface Friction Related to Soil Crushability. *Geomechanics, Testing, Modeling and Simulation. ASCE. GSP* 143: 627–641.

Prediction and Simulation Methods for Geohazard Mitigation – Oka, Murakami & Kimoto (eds)
© 2009 Taylor & Francis Group, London, ISBN 978-0-415-80482-0

Defining and monitoring of landslide boundaries using fiber optic systems

M. Iten, A. Schmid, D. Hauswirth & A.M. Puzrin
Swiss Federal Institute of Technology, Zurich, Switzerland

ABSTRACT: Defining and monitoring of the landslide boundaries is essential for the landslide understanding, analysis and stabilization. It is, however, a rather challenging task in both urban and rural areas. Distributed fiber optic sensors are offering new possibilities in the field of geotechnical monitoring. This paper describes three novel fiber optic landslide boundary monitoring systems, which have been designed, put in place and tested on several locations around St. Moritz, Switzerland.

1 INTRODUCTION

1.1 *Motivation*

Differential soil displacements initiated by creeping landslides can cause immense problems by damaging infrastructure and buildings in the sliding area. Moreover, special construction and reinforcement requirements, or even total halt of construction within a landslide area may be demanded by local construction laws. In some cases it is, therefore, of crucial importance to determine the exact position of the boundary between the landslide and the stable part of the slope. For a comprehensive analysis, this boundary needs to be identified on the surface, as well as in the subsoil.

Traditional monitoring techniques for this problem include geodetic measurements and inclinometers. Geodetic measurements can identify the boundary on the surface, but not necessarily with high precision. Additionally, depending on the amount of required measurement points and survey frequency, this can be a costly campaign. Inclinometers serve for the detection of this boundary underneath the surface, called the sliding surface. Once an inclinometer pipe is sheared, a conventional inclinometer probe can not be inserted anymore and the inclinometer will no longer produce results.

This paper is an attempt to outline new landslide monitoring techniques by means of continuous strain measurements in optical fibers embedded into an asphalt road, an old inclinometer pipe, or even directly into the soil. The necessary technology of measuring continuous longitudinal strain in an optical fiber is based on Brillouin Optical Time Domain Analysis (BOTDA), nowadays commercially available. It offers a great potential for application in the geotechnical field.

1.2 *Projects*

In all the projects presented in this paper, the goal is to determine the boundary between the landslide and the stable part. By performing optical strain measurements along the sensor cable, the transition zone between the sliding and the stable parts can be identified.

The first system, an asphalt road-embedded sensor cable, serves for the evaluation of such a boundary in an urban region. A road, which intersects this boundary, can be seen as a large-scale strain gauge. Commencing in 2006, up to date three such road-embedded sensor systems have been integrated and tested in the field.

For the boundary identification in an area where no road or other infrastructure exists, to which the fiber cable could be attached, a soil-embedded "micro-anchor"-cable system has been developed. The principle of this second system is that a cable fixed to "micro-anchors" buried in soil experiences the same movement than the soil around it. Laboratory testing of system parts started in 2007 and a first field integration of the novel specially developed cables and anchors took place in July 2008.

The third monitoring system takes advantage of old, out of service, inclinometer pipes. In order to continue using such pipes, a fiber cable is placed inside and the pipe is filled with grout. The current sliding surface can then be identified and displacements on this surface back calculated. Such a fiber optics equipped inclinometer has been installed on site in July 2008.

1.3 *Location*

The two creeping landslides monitored in this work are located in the heart of the renowned mountain

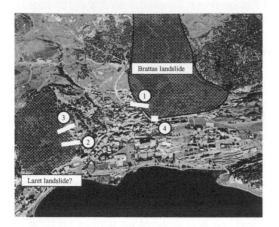

Figure 1. A view of the lake and the town of St. Moritz from the South. The sensors installed are as follows: 1 and 2, road-embedded sensors, 3 soil-embedded sensor system, 4 fiber optics inclinometer.

resort town of St. Moritz, Switzerland (Fig. 1). The displacements of the Brattas landslide have been monitored by geodetic measurement techniques for over 100 years. The Laret landslide, on the other hand, has not been recognized until recently, and it is still discussed whether it even represents a landslide or just localized displacements.

2 DISTRIBUTED FIBER OPTIC SENSING

2.1 *Brillouin optical time domain analysis*

Distributed temperature and strain sensing along an optical fiber using the effect of Brillouin scattering was proposed by Horiguchi et al. (1990). It has since evolved towards a high performance technology that can achieve 1 meter spatial resolution over long fiber lengths with absolute strain measurements in the range of a few $\mu\varepsilon$ (1 microstrain = 10^{-4}%).

Spontaneous Brillouin scattering occurs when a portion of light guided through a silicia fiber is backscattered by a nonlinear interaction with thermally excited acoustic waves. The scattered light undergoes a frequency shift. This frequency shift depends, among others, on the acoustic wave velocity, which is directly related to the strain and temperature dependent medium density.

In the more refined stimulated Brillouin scattering configuration, two counter-propagating lightwaves, at different frequencies, interact via stimulated acoustic waves. The use of stimulated Brillouin scattering for distributed fiber measurements was demonstrated by Horiguchi et al. (1989) and named as Brillouin optical time domain analysis (BOTDA).

The Brillouin backscattered light is recorded in the time domain to obtain information of the scattering along the fiber and the frequency shift of the signal is analyzed and converted into strain and temperature data. As the frequency shift is dependent on strain and temperature change in the cable, a temperature reference measurement is needed in order to separate the two components.

Today's commercially available distributed Brillouin scattering measurement units can be operated by an experienced engineer without profound knowledge of the physics behind the phenomenon. For the projects presented in this paper, a BOTDA measurement unit from Omnisens, Switzerland was utilized. It is capable of measuring the strain value with a resolution of 2 $\mu\varepsilon$ continuously along the optical fiber of up to 30 km length for any minimally 1 m long section of this fiber (Omnisens 2007).

2.2 *Design and selection of fiber optic cables*

Fiber optic cables for integration into different environments have to comply with several requirements, such as being strong enough to withstand harsh installation conditions, transmitting strain applied on the jacket without loss to the fiber core, allowing unproblematic handling and offering flexible adjustment to project modifications.

Specialty fiber optics strain and temperature sensing cables can be found on the market, but they appear to have a series of handicaps varying from high signal loss (attenuation), tricky handling and inflexibility in project alterations to long production and delivery

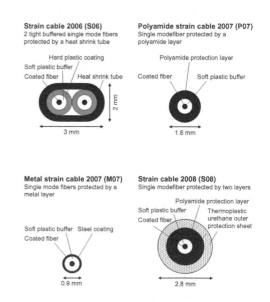

Figure 2. Cross section of strain sensing cables used in this study.

Table 1. Main characteristics of custom produced fiber optic sensors used in the projects described in this paper.

Sensor	Robustness	Handling	Attenuation	Price
S06	– – –	–	– – –	+ + +
P07	+	+ +	+	–
M07	+ +	+	+	–
S08	+ +	+	+	– –

+ advantage.
− limitation.

time and high prices. Therefore, it was chosen to custom produce strain sensing cables for the projects in this paper. The first custom sensor has been produced in 2006 at the institute's laboratory, while the later versions where developed in a joint project with the cable manufacturer Brugg Cables. The different sensor cables are displayed in Figure 2 and their main characteristics are qualitatively described in Table 1.

2.3 Laboratory testing of cables

The strain response of the fiber cables used in this study was separately tested in the laboratory. This has been done using a fiber strain calibration device which allows for fixation and straining of the fiber at known values and reference measurements thereof with Brillouin technology. As a result, the characteristics of all the cables, such as Brillouin shift, fiber slip and temperature influence could be specified and field readings can be, if necessary, adjusted to these values.

3 ASPHALT ROAD-EMBEDDED FIBER OPTIC SENSOR

3.1 Site description

The creeping Brattas landslide (Fig. 1) divides the heavily populated hillside into a stable part (outside the landslide boundary) and a moving part (landslide). An asphalt road crosses the landslide boundary in the south-western region of the landslide. The road is surrounded by buildings on both sides. No cracks have been observed in asphalt, indicating that the boundary shear zone is sufficiently wide for the asphalt to absorb deformations without cracking. To locate the boundary shear zone, it was decided to instrument the road with the fiber optic cable (Site 1 in Fig. 1).

3.2 Installation of sensor

In October 2006, a 89 m long, 10 mm wide and 70 mm deep trench was cut along the road in the asphalt. The sensor cable S06 was then attached to the bottom of the trench by clips at 2 m length intervals. This procedure allowed pre-straining of the cable over the

whole length. Subsequently, a two-component epoxy adhesive fixed the cable definitely at 1 meter intervals to the asphalt at the trench bottom. On top of the strain cable, a temperature compensation cable was loosely positioned and the whole trench was filled with an elastic cold sealing compound. The detailed installation procedure is described in Iten et al. (2008).

3.3 Monitoring process

Strain and temperature measurements were performed the day following the sensor installation and thereafter every couple of months. There was no need for continuous monitoring, because the landslide is slow moving (creeping) and no large-scale movements have been observed. The monitoring was performed successfully for 7 months until the sensor broke.

3.4 Monitoring results

The strain change in the fiber along the road in the 7 months of monitoring is shown in Figure 3. It is clearly recognizable that in a 15 m long section (between 30 and 45 meters from the upper shaft), strain increased by about 1000 $\mu\varepsilon$, while in the other parts of the fiber, strain change stays within a 500 $\mu\varepsilon$ band around the zero line.

Temperature measurements could not be successfully performed, as the signal loss was exceeding the required level for measuring with BOTDA. Therefore, temperature compensation is not included in the results of the strain measurements.

3.5 Data interpretation

For the data interpretation of the measured strain in the fiber, a simple model can be considered (Fig. 4). For correct interpretation, the angle between the sensor and the movement needs to be specified. Visual observations of damaged structures, geodetical data, as well as the location of the strained section lead to the assumption, that this angle is around 45°.

In order to achieve a strain of about 1000 $\mu\varepsilon$, over the transition zone of 15 m, the landslide movement should be about 20 mm. This displacement is of the order of magnitude of the yearly displacement geodetically recorded in this area.

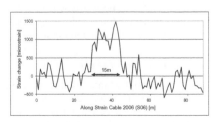

Figure 3. Strain change along road-embedded S06 sensor.

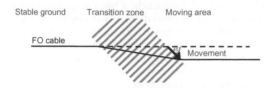

Figure 4. Simple model for movement interpretation from strain data.

3.6 *Application of asphalt road-embedded sensor system in other projects*

Following the cable breaking of the S06, a new set of sensors was installed on the same location in summer 2007. This time, a metal protected strain cable (M07), as well as a polyamide protected strain cable (P07) (Fig. 2), were embedded into the road. In addition, in 2008 an asphalt road crossing the Laret landslide boundary (Site 2 in Fig. 1) was equipped by 47 m of the P07 cables. The monitoring of these sensors is still continuing and is beyond the scope of this paper.

4 SOIL-EMBEDDED "MICRO-ANCHOR"—CABLE SYSTEM

4.1 *Site description*

Uphill from the road-embedded sensor on the Laret landslide (Site 2 in Fig. 1), a soil-embedded sensing system was placed beneath a hiking path (Site 3 in Fig. 1). Downhill from that path, and above the road-embedded sensor, a building was severely damaged in 2006 due to differential displacements. At that point, concerns began to rise that a creeping landslide may exist at this location. It is assumed that the landslide is almost not moving under normal conditions, but may be triggered by a nearby excavation, undertaken in 2006 for the construction of a new building.

4.2 *Sensor system*

The soil-embedded sensor system consists of the two parts: "micro-anchors" and optical cable. The purpose of the "micro-anchors" is to connect the cable to the soil at the specified fixation points, so that the cable experiences the same movement than the soil around it, instead of the soil just flowing around the cable. Such a system is sketched in Figure 5.

4.3 *Laboratory testing of sensor system*

In order to dimension the "micro-anchors", a pullout box was used (Fig. 6). The box is 2 m long, 0.1 m wide and 0.2 m deep and has a step motor at the front. For the tests, the box was filled with a poorly graded sand up to the half of its height. The sensor was then placed

and the box was entirely filled with sand. On top of this sand, it was possible to add additional weight simulating different sensor depths. Three different tests were carried out: the cable failure test, the anchor failure test and the anchor interaction test (Fig. 7).

The cable failure test allows for the measurement of cable pullout resistance, and therefore, the friction between the cable and the sand. The purpose of the anchor failure test is to determine the anchor bearing capacity (failure load), as well as the design load (before the anchor starts loosing contact with the soil). The anchor interaction test determines the minimal

Figure 5. Soil-embedded sensor system.

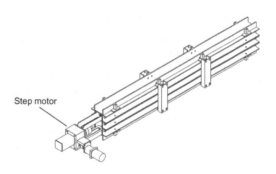

Figure 6. Pullout box.

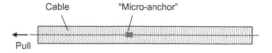

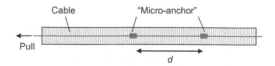

Figure 7. Setup of the laboratory tests with the pullout box.

Figure 8. The trench cut in the hiking path with the sensor system already embedded.

distance d between two identical anchors, at which they do not affect each others performance.

In advance of the field installation, these tests were run in order to design the system components properly. The cable S08 (Fig. 2) in connection with a $40 \times 40 \times 40$ mm anchor was subsequently chosen for field installation.

4.4 Installation of the sensor system

For the installation of the sensor system, a 80 m long and approximately 0.4 m deep trench was cut in the path. The location of the trench was chosen so that part of the trench should be on stable ground while the rest is in the creeping zone. The bottom of the trench was then filled with compacted sand, the sensor was slightly pre-strained, placed and covered with sand again. In addition, a temperature compensation cable as well as a strain cable without anchors was put in the trench. A picture of the installation is shown in Figure 8.

4.5 Monitoring and outlook

Zero readings of the strain were taken the day following the implementation in July 2008. First results from strain monitoring are expected for summer 2009.

5 REACTIVATION OF OLD INCLINOMETER PIPES

5.1 Site description

Close to the compression zone and lower boundary of the Brattas landslide, an inclinometer pipe was installed in 1982 for the localization of the sliding surface and monitoring of displacements (Site 4 in Fig. 1). The inclinometer pipe has a standard 71 mm diameter, and goes 14.75 m deep into the ground. Since 1987, the inclinometer can not be monitored

anymore because the conventional probe does not go all the way through the pipe. For that reason, it was decided to reactivate this inclinometer by installing a fiber optic sensor.

5.2 Installation of sensor

The sensor consists of the P07 cable, which was inserted into the inclinometer to 14 m depth and slightly prestrained. The whole pipe was then filled with a cement-bentonite grout backfill. This grout is intended for the overall bonding of the fiber along the pipe.

5.3 Monitoring process

In July 2008, after hardening of the grout, zero readings of the strain in the fiber were taken. In October, the strain was measured again and monitoring will continue every few months.

5.4 First monitoring results

The strain change along the cable between July and October 2009 is displayed in Figure 9. It can be clearly seen, that between 5.5 m and 7.5 m, the strain in the fiber increased by up to 400 $\mu\varepsilon$, while in the other parts, strain increase stays about ± 100 $\mu\varepsilon$ around 0. The largest strain increase was measured at 6.2 m with 410 $\mu\varepsilon$.

5.5 Data interpretation and outlook

Monitoring data from 1982 to 1984 indicate that the sliding surface lies between 5.8 m and 6.7 m. By looking at the optically measured strain in the fiber, one can see that principal displacements must occur between 6.2 m and 6.8 m. The measured strain over the 1 meter thick sliding surface suggests a monthly horizontal displacement of about 1 mm, which is in the range measured in 1982–1987. The optical data series at this site consists of just two measurements, one in July and the other one in October. Therefore, it would be too early to give a comprehensive validation of this sensor

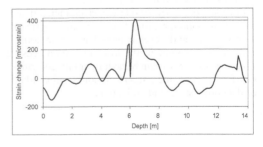

Figure 9. Average strain change in the 3 months period following the sensor installation in the inclinometer.

system. Nevertheless, the agreement between the two methods strongly supports applicability of the fiber optics sensors in old inclinometer pipes.

6 DISCUSSION

6.1 Cable selection

In any fiber optic strain sensing project, the selection of the adequate cable is a key issue. The main concern is usually put on the protection and on the strain transfer from the outside sheath of the cable to the optical fiber. Cable manufacturers are increasingly adjusting to the demands of the fiber sensing community and can offer custom made strain sensor cables. Shortly, such cables should be available of the shelf.

6.2 Attachment of cable to the structure / ground

For the attachment of a sensor cable to a structure or to the ground, basically two methods exist: continuous bonding and point fixation. It is a difficult task to determine the best fixation method and, especially, to choose a procedure which allows to achieve a durable and trustworthy fixation. The options shown in this paper have demonstrated a good performance.

6.3 BOTDA technology vs. traditional technologies

The possibility to measure deformation continuously along a fiber optic cable, presents several advantages over traditional technologies such as geodetic measurements or inclinometers. First, it offers distributed measurements of strain over long distances and therefore, hundreds and thousands of measuring points. This is a clear advantage when soil stability monitoring extends over long distances (e.g. for ground movement detection along roads, train tracks, power lines or pipelines). Second, measurements can be done remotely from the site, as the cable can extend to the instrument placed in an office, and therefore, a large area can be monitored from one source. Other advantages of fiber optic strain sensing include its immunity to electromagnetic hazards (lightning), water resistance and its long lifetime.

In order to deploy BOTDA measurements meaningfully to monitor ground movement, a series of improvements and developments have to be considered, such as commercially available strain sensing cables with known strain behavior parameters, guidelines for structure integration techniques and standards for data processing and evaluation.

7 CONCLUSIONS

The design, implementation and testing of three novel fiber optic landslide boundary monitoring systems using BOTDA technology has been demonstrated. Where the time frame allowed, measurements have been taken and the obtained results were in good agreement with traditional monitoring techniques. This validation leads to the conclusion that distributed BOTDA sensing has the potential to become a widely accepted new tool in landslide monitoring, and, in some cases, might even outperform traditional techniques. At present, until methods and standards in this field are established and reliable, combination with traditional measurements is necessary.

ACKNOWLEDGMENTS

The authors would like to thank the cable manufacturer and "micro-anchor" producer BRUGG CABLES and the measurement unit provider OMNISENS for their collaboration and advice, VSS/ASTRA (the Swiss federal department of transportation) as well as to CTI (the Swiss innovation promotion agency) for funding. In addition, we are grateful to the authorities of the municipality of St. Moritz for the test sites and support during field measurements.

REFERENCES

Horiguchi, T., Kurashima, T. & Tateda, M. 1990. A technique to measure distributed strain in optical fibers. In Photonics Technol. Lett., 2: 352–354.

Horiguchi, T. & Tateda, M. 1989. Optical-fibre-attenuation investigation using stimulated Brillouin scattering between a pulse and a continuous wave. In Optic Letters, 14: 408–410.

Iten, M., Puzrin., A.M. & Schmid, A. 2008. Landslide monitoring using a road-embedded optical fiber sensor. In Smart Sensor Phenomena, Technology, Networks, and Systems 2008; Proceedings of SPIE 6933.

Omnisens, S.A. 2007. STA100/200 Series, Fiber Optic Distributed Temperature and Strain Analyzer; User Manual.

Dynamic interaction between pile and reinforced soil structure – Piled Geo-wall –

T. Hara, S. Tsuji, A. Yashima & K. Sawada
Gifu University, Gifu, Japan

N. Tatta
Maeda Kosen Co., Ltd., Fukui, Japan

ABSTRACT: This paper proposes a new application of pile foundation to reinforced soil structure in order to improve horizontal resistance of the structure. The effectiveness of the application and dynamic interaction between pile and reinforced soil structure was confirmed from the result of a dynamic centrifuge (50G) model test, and a design method of the reinforced soil structure with using piles (Piled Geo-wall) was proposed through the simulation of the test by two dimensional dynamic FEM analysis.

1 INTRODUCTION

High ductility of soil structure reinforced by geogrid is well known and it is also possible to build independent soil structure as a retaining wall. The independent reinforced soil structure (Geo-wall) can be applied to countermeasures against collapse of embankments built on slope, or rockfall impacts or etc., as reasonable one, because it can be constructed by using soil prepared at the construction site. Hence, the application of Geo-wall, such as a countermeasure wall against rockfall impacts as shown Figure 1 and Photograph 1, has been increased.

The foundation of Geo-wall, however, has been confined to spread foundation, thus wide Geo-wall is designed and the application to narrow construction site, such as on a slope, is too difficult. If the stability of narrow width Geo-wall was assured, rational and economical geo-structure without expensive temporally bridge to convey materials of structure even though construction site is on a slope can be achieved.

Therefore, the application of pile foundation to Geo-wall, "Piled Geo-wall", as shown in Figure 2, has been studied from a dynamic centrifuge (25G) test and its numerical simulations.

In this study, the assumption concerning interaction between pile and Geo-wall that reinforced soil structure with high ductility can transmit the lateral force to the pile foundation, as shown in Figure 3, is confirmed, and a design method that can reproduce the interaction behavior is proposed. The interaction behavior between pile and Geo-wall and its numerical simulation as well as the results of the dynamic centrifuge model test are described in this paper.

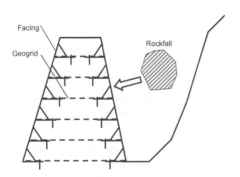

Figure 1. Counter wall against rockfall impacts.

Photograph 1. Protective wall against rockfall impacts.

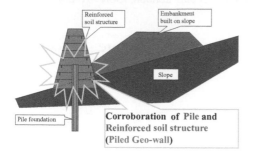

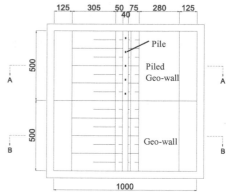

125 305 50 75 280 125
40

Pile

Piled
Geo-wall

Geo-wall

1000

(a) top view.

Figure 2. Image of Piled Geo-wall.

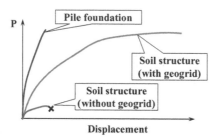

P

Pile foundation

**Soil structure
(with geogrid)**

**Soil structure
(without geogrid)**

Displacement

Figure 3. Assumption of interaction between pile and Geo-wall.

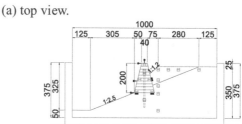

(b) A-A section.

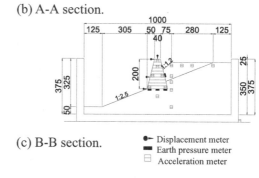

(c) B-B section.

●— Displacement meter
■ Earth pressure meter
▯ Acceleration meter

Figure 4. Experiment model (1/25 scale model).

2 AN EXPERIMENT OF PILED GEO-WALL

2.1 Summary

A dynamic centrifuge test (25G) was carried out to confirm the effectiveness of application of pile foundation to Geo-wall, and dynamic interaction between pile and Geo-wall. In this test, the difference of the behaviors between a piled Geo-wall and normal one (without pile) as a seismic countermeasure against deformation of road embankment built on slope, as shown Figure 4, was compared. Tables 1 and 2 shows geotechnical and structural parameters adopted in this test.

Where, the material, which has equivalent initial elastic modulus with actual geogrid as shown in Figure 5, was adopted to geogrid in the test model. Geo-wall was built at every step as same as actual Geo-wall execution as shown in Figure 6. Furthermore, longitudinal additional geogrid, in particular, is had in place at every layer, as shown in Figure 7, in order to transmit the load to pile from Geo-wall body smoothly. Figure 8 shows input earthquake wave in this test.

2.2 Results

2.2.1 Effectiveness of pile foundation

Photograph 2 shows the state of road surface after shaking. And residual deformation and transition of

Table 1. Geotechnical parameters*

	Elastic modulus	Cohesion	Friction angle
	kN/m^2	kN/m^2	degree
Slope	3.26×10^5	–	–
Embankment	3.0×10^4	0	40
Geo-wall	3.0×10^4	0	40

* converted value to actual scale.

displacement of the Geo-walls, transition of subgrade reaction under the Geo-walls, transition of earth pressure at the back of the Geo-walls, and distribution of the strain in depth occurring in the pile are shown in Figure 9 to Figure 13 respectively. According to the

Table 2.   Structural parameters*

	Elastic modulus	sectional area	Moment of inertia
	kN/m^2	m^2	m^4
Pile	2.0×10^8	4.79×10^{-3}	2.04×10^{-4}
Geogrid	8.0×10^5	1.0×10^{-3}	–

* converted value to actual scale.

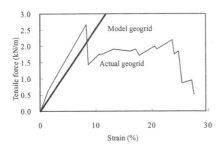

Figure 5.   Tensile stiffness of geogrid.

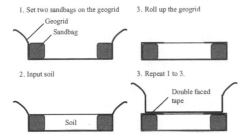

Figure 6.   Procedure for building Geo-wall.

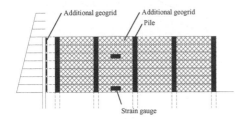

Figure 7.   Additional geogrid to unify piles and Geo-wall.

result, the effectiveness of the application of the pile to Geo-wall was confirmed as follows;

a. the load of embankment is transmitted to the piles (Figure 13)
b. as the above result, the Piled Geo-wall can receive larger earth pressure than the normal one without pile (Figure 12)

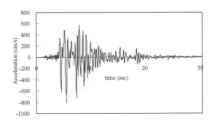

Figure 8.   Input wave.

Photograph 2.   State of road surface after shaking.

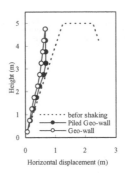

Figure 9.   Deformation of Geo-wall.

c. and, response and residual displacement of Piled Geo-wall are smaller than the normal one without pile (Figure 9 to 11)
d. conclusively, the deformation of the road surface can be reduced (Photograph 2)

By the way, the cause of that subgrade reaction at point C of normal Geo-wall is smaller than the one at point B is considered that local collapse was occurred in the back-bottom of the Geo-wall. Furthermore, it was also confirmed that the displacement of the top part (pile un-inserted) of the Piled Geo-wall is large.

With respect to the fact that the earth pressure acting on back of Piled Geo-wall is larger than the one of the

459

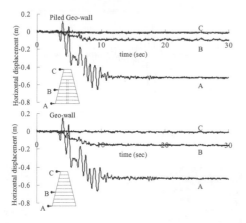

Figure 10.   Transition of the displacements of Geo-wall.

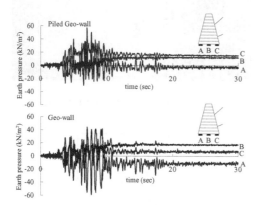

Figure 11.   Transition of subgrade reaction under Geo-wall.

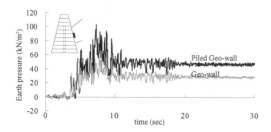

Figure 12.   Transition of earth pressure on back of Geo-wall.

normal Geo-wall (without pile), Figure 14 shows the comparison between the locations of the crack occurring on the road surface and Coulomb's failure line. According to the result, it was considered that the behaviors of back embankment of Piled Geo-wall is similar with passive behavior respect to one of normal Geo-wall (without pile), which is active behavior. The effectiveness of pile foundation to improve lateral

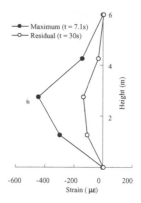

Figure 13.   Strain occurring in the pile.

resistance of Geo-wall can be also confirmed from the view point.

Figure 15 shows the transition of the strain occurring in the longitudinal additional geogrid. According to the result, it was confirmed that the geogrid is effectiveness to transmit the load of the back embankment to the piles from the view point that large strain occurs in the geogrid.

2.2.2 *Dynamic interaction between pile and Geo-wall*

The distribution of the displacement of the pile and Geo-wall in depth, and transition of the displacement of the pile and Geo-walls are shown in Figures 16 and 17 respectively. According to the result, because

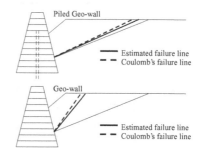

Figure 14.   Presumption of slide surface.

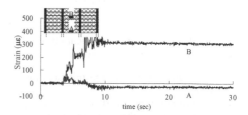

Figure 15.   Strain occurring in the additional geogrid.

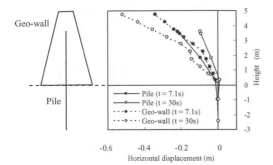

Figure 16. Displacement distribution of pile and Geo-wall.

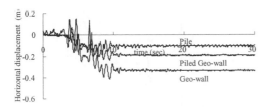

Figure 17. Transition of displacement of pile and Geo-wall.

of high ductility of reinforced soil structure, relative deformation between pile and Geo-wall is not small though, it was considered that Geo-wall can transmit the force to pile foundation predictably.

3 NUMERICAL SIMULATION

3.1 Simulation model

3.1.1 Analysis model

Two dimensional dynamic FEM, UWLC (Lee 2000), is adopted for simulation of the experiment. Figure 18 shows FEM mesh and Tables 3 and 4 shows geotechnical and structural parameters. Where, because piles are three dimensional structure, the three dimensional effect is reflected to flexural stiffness of the piles from the view point of load apportionment.

3.1.2 Three dimensional effect of pile foundation

Three dimensional effect for modeling of piles as a continuous wall in two dimensional analysis is studied from the comparison of the behavior between a static three dimensional FEM and a static two dimensional FEM analyses. Figure 19 shows the three dimensional FEM mesh.

At first, in order to confirm the reproducibility of the three dimensional FEM with respect to dynamic three dimensional behavior of pile and Geo-wall, below mentioned analyses were carried out.

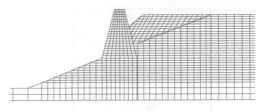

Figure 18. FEM mesh.

Table 3. Geotechnical parameters

	Elastic modulus	Cohesion	Friction angle
	kN/m^2	kN/m^2	degree
Slope	3.26×10^5	–	–
Embankment	3.0×10^4	0	40
Geo-wall	3.0×10^4	0	40

Table 4. Structural parameters

	Elastic modulus	sectional area	Moment of inertia
	kN/m^2	m^2	m^4/m
Pile	2.0×10^8	4.79×10^{-3}	$*4.08 \times 10^{-5}$
Geogrid	8.0×10^5	1.0×10^{-3}	–

* the value, which was considered 3D effect.

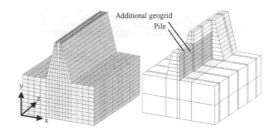

Figure 19. Three dimensional analysis model.

a. Inertia force of the Piled Geo-wall and earth pressure of the road embankment were loaded on the lateral surface of the Geo-wall
b. The forces were adjusted as the stresses occurring in the pile can be reproduced the maximum stress distribution observed in the experiment as shown in Figure 20.
c. The strain occurring in the longitudinal additional geogrid obtained from the analysis was compared with observed one.

Conclusively, the strain of the additional geogrid, which is obtained from the analysis, reproduced

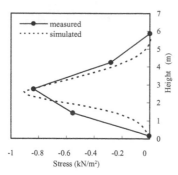

Figure 20. Reproduction of maximum stress occurring in pile.

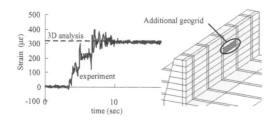

Figure 21. Comparison with strain of additional geogrid between experiment and 3D analysis.

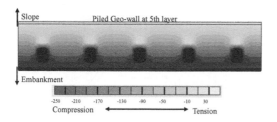

Figure 22. Internal stress state of piled Geo-wall.

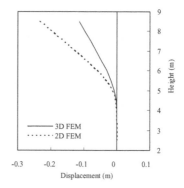

Figure 23. Pile deformation obtained from 3D and 2D FEM.

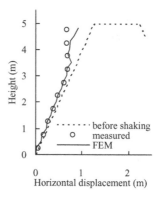

Figure 24. Residual deformation of Geo-wall.

observed one from the test as shown in Figure 21, and the validity of the three dimensional analysis to study three dimensional effect for modeling of the piles in two dimensional analysis could be confirmed. Figure 22 shows horizontal stress distribution obtained from the three dimensional FEM analysis at the section, where the strain of longitudinal additional geogrid was observed. The state of three dimensional lateral resistances of pile and Geo-wall can be confirmed from the result.

Three dimensional effects of piles for modeling of piles as a continuous wall in two dimensional FEM analysis is estimated from the difference of pile displacements obtained from three dimensional and two dimensional FEM analyses. In where, each pile was modeled as an independent pile in three dimensional

FEM analysis, and the piles were modeled as a continuous wall in two dimensional FEM analysis. Figure 23 shows the difference of the pile displacements.

According to the result, the displacement of the pile obtained from the two dimensional FEM analysis is about two times of the three dimensional one, hence the pile displacement obtained from three dimensional FEM analysis can be reproduced by two dimensional FEM analysis with using about half flexural stiffness of the pile as a three dimensional effect in the case of the experiment condition.

This three dimensional effect for modeling of piles in two dimensional analysis, however, depends on the condition of Piled Geo-wall, such as alignment and flexural stiffness of the piles, alignment and tensile stiffness of the geogrid, the materials adopted to the Geo-wall. So the effect has to be estimated with considering the condition in the design of Piled Geo-wall.

3.2 *Result and reproducibility*

Comparisons of the analysis results and observed ones with respect to residual deformation of Geo-walls

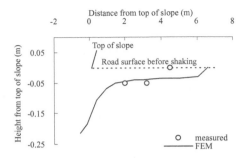

Figure 25. Residual deformation of road surface.

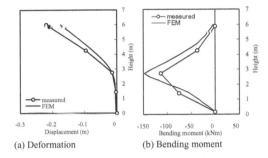

(a) Deformation   (b) Bending moment

Figure 26. Maximum response deformation and bending moment of pile.

and road surface, maximum response deformation and bending moment of pile are shown in Figure 24 to 26 respectively. According to the results, because the analysis results comparatively good reproduce the results observed from the experiment, it is considered that Piled Geo-wall can be designed by using the analysis approach proposed in this study.

4 CONCLUSION

This paper can be concluded as follows;

– Reasonability with respect to development of narrow width Geo-wall, from the view point that economical Geo-wall can be built on shallow construction site such as on slop, was described.

– For the development of the Geo-wall, a hypothesis that pile foundation can be applied to high ductile Geo-wall in order to improve lateral resistance was established.
– The validity of the hypothesis, more specifically, effectiveness of application of pile foundation to Geo-wall was confirmed predictably thorough an experiment, a dynamic centrifuge model test (25G), which carried out in this study.
– The behaviors obtained from the experiment were simulated by two dimensional dynamic FEM analysis.

  • Evaluation method of three dimensional effect of piles with respect to modeling of the piles as a wall in two dimensional analysis was proposed.
  • Two dimensional dynamic FEM analysis, with considering the three dimensional effect of piles, comparatively good reproduced the observed behaviors of the Piled Geo-wall from the experiment

– It was considered that reasonable Piled Geo-wall can be achieved by using proposed design approach in this study.

ACKNOWLEDGEMENT

The research reported herein was supported financially by The Japan Iron and Steel Federation. This support is gratefully acknowledged.

REFERENCES

Lee, G. 2000. Development of a New Finite Element Program for Liquefaction Analysis of Soil and Its Application to Seismic Behavior of Embankments on Sandy Ground, Ph.D paper of Gunma University.

Prediction and Simulation Methods for Geohazard Mitigation – Oka, Murakami & Kimoto (eds)
© 2009 Taylor & Francis Group, London, ISBN 978-0-415-80482-0

Upgrade of existing stone-guard fence with using high-energy absorption net

S. Tsuji, T. Hara, A. Yashima & K. Sawada
Gifu University, Gifu, Japan

M. Yoshida
Maeda Kosen Co., Ltd., Fukui, Japan

ABSTRACT: This paper proposes a new technique with using high-energy absorption net in order to upgrade the performance of existing stone guard fences. The technique is proposed from case studies of FEM analysis that can reproduce results of an experiment with using an actual target net. Summary of the experiment, reproducibility of the analysis using in this study with respect to the results of the experiment and effectiveness of the proposed method as well as necessity to upgrade performance of existing stone guard fences are described in this paper.

1 INTRODUCTION

A great number of stone guard fences that protect the traffic of mountainous roads from rockfall impacts have been constructed so far in Japan.

Most of the existing fences, however, has not satisfied the performance requirements with respect to absorbing energy of rockfall impacts because the fence was originally designed by a target design rockfall impact. There are many cases that a new larger stone than the initial design target stone scale was often found during the periodic site investigation. This means that a lot of mountainous roads are in peril of rockfall impacts. Therefore a quick execution of the treatment to upgrade the performance of existing stone guard fences is required. At the same time, expensive cost is estimated if the insufficient fences are replaced to new ones. Accordingly, development of a reasonable upgrading technique for the existing fences has been required.

Therefore, a study on a reasonable (i.e. rational and economic) technique to upgrade the performance of the existing fences, which uses a high-energy absorption net, with respect to larger rockfall impact than the initial design one has been conducted by authors. In this study, an experiment with using the actual target net was carried out on site to confirm the absorbability of rockfall impacts, and a new application method of the high energy absorption net to upgrade the existing fences is proposed through FEM analysis that can reproduce the results of the experiment.

2 TARGET EXISTING STONE-GUARD FENCE

Photograph 1 and Figure 1 show the target stone guard fence in this study. This type of fence is one of popular stone-guard facilities in Japan. The design absorbing rockfall energy of the target fence is 56 kJ. In this study, a reasonable technique to upgrade of the performance with respect to the absorbing energy of this fence is proposed.

Photograph 1. Target existing stone-guard fence.

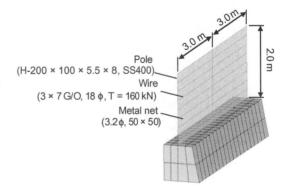

Figure 1. Target existing stone—guard fence.

3 PROCEDURE OF THIS STUDY

3.1 Aim of this study

The aim of this study is a proposal of a technique to upgrade performance of the target fence with respect to absorption of rockfall impact, more specifically, the view point of this study is the behavior of the fence that received the design rockfall impact, which is adopted as a criterion measure to evaluate the validity of the proposed technique. And the technique, which the behavior of the fence received larger rockfall impact than the design one can be restricted smaller than the criterion measure, is studied. Therefore, from the view point of knowing that the response behavior of the fence is larger or smaller than the criterion measure, a simple elastic analysis with considering equivalent stiffness based on equivalent energy respect to a given plastic behavior is adopted in this study.

3.2 Procedure of this study

Figure 2 shows the procedure of this study. The content is described as follows;

I. Establishment of criterion measure: the response behavior of the target existing fence respect to design rockfall impact is analyzed, and the behavior is established as a criterion measure.

II. Estimation of the response behavior of the fence respect to a new target stone: in where, the necessity to upgrade performance of the existing fence is confirmed from the response behavior of the fence respect to larger stone than the design one. (it must exceed the criterion measure)

III. Proposal of a technique: a technique, which uses high-energy absorption net, to reduce the behavior of the fence to smaller one than the criterion measure is proposed.

    i. Modeling of the high-energy absorption net: the modeling of the target net on the analysis in this study is established from an experiment with using actual net on site and its simulation.

    ii. Effectiveness of the target technique: it is confirmed that the response behavior of the existing fence respect to the new target stone can be reduced to smaller one than the criterion measure.

4 PERFORMANCE OF THE EXISTING FENCE

4.1 Design performance of the existing fence

The behavior of the target existing fence that received the design rockfall impact is estimated by using LS-DYNA, at first, as a criterion measure to evaluate lack of performance of the existing fence and effectiveness of the technique proposed in this study, when they received a larger rockfall impact than the design one. An elastic analysis with considering equivalent stiffness based on equivalent energy respect to a given plastic behavior was adopted to estimate the behavior of criterion measure in this study. Figure 3 shows the analysis model and structural parameters adopted in this analysis were shown in Table 1.

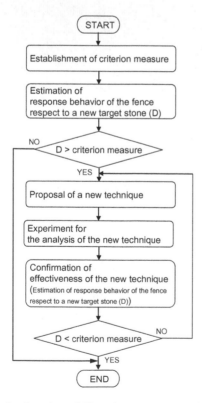

Figure 2.   Procedure of this study.

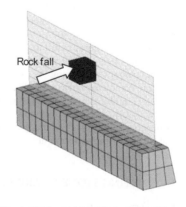

Figure 3.   Analysis model of the existing fence respect to the design rockfall impact.

Table 1. Structural parameters.

	Elastic modulus (kN/m^2)	Sectional area (m^2)	Moment of inertia (m^4)
Metal net*	–	–	–
Center pole	3.32×10^6	2.72×10^{-3}	1.84×10^{-5}
Side Pole	6.56×10^6	2.72×10^{-3}	1.84×10^{-5}
Wire rope	2.00×10^7	1.29×10^{-4}	–

* Energy absorption is considered in the rockfall impact.

In where, the equivalent stiffness based on equivalent energy respect to a given plastic behavior were established as shown in Figure 4 by the assumption of the existing fence deformation concurrently with energy absorption on the present design (JRA 2000) shown in Figure 5. Although, there are two kinds of assumptions of the existing deformation, which depend on the place where the stone collides, more specifically, cases of collision to net as shown in Figure 5 (a) and collision to pole directly as shown in Figure 5 (b), the latter case was focused on in this study.

The design rockfall energy in this analysis is established by that absorption energy of un-modeled wire and metal net that is difficult to model are taken from the design rockfall energy, as described in Equation 1, because the extent of three poles was modeled in this analysis as shown in Figure 3.

$$ED = EDP - EWE - EN$$
$$= 71 - 9 - 25 = 37 \text{(kJ)} \qquad (1)$$

where ED = Design absorption energy of the existing fence in this analysis (kJ); EDP = Design absorption energy of the existing fence in the present design (kJ); EWE = Design absorption energy of the end wires in the present design (kJ); and EN = Design absorption energy of the metal net in the present design (kJ), which is assumed 25 kJ in the present design.

Figure 6 shows the deformation of the target existing fence respect to the design rockfall impact. And the behaviors, deformation and bending moment of the center poll and distribution of tensile forces occurring in the wire, are shown in Figures 7 and 8 respectively. The behaviors are assumed as the limit state of the fence in this fence on the present design, and it is adopted as a criterion measure to evaluate lack of performance of the existing target fence respect to larger rockfall impact than the design one and effectiveness of the technique that is proposed in this study.

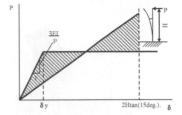

(a) Central pole

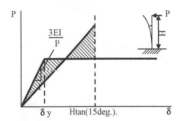

(b) Side poles

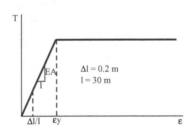

(c) Wires

Figure 4. Deformation characteristics of structural members.

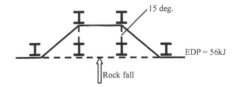

(a) Collision to net

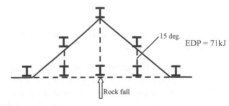

(b) Collision to pole

Figure 5. Assumption of the existing fence deformation concurrently with energy absorption on the present design.

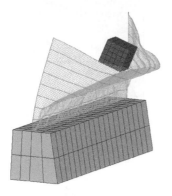

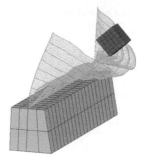

Figure 6.   Deformation of the existing fence respect to the design load.

Figure 9.   Deformation of the existing fence respect to a new target impact.

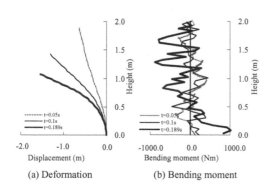

(a) Deformation

(b) Bending moment

Figure 7.   Behaviors of the Central pole.

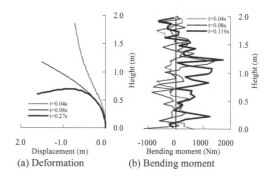

(a) Deformation

(b) Bending moment

Figure 10.   Behaviors of the Central pole.

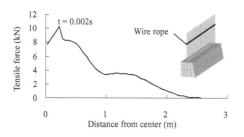

Figure 8.   Distribution of tensile forces occurring in wire.

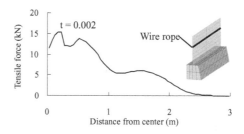

Figure 11.   Tensile force distribution occurring in wire.

4.2   *The response of the existing fence respect to a new target impact*

Figure 9 shows the deformation of the existing target fence respect to a new target rockfall impact, and the behaviors of the central pole and wire are shown in Figures 10 and 11 respectively. In where, the new target rockfall impact is 135 kJ (i.e. input energy estimated by Eq. 1 is 98 kJ). According to the results, because the behaviors of the existing fence are larger than the criterion measure, it is assumed that the performance of the fence cannot satisfy the requirements respect to the new target rockfall impact.

A reasonable technique to reduce the behaviors of existing fence to smaller ones than the criterion measure is proposed.

5   PROPOSAL OF A REASONABLE UPGRADE TECHNIQUE OF THE EXISTING FENCE

A new technique with using high-energy absorption net as shown Figure 12 is proposed in this study. This technique is that the new rockfall impact affected to the existing fence is reduced to smaller one than the initial design rockfall impact with absorbing the energy by both of high-energy absorption net and stay rope before collision of the new target stone to the existing fence.

5.1 Experiment of the target net

5.1.1 Summary
An experiment with using actual target net on site carried out (Yoshida et al. 2008), in order to confirm the absorbing ability of rockfall impact and model the characteristics of the net in the analysis using this study. Figure 13 shows the summary of the experiment and measurement system.

5.1.2 Results
Photograph 2 shows the maximum deformation of the net occurred concurrently with absorption the impact energy of the plumb bob. Transition of acceleration of the plumb bob and tensile strain occurring in the stay rope after collision of the plumb bob to the net are shown in Figures 15 and 16 respectively. According to the results, the absorbing ability of the rockfall impact energy of the net was confirmed from the fact that the plumb bob can be received by only the net and stay rope.

5.1.3 Simulation
The result of the experiment was simulated by the analysis using in this study with considering the deformation characteristics of the net and stay rope. Structural parameters adopted in this analysis are shown in Table 2.

Figure 14 shows analyzed deformation of the net fence. Comparisons with analysis results and observed ones concerning transition of acceleration of the plumb bob and tensile forces occurring in the stay rope after collision of the plumb bob to the net, and the maximum deformation of the net occurred concurrently with absorbing the energy are shown in Figures 15 to 17 respectively. According to the results, because the analysis results could comparatively good reproduce the observed ones, it was considered that the modeling of the net can apply to the analysis model to evaluate effectiveness of proposed technique.

Photograph 2.   Deformation of the net.

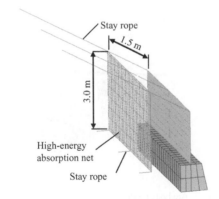

Figure 12.   A new technique with using target net.

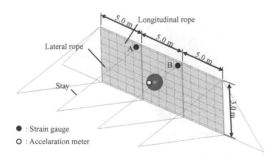

Figure 13.   Summary of the experiment and measurements.

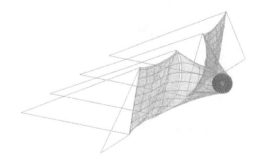

Figure 14.   Analyzed deformation of the net fence.

Table 2.   Structural parameters.

	Elastic modulus kN/m^2	Sectional area m^2
Rope making up net	2.36×10^8	6.36×10^{-6}
Longitudinal rope	2.78×10^9	1.20×10^{-4}
Lateral rope	1.69×10^8	3.14×10^{-4}
Stay	2.78×10^9	1.20×10^{-4}

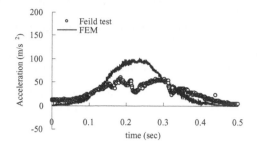

Figure 15. Transition of acceleration of the plumb bob.

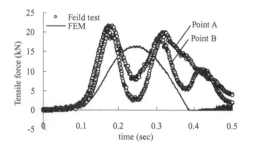

Figure 16. Transition of tensile forces occurring in the stay.

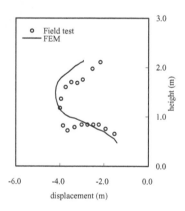

Figure 17. Deformation of the net.

collision of the stone. The behaviors, deformation and bending moment of center pole and transition of strain occurring in the stay rope and the wire of the fence, are shown in Figures 20 to 22 respectively. According

Table 3. Structural parameters.

	Elastic modulus kN/m²	sectional area m²	Moment of inertia m⁴
Metal net*	–	–	–
Center pole	3.32×10^6	2.72×10^{-3}	1.84×10^{-5}
Side pole	6.56×10^6	2.72×10^{-3}	1.84×10^{-5}
Wire rope	2.00×10^7	1.29×10^{-4}	–

	Elastic modulus kN/m²	sectional area m²
Rope making up net	2.36×10^8	6.36×10^{-6}
Longitudinal rope	2.78×10^9	1.20×10^{-4}
Lateral rope	1.69×10^8	3.14×10^{-4}
Stay	2.78×10^9	1.20×10^{-4}

* Energy absorption is considered in the rockfall impact.

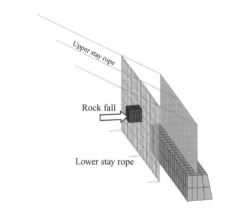

Figure 18. Analysis model.

5.2 Effectiveness of the proposed technique

5.2.1 Analysis model
Figure 18 shows the analysis model to estimate the effectiveness of the proposed technique for upgrade of the existing fence. Structural parameters adopted in this analysis are shown in Table 3.

5.2.2 Analysis results
Figure 19 shows the deformation of the net before collision of the target stone to the existing fence and, the maximum deformation of the existing fence after

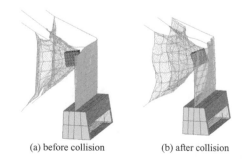

(a) before collision   (b) after collision

Figure 19. Deformation of the existing fence with proposed technique.

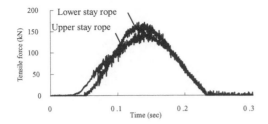

Figure 20. Transition of tensile force occurring in stay rope.

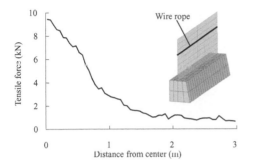

Figure 21. Tensile force distribution occurring in wire.

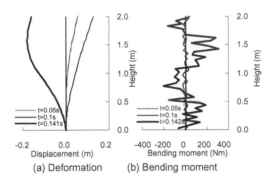

(a) Deformation   (b) Bending moment

Figure 22. Behaviors of the center pole.

to the results, from the view point that the behaviors of the existing fence are smaller than the criterion measure, the effectiveness of the proposed technique can be confirmed.

6 CONCLUSION

This paper can be concluded as follows;

– There are many existing stone-guard fences, which are lack of performance for absorbing ability of rockfall impact from the discovery of new larger stone than original design one during periodic site investigation, and early development of reasonable technique for upgrade of the fences have been required.
– The behaviors of an existing fence respect to the original design rockfall impact were estimated by an equivalent elastic analysis, and the results were established as a criterion measure to evaluate the effectiveness of a technique to reduce the transmitted energy of a stone to the existing fence.
– It was confirmed that the behaviors of existing fence respect to a new target rockfall impact that is larger than original one exceed the criterion measure.
– A new technique with using high-energy absorption net, which can reduce the transmitted energy of a stone to the existing fence, was proposed.
– An experiment of the actual target net on a site in order to confirm absorbing ability of rockfall impact and modeling of the net in the analysis was carried out.
– Effectiveness of the proposed method, the behaviors of the existing fence respect to a new rockfall impact that is larger than the original design one can be reduced to smaller than the criterion measure, was confirmed.

REFERENCES

Japan Road Association (JRA) 2000. Guideline for rockfall counter measure: Maruzen.
Yoshida, M. Kobayashi, H. Arakawa, M. & Okumura, H. 2008. Impact resistance test for rockfall preventing fence using fiber net and rope, Geosynthetics Engineering Journal (Japan), No. 23: 113–118, Japan branch of International Geosynthetics Society.

Prediction and Simulation Methods for Geohazard Mitigation – Oka, Murakami & Kimoto (eds)
© 2009 Taylor & Francis Group, London, ISBN 978-0-415-80482-0

Damage detection and health monitoring of buried concrete pipelines

A.S. Bradshaw, G. daSilva & M.T. McCue
Merrimack College, North Andover, USA

J. Kim, S.S. Nadukuru, J. Lynch & R.L. Michalowski
University of Michigan, Ann Arbor, USA

M. Pour-Ghaz & J. Weiss
Purdue University, West Lafayette, USA

R.A. Green
Virginia Tech, Blacksburg, USA

ABSTRACT: Rapid assessment of damage to buried pipelines from earthquake ground faulting is a crucial component to quickly plan repair efforts. This paper briefly reviews sensor technologies currently used for monitoring the health (i.e. assessing damage) of buried concrete pipelines. This paper also reports on the first of a four-year study aimed at developing rapid, reliable, and cost-effective sensing systems for health monitoring of buried concrete pipelines. The study includes testing of buried concrete pipelines in a large-scale facility that is capable of simulating earthquake ground faulting. Two modes of failure were identified in the first pipeline test, which were compression and bending at the pipeline joints closest to the fault line. As a result future research aimed at advancing sensing technology will likely focus on the behavior of the joints.

1 INTRODUCTION

Assessment of damage to lifelines after natural disasters, such as earthquakes, is crucial for management of an effective emergency response. Of particular importance is the water supply system because water is an important survival resource; even minor damage to water pipelines may result in contamination and epidemic outbreaks. Water pipelines are considered one of the most vulnerable systems to damage from earthquakes (Eidinger 1996). In particular, pipelines in the vicinity of permanent ground displacements are most vulnerable to damage.

The assessment of the condition or health of pipeline systems is very difficult given that they are typically buried and therefore not readily accessible for visual inspection. In urban settings the water system is just one of a number of underground utilities, making access for inspection or repair even more difficult. Furthermore, the need for earthwork and heavy equipment to expose buried pipelines makes a rapid emergency response unfeasible. Hence, there is a clear need for systems which can rapidly assess the health of a pipeline after an earthquake. For this reason the Authors are currently participating in a four-year study aimed at developing sensing technologies for health monitoring (i.e. assessing damage)

of buried concrete water pipelines. Though damage to buried pipelines can be caused by seismic waves (Barenberg 1988, O'Rourke & Ayala 1993), this study focuses on damage attributed to permanent ground displacement, which is often more severe (O'Rourke 2005).

This paper reviews some of the current technologies used for damage detection of concrete pipelines. This paper also reports on the first of four large-scale tests that was performed on a buried segmental concrete pipeline at the Large Displacement Facility at Cornell University (part of the Network for Earthquake Engineering Simulation). The test facility is capable of simulating permanent ground displacements that can occur in regions subjected to earthquakes.

2 SENSING TECHNOLOGIES FOR PIPELINES

The three basic types of sensing technologies for concrete pipelines include: (1) internal sensing, (2) fiberoptic sensing, and (3) remote sensing. Internal sensing involves inspecting the pipeline walls using technology deployed inside the pipe. These include remote video cameras or ultrasonic transducers. Ultrasonic transducers are used to map the thickness of the pipe walls by transmitting an acoustic wave and recording the travel time of the waves that are reflected at the interior and exterior surface of the pipe.

Fiber optic sensors, such as the Fiber Bragg Gratings (FBG), have been used for structural sensing since the 1970s. The FBG strain sensor consists of a traditional silicon glass fiber, upon which a Bragg grating is etched (Tennyson 2003). The sensor works by measuring the optical wavelength, which changes linearly with strain. Recently in Italy, a 500 meter stretch of pipeline was instrumented with FBG strain sensors to monitor its strain response to landslides (Inaudi & Glisic 2005). The application of fiber-optic sensing technology for pipelines is typically cost prohibitive, with a typical system ranging from $20k to $100k (Bergmeister 2000).

Remote sensing technologies include Infrared Thermography Systems (ITS) and Ground Penetrating Radar (GPR). Infrared technology indirectly detects damage by detecting leaks in the pipeline, which show up as temperature anomalies. ITS has been successfully used to detect pipeline leaks and poor backfill conditions (Inagaki & Okamoto 1997). GPR uses electromagnetic wave energy to map the conditions below the ground surface by measuring the reflections that occur at discontinuities between soil strata and soil pipe interfaces. However, the image quality is usually poor and requires a fair amount of user judgment to interpret (Hayakawa & Kawanaka 1998).

There is currently a need for reliable and cost-effective sensing systems that can rapidly assess the health of buried pipelines. Development of new technologies requires a test facility that can simulate the conditions that a pipe might experience during ground faulting. Given the difficulty in scaling experiments, it is desired to test as close as possible to full-scale. To this end, this study utilized the large-scale testing facility at Cornell University, which can simulate permanent ground displacements associated with earthquake ground faulting.

3 LARGE-SCALE PIPELINE TESTING

Pipeline testing for the four-year study is being performed in the large-scale pipeline test facility at Cornell University. The facility is one of the Network for Earthquake Engineering Simulation (NEES) equipment sites which are located throughout the United States. Previous tests at this facility included pipelines made of materials other than concrete (e.g. high-density polyethylene) (O'Rourke et al. 2008). The study described herein is the first such study performed on segmental concrete pipes. This section describes some preliminary results of the first pipeline test performed in June 2008. The objective of this first test was to observe the modes of failure in order to direct future research.

3.1 Test facility

The pipeline test basin, shown in Figure 1, consists of a 11.7-meter long by 3.4-meter wide by 1.9-meter deep box, having a transverse fault oriented at an angle of 65 degrees relative to the longitudinal axis of the basin. The basin is able to accommodate a pipeline that is fixed to the ends of the test basin and buried with granular backfill. To simulate earthquake induced permanent ground displacement, one half of the basin is moved laterally parallel to the fault line using two large hydraulic actuators placed between the basin and a reaction wall, while the other half of the basin remains stationary. The box can be displaced in either direction, causing the pipeline to be put in axial compression or tension.

Figure 1. Pipeline test basin at Cornell University.

Figure 2. Typical bell-and-spigot connection between two pipe segments. The three potentiometers shown inside the pipe were used to measure displacement and rotation at the joint.

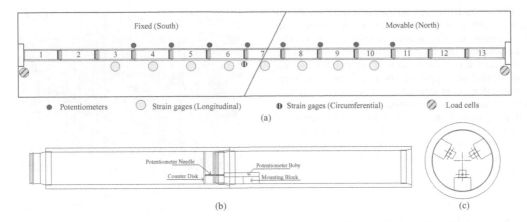

Figure 3. Instrumented sensors: (a) three types of sensors (potentiometers, strain gages and load cells) and their instrumentation location; (b) schematic of a potentiometer installation at a typical joint; (c) cross-sectional view of a potentiometer set at a joint.

3.2 *Test preparation*

Pipe segments used for the experimental program were manufactured according to AWWA C300 (AWWA 2004). This standard specifies requirements for concrete pressure pipes with internal diameter of 76 to 365 cm. Since there was some concern in the first test of exceeding the load capacity of the test basin, pipe segments were linearly scaled down to 15.24 cm inner diameter resulting in 19.2 cm outer diameter.

Fifteen 90-centimeter long pipe segments were fabricated at the Materials Engineering Lab at Purdue University. The pipe segments consist of a thin-walled (0.8 mm) steel tube covered by a concrete shell with steel rebar. Thirteen pipe segments were required to span the entire length of the test basin. The segments were connected by bell-and-spigot connections, shown in Figure 2. The segments were placed such that the center segment bisected the fault plane.

The pipeline was instrumented with three different types of sensors whose locations are shown in Figure 3. Sixteen surface-mounted strain gages were attached to the exterior of the pipeline at various locations. Four sets of three potentiometers were mounted on the interior of pipe to measure rotation at the four joints closest to the fault plane. Six load cells were used to measure axial forces at the ends of the pipeline.

The pipeline was assembled inside the test basin working from one side of the basin to the other. Once the entire pipeline was assembled (Figure 4), the gaps between segments were grouted using mortar with 50% aggregate and a water-cement ratio of 0.5 produced using ASTM Type I/II Portland cement. Given the narrow joints, a 0.5% high range water reducer (by weight of cement) was used as needed to improve the flowability of the grout. Plastic sheets or "diapers"

Figure 4. Assembled pipeline segments prior to joint grouting. Grout was poured in the annular space shown at each joint location.

wrapped around the joints kept the grout in place during pouring and also served to prevent moisture loss during curing.

The pipeline was buried beneath about 115 cm of granular soil, compacted in roughly 20-cm thick lifts. The density and water content of each lift was measured with a nuclear density gauge. The backfill soil was classified as poorly-graded sand (SP), per ASTM D 2487 (ASTM 2002). The optimum water content was about 9% and the maximum dry density was 2.1 Mg/m^3, as determined by a modified Proctor test, ASTM D 1557 (ASTM 2002). Direct simple shear tests (see O'Rourke et al. 2008) indicate that the effective friction angle of the backfill soil is between 39° and 40° when prepared to dry densities within the range of 1.58 Mg/m^3 to 1.61 Mg/m^3.

The in situ dry density and water content are shown in the depth profile in Figure 5. Each data point in the figure represents the average value for the lift. The data indicate relatively consistent properties within the soil placed above the bottom of the pipeline, which was located at a depth of about 115 cm. The average dry density in this zone was about 1.67 Mg/m^3, and the average water content was about 5%. These values indicated an average relative compaction of about 80%, relative to the modified Proctor test. The soil located below the pipeline had a higher density due to foot traffic while the pipeline was being assembled.

3.3 *Preliminary results*

Earthquake induced permanent ground displacement was simulated by displacing one half of the test basin parallel to the strike of the fault plane. The box was configured to put the pipeline in compression as half of the basin was slowly displaced. The box was displaced at a constant rate of 0.5 cm/s to a maximum displacement of 1.22 m. The test was paused at 15-cm intervals for a period of 60 seconds. This was done to allow more controlled observation of the test and to investigate possible stress relaxation phenomena. The ground deformation at the completion of the test is shown in Figure 6.

After the test was completed, the pipeline was carefully excavated so as not to disturb its deformed condition. Observations of the deformed pipe indicated that most of the pipeline damage was concentrated in the two joints on both sides of the fault plane, with cracks propagating away from the damaged joints.

Two modes of joint failure were observed: compression and bending. During the initial phases of

Figure 6. Ground faulting observed at the end of the test. The gridlines shown are spaced at approximately 10 cm.

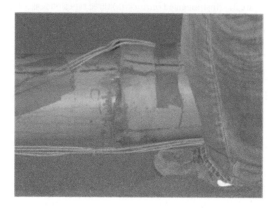

Figure 7. Compressive telescoping observed in one of the bell-and-spigot joints adjacent to the fault plane.

the test, large compressive loads caused the spigot to telescope into the bell as shown in Figure 7. However, these axial forces were not detected in the load cells at the ends of the pipeline. Therefore, it is hypothesized that the compressive loads generated in the mid-section of the pipeline were reacted by the shear forces induced at the pipe-soil interface. The telescoping-type failures occurred between the second and third pipe segments, on either side of the fault plane.

In the bending mode of failure, the un-reinforced grouted joints underwent initial micro-cracking which ultimately lead to coalesces of micro-cracks and macro-cracking and spalling/crushing of grout. The large deformations and rotation observed at the joints adjacent to the fault, shown in Figure 8, relieved any axial compressive loads that were initially developed within the pipeline.

The two failure modes are also confirmed by the data obtained from the sensors instrumented on the pipeline. Figure 9 represents the calculation of joint displacement and rotation using data obtained from

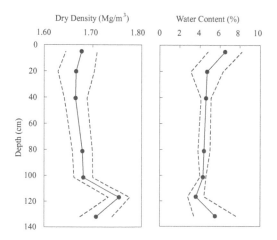

Figure 5. Profile of dry density and water content of the backfill soil averaged within each lift. The dashed lines indicate ± one standard deviation.

Figure 8. Pipeline deformation observed near the fault. Note that the fault plane ran through the middle of the pipe segment shown in the center of the figure.

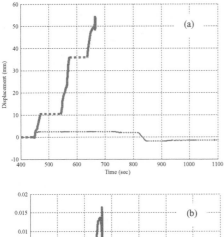

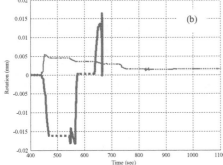

Figure 9. (a) Joint displacements and (b) rotations obtained from the potentiometers at the intersection of pipe segments. The continuous line corresponds to the response during displacement while the dotted line is with the actuator off. The thick line corresponds to the joint between pipe segments 6 and 7 (see Figure 3) while the thin line corresponds to the joint between segments 3 and 4.

the three potentiometers installed at two of the instrumented joints. Specifically, the thick line corresponds to joint behavior between segments 6 and 7 (see Figure 3) while the thin line corresponds to the joint

between segments 3 and 4. The joint between segments 6 and 7 experienced large displacements and rotations during the test, suggesting plastic behavior at the joint. In fact, at the end of the third actuated displacement (650 seconds into the record of Figure 9) the joint movement exceeded the maximum sensing range of the potentiometers. This is consistent with extreme rigid body motion of pipe segment 7; during post-testing excavation, this segment was later found to be totally separated from segments 6 and 8 (Figure 8).

It is also interesting to consider the response of the joint between segments 3 and 4. Under the initial fault displacement, the joint is seen to compress, consistent with telescopic compression due to the faulting. However, once segment 7 rotated free of segments 6 and 8, relaxation of the joint between segments 3 and 4 is evident in the displacement record (Figure 9a).

4 SENSOR DEVELOPMENT

The first pipeline test provided significant information on the nature and the most prominent types of damage due to faulting of segmented concrete pipelines. In the pipeline segments closest to the fault, the joints and the areas immediately adjacent to joints were damaged (see Figure 8). This leads to the conclusion that sensing in the joint areas will be most effective. Hence, sensors will be designed to monitor damage in joints and in areas adjacent to joints. Currently two groups of sensing techniques are being developed: (1) electrical sensing, and (2) Acoustic Emission (AE) sensing.

Electrical sensing techniques utilize highly conductive materials typically in the form of a graphite-filled epoxy which is applied to the joint area, or an electrically conductive grout used in the joint. Damage is detected by measuring changes in the electrical resistance of these materials over predefined time intervals. This technique has been utilized in the detection of shrinkage cracking in cement-based composites (Pour-Ghaz & Weiss, in prep.).

Acoustic emission techniques evaluate damage through analysis of acoustic waves using either active or passive methods. Continued development will focus on passive methods where acoustic waves that are generated from damage are captured and the energy, amplitude and duration of the captured waves are evaluated and related to the type of damage (Kim & Weiss 2003).

5 CONCLUSIONS

There is a need for rapid assessment of damage and health monitoring of buried concrete pipelines after earthquakes. Currently available sensing systems are either slow, unreliable, or cost prohibitive. For this reason, development of rapid, reliable and cost-effective systems is needed. In order to focus future sensor system development, a test was performed on a segmental concrete pipeline to investigate possible failure modes. The large-scale test represented the conditions that a buried pipeline will likely experience during earthquake-induced permanent ground displacement. Two modes of failure were observed, which were telescoping of the bell-and-spigot connections and plastic bending at the joints. Most of the damage was confined to the four joints closest to the fault plane. Accordingly, future efforts in this study will include methods that focus on behavior of these joints.

ACKNOWLEDGEMENTS

This research was sponsored by the National Science Foundation under the NEES Program (Grant CMMI-0724022). We would like to thank the Cornell staff for helping with the experiment. In particular, we acknowledge the help of Mr. Tim Bond, manager of operations of the Harry E. Bovay Jr. Civil Infrastructure Laboratory Complex at Cornell University, and Mr. Joe Chipalowski, the manager of Cornell's NEES Equipment Site. We also thank Professor Tom O'Rourke for his participation. Additional thanks go to Ms. Qinge Ma, and Mr. John Davis their support, and to Cornell graduate students Nathan Olson and Jeremiah Jezerski for sharing their expertise during preparation of the pipeline test.

REFERENCES

ASTM 2002. *Annual Book of ASTM Standards 2002. Section Four, Construction. Volume 04.08. Soil and Rock (I): D420–D5779*. West Conshohocken: ASTM International.

AWWA 2004. *AWWA C300 Standard for Reinforced concrete Pressure Pipe, Steel Cylinder Type, for Water and Other Liquids*. Denver: American Water Works Association.

Barenberg, M.E. 1988. Correlation of pipeline damage with ground motions. *Journal of Geotechnical Engineering* 114: 706–711.

Bergmeister, K. 2000. Maintenance and repairs- an expert opinion. *Freyssinet Magazine* 10(1): 4–5.

Eidinger, J. 1996. Lifeline Considerations and Fire Potential. In D. Eagling (ed.), *Seismic Safety Manual*: Livermore: Lawrence Livermore National Laboratory.

Hayakawa, H. & Kawanaka, A. 1998. Radar imaging of underground pipes by automated estimation of velocity distribution versus depth. *Journal of Applied Geophysics* 40(1–3): 47–48.

Inagaki, T. & Okamoto, Y. 1997. Diagnosis of the leakage point on a structure surface using infrared thermography in near ambient conditions. *NDT&E International* 30(3): 135–142.

Inaudi, D. & Glisic, B. 2005. Field applications of fiber optic strain and temperature monitoring systems. *Optoelectronic sensor-based monitoring in geo-engineering; Proc. intern. workshop.*, Nanjing, China, 23–24 November, 2005.

Kim, B. & Weiss, W.J. 2003. Using Acoustic Emission to Quantify Damage in Restrained Fiber Reinforced Cement Mortars. *Cement and Concrete Research* 33(2): 207–214.

Pour-Ghaz, M. & Weiss, W.J. in prep. Utilizing A.C. Impedance Spectroscopy for Characterizing Damage in Cement Based Composites.

O'Rourke, T.D. 2005. Soil-structure interaction under extreme loading conditions. *13th Spencer J. Buchanan lecture; Texas A&M University, College Station, Texas, USA, 18 November 2005*.

O'Rourke, M.J. & Ayala, G. 1993. Pipeline damage due to wave propagation, *Journal of Geotechnical Engineering*, 119: 1490–1498.

O'Rourke, T.D., Jezerski, J.M., Olson, N.A., Bonneau, A.L., Palmer, M.C., Stewart, H.E., O'Rourke, M.J. & Abdoun, T. 2008. Geotechnics of pipeline system response to earthquakes. In D. Zeng, M.T. Manzari, D.R. Hiltunen (eds), *Geotechnical earthquake engineering and soil dynamics IV; Proc., Sacramento, CA, 18–22 May 2008*. Reston: American Society of Civil Engineers.

Tennyson, R.C. 2003. Fiber optic sensing for civil structures. *Caltrans/UCSD workshop on structural health monitoring and diagnostics of bridge infrastructure; Proc., San Diego, CA, USA, 7–8 March 2003*.

Prediction and Simulation Methods for Geohazard Mitigation – Oka, Murakami & Kimoto (eds)
© 2009 Taylor & Francis Group, London, ISBN 978-0-415-80482-0

Inclinodeformometer: A novel device for measuring earth pressure in creeping landslides

M.V. Schwager, A.M. Schmid & A.M. Puzrin
Swiss Federal Institute of Technology, Zurich, Switzerland

ABSTRACT: The inclinodeformometer (IDM) is a novel device to measure changes of earth pressure in a sliding layer of a creeping landslide. The device makes use of the existing and widely used technology of the inclinometer measurements. The change of earth pressures in the sliding layer leads to the changes in the inclinometer pipe shape and dimensions. If these changes are carefully measured, the pressure change can be backcalculated, from a solution of a boundary value problem with properly described constitutive behaviors of the pipe and the surrounding soil. An advantage of the inclinodeformometer is that it does not require any additional infrastructure than standard inclinometer pipes, which are being installed anyway for landslide monitoring. Furthermore, these pipes can be used for pressure change measurements in the sliding layer long after they were sheared and became unsuitable for inclinometer measurements. Full scale laboratory tests performed in a 2 m high calibration chamber demonstrated that the pressure measurement accuracy can be as high as 3 kPa. Initial field measurements performed on the St. Moritz landslide confirmed significant stress anisotropy in the compression zone of this constrained creeping landslide.

1 INTRODUCTION

1.1 Motivation

Information about the earth pressure changes in a sliding layer of a creeping landslide is a crucial component for understanding, analysis and stabilization of the creeping landslides.

This information is especially important for constrained landslides, where the pressures in the compression zone could reach the passive pressure and lead to a catastrophic failure. Combining the measured increase in pressure with geodetic measurements allows for predicting whether and when the constrained landslide will fail. The procedure for this long-term inverse stability analysis was proposed by Puzrin & Sterba (2006).

In general the combination of the displacements and the changes in pressure observations is very promising for the analysis of landslides. Knowing the pressure changes and displacements in two sections of a long thin sliding layer (Fig. 1) allows for backcalculation of the average shear strength on the sliding surface and average lateral stiffness of the soil in the sliding layer.

Unfortunately, measuring the earth pressures is also one of the most challenging problems in the geotechnical monitoring.

1.2 The concept

The inclinodeformometer (IDM) is a new device allowing for backcalculation of the changes of earth pressure using a two step procedure. In the first step, IDM measures the change in dimensions of an inclinometer pipe in the sliding layer. The change in shape is assumed to be caused by the changes in the surrounding stress field. Therefore, in the second step, the measured deformations are used to backcalculate the change in pressure via inverse analysis of the corresponding boundary value problem of a plastic pile surrounded by soil under a changing stress state.

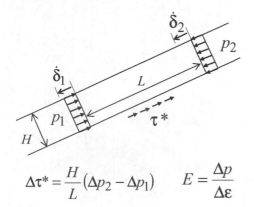

$$\Delta\tau^* = \frac{H}{L}\left(\Delta p_2 - \Delta p_1\right) \qquad E = \frac{\Delta p}{\Delta\varepsilon}$$

Figure 1. Backcalculation of the average shear strength on the sliding surface and average lateral stiffness of soil.

2 MEASURING THE PIPE DEFORMATIONS

2.1 Principle of measurement

The inlclindeformometer makes use of the existing and widely used technology of the inclinometer measurements. The IDM probe is being lowered down the depth of the pipe on three wheels guided along the channels of the inclinometer pipe (Fig. 2). Continuous diameter measurements in two perpendicular directions can be taken.

The upper and the lower wheels are rolling in the same channel. These wheels are fixed to the probe. The middle wheel is connected via a lever with two springs, so that it can be pressed against the opposite channel. Change in the diameter of the pipe leads to change of the position of the middle wheel in respect to the probe. There are two tilt sensors (VTI Technologies, 2006) detecting the relative inclination between the probe and the lever of the middle wheel. One sensor is located on the top of the probe, another on the middle wheel (Fig. 2).

In addition to the two tilt sensors in the plane of the measured diameter there is another tilt sensor measuring in a perpendicular direction out of this plane. This sensor is used for correction of the measurements (see Section 2.2). Above the top wheel there is a pressure cell to measure the water pressure in the inclinometer pipe.

At the top of the borehole the cable on which the device is hanging goes around a wheel. An incremtental rotation sensor (Wachendorff, 2008) measures the wheel rotation which determines the depth of

the device in the inclinometer pipe. As the device is lowered down into the inclinometer pipe, all the sensor measurements get saved on the computer for the corresponding position in the pipe.

The IDM is built for the two most common diameters of inclinometer pipes in Switzerland: 71 and 84 mm. The device can be easily switched between the different inclinometer pipe diameters.

2.2 Correction for the pipe inclination

The measured diameter D inside of the pipe is a function of the two angels α_W and α_P measured at the middle wheel and at the probe:

$$D = d + X + \sin(\alpha_W - \alpha_P) \cdot Y \qquad (1)$$

X, Y and d are constants depending on geometry.

The measurements of α_W and α_P are not independent of the inclination β of the device out of the plane. Assuming that the sensors are giving the true $(\alpha_W - \alpha_P)$ value at $\beta = 0°$ there is an error occurring on the tilt measurements, when β is different from $0°$. Because the diameter is just a function of the difference $(\alpha_W - \alpha_P)$, it is sufficient to describe the error Δ affecting this difference. This error can be found by calibration measurements on a biaxial inclinable table (Fig. 3).

The error occurring due to the device inclination out of the plane can be described as a function of α_W,

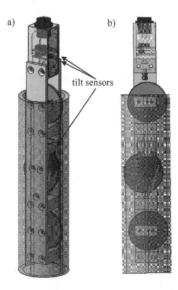

Figure 2. The inclinodeformometer in the inclinometer pipe: a) The complete device, b) The device without the front panel.

Figure 3. The inclinodeformometer on a biaxial inclinable table for calibration measurements.

480

α_T and β as follows:

$$\Delta(\alpha_W - \alpha_P) = (A_1\alpha_W + A_2\alpha_P + A_3)\beta^2$$
$$+(C_1\alpha_W + C_2\alpha_P + C_3)\beta \qquad (2)$$

A_1, A_2, A_3, C_1, C_2 and C_3 are constants derived by a regression analysis of the calibration measurements. Correcting the difference $(\alpha_W - \alpha_P)$ by the error function leads to the corrected diameter D_{cor}:

$$D_{cor} = d + X + \sin(\alpha_W - \alpha_P - \Delta) \cdot Y \qquad (3)$$

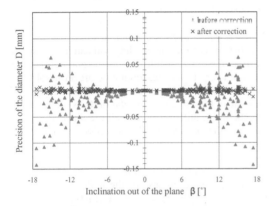

Figure 4. The precision of the diameter D compared to the measurement at $\beta = 0°$.

Figure 5. The reference gauges for both types of inclinometer pipes.

By using the error function from Equation 2, the corrected measurements of the diameter D reach the precision of ± 2 micrometers within $\beta = \pm 4°$, compared to the measurement at $\beta = 0°$ (Fig. 4).

2.3 Calibration of the device

The aim of IDM is it to have measurements of the diameters of the inclinometer pipes over a period of several years so that the change in shape can be computed. It is therefore important to avoid the influence of a possible shift of the device reference. For this reason the field measurements of different years should be preceded by the calibration in a constant diameter high precision reference gauge. There are two different reference gauges for both types of inclinometer pipes (Fig. 5).

3 BACKCALCULATING PRESSURES

3.1 Boundary value problem

The pressure change can be backcalculated, from a solution of a boundary value problem. The boundary value problem is given by a horizontal crossection of the pipe and the surrounding soil in plane stress (Fig. 6). The boundary conditions are static: the two principle earth pressures. The major principle stress direction is assumed to coincide with the direction of displacement vector which is known from the conventional inclinometer measurements.

The pressure increments can be backcalculated, from the measured changes in pipe diameters, provided the stiffness of the soil and the stiffness of the pipe in this range of stresses are known.

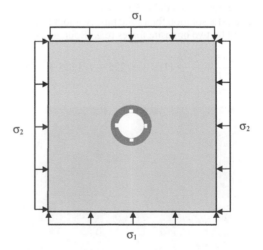

Figure 6. Plain strain model in case of principle stresses parallel to the axes of the pipe.

The measured diameters are not only affected by the earth pressure. Also the bending of the inclinometer pipe produced by the movement of the landslide causes changes in diameter. This influence has to be corrected before modeling in the plain stress problem.

3.2 Effects of stiffness

The stiffness of the pipe and the soil affect the result of the backcalculation significantly. Therefore, it is very important to describe the constitutive behaviors of the pipe and the surrounding soil in an appropriate way.

3.2.1 Stiffness of the inclinometer pipe

In case of the inclinometer pipe the short term Young's modulus for fast loading could be found by compression tests. The pipe was loaded by a linear distributed load in a purpose-built test apparatus (Fig. 7). The deformations had been measured for several angles between the direction of the force and the direction of the channels in the pipe.

The tangent stiffness of the pipe k is defined as the ratio between the increment of force f divided by the increment of displacement u (or, more precisely, a half of the displacement due to the symmetry of the setup):

$$k = \frac{\Delta f}{(\Delta u)/2} \qquad (4)$$

The stiffness is strongly dependent on the angel between the direction of force and the direction of the channels in the pipe, because they soften the pipe cross-section. The highest stiffness is achieved at the angle of 45° between the force direction and the channels (Fig. 8). In this configuration there are hardly any bending moments acting in the area of the channels, where the bending stiffness is reduced a lot.

The Young's modulus E is related to k by an analytical elastic solution for a solid pipe (without channels) loaded by two opposite forces (e.g., Bouma, 1993):

$$E = \frac{2kr}{\pi A} + \frac{2r^3 k}{\pi I} \, (1/9 + 1/225 + 1/1225 + \cdots) \quad (5)$$

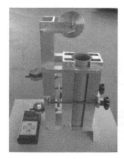

Figure 7. Compression test on the inclinometer pipe.

where r = middle radius; A = area of the pipe section; and I = moment of inertia of the pipe section

Using Equation 5 with the stiffness k from the 45° constellation (Fig. 7) leads to the immediate Young's modulus of E = 2850 MPa for fast loading. In fact this value is a lower bound because the formula is made for pipes without channels. Nevertheless, when used for the modeling of this test with the finite element program ABAQUS, it appeared to reproduce the results for a channeled pipe with high accuracy.

For backcalculations of pressures in creeping landslides, however, the long-term modulus is of much bigger concern then the immediate modulus for fast loading. The viscous behaviour of the pipe has to be considered. The long term creep and relaxation test are currently being carried out.

3.2.2 Stiffness of the surrounding soil

The stiffness of the surrounding soil can be measured by dilatometer tests while drilling the borehole for the inclinometer. If no measurements are available the stiffness of the soil can be estimated from results of laboratory tests (e.g., consolidation tests). The major effects on the stiffness of the soil are the loading history and the nonlinearity due to stress dependency of stiffness.

3.2.3 Linear elastic solution

In the axisymmetric case of the hydrostatic state of stress, where $\sigma_1 = \sigma_2 = \sigma$, there exists an analytical elastic solution relating the acting free field stress σ and the radial deformation u:

$$\sigma = u \left(\frac{E_S}{R(1 + \upsilon_S)} + \frac{E_P t}{(R - t/2)^2} \right) \qquad (6)$$

where E_S = Young's modulus of the surrounding soil; υ_S = Poisson ratio of the soil; E_P = Young's modulus of the inclinometer pipe; R = external radius of the pipe; and t = thickness of the pipe.

Equation 6 clearly shows the effects of the both the pipe and the soil stiffness, but does not consider any channels. This formula can be used as a benchmark for the finite element model.

3.3 Inverse analysis

In reality, however, the soil is visco-elasto-plastic and the pipe is visco-elastic with geometric non-linearity due to the channels. Therefore, the forward boundary value problem has to be solved using numerical analysis. The backcalculation of pressures is performed in two steps: first, a finite element program computes the deformations caused by the trial stresses. An inverse analysis algorithm then solves the optimization problem to minimize the objective function F (the sum of squared errors between the measured and the computed pipe deformations) by changing the trial stresses σ_1 and σ_2.

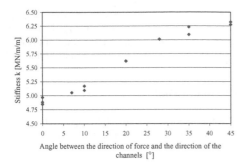

Figure 8. Stiffness k of the pipe as a function of the direction of force.

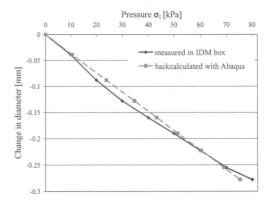

Figure 11. The comparison between the backcalculated and applied pressures in the IDM box experiments; $\sigma_2 = 0.6\,\sigma_1$.

Figure 8 shows F as a function of σ_1 and σ_2. The Young's moduli of the soil and the pipe were set to 50 MPa and 2850 MPa, respectively. The poisson ratios were assumed to 0.3 and 0.25. The minimum of the objective function is located at $\sigma_1 = 100$ kPa and $\sigma_2 = 60$ kPa.

It can be seen in Fig. 9, that the gradient is quite low along the objection function "valley". This means that a variety of states of stress may lead to similar pipe deformations. This can be explained by the fact, that bending moments cause much bigger deformations of the pipe then normal forces, and different combinations of the stress ratio σ_2/σ_1 and the average stress $(\sigma_1 + \sigma_2)/2$ can produce the same bending

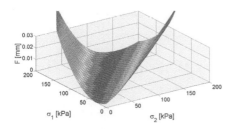

Figure 9. Objective function for inverse analysis.

Figure 10. The IDM box.

moments at different levels of compression of the pipe. From Fig. 9 it follows that variation in σ_2/σ_1 at a fixed $(\sigma_1 + \sigma_2)/2$ produces larger pipe deformation than other way round. This makes backcalculation of stress increments more challenging.

3.4 Validation: IDM box

For validation of the backcalculation full scale laboratory tests were performed in a 2 m high calibration chamber (IDM-box) with a cross-section of 40 by 40 cm. Each of the 4 vertical walls of the chamber is equipped with a pressure membrane allowing for application of 2 independent principle horizontal stresses. The chamber is filled with sand and the inclinometer pipe is fixed in the middle of the chamber. Increase in principle stresses results in the deformations in the pipe which are measured using the inclinodeformometer.

Validation with finite elements showed that the dimensions of 40 by 40 cm, when compared with a free field solution, are sufficient to avoid the effect of the boundaries.

Preliminary measurements in the calibration chamber demonstrated an accuracy in pressure measurements of about ±5 kPa. The comparison between applied and backcalculated pressures is given in Fig. 11.

4 INITIAL FIELD MEASUREMENTS

Initial field measurements performed in the compression zone of the St. Moritz landslide confirmed significant stress anisotropy (Fig. 12). The A-axis coincides with the direction of the landslide velocity, B-axis is perpendicular to it. The pipe diameters are averaged every 3 meters—e.g., within each continuous pipe section between the installation joints. The measurements consistently demonstrate an elliptical pipe shape with a smaller diameter parallel to the landslide velocity. The difference between the pipe

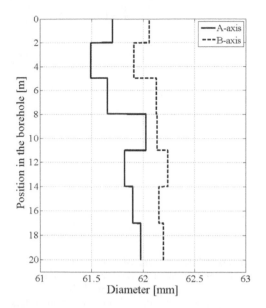

Figure 12. Initial measurements of the sheared inclinometer pipe diameter in the compression zone of the St. Moritz landslide.

sections is most likely due to the variation in the initial pipe diameters. In 2009 the measurements of the diameter changes will be taken and the earthpressure changes backcalculated.

5 CONCLUSIONS

The research on the novel inclinodeformometer device is in its preliminary stages. Nevertheless it looks promising, due to simplicity and accuracy of measurements. In addition, it does not require any additional infrastructure than standard inclinometer pipes, which are being installed anyway for landslide monitoring. Furthermore, these pipes can be used for pressure change measurements in the sliding layer long after they were sheared and became unsuitable for inclinometer measurements. Backcalculation of pressures is a challenging task, which requires additional study.

ACKNOWLEDGMENTS

The work has been partially supported by the ASTRA/VSS grant VSS 2005/502 "Landslide-Road-Interaction".

REFERENCES

Bouma, A.L. 1993. *Mechanik schlanker Tragwerke. Ausgewählte Beispiele in der Praxis.* Berlin: Springerverlag.
Puzrin, A.M., & Sterba, I. 2006. Inverse long-term stability analysis of a constrained landslide. *Géotechnique* 56(7): 483–489.
VTI Technologies Oy. 2006. The SCA103T Differential Inclinometer Series; Data Sheet.
Wachendorff Automation GmbH. 2008. Drehgeber WDG 58E; Datenblatt.

Prediction and Simulation Methods for Geohazard Mitigation – Oka, Murakami & Kimoto (eds)
© 2009 Taylor & Francis Group, London, ISBN 978-0-415-80482-0

Monitoring of model slope failure tests using Amplitude Domain Reflectometry and Tensiometer methods

S. Shimobe
Nihon University, Funabashi, Japan

N. Ujihira
Graduate School, Nihon University, Tokyo, Japan

ABSTRACT: This study describes the practicability of ADR (Amplitude Domain Reflectometry) soil moisture meter and its application to the monitoring system using the Tensiometer together for soil column and model slope failure tests. The laboratory tests were carried out in both a sandy soil and volcanic cohesive soil by the rainfall infiltration etc. As a result, the validity of this monitoring system using the ADR-Tensiometer simultaneous measurement methods was confirmed and the soil water content and pressure head distributions within the soil layers were capable of clarifying in real-time. In the model slope failure tests by rainfall, the some characteristic changes in volumetric water content and pressure head were observed just before the partial and general surface layer failures of slopes on both the two soils, respectively. These trends seem to indicate the presence of each phase pattern in relation to the slope failure process by rainfall.

1 INTRODUCTION

The natural disasters such as storm, flood, earthquake and volcanic damages often generate in Japan. It is due to the geographical and geological environments that the Japanese country has fateful ones. Recently, in particular, the geotechnical hazards of sediment disasters (debris flow, earth flow and landslide) by heavy rainfall are a more serious problem and the various disaster prevention works have been carried out. However, the slope failure mechanism as sediment disasters by rainfall has not been clarified enough yet, in relation to the changes in soil water content and pore water pressure with time during rainfall.

On the other hand, the standard testing method for soil water content is the oven drying method, widely used in the world. However, the oven drying method has the non-efficient defect with the long measurement time. Therefore, the new simplified and quick measuring technique for soil water content is strongly desired, irrespective of the laboratory and the in-situ.

Recently, as the soil water measurement called the soil moisture meter, TDR (Time Domain Reflectometry), FDR (Frequency Domain Reflectometry) and ADR (Amplitude Domain Reflectometry) are listed. Every one of the fundamental measurement principles is to obtain the dielectric constant of soils using the electromagnetic wave and to calculate indirectly the soil water content by their calibrations. As the ADR method is cheap and easy with operation compared with the TDR and FDR methods, its applicability is expected as a simplified and quick measurement method of soil water content (Shimobe 2004). Furthermore, as its engineering application to monitoring technique, the simultaneous monitoring system for the soil water content—pressure head profiling using the Micro-Tensiometer probes together at different depths within the soil layers is a promising new method (Shimobe & Ujihira 2008).

This paper mainly describes the practicability of ADR soil moisture meter and the validity of ADR-Tensiometer simultaneous monitoring system for model slope failure tests by rainfall.

2 TESTING DEVICE AND METHOD

2.1 Samples used

The samples used in this study are the sandy soil, Narita sand (soil classification: S-F) and volcanic cohesive soil, Kanto loam (ditto: VH_2). The physical properties of these soils are listed in Table 1.

2.2 ADR probe and calibration

ADR soil moisture meter (Delta-T Devices: ML2x) shown in Figure 1 is a device to obtain the dielectric constant ε of soils in terms of its output voltage. Hereinafter, we abbreviated this sensor the ADR probe. In carrying out the preliminary test actually, we must prepare for the calibration curve between output voltage V_{wet} and volumetric water content θ_w

Table 1. Physical properties of soil samples used.

Sample name	Particle density (ρ_s g/cm^3)	Fine content (F_c%)	Liquid limit * (W_L%)	Plasticity index* (I_p)
Narita sand	2.668	35	–	–
Kanto loam	2.761	77	126.9	41

* Values obtained from extended fall cone test (e.g. Shimobe 2000).

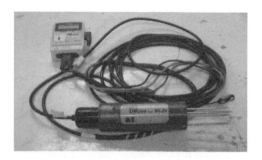

Figure 1. ADR soil moisture meter (ADR probe).

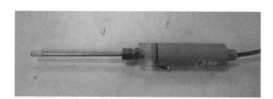

Figure 2. Micro-Tensiometer (MT probe).

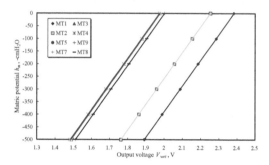

Figure 3. Relationship between output voltage and matric potential for MT probes.

for a soil in advance, and can obtain the volumetric water content by substituting the measured values to the calibration curve.

In this test, we prepared for the twelve ADR probes, packed uniformly the soil with prescribed water content into the mould for specimen preparation and carried out the semi-dynamic compaction using the Harvard miniature compactor. Thereafter, we inserted the rod portion of ADR probe in three points on the compacted specimen surface and obtained the average of output voltage. Furthermore, we determined the water content w and volumetric water content based on the oven drying method by JIS (Japanese Industrial Standard) and made up the calibration curve for a soil.

2.3 Soil column test

In soil column test, the rainfall infiltration tests of two soils using artificial rainfall apparatus were performed under the average rainfall intensity 20 mm/h and continuous rainfall 8 hours. The water content and suction (i.e. matric potential) of soils in rainfall infiltration-drainage processes were simultaneously measured at set time intervals by the ADR probes and the Micro-Tensiometers, respectively. Figure 2 shows the Micro-Tensiometer (Sankeirika: SK5500-M6) used in this experiment. Hereinafter, we abbreviated this sensor the MT probe and carried out the calibrations in advance of the test.

The calibration results of these probes are shown in terms of the relationship between output voltage and matric potential in Figure 3. From this figure, it was seen that all the relationships were linear but

the ones had the individual MT probe's characteristics. The soil column test of Narita sand during the rainfall infiltration is shown in Figure 4.

2.4 Slope failure test

In slope failure test by rainfall, the model slope with inclined angle of 33° was made into the experimental soil box (inner size: 80 width, 60 height and 120 length in cm) using the each soil which was set in the prescribed compactive state and the failure test by artificial rainfall was carried out respectively. Moreover, the eleven ADR probes and six Micro-Tensiometers were buried at the prescribed positions inside the slope, and the measurements of all probes were performed in real-time until the slope reached to failure under the prescribed average rainfall intensity and continuous rainfall. Here, these rainfall conditions set 20 mm/h and 7 hours in Narita sand and 40 mm/h and 30 hours in Kanto loam, respectively. Figure 5 shows the soil box for slope failure test.

Figure 4.   Soil column test of Narita sand.

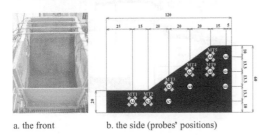

a. the front   b. the side (probes' positions)

Figure 5.   Soil box for slope failure test.

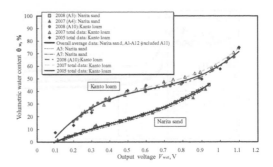

Figure 6.   Relationship between output voltage and volumetric water content of two sample soils used.

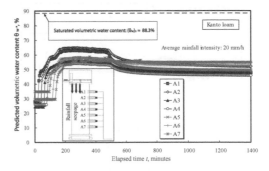

Figure 7.   Relationship between elapsed time and predicted volumetric water content of Kanto loam in soil column test with rainfall infiltration-drainage processes.

3   TEST RESULT AND DISCUSSION

3.1   Water content measurement by ADR method

Figure 6 shows the relationship between output voltage and volumetric water content of two sample soils used. As a result, the calibration curve of an ADR probe selected randomly this time on each soil was almost similar to the overall average curve of 11 ADR probes on the soil. The differences between these probes settled within ±3% in absolute error and also within ±5% in relative error on the two soils, respectively. In addition, the Kanto loam of volcanic cohesive soil indicated the higher volumetric water content than the Narita sand of sandy soil compared with the same output voltages. This is a characteristic of volcanic soils with aggregate structure. Moreover, these results were in good agreement with the ones suggested by Shimobe (2004).

3.2   Soil column test

Figure 7 shows the relationship between elapsed time t and volumetric water content θ_w^* predicted by ADR

probes' calibration curves of Kanto loam in rainfall infiltration-drainage processes. As a result, the volumetric water content generated in turn from the probe A1 in upper portion of column and the values became higher toward the probes of lower portions. Though these values did not reach to the saturated volumetric water content $((\theta_w)_s = 88.3\%)$ of this sample, the differences in values of θ_w^* were seen due to the depth of column and thereafter each θ_w^*-value almost became constant. In drainage process after the end of rainfall, the reduction in volumetric water content was higher as to the upper portion of column.

On the other hand, similarly the relationship between elapsed time and predicted volumetric water content of Narita sand in soil column test is shown in Figure 8. From this figure, the volumetric water content behaviours of Narita sand with elapsed time were, on the whole, the same compared to Kanto loam but those in drainage process after continuous rainfall were the remarkable reduction in volumetric water content. In addition, the probes A7 and A6 in lower portion of column reached to the saturated volumetric water content $((\theta_w)_s = 38\%)$ of this sample. Figure 9 shows the

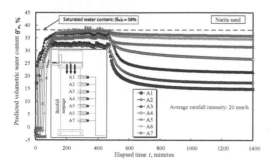

Figure 8. Relationship between elapsed time and predicted volumetric water content of Narita sand in soil column test with rainfall infiltration-drainage processes.

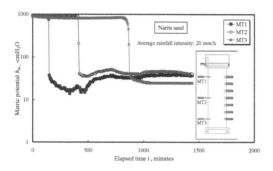

Figure 9. Relationship between elapsed time and matric potential of Narita sand in soil column test with rainfall infiltration-drainage processes.

relationship between elapsed time and matric potential h_m of Narita sand in soil column test with rainfall infiltration-drainage processes. From this figure, it was seen that the matric potential behaviours almost corresponded with the results of Figure 8.

3.3 Model slope failure test by rainfall

Figure 10 shows the relationship between elapsed time and predicted volumetric water content of Kanto loam in model slope failure test by rainfall. As a result, from the overall point of view, the volumetric water contents in probes A3 and A4 near the surface layer of slope generated most quickly. It was monitored the appearances that the ADR probes near the surface layer indicated the quick detection response of water content and their responses of the ones with deeply inserted positions were late. After 1290 minutes, the probe A3 had exposed perfectly.

Furthermore, in order to investigate this failure process in detail, the relationship between the volumetric water content behaviours of each ADR probe set up into surface layer and the phases of slope failure for Kanto loam was presented in Figure 11. From this figure, it was found that the phases of slope failure

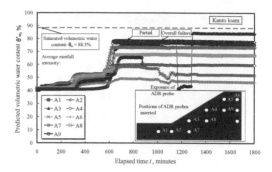

Figure 10. Relationship between elapsed time and predicted volumetric water content of Kanto loam in model slope failure test by rainfall infiltration.

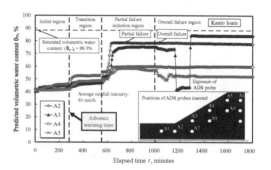

Figure 11. Relationship between volumetric water content behaviours of surface layer and phases of slope failure for Kanto loam.

were roughly divided into four stages, and that these were consisted of the first stage which no change in volumetric water content was almost seen (initial region), the second one which volumetric water content began to generate (transition region), the third one which volumetric water content generated suddenly and the partial failure of surface layer began almost simultaneously (initiation region of partial failure), the final one which volumetric water content remained constant and overall the surface layer failed (overall failure region). Here, we provisionally defined the time which plunged into the transition region an advance warning time for failure. Then again, the volumetric water content became higher toward the slope toe after the partial failure initiation of the third stage. This is considered for the reason why the water content is easy to concentrate toward the slope toe.

On the other hand, since the Micro-Tensiometers were buried together at the same positions as the ADR probes to measure the internal suction of slope in its failure process, the relationship between elapsed time and matric potential of Kanto loam in model slope failure test by rainfall infiltration is shown in Figure 12. From this figure, all the Micro-Tensiometers responded immediately after the initiation of rainfall and those (probes MT4 & MT2) near the slope shoulder and slope toe quickly responded compared with the other ones. This was considered for the reason why the water by rainfall infiltrated from the slope shoulder, the water movement along the slope caused and it concentrated in the slope toe. Then, the matric

potentials of all these became almost zero at about six hundreds minutes (ten hours) and subsequently the zero-value was held constantly with elapsed time. These matric potential behaviours presented a piece of evidence which supported the phase process to slope failure.

Next, to compare the difference between Kanto loam and Narita sand as regards the phases of slope

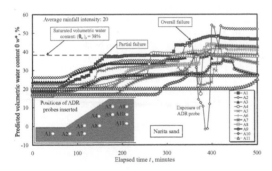

Figure 12. Relationship between elapsed time and matric potential of Kanto loam in model slope failure test by rainfall infiltration.

a. Start of rainfall.　　　　　b. After 750 minutes (partial failure)

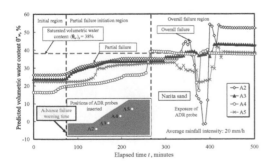

Figure 13. Relationship between elapsed time and predicted volumetric water content of Narita sand in model slope failure test by rainfall infiltration.

c. After 990 minutes (initiation of overall failure)

Figure 14. Relationship between volumetric water content behaviours of surface layer and phases of slope failure for Narita sand.

d. After 1290 minutes (exposure of ADR probe)

Figure 15a, b, c, d. Slope failure stage on Kanto loam.

failure, the relationship between the volumetric water content behaviours of each ADR probe inserted within Narita sand model slope and the phases of slope failure for this sand was indicated in Figure 13 and Figure 14. According to these figures, the overall behaviour of Narita sand apparently differed from Kanto loam and also it was found that the phases of slope failure for the sand were roughly divided into three stages. In case of Narita sand, the above-mentioned transition region connected to the initiation one of partial failure did not exist. It is very interesting that the different phase of slope failure due to the soil type was found out.

Furthermore, as an example, the photographs of slope failure stage on Kanto loam are shown in Figure 15a, b, c, d. Here, in Figure 15d the ADR probe was seen to be exposed From these photographs, it can be seen that the validity of this monitoring system using the ADR-Tensiometer simultaneous measurement methods was confirmed.

4 CONCLUSIONS

The conclusions obtained within the range of this study are as follows:

1. From the calibration test results of water content using Narita sand and Kanto loam, the reproducibility of these calibration curves is good and also the errors between ADR probes on volumetric water content are within $\pm3\%$ in absolute error and within $\pm5\%$ in relative error. Therefore, the practicability of ADR method is high.
2. In soil column test, using the ADR and Micro-Tensiometer probes together, the water content and suction at every depth within column are capable of monitoring in rainfall infiltration.
3. In model slope failure test, using the ADR and Micro-Tensiometer probes together as well as the soil column test, the water content and suction within slope are capable of monitoring in rainfall infiltration. Therefore, this monitoring system in real-time can detect the slope failure in advance and is expected to the establishment of detection technology for the mitigation of geotechnical hazards.

REFERENCES

Delta-T Devices Ltd. 1999. *ThetaProbe Soil Moisture Sensor Type ML2x User manual ML2x-UM-1.21*. Cambridge: Delta-T Devices Ltd.
Sankeirika Inc. 2004. *User manual for instruction of Micro-Tensiometer SK5500*. Tokyo: Sankeirika Inc.
Shimobe, S. 2000. Correlations among liquidity index, undrained shear strength and fall cone penetration of fine-grained soils. *Proc. the International Symposium on Coastal Geotechnical Engineering in Practice 1, Yokohama, 20–22 September 2000: 141–146*. Rotterdam: Balkema.
Shimobe, S. 2004. Applicability of ADR soil moisture meter and its engineering utilization. *Proc. the 49th Geotechnical Symposium, Tokyo, 18–19 November 2004: 39–46. Tokyo*: The Japanese Geotechnical Society.
Shimobe, S. & Ujihira, N. 2008. Monitoring of model slope failure test using ADR and tensiometer methods. *Proc.the 43rd Japanese National Conference on Geotechnical Engineering, Hiroshima, 9–11 July 2008: 1945–1946. Tokyo*: The Japanese Geotechnical Society.

Prediction and Simulation Methods for Geohazard Mitigation – Oka, Murakami & Kimoto (eds)
© 2009 Taylor & Francis Group, London, ISBN 978-0-415-80482-0

Prediction on volume of landslide in Shih-Men reservoir watershed in Taiwan from field investigation and historical terrain migration information

B.-H. Ku, C.-T. Cheng, S.-Y. Chi, C.-Y. Hsiao & B.-S. Lin
Sinotech Engineering Consultants, Inc, Taiwan

ABSTRACT: Shih-Men Reservoir is an important infrastructure that provides the potable water to inhabitants living in northern Taiwan. Recently, Shih-Men Reservoir suffered from sediment due to landslides and soil erosion in headstream watershed and reduced its service ability and performance by degrees. Those disasters generally triggered by heavy rainfalls and typhoons. In past years, Governments had paid much attention to mitigate the great deal of sediment and monitor its variation during typhoon season. The sediment variation of Shih-Men Reservoir in past years was heavily influenced by typhoons. Compared with other events, the fatal sediment yields of landslides and debris flows induced by Aere typhoon (2004/08/23~08/26) which reach higher quantity and brought enormous rainfall intensity in the meanwhile. This study would perform some field investigation to measure soil erosion and estimate the thickness of weathering zone on slope for evaluating the sediment yields. Then, using satellite images with several events is to acquire historical terrain migration information. Besides, the relationship between landslide and rainfall was discussed and analysed by the field investigation and remote measurement. It could discuss the sediment transportation and process during tyhoon events and validated with field investigation data and GIS Information, and realize the relationship between rainfall and landslide for mitigating the disasters in Shih-Men Reservoir watershed.

1 INTRODUCTION

Shih-Men Reservoir is an important infrastructure that provides the potable water to inhabitants living in northern Taiwan, as shown in Figure 1. The Shin-Men Reservoir was built in July 1956, and completed in June 1964. The type of Shih-Men Reservoir dam is earth rockfill dam with effective water storage for 25188000 cubic meters. The function of dam contains irrigation hydropower generation, public water supply, flood prevention, and sightseeing.

It is widely recognized a large number of shallow landslides in Taiwan occurred as a result of heavy rainfall inducing pore pressure increases in the soil layer. Typhoon events with long duration, high intensity rainfall usually triggered shallow, rapidly moving landslides resulting in severe property damages and casualties in the Shih-Men reservoir watershed along the recent year. A huge amount of sediment form landslide and erosion was carried to the Shin-Men reservoir section during these events as shown in Figure 2.

Thus the sediment concentration increased and resulted in the lack of potable water. In order to maintain the working capability and extend the service life of shih-men reservoir, the government put great emphasis on strategies of monitoring sediment variation. For this reason, several kinds of field tests were applied in the Shin-Men watershed. Besides, satellite images can provide much useful information for the section that difficult to perform field tests. The drift of this article illustrated facing problems with the viewpoint of watershed. Then, the relationship between sediment source and influence was discussed and analysed by the field investigation and remote sensing.

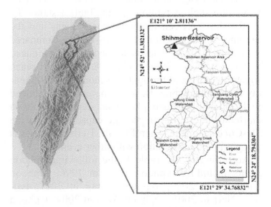

Figure 1. The location of Shih-Men Reservoir Watershed in the northern Taiwan.

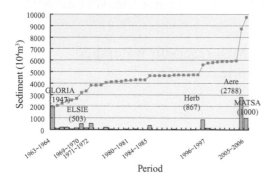

Figure 2. The sediment deposition of Shih-Men Reservoir in past years.

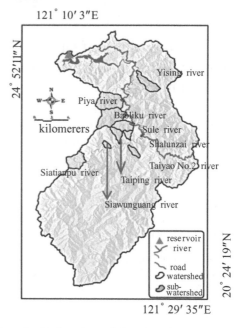

Figure 3. Study region and nine tributaries.

1.1 Study region

The study region includes a major river, named Da-Han river. The Da-Han River stretch extends 94.01 km from upstream to the end. There are many tributaries along the Da-Han river. In this research, about nine tributaries were selected to be treated as study region, as shown in figure 3. The area of our study region is 763.529 km². Check the past typhoon events, the geohazards were insignificant until the Chi-Chi earthquake hit Taiwan in 1999. Accoring to this phenomenon, the following discussion is mainly focused on the events after 1999 in the study region.

1.2 Major typhoon events in the study region

The study region is abundant rainfall, and the water resources are profuse. The annual rainfall is about 2,400 to 2,500 mm in average. The rainy seasons begin in June and the most of rain increased in June to August. There is always heavy rainfall, more than 300 mm per day in the typhoon season. The dry seasons begin in October till the next June. Sediment usually increases with volume of flow triggered by heavy rain such as typhoon event. Table 1 shows the relationship between major typhoon events and the increment of sediment in the study region. Compared with other events, the newly applied sediment induced by Aere typhoon (2004/08/23~08/26) would reach higher quantity and brought enormous rainfall intensity in the meanwhile. Besides, the sediment induced by Herb typhoon also played an important role in shortening the reservoir storage capacity before the Chi-Chi earthquake. Therefore, the Herb typhoon was brought into consideration with other events.

According to the Table 1, the following typhoon events including typhoon Herb in 1996, typhoon Xangsane in 2000, typhoon Toraji in 2001, typhoon Nari in 2001, typhoon Aere in 2004, typhoon Haitang in 2005, typhoon Matsa in 2005, typhoon Talim in 2005, typhoon Longwang in 2005

and typhoon Krosa in 2007 have hit the study region. Besides, the Chi-Chi earthquake on 21 September 1999 is the largest earthquake in recent years. The ground motion of it was more than 0.1 g and caused some slope failures or cracks in the study region. These typhoon events represent a large-scale experiment that provides much information about the impact of heavy rainfall on landslide and the transformation of sediment yields.

1.3 Remote sensing data

Remote sensing data including aerial photo and satellite image of Spot 1~5 satellite and FORMOSAT-II satellite. In this research, remote sensing data include aerial photos of two stages and satellite images of fourteen temporal stages which are the stage before and after typhoon events listed in Table 1. Aerial photos were utilized for field investigation site chosen in advance and satellite images were utilized for landslide mapping. Subsequently, the landslide information acquired from satellite images can contribute to the calculation of landslide area.

2 MEASUREMENT

For Governments have to pay much attention to a great deal of sediment and need to expand monitoring of sediment variation during typhoon season, This research plans to perform several kinds of field tests, such as

Table 1. Comparison with Sediment and typhoon events in Shih-Men reservoir.

Period	Name of typhoon	Peak flow (cms)	Average rainfall (mm)	Maximum rainfall (mm/hr)	Increment of sediment (10^3m^3)
1996	HERB	6,363	715	82	8,670
2000	XANG-SANE	2,129	320	38	−3,622
2001	TORAJI	63	58	23	
2001	NARI	4,123	1004	88	
2004	AERE	8,594	967	88	27,884
2005	HAI-TANG	3,199	688	68.5	2,332
2005	MATSA	5,322	819	49	
2005	TALIM	3,486	362	64	
2005	LONG-WANG	1,310	202	44	
2007	KROSA	4,597	623	56	9,624

shallow soil depth test, erosion pin, Ground LiDAR (LiDAR, Light Detection And Ranging), Airborne LiDAR in the study Region.

2.1 Field investigation

LiDAR is an optical remote sensing technology that measures properties of scattered light to find range and other information of a distant target. The prevalent method to determine distance to an object or surface is to use laser pulses. Like the similar radar technology, which uses radio waves instead of light, the range to an object is determined by measuring the time delay between transmission of a pulse and detection of the reflected signal. LIDAR technology has applications in geography, geology, seismology and remote sensing.

LiDAR efficiently produces high definition, high accuracy terrain, elevation models. In densely vegetated areas, LiDAR routinely provides a more rapid and detailed terrain model than photogrammetry for broadscale projects. Use the two stage of LiDAR result, the variation of terrain can be identified easily. This research is expected to perform twice this year. First time of Ground LiDAR and Airborne LiDAR were finished in the early of August, as shown in Figure 4. But the Second time of LiDAR measurement will be conducted in the end of November. Using Ground LiDAR and Airborne LiDAR must be easy to obtained the volume of sediment yields, especially in river channel and slope failure area crossover the typhoon season in study region.

Shallow soil depth test was conducted to acquire the shallow landslide depth. Soil depth was measured by screwing the hand auger or machine auger. The test performed vertically in to the ground until it wouldn't go further, assuming that auger had reached the bottom of soil layer or bedrock. The soil depth was deduced from the length of the auger remaining above ground after vegetation had been cleared away. This measurement was carried out at each chosen site in the

(a)Real Site image   (b)Virtual Site from LiDAR Data

Figure 4. The LiDAR result of Tai-Ping river in the early of August.

sub-watershed. The auger length (2.0 m) limited the upper boundary on the maximum soil depth that could be measured. The precision of the measurement was 0.05 m. Surface boulders were avoided and usually underestimated. In this case, the second measurement should be taken nearby.

2.2 Results of field investigation and analysis

There are several factors that are important in the estimation of shallow landslide volume, including geological factors, topographic factors, soil properties, the hydrological system, meteorology, vegetation, seismism, and land management. Two parameters that are particularly important in the limit equilibrium method for the infinite slope factor of safety (the standard basis for modeling shallow landslide occurrence) are the slope angle and the soil depth. The experiential equation about slope angle and soil depth was presented by Khazai and Sitar (2000), as shown in Table 2. In this research, thirty-two location were selected to performed shallow soil depth test in nine tributaries of study region. The chosen sites contain different geologic deformation in whole Shih-Men watershed, as shown in Table 3.

In order to estimate relationship between the slope angle and shallow landslide depth, shallow soil depth test were performed by hand auger in nine tributaries of this research. CGS (2008) had performed tests in

Table 2. The experiential equation about slope angle and soil depth presented by Khazai and Sitar (2000).

Slope angle (degree)	Average landslide depth (m)
<30	2.0
30~40	1.5
40~60	1.0
>60	0.5

Table 3. The distribution of investigation sites among different geologic formation.

No	Geologic formation	This research	*CGS (2008)	S.H. Jhong (2008)
1	Paling	10	26	25
2	Mushan	8	14	13
3	Peiliao	3	0	0
4	Taliao	6	6	0
5	Talu Shale	4	0	0
6	Shihti	1	2	0
7	Terrace	0	3	0
8	Szeleng Sandstone	0	3	0
9	Nanchuang	0	6	0
10	Hsitsun	0	7	0
	sum	32	67	38

* CGS means Central Geological Survey, Taiwan.

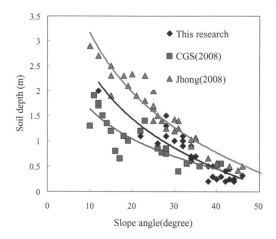

Figure 5. The regression curve with each case of shallow soil depth test.

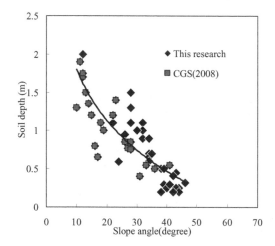

Figure 6. The suggested regression in this research included the major river and the tributaries in Shih-Men reservoir.

sixty-seven locations by machine auger along the Da-Han river and Jhong (2008) had performed tests in thirty-eight locations by hand auger in Pi-Ya river, which is one tributary of Da-Han river included in this study region. Plot three data set of shallow soil depth test in one graph and make regression, as shown in Figure 5. Herein, on the aspect of same slope angle, the shallow soil depth from CGS (2008) regression was low-estimated and Jhong (2008) regression was high-estimated. The difference in results could be influenced by measurement sites.

For regression curve could be applied in all Shih-Men reservoir and included the major river and all of the tributaries, the test data of CGS(2008) and this research were combined together to fit the relationship between slope angle and shallow soil depth, as shown in Figure 6.

The equation of the relationship between slop angle and soil depth is determined by

$$y = -0.9596 \ln(x) + 4.0072 \quad R^2 = 0.7214 \quad (1)$$

where y = shallow soil depth; and x = slope angle. In order to use the above equation, the shallow soil depth tests of chosen sites need to be conducted prior to the analysis. Refer to the slope angle, the shallow soil depth could be obtained from Eq. (1). Following studies are focusing on the volume of shallow landslide.

2.3 Estimation of potential landslide volume

Four kinds of empirical equations, which included Khazai (2000), NCDR (National Science and Technology Center for Disaster Reduction, Taiwan) method, CGS area method and Eq. (1) of this research, were adopted to calculate the volume of potential landslide. Take the Su-le river sub-watershed for example, the potential landslide volume were calculated, as shown in Figure 7. As indicated in the figure 7, the volume obtained from Eq. (1) was the lowest and the results of

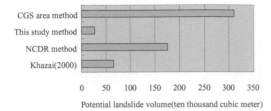

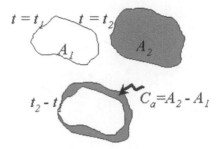

Figure 7. The potential landslide volume of four kinds empirical equations for Sule river sub-watershed.

Figure 8. The illustration of Uchihugi's empirical equation (1971).

the other methods were high-estimated. Among these methods, potential landslide area were the same, and the differences were acquired form the estimation of shallow soil depth. For the shallow soil depth tests were actually performed in the study region, the estimation of potential shallow landslide volume form Eq. (1) was more reliable in Shih-Men Reservoir watershed.

3 ESTIMAT NEW LANDSLIDE INDUCED BY RAINFALL

In view of watershed management, the typhoon events play an important role induced new landslide increment. In order to estimate the loss of sediment yields caused by heavy rainfall in each sub-watershed of Da-Han river, Uchihugi's empirical equation was adopted for predicting the new landslide increment rate of each sub-watershed. The illustration of Uchihugi's empirical equation was shown in Figure 8 and described as following Equation,

$$Y = \frac{C_a}{a} = K \times 10^{-6}(R - r)^2 \qquad (2)$$

where Y = increment rate of new landslide, C_a = the sum of new landslide area, a = sub − watershed area, K = specific coefficient for actual region, R = accumulated rainfall, r = critical rainfall induced landslide.

3.1 Characteristic of landslide region

In order to decide the new landslide increment rate of each sub-watershed, remote sensing data including aerial photos of two stages and satellite images of four-teen temporal stages, which are the stage before and after typhoon events, were used for landslide mapping. Nevertheless, the critical rainfall induced landslide should be taken into consideration when utilizing the Uchihugi's Equation. According to the critical rain-fall, which 300 mm was adopted in Fusing township of Taoyuan county by Taiwan government, four satellite images included typhoon Herb, typhoon Nari, typhoon Aere and typhoon Matsa were selected to analysis the

Table 4. The relationship between new landslide increment rate and typhoon events in nine tributaries (unit: %).

Name	Herb	Nari	Aere	Matsa
Yising river	0.46	0.10	0.20	0.06
Piya river	0.18	0.11	1.16	0.08
Baoliku river	0.02	0.07	0.80	0.52
Sule river	0.00	0.71	3.52	1.68
Shalunzai river	0.24	0.00	3.26	3.89
Taiping river	1.53	0.54	5.09	2.52
Siawunguang river	0.00	0.94	9.03	5.25
Taiyao No. 2 river	1.69	0.27	0.96	0.91
Siatianpu river	0.03	0.06	1.39	1.03

New landslide increment rate

new landslide increment rate of nine sub-watershed, as shown in Table 4.

3.2 Analysis of rainfall characteristics

To explore the relationship between landslide and rain-fall, analysis of rainfall characteristic was conducted in this research. Precipitation data from sixteen rainfall stations, which recorded data were all over 25 years, operated by Water Resource Agency of Taiwan in the study region were all adopted for the rainfall analysis. For the coincidence of rainfall characteristic and representative of average rainfall, Thiessen polygons method was utilized to set the influence region of each rainfall stations, as shown in Figure 9. Besides, the results of the frequency analysis for 1 day rainfall were presented as the return period of rainfall, as shown in Table 5.

3.3 Estimate new landslide area and volume

According to the results of new landslide increment rate obtained from satellite images and rainfall of nine tributaries obtained by Thiessen polygons method, Uchihugi's equation was transformed as following Equation (TPC, 2006),

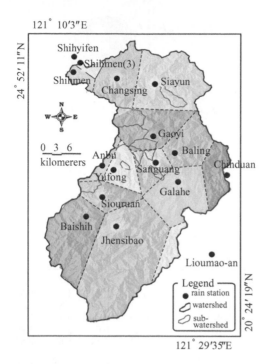

121° 10′3″E

24° 52′11″N

20° 24′19″N

121° 29′35″E

Figure 9. The influence region of each rainfall station by Thiessen polygons method.

Table 5. The result of frequency analysis for 1 day rainfall (unit:mm).

Watershed name	Return period of rainfall					
	5 years	10 years	20 years	25 years	50 years	100 years
Yising river	429	505	568	586	637	681
Piya river	453	548	632	657	732	803
Baoliku river	485	588	679	707	789	867
Sule river	477	578	668	696	777	854
Shalunzai river	432	540	644	677	778	879
Taiping river	449	561	668	702	806	910
Siawunguang river	544	694	837	883	1023	1162
Taiyao river	423	530	632	664	764	863
Siatianpu river	452	569	681	716	826	934

$$Y = \frac{C_a}{a} = A + K \times 10^{-8}(R - r)^2 \qquad (3)$$

where Y, C_a, a, K and R were the same with Eq. (2) and A = specific coefficient for different tributary. Eq. (3) was applied to calculate the coefficient of K and A in nine tributaries, as shown in Figure 10. Utilized the results of Figure 10 and Table 5, the new landslide

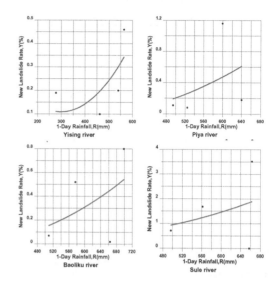

Figure 10. The relationship between increment rate of new landslide and 1 Day rainfall.

areas for various return period rainfall could be estimated by Eq. (3). The volume of new landslide, which induced by different return period of rainfall, could be expected by the depth utilized Eq. (1) from average slope angle and new landslide area utilized Eq. (3) from different return period of rainfall, as shown in Table 6.

4 CONCLUSIONS

Shih-Men Reservoir suffered from sediment mainly due to landslides in headstream watershed and reduced its service ability and performance by degrees. Shallow landslide is the most common landslide on steep slope in this region of Shin-Men reservoir. Those disasters generally triggered by heavy rainfalls and typhoons. For the estimation of volume of shallow landslide, shallow soil depth tests and LiDAR technique were performed in the Shih-Men region. Aerial photos were utilized for field investigation site chosen in advance and satellite images were utilized for landslide mapping. Finding from this study are described as following.

Based on the results of shallow soil depth tests in the study region, the equation of the relationship between slope angle and soil depth was determined and able to estimate the depth by slope angle in Shih-Men reservoir watershed.

Refer to the concept of Uchihugi's equation, four typhoon satellite images included typhoon Herb, typhoon Nari, typhoon Aere and typhoon Matsa were selected to analysis the new landslide increment rate

Table 6. New landslide volume for various return period rainfall (unit: m³).

Water-shed name	New landslide area for various Return period of rainfall					
	5 years	10 years	20 years	25 years	50 years	100 years
Yising river	6388	9675	13525	14830	18926	23015
Piya river	8673	24267	44226	51416	75848	103154
Baoliku river	9403	22019	37833	43481	62639	83961
Sule river	30484	47609	69230	77000	103300	132679
Shalunzai river	9168	17056	28909	33553	50512	71308
Taiping river	12400	39664	79741	95332	151936	221069
Siawunguang river	5864	21794	44075	52578	83123	119941
Taiyao river	11424	12507	14162	14810	17194	20128
Siatianpu river	6454	18705	36849	43905	69577	100983

of nine sub-watershed. The new landslide area for various return period of rainfall could be estimated. The volume of new landslide, which induced by different return period of rainfall, could be expected by the depth from average slope angle and area of new land-slide induced by various rainfall. Furthermore, results of the analysis from the above could be validated by the divergence analysis from two stages of LiDAR measurement in the future.

REFERENCES

Central Geology Survey (CGS) 2008. Project of evaluation and investigation about hydrology and geology effecting on slope stabilization (in Chinese).

Khazai, B. & Sitar, N. 2000. Assessment of Seismic Slope Stability Using GIS Modeling, *Geographic Information Sciences*, 6(2): 121–128.

Taiwan Power Company (TPC) 2006. Investigation and Strategy Study of Landslide and Debris flows in Da-Chia river Basin Between Te-Chi and Ma-An (in Chinese).

Tan, C.H., Ku, C.Y., Chi, S.Y., Chen, Y.H., Fei, L.Y., Lee, C.F. & Su, T.W. 2008. Assessment of Regional Rainfall-induced Landslides Using 3S-based Hydro-geological Model, *10th Landslides and Engineered Slopes*; *Proc. intern. symp., 30 June–4 July, 2008*. Xi'an: China.

Extreme rainfall for slope stability evaluation

N. Gofar, M.L. Lee & A. Kasim
Universiti Teknologi Malaysia, Johor, Malaysia

ABSTRACT: Slopes in tropical region are prone to frequent rainfall-induced failures; hence the evaluation of slope stability has to consider the rainfall characteristics and other factors related to rainfall infiltration. This paper demonstrates the results of numerical simulation using Seep/W for the evaluation of rainfall-induced slope instability at five selected sites at Johor Bahru, Malaysia. The slopes were assigned with 1-day, 2-day, 3-day, 5-day, 7-day, 14-day and 30-day extreme rainfall of ten-year return period. A suction envelope was obtained for each site and slope stability analysis was performed for the corresponding suction envelope. The results show that the reduction in factor of safety is relatively more significant for fine-grained soil. The finding was verified by field observation after the slopes were subjected to a series of intense rainfalls in December 2006. The application of extreme rainfall in slope stability analysis gives a good prediction of rainfall-induced slope failure.

1 INTRODUCTION

Slope failures in tropical region, particularly Malaysia are commonly triggered by prolonged yet intense rainfalls during monsoon seasons. The mechanism of the slope failure is as follow: the prolonged rainfall infiltration reduces matric suction of soil which in turn decreases the soil shear strength, and subsequently triggers the slope failure (Li *et al.*, 2005). It is thus essential to consider the rainfall characteristics and the suction profile of the soil slope for a more comprehensive evaluation of slope stability.

Studies relating the slope stability analysis with the extreme rainfall have been reported by a few researchers. Ng & Shi (1998) carried out a numerical simulation to study the influence of rainfall infiltration on the slope stability in Hong Kong. The rainfall applied in their study was based on the 10-year return period extreme rainfalls analyzed by Lam & Leung (1994). They suggested that the critical rainfall duration between three and seven days is required to cause failure. Pradel & Raad (1993) carried out a study on the equations governing the seepage and rainfall data in Southern California. They suggested that there was a threshold permeability of soil corresponding to an extreme rainfall. By applying a 50-year return period extreme rainfall, they found that the soils with permeability greater than 10^{-4} m/s will never become saturated; hence the surficial instability will not be developed.

It can be inferred from the previous studies that several attempts have been made to integrate the extreme rainfall into the slope stability analysis. In fact, the statistical prediction of extreme rainfall has played an integral role in the flood risk estimation and flood protection management. Development of the model for extreme rainfall application to slope stability analysis was presented by Gofar & Lee (2008).

The main objectives of this paper are to demonstrate the application of extreme rainfall in the slope stability evaluation and to prove its practicability by comparing the evaluation results with the observations of actual slope condition.

2 METHODOLOGY

Five sloping sites in the region of Johor Bahru, Malaysia (Figure 1) were selected for slope stability analysis and field observation. The slopes are made of different types of soil and the inclination of the slopes varies from 21° to 40° (Figure 2).

Intensity-Duration-Frequency (IDF) curve of Johor Bahru area (Figure 3) was used for the slope stability analysis. The 1-day, 2-day, 3-day, 5-day, 7-day, 14-day and 30-day extreme rainfalls were obtained from statistical analysis carried out by Gofar & Lee (2008). The properties of the soils retrieved from the study sites were determined through a series of laboratory tests and the results are summarized in Table 1. Figures 4, 5, and 6 shows the particle size distribution (PSD), soil water characteristic curve (SWCC) and hydraulic conductivity function of the soils.

Seep/W (GEO-SLOPE International Ltd., 2004a) was used to perform transient seepage analysis for each of the study sites. Prior to the analysis, the initial condition for each site was simulated based on the field suction measurements during dry period. In general, these suctions can be approximated to the suction corresponding to residual water content (Gofar *et al.*, 2007). Subsequently, the suction profiles generated from the seepage analyses were integrated into

Table 1. Properties of soil at the selected study sites.

	Site 1	Site 2	Site 3	Site 4	Site 5
Composition					
Gravel (%)	48	21	7	1	1
Sand (%)	15	39	37	26	16
Silt (%)	20	13	40	64	63
Clay (%)	17	27	16	9	20
LL (%)	53.2	64.8	61.9	60.2	64.6
PL (%)	35.5	42.2	41.9	45.4	40.0
PI	17.7	22.6	20.0	14.8	24.6
Classification BSCS	GMH	SMH	MHS	MHS	MHS
ρ_b (kg/m^3)	1805	1833	1724	1550	1704
Natural MC (%)	24.3	29.1	35.0	42.5	27.3
Saturated Permeability (m/s)	1.23×10^{-5}	1.44×10^{-5}	3.25×10^{-8}	8.36×10^{-8}	2.22×10^{-8}
Shear Strength,					
c'(kPa)	3.3	6.3	2.6	2.1	5.7
$\phi'(^{\circ})$	39.5	28.4	24.6	20.4	22.3

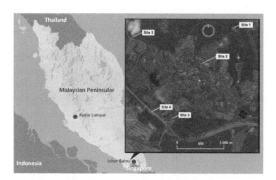

Figure 1. Locations of the selected study sites.

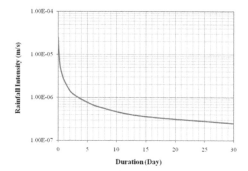

Figure 3. Intensity-Duration-Frequency (IDF) curve of Johor Bahru, Malaysia.

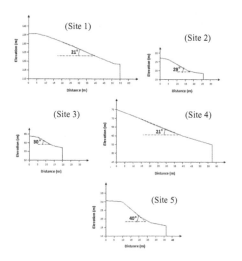

Figure 2. Geometry of the slopes.

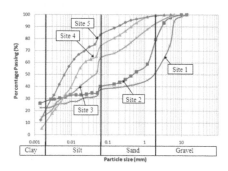

Figure 4. Particle size distribution.

Slope/W (GEO-SLOPE International Ltd., 2004b) for the slope stability analysis.

Field observation was made after the southern part of the Malaysian Peninsular experienced excessive rainfall for four continuous days in December 2006.

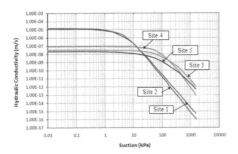

Figure 5. Soil Water Characteristic curves.

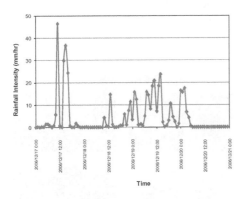

Figure 6. Hydraulic conductivity curves.

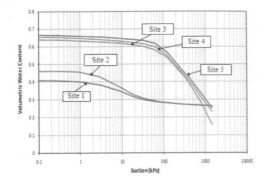

Figure 7. Hourly Rainfall Recorded at Johor Bahru from 17 to 20 December 2006.

As shown in Figure 7, the hourly rainfall measured at Johor Bahru within these four days was 450.4 mm, which was much higher than the average monthly rainfall of the Malayisan Peninsular (250.2 mm). Several slope failure incidents were reported following the rainfall events. The incident provided a unique opportunity to validate and evaluate the practicability of the model developed for slope stability analysis using extreme rainfall (Gofar & Lee, 2008).

3 RESULTS AND ANALYSIS

Figure 8 shows the suction profiles developed in soil mass at the selected sites computed by Seep/W. These suction envelopes indicate the worst suction condition that may occur in the corresponding soil slope under extreme rainfall. The figure indicates that there are two distinctive types of soil response to rainfall infiltration. Suction envelopes for Site 1 and 2 are formed by all the extreme rainfalls while the suction envelopes for Site 3, 4 and 5 were formed by the suction profile caused by the 30-day extreme rainfall only. The reason for this observation was that the slopes at Site 3, 4, and 5 consisted of fine-grained soils with the saturated permeability lower than the intensity of extreme rainfalls. Thus, the infiltration was controlled by the saturated permeability and the extreme rainfall with the longest duration would definitely result in the worst suction profile.

Figure 9 shows the factor of safety computed from the initial suction condition (dry condition) and the suction envelope (extreme rainfall condition). Apparently, the differences in the factor of safety between the two suction conditions were relatively less significant for Site 1 and 2 as compared with Site 3, 4, and 5. This was because the initial suction existed in the coarse-grained soil was considerably low (20 kPa to 30 kPa),

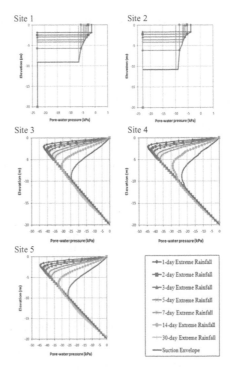

Figure 8. Suction envelopes as the result of extreme rainfalls.

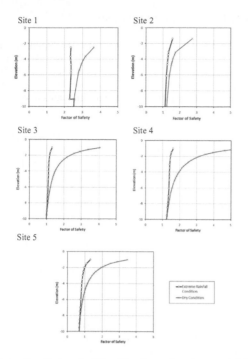

Figure 9. Factor of safety of the slopes for extreme rainfall condition and dry condition.

indicating the shear strength properties of the soil play a more crucial role than the suction towards the stability of the slope.

Conversely, greater variations in the factor of safety were observed for the slopes consisting of fine-grained soils (i.e. Site 3, 4 and 5). The huge differences between the initial suction condition (50 kPa) and the critical suction condition (0 kPa) should explain the result.

From the factor of safety chart, it can be noted that the influence of rainfall infiltration on the stability of the slope of fine-grained soil was particularly significant at shallow depth (i.e. within top 5 m). In other words, the potential failure for a slope consisting of fine-grained soil should be a shallow to intermediate-depth failure. This result contested the findings from previous studies which suggested that the failure of fine-grained soil should be comparatively deep-seated due to high cohesion.

4 DISCUSSION

The stability of a slope is governed by several parameters including slope geometry, shear strength properties and the suction profile of soil. Slope stability analysis is a way of assessing the factor of safety of the slope by assuming that the ultimate magnitudes for these parameters are known. In reality, however,

these parameters are exposed to uncertainties due to the inhomogeneity of soil and slope environment.

The use of extreme rainfall in the evaluation of slope stability could minimize the uncertainties associated with the conventional slope stability analysis through a statistical approach. It is not a method to obtain the ultimate factor of safety of the slope, but to compare the relative factor of safety between dry and extreme rainfall condition. It offers an advantage of better understanding of the mechanism of rainfall-induced slope failure.

In the present study, the factors of safety for the five selected study sites were computed for both dry and extreme rainfall conditions (Figure 9). Consistencies were revealed through the comparison between the analysis results and the slope conditions observed at the actual sites (Figure 10) under extreme rainfall condition.

For Site 1 and 2, the reductions in the factor of safety as the result of extreme rainfall were relatively less significant, thus only minor surface erosions were observed at the actual slopes. On the other hand, the reductions in factor of safety as the result of extreme rainfall for Site 3, 4 and 5 were significant, particularly within top 5 m of the soil slopes. The observations at the actual slopes confirmed these findings with the failure plane of these slopes are 2 to 3 m depths. It is noted that the factor of safety obtained for Site 4 was slightly higher than one but still the slope failed. This may be due to the inconsistency of the actual shear strength of the soil with the value used in the analysis. Nonetheless, the analysis showed that

Figure 10. Slope conditions at the selected sites under prolonged intense rainfall in December 2006.

the effect of extreme rainfall on this slope is significant. In this study, the mechanism of rainfall-induced slope failure was clearly revealed through numerical analysis as well as field observation. The integration of extreme rainfall in rainfall induced slope stability analysis provides a tool to analyse the susceptibility of any slope to rainfall infiltration.

5 CONCLUSIONS

The following conclusions can be drawn from the slope stability evaluation using extreme rainfalls for five selected sloping sites in Johor Bahru, Malaysia:

1. The extreme rainfall was integrated successfully into the slope stability evaluation to serve as an alternative approach to the conventional method of slope stability analysis. The field observation proved the viability of the method.
2. Despite the fact that the ultimate factor of safety of a slope is governed by several parameters including slope geometry, shear strength properties and suction profile of soil, the use of extreme rainfall in the slope stability evaluation could compare the relative factor of safety during dry and extreme rainfall conditions, hence provides an insight to the mechanism of rainfall-induced slope failure.
3. The potential depth of failure plane can be predicted from the slope stability evaluation using extreme rainfall. In general, the potential failure plane for fine-grained soil is within 5 m depths. The effect of extreme rainfall on the stability of coarse-grained soil slope is less significant.

REFERENCES

GEO-SLOPE International Ltd. 2004a. *Seepage Modeling with SEEP/W*. Calgary, Alta., Canada.
GEO-SLOPE International Ltd. 2004b. *Stability Modeling with SLOPE/W*. Calgary, Alta., Canada.
Gofar, N. & Lee, M.L. 2008. Extreme Rainfall Characteristics for Surface Slope Stability in the Malaysian Peninsular. *Journal of Assessment and Management of Risk for Engineered Systems and Geohazards (Georisk)*, Taylor and Francis. 2(2): 65–78.
Gofar, N., Lee, M.L. & Kassim, A. 2007. Stability of Unsaturated Slopes Subjected to Rainfall Infiltration. *Proceedings of the Fourth International Conference on Disaster Prevention and Rehabilitation, Semarang, Indonesia 10–11 September 2007*: 158–167.
Lam, C.C. & Leung, Y.K. 1995. Extreme Rainfall Statistics and Design Rainstorm Profiles at Selected Locations in Hong Kong. *Technical Note No. 86, Royal Observatory*. Hong Kong.
Li, A.G., Tham, L.G., Yue, G.Q., Lee, C.F. & Law, K.T. 2005. Comparison of Field and Laboratory Soil-Water Characteristic Curves. *Journal of Geotechnical and Geoenvironmental Engineering, ASCE*. 131(9): 1176–1180.
Ng, C.W.W. & Shi, Q. 1998. A Numerical Investigation of the Stability of Unsaturated Soil Slopes Subjected to Transient Seepage. *Computer & Geotechnics* 22(1): 1–28.
Pradel, D. & Raad, G. 1993. Effect of Permeability on Surficial Stability of Homogeneous Slopes. *Journal of Geotechnical Engineering, ASCE*. 119(2): 315–332.

Evaluation of existing prediction methods, performance-based design methods, risk analysis and the management of mitigation programs

Prediction and Simulation Methods for Geohazard Mitigation – Oka, Murakami & Kimoto (eds)
© 2009 Taylor & Francis Group, London, ISBN 978-0-415-80482-0

Risk assessment for hydraulic design associated with the uncertainty of rainfall

C.-M. Wang & C.-L. Shieh

National Cheng Kung University, Tainan, Taiwan

ABSTRACT: The purpose of this paper is to propose a method that incorporates the uncertainties of the depth and the duration of rainfall into the risk assessment for hydraulic design. In this paper, the risk of a hydraulic system is defined as the probability of failure of the hydrological system. The Hasofer-Lind reliability index (HLRI), which is a popular index for risk assessment, is used to improve the computation efficiency. The evaluation of the HLRI can be transformed into a constrained optimization problem. To solve the constrained optimization problem, modified simple genetic algorithms (SGA), which combine the penalty function and the constraint handling technique proposed by Deb (2000), is developed. The proposed method can produce the relation of the probability of failure versus the central safety factor. Based on the relation, a comprehensive benefit-cost analysis can be performed. The optimal alternative can be selected according to the result of the benefit-cost analysis. The proposed method provides an aid for performing the benefit-cost analysis.

1 INTRODUCTION

Design of a hydraulic system is the procedure to determine the impact of extreme events and then to select the proper value of the design variable of a hydraulic system. Therefore, one typical consideration for designing a hydraulic system is the recurrence interval of extreme events. The recurrence interval of extreme events implies the probability of occurrence of extreme events. However, in such a way, uncertainties associated with the hydraulic system, for example the uncertainty of rainfall, are difficult to be considered in the design of the hydraulic system (Chow et al. 1988). A better way to design a hydraulic system is to incorporate the uncertainties into the design consideration. It is therefore crucial to evaluate the risk resulted from the uncertainties of the system.

The risk of a system, in this paper, is defined as the probability of the system that fails under a specific condition. The failure of a system represents that the loading of the system exceeds the design capacity of the system, and the specific condition represents the design capacity of the system. Obviously, the risk of the system changes with the changing of the design capacity. In the hydraulic system, the design discharge is a common choice as the design variable. Under a specific design discharge, the probability of failure of a hydrological system can be seen as the risk of the system.

As aforementioned, it is difficult for the conventional hydraulic design to take uncertainties of the system into account. A feasible method, for example, to incorporate the uncertainties into the risk analysis

of the hydraulic design is the Monte Carol method (Apel et al. 2004). However, the risk analysis, that is performed using the Monte Carol method, is time consuming. To improve the efficiency of the risk analysis associated with uncertainties, the Hasofer-Lind reliability index (Hasofer & Lind, 1974), which is a popular index to evaluate the reliability of a system (Tandjiria et al. 2000), is an option. Therefore, the Hasofer-Lind reliability index (abbreviated as HLRI hereafter) is employed in this paper to evaluate the risk of a hydraulic system associated with uncertainties.

The value of the design discharge is usually the basis for a hydraulic system to arrange appropriate facilities and to determine the dimensions of the facilities. The design discharge is also the basis to evaluate the corresponding risk of the hydraulic system. The maximum water conveying capacity of the hydraulic system is the implication of the design discharge. One method to determine the design discharge for a hydraulic system is to employ the rainfall-runoff modeling. This method is employed in this paper.

The method for the determination of the design discharge through the rainfall-runoff modeling is simple. The first step is to determine the design storm. Then a well-calibrated rainfall-runoff model is used to calculate the hydrograph resulted from the design storm. The peak discharge of the resulted hydrograph is the design discharge. Apparently, the determination of a proper design discharge depends on a well rainfall-runoff model.

There are many rainfall-runoff models. For this paper, a suitable rainfall-runoff model should well represent the rainfall-runoff process of the study watershed, and should be able to be used to determine the design discharge. There is a requirement for the rainfall-runoff models when they are used to estimate the design discharge. The requirement is that the rainfall-runoff models must use only the rainfall as input data. The artificial neural networks (ANN) based rainfall-runoff models are not appropriate for estimating the design discharge, because they need previous discharge as the input (Lin & Wang 2007a). A possible choice for determining the design discharge, for example, is the HEC-HMS, which is a model of public domain. The HEC-HMS is capable of determining the design discharge of a hydraulic system, because it does not need the previous discharge records as input. When the HEC-HMS is well calibrated, it can properly represent the rainfall-runoff process of the study watershed. To employ the HEC-HMS, in other words, the model calibration must be done. However, the job of the model calibration may take a long time, and thus may increase the burden of engineers.

In this paper, the NCUC (nonlinear computational units cascaded) model (Lin & Wang 2007a, b) is selected for modeling the rainfall-runoff process, because it does not require the runoff records as input. Moreover, the NCUC model is embedded with an automated calibration method that facilitates engineers to calibrate the model parameters. Therefore the burden of engineers can be eased, and the modeling accuracy can be improved.

The evaluation of the HLRI can be transformed into a constrained optimization problem. Description of the evaluation of the HLRI is given in section 2 of this paper. Since the evaluation of the HLRI is a constrained optimization problem, an effective method to solve the problem is necessary. A popular method to solve the constrained optimization problem is the genetic algorithms (GAs) (Goldberg 1989). However, to handle the constraints, the GAs should be slightly modified and fine tuned. A typical technique to handle the constraints using the GAs is the penalty function. The mechanism of the penalty function is to give a penalty to the fitness score when the solution violates the constraints. A more intuitive method to handle the constraints using the GAs is proposed by Deb (2000) (referred to as the Deb's method herein). In this paper, the penalty function and the Deb's method is combined to estimate of the HLRI.

The objective of this paper is to propose a method to analyze the risk of a hydraulic system. The uncertainties which affect the risk are considered in the risk analysis. The HLRI is the basis in this paper to estimate the risk of a hydraulic system. The GAs, which are associated with the penalty function and the Deb's method, is adopted to solve the optimization problem with constraints resulted from the estimation of the HLRI. The rainfall-runoff modeling is performed using the NCUC model. A case study of the Tsengwen

reservoir is also given in this paper to illustrate the performance of the proposed method.

2 METHOD

2.1 The Hasofer-Lind reliability index

The HLRI is a popular method to evaluate risk of a system (Mailhot & Villeneuve 2003; Paik 2008; Tandjiria et al. 2000). The HLRI can be expressed by the following equations:

$$\beta = \min_{X \in F} \sqrt{(X - \mu)^T C^{-1}(X - \mu)} \tag{1}$$

where β is the shortest distance between the origin and the limit state boundary, F; X is the vector of random variables that are considered the source of uncertainties; μ is the vector of mean values of X, and C is the covariance matrix. The reliability, r, of the system can be expressed as:

$$r = \Phi(\beta) \tag{2}$$

where $\Phi(\cdot)$ is the cumulative density function of normal distribution. The risk of the system, i.e. the probability of failure (P_f), is

$$P_f = 1 - r \tag{3}$$

From the above explanation, it can be found that the point for the evaluation of the HLRI is to find the β that the X is exactly on the limit state boundary.

2.2 Problem formulation

The design variable for the hydraulic system in this paper is the design discharge, which is a common choice for designing a hydraulic system. The design discharge is determined by the aforementioned method. When the discharge that is conveyed in the hydraulic system is larger than the design discharge, the hydraulic system fails. Factors that affect the discharge are many. To keep the balance between the actual complexity of a hydraulic system and the feasibility for solving the problem, the factors that affect the discharge are considered the total rainfall depth and the duration of the rainfall. The covariance between the total rainfall depth and the duration of the rainfall is neglected, and hence the C in Eq. (1) is a unity matrix.

The evaluation of the risk of a hydraulic system can then be illustrated using the following equations, which is a constrained optimization problem:

minimize d (4)

where d is calculated by

$$d = \sqrt{K_p^2 + K_t^2} \tag{5}$$

subject to

$$SF = 1 \qquad (6)$$

where d is the distance in the standardized space; SF is the safety factor of the hydraulic system. The eq. (6) represents the limit state boundary of the hydraulic system. The definition of K_p and K_t are as follows:

$$x_p = \mu_p + K_p \sigma_p \qquad (7)$$

$$x_t = \mu_t + K_t \sigma_t \qquad (8)$$

where x_p is the rainfall depth; μ_p is the mean value of the rainfall depth; σ_p is the variance of the rainfall depth; K_p is a random variable whose range is $(-7, 7)$; x_t is the duration of rainfall; μ_t is the mean value of the rainfall duration; σ_t is the variance of the duration of rainfall; K_t is a random variable whose range is $(-1, 1)$. When d is minimized, d equals to β.

The calculation of SF uses the following equation:

$$SF = \frac{Q_D}{Q_L} \qquad (9)$$

where the Q_D is the design discharge of the hydraulic system and is the capacity of the hydraulic system as well; Q_L is the actual discharge of hydraulic system and is also the loading of the hydraulic system.

In this paper, the Q_L is calculated by:

$$P = \text{TriH}\left(x_p, x_t\right) \qquad (10)$$

$$Q_L = \max\left(\text{NCUC}\left(P\right)\right) \qquad (11)$$

where P is the hyetograph derived using the triangular hyetograph method (Chow et al. 1988); Q_L is the peak discharge derived using the NCUC model; Tri($\cdot$) represents the triangular hyetograph method; NCUC($\cdot$) represents the NCUC model.

The above equations (4) to (6) are an alternative representation of the HLRI. It is easier to evaluate the P_f using the equations (4) to (6). Readers can refer to Hasofer & Lind (1974) for more information.

2.3 The NCUC model

The NCUC model, which can model the rainfall-runoff process, is developed by Lin and Wang (2007a, b). Unlike the ANN based rainfall-runoff models, the NCUC model can accurately model the rainfall-runoff process without manual system identification. Furthermore, the NCUC model is embedded with an automated calibration method, which facilitates engineers to acquire appropriate parameters in a short time and to improve the accuracy of the modeling.

The building block of the NCUC model is the non-linear computational unit (NCU). The NCUC model consists of several NCUs. The pattern of the NCUC model can be arbitrarily adjusted in order to meet the characteristics of the watershed and the requirement of engineers. The automated calibration method can be applied to any pattern of the NCUC model without any manual adjustment. For more information regarding the NCUC model, please refer to Lin and Wang (2007a, b).

2.4 Genetic algorithms

"Survival of the fittest" is the key concept of the GAs. There are many different kinds of GAs. In this paper, the simple genetic algorithms (SGA) (Goldberg 1989) are sufficient to solve the problem in hand and are hence employed in this paper. More details of the SGA can be found in Goldberg (1989) and only a brief explanation is given herein.

The SGA consists of three major operators: the selection, the crossover, and the mutation. At each generation of SGA, these three operators execute sequentially to manipulate the chromosomes. The chromosomes are representations of solutions to the problem.

The selection operator selects better solutions according to the fitness score which is evaluated using the fitness function. The fitness score is the driven force of the SGA. After the selection, the crossover operator is used to generate new solutions, which are potentially better. Following the crossover, the mutation operator randomly shifts the chromosomes with a very small probability to direct the searching to a possibly new direction.

When one generation ends, another new generation immediately begins. The whole process of the SGA completes till the optimal solution is found.

2.5 Constraint handling

The SGA does not have the ability to handle the constraints within the optimization. A slight modification of the SGA is necessary to handle the constraints. Typically, in the SGA constraints is handled using the penalty function method, which is implemented in the evaluation of the fitness score in this paper. The penalty function method is an intuitional and easily-implemented one. However, the penalty parameters in the penalty function method are difficult to be determined (Deb 2000).

The Deb's method, which eliminates the requirement of estimating the penalty parameters, is utilized in this paper to handle the constraints. The Deb's method involves three criteria (Deb 2000):

1. "Any feasible solution is preferred to any infeasible solution.
2. Among two feasible solutions, the one having better objective function value is preferred.
3. Among two infeasible solutions, the one having smaller constraint violation is preferred."

The Deb's method can be easily implemented in the selection operator of the SGA. Using the Deb's method to solve the constrained optimization problem, the penalty parameters are not required and thus the difficulty of estimating the penalty parameters is avoided.

In this paper, the penalty function method and the Deb's method are combined to derive the optimal estimate of the HLRI.

3 CASE STUDY

3.1 Study area

The study area is the upstream area of the watershed of the Tsengwen creek. There is a reservoir, namely the Tsengwen reservoir, on the Tsengwen creek. The capacity of the Tsengwen reservoir is the largest in Taiwan. The Tsengwen reservoir is a multipurpose one. The purposes of the Tsengwen reservoir include water supply, power generation, irrigation, flood mitigation and entertainment. Therefore, the Tsengwen reservoir is an important reservoir in Taiwan. The study area is an upstream watershed within the watershed of the Tsengwen reservoir. There is a sabo dam on the outlet of the study area. A gauging station, which measures the water level, is also installed on the sabo dam.

The location of the study area is shown in Figure 1. The area of the study area is 280.4 km^2. There are five rain gauges in the study area. According to the handbook of hydrological design (Water resource agency, Ministry of economic affairs, Taiwan, 2000), the duration and the rainfall depth of the design storm of the study area are 48 hours and 1208 mm, respectively. The design storm of the study area is shown in Figure 2.

The variance of the rainfall depth, σ_p, is assumed 0.1 μ_p, and the variance of the rainfall duration, σ_t, is assumed 0.1 μ_t.

3.2 Setup of the constrained optimization method

The coding of the chromosomes used by the SGA in this paper is the real coding (Michalewicz 1996). Two parameter, i.e., the K_p and K_t in equations (7) and (8) respectively, constitutes the chromosomes. With the real coding, the efficiency of the SGA is increased, and hence is adopted by Deb (2000) as well.

As explained above, a proper choice of the fitness function is critical. In this paper, the Eq. (5) is used as the fitness function to evaluate the fitness score of each chromosome.

The constraint of the problem in hand is the Eq. (6). However, the constraint is strict. It is perhaps impossible to find the optimal solution that exactly conforms to the constraint under certain μ_p and μ_t. A feasible constraint of the problem is:

$$1 - \varepsilon \leq SF \leq 1 + \varepsilon \qquad (12)$$

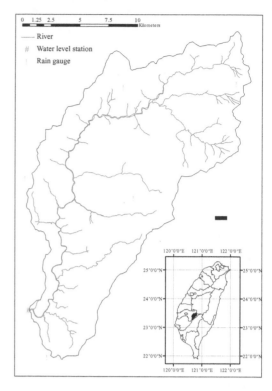

Figure 1. The location of the study area.

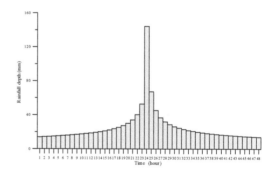

Figure 2. The design storm of the study area.

where ε is the tolerance. In this paper, the tolerance is 0.001. That is the SFs between 0.999 and 1.001 conform to the constraint.

The setup of the SGA is: the population size is 30, the number of generations is 1000, the probability of mutation is 0.001, and the probability of crossover is 0.8.

3.3 Rainfall-runoff modeling

To appropriately model the rainfall-runoff process of the study area, the NCUC model is applied. The pattern of the NCUC model used in this paper is {D3, D3, D3}. Regarding the notation, {D3, D3, D3}, please refer to Lin & Wang (2007a).

The NCUC model is constructed using two events, which are used for the calibration and the validation, respectively. The storm event, which occurred at June, 9, 2006, is used for calibrating the NCUC model; the Typhoon Bilis, which occurred at July, 12, 2006, is used for validation. The result of the calibration is drawn in Figure 3. The corresponding coefficient of efficiency (Nash & Sutcliffe, 1970) is 0.97, which indicates the good performance of the NCUC model. The result of validation is depicted in Figure 4. The coefficient of efficiency during the validation is 0.93. The result of validation implies that the NCUC model can well model the rainfall-runoff process of the study area.

The design discharge is determined by the NCUC model whose input is the design storm shown in Figure 2. The result shows that the design discharge is 3211.5 m³/s.

The NCUC model obtained here is also used to calculate the hydrograph that resulted from the presumed hyetograph. As aforementioned, the hyetograph is generated using the triangular hyetograph method.

The peak discharge of the resulted hydrograph is then abstracted. The SF for the resulted peak discharge is calculated using the Eq. (9).

3.4 Result

The ultimate objective of this paper is to derive the probability of failure associated with the uncertainty of rainfall for each system loading. Since the uncertainty of the depth of rainfall is more significant than that of the duration of rainfall, the probability of failure that corresponds to the mean values of the depth of rainfall is considered. The uncertainty of the duration of rainfall is also considered in this paper, but the uncertainty of the mean value of the duration of rainfall is not considered.

In this paper, to clearly represent the probability of failure for each system loading, the relation of the probability of failure versus the central safety factor is drawn in Figure 5. The definition of the central safety factor (CSF) is

$$CSF = \frac{P_D}{\mu_P} \qquad (13)$$

where P_D is the depth of the design storm; μ_p is the mean value of the depth of the rainfall that results in the loading (Q_L) of the hydraulic system. As aforementioned, P_D is 1208 mm. The μ_p ranges from 800 mm to 1600 mm with an increment 200 mm.

The relation of the probability of failure versus the central safety factor can be fitted by

$$\frac{1}{1 + \left(\frac{1}{CSF}\right)^a} \qquad (14)$$

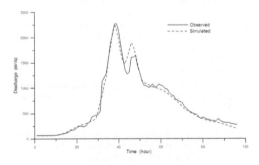

Figure 3. Comparison of the simulated and observed discharge during the calibration.

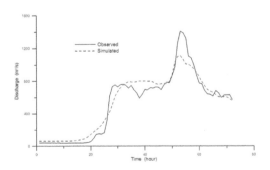

Figure 4. Comparison of the simulated and observed discharge during the validation.

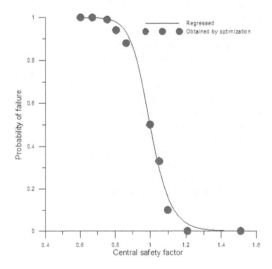

Figure 5. The relation of the probability of failure versus the central safety factor.

where a is a coefficient to be determined. The coefficient, a, can be determined from the red points in Figure 5. The value of a is -17.55 and the R^2 (coefficient of determination) is 0.9955. The value of R^2 shows that the result of the fitting is well.

3.5 Discussion

From Figure 5, it can be found that the probability of failure decreases with the decreasing of the mean value of the depth of rainfall. When the depth of rainfall equals to that of the design storm, the probability of failure is 0.5.

Figure 5 provides important information for engineers who design the hydraulic system. According to Figure 5, engineers can know the probability of failure under a specific depth of rainfall. The construction costs and benefit of hydraulic systems that resist the peak discharges caused by different magnitudes of storms are different. The design considerations for hydraulic systems with different values of the design variable are obvious different as well. With the information provided by Figure 5, a comprehensive benefit-cost analysis of the alternatives of the hydraulic system can be obtained. With the result of the benefit-cost analysis, engineers can select the optimal alternative among several feasible alternatives. Meanwhile, the uncertainties of the depth and the duration of rainfall can be incorporated in the benefit-cost analysis.

4 CONCLUSION

A risk assessment, which incorporates the uncertainties of the depth and the duration of rainfall, for a hydraulic system is proposed in this paper. The proposed method eliminates the requirement of the Monte Carol simulation to evaluate the risk. Therefore, the computation burden is eased. Moreover, unlike the conventional method for hydraulic design, the uncertainties that affect the design discharge are considered in the proposed method. The probability of failure versus the central safety factor can be derived from of the proposed method. According to the information, a comprehensive benefit-cost analysis can performed. Engineers can select the optimal alternative through the benefit-cost analysis. It can be concluded that the proposed method can estimate the risk with slight computation effort and can provide an aid for engineers to conduct the benefit-cost analysis.

REFERENCES

Apel, H., Thieken, A.H., Merz, B., Blöschl, G. 2004. Flood risk assessment and associated uncertainty. *Natural Hazards and Earth System Sciences* 4:295–308.
Chow, V.T. Maidment, D.R. Larry, W.M. 1988. *Applied Hydrology*. Singapore:McGraw-Hill.
Deb, K. 2000. An efficient constraint handling method for genetic algorithms. *Computer Methods in Applied Mechanics and engineering* 186:311–338.
Goldberg, D.E. 1989. *Genetic Algorithms in Search, Optimization and Machine Learning*. Addison-Wesley.
Lin, G.F., Wang, C.M., 2007a. A nonlinear rainfall-runoff model embedded with an automated calibration method. Part 1. The model. Journal of Hydrology 341(3–4): 186–195.
Lin, G.F., Wang, C.M., 2007b. A nonlinear rainfall-runoff model embedded with an automated calibration method. Part 2: The automated calibration method. Journal of Hydrology 341(3–4): 196–206.
Hasofer A.M., Lind N. 1974. An exact and invariant first-order reliability format. *Journal of Engineering Mechanics ASCE* 100(1): 111–121.
Mailhot A., Villeneuve, J.P., 2003. Mean-value second-order uncertainty analysis method: application to water quality modelling. *Advances in Water Resources* 26(5): 491–499.
Michalewicz, Z., 1996. *Genetic algorithms + data structures = evolution programs*. Springer-Verlag, New York.
Nash J.E., Sutcliffe J.V., 1970. River flow forecasting through conceptual models. Part 1: A discussion of principles. *Journal of Hydrology* 10(3): 282–290.
Paik K., 2008. Analytical derivation of reservoir routing and hydrological risk evaluation of detention basins. *Journal of Hydrology* 352(1–2): 191–201.
Plate, E.J. 2002. Flood risk and flood management. *Journal of Hydrology* 267:2–11.
Tandjiria V., Teh C.I., Low B.K. 2000. Reliability analysis of laterally loaded piles using response. *Structural Safety* 22:335–355.
Surface methods
Water resource agency, Ministry of economic affairs, Taiwan. 2000. Handbook of hydrological design. Water resource agency, Ministry of economic affairs.

Prediction and Simulation Methods for Geohazard Mitigation – Oka, Murakami & Kimoto (eds)
© 2009 Taylor & Francis Group, London, ISBN 978-0-415-80482-0

Geotechnical risk assessment of highly weathered slopes using seismic refraction technique

M.U. Qureshi, I. Towhata, S. Yamada, M. Aziz & S. Aoyama
The University of Tokyo, Tokyo, Japan

ABSTRACT: A huge number of slope failures have been reported following the 2005 Kashmir earthquake in Pakistan of magnitude 7.6. Thorough examination of disturbed slopes indicated that the rock is extensively weathered and lost its shear strength. Mechanism of weathering is repeated changes in temperature and moisture. Authors set to think about the geotechnical problems in near future from those slopes with a global approach and demystify the problem by delineating the thickness and mechanical properties of weathered layer exposed by slopes. Seismic refraction survey and intrusive tests indicated that the weathered layer has S-wave velocity range of 150–300 m/s and thickness is 1 m or its roundabouts. The in-situ information on geometry and mechanical properties of slope is believed to be a useful tool for the slope stability analysis.

1 INTRODUCTION

During the 2005 Kashmir earthquake of magnitude 7.6, a huge number of landslides possibly close to 1000 were triggered. Those slope failures had drastically affected the communities and infrastructures in surrounding steep mountain valleys. Landslide remains the largest threat to the community, especially during the monsoon season in July and August. Major risks to the human life include slope failures in seismically disturbed slopes and later effects of the failed deposited material. Large sized cones of the debris are still an unresolved threat (Fig. 1), which can be seen in the limestone slope failure on the bank of River Neelum in north of Muzaffarabad, Pakistan.

Examining the disturbed slopes indicated that the rock is extensively weathered and lost its shear strength. Mechanism of weathering are repeated temperature and moisture changes during summer and winter or during day and night, freezing and thawing of infiltered rain water and joints created by the tectonic forces. Weathering is believed to be accountable for the topographical changes, degradation of mechanical properties, surface deterioration and erosion. Fully or partially disturbed slopes are still and unknown geotechnical risk to the inhabitants living in the vicinity of those slopes. Hazardous slope movements due to weathering can result from one or more of two causes: change in the mechanical properties or the change in geometry of the slopes.

Change in mechanical properties is associated with the process of ageing, which is the degradation of the geomaterials with time due to environmental agents. On the other hand change in geometry corresponds to the extent of loose/weathered geomaterials exposed by the slope, removal of support (erosion of geomaterial by rain or wind and excavation due to human activity) and increase in load due to rain water infiltration.

Nearly three years have passed, but still rehabilitation activities are lacking the wide scope of geotechnical risk assessment and respective remedial measures. Authors believed that the seismically disturbed slopes will be a long term geotechnical risk on the surroundings, unless stabilized. As a matter of fact the identification of those potentially unstable slopes takes precedence if long term geotechnical risk assessment is in question.

Generally researchers are interested in the slope failures which have directly affected the community and infrastructure immediately after any event like earthquake or rainfall. In Figure 1, the slope failure which occurred in 2005 Kashmir earthquake dammed the River Neelum and blasting was done to streamline the river flow. Huge cones retaining the loose geomaterial at the toe of the slope failure were not given the importance at that time. But an increase in

Figure 1. Earthquake triggered slope failure along the River Neelum in north of Muzaffarabad, Pakistan.

sediment load has been observed in the lake of Mangla dam of which River Neelum is one of the tributaries. Including other problems, there are many case histories which reveal the facts pointed out by the authors.

2 CASE HISTORIES

Before presenting the experimentation, authors would like to present some case histories which have been observed during their reconnaissance visits to the earthquake affected topographies. Motivation to this research was accredited to the slope failure problems in developing countries with limited capital for geotechnical risk assessment. Figure 1 shows a slope failure in highly weathered limestone formation in north of Muzaffarabad, Pakistan after 2005 Kashmir earthquake. A quite similar example of continuous slope failures due to earthquake can be seen in Figure 2 taken during the reconnaissance after 2008 Wenchuan earthquake in China. In both the cases temporary restorations have been made to reconstruct the transportation conditions. For the long term stability issues permanent measures are required with intelligent understanding of the disturbed and failed slopes.

The extent of damage to the slope is far deeper than what it looks like in most of the cases and environmental agents including rain, wind and temperature exposes the disturbed slopes with time. The span of time may vary up to several hundreds of years for the stabilization of disturbed slopes. Figure 3 illustrates the erosion control works to stop the debris from Ohya slide in Shizoka Prefecture, Japan which is said to have been caused by the 1707 Hoei gigantic earthquake. Since then, the destabilized slope has been producing debris flow at heavy rainfalls frequently. Three centuries have been passed but the debris problem has not yet been solved with their secondary effects in the

Figure 3. Erosion control works in vicinity of Ohya slide. Shizuoka Prefecture, Japan.

Figure 4. Debris control works along the tributary of Tachia River, Taichung, Taiwan.

form of raising bed of Abe River and scouring at the bridge piers.

Similar situation has been observed in the Taichung County of Taiwan after a decade of 1999 Chichi earthquake, due to the debris flow from seismically disturbed slopes. The extremely high rate of rainfall contributes more towards the amount of debris flow.

Figure 4 gives an interesting illustration of the situation after a decade of earthquake. As a matter of fact, similar situation is expected to the disturbed topography of the earthquake affected areas in Pakistan and China. So lesson learnt from the Ohya slide and situation in Taiwan are helpful in long term geotechnical risk assessment.

3 LITERATURE REVIEW

In the past, many studies were conducted to investigate the weathering potential of rocks in relation to the slope stability problems, but still the understanding is not

Figure 2. Slope failures along a river valley in epicentral town of Ying Xiu, China.

reproducible. The approaches adopted for the study were of narrow scope for their application. Weathering studies, in practical assessment of slope failures, helps to understand what stage in the process the landscape has reached. In the field of geotechnical engineering, recognition of the weathered conditions of the slope is a critical issue to be faced in the process of evaluating the slope failure hazards. So, a three dimensional architecture of weathering is significant in this regard but difficult to predict.

Borrelli et al. (2006) studied the methodology of the reconnaissance surveys of weathering grade, which is then related to the slope instability for plutonic and metamorphic rocks in Acri, Italy. To achieve this aim reconnaissance procedure included the observations and index tests (discoloration, sound when struck with geological hammer, effect of the point geological pick, breaking with hands, rebound of Schmidt hammer and grain size analysis) carried out at check sites. The weathering grade was confirmed by the quantitative measurements of weathering. Reconnaissance mapping was carried out at 1:50,000 scale and was checked by detailed mapping at 1:10,000 scale.

Shang et al. (2004) estimated the weathering rate to be 41.2–52.5% of the volume loss per year in Southeast of Tibet, which was believed to be formed by three ways, i.e. weathering, avalanching and rock falling, and sliding. For calculation of weathering rate, one grandiorite block exposed in a slightly dipping angle was chosen to estimate the volume change due to surface deterioration for one year.

Hachinohe et al. (1999) conducted the study on sandstone and mudstone samples from Boso Peninsula, Japan and change in needle penetration hardness, pore size distribution and mineralogical and chemical compositions were observed with depth. Mudstone was thought to weather faster than the sandstone. The studies discussed lack the global approach of the in-situ state of the geomaterials but guide the authors to think about more global and generalized approach, which eliminates the limitations of previous studies.

Exclusively, field observations or focused measurements of volume change for limited time are reliable in the sense of local application, but helps to generate more sophisticated approaches. A review of various studies has been made, which makes significant contributions to bringing together the ideas of engineering, soil and rock mechanics and geomorphology. Techniques used in these studies for description and analysis are by no means uniform, and if standardization is desirable there is still a long way to go.

4 METHODOLOGY

The authors set out a challenge to investigate the weathering potential of the soft/jointed rocks in the field focusing on the in-situ geotechnical properties and geometry of the slopes. Field tests were performed with the objective to delineate the thickness and in-situ mechanical properties of surface weathered layer. To avoid the local approach such as drilling and for time and cost saving, authors suggested adopting the geophysical exploration. Because the scope of investigations with respect to the extent was anticipated to be shallow, seismic refraction technique was thought to be the best among other non-intrusive geophysical techniques. Since the reliability of seismic refraction had to be confirmed, so Swedish weight sounding and portable dynamic cone penetration tests were perform wherever applicable.

4.1 Seismic refraction survey

The setup for the seismic refraction survey can be seen in Figure 5. The impulsive source used was a steel plate hit by a wooden hammer to produce Rayleigh waves. The spread of receivers was 4.5 Hz vertical geophones. Ten shots at each end of the spread were recorded in the mobile data logger. The quality of data was checked at the site and in case of abnormal data the test was repeated. Data was transferred to the computer to perform the refraction analysis to judge the depth and seismic velocity of two layered model of strata. Intercept time method (Palmer D., 1986) was used to generate the seismic refraction profiles from the field data.

4.1.1 Assumptions
A two layer model of the soil stratum was employed, assuming;

1. Seismic velocity of each layer of soil stratum is uniform and isotropic.
2. The layers of the strata are bounded by plane dipping interfaces.
3. The velocity increases with depth so that critical refraction can occur at each interface.
4. Layer should be of sufficient thickness to be judged by travel time data.

Figure 5. Performance of seismic refraction survey at Yokosuka, Japan.

4.2 Swedish weight sounding (SWS) test

Swedish weight sounding test is a portable method of ground investigation. The equipment is illustrated in Figure 6a. The whole setup is rotated manually and the number of rotations (180-degree rotation) needed for 1 m of penetration is counted and recorded as N_{sw}.

Inada (1960) proposed the empirical co-relations between N_{sw} and SPT-N

For sand,
$$N = 0.02 \ W_{sw} \text{ (kg) and N} = 2 + 0.067N_{sw} \quad (1)$$

For cohesive soil,
$$N = 0.03 \ W_{sw} \text{ (kg) and N} = 3 + 0.05N_{sw} \quad (2)$$

Further the SPT-N value can be empirically co-related with the field shear wave velocity V_s. Among various correlations, those employed by Japanese Highway Bridge Design Code appear the easiest to memorize.

For sand, $\quad V_s = 80N^{1/3} \text{(m/s)} \quad (3)$

For clay, $\quad V_s = 100N^{1/3} \text{(m/s)} \quad (4)$

4.3 Portable dynamic cone penetration (PDCP) test

In the site where mixture of gravels and loose soil is anticipated, dynamic cone penetration is one of the best choices for the confirmation of seismic refraction survey. The arrangement of the test can be seen in Figure 6b, consisting of metallic cone having a cone angle of 60-degrees, fitted to a metallic rod of 16 mm diameter and 1 m length. A knocking head is screwed to the other end of rod followed by another rod which guides a 5 kg hammer with a 50 cm height of fall. Number of drops N_d for 10 cm of penetration is recorded.

N_d can be co-related to the SPT-N by using the following empirical relations;

$$N_d = 1 \sim 3N \quad (5)$$

(Okubo equation for normal/mild slope)

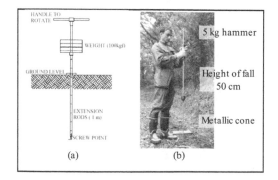

Figure 6. (a) Swedish weight sounding equipment, (b) Portable dynamic cone penetration test equipment.

$$N_d = 1.5N \quad (6)$$

(Ogawa equation for steep slopes and for $N_d < 20$)

5 FIELD EXPERIMENTS

5.1 Seismic refraction survey

For the exploration on the thickness and in-situ mechanical properties, seismic refraction surveys were conducted at various topographies of soil and rock formations. The tests were conducted at Tokyo and Yokosuka in Japan and at Muzaffarabad, Pakistan. In Tokyo (Hongo-campus) the tests were performed on the plane topography covered with grass as shown in Figure 7. The seismic refraction exploration was further extended to actual slopes in Muzaffarabad, Pakistan and Yokosuka Japan, as shown in Figures 8 & 9, respectively. Muzaffarabad site was included in the Murree formation of Miocene age consisting of medium to thickly bedded, graded fluvial sandstone and flood plain siltstone and clay with occasional limestone and inter-formational conglomerate. The site at Yokosuka belongs to the Zushi formation of Miocene age consisting of alternating sandstone and mudstone having sand content below 40%. It has been observed that the loose/weathered layer is of shallow extent having thickness of 1m or its roundabout.

The shear wave velocities calculated form the seismic refraction analysis, for the weathered layer, were in range of 200–300 m/s in case of rocks and 100–150 m/s in case of soil. If factually interpreted, vegetation cover also gives some idea about the weathered layer thickness, i.e. slopes covered by grass have shallow thickness of weathered layer as compared to the one covered by trees. But the cover of trees also suggests that the slope is stable for a long time depending upon the life of trees.

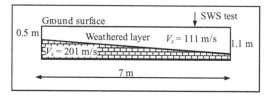

Figure 7. Seismic refraction profile at Hongo-campus, Tokyo, Japan.

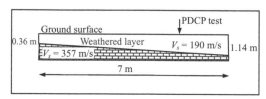

Figure 8. Seismic refraction profile at Muzaffarabad, Pakistan.

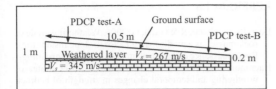

Figure 9. Seismic refraction profile at Yokosuka, Japan.

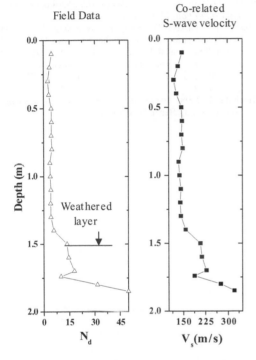

Figure 10. Portable dynamic cone penetration test at Muzaffarabad, Pakistan.

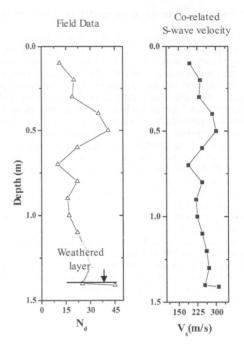

Figure 11. Portable dynamic cone penetration tests at Yokosuka-A, Japan.

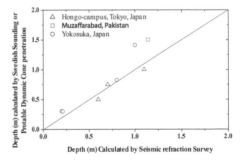

Figure 12. Validation of weathered layer thickness.

5.2 Intrusive tests

In case of exploration in soil, Swedish weight sounding tests were performed and intrusive exploration in weathered rock slopes was done by portable dynamic cone penetration tests. The test results from portable dynamic cone penetration tests and their co-relations to the S-wave velocity for the sites in Muzaffarabad and Yokosuka are presented in Figures 10 & 11. Equations 1–6 have been used for co-relating the N_{sw} or N_d to the S-wave velocity.

5.3 Validation of weathered layer thickness

A good agreement can be see in Figure 12 in which the weathered layer thicknesses calculated by the intrusive and non-intrusive tests have been plotted. Authors believed that elucidated range of thickness of the weathered layer requires more exploration for robust conclusion.

6 CONCLUSIONS AND RECOMMENDATIONS

Field and investigations suggested that the thickness of weathered layer was of shallow extent ranges from 0.5 to 1.5 meter as detected by the dynamic cone penetration tests, Swedish weight sounding tests or seismic refraction profiles. The field shear wave velocity which can be associated with the weathered geomaterial ranges from 150–300 m/s.

Authors believe that the elucidated range of thickness and the shear wave velocity of the weathered geomaterial require more exploration for robust conclusions. Whilst selecting the adopted methodology the simplicity in interpretation and economy in performance was given with the foremost importance so that the benefit

517

of this research could be successfully extended to the developing countries.

Authors concluded that the slopes having the weathered layer thickness greater than the evaluated range or shear wave velocity lower than the demystified range are potentially unstable. As a matter of fact for highly jointed slopes, exposures should be protected against the rain water infiltration by smearing with impermeable layer on surface. For the earthwork operations at potentially unstable slopes careful attention should be given to develop the design criteria considering the aspects pointed out by the authors. Improving the slope angle, reducing the height of slope (benching) and rock bolting are some of the recommendations for such areas, but in case of developing countries with limited resources identification of potentially unstable slopes can also contribute towards the geotechnical risk assessment and mitigation. The in-situ information on the thickness and mechanical properties of weathered layer is useful tool for the slope failure risk assessment, which is determined by seismic refraction survey and been confirmed by the field intrusive tests and laboratory explorations. In future more exploration on the thickness and in-situ mechanical properties of weathered geomaterials will of interest to the authors.

REFERENCES

Borrelli, L., Greco, R & Gulla, G. 2006. Weathering grade of rock masses as a predisposing factor to slope instabilities: Reconnaissance and control procedures. *Geomorphology.*

Hachinohe, S., Hiraki, N., & Suzuki, T. 1999. Rates of weathering and temporal changes in strength of bedrock of marine terraces in Boso Peninsula, Japan. *Engineering Geology*, 55: 29–43.

Inada. 1960. Interpretation of Swedish sounding test results: 13–18. (in Japanese)

Palmer, D. 1986. *Refraction Seismics, Handbook of geophysical exploration*, 13, Geophysical press.

Shang, Y.J., Park, H.D., Yang, Z.F. & Zhang, L.Q. 2004 Debris formation due to weathering, avalanching and rock falling, landsliding in Se Tibet. *SINOROCK2004 Symposium; International Journal of Rock Mechanics and Mining Sciences* 41(3), Paper 3B 12, CD-ROM © Elsevier Ltd.

Towhata, I. 2008. *Geotechnical Earthquake Engineering*, Spinger-Verlag Berlin Heidelberg.

Prediction and Simulation Methods for Geohazard Mitigation – Oka, Murakami & Kimoto (eds)
© 2009 Taylor & Francis Group, London, ISBN 978-0-415-80482-0

Bridging satellite monitoring and characterization of subsurface flow: With a case of Horonobe underground research laboratory

Q. Li, K. Ito, Y. Tomishima & Y. Seki

National Institute of Advanced Industrial Science and Technology (AIST), Tsukuba, Japan

ABSTRACT: In this paper, we introduce the methodology to link satellite monitoring to characterization of deep groundwater flow. The JAEA's Horonobe URL (underground research laboratory) is considered as an experimental site. The corner reflector based permanent scatterer interferometric synthetic aperture radar (CR-PS-InSAR) technology is firstly used to monitor the earth's surface deformation, strongly associated with subsurface flow, of an URL during its construction. The proposed modeling methodology of coupled inversion is promising.

1 INTRODUCTION

In hydrogeology, accurate predication of groundwater flow and solute transport relies on detailed knowledge of spatial distribution of hydraulic parameters. However, the subsurface structure is naturally heterogeneous at different scales. To obtain detailed spatially distribution of hydraulic parameters for a field problem, a direct measurement method usually requires a large number of measurements at many different locations, which is high costly and impractical (Cai & Yeh 2008). On the other hand, direct aquifer responses (i.e., hydraulic head) and indirect aquifer responses (i.e., surface deformation) are relatively inexpensive and easy to measure in practice. Theoretically, these responses can be used to inverse the characterization of deep underground such as the spatial distribution of hydraulic parameters. However, the detailed aquifer characterization using responses requires large number of measurements, therefore, requires cost-effective data collecting techniques (Li et al. 2009). Up to now, some methods using direct or indirect aquifer responses have been conducted to characterize aquifer parameters (de Barros & Rubin 2008, Gish et al. 2002, Ito et al. 2004, Vasco et al. 2002, Zhu & Yeh 2005). As a novel data measuring technique, the satellite InSAR observation method is not far mature with application to characterization of subsurface flow, though latest satellite technologies have greatly improved the development of this new method. In this paper, we firstly use CR-PS-InSAR technique to infer the earth's surface deformation, and firstly use these measurements as indirect subsurface responses to conduct hydrogeological site characterization (Li & Ito 2008). The most innovation of our research program is to build a flexible modeling system to bridge CR-PS-InSAR measurements (and/or tilt data) with subsurface parameters (i.e., head and/or permeability) via a strong coupled inverse analysis.

JAEA's Horonobe URL site is chosen as an experimental field to implement our plan. As the product of the first stage of this plan, it is addressed in this paper. Section 2 introduces the Horonobe URL site. The methodology is addressed in Section 3. The preliminary results obtained from small amount of data in the early stage are concluded in Section 4.

2 HORONOBE URL SITE

The Japan Atomic Energy Agency (JAEA) is undertaking geologic investigations at Horonobe-cho, in northwestern Hokkaido, for the potential development of an underground research laboratory for research into geologic disposal of high-level radioactive wastes (HLW) in Neogene sedimentary host rocks. The Horonobe URL (Fig. 1) is being constructed toward a depth of approximately 500 m in a basined area (Kurikami et al. 2005). The research and development programs for this URL site include geologic and hydrochemical investigations of the deep geologic environment, developing techniques for planning and constructing repository facilities, and developing methodologies to confirm reliability of technologies for HLW disposal and to advance safety performance assessment. The whole Horonobe URL project is planned over about 20 years in three phases as follows:

Phase 1: Research activities directed from the ground surface including borehole investigations (for up to six years);
Phase 2: Investigations in construction of the underground facilities (for up to six years);

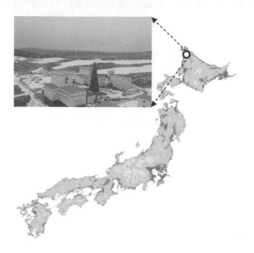

Figure 1. Site location of Horonobe URL. The DEM map of Japan was generated from SRTM data using ESRI ArcGIS. Photograph was taken on Nov. 30, 2008.

Phase 3: Investigations using specially excavated experimental facilities (for 9 to 11 years).

Phase 1 of the investigations began in 2001 and closed in 2005. Currently, Phase 2 of the construction of underground facilities is progressing.

3 METHODOLOGY

3.1 *PS-InSAR*

With the advancement of satellite and radar technologies, the satellite InSAR was recognized as powerful technique for indirect measurement of earth's surface changes since the early 1990s. Spaceborne SAR sensors have obvious merits, such as cloud-free and Day-and-Night observing ability, rather than optical sensors. While InSAR has the power to detect precise change of the earth's topography and surface deformation, it does have limitations. They mainly include temporal and geometrical decorrelation (low SNR in the phase change estimate), and variable tropospheric water vapor, which can generate variable phase delay due to the impact of water vapor on the propagation speed of microwave signals. Under such circumstances, the corresponding phase changes can be misinterpreted as surface change. The effects can be quite large, especially in tropical and subtropical regions. In tropical regions, up to 10 cm of variable path delay over several weeks has been reported (Dixon et al. 1991).

In the late 1990s, the multi-image Permanent Scatterer (PS) technique was introduced in an innovative way to deal with the aforementioned problems encountered in traditional InSAR processing (Ferretti et al. 1999). PS-InSAR exploits several characteristics of radar scattering and atmospheric decorrelation to measure surface displacement in otherwise non-optimum conditions. Atmospheric phase contributions are spatially correlated within a single SAR scene, but tend to be uncorrelated on time scales of days to weeks. Conversely, surface motion is usually strongly correlated in time. An example is surface subsidence caused by the underground excavation, which is usually steady over periods of months and sometimes years. Thus, atmospheric effects can be estimated and removed by combining data from long time series of SAR images, averaging out the temporal fluctuations. Radar scatterers that are only slightly affected by temporal and geometrical decorrelation are used, allowing exploitation of all available images regardless of imaging geometry. In this sense the scatterers are "permanent", i.e., persistent over many satellite revolutions (Kampes 2006). The detailed algorithm for our PS-InSAR processing will be addressed in another article.

3.2 *Corner reflector (CR)*

To improve the radar signal return to the spaceborne SAR sensors, corner reflectors are usually installed in practice. The reflectors are trihedral shaped and made of aluminum (Fig. 2). The trihedral design ensures that the radar signal is returned exactly in the incident direction and with the same polarity. The size of the corer reflector is proportional with the quality of the signal strength and implicitly with the quality of the measurement. The minimum size of the reflectors is a function of the SAR sensor wavelength and of the expected strength of the natural radar targets (e.g., rocks, buildings). The corner reflector signal should dominate all the other reflections located in the immediate vicinity. The orientation of the corner reflector is perpendicular to the radar line of sight. This needs a very delicate design and installation (Froese et al. 2008). The total five corner reflectors

Figure 2. Photo of a corner reflector installed near the borehole HDB-8.

installed in the Horonobe URL site were designed to properly work with Canada RADARSAT-2 by Japan ImageONE Co., Ltd.

One of our research aims is to exploit the temporal and spatial characteristics of interferometric signatures collected from CR point targets to accurately map surface deformation histories and terrain heights of the Horonobe URL site, especially around the boreholes. Technically speaking in a word, the CR-PS-InSAR analysis can be summarized as an iterative improvement of the model parameters to achieve an optimal match to the observed interferometric phases. The detailed results will be reported in our latter publications.

3.3 Surface subsidence

The deep underground excavation, such as construction of URLs, usually causes long-term surface subsidence. This kind of subsidence is usually groundwater-related. The excavation of shafts and tunnels requires drainage. The drained geomaterials consolidate as a result of increased effective stress. In general, any deformation of the earth's surface has been viewed as deleterious, an undesirable side effect of the extraction (e.g., oil and gas) or injection (e.g., CO_2) of fluids (Li et al. 2006). Correspondingly, much effort has been devoted to understanding and predicting the consolidation and compaction induced by increased effective stress of geomaterials. The consolidation process has two major matured solving methods. Biot's theory is used to solve coupled single phase consolidation problems (Biot 1941). Lewis & Schrefler (1998) developed Biot's theory to solve uncoupled multiphase consolidation problems. If interested, the detailed background of theory and the numerical solutions of partial differential equations can be referred in the books (Lewis & Schrefler 1998, Wang 2000).

The processes of deformation and flow associated with subsidences are typically viewed in isolation, although a coherent thread links the two processes. Isolated description of individual events (e.g., deformation) can be justified only when the single event becomes relatively dominant in isolation. As individual processes, momentum must be conserved for deformation, and mass or energy must be conserved for flow. If the systems are coupled, the individual conservation laws must be jointly and simultaneously satisfied for such multiple processes. (Bai & Elsworth 2000).

3.4 Coupled inversion

Technically speaking, most geoscience problems can be regarded as inverse analyses. The subsurface characterization of flow associated with surface deformation is such an inverse problem. Although many investigations have focused on the mechanism and the

numerical solution of consolidation and compaction (Chilingarian et al. 1995, Lewis & Schrefler 1998, Wang 2000), littler research has focused on the utility of InSAR inferred surface deformation in furthering the understanding of the characterization of flow and transport in the deep subsurface (Galloway & Hoffmann 2007, Massonnet et al. 1997, Vasco et al. 2001, Vasco et al. 2008). In this paper, we propose a CR-PS-InSAR based coupled inverse modeling to study the subsurface flow related with surface deformation. The starting point of the mathematical-physical formulations is Biot's equations (Biot 1941, Biot 1956), which well established the relationship between surface deformations and subsurface flow changes. It should be noted that there may be departures from ideal poroelasticity, and many modifications of Biot's theory have been proposed (Wang 2000). However, Biot's equations (Biot 1941) are still the easy to use and fast-to-implement formulations for most investigations of the deformation over a poroelastic target, e.g., reservoir.

Our primary goal is the integration of CR-PS-InSAR based surface deformation data and hydromechanical coupling solution of Biot's poroelasticity to map the subsurface flow with providing new insights into underground stabilities during the construction of Horonobe URL. The flowchart of the integration is depicted in Figure 3. The primary quantities of interest of the coupling equations are the displacement, $\mathbf{U}$, and the pore pressure, P (Fig. 3). The primary output of the CR-PS-InSAR processing module is the surface deformation (Fig. 3). The governing equations, which are often adopted in practice, of a deforming poroelastic medium are the results of Rice & Cleary's study (Rice & Cleary 1976).

The general constitutive relations for an isotropic poroelastic medium can be expressed by the following equations (Biot 1941, Rice & Cleary 1976):

$$2G\varepsilon_{ij} = \sigma_{ij} - \frac{\nu}{1+\nu}\sigma_{kk}\delta_{ij} + \frac{3(\nu_u - \nu)}{B(1+\nu)(1+\nu_u)}p\delta_{ij}$$

(1)

$$m = m_0 + \frac{3\rho_0(\nu_u - \nu)}{2GB(1+\nu)(1+\nu_u)}\left(\sigma_{kk} + \frac{3}{B}p\right)$$ (2)

which involve four elastic constants: G (shear modulus), ν (Poisson ratio), B (Skemptons coefficient), ν_u (undrained Poisson ratio). p is the pore pressure. σ_{ij} is the total stress. ε_{ij} is the strain. m is the pore fluid mass per unit volume. ρ_0 is the fluid density. The detailed discussion of these material constraints are commented by Rice & Cleary (1976).

In this research, we are interested in quasi-static deformation in which we may neglect inertial terms in the equation of equilibrium. Then, we get the stresses in terms of displacements and the change in pore fluid

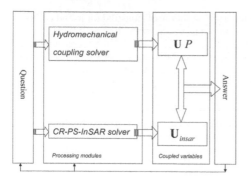

Figure 3.   Flowchart of coupled inversion modeling system.

mass per unit volume by solving (1) and (2) for σ_{ij} in terms of ε_{ij} and Δm (i.e., $m - m_0$).

$$\sigma_{ij} = G\left(\frac{\partial u_i}{\partial x_j} + \frac{\partial u_j}{\partial x_i}\right) + \lambda_u \frac{\partial u_k}{\partial x_k}\delta_{ij} - \frac{BK_u}{\rho_0}\Delta m \delta_{ij} \quad (3)$$

where λ_u is the undrained Lame constant, and K_u is the undrained bulk modulus. Conclusively, we obtain a partial differential equation for the displacements (Vasco et al 2002).

$$\frac{\partial}{\partial x_j}\left[G\left(\frac{\partial u_i}{\partial x_j} + \frac{\partial u_j}{\partial x_i}\right)\right] + \frac{\partial}{\partial x_j}\left(\lambda_u \frac{\partial u_k}{\partial x_k}\right)\delta_{ij}$$

$$= \frac{\partial}{\partial x_j}\left(\frac{BK_u}{\rho_0}\Delta m\right)\delta_{ij} \quad (4)$$

On the other hand, we assumed that the constitutive relations governing pore fluid diffusion obey Darcy's law. Given that the initial pore fluid mass per unit volume is constant and assumed that the movement of the solid phase is much smaller than the motion of the pore fluid, the conservation of the pore fluid mass can be described by the following equation.

$$\nabla \cdot \left(\rho_0 \frac{k}{\mu_f}\nabla p\right) + \frac{\partial \Delta m}{\partial t} = \rho_0 Q \quad (5)$$

where k is the permeability, μ_f is the fluid viscosity, and Q is a source term. This equation shows the relationship between the pressure and the time change of the pore fluid mass per unit volume. Note that for steady state conditions, the time derivative vanishes and Δm does not appear explicitly in (5). This is in accordance with the observation that coupling only occurs under time-transient conditions (Lewis & Schrefler 1998, Vasco et al 2001).

As noted above, one of our goals is to build a strong inversion modeling system to integrate the satellite observations and other measurements in order to characterize the properties, such as permeability, of subsurface reservoirs. By (2) and (4), we can estimate Δm, the change in fluid mass per unit volume in the

subsurface. Then, by solving (2) for the pore pressure, we can map our estimated Δm into subsurface pressure. On the basis of this derived pressure variation, we may use (5) to infer variations in permeability. Because Δm enters (4) on the right-hand-side as a source term, the inverse problem is linear. The solution of the inverse problem for Δm can be described as follows:

$$u_i(\mathbf{x}, t) = \frac{B}{\rho_0}\int_V g_i(\mathbf{x}, \mathbf{y})\Delta m(\mathbf{y}, t)dV \quad (6)$$

in which $g_i(\mathbf{x}, \mathbf{y})$ is the Green's function solution of (4), i.e., $g_i(\mathbf{x}, \mathbf{y})$ is the solution of (4) for a delta function source term, located as position $\mathbf{y}$, and $g_i(\mathbf{x}, \mathbf{y})$ may be derived analytically for simple models such as a homogeneous half-space. Alternatively, we many appropriate $g_i(\mathbf{x}, \mathbf{y})$ for arbitrary poroelastic media by solving (4) using a numerial technique such as finite differences (Vasco et al 2002) or finite elements (Li & Ito 2008).

Furthermore, the basic datum in InSAR is the change in range over some time interval. If the earth's surface deforms during this period, the accumulated displacements of the imaging elements are projected onto the range vector which points toward the satellite. Thus, the change in distance to the satellite, $\Delta \chi$, that we seek can be written as:

$$\Delta \chi(\mathbf{x}) = u_i \cdot l_i \quad (7)$$

in which l_i is a unit range vector. With the similar numerical processing, we can get the Green's function solution of $\Delta \chi$ as follows:

$$\Delta \chi(\mathbf{x}) = \int_V r(\mathbf{x}, \mathbf{y})\psi(\mathbf{y})d\mathbf{y} \quad (8)$$

in which

$$r(\mathbf{x}, \mathbf{y}) = l_i g_i(\mathbf{x}, \mathbf{y}) \quad (9)$$

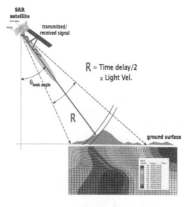

Figure 4.   Illustration of coupled inversion analysis of subsurface flow and pore volume change.

is the projection of the displacement Green's functions onto the range vector. The components of the range vector are known from the geometry of the satellite orbit. $\psi(\mathbf{y})$ is stress-free volume strain. The numerical processing of the aforementioned inverse problems can be treated as an optimization problem. The detailed solution of this optimization problem will be discussed elsewhere.

Conclusively, the solution of CR-PS-InSAR based coupled inversion of poroelastic reservoirs can be simply envisioned in Figure 4.

4 CONCLUSION

The modeling methodology is proposed in characterizing subsurface flow problems with CR-PS-InSAR observations during construction of the Horonobe URL for HLW disposal research. The coupled inversion, other than direct hydraulic inversion, of subsurface flow and surface deformation can easily incorporate multiple coupling effects into history matching during evolvement of natural systems. The potential applications of this integrated modeling system are envisioned to cover three major geoscience subjects as follows: 1) assessment of subsurface flow and parameter identification of aquifers, 2) site assessment of nuclear repositories, 3) reservoir monitoring and characterization, such as the new cases of geological sequestration of CO_2.

REFERENCES

Bai, M. & Elsworth, D. 2000. *Coupled processes in subsurface deformation, flow and transport*. Reston, VA, USA: ASCE.
Biot, M.A. 1941. General theory of three-dimensional consolidation. *Journal of Applied Physics* 12(2): 155–164.
Biot, M.A. 1956. Theory of deformation of a porous viscoelastic anisotropic solid. *Journal of Applied Physics* 27(5): 459–467.
Cai, X. & Yeh, T.-C.J. (eds.) 2008. *Quantitative information fusion for hydrological sciences*, New York: Springer.
Chilingarian, G.V., et al. (eds.) 1995. *Subsidence due to fluid withdrawal*, Amsterdam: Elsevier.
de Barros, F.P.J. & Rubin, Y. 2008. A risk-driven approach for subsurface site characterization. *Water Resources Research* 44: W01414.
Dixon, T.H., et al. 1991. First epoch geodetic measurements with the global positioning system across the northern Caribbean plate boundary zone. *Journal of Geophysical Research* 96(B2): 2397–2415.
Ferretti, A., et al. 1999. Persistent scatterers in SAR interferometry. *IEEE 1999 International Geoscience and Remote Sensing Symposium: Remote Sensing of the System Earth—A Challenge for the 21st Century, Congress Centrum Hamburg, Germany, 28 June–2 July, 1999*.
Froese, C.R., et al. 2008. Characterizing complex deep-seated landslide deformation using corner reflector InSAR (CR-InSAR): Little smoky landslide, Alberta. *Proceedings of the 4th Canadian Conference on Geohazards: From Causes to Management, University Laval, Quebec City, Quebec, Canada, May 20–24, 2008.*
Galloway, D.L. & Hoffmann, J. 2007. The application of satellite differential SAR interferometry-derived ground displacements in hydrogeology. *Hydrogeology Journal* 15(1): 133–154.
Gish, T.J., et al. 2002. Evaluating use of ground-penetrating radar for identifying subsurface flow pathways. *Soil Science Society of America Journal* 66(5): 1620–1629.
Ito, K., et al. 2004. Hydrogeological characterization of sedimentary rocks with numerical inversion using vertical hydraulic head distribution: An application to Horonobe site. *Journal of the Japan Society of Engineering Geology* 45(3): 125–134.
Kampes, B.M. 2006. *Radar interferometry: Persistent scatterer technique*. Dordrecht, The Netherlands: Springer.
Kurikami, H., et al. 2005. Analytical study on groundwater flow characteristics of sedimentary rocks in Horonobe. *Abstracts of Japan Earth and Planetary Science Joint Meeting 2005, International Convention Complex, Chiba Prefecture, Japan, May 22–26, 2005.*
Lewis, R.W. & Schrefler, B.A. 1998. *The finite element method in the static and dynamic deformation and consolidation of porous media*. London: Wiley.
Li, Q. & Ito, K. 2008. An integrated thermal-hydraulic-mechanical-chemical-biological multiscale and multi-physics coupled system with applications to geological disposal problems. *AIST Symposium: The Annual Conference of GREEN 2008, Akihabara Convention Hall, Tokyo, Japan, November 20, 2008.*
Li, Q., et al. 2006. Thermo-hydro-mechanical modeling of CO_2 sequestration system around fault environment. *Pure and Applied Geophysics* 163(11–12): 2585–2593.
Li, Q., et al. 2009. Coupling and fusion in modern geosciences. *Data Science Journal* (in press).
Massonnet, D., et al. 1997. Land subsidence caused by the East Mesa geothermal field, California, observed using SAR interferometry. *Geophysical Research Letters* 24(8): 901–904.
Rice, J.R. & Cleary, M.P. 1976. Some basic stress diffusion solutions for fluid-saturated elastic porous media with compressible constituents. *Reviews of Geophysics and Space Physics* 14(2): 227–241.
Vasco, D.W., et al. 2001. A coupled inversion of pressure and surface displacement. *Water Resources Research* 37(12): 3071–3089.
Vasco, D.W., et al. 2002. Geodetic imaging: Reservoir monitoring using satellite interferometry. *Geophysical Journal International* 149(3): 555–571.
Vasco, D.W., et al. 2008. Reservoir monitoring and characterization using satellite geodetic data: Interferometric synthetic aperture radar observations from the Krechba field, Algeria. *Geophysics* 73(6): WA113–WA122.
Wang, H.F. 2000. *Theory of linear poroelasticity with applications to geomechanics and hydrogeology*. Princeton: Princeton University Press.
Zhu, J.F. & Yeh, T.-C.J. 2005. Characterization of aquifer heterogeneity using transient hydraulic tomography. *Water Resources Research* 41(7): 10.

Prediction and Simulation Methods for Geohazard Mitigation – Oka, Murakami & Kimoto (eds)
© 2009 Taylor & Francis Group, London, ISBN 978-0-415-80482-0

Validation of a numerical model for the fault-rupture propagation through sand deposit

M. Rokonuzzaman & T. Sakai
Mie University, Mie, Japan

A. El Nahas
Haskoning UK Ltd., UK

ABSTRACT: Over the last few decades the researchers were mainly concentrated towards the understanding of the mechanisms due to the earthquake shaking. The effect of the permanent offset along the fault ruptures on the overlaying deposits and the structures on or near to it in some recent earthquake events has emerged the importance of proper understanding these phenomena. In this study a sophisticated numerical technique was validated in the process of (a) calibration of the constitutive model parameters from model direct shear tests (b) comparing its predictions with centrifuge experiments for normal and reverse faults in free field condition in terms of normalized vertical displacements profile of the ground surface, minimum relative vertical base displacement for the rupture to reach the ground, the average dip angle propagated into the soil as well as the horizontal extent of the deformed surface ground, and, then, the increased confidence on the proposed model inspired (c) to apply it in understanding the mechanism for the interaction problems of normal fault and rigid raft foundation.

1 INTRODUCTION

The movements on the fault surface in the dip direction (Dip Slip) cause permanent soil deformations in the overlaying alluvium. Besides the earthquakes shaking that might occur due to the dip slip, the resulting permanent soil displacements in the overlaying alluvium, subject the geotechnical structures in and above the alluvium to additional stresses. The resulting strains or displacements for the structure may violate its serviceability requirements, or worse, cause structural collapse. To provide design guidelines for the facilities in the seismically active regions requires a better understanding of the sequence of failure events during fault rupture, the associated ground deformation pattern, as well as the interaction of an existing structure with these mechanisms and soil deformation patterns.

The main goals of the past researches were to identify the failure pattern in the alluvium and the height of fault rupture in the model ground, as well as to find general criterion for surface faulting (Bray et al. 1994a, b & Scott et al. 1974). The field case studies investigated the failure mechanisms in the soil during a fault rupture event and the effects of the resulting ground movements and deformations, on the exiting structures across or near the fault rupture plane (Anastasopoulos 2005). Soil behavior after failure has been shown to play a major role in problems related to shear-band formation and propagation.

Scott and Schoustra 1974, applying the FE method in combination with elastic-perfectly plastic constitutive soil model with Mohr-Coulomb failure criterion, produced results contradicting both reality and experiments. In contrast, Bray et al. 1994a, b also utilizing the FE method, but with a hyperbolic non-linear elastic constitutive law achieved satisfactory agreement with small-scale tests. Most recently, Anastasopoulos et al. 2007, has employed similar kind of constitutive model with isotropic strain softening- linearly reducing the mobilized frictional angle and dilation angle with the increase of octahedral plastic shear strain. Walters & Thomas 1982 performed sandbox experiment and conducted numerical simulation of their experiment by FEM. They found that non-associated flow rule and proper strain softening were essential in localization of rupture. In their FE analyses, rupture propagated through the sand and broke the ground surface with only a fraction of the displacement observed in experiments. Tani 1994 performed sandbox tests and FE analyses. He showed the importance of proper modeling the discontinuous behavior of failure surface in analyzing the post failure process as well as the process before rupture. The previous numerical models used for the analysis of fault rupture propagation through overlaying soil beds were not accurate in modeling shear band effect, mesh size effect and confining pressure effect, which shows the necessity for a sophisticated numerical model.

The current study centrifuge modeling of dip fault events (reverse and normal) for a dip angle of 60° in free field condition, as well as, light and heavy rigid strong raft foundation placed on the normal faults were conducted and used to validate a sophisticated numerical modeling, comparing the observed failure mechanisms from tests and numerical results.

2 TESTING PROCEEDURES

For the experiment of fault rupture propagation through Fontainebleau sand ($D_{50} = 0.24$ mm, $U_0 = 1.33$, $G_s = 2.59$, $e_{max} = 0.833$, $e_{min} = 0.55$, fines content = 0%) deposit, University of Dundee's beam centrifuge was used. The strong box internal model dimensions are 800 mm × 500 mm × 500 mm (Figure 1), with a front and back transparent Perspex plates, through which the models were monitored during the tests. Two hydraulic cylinders were used to push the hanging or right part up or down to simulate reverse and normal faulting. A central guidance (G) and three wedges (A_1–A_3) were used to guide the imposed displacement at the desired dip angle (60°, Figure 1). Sand was pluviated in the strong box on 20–30 mm thick layers to fill up to 217 mm. On top of each layer, a line of dyed sand was laid behind each Perspex wall to clearly visualize the shear bands. The corner and internal cans were placed to verify the sand unit weights inside the strong box and near the edges at the bottom and in the middle of the model ground. A series of digital images were taken for the displaced model ground after each stepwise fault dip slip of about 0.5 to 1.5 mm till to the total machine allowable dip slip or maximum vertical component of base dislocation, h_{max}, shown in Table 1. In addition, linearly variable differential transformers (LVDTs) were used to monitor the vertical settlements of the model ground surface, and the vertical component of the base dislocation (h, and all definitions of the physical model used in this work

Table 1. Prototype dimensions and basic parameters for the centrifuge models (centrifuge acceleration = 115 g).

Test name (fault type)	D_r (%)	H (m)	L (m)	W (m)	h_{max} (m)	q (kPa)
Test 8 (Reverse)	60.9	25.3	75.9	24.15	2.56	0
Test12_R2 (Normal)	60.2	24.6	75.9	24.15	−3.15	0*
Test15 (Normal)	59.2	24.9	75.9	24.15	−3.15	37**
Test14_R (Normal)	62.5	24.6	75.9	24.15	−2.89	91**

*, ** fault rupture in free field and 10 m wide strip foundation centered at the scarp position in the free field test, respectively.

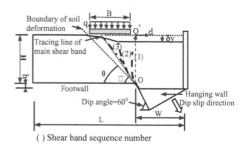

( ) Shear band sequence number

Figure 2. Definitions of the physical model with raft (normal fault with raft above the ground surface, not to scale).

are shown in Figure 2). Two model strip footings were used (shown by R in Figure 1) to understand the fault and structure interaction. Each had dimensions of 87 mm × 500 mm × 10 mm, with sand paper sheets (No. 100) glued on their bottom surfaces to create rough base condition. In Tests 14_R and 15, the footings were made of steel and aluminum, respectively, to have foundation bearing pressures (q, Figure 2) of 91 and 37 kPa, respectively. The prototype dimensions and parameters used in the all experiments are given in Table 1. After the model preparations, the strong box was mounted on the centrifuge, and the centrifuge was spun to 115 g-level in this study. Next, the fault movements were made stepwise. The detailed experimental set up and procedures can be found in El Nahas et al. 2007.

3 NUMERICAL MODEL

3.1 Formulations

This FE model uses an elasto-plastic framework with non-associated flow rule and strain-hardening/softening law. A yield function corresponding to the Mohr-Coulomb model and a plastic potential

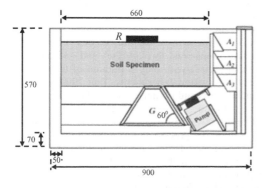

Figure 1. Basic dimensions of the experimental apparatus installed in the Dundee University centrifuge (all dimensions are in mm, not to scale).

function geometrically represented by the Drucker-Prager model are employed. The modeling of the materials having softening properties is full of serious difficulties both in modeling strain localization and from the view point of numerical analysis. The straightforward use of the material softening model in a classical continuum, generally, does not result in a well-posed problem. The standard finite element solution of strain localization in a rate-dependent material result in solutions those are strongly mesh-sensitive. Here, objectively, the shear band effect is introduced into the constitutive equation. An explicit dynamic relaxation method combined with generalized return-mapping algorithm is applied to the integration algorithms. The details of the numerical model are referred in Rokonuzzaman et al. 2008a due to space limitation.

3.2 Calibration

To use the numerical model for the analyses of the fault problems, it is necessary to calibrate the material parameters for hardening-softening (ε_f, ε_r and m) and stress-dilatancy relationship (ε_d and β) of the incorporated constitutive model. For this purpose, the model experimental data from direct shear (DS) tests are used, as DS tests closely mimic the shearing in the faults. In Rokonuzzaman et al. 2008a, the calibration process of the proposed numerical model is discussed in details. The calibrated material parameters are given in the Table 2.

3.3 FE discretization and analysis techniques of fault problems

The elements employed for the analyses are four noded quadrilateral Lagrange type with reduced integration. Linear elastic elements are used for the footing with Young's modulus of 2.1×10^4 MPa and Poisson's ratio of 0.3. No particular interface elements are used in this study, since no displacement discontinuity was observed between the rough footing base and sand in the experiment (Rokonuzzaman et al.

Table 2. Material parameters of the numerical model.

Density (kN/m³)	15.57
Initial void ratio (e_0)	0.64
Initial earth pressure coefficient (K_0)	0.5
Coefficient of shear modulus (G_0, kPa)	50
Residual friction angle (ϕ_r :⁰)	30.2
Poission's ratio (V)	0.3
Shear band thickness	3.84
(SB: mm, model scale)	
ε_f	0.2
ε_r	0.6
ε_d	0.3
m	0.8

2008c). The finite element discretization is displayed in Figure 3. It refers to a uniform soil deposit of thickness, H, at the base of which a reverse fault, dipping at angle of 60° (measured from the horizontal), ruptures and produces upward displacement, with a vertical component, h. Following the recommendation of Bray 1990 and to minimize undesired boundary effects, the width, L, of the FE model was set equal to 4 H. The discretization is finer in the central part of the model with the quadrilateral elements than those at the two edges where limited deformation is expected. The differential quasi-elastic displacement is applied to the right part of the model (hanging-wall) in small consecutive increments (upward for reverse faulting and downward for normal faulting).

The all experiments were conducted at 115 g centrifuge acceleration. The sand ($D_{50} = 0.24$ mm), modeled in the centrifuge sand box, corresponds to a prototype material with mean particle size diameter equals to nD_{50} (where, n is scale factor). So, the shear band thickness used for 115 g (n = 115) model tests FE analyses is 115×16 d$_{50}$ (≈ 441.5 mm).

3.4 Issue of mesh-size effects

Such a numerical model incorporating hardening-softening model must be verified before to apply to the real world fault problems, as strain softening makes the analysis sensitive to mesh size. For this purpose, finite elements of sizes: 1 m × 1 m (width × height), 1.5 m × 1.5 m, 2 m × 2 m, respectively, are used in the central part. At the two edges, 2 m × 1 m, 3 m × 1.5 m, 4 m × 2 m, respectively, are used (Figure 3). For the all numerical analyses in this study, the used model parameters are shown in Table 2. The results are compared with an arbitrary set experimental result (Test 8). Figure 4 shows that the vertical displacements (δy) are mesh size independent (where, d is measured from the point of application of the base dislocation, Figure 2). The orientation of the progressive path is less affected by the mesh size: reducing the size of mesh leads slight shifting of out cropping location towards the foot wall (Rokonuzzaman et al. 2008a, b). So, the inclusion of shear band effect into the constitutive relation makes the numerical model insensitive to mesh size.

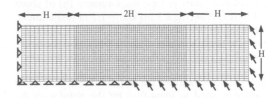

Figure 3. Finite element mesh and boundary conditions (reverse fault in free field).

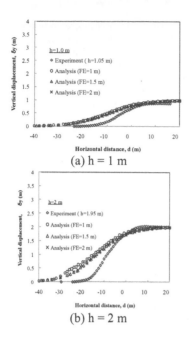

(a) h = 1 m

(b) h = 2 m

Figure 4. Analysis of sensitivity of mesh density (compared with Test 8) for vertical displacements of ground surface.

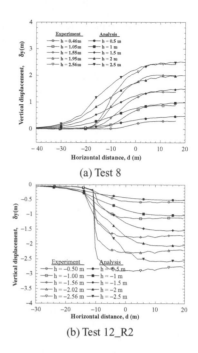

(a) Test 8

(b) Test 12_R2

Figure 5. Comparison of experimental and numerical normalized vertical displacements of ground surface for (a) Test 8 and (b) Test 12_R2.

4 RESULTS AND DISCUSSIONS

4.1 Faults in free field (Test 8 & Test12_R2)

Figures 5a, b show that the experimental normalized vertical displacements on the ground for Test 8 (reverse fault) and 12_R2 (normal fault) are closely predicted by the FE analyses for absolute base dislocations: h = 0.5 to 2.5 m, respectively. The progressive development of failure mechanism in the model ground during faulting in free field are discussed in Rokonuzzaman et al. 2008a, b. Here, the model ground images and the corresponding FE deformed mesh with average shear strain plot after failure are shown in Figures 6a, b (Test 8) and c, d (Test 12_R2), showing the sequence number of shear band formation. The shear bands are drawn by naked eye observation from the deformed colored sand layers. The strain localization in narrow shear bands start after h = 1.2 m and h = −0.4 m in reverse and normal fault, respectively. The fault rupture propagated along the full depth of the model ground after vertical base displacement (h) of about 2.8 m (experiment: 2.5 m) and −1.1 m (experiment: −1.0 m) in reverse and normal fault, respectively. Now, for the calculation of the inclination of the fault's main slip surface in the model ground to the horizontal (Δ, Figure 2), an operative definition (Yilmaz et al. 2007) of the location of the fault rupture at the ground surface is used to identify it by the point with the maximum absolute value of the second derivative of the vertical displacement along the horizontal direction. The average dip angles, Δs, are 55° (experiment: 60°) and 61° (experiment: 65°) in reverse and normal fault, respectively. Additionally, outside the main shear band, some soil deformation took place, and that led to ground surface inclination, as illustrated in Figure 6. The inclinations of the line between the faults rupture point and the ends of the soil deformations at ground surface at the end of each test (θ_{max}, Figure 2) are 35° (experiment: 44°) and 47° (experiment: 56.1°) in reverse and normal fault, respectively. The above discussed results for the normal fault (Test12_R2) and reverse fault (Test 8), collected from Rokonuzzaman et al. 2008a, b are summarized in Table 3 to compare the failure mechanisms. It shows: (a) the shear band outcrops faster in normal faults than that in reverse faults, (b) the inclination of the main shear band to the horizontal is equal or less than dip angle in reverse faults but larger in normal faults, and (c) the extension of the soil deformation on the ground surface in the normal faults is less than that in reverse faults.

4.2 Faults with raft foundation (Test15 & Test14_R)

In this present study, ordinary sand elements are used adjacent to the contact surface, following the method that is described in Kotake et al. 1999. It is shown that

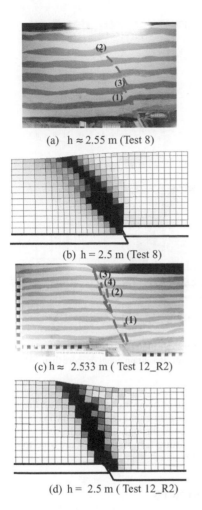

(a) h ≈ 2.55 m (Test 8)

(b) h = 2.5 m (Test 8)

(c) h ≈ 2.533 m (Test 12_R2)

(d) h = 2.5 m (Test 12_R2)

Figure 6. Comparison of experimental ground images and numerical shear strain plots on deformed mesh for Test 8 (a and b) and Test 12_R2 (c and d) (the darkest color indicates shear strain equal or more than 20%).

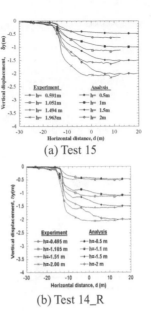

(a) Test 15

(b) Test 14_R

Figure 7. Comparison of experimental and numerical vertical displacements of ground surface for (a) Test 15 with light raft and (b) Test 14_R with heavy raft.

Table 3. Vertical base displacement, propagated dip angle and extension of soil deformation when fault rupture reached the ground surface.

Test name	h (m) Exp.	h (m) FEM	Δ (°) Exp.	Δ (°) FEM	θ_{max} (°) Exp.	θ_{max} (°) FEM
Test 8	2.5*	2.8	60	55	44	35
Test 12_R2	−1.0	−1.1	65	61	56.1	47
Test 15	−1.0	−1.1	63	57.8	54	48.8
Test 14_R	−0.9	−1.1	62.8	57.3	58.2	49.2

*The shear band propagated up to 93.1% of the soil depth (H) at h = h_{max} in the centrifuge, having some lighting problems near ground surface.

interface elements with friction angle of 6° closely predict the footing tilting and others (Rokonuzzaman et al. 2008c). Such interface elements are also not susceptible to thickness due to the shear bands included in the constitutive mode and will be used in the all analyses. The rafts (B = 10 m, Figure 2) are centered at the scarp position confirmed in the free field test (Test12_R2). Figure 7 shows that the profiles of vertical displacements on the ground surface from experiment for base dislocations, h = −0.5 to −2.0 m, are, satisfactorily, predicted by the numerical model in Tests 15 and 14_R with light (q = 37 kPa) and heavy (q = 91 kPa) rigid raft, respectively. Figure 8 show some characteristic ground photographs taken during the experiments and the numerically obtained deformed FE mesh with

shear strain distribution. The rafts (light or heavy) do not affect the amount of base dislocation required for the main shear band to outcrop on the ground surface as in all the cases the shear bands cross the model ground after h = −1.1 m. However, the foundations divert the shear band so as to reach the ground surface outside. In the both cases, a shear band radiates downward from the ground surface and the corner of the raft, and join the another one coming upward from the fault to make the main shear band outcrops, see Figures 8a, b. This phenomenon is more clearly observed in fault with heavier foundation, see Figures 8b, c. The deviation of the main shear band may prevent damages to the structure to some extent. The average dip angles,

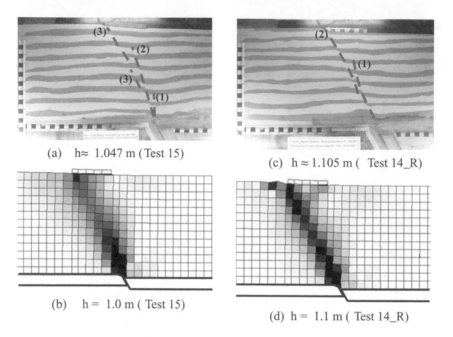

(a) h≈ 1.047 m (Test 15)

(c) h ≈ 1.105 m (Test 14_R)

(b) h = 1.0 m (Test 15)

(d) h = 1.1 m (Test 14_R)

Figure. 8   Comparison of experimental ground images and numerical shear strain plots on deformed mesh for Test 15 (a and b) with light raft and for Test 14_R (c and d) with heavy raft (darkest region indicates strain equal or more than 20%).

Δs, are 57.8° (experiment: 63°) and 57.3° (experiment: 62.8°) with light and heavy raft, respectively. The average dip angles at the presence of foundations are less than that in free field. The horizontal extents of the deformed surface ground (θ_{max}) are 48.8° (experiment: 54°) and 49.2° (experiment: 58.2°) with light and heavy raft foundation, respectively. The results discussed above are summarized in Table 3.

5   CONCLUSIONS

This paper has validated a numerical technique for predicting the failure mechanisms of reverse and normal faults in free field condition and normal fault-raft foundation interaction, comparing the numerical results and extensive experimental results and the following conclusions have been drawn:

1. in the reverse fault rupture event, the strain localization in narrow shear bands start slow and the main shear band outcrop also later than that in a normal fault rupture events.
2. the horizontal extent of the associated soil deformations in the reverse faults is more than that associated with the normal fault rupture events.
3. in the normal fault rupture event, the inclination of that main shear band to the horizontal is more than the dip angle, while it is equal or less in reverse faults.

4. the light or heavy raft does not affect the amount of fault displacement required to propagate the main shear band through the sand deposit.
5. rigid strong raft foundations are able to divert the fault rupture shear bands.

REFERENCES

Anastasopoulos, I., Gazetas, G., Bransby, M.F., Davies, M.C.R. & El Nahas, A. 2007. Fault rupture propagation through sand: finite element analysis and validation through centrifuge experiments. *J. Geotech. Geoenv. Eng.* 133(8): 943–958.

Anastasopoulos, I. 2005. Behaviour of foundations over surface fault rupture: analysis of case histories from the Izmit (1999) earthquake. *Proc. of the 16th International Conference on Soil Mechanics and Geotechnical*, Japan, 2623–2626.

Bray, J.D., Seed, R.B., Cluff, L.S. & Seed, H.B. 1994a. Earthquake fault rupture propagation through soil. *J. Geotech. Eng.* 120(3): 543–561.

Bray, J.D., Seed, R.B. & Seed, H.B. 1994b. Analysis of earthquake fault ruptures propagation through cohesive soil. *J. Geotech. Eng.* 120(3): 562–580.

Bray, J.D. 1990. *The effects of tectonic movements on stresses and deformations in earth embankments.* Ph.D. thesis, University of California, Berkeley.

El Nahas, A., Bransby, M.F. & Davies, M.C.R. 2006. Centrifuge modelling of the interaction between normal fault rupture and rigid, strong raft foundations. *Proc. International Conference on Physical Modelling in Geotechnics*, Hong Kong, August, 317–323.

Kotake, N., Tatsuoka, F., Tanaka, T., Siddiquee, M.S.A. & Yamauchi, H. 1999. An insight into the failure of reinforced sand in plane strain compression by FEM simulation. *Soils and Foundations* 39(5): 103–130.

Rokonuzzaman, M., Sakai, T., El Nahas, A., Tanaka, T. & Hossain, M.Z. 2008a. Experimental validation of a numerical model: reverse fault rupture propagation through sand. *Submitted to JSCE*.

Rokonuzzaman, M., Sakai, T., El Nahas, A., Tanaka, T. & Hossain, M.Z. 2008b. Experimental validation of a numerical model: normal fault rupture propagation through sand deposit. *Submitted to JSCE*.

Rokonuzzaman, M., Sakai, T., El Nahas, A., Tanaka, T. & Hossain, M.Z. 2008c. Centrifuge and numerical modeling of normal fault rupture-sand deposit-raft foundation interaction, *Submitted to JSCE*.

Scott, R.F. & Schoustra, J.J. 1974. Nuclear power plant sitting on deep alluvium. *J. Geotech. Eng.* 100(4): 449–459.

Tani, K., Ueta, K. & Onizuka, N. 1994. Scale effect of Quaternary ground deformation observed in model tests of vertical fault. *Proc. of 29th Japan National Conference on SMFE*, Japan, 1359–1362 (In Japanese).

Walters, J.V. & Thomas, J.N. 1982. Shear zone development in granular materials. *Proc. 4th Intl Conf. Num Methods in Geomech.*, Canada, 263–274.

Yilmaz, M.T. & Paolucci, R. 2007. Earthquake fault rupture-shallow foundation interaction in undrained soils: a simplified analytical approach. *Earthquake Eng. Struct. Dyn.* 36: 101–118.

Prediction and Simulation Methods for Geohazard Mitigation – Oka, Murakami & Kimoto (eds)
© 2009 Taylor & Francis Group, London, ISBN 978-0-415-80482-0

A research on the quantitative evaluation of slope stability during rainfall

K. Sako
Ritsumeikan University, Shiga, Japan

R. Kitamura
Kagoshima University, Kagoshima, Japan

T. Satomi & R. Fukagawa
Ritsumeikan University, Shiga, Japan

ABSTRACT: A lot of lives, infrastructures and heritages were lost due to slope failure. Therefore, it is necessary to construct a prevention system in order to predict the probability of slope failure due to heavy rainfall. We have proposed the research strategy of the prevention system, which consists of the laboratory soil tests on disturbed and undisturbed samples, the numerical simulations, seepage and failure tests, the filed measurements of suction, the temperature and rainfall, and the some in-situ tests for identification of geological and geotechnical characteristics of slope. In this paper, numerical simulation for the evaluation of slope instability due to rainfall will be presented. Furthermore, experiments are also conducted in this paper in order to validate the simulation models. The results of this paper have shown that the stability of the slope during rainfall can be well evaluated by using our method.

1 INTRODUCTION

In 2004, there were 10 typhoons hit Japan, and many natural disasters had been occurred due to heavy rainfall. As the result, a lot of lives, infrastructures and heritages were lost due to slope failure. Therefore, it is necessary to construct a prevention system in order to predict the probability of slope failure due to heavy rainfall.

It is basically considered that the slope failures occurred during heavy rainfall are mainly caused by the increase of soil mass, the decrease in apparent cohesion due to the increase of suction in unsaturated soil, and the increase of underground water level. Therefore, it is very important to consider the seepage properties and shear strength parameters of unsaturated soil in order to quantitatively evaluate the slope stability.

We have proposed the research strategy of the prevention system, which consists of the laboratory soil tests on disturbed and undisturbed samples, the numerical simulations, seepage and failure tests, the filed measurements of suction, the temperature and rainfall, and the some in-situ tests for identification of geological and geotechnical characteristics of slope.

In this paper, numerical simulation for the evaluation of slope instability due to rainfall will be presented. Numerical models for void and apparent cohesion, which are formulated based on some mechanical and probabilistic theories, will be described through out this paper. The void model is used to obtain the soil-water characteristic curve and the relationship between unsaturated-saturated permeability coefficient and degree of saturation. On the other hand, the apparent cohesion model is used to calculate the change in apparent cohesion with respect to the change in degree of saturation. The results obtained from the models are then applied to the 2-D unsaturated-saturated seepage analysis (FEM) and slope stability analysis (Janbu's method). Furthermore, experiments are also conducted in this paper in order to validate the simulation models. The results of this paper have shown that the stability of the slope during rainfall can be well evaluated by using our method.

2 PREDICTION SYSTEM FOR SLOPE FAILURE DUE TO RAINFALL

Figure 1 shows the research strategy of the prevention system for slope failures due to heavy rainfall proposed by Kitamura et al. (Kitamura et al. 2002). This strategy is mainly composed of five items: 1) laboratory soil tests on disturbed and undisturbed samples, 2) numerical simulations, 3) soil tank tests of seepage and failure, 4) Field measurement of suction, temperature and precipitation and 5) some in-situ tests for identification of geological and geotechnical characteristics of slope.

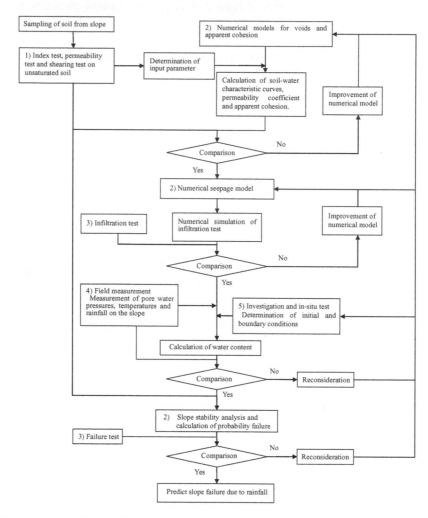

Figure 1. Research strategy of the prediction system for slope failure due to rainfall.

The purposes of each item will be explained as follows:

1. Results of laboratory soil tests are used as the input parameters for numerical models and also used to validate the simulation.
2. The numerical simulations are important items for the prevention system. These items consist of a numerical model for voids, a numerical model for apparent cohesion, a 2-D unsaturated saturated seepage analysis, and a slope stability analysis. The use of numerical model for voids (Kitamura et al. 1998) computed the unsaturated seepage characteristics of soil, i.e. the soil-water characteristic curve and the relationship between degree of saturation and unsaturated saturated permeability coefficient. On the other hand, the use of

numerical model for apparent cohesion (Sako et al. 2001) computed change in apparent cohesion due to change in water content. This change is one of the main causes of slope failures due to rainfall. The Finite Element Method is used for the numerical seepage model. Results of numerical model for voids are employed in the above numerical seepage model. The Janbu method is used for the slope stability analysis, and a probability of failure (Sako et al. 2003) is also implemented as the new estimation method for slope stability in the rainfall.

3. The soil tank tests of seepage and failure are conducted to validate numerical scheme by means of the above-mentioned models in the soil tank where the initial and boundary conditions can be easily controlled.

4. The data of field measurement, which are pore water pressure, rainfall and temperature, are used to determine the initial and boundary conditions for 2-D unsaturated-saturated seepage analysis and to discuss about the validity of numerical models.

5. Some in-situ tests such as CPT and boring are carried out to identify the geological and geotechnical characteristics of the slope. The calculation domain, the structure of soil layers and the potential slip plane are determined based on the data obtained from some in-situ tests.

3 NUMERICAL MODELS

3.1 Numerical model for voids

Kitamura et al. proposed a numerical model to simulate the unsaturated seepage characteristics of soil (Kitamura et al. 1998). Figure 2a shows the imaged soil particles in the small element. This situation can be modeled as shown in Figure 2b, i.e., voids are modeled by pipes and soil particles are modeled by other impermeable parts. Based on mechanical and probabilistic considerations in Figure 2b, the void ratio (e), volumetric water content (W_v), suction (s_u) and unsaturated-saturated permeability coefficient (k) can be derived as follows:

$$e = \int_0^\infty \int_{-\frac{\pi}{2}}^{\frac{\pi}{2}} \frac{V_P}{V - V_P} \cdot P_d(D) \cdot P_c(\theta) d\theta dD \quad (1)$$

$$W_v = \frac{e(d)}{1+e} = \frac{1}{1+e} \int_0^d \int_{-\frac{\pi}{2}}^{\frac{\pi}{2}} \frac{V_p}{V - V_p} \cdot P_d(D)$$
$$\times P_c(\theta) d\theta dD \quad (2)$$

$$s_u = \gamma_w \cdot h_c = \frac{4 \cdot T_s \cdot \cos \alpha}{d} \quad (3)$$

$$k = \int_0^d \int_{-\frac{\pi}{2}}^{\frac{\pi}{2}} \frac{\gamma_w \cdot D^3 \cdot \pi \cdot \sin \theta}{128 \cdot \mu \cdot \left[\frac{D}{\sin \theta} + \frac{D_h}{\tan \theta} \right]}$$
$$\times P_d(D) \cdot P_c(\theta) d\theta dD \quad (4)$$

where, V is the volume of element, V_p is the volume of pipe, D is the diameter of pipe, θ is the inclination angle of pipe, d is the maximum diameter of pipe filled with water and D_h is height of element as shown in Figure 2b. $P_d(D)$ is the probability density function of D, $P_c(\theta)$ is the probability density functions of θ, γ_w is the unit weight of water, μ is the viscous coefficient of water, T_s is the surface tension of water, h_c is the height of water column due to surface tension and α is the contact angle between pipe and water.

Using the above model associated with some simple parameters such as: grain size distribution curve, density of soil particles, void ratio, viscous coefficient of pore water, and surface tension of pore water, the soil-water characteristic curves and the relationship between the unsaturated permeability coefficient and the degree of saturation can be calculated.

3.2 Numerical model for apparent cohesion

Sako et al. proposed a numerical model for apparent cohesion in order to obtain the relationship degree of saturation and apparent cohesion (Sako et al. 2001). Figure 3 shows two adjacent particles with pore water at the contact point. The inter-particle force between two adjacent particles is generated by the surface tension of pore water. The inter-particle force F_i can be expressed by the following equation.

$$F_i = 2\pi r' T_s + \pi r'^2 s_u \quad (5)$$

where, F_i is the inter-particle force, T_s is the surface tension, s_u is suction ($= u_a - u_w$), r' is the radius of meniscus, a is the radius of curvature of meniscus.

Equation 5 includes the suction in the second term of right side therefore the volumetric water content in soil mass must be estimated to obtain the interparticle force. Based on mechanical and probabilistic consideration, the apparent cohesion due to

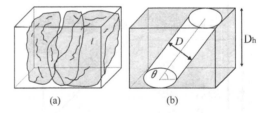

(a)   (b)

Figure 2. Modeling of particles and void in soil mass: (a) a container with a few soil particles; (b) Pipe and the other impermeable parts.

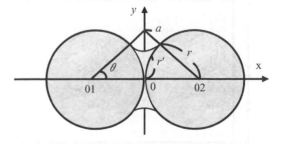

Figure 3. Inter-particle force between two particles due to surface tension.

inter-particle force c_1 is obtained using the following equation.

$$c_1 = \frac{\pi}{\pi - 2} \cdot \overrightarrow{F_i} \cdot Nc \cdot \tan \phi \qquad (6)$$

where Nc is total number of contact points per unit area, ϕ is internal friction angle.

The apparent cohesion, c, obtained by shear test includes some components due to interlocking, physical and chemical action, and the inter-particle force (or suction and surface tension):

$$c = c_0 + c_1 \qquad (7)$$

where, c_0 is the apparent cohesion that included some components due to interlocking, physical and chemical action, and c_1 is calculated from Eq. 6.

4 NUMERICAL SIMULATIONS

4.1 Input data of numerical simulations

In our research, a soil tank test on sandy soil called Masa-soil was carried out to investigate the failure mechanism of an unsaturated slope due to the increase in degree of saturation. Figure 4 shows the schema of laboratory soil tank test. The numerical simulations using 2D unsaturated-saturated seepage analysis and slope stability analysis is then performed to compare with the experiment in order to prove the validity of numerical simulations at the model scale.

Figure 5 shows the analytical area for 2-D unsaturated-saturated seepage analysis and slope stability analysis (Janbu method), which corresponds to the slope prepared for the laboratory soil tank test. Circle points in Fig. 5 show the installation positions of tensiometers. The conditions of experiment are listed in Table 1. Table 2 and Figure 6 are input parameters of numerical models for voids and apparent cohesion.

4.2 Results of numerical models for voids and apparent cohesion

Using the parameters as shown in Table 2 and Figure 6, the Soil water characteristic curve (SWCC) and the relationship between degree of saturation and unsaturated-saturated permeability coefficient can be obtained as shown in Fig. 7. The dots in Fig. 7 are obtained from laboratory tests, while the lines show the results calculated using the input parameters in

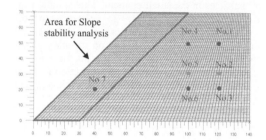

Figure 5.   Analytical area.

Table 1.   Condition of experiment.

Rainfall	50 mm/h	
Boundary condition	Top	Drainage
	Slope	Drainage
	Bottom	Undrainage
	Back	Undrainage
Embankment condition	Water content(%)	5
	Total density (g/cm³)	1.6
	Angle of slope (deg.)	45

Table 2.   Input parameters of numerical simulations.

Sample	Masa soil
Density of soil particles (g/cm³)	2.63
Surface tention (N/m)	73.48×10^{-3}
Viscosity coefficient (Pa · s)	1.138×10^{-3}
Viod ratio	0.75
Grain size distribution	Figure 6
Apparent cohesion (kPa) (Sr = 10.8%)	5.63
Internal friction (deg.)	18.9

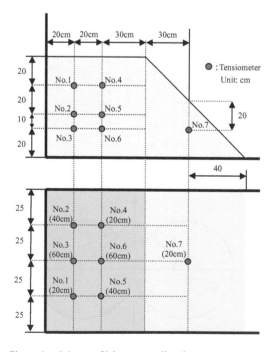

Figure 4.   Schema of laboratory soil tank test.

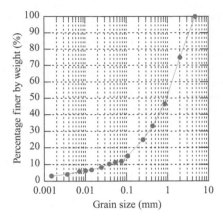

Figure 6.   Grain size distribution.

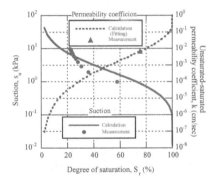

Figure 7.   SWCC and Relationship between S_r and k.

Table 2. It is found that the calculations of SWCC show the decrease in the suction with the increase in degree of saturation. However, the calculations of SWCC are larger than those of the measurements at the same degree of saturation. On the other hand, an improved method (Sako *et al.* 2006) is used to calculate the unsaturated-saturated permeability coefficient because they are overestimated as compared with the measurements. These results are shown as broken line in Fig. 7.

Substituting the results in Fig. 7 into Eq. 5, the relationship between the apparent cohesion and the degree of saturation are derived as shown in Fig. 8. In Fig. 8, c_0 is equal to 5.63 kPa. It is found from this figure that the apparent cohesion is decreased with the increase in the degree of saturation. However, it is difficult to prove the validity of this relation by means of soil tests since the change in apparent cohesion with the degree of saturation is extremely small.

4.3   Results of seepage analysis and slope stability analysis

These results, Figs. 7 and 8, are used to the 2-D seepage analysis and slope stability analysis. The seepage analysis was carried out by using the Finite Element Method (FEM). In the experiment, 50 mm/hour of rainfall were supplied to the soil tank. However, the amounts of rainfall per hour at each area in the soil tank were measured as shown in Figure 9. Therefore, the average values at the each area were used as boundary condition in this calculation.

Figure 10a, b shows the change in negative pore water pressure with time obtained from the numerical simulation and experiment. It is found from these figures that the changes in calculation results are later than the change in experiment results in all the measurement points. The great difference is especially seen at the data of No. 2 and No. 5 that lies to 40 cm in depth. These problems should be investigated more based on the various soil tests on unsaturated soil and then the numerical simulation should be improved.

The slope stability analysis was carried out by using the results of seepage analysis and the relationship between apparent cohesion and degree of saturation.

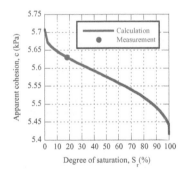

Figure 8.   Relationship between apparent cohesion and degree of saturation.

	(cm)						
	0	20	40	60	80	100	120	140

	66	58	32	32	32	26	18	X
	34	44	18	28	34	28	16	Back
Slope side	56	48	24	28	34	36	40	side
Rainfall per hour (mm/hour)	34	24	30	16	34	60	40	
	20	34	28	22	26	46	48	

Average rainfall (mm/hour)	42	41.6	26.4	25.2	32	39.2	32.4

Figure 9.   Amount of rainfall per hour at each area in the soil tank.

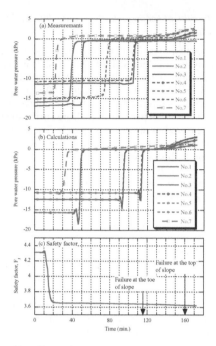

Figure 10. Change in pore water pressure and safety factor with time.

Figure 10c shows the change in safety factor with time and the time that the slope failures occurred. In this soil tank test, firstly, the slope failure occurred at the toe of slope. And then, the slope failure has progressed from the toe to the top of slope. In the calculation of slope stability analysis, the Janbu method was employed. It is understood that the safety factor decreases with time and this tendency corresponds to the increase in the degree of saturation. However, the values of safety factor are still very large when the slope collapsed. This result can be explained as follows. In the experiment, the slope failure has progressed from the toe to the top of slope. Therefore, it is considered that the safety factor was very large because the slope stability analysis method was not expressible well for such slope failure behavior. And then, the variation of the apparent cohesion due to rainfall that obtained by numerical models was considerably little. Therefore, it is necessary to improve the numerical models and to check the validity of them.

5   CONCLUSIONS

In this paper, the prediction system for the slope failure during heavy rainfall was presented. And then, the numerical simulations were explained and carried out for the result of laboratory soil tank tests. The results of our research are summarized as follows:

1. The prediction system employed the field monitoring system, the numerical simulations, the laboratory soil tests and the in-situ tests. Herein, the field monitoring system was used to measure the suction, rainfall, temperature, while numerical models were used to calculate seepage flow behavior as well as change in slope stability during rainfall. Experiments were also conducted to validate the proposed numerical models.

2. Using the Numerical models, the safety factor of the slope was decreased as the degree of saturation increases. However, the values of safety factor are still very large when the slope collapsed. We suggested that this overestimated safety factor is because of the incapability of the slope stability analysis method in describing the slope failure mechanism obtained in this paper, i.e. the slope failure progressed from the toe of the slope to the upper surface. And then, the variation of the apparent cohesion due to rainfall that obtained by the numerical models is considerably little. Therefore, it is necessary to improve the numerical models and to check the validity of them.

3. Further studies on this slope failure mechanism are in progress in our research group. For the near future, validation of the proposed simulation model for full stress scale problem will be obtained by comparing simulation results with field measurement data. We would like to make a slope failure prevention system in which we plan to use the combination between the simulation model and the field measurement data. Therefore, the proposed model for simulation of slope failure in the current study will play an important role in our process.

REFERENCES

Kitamura, R., Sako, K. & Matsuo, K. 2002. Research strategy on unsaturated soil for prediction of slope failure due to heavy rainfall. *Proc. 2nd World Engineering Congress, Engineering Innovation and Sustainability: Global Challenges and Issues*: 196–202.

Kitamura, R., Fukuhara, S., Uemura, K. Kisanuki, G. & Seyama, M. 1998. A numerical model for seepage through unsaturated soil. *Soils and Foundations, Vol.38, No.4*: 261–265.

Sako, K., Kitamura, R. & Yamada, M. 2001. A consideration on effective cohesion of unsaturated sandy soil. *Proc. of the Fourth International Conference on Micromechanics of Granular Media, Powders and Grains 2001*: 39–42.

Sako, K., Araki, K. & Kitamura, R. 2003. Research on probabilistic estimation of shear strength parameters for slope stability analysis. *Proc. of The Sino-Japanese Symposium on Geotechnical Engineering, Geotechnical Engineering in Urban Construction*: 320–325.

Sako, K. & Kitamura, R. 2006. A practical numerical model for seepage behavior of unsaturated soil. *Soils and Foundations* 46(5): 595–604.

Prediction and Simulation Methods for Geohazard Mitigation – Oka, Murakami & Kimoto (eds)
© 2009 Taylor & Francis Group, London, ISBN 978-0-415-80482-0

Prediction on landslides induced by heavy rainfalls and typhoons for Shih-Men watershed in Taiwan using SHALSTAB program

B.-S. Lin, C.-K. Hsu, W.-Y. Leung, C.-W. Kao & C.-T. Cheng
Sinotech Engineering Consults, Inc, Taiwan

J.-C. Lian
Soil and Water Conservation Bureau, Council of Agriculture, Taiwan

B.-S. Lin
National Chiao Tung University, Taiwan

ABSTRACT: In recent years, extreme weather made particular typhoons and heavy rainfalls under global warming effect, which caused enormous landslides and deadly debris flows in Shih-Men Watershed. Those disasters brought a great deal of sediment to a river course and silt up the some Sabo dams. Over flowing sediment would block up the route via downtown. Due to fine content of sediment, nephelometric turbidity unit (NTU) would grow up so highly that potable water could not be supplied to drink. Since the Chi-Chi earthquake, the main geological structure of Shih-Men Watershed became fragile and porous. After that event, the water table was apt to get rising highly when heavy rainfall infiltrated into slopes. In the meanwhile, the effective stresses of soil particles would be reduce and approach to be zero, and it would cause several shallow landslides over this region. In view of that, this study would apply hydrology model incorporated with infinite slope theory to analyze large-scale slope stability using SHALSTAB program. Prior to analysis, some soil parameters need to be determined using iterative procedure in rational range. Besides, soil depth was sensitive to the results in connection with terrain curvature and slope angle. It should be obtained from field investigation using hand auger and spade or based on empirical equations. Hydraulic conductivity and infiltration rate should be also considered within. Through the above procedure, it could help us realizing the sediment sources, distinguishing hazard-induced factors, and spatial distribution of slope failure potential. Moreover, engineers might find out how to mitigate the nature hazards and retrofit the hydraulic structure during the typhoon seasons

1 INTRODUCTION

At present, the sediment became a serious problem to Shih-Men watershed, where its main river was called Ta-Han River. Since 1963, Shih-Men dam have been constructed mainly to provide the potable water to population of downstream area. It could also generate hydraulic power to industry and factories for promoting economic development. In 2004, Typhoon AERE assaulted northern Taiwan as shown in Figure 1, which its maximum rainfall intensity was 90 mm and accumulated up close to 1,600 mm recorded by the nearby rainfall gauge station, and it exceeded 200 year of return period. It caused many landslides and debris flows around the watershed. Accompanied with a series of landslides, some routes and roads had been interrupted to pass along. In the meantime, a mountain of rocks and soils had been taken into the main of Ta-Han River and became sediment. When the sediments had transported by the river flow from upstream to downstream and it had directly dropped to the Shih-Men dam. They made water of

the main river more turbid because fine content was the major component, which could not rapidly sink in the short term so that NTU excited up to high value. This critical situation could make no potable water to

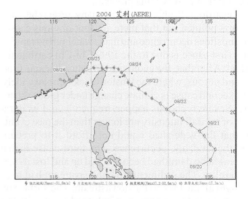

Figure 1. Moving Path of Typhoon AERE (from Central Weather Bureau in Taiwan).

drink about half of month. After Typhoon AERE, our government and research center had paid more attention and started to renovate the main and branch of Ta-Han River. Besides, they also wanted to find out which causes made the disasters happen.

This study adopted the large-scale slope with hydrology model to analyze and predict the spatial distribution of landslides. Prior to analysis, site investigation should been conducted to get the field data. Soil depth would be an important parameter to analyze slope stabilization. It would be related closely to slope angle and its relationship would be described clearly in the following paragraphs. Thus, it could find out landslide-induced factor and understand the composition of sediment through this procedure.

2 LOCATION AND ITS BACKGROUND

Shih-Men watershed is the third biggest one of Taiwan Island. It is abundant in water resource and woods. The landforms increase by degree from the northwest to the southeast. There are three administrative counties within. All the people living there have engaged in farming. The traffic is based on the Route No. 7 and some roads to communicate with downtown and the other places. Its location map has been shown in Figure 2. In 1963, Shih-Men dam was to constructed as multi-purpose hydraulic structure. At the first year, Typhoon GLORIA attacked this region and brought a lot of sediment to Shih-Men dam. The volume approached to $1,943 \times 10^4$ m³, which was too huge to reduce the storage of water substantially. And then, a series of typhoon events gave the same damage to the dam. Figure 3 shows the sediment history of Shih-Men Dam in different typhoon events. One could found that until Typhoon AERE, the extreme heavy rainfall made the sediment problem serious again and it directly threatened the people's livelihood and wealth. It also inflicted severe damages on economy. Afterwards, the government undertook some efforts to find the solution to solve this problem.

Regarding Typhoon AERE, many researches and reports pointed out that there are four obvious causes to landslides occurrence and make water turbid rapidly. The first cause is due to the 1999 Chi-Chi earthquake effect. After that, the geological structure became fragile and porous, where the tension cracks spread widely. The second cause is climate change under globe warming. Severe typhoons happen frequently and its duration of rainfall intensity are longer than the past events so that the watershed could not afford it to precipitate the landslide occurrence. The third cause is that several Sabo dams had been blocked up and lost its serviceability. The sediment from the upstream had been brought all the time to the downstream. The surface on riverbed need to approach in balanced level. The last cause is that the permanent sediment in the bottom of

Figure 2. Location map of Shih-Men watershed.

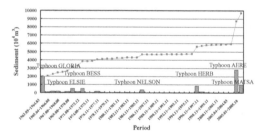

Figure 3. Sediment History of Shih-Men Dam in different typhoon events.

dam had been slumped off to be turbid at high flow rate especially in typhoon seasons. One could notice all of the above to realize the real reasons and distinguish the sources from the sediment for management of watershed.

3 STUDY AREA

Nowadays, the geology of Shih-Men watershed still tends to young and fresh formation by geologist investigation. There are some faults cross this region, where have a few ground motions once in a while. After Typhoon AERE, Soil and Water Conservation Bureau had started to some remediation strategies, such as Sabo engineering, drainage facilities and plant covers. Field survey indicated that Suler sub-watershed (see Fig. 4), whose area is approximately 585.5 ha, should have the priority to manage and renovate. Figure 5 shows that there were vast landslide area existing in the upstream and numerous shallow landslides in downstream. In view of the above points, this study would take Suler sub-watershed for a case study to analyze the potential of landslides using SHALSTAB program. After that, those results would validate with landslide interpretation from a satellite images for classifying the accuracy. About the geologic condition and its landslide distribution are introduced in the following passages.

3.1 Geologic condition

The geology of study area includes Paling and Mushan formation (CGS, 2008). Regarding area distribution, the Paling formation has 65 percentages and the other is 35 percentages (see Fig. 5). The Paling formation mainly consists of stiff dark-gray shale and gray slate and mixed up muddy shallow sandstone. That formation has almost intact cleavage, which is liable to fracture under crushing pressure and external forces. About Mushan formation, the principal component

(a) Aerial photo of Suler sub-watershed in upstream

(b) Shallow landslide near Route No. 7

Figure 5.   Landslide failure photos in Suler sub-watershed.

is grained and fine sandstone. A little gray shale or peats blend inside. The layer between sandstone and shale had weakness under weathering action to trigger landslides easily. This geologic condition would make the slopes unstable under heavy rainfall infiltration and strong earthquake shaking.

3.2 Spatial distribution of landslides

Since the 1999 Chi-Chi earthquake, the sub-watershed affected by the seismic motions had remained certain of landslide area. Through landslide interpretation from satellite images, the exposed area is about 1.29 ha. In the same way, the authors also collected the other events from 1999 to 2004. Figure 6 shows the landslides area change of the sub-watershed in his historical disasters including Typhoon XANGSANE, Typhoon TORAJI, Typhoon NARI, and Typhoon AERE. It is deserved to be mentioned that landslide area induced by Typhoon AERE is about 21.83 ha and increase rapidly 16.48 ha compared with Typhoon TORAJI. Its landslide ratio is highest of all the events closed to 3.7%. The spatial landslide distribution had been drawn in Figure 7. Almost landslides happened adjacent to the river course because of higher water table and tendency to be saturated. When heavy rainfalls and typhoons is coming, those places should be careful to monitor and be admonished at first for effective mitigation.

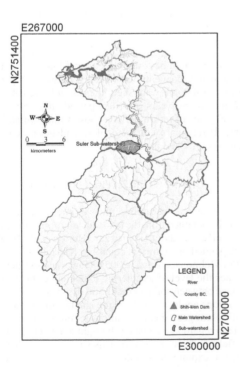

Figure 4.   Location map of Suler sub-watershed.

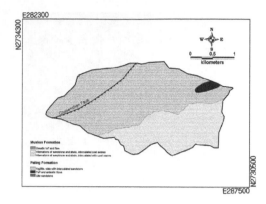

Figure 6.　Geologic map for Suler watershed.

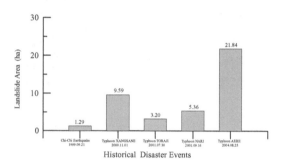

Historical Disaster Events

Figure 7.　Bar chart of landslide area change in historical disaster events.

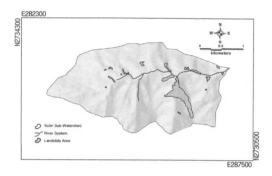

Figure 8.　Spatial disribution of landslide induced by Typhoon AERE.

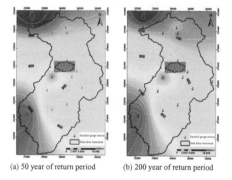

(a) 50 year of return period　　(b) 200 year of return period

Figure 9.　Annual average rainfall intensity in Shi-Men watershed.

4　FREQUENCY ANALYSIS OF RAINFALL INTENSIITY

For the consistence of annual average rainfall intensity, Thiessen polygons method has been used to estimate them in practice. Rainfall intensity data should use over twenty years of rainfall histories to do frequency analysis for high reliability. This study adopted the annual maximum series method and selected the one day of torrential rain to analyze in different return periods. Besides, frequency analysis is to be considered six probability distribution involved with the normal distribution, two-parameters lognormal distribution (LN2), three-parameters lognormal distribution (LN3), Pearson type III distribution (PT3), Log-Pearson type III distribution (LPT3) and extreme value type I Distribution (EV1). The optimum distribution would be chosen from all of the results based on the sum of square due to error (SSE) and standard error (SE). Figure 8 presents spatial distribution of annual average intensity suffered torrential rain respectively in 50 years and 200 years of return period by the means of kriging interpolation

method. It could be found that the predominant rainfall intensity concentrated in the southwestern part. Around Suler sub-watershed, the rainfall intensity is 750 mm on 50 years of return period, and 950 mm on 200 years of return period. Those values would be applied to compare with the shallow landslide potential from SHALSTAB program.

5　SITE CHARACTERIZATION OF SOIL DEPTH

In the previous researches, the volume of landslide could be obtained from the landslide multiplied by specific soil depth about 1.5 m to 2 m (Burton *et al.*, 1998) or depended on empirical equation (Khazai and Sitar, 2000; Delmonaco *et al.*, 2003). However, Soil depth implies the level of weathering effect and critical slip surface between soil and rock. Soil depth is associated with slope angle, slope direction and terrain curvature in different geological condition. It could not be possible to regard as a constant. It ought to be determined by field exploration in situ. This study is to use hand auger to explore the soil depths in study area. Exploring standard of operation is presented in Figure 10.

According to the above operation, SWCB (2008) and CGS (2008) had been to Shih-Men watershed and got the soil depths in various slope angles as shown in Figure 11. The relationship between soil depths and slope angle could be generated by regression method. The equation could be expressed as follows,

$$Y = -0.9596 \times \ln(X) + 4.007 \quad R^2 = 0.7214 \quad (1)$$

where Y is the soil depth and X is the slope angle. When analyzing the potential of slope failure, it could be estimated the depths of soil based on the above equation into the numerical model such as SHALSTAB program. The precision of parameter would be a key to affect the distribution of landslide area for the given area.

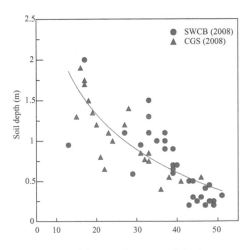

(a) spade arrangement   (b) use the hand auger

(c) check if the bedrock   d) record the soil depth

Figure 10.   Schematic photos for exploration of soil depth.

6   SHALSLAB PROGRAM

6.1   *Theory*

The SHALSTAB program had been developed in 1998 by Montgomery and Dietrich (1994). It aimed to map the potential of shallow slope instability using infinite slope theory with a steady state hydrological model (Salciarini and Conversini, 2007). This program could be also executed and demonstrated on a GIS-based platform. By the topographic element derived from DEM data, one can divide several cells as 5 m or 30 m size for user's default. According to the soil properties and geometric condition, each cell unit could be figured out all the potential and distribution of landslide. The governing equation would be written as follows,

$$\frac{q}{T} = \frac{\sin\alpha}{A/b}\left[\frac{c}{\gamma_w Z \cos^2\alpha \tan\phi}\right] + \frac{\gamma_s}{\gamma_w}\left[1 - \frac{\tan\alpha}{\tan\phi}\right] \quad (2)$$

where q = the rainfall intensity, T = the soil transmissivity, α = the slope angle, A = the drainage area, b = the overflow boundary length, c = the soil cohesion, ϕ = the soil frictional angle, γ_s and γ_w are the bulk density of soil and water respectively.

6.2   *Revised input data*

In the original SHALSTAB code, the soil depth is assumed to be a constant. It is not reasonable for practical application. The soil depth varies in arbitrary region and depends on geographic and hydrologic factor. For the purpose to fit the field situation, one could utilize terrain algorithm to calculate the average slope angle in grid-based cell. Then, the soil depth could be obtained, which was based on Eq. (1). Figure 13 presents the spatial distribution of soil depth. Almost soil depth lies in 1m to 1.5 m and fewer

Figure 11.   Plot of slope angle versus soil depth.

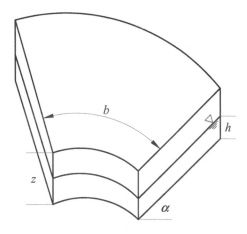

Figure 12.   Schematic layout of cell unit for spatial pattern of wetness (after Montgomery and Dietrich, 1994).

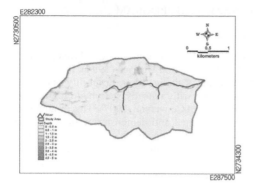

Figure 13. Soil depth distribution map.

Table 1. Soil properties for the study area.

Formation Name	γ_s N/m^3	c N/m^2	ϕ degree
Baling formation	22	2950	33
Mushan formation	22	2600	32

Table 2. Classification error matrix.

Reality	Prediction Non-landslide area	Landslide area
Non-landslide area	N_1	N_2
Landslide area	N_3	N_4

portions are over 3 m to 3.5 m. Thus, this study had changed and revised the input format of SHALSTAB code to make the parameter of soil depth variable. Once, the distribution of soil depth had been generated as input data and supposed to promote the results of accuracy.

7 PREDICTION ON LANDSLIDE POTENTIAL

This paragraph would discuss the results of landslide potential predicting by SHALSTAB program analysis. Compared with the results about the parameter of soil depth, it would be observed the difference and accuracy. Those results would also be demonstrated in a GIS platform to display its distribution.

7.1 Selection of parameters

To predict the potential of landslide, those parameters should be determined in a rational range. When starting to simulate the potential of landslide, the preparation of parameters would be established including soil properties and geometry condition. Soil properties could be referred from the borehole data in the field. Those data to the two formation of this study area list in Table 1. On the other side, the geometry condition consists of several 5 m × 5 m grids based on the digital elevation model.

7.2 Classification error matrix

Classification error matrix usually applied to explain the accuracy between predicting and real situation in the remote sensing field (Lillesand and Kiefer, 2000). Depending on the concept, the accuracy would be defined as three types including overall accuracy, producer's accuracy and user's accuracy. Overall accuracy

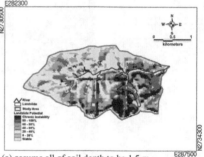

(a) assume all of soil depth to be 1.5 m

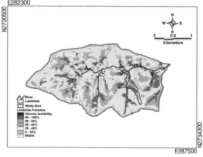

(b) various soil depth depending on slope angle

Figure 14. Landslide potential of Suler watershed.

is the principle discriminat expressed as follows (see table 2),

$$\text{Overall accuracy} = (N_1+N_4)/\Sigma N_i \qquad (3)$$

where N_i means the matching events. Through Eq. (3), one could validated and illustrate the accuracy of prediction on landslide potential.

7.3 Predicting results

This case study conducted two different assumptions of soil depth to simulate the landslide potential. Figure 14 shows the spatial distribution of landslide potential by degrees. Based on the hydrologic ratio (q/T) referred to Eq. (1), the potential of landslide could be divided into seven categories. Fig. 14(a) indicated the landslide might probably happen around 69% of the study area at assumption of a constant of soil depth, and only 31% were under stable situation. To calculate the above overall accuracy, its value simply approached to 35%. On the contrary, when employing various soil depths, the predicting results of overall accuracy became 52.2% and had increase of 17.2% obviously as shown in Fig. 14(b). Therefore, in case of a uniform soil depth, the topographic result would over-predict the potential of landslide and misunderstand the occurring position of that.

8 CONCLUDING REMARK

By the correlation of soil depth and slope angle, it will help us to effectively solve the complexity of soil stratum and build up the spatial distribution throughout the given area. Then, it also revises analytic procedure, promotes the overall accuracy and reduces the runoff error of simulation. Essentially, SHALSTAB program only depends on the hydrologic ratio (q/T) to classify different potential levels, but it can not completely point out where it will cause failure. In the meantime, the soil cohesion and frictional angle should be variable and randomize. Consequently, it is supposed to represent the safety of factor involving soil profile and easily apply in engineering design.

REFERENCES

Central Geology Survey (CGS) 2008. Project of evaluation and investigation about hydrology and geology effecting on slope stabilization (in Chinese).

Soil Water Conservation Bureau (SWCB) 2008. Study on high quality measurement and terrain historical migration in Shih-Men watershed (in Chinese).

Burton, A., Arkell, T.J. & Bathrust, J.C. 1998. Field Variability of Landslide model parameters. *Environmental Geology* 35: 100–114.

Khazai, B. & Sitar, N. 2000. Assessment of Seismic Slope Stability Using GIS Modeling. *Geographic Information Sciences* 6(2): 121–128.

Delmonaco, G., Leoni, G., Margottini, C., Puglisi, C. & Spizzichino, D. 2003. Large Scale Debris-flow Hazard Assessment: a Geotechnical Approach and GIS Modeling, *Nature Hazards Earth System Science* 3: 443–455.

Montgomery, D.R. & Dietrich, W.E. 1994. A physically-based model for the topographic control on shallow landsliding. *Water Resource Research* 30: 1153–1171.

Salciarini, D. & Conversini, P. 2007. A comparison between analytic approaches to model rainfall-induced development of shallow landslides in the central Apennine of Italy. In Mcinnes, Jakeways, Fairbank & Mathie (eds.), *Landslide and Climate Change*: 185–193. Taylor & Francis Group.

Lillesand, T.M. & Kiefer, R.W. 2000. Remote sensing and image interpretation. *Wiley & Sons*, New York.

Risk evaluation and reliability-based design of earth-fill dams

S. Nishimura, A. Murakami & K. Fujisawa
Okayama University, Okayama, Japan

ABSTRACT: A reliability-based design for the improvement of earth-fill dams is discussed in this study, and the effect of improving of the embankment is evaluated in relation to the safety of the embankment against earthquakes. An inclined core is considered here as the means of improvement for the earth-fill embankment. To evaluate the stability of earth-fill dams, the elasto-viscoplastic finite element method (EVP-FEM) is used. Then, combining the Monte Carlo method with the EVP-FEM, the expected damage costs are calculated. Finally, the effect of this improvement is evaluated by comparing the total costs before and after the restoration of the embankment.

1 INTRODUCTION

There are many earth-fill dams for farm ponds in Japan, particularly in the Setouchi region, which is the area surrounding the Inland Sea of Japan. Some of them are getting old and decrepit, and have weakened. Every year, a number of them are damaged by heavy rains and earthquakes, and in a few worst cases, the dams are completely ruined. To mitigate such disasters, improvement works are conducted on the most decrepit earth-fill dams. Since there is a recent demand for low-cost improvements, the development of a design method for optimum improvement works at a low cost is the final objective of this research. A reliability-based design method is introduced here in response to this demand.

To evaluate the stability of earth-fill dams, the elasto-viscoplastic finite element method is used as the stability analysis method in this study. With the finite element method, structures which have complicated boundaries or heterogeneity of the material parameters can be analyzed easily. The performance-based design will require a displacement evaluation of the embankment during the earthquakes. To this point, FEM is a very convenient tool for calculating the embankment displacement and for keeping consistency with the stability analysis, although this paper addresses only the stability analysis.

The Monte Carlo method is combined with the finite element method. The strength parameters, namely, cohesion c and internal friction angle ϕ, are considered to be the probabilistic variables in this research. The risk to earth fill-dams is evaluated from the viewpoints of damage loss and the stability of the embankment.

Many works dealing with reliability analyses for geotechnical problems are based on the finite element method combined with the Monte Carlo method. Fenton & Griffiths (2002) and Griffiths & Fenton (2005) analyzed the settlement of the footing on the ground considering the random field of Young's modulus. Griffiths et al. (2002) calculated the bearing capacity considering the random field of the undrained shear strength with the elasto-plastic finite element method. Bakker (2005) analyzed the stability of a dike with the elasto-plastic model based on the random field of the undrained shear strength.

In this study, the effect of improving the embankment is evaluated in terms of the safety of the embankment against earthquakes. An inclined core is considered here as the means of improvement for an earth-fill embankment. In this research, the spatial distribution and the spatial correlation structure of the strength of the earth-fill materials are firstly discussed. For this task, mainly Swedish weight sounding tests are conducted on several earth-fill dams. Based on the test results, several statistical models for the spatial distribution of the strength parameters are identified. The elasto-viscoplastic finite element analysis is applied for the stability estimation. Then, combining FEM with the Monte Carlo method, the variability of the damage level is evaluated. Strength parameters c and ϕ are treated as the probabilistic variables. The spatial correlations of these parameters are assumed from the sounding test results. Finally, the total costs before and after the improvement work are compared, and the effect of the improvement work is evaluated as the difference between the expected total costs.

2 NUMERICAL METHOD

To evaluate the stability of the earth-fill dams in this study, the elasto-viscoplastic finite element method (EVP-FEM) (Owen & Hinton 1980) is used as the analytical tool. With the finite element method, structures which have complicated boundaries or heterogeneity of the material parameters can be easily analyzed.

The elasto-plastic model is used as the governing equation here. The model follows the Mohr-Coulomb failure criterion given as:

$$\tau_f = c' + (\sigma - u)\tan\phi' = c' + \sigma'\tan\phi' \quad (1)$$

in which τ_f is the shear strength, c' is the effective cohesion, σ is the normal stress, u is the pore water pressure, and ϕ' is the effective internal friction angle. Since the non-associated flow rule is employed here, the plastic potential Q is defined as:

$$Q = k + \sigma'\tan\psi \quad (2)$$

in which k is a constant not affecting the analysis, and ψ is the dilatancy angle. Based on the over stress model, the viscoplastic strain rate is obtained from the following equation.

$$\dot{\varepsilon}_{vp} = \lambda \langle\Phi(F)\rangle \frac{\partial Q}{\partial T} \quad (3)$$

$$\langle\Phi(F)\rangle = 0 \quad (F < 0)$$
$$\langle\Phi(F)\rangle = F \quad (F \geq 0) \quad (4)$$

$$F = \tau - \tau_f \quad (5)$$

in which λ is the flow parameter, $\Phi(F)$ is the flow function, F is the yield function, τ is the shear stress, and T is the stress tensor.

Although the time domain is addressed in the EVP model, the viscosity is hypothetical in this research. It is just used as a tool to create equivalence between the stress and the applied load.

The failure of the fill embankment is judged from the equivalent viscoplastic stain rate given by Equation (6) in the EVP-FEM, namely,

$$\bar{\dot{\varepsilon}}_{vp}^n = \sqrt{\frac{2}{3}\left\{(\dot{\varepsilon}_{ij})_{vp}^n (\dot{\varepsilon}_{ij})_{vp}^n\right\}^{1/2}} \quad (6)$$

in which n is the time step number, and is the viscoplastic strain rate of the i-j component. Parameter TSR is defined as follows:

$$TSR = \sum_{A.G.P} \bar{\dot{\varepsilon}}_{vp}^n \quad (7)$$

in which $\bar{\dot{\varepsilon}}_{vp}^n$ is the equivalent viscoplastic strain rate at time step n and TSR is the sum of $\bar{\dot{\varepsilon}}_{vp}^n$ at all the Gauss points.

The overall failure of the earth-fill embankment is defined as:

$$TSR \geq TOLER \quad (8)$$

in which $TOLER$ is the critical value for TSR. A detailed definition of the failure mode corresponding to $TOLER$ is given in Section 5.2.

Furthermore, the damage level of the embankment is defined here as parameter RSR. The upper line of Equation (9) presents the local failure, while the lower line corresponds to the overall failure.

$$RSR = \begin{cases} TSR/TOLER & (TSR \leq TOLER) \\ 1 & (TSR > TOLER) \end{cases} \quad (9)$$

Since an evaluation of the pore water pressure is very important in the analysis of earth-fill dams, the water-soil coupling method is employed here. In other words, the equation for Darcy's law is solved simultaneously with the elasto-viscoplastic deformation.

3 STATISTICAL MODEL FOR THE SOIL STRENGTH

3.1 Determination method

The soil parameters which are obtained from tests are defined here as $\Xi = (\Xi_1, \Xi_2, \ldots \Xi_M)$. The symbol M signifies the number of test points. Vector Ξ is considered as a realization of the random vector $\xi = (\xi_1, \xi_2, \ldots, \xi_M)$. If the variables $\xi_1, \xi_2, \ldots, \xi_M$ constitute the M—variate normal distribution, the probability density function of ψ can then be given by the following equation.

$$f_\Xi(\xi) = (2\pi)^{-M/2}|C|^{-1/2}$$
$$\times \exp\left\{-\frac{1}{2}(\xi-\mu)^t C(\xi-\mu)\right\} \quad (10)$$

in which $\mu = (\mu_1, \mu_2, \ldots, \mu_M)$ is the mean vector of random function $\xi = (\xi_1, \xi_2, \ldots, \xi_M)$ and is assumed to be the following regression function. In this research, a 2-D statistical model is considered, namely, the horizontal coordinate x, which is parallel to the embankment axis, and the vertical coordinate z are introduced here, while the other horizontal coordinate y, which is perpendicular to the embankment axis, is disregarded.

$$\mu_k = a_0 + a_1 x_k + a_2 z_k + a_3 x_k^2 + a_4 z_k^2 + a_5 x_k z_k \quad (11)$$

in which (x_k, z_k) means the coordinate corresponding to the position of the parameter ψ_k, and a_0, a_1, a_2, a_3, a_4, and a_5 are the regression coefficients.

C is the $M \times M$ covariance matrix, which is selected from the following four types in this study.

$$C = [C_{ij}]$$

$$= \begin{cases} \sigma^2 \exp\left(-|x_i - x_j|/l_x - |z_i - z_j|/l_z\right) & \text{(a)} \\ \sigma^2 \exp\left\{-(x_i - x_j)^2/l_x^2 - (z_i - z_j)^2/l_z^2\right\} & \text{(b)} \\ \sigma^2 \exp\left\{-\sqrt{(x_i - x_j)^2/l_x^2 + (z_i - z_j)^2/l_z^2}\right\} & \text{(c)} \\ N_e\sigma^2 \exp\left(-|x_i - x_j|/l_x - |z_i - z_j|/l_z\right) & \text{(d)} \end{cases}$$

$$i,j = 1,2,\ldots,M$$

$$\begin{cases} N_e = 1 \ (i = j) \\ N_e \leq 1 \ (i \neq j) \end{cases} \tag{12}$$

in which the symbol $[\ ]$ signifies a matrix, σ is the standard deviation, and l_x and l_z are the correlation lengths for x and z directions, respectively. Parameter N_e is the nugget effect. The Akaike's Information Criterion, AIC (Akaike 1974) is defined by Equation (13), considering the logarithmic likelihood.

$$\text{AIC} = -2 \cdot \max\left\{\ln f_\Xi\left(\Xi\right)\right\} + 2L \tag{13}$$

$$= M \ln 2\pi + \min\left\{\ln|C| + (\Xi - \mu)^t C^{-1} (\Xi - \mu)\right\} + 2L$$

in which L is the number of unknown parameters included in Equation (13). By minimizing AIC (MAIC), the regression coefficients of the mean function, the number of regression coefficients, the standard deviation, σ, a type of the covariance function, the nugget effect parameter, and the correlation lengths are determined.

3.2 Example of the strength spatial distribution

In order to identify the spatial correlation structures of soil parameters, high density samplings are required. However, since sampling data are not usually sufficient for identifying the spatial correlation structures, sounding tests are convenient. As an example, the results of Swedish weight sounding (SS) tests are shown in Figure 1. Figure 1(a) exhibits the locations of the testing points on an embankment. Figure 1(b) shows the SPT N-value distribution on the longitudinal section along an embankment axis. The N-values are transformed from the Swedish weight sounding test results based on Equation (16) (Inada 1960), namely,

$$N = 0.002W_{SW} + 0.067N_{SW} \tag{14}$$

in which N is the SPT N-value, N_{SW} is the number of half rotations, and W_{SW} is the load applied in the Swedish weight sounding tests.

As the optimum covariance function, Equation (12d) is selected. The determined parameters of the

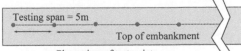

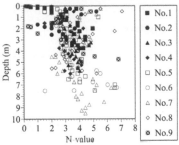

(a) Plane view of embankment top and SS test points.

(b) Spatial distribution of N-values.

Figure 1. Results of SS tests.

function are as follows:

$$\mu = 2.44 - 0.315z$$

$$C_{ij} = N_e \cdot 1.26^2 \exp\left(-|x_i - x_j|/9.3 - |z_i - z_j|/0.46\right)$$

$$\begin{cases} N_e = 1.0 \ (i = j) \\ N_e = 0.77 \ (i \neq j) \end{cases}$$

It can be seen that the horizontal correlation length is several times the vertical one, and this is a reasonable tendency.

4 RELIABLITY-BASED DESIGN METHOD

Ordinarily, risk is evaluated as the expected cost of failure, namely,

$$C_F = P_f C_f \tag{15}$$

in which P_f is the probability of failure and C_f is the failure loss when an earth-fill embankment is completely ruined.

In the case of earth structures, it is difficult to determine the probability of failure, since the boundary between the stable and the unstable states is not clear. If we apply the circular slip surface method, the overall failure of a structure can be defined, but only the overall failure. Since local failures can also be defined when using the finite element method, the expected damage loss, shown in Equation (16), which considers the damage level, is proposed for the risk assessment in this research.

$$C_F = E[RSR] C_f \tag{16}$$

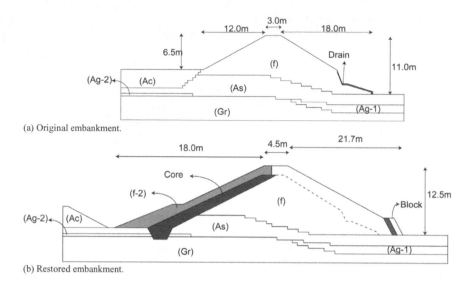

(a) Original embankment.

(b) Restored embankment.

Figure 2. Cross sections of earth-fill embankments.

The total cost is derived as:

$$C_T = C_I + C_F \qquad (17)$$

in which C_I is the initial cost. It is presented as the restoration cost in this research. By minimizing the total cost, an optimum design can be determined within the general reliability design.

5 ANALYSIS OF A FILL-EMBANKMENT

5.1 Embankment model

In this study, the stability of an earth-fill embankment, before and after restoration, is compared. The cross sections of the embankment are shown in Figure 2. Figure 2(a) shows the original embankment, while Figure 2(b) presents the restored embankment for which inclined cores have been installed. Table 1 shows the fill and the ground materials, in which the material symbols correspond to the material zones shown in Figure 2. The parameters are determined from the SPT N-values, soil sampling test results and in-situ permeability tests. Consistency tests and $\overline{CU}$ triaxial compression tests are conducted as the laboratory soil tests. The internal friction angle is determined from the Ohsaki's equation, $\phi = \sqrt{20N} + 15$ (Hatanaka et al. 1996), while the cohesion is assumed based on the soil sampling tests. The effective cohesion and the effective internal friction angle are dealt with as probabilistic parameters. The coefficients of the variations are assumed to be 0.2 for c' and 0.1 for ϕ'. These values are supposed to be realistic, referring to the previously published report (e.g., Phoon & Kulhawy

1999a, b). If a large number of shear tests is conducted, the coefficients of variation can be determined as corresponding exactly to those at the analytical sites.

5.2 Seepage analysis

Although the coupling method is employed in this research, a seepage analysis is firstly conducted to avoid an evaluation of the behavior in the unsaturated zone. As the first step of the analysis, a saturated-unsaturated seepage analysis (Nishigaki 2001) is conducted to distinguish the saturated and the unsaturated zones. Secondly, a stability analysis is performed with the coupling method to estimate the pore water pressure behavior in the saturated zone, while the negative pore pressure in the unsaturated zone is not considered in the analysis.

Intrinsically, the variabilities of the permeability and the characteristic curve of the unsaturated soil should be considered in the analysis. Since the negative pore water pressure is not considered in Equation (1), namely, the pore water pressure is considered in the saturated zone, the parameters in the seepage analysis do not seriously affect the stability of the embankment in this analysis. This is the assumption for the design on the safer side.

5.3 Stability analysis

The stability analysis is conducted by combining EVP-FEM with the Monte Carlo method. The Monte Carlo method is applied for the probabilistic parameters, namely, cohesion c' and internal friction angle ϕ',

Table 1. Material properties of earth-fill dam and ground.

Materials	Unit weight (kN/m³)	Cohesion (kPa)	Internal friction angle (°)	Permeability (m/d)	Young's modulus (kPa)	Poisson's ratio	Dilatancy angle (°)
f	20.0	12.45	$\phi = \sqrt{20N} + 15$	2.88×10^{-1}	$2,800N$[1]	0.3	0.0
As	18.0	38.2	24.4	1.18×10^{-1}	12.0×10^3	0.3	0.0
Ac	15.0	12.5	19.5	1.47×10^{-1}	28.0×10^2	0.3	0.0
Ag-1	18.0	10.0	36.4	9.50×10^{-2}	66.4×10^4	0.3	6.4
Ag-2	18.0	10.0	50.0	9.50×10^{-2}	19.0×10^4	0.3	20
Gr	18.0	755.0	50.0	2.88×10^{-1}	28.0×10^4	0.3	20
Core	17.6	20.0	27.0	1.00×10^{-4}	28.9×10^2	0.3	0.0
f-2	17.1	19.0	26.0	8.76×10^{-2}	28.9×10^2	0.3	0.0
Drain	19.0	20.0	27.0	8.0	10.0×10^6	0.3	0.0
Block	24.0	3.0×10^8	38.0	1.00×10^{-6}	22.6×10^6	0.2	8.0

1) Japan Road Association (2002).

and the calculation is repeated 1,000 times. A random number, based on the random field model determined in Section 3.2, is assigned to each finite element. In fact, the number of iterations is not sufficient to obtain an accurate $E[RSR]$. As the next step of this study, the calculation efficiency must be raised.

The earthquake load is considered here as the load factor, and the design earthquake intensity is set to be 0.15. Figure 3 exhibits an example of the convergence process for the EVP analysis. The figure does not show the probabilistic analysis results, but the deterministic analysis results for the expected values of the strength parameters. The two lines correspond to safety factors 1.23 and 1.04. The former is for the overall failure, and the latter case is for the safer case of no failure, since the equivalent elasto-viscoplastic strain rate vanishes after 580 iterations. According to these results, 1,000 iterations are judged to be enough for the convergence of the EVP analysis. The safety factors, which are obtained based on the shear strength reduction method (Ugai 1990), are shown.

Figure 4 shows the results of the stability analysis. Figures 4(a) and (b) correspond to the original and the restored states, respectively. These figures are also derived from the deterministic analysis; they correspond to the overall failure mode, which shows a circular slip surface. The TSR values of this failure mode are defined as the critical values of TSR and $TOLER$. It can be seen that the safety factor value increases from 1.12 to 1.23 with the improvement of the fill materials.

5.4 *Reliability analysis*

In this study, expected damage cost C_F is calculated with Equation (18), using the Monte Carlo method, while initial cost C_I is introduced as the restoration cost. The total costs derived from Equation (17), for

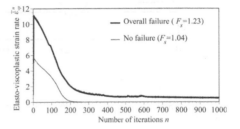

Figure 3. Convergence process of EVP analysis for the restored embankment.

Table 2. Results of reliability analysis (Unit 1,000 JPY).

Embankment	Intial cost C_I	Failure loss C_f	$E[RSR]$	Total cost C_T
Original	0	1,174,769	0.095	111,603
Restored	87,500	1,174,769	0.0143	104,299

the original and the restored embankments, are shown in Table 2. Failure loss means the damage to houses, farms, and other agricultural facilities for cases in which the fill embankments have collapsed.

The expected values for RSR are evaluated based on the Monte Carlo method. According to Table 2, $E[RSR]$ is reduced drastically due to the improvement of the embankment. The total cost is also reduced due to this factor.

Comparing the total costs of the original and the restored states, the profit procured by the improvement of the earth-fill dam, in terms of the safety of the embankment against earthquakes with a design earthquake intensity of 0.15, is evaluated as 111,603–104,299 = 7,304 (Unit: 1,000 JPY).

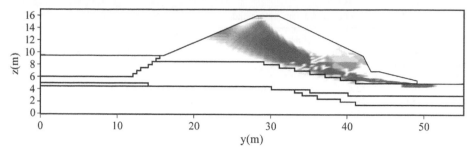

(a) Original embankment.

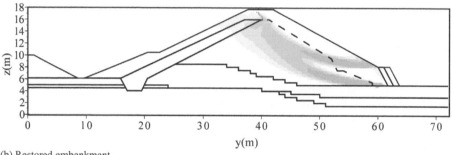

(b) Restored embankment.

Figure 4. Distribution of viscoplastic strain for expected values of strength parameters and safety factors. (Design earthquake intensity = 0.15).

6 CONCLUSIONS

1. The spatial distribution model for the strength of earth-fill dams has been determined based on sounding test results through the use of MAIC based on the sounding test results.
2. A reliability analysis has been conducted with the elasto-viscoplastic FEM combined with the Monte Carlo simulation method.
3. A reliability-based design has been conducted for the improvement work of earth-fill dams. Making a comparison between the total costs of the original and the restored states of the embankment, the profit procured by the improvement has been evaluated against the risk of damage brought about by earthquakes.

REFERENCES

Akaike, H. 1974. A new look at the statistical model identification, *IEEE Trans. on Automatic Control*, AC-19(6): 716–723.
Bakker, H.L. 2005. Failure probability of river dikes strengthened with structural elements, *Proc. of 16th ICSMGE*, 1: 1845–1848.
Fenton, G.A. & Griffiths, D. V. 2002. Probabilistic foundation settlement on spatial random soil, *Journal of Geotechnical and Geoenvironmental Engineering*, 128(5): 381–391.

Griffiths, D.V., Fenton, G.A. & Manoharan N. 2002. Bearing capacity of rough rigid strip footing on cohesive soil: probabilistic study, *Journal of Geotechnical and Geoenvironmental Engineering*, 128(9): 743–755.
Griffiths, D.V. & Fenton, G.A. 2005. Probabilistic settlement analysis of rectangular footings, *Proc. of 16th ICSMGE*, 1: 1845–1848.
Hatanaka, M. & Uchida, A. 1996. Empirical correlation between penetration resistance and internal friction angle of sandy soils, *Soils and Foundations*, 36(4): 1–9.
Inada, M. 1960. Usage of Swedish weight sounding results, *Tsuchi-to-Kiso, J. of JSSMGE*, 8(1): 13–18 (in Japanese).
Japan Road Association 2002. *Specification for highway bridge, Part IV*, Tokyo: Maruzen; 1990 (in Japanese).
Nishigaki, M. 2001. *AC-UNSAF3D User's Manual*. (in Japanese).
Owen, D.R.J. & Hinton, E. 1980. *Finite elements in plasticity: Theory and Practice*, Pineridge Press, UK.
Phoon, K.-K. & Kulhawy F.H. 1999a. Characterization of geotechnical variability, *Can. Geotech. J.*, 36: 612–624.
Phoon, K.-K. & Kulhawy F.H. 1999b. Evaluation of geotechnical property variability, *Can. Geotech. J.*, 36: 625–639.
Ugai, K. 1990. Availability of shear strength reduction method in stability analysis, *Tsuchi-to-Kiso, J. of JSSMGE*, 38(1): 67–72 (in Japanese).

Study on road slope disaster prevention integrated management system

N. Sekiguchi
Pacific Consultants Co., Ltd, Osaka, Japan

H. Ohtsu
Kyoto University, Kyoto, Japan

T. Yasuda
Pacific Consultants Co., Ltd, Osaka, Japan

ABSTRACT: In order to promote infrastructure management that takes into consideration the challenging social and economic conditions and the occurrence of many natural disasters in recent years, it is necessary to pay attention to disaster prevention and mitigation, which is currently one of the major public concerns, and explore infrastructure management possibilities from the viewpoints of risk management and asset management. Paying attention to rain, whose frequency is relatively high among the natural hazards involved in road slope disaster prevention, this study proposes a slope risk evaluation system implemented on a geological information system (GIS) platform as the base system of a road slope disaster prevention integrated management system that integrates asset management with risk management. This study then shows an example of the application of the proposed system to slopes in a real road network. This study also shows the direction of efforts toward the development of a road slope disaster prevention integrated management system.

1 INTRODUCTION

Focusing on rain hazards, which occur relatively frequently, among natural hazards that are of interest from the viewpoint of road slope disaster prevention, this study proposes a Slope Risk Evaluation System as a method (system) for integrated management of road slope disaster prevention efforts such as integrating slope data and disaster information into a geological information system (GIS), predicting when and where a slope failure occurs, estimating the resultant economic loss (road slope risk) quantitatively, prioritizing the measures to betaken, and taking timely measures for slope failure risk areas. The proposed method makes it possible to automatically acquire slope data (e.g., slope height, average gradient, collapsible soil layer thickness) from three-dimensional terrain data, calculate the annual failure probability of individual slopes during wet weather and perform a risk evaluation of a particular road network based on information on economic losses due to detours. The method is expected to contribute to quantitative evaluation in road slope management.

In this study, the concept of risk (Σ, probability of failure × the amount of loss) is introduced into LCC analysis and project prioritization for the management and operation of infrastructure. By so doing, this study indicates a direction for the development of a system that is capable of providing information reflecting such risks as cost and benefit uncertainty for decision making and an integrated management system for road slope disaster prevention that integrates asset management with risk management.

2 ROAD SLOPE DISASTER PREVENTION: PRESENT STATE AND PROBLEMS

Examination of the causes of road slope disaster prevention reveals that rain-induced disasters are conspicuous among the natural disasters in recent years. There is a growing need, therefore, for rainfall prediction (use of meteorology) in order to organize effective measures against rain hazards and minimize damage. In the area of nonstructural measures designed to minimize damage by providing information before and after the occurrence of a slope disaster, traffic control based on continuous rainfall and other criteria is practiced at present. Because of ambiguity of the rationale behind traffic control criteria, however, there is a need for the establishment of new traffic control criteria based on the rain seepage characteristics (water level in the slope, moisture content) of slopes, the provision of early warning information, and assistance for evacuation activities. In the area of structural measures designed to increase slope stability by taking slope control measures, various measures based on slope failure risk evaluation is currently practiced.

There is a need, however, for such measures as project prioritization based on the evaluation of social loss due to detour in the event of a slope failure and the evaluation of the risk of unnecessary traffic control.

Thus, in the field of slope disaster prevention, there is a pressing need for efficient and effective road slope management based on the evaluation of rain hazard risks. To this end, the development of an integrated management system for road slope disaster prevention (described later) is essential.

3 ABOUT SLOPE RISK EVALUATION SYSTEM

In the area of road slope disaster prevention, the authors (Ohtsu et al. 2001, Ohtsu et al. 2002, Takahashi et al. 2003, Togo et al. 2004) have conducted studies on risk management methods based on the evaluation of slope-failure-induced losses necessary for the strengthening of roadside slopes. On the basis of the research findings of the authors, this study proposes the Slope Risk Evaluation System, a GIS-based package of analysis software for the automatic acquisition of slope data, the slope-by-slope calculation of the probability of slope failure during wet weather, and road network risk evaluation based on losses due to detours, as a tool for assisting in risk-evaluation-based decision making based on the evaluation of socioeconomic losses caused by rain-induced slope failures.

3.1 Considerations in developing the slope risk evaluation system

Slopes to be managed by road slope managers abound along the road sections to be managed. For some road networks where road safety inspection has been conducted, a relatively complete set of information is available about the characteristics of the inspected slopes. For most slopes, however, slope data and relevant basic data such as the area and gradient of each slope are not fully available, and there are many slopes that have not even been inspected.

The key to success in road slope management, therefore, is the ability to prepare road data and relevant basic data for a number of slopes efficiently and provide quantitative data needed to prioritize inspection needs and measures to be taken. Considerations in developing the Slope Risk Evaluation System, therefore, are as follows.

3.1.1 Creating a system capable of evaluating a road network

Road slopes do not perform their functions independently. Instead, they perform their functions to create a safe and comfortable road network. In road slope management, therefore, it is necessary to create a system capable of evaluating the overall slope risk of a road network.

3.1.2 System capable of generating basic data efficiently

Road administrators need to manage many slopes, and it is a time- and money-consuming task to prepare slope data and relevant basic data. It is therefore necessary to create a system that can efficiently generate basic data to be used for risk evaluation. To be more specific, it is desirable that simple tasks that do not require engineering judgment be simplified as much as possible.

3.1.3 System capable of providing information necessary for decision making

Priorities of inspection needs and control measures derived from the Slope Risk Evaluation System can be used as quantitative criteria for decision making based on an evaluation method. These priorities, however, are by no means final decisions. A system that makes more rational decision making possible needs to be created by adding the road administrator's engineering judgment to the output of the Slope Risk Evaluation System.

3.1.4 Creating a system making full use of the advantages of GIS

In order to meet these system performance requirements, it is necessary to create a system that makes effective use of the functions described below, taking advantage of the use of a geographic information system as a platform.

1. Use of layering capability
   A system capable of providing a variety of quantitative information that can be used for decision making is created by using the ability to layer a number of information attributes, different types of information such as data on landforms, road networks, slope area, the zone of influence of slope failure, slope risk, aerial image data, vegetation maps and geological maps.
2. Use of network analysis capability
   By using network analysis capability, a system capable of performing detour search and road network slope risk evaluation is created.
3. Use of thematic mapping capability
   Thematic mapping capability is used to create a system with excellent visual presentation capability.

3.2 Configuration of slope risk evaluation system

Figure 1 shows the configuration of the Slope Risk Evaluation System proposed in this study.

The system consists mainly of a slope data acquisition system, which automatically generates slope cross

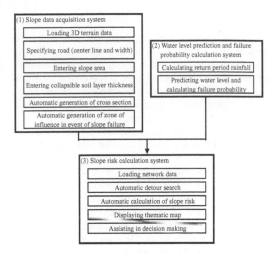

Figure 1.   Configuration of Slope Risk Evaluation System.

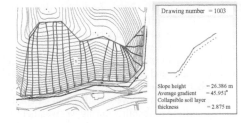

Figure 2.   Slope extraction.

sections from 3D terrain data, a water level prediction and failure probability calculation system, which estimates groundwater level during wet weather and calculates conditional failure probability and annual failure probability, and a slope risk calculation system, which performs detour search and automatically calculates slope risk. It is also possible to perform integrated management and analysis of slope-related spatial information, visualize information, and perform sophisticated information processing.

3.3   Slope data acquisition system

For road networks where road safety inspection has been conducted, a relatively complete set of information on slope characteristics at the inspection sites is available. For a majority of slopes, however, slope data and relevant basic data such as the area and gradient of each slope are not fully available. It is necessary, therefore, to find ways to generate these basic data efficiently.

The slope data acquisition system is designed to automatically generate slope data, such as the height, average gradient and collapsible soil layer thickness of slopes adjacent to the roads to be managed, from 3D terrain data.

The first step is to enter road details (road center line, width, starting point kilometer marker, etc.) so as to evaluate the slope risk of the road network concerned. By so doing, the road center line and the kilometer marker are automatically shown on the map so that thereafter data can be managed on the basis of kilometer points.

The next step is to specify the slope area to extract slopes along the road to be managed. By specifying the slope area on the map, the system can be used to automatically generate cross sections (slope height, average gradient, collapsible soil layer thickness) at 10-meter intervals by using 3D terrain data (TIN data) (Figs. 2).

The thickness of the collapsible soil layer is automatically set by a digital terrain model (TIN data) from nearby boring data. The method of entering boring data can be chosen from two options: either entering them by specifying a location on the map or entering them by specifying coordinates.

The system can also manage slope attribute data (e.g., disaster prevention check sheet number, disaster prevention check sheet evaluation) from disaster prevention check sheets, etc., as basic data on individual slopes.

After cross sections are generated, slope data (slope height, average gradient, collapsible soil layer thickness) and slope attribute data are compared, and a cross section that is deemed most dangerous in the slope zone and that has the steepest gradient, the greatest height difference, the greatest collapsible soil layer thickness, etc., is selected as a cross section representative of the slope. The cross section data thus extracted are handed over to the water level prediction and failure probability calculation system.

If the extent of the slope region that is likely to collapse because of rain is entered, the zone of influence in the event of collapse is automatically mapped. Information on the zone of influence thus determined is related to residential map data so that it can be used for network evaluation performed by the Slope Risk Evaluation System.

The volume of collapsed soil is calculated from Eq. (1):

Collapsed soil volume V_0

$= W_0$ (collapsed slope width)

$\times L_0$ (collapsed slope length)

$\times d$ (soil layer thickness)   (1)

3.4   Water level prediction and failure probability calculation system

The water level prediction and failure probability calculation system calculates the annual rain-induced failure probability of slopes from the slope data obtained by the GIS (Takahashi et al. 2003).

This system consists of the return period rainfall calculation system, which calculates return period rainfall from Japan Meteorological Agency's rainfall observation data, and the water level prediction and failure probability calculation system, which predicts groundwater level during wet weather and calculates conditional failure probability during wet weather and annual failure probability.

The return period rainfall calculation system basically uses Japan Meteorological Agency's local rainfall observation data and is capable of processing up to five rainfall patterns for a slope so that the most dangerous rainfall pattern can be identified by evaluating the effects of different rainfall patterns on the failure probability of a slope.

The water level prediction and failure probability calculation system predicts the groundwater level in a slope during wet weather by using a three-stage one-dimensional tank model. In the calculation method used by the system, a performance function assisted by the stability analysis method for a large slope shown in Eq. (2) is used, and strength coefficients ($\tan \phi d$, Cd) are used as random variables. For each of their coefficients of variation (standard deviation/mean value), the failure probability P_f during wet weather is calculated by using the first-order second-moment method (FOSM).

$$Q = \left(1 - \frac{\gamma_w H_w}{\gamma H}\right) \cdot \frac{\tan \phi d}{\tan \alpha} + \frac{Cd}{\gamma H} \cdot \frac{1}{\sin \alpha \cos \alpha} - 1 \quad (2)$$

where γ_w is the unit weight of water; γ, the unit weight of soil; ϕd, the internal friction angle of soil; Cd, cohesion; α, the inclination angle of an infinite slope; H, the thickness of the collapsible soil layer; and H_w, water depth in the collapsible soil calculated by the water level prediction and failure probability calculation system.

The variability of the strength coefficients applied to the performance function is modeled as a normal distribution by setting the present factor of safety and then back-calculating the internal friction angle (ϕd) and cohesion (Cd).

The annual failure probability of each slope during wet weather is formulated by incorporating a wide range of rainfall events ranging from relatively frequent minor rainfall events to very rare heavy rainfall events in the form of the annual average. As shown in Eq. (3), convolution is performed by weighting the conditional failure probability $p_f(\alpha)$ obtained at each level of rainfall with the possibility of occurrence of that level of rainfall in one year .

$$P_a = \int_0^\infty p_f(\alpha) \cdot \frac{dP(\alpha)}{d\alpha} \cdot d\alpha \quad (3)$$

where $P(\alpha)$ is the probability of occurrence of rainfall; $p_f(\alpha)$, conditional failure probability; and Pa, annual failure probability.

3.5 Slope Risk Calculation System

The Slope Risk Calculation System calculates the annual risk of slopes in a road network reflecting social losses by using the annual failure probability of each slope calculated by the slope data acquisition system and the water level prediction and failure probability calculation system and performing automatic searches for detours in the event of slope failure.

The first step in detour search is to set unit quantities necessary for risk calculation. The unit quantities to be preset include human loss, building loss and road restoration cost used to calculate manager loss and user loss (described later). Road attributes (e.g., travel speed, traffic volume) are also preset.

Detour search is performed to extract road networks by using the GIS and find a detour that minimizes the travel time for each route. Data such as the length of the detour thus found and detour time are used to calculate the annual risk of each slope.

The annual risk of each slope is calculated on the basis of the annual failure probability (Pa) and the amount of loss (Ci).

Annual risk Ra = Annual failure probability Pa

$\times$ Amount of loss Ci (4)

The amount of loss consists of manager loss, which is a direct loss, and user loss due to the use of detours, etc., which is an indirect loss. The proposed system assumes a damage scenario due to slope failure (damage to passing vehicles caused by the collapsed soil) and calculates the amount of loss due to that slope failure (Fig. 3).

Manager loss = Loss to passing vehicles and

people onboard ($C1$)

+ Cost for collapsed soil removal

and restoration ($C2$) (5)

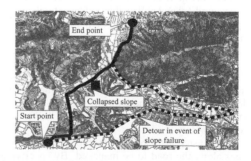

Figure 3. Detour search.

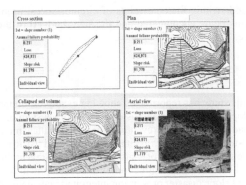

Figure 4. Decision support.

User loss $(C3) =$ Detour loss

$$(\text{time cost} + \text{travel cost}) \quad (6)$$

Amount of loss $(Ci) = C1 + C2 + C3 \quad (7)$

The amount of loss in each category is calculated by the methods shown in a previous study (Slope Disaster Prevention Research Committee 2006), but the effect of collapsed soil on housing in the event of slope failure (building loss) mentioned in connection with the slope data acquisition system is also taken into consideration.

Thus, the proposed system makes it possible to quantitatively evaluate the priority of the implementation of slope inspection or control projects according to the annual risk of each slope. As a final stage of decision making assistance, however, the system has a "decision making" stage at which engineering judgment can be made.

At this stage, first the high-priority slopes identified by the Slope Risk Calculation System are narrowed down to three highest-ranking slopes. By showing data relevant to those three slopes such as calculation results, cross sectional views, plan views, aerial images and slope data simultaneously on the display screen, the system assists engineers in making final judgments (Fig. 4).

4 EXAMPLE OF ANALYSIS BY SLOPE RISK EVALUATION SYSTEM

This section considers the applicability of the Slope Risk Evaluation System by using a real topographic map. In the example of risk calculation described here, the case in which the rainfall pattern and the items used in annual risk evaluation are varied is verified.

4.1 Annual failure probability and annual risk

The annual failure probability, the amount of loss and annual risk thus obtained for each slope are shown in Figure 5 to Figure 7.

As shown, Slope No. 2 shows the highest value (about 0.25) of annual failure probability. From the viewpoint of the conventional concept of slope failure risk level, control works for Slope No. 2 should be given the highest priority. However, from the viewpoint of the annual risk taking into consideration the amount of slope-failure-induced loss based on detour search, Slope No. 1, which shows the greatest annual risk in the road network, should be selected as the highest-priority slope.

4.2 Rainfall pattern and annual failure probability

Figure 8 shows the effects of the rainfall pattern (early concentration type, central concentration type or late concentration type) on the results of annual failure probability calculation. As shown, the annual failure probability for the late concentration type rainfall tends to be highest (i.e., least safe).

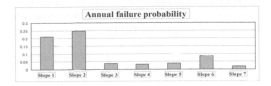

Figure 5. Annual failure probability.

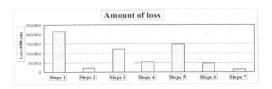

Figure 6. Amount of loss.

Figure 7. Annual risk.

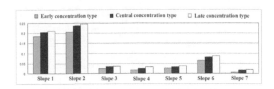

Figure 8. Rainfall pattern and annual failure probability.

5 DIRECTION OF DEVELOPMENT OF ROAD SLOPE DISASTER PREVENTION MANAGEMENT SYSTEM

5.1 *Constituent technologies supporting road slope disaster prevention*

The constituent technologies that support road slope disaster prevention include the following: slope stability evaluation and prediction technology (stability analysis model, hazard model, slope groundwater level and moisture content prediction model, control works deterioration prediction model, back analysis model, failure probability calculation, etc.), management technology (management target setting and risk evaluation, risk analysis and loss evaluation, LCC calculation, project prioritization, information disclosure and accountability, traffic control management, etc.), spatial information processing system technology (digital map data acquisition and terrain analysis, slope model information acquisition, GIS, etc.), and monitoring technology (e.g., slope groundwater level and moisture content monitoring technology).

5.2 *Direction of development of road slope disaster prevention integrated management system*

Figure 9 illustrates the direction of the development of a management system that can be realized by adding different levels of management to a combination of some of the constituent technologies mentioned earlier. Systems that can be realized at initial stages of management may range from a slope risk evaluation system that performs such tasks as the automatic acquisition of slope data from 3D terrain data, the calculation of the annual failure probability of slopes during wet weather and the risk evaluation of road networks based on detour losses to an asset management decision support system as an integrated management system designed to perform project priority evaluation based on life cycle cost calculated on the basis of control works deterioration prediction over a medium to long period of time. It is important to develop methods (tools) for providing information so as to enable road slope managers to respond flexibly to management needs within the constraints of circumstances, problems being encountered and the size (budget) of the local government concerned.

6 CONCLUSIONS

This study has reported, by showing examples, on the Slope Risk Evaluation System as the core system of a road slope disaster prevention integrated management system that integrates asset management and risk management in the area of road slope disaster prevention, and has indicated a direction of the development of a road slope disaster prevention integrated management system.

In view of the direction of the development of a road slope disaster prevention integrated management system discussed in the latter part of this paper, the authors intend to extend and upgrade the system to a road slope asset management decision support system in order to realize integrated management of road slope disaster prevention efforts.

REFERENCES

Ohtsu, H., Ohnishi, Y., Mizutani, M. & Ito, M. 2001. The proposal of the methodology associated with decision-making of reinforcement of slopes based on cost-benefit-analysis. Proceedings of the Japan Society of Civil Engineers.

Ohtsu, H., Ohnishi, Y., Nishiyama, S. & Takeyama, Y. 2002. The investigation of risk assessment of rock slopes considering the socioeconomic loss due to rock fall. Proceedings of the Japan Society of Civil Engineers.

Takahashi, K., Ohtsu, H. & Ohnishi, Y. 2003. Research on the slope risk evaluation in consideration of groundwater behavior using the storage tank model. Soils and Foundations.

Togo, S., Ohtsu, H. & Ohnishi, Y. 2004. A study on strategic slope asset management using GIS. Journal of Construction Management.

Slope Disaster Prevention Research Committee, Kinki Chapter, Japan Civil Engineering Consultants Association. 2006. Introducing the concept of deterioration into slope stability evaluation (in Japanese).

Figure 9. Direction of development of road slope disaster prevention integrated management system.

Prediction and Simulation Methods for Geohazard Mitigation – Oka, Murakami & Kimoto (eds)
© 2009 Taylor & Francis Group, London, ISBN 978-0-415-80482-0

An empirical study on safety management for river levee systems

K. Fukunari
Fukken Co., Ltd., Tokyo, Japan

M. Miyamoto & K. Yoshikawa
Nihon University, Chiba, Japan

ABSTRACT: In Japan, approximately half the population and three quarters of property is located on flood plains where they are at risk from river flooding. However, there is next to no research into managing the continuous levee systems. This paper reports on breaches of the levees along the Tone River system over the past 80 years and presents empirical information on the causes for the breaches. The most common reason for levee breaches is overtopping water; however, problems have occurred with breaches caused by leakage around sluice pipes in levees. Based on this survey, the paper clearly indicates the functional and control limitations of managing the historical levees that are still in use today and proposes ways to manage the levee systems in the future, keeping in mind excess flooding and the added factors of damage and risk management.

1 GENERAL INSTRUCTIONS

A river levee is a basic means of flood control that has been successively built up and repaired over an extended period based on experiences of the damage caused by flooding. Once a levee is breached, the inundated area suffers considerable damage. The Tone River, one of the major rivers in Japan, has been the heart of many major floods.

This paper systematically organizes actual cases of floods and levee breaches involving the Tone River over the past eighty years or so to elucidate the mechanism and causes of individual cases of collapse. From this viewpoint, this paper examines a range of cases of levee breaches along the Tone River over an extended period. It takes cases of levee overtopping or collapse based on the review mentioned above, irrespective of the design flood scale or whether the flood capacity of the existing river levee is exceeded. The paper then recommends the optimum means of river levee development and management with a view to minimizing the damage caused by flooding.

2 CHARACTERISTICS OF LEVEES FOR LARGE-SCALE RIVERS

Along the Tone River, work to improve the river has been conducted incrementally since the Meiji period. Even today, means of flood control are being planned and constructed to address the ultimate long-term goal of restricting the incidence of flooding so that it becomes an event that occurs once every two hundred years. However, the current safety level for flood

control has failed to meet the objectives envisioned in the long-term plan. What is worse, the safety level is not uniform as the river levee has several characteristics.

In other words, the constitution of the existing river levees and the scale and characteristics the flooding that serve as exogenic agents are so diverse as to render it impossible to verify that the level of safety is uniform throughout their entire length. This explains why the aforementioned factors cause breaches in the continuous levee along the Tone River at positions where the safety level is poorest.

River levees have the characteristics mentioned above. In controlling the safety of the river levee, it is difficult to apply any simple model or approach. It is significant, however, to examine actual levee breaches that have occurred in the past.

3 LEVEE BREACHES THAT HAVE OCCURRED IN THE TONE RIVER SYSTEM

3.1 *Disasters during and before the Meiji period*

In the Edo Period, flood control along the Tone River depended largely on the retarding function of lakes and ponds, as well as on open levees. Levees were constructed to protect the land. They were small and built using manual labor. During the period, the Tone River was responsible for several major floods where river levels exceeded the capacity at that time. These include the flood of 1742 and the flood of 1786, the latter of which was partly due to an elevation in the riverbed after the eruption of the volcano, Mt. Asama, three years earlier.

In the Meiji Period, a major flood occurred in August 1910. The levee was breached at several points near the upper reaches of the river near the limit between the mountainous section and the flood plain, which are now found in Honjo, Fukaya and Menuma-machi in Saitama Prefecture. The flood caused massive damage.

3.2 Cases of levee breaches caused by flooding during and after the Showa period

Figure 1 illustrates the failure points in the levees in the Tone River System over the past eighty years, from the start of the Showa Period until today. There are many more points where the levee was damaged without being breached.

Many of the cases where the levee was breached and the floodwaters submerged the floodplain have actually resulted from levee overtopping. Other cases of breaches were due either to water leaks around a structure that crosses the levee at a level lower than the levee or to water leaks in the ordinary part of the levee (Table 1).

3.3 Level breaching by typhoon kathleen in 1947

The flood in September 1947 was triggered by overtopping at the village of Higashi-mura in Saitama Prefecture on the right bank of the Tone River. It was caused by Typhoon Kathleen. In Higashi-mura in particular, overtopping led to levee breaching on the right bank. The floodwaters floewed through Saitama Prefecture, Katsushika-ku and Edogawa-ku in Tokyo and

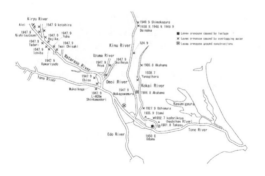

Figure 1. Points at which the levee has been breached along the Tone River in recent years.

Table 1. Number of points at which the levee has been breached in the past eighty years.

Cause	Number of Points
Levee overtopping	28
Leaks around a structure	3
Leaks in the ordinary levee	1

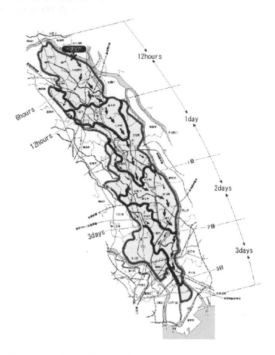

Figure 2. Levee breach due to overtopping at Higashi-mura, Saitama Prefecture on the right bank of the Tone River (September 1947 flood).

reached the Tokyo Bay five days later (Figure 2). It left approximately 70,000 houses under water, 70% of which were inundated above floor level. Within Tokyo, at least 80% of the affected houses were inundated above floor level. The depth of the floodwaters was considerable throughout the whole area. While the breach of the levee caused the river water to overflow in the upstream area, the water level in the downstream section was below the peak level marked in 1941. As a result, flood damage in the lower reaches was relatively limited.

4 IDENTIFYING THE CAUSE OF THE BREACH

4.1 Approach towards the cause of the breach

4.1.1 Levee design and safety limit (control responsibility limit)

The Government Ordinance for the Structural Standard for River Administration Facilities stipulates that a levee, integrated with bank protection, groins and other similar facilities, shall be designed to have a safe structure when subject to the normal flow of water at or below the design flood level and that it shall prevent flooding at or below the design flood level. The height of the river levee is calculated by adding a freeboard to the design flood level. At this height, the cross section of the levee is required to have a crest width commensurate with the river size and a fixed levee gradient. Figure 3 indicates the relationships between these elements.

4.1.2 Safety of the levee system
The levee is considerably long as it runs beside the river. The levee must therefore be regarded as a line and a system that is formed in the longitudinal direction. Any levee breach or overtopping has a massive impact on the propensity for flooding along the whole section of the river. For instance, it may lower the water level in other sections. River levee safety has to be assessed from the perspective of regarding the levee as a continuous longitudinal unit.

4.1.3 Safety from the viewpoint of damage
River levees are generally planned and designed by assuming a certain flood scale. For this reason, many cases of flooding that exceed the presumed scale are treated as events beyond the limit of responsibility of river management. It is imperative to also study a means of restricting and discouraging land use in the flood plain, as well as risk management.

4.2 Identifying the cause of the breach

Many cases of level breaches are caused by overtopping. Other cases of breaches include those stemming from leakage around a sluice pipe or other structure and those due to permeation without overtopping. Past cases of breaches caused by overtopping on the Tone River include some cases that were caused by a combination of overtopping and weakened levee body as a result of the permeation of rainwater and river water. Four causes of levee breaches are examined below.

4.2.1 Levee breaches caused by overtopping
From the standpoint of river management, any case of breaching following a flood that exceeds the design flood level where water flows over the levee crest is beyond the responsibility of river management.

After studying the process leading up to breaching after overtopping, it has been found that, in many cases, the depth of overflow reaches 50 to 60 centimeters several hours after the start of overtopping and that the overflowing water erodes and scours the back slope of the levee, which finally leads to levee failure. However, there are many other cases in which overtopping did not result in levee failure (Photograph 1).

4.2.2 Levee breaches caused by leaks around a structure without overtopping
Even when the water level does not exceed the design flood level or in any situation where overtopping does not occur, levee breaching may arise from leakage around a sluice pipe in the levee body or any other structure that penetrates the levee. Among other factors, any structure on unstable ground creates a state of unequal settlement with the earth levee that surrounds it. It is likely that cavities will form to create water paths underneath or around the structure. They serve as a cause of levee breaching as they give rise to the discharge of sand and soil, causing the earth that constitutes the levee body to collapse. Sluice pipes supported by piles were constructed during a certain period of time after a particular era, and they require attention.

4.2.3 Levee breaches caused by leakage in the general part of the levee without overtopping
In the event of a flood, it is normal to see levee leakage at several points in the levee section. Levee damage caused by a leak takes different forms. One example is the phenomenon called boiling or piping, in which water and soil spout up by a leak from the levee foundation and accumulate near the outlet. Another example is where the levee slope slides down, resulting in failure, after rainwater and the river permeate the levee body to weaken it and reduce its stability.

As far as the Tone River System is concerned, this phenomenon has attracted considerable attention since the time of the 1982 flood. Figure 4 illustrates the sites where leaks took place along the Tone River in the 1982 flood. It is unknown whether leaks like these would lead directly to a levee breach, but it is important to understand a phenomenon like this that takes place at the time of flooding and to carry out leakage prevention measures.

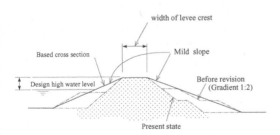

Figure 3.   Relationship between the structure and the height of a levee.

Photo 1.   The Kokai River flood in 1986 caused by overtopping.

Figure 4. Sites where the levee was damaged or where foundation leakage occurred after leakage from the Tone River and the Edo River in the 1982 flood.

Where there is a levee breach following a leak from the ordinary part of the levee, it is believed that the levee body has become wetted by rainwater and river water and collapses after a decline in the stability of the levee wall. It is unclear whether piping caused by a leak from the levee foundation alone develops into a levee breach. It is possible that a levee breach without overtopping may occur after a combination of piping and sliding.

4.2.4 Levee breaches due to erosion

In the Tone River System, there have been no cases of levee breaching due to erosion or scouring caused by the flow of floodwaters. If the scope of the case studies is expanded to the Kanto region, one such case is found, which was caused by a flow diverted at the Shukugawara Weir on the Tama River during the 1974 flood. Erosion can cause levee breaches in fast-flowing streams or where there is any structure across the river that disturbs the flood flow.

5 LEVEE SAFETY CONTROL

While construction and reinforcement of levees are basic to levee safety control, it is also vital to take measures to protect levees and preserve the construction that has been performed up to that time. A report on the construction and reinforcement of river levees will be delivered in the future. However, some actions other than levee construction are discussed below.

5.1 Safety control to prevent flooding through levee protection measures

It is presumed that there existed some cases in which measures to protect levees were taken to prevent levee breaches and to reduce damage. When Typhoon No. 10 (Typhoon Bess) struck the region in 1982, there was a major flood in the upper reaches of the Tone River System. The magnitude of the typhoon was the largest after that of the 1947 flood. The levee protection measures helped prevent a levee breach in this case.

5.2 Daily safety control of levee systems

It is critical to obtain information about the traces of water levels in past floods and its disparities with the levee height as well as the flood characteristics on the site. Another key task to prevent leakage around structures is to use cut-off sheet piles and to conduct daily checks of the cavities.

5.3 Safety control aimed at mitigating flood damage

5.3.1 Calculation of the breach probability

The authors have formulated the probability of a breach considering the levee's flood control capacity and the level of flooding, which constitute causes of failure in the levee system.

$$P = f(H, E, M, N, F) \tag{1}$$

$$P' = \int_0^L f(H, E, M, N, F)dL \tag{2}$$

Where, P is the local breach probability, P' is the longitudinal breach probability, H is the levee height, E is the infiltration resistance, M is the structure that crosses the levee, N is the channel condition, F is the water level or the flow rate and L is the longitudinal length.

This equation calculates the possibility of a levee breach at a specific point. Given that the flood scale is nearly uniform at different locations before a levee breach takes place, the breach depend greatly on the factors underlying the capacity of the levee to control floods. Equation (2) ingrates the breaching probability P with respect to the levee's longitudinal length L to compute the possibility or probability of levee branching in longitudinal terms, P'.

With these two equations, it is possible to examine the risk of failure in a levee system in both local and longitudinal terms.

5.3.2 Calculation of damage according to the potential for damage

For the purpose of levee development and management, it is requisite to identify the damage in an area that may turn into a flood plain. The authors have formulated the damage considering the potential for damage in addition to the flood control capacity of the levee and the magnitude of the flood. The following shows the results of the formulation.

$$D = D(F, H, E, M, N, S) \tag{3}$$

$$D' = \int_0^L DdL \tag{4}$$

$$\overline{D} = \int_0^\infty P(F) \cdot D'dF \tag{5}$$

Where, D is the damage, D' is the damage when calculated on an intermittent basis, $\overline{D}$ is the annual average damage, H is the levee height, E is the infiltration resistance, M is a structure that crosses the levee, N is the channel condition (geotechnical factor), F is the water level or the flow rate, L is the longitudinal length and S is the damage potential.

Using the formulation mentioned above, it is possible to calculate intermittent damage by computing the flood control capacity of the levee, the flood scale and the damage potential and by considering the longitudinal length. In addition, the annual average damage may be used in the study by taking account of the breach probability P calculated from Equation (1) to the damage D and the flood scale F.

6 DEVELOPING A SYSTEM OF RIVER LEVEE CONTROL TECHNIQUES

Because of the recent climatic changes accompanying global warming, variations in rainfall have become greater and the trend is toward increases in the magnitude and frequency of heavy rainfall. Therefore, in the years to come, it will be difficult to guarantee the reliability of current flood controls at many rivers and there is potential for increased inundation and flooding.

However, if we consider the financial restraints, it will be difficult to respond with nothing but flood control measures that are based on past thinking. Flood control plans for future large-scale basins must be defined by either line or plane, and subjected to comprehensive scrutiny including land use and town planning, schemes for residents and crisis management.

River levees, the most fundamental of flood control facilities, are no exception. In order to guarantee a degree of reliability of the flood controls while minimizing damage caused by river basin submergence, it will be necessary to 'develop a system of river levee control techniques' that perceives river levees as continuous in a longitudinal direction, and which include responses to a variety of external forces including excess flooding.

6.1 *Longitudinal safety assessment of levees as river systems*

In terms of river improvements with regard to existing external force (submergence), it is unclear when and at what point the levees will be breached and flood. This is the most dangerous situation from the viewpoint of crisis management and it is an urgent matter to clarify the risk of levees breaching in response to excess flooding. Consequently, it is necessary to take the perspective of hydrology to analyze the possibility of levees breaching in response to various external forces (inundation), and to urgently clarify at what points there are high risks of levees breaching with respect to inundation.

6.2 *Longitudinal safety assessment of the levees at tone river*

Figure 5 shows the water levels that were generated by actual flooding below the main bed of the Tone River as it was in 1982. Viewed longitudinally, the actual maximum water level is not uniform, but it is clear that there are points where the level is relatively high with respect to the height of the levee and design flood levels.

In the future, it is projected that various magnitudes of flooding, including excess flooding, will occur. Therefore, with respect to actual controls of present river levees and to reduce damage to the river basin as a whole with regard to various magnitudes of flooding, it is important from the viewpoint of crisis management to, for example, specify in advance the points where overtopping, which is the greatest cause of levee breaches, will occur, and to investigate how to avoid levee breaches even if overflow is permitted.

Figure 6, 7 are examples of estimating the points where overflow from the river levee may be projected in case of various magnitudes of flooding at the Tone River.

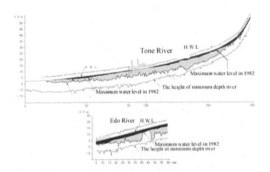

Figure 5. The longitudinal relationship between water levels, levee crown height and the height of low-lying inland areas in the case of flood.

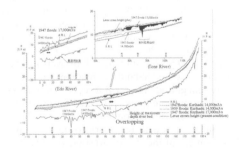

Figure 6. Estimates of points where levee overtopping will occur in case of flooding of large magnitudes.

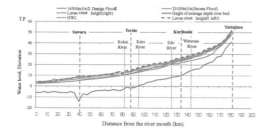

Figure 7. Estimates of flood depths in the present river channel.

6.3 Proposals for levee control techniques at the tone river from the viewpoint of damage reduction

By calculating projected flood runoff in response to flooding magnitude in case of overtopping or breaches at a cross-section of the Tone River levees, it is possible to find the depth of inundation in the projected areas of flooding. Then, we calculate the extent of the damage D by multiplying the damage level determined by the depth of the inundation by the damage potential S in the projected flood areas. Based on present topographical conditions and damage potential, we reproduced calculations in case of the levees breaching on the right bank of the Tone River at Otonemachi in Saitama Prefecture by flooding on the magnitude caused by Typhoon Kathleen in 1947. The results indicate approximately 2.6 million victims, damage to approximately one million households and 110,000 businesses with the extent of the damage reaching approximately 33 trillion yen.

Incidentally, in order to assess the safety of river levees, it is necessary to consider the risk of overtopping or levee breach P. The risk of overflow or levee breach P is, as equation (1) indicates, determined by the magnitude of the discharge (external forces) F, the height of the levee (H), the infiltration resistance E, the presence or absence of structures that transect the levee M, and the conditions of the river channel N. Consequently, to calculate the annual average damage extent D, we project various external forces (inundation) F according to event probability and we use the damage potential S and the risk of overflow or breaches at a cross-section of the river levees P as determined by the above-mentioned H, E, M and N. That is to say, by assessing the risk of overtopping or breaches P, we understand the relationship between discharge magnitude (external forces) F and damage extent D at multiple points projected for overflow or breaching when viewed longitudinally. At such time, for example, it is also necessary to do a longitudinal assessment of the risk of overflow or breaches in case of a defined magnitude of discharge (external forces) F since downstream flood discharge generally

decreases in case of overtopping or breaching of levees at the headwater.

In short, in response to various discharge magnitudes (external forces) and with a view to reducing damage to the river basin as a whole, it is necessary to propose ways to control the river levees as a continuous system by means of a longitudinal understanding of the relationship between "the extent of damage to the river basin as a whole" and "the size of the risk of overtopping or levee breaches."

6.4 Establishing flexible flood control risk for river channel systems

Since the lower reaches, where urban areas are in close vicinity, generally suffer enormous damage compared to the upper reaches if a disaster occurs, flood control risks for the river basin should not be uniform. That is to say, with regards to maintaining consecutive levees, (1) consider a systematic arrangement of flood control basins in order to keep water levels low at the time of downstream flooding. (2) when it is projected that damage due to flooding will be comparatively small even in protected lowlands, make the height of the levees lower than those upstream and downstream, and shift toward land use wherever there is promise of a retarding effect. (3) depending on the discharge magnitude, specify places where overtopping is projected in advance. (4) since super-levees are not breached even with overtopping, make the height of the levees lower than those upstream and downstream, and maintain the levees at the height where overtopping is projected

In this way, river improvement plans for river basins should establish flood control levels according to damage potential at the time of a disaster. That is to say, from the viewpoint of reducing flood damage to the river basin as a whole, a system of levee maintenance and control techniques should be constructed that differentiates between degrees of safety, such as levee height, upstream and downstream and on either river embankment.

7 CONCLUSIONS

Continuous levees fulfill their function when upstream and downstream areas and the right and left river embankments are uniform. To control the safety of levees with the unique technical qualities of historical structures, the most important data are the results of verifying the facts of flooding experienced over a comparatively long time span for the first time since modern flood controls were implemented.

Most cases of levee breaches on the Tone River in the past were caused by overtopping, making it clear that a response to overtopping is most important to reduce flood damage. Future research topics include devising ways to reduce damage in the whole river basin by means of projecting non-breaching over-topping from specific points and utilizing the cross-sectional continuity of the levee system. In addition, from the viewpoint of devising ways to reduce damage to the river basin as a whole, the discourse on levees must be studied including crisis management responses and regulations and shifts in land use.

With respect to these issues, the present paper is an empirical report based on first stage results. In the future, we plan to conduct specific simulations associated with damage extent and the risk of levee breaches or overtopping in the Tone River. We plan to report on levee control discourse as a system for devising ways to reduce damage while also incorporating the viewpoints of land use and risk management in response to various projected discharge magnitudes (external forces).

ACKNOWLEDGEMENT

This study presents a part of the results of the study on 'Developing the management methods for River Levee system (Katsuhide Yoshikawa) which was supported by a grant for scientific research from the Ministry of Education, Culture, Sports, Science and Technology.

REFERENCE

K. Yoshikawa, 2008. Urban Planning with Basin Management, Kajima Institute Publishing Co., Ltd.

Case records of geohazards and mitigation projects

Prediction and Simulation Methods for Geohazard Mitigation – Oka, Murakami & Kimoto (eds)
© 2009 Taylor & Francis Group, London, ISBN 978-0-415-80482-0

Characterization of Uemachi-fault in Osaka against displacement induced disaster

Y. Iwasaki
Geo-Research Institute. Osaka, Japan

ABSTRACT: Osaka has been developed in a basin area where the boundaries are defined as active faults. The highland of Uemachi area is higher than the surrounding lowlands by about 20 m. The boundary between the Uemachi area and the western lowland was identified as Uemachi fault. Since the shape of the surface ground deformation is the same displacement pattern as in the past, it is necessary to develop characterization of fault deformation. Seismic reflection surveys across the fault were studied to characterized fault. Deformation of layers caused by faulting are found to fit well with a curve of hyperbolic tangent that was characterized by parameters of reference width and throw. This paper describes characteristics of typical types of the Uemachi faults based upon the results of seismic as well as boring results.

1 INTRODUCTION

Topologically, Osaka basin is divided into three zones of west lowland, east lowland, and Uemachi Upland as shown in Figure 1. The ground surface of Uemachi-Upland is about 30 m higher than the surrounding lowlands. The boundary between the West Lowland and the Uemachi Upland is a formed straight cliff with the NS direction. Since these topological features, the cliff was historically considered as a fault and has been named as Uemachi-Fault.

Figure 1. Osaka Basin.

Various lifelines are located above the fault and necessary to prepare displacement induced problems. To estimate deformation of ground by fault movement, this paper describes characterization of fault deformation.

2 UNDERGROUND STRUCTURE OF UEMACHI-FAULT

2.1 Topological study

In the late of 1960', a subway excavation in the Uemachi Upland revealed the geological underground structures of the Uemachi fault. The subway location is shown in Figure 3 denoted as Namba SW and the vertical geological section along the subway line is shown in Figure 2. It should be noticed that the horizontal position of the fault is about 0.75 km west of the topological fault.

2.2 Seismic reflection study

To overcome the shortage of geomorphology, we have introduced seismic reflection to study deep underground structure since 1980.

Some of these surveyed lines are shown in Osaka in Figure 3. The Uemachi faults consist of many segments that have been induced by basically reverse movement of base rock with dipping east and strike of NS direction. The base rock was confirmed to exist at about 900–1500 m in the west Osaka and 600–900 m in the east Uemachi upland. Active faults are shown in red bold lines in Figure 3.

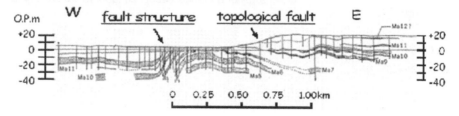

Figure 2. Underground structure of Uemachi Fault along subway line (Ikebe, Iwatsu & Takenaka, 1970).

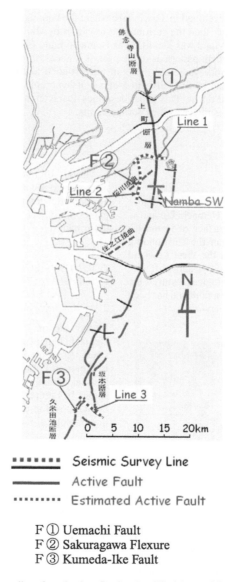

Seismic Survey Line
——————— Active Fault
·········· Estimated Active Fault

F ① Uemachi Fault
F ② Sakuragawa Flexure
F ③ Kumeda-Ike Fault

Figure 3. Uemachi Faults and survey lines for seismic reflection (modified from Ichihara, 2001).

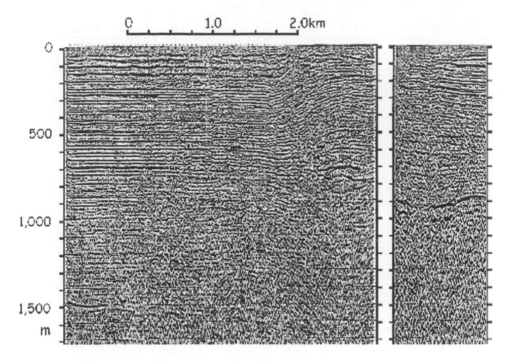

Figure 4.   Seismic Reflection Result (Line-1) (Uemachi-Fault/Nakanoshima-line).

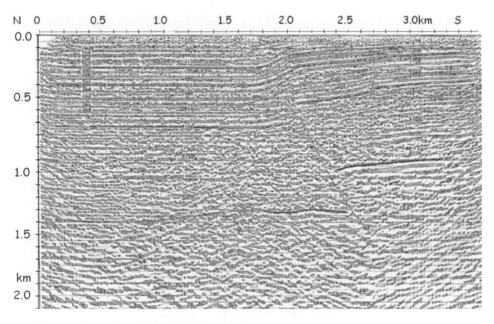

Figure 5.   Seismic Reflection Result (Line-2) (Sakuragawa Flexure).

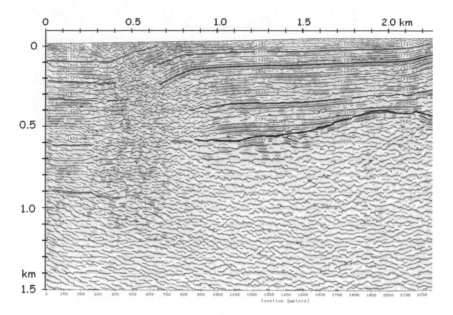

Figure 6. Seismic Reflection Result (Line-3) Kumeda Ike Fault).

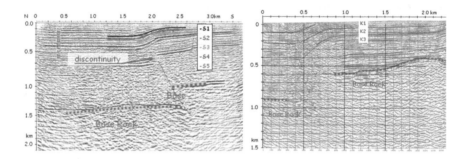

Figure 7. Deformation of sedimentary layers above faulting.

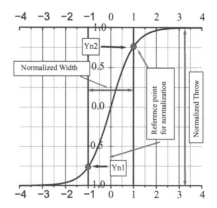

Figure 8. Curve of hyperbolic tangent.

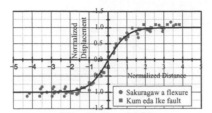

Figure 9. Normalized plots of fault deformation of layers.

Table 1. Normalization parameters of deformation of layers faulting.

Yn1	Yn2	R. Distance	Throw
Sakuragawa F.			
220	120	325	140
255	145	305	160
320	205	310	150
390	250	310	200
450	295	320	210
Kumeda Ike F.			
400	240	100	140
298	158	80	150

Yn1, Yn2: Depth of normalization point of layer
R. Distance: Reference distance for nomarlization
Throw: Vertical component of fault deformation.

Seismic reflection survey lines are shown in black lines. Among these seismic survey lines three seismic survey lines of Line 1, 2, and 3 are shown in this study with dotted bold blue lines. The seismic survey Line 1 is for Uemachi-faults along Nakanoshima shown in Figure 4. The seismic survey Line-2 is along N-S direction crossing Sakuragawa Flexure shown in Fig. 5. The seismic survey Line-3 is shown in Fig. 6 for Kumeda-Ike fault.

2.3 Characterization of fault deformation

Among these seismic sections, deformations of each soil layers caused by fault movement are clearly identified for Sakuragawa Flexure in Line No. 2.

Deformation of the seismic layers of the seismic reflection result was shown in Fig. 7 and digitized and plotted in Fig. 8 for Lines 2 and 3.

In Line-1, it was difficult to identify the deformation of sedimentary layers across the Uemachi-fault, which is referred later in this paper. These seismic surveys are compiled by Ichihara, 2001.

2.4 Normalization of deformation of the layered sediment by faulting

Deformation of sedimentary layers above the fault shown in Figure 7 may be expressed by normalized hyperbolic function. Figure 8 shows a curve of hyperbolic tangent and two reference points for normalization. We need two parameters for normalization of deformation curves to fit as hyperbolic tangent curve. Two points are selected as reference points that are defined on the curve at $(x, y) = (-1.0, -0.762)$ and $(1.0, 0.762)$ as shown in Figure 8. Additional parameter of vertical deformation may be added as throw.

Table 1 shows parameters for normalization for Sakuragawa F. and Kumeda Ike F.

Normalized deformation curve may be obtained by normalizing horizontal distance by reference width and depth by reference throw or vertical distance between the reference points.

In Figure 9, normalized deformation plots are shown for Sakuragawa F. and Kumeda Ike F, where the deformations are normalized by two reference points. Error may be evaluated by the difference of the vertical plot and the hyperbolic curve. Errors based upon difference of the normalized plots at both ends from the hyperbolic curve suggests about within 20% of the throw for the layer.

2.5 Continuity of deformations of layered sediment by faulting

Dip faulting at the base rock causes deformations of the sedimentary layers above the fault.

The sedimentary layers just above the fault became discontinued due to large displacement.

Those surface layers with thick sediments may keep their continuity of the formations.

The deformations in the surface layer are basically shear mode at the center as well as additional bending mode at the end of the fault. A parameter of Deformation Ratio, D.R., in a reversed fault is defined as ratio of reference fault throw to reference width as shown in Fig. 10. If the deformation is regarded as simple shear mode, the DR is shear strain that has been accumulated by successive fault displacement in the past.

The DR is plotted against ratios of base rock throw to averaged depth distance in Figure 11 for Skuragawa and Kumeda Ike Fault.

The plotted DR is divided into two groups of continuous and discontinuous ones. The continuous group

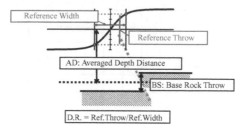

Figure 10. Definition of Deformation Ratio: D.R. in a reversed fault.

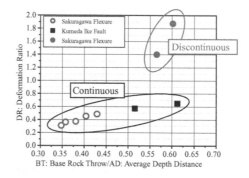

BT: Base Rock Throw/AD: Average Depth Distance

Figure 11. Deformation Ratio against Averaged depth distance.

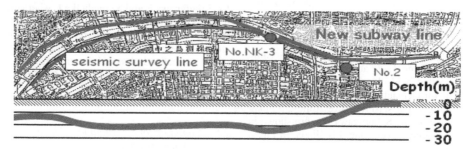

Figure 12. Plan and vertical section of new subway line by Keihan Electric Railway Co.

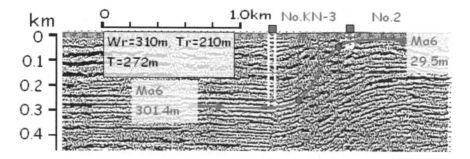

Figure 13. Characterization of Uemachi F. based upon boring data and seismic profile.

increases from 0.25 to 0.7 with depth distance. For discontinuous group the DR is larger than 1.0.

3 NAKANOSHIMA LINE ABOVE UEMACHI FAULT 4 CONCLUSIONS

A new subway line was planned above the Uemachi fault as shown in Figure 12, whose seismic reflection lines are difficult to identify continuously within the faulting range. To determine fault parameter, two borings of No.2 and No.KN-3 were performed to identify the same geological layer at both sides of the fault.

Marine clay deposits of Ma6 were identified at depth 301.4 m and 29.5 m in these boring samples. It is rather easy to find position and scale of the hyperbolic tangent curve that fits with the estimated deformation line of Ma6 as shown in Figure 13. It was found the reference width as Wr = 310 m and reference throw Tr = 210 m. Displacement of the Uemachi Fault by one event of the fault movement is estimated as 4 m. Fault displacement by the next movement is considered to have the same fault parameters except vertical movement.

4 CONCLUSIONS

Characterization of dip slip fault based upon seismic reflection survey was shown for Uemachi Fault in Osaka for the first step. There are more complex structures of fault that should be studied in the future.

The present conclusions obtained are as follows,

1. It is necessary to estimate deformation of the surface and underground by fault to prepare countermeasure against fault displacement.
2. Geomorphologic study does not provide enough information of the ground deformation.
3. Seismic reflection survey is the most useful information that gives deformation of sedimentary layers caused by faulting.
4. Deformation caused by faulting is found to fit well with a curve of hyperbolic tangent.
5. Characterization of the curve may be defined by two reference points and parameters of reference width and reference throw as well as throw of the vertical dip displacement of the layer.
6. When seismic profile does not show continuous image of layers, boring study is useful to obtain corresponding layers between opposite sides of the fault.

REFERENCES

Ichihara, M. 2001. Uemachi Fault and Uemachi upland zone, Urban Kubota 39: 24–27, Osaka, Kubota.
Ikebe, N., Iwatsu, J., and Takenaka, J., 1970. Quaternary geology of Osaka with special reference to land subsidence., Jour. Geoscience Osaka City Univ., 13: 39–98, Osaka.

Prediction and Simulation Methods for Geohazard Mitigation – Oka, Murakami & Kimoto (eds)
© *2009 Taylor & Francis Group, London, ISBN 978-0-415-80482-0*

Liquefaction potential assessment at Laem Chabang port, Thailand

P. Pongvithayapanu
Kasetsart University, Si Racha, Thailand.

ABSTRACT: Laem Chabang Port was developed in the late 1980s to facilitate Thailand's long-term economic and trade growth. Laem Chabang Port locates at Chonburi province in the upper Gulf of Thailand, 110 km of southeast of Bangkok. The ranking position of Laem Chabang Port in the World Top Container Port had risen up from the top 23rd in 1999 to top 19th in 2004. Similar to many of the ports around the world, some areas of Leam Chabang port were built from backfill materials which are highly suspected to soil liquefaction phenomena from moderate to strong earthquakes. Leam Chabang port situates near two major active faults in Thailand which can trigger an earthquake of magnitude $M_w = 8.0$. The liquefaction potential assessment based on simplified analysis is adopted in the study. The major input parameters in the analysis consist of SPT value, Peak Ground Acceleration (*PGA*) and soil density. From the soil characteristics of the studied area, we found that some layers of backfill soils have very low SPT N-value, i.e. below 10, which can be easily liquefied under strong earthquake. The outcome of the assessment divulges that Leam Chabang port is suspected to liquefaction in the backfill soil layers at the depth -6.00 to -10.00 m below ground surface if subjected to strong shaking from potential active faults near the port.

1 INTRODUCTION

1.1 *Background of Laem Chabang port*

Laem Chabang Port is one of the highest (trading) growth rates in the world and also one of the top deep-sea ports in Southeast Asia, positioned as the most efficient gateway to Thailand and the greater Indochina region. Laem Chabang Port is approximately 110 km southeast of Bangkok, in Tungsukhla, Sriracha and Banglamung district of Chonburi province, Thailand. Laem Chabang port presently operates 11 terminals to accommodate container ships, bulk carriers, pure car carriers and passenger liners of up to 120,000 displacement tons. Estimated capacity of the port will increase to 5 million TEU in 2010. The developments of Laem Chabang Port consist of 3 phases:

- The construction period of the first phase had been started from 1987 to 1991, the backfill (reclaimed land) of this phase can be shown in the area under the solid line in Figure 1.
- The second phase construction period had been started from 1997 to 2000, the backfill land area shown in Figure 1 enclosed by dotted line.
- The third phase will be commenced in the near future, i.e. about 2010.

The elevation of the backfill ground surface, with respect to the dredging seabed, is approximately 14–16 m.

1.2 *Backfill Soil Characteristics at Site*

The locations of the boring are around the shadowed area displayed in Figure 1. From the boring log and soil profiles we can see that the types of backfill soil in the area are mostly silty sand (SM) and clayey sand (SC) with very loose to medium conditions. The SPT value, from top of the backfill ground surface to seabed, varies from about 5–28 blows/ft. Below the seabed up to 25 m deep, the SPT value increases drastically, i.e. 35 blows/ft. to 50 blows/25 cm. The ground water level measured 24 hours after completion of the borings is approximately -0.8 m below the backfill ground surface. The natural water content is averagely 20%. The

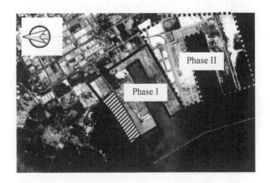

Figure 1. The bird eye view of the reclaimed land of Laem Chabang port in Phase I and II.

complete generalized backfill soil profile, with SPT N-value and soil density, can be shown in Figure 2.

1.3 Earthquake activities in Thailand

The earthquake activities which an epicenter is in Thailand are somewhat low and small to moderate in magnitudes. Some events had been documented in the western and northern of the country. The moderate earthquake, 5.9 in Richter scale, shook Kanchanaburi in 1983 and 5.2 Richter earthquake hit Chiang Mai in 1995. Thailand also faces numerous small to moderate earthquakes which an epicenter is in the neighbor countries such as Burma and Laos. In 2006, a number of earthquakes could be detected by the Thai Meteorological Department (TMD) in Prachuab Khiri Khan and Petch Buri ranged between 3.7 to 5.6 Richter scale. These events were triggered by TaNao Si fault in Burma. In 2007, the strong earthquake of 6.3 Richter, activated by a fault in Laos, struck Chiang Rai and a few damages to houses and temples could be observed. Up to date, identified by the Department of Mineral Resources (DMR), Thailand totally consists of 15 active faults. Some major faults, shown in Figure 3, are in the north, west and south, i.e. Mae Chan fault, Moei fault, Si Sawat fault, Three Pagodas fault, Ranong fault and Khlong Marui fault. Table 1 details the fault parameters, i.e. rupture length and slip rate, of these active faults reported by Peterson

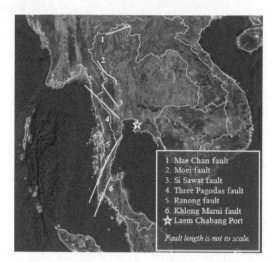

Figure 3.   Map of major fault sources in Thailand.

Table 1.   Parameters of major faults in Thailand.

Faults	Rupture Length, L (km)	Moment Magnitude (M_w)*	Slip rate (mm/year)
Mae Chan	154	7.6	0.70
Moei	226	7.8	0.36
Si Sawat	209	7.8	0.60
Three Pagodas	380	8.0	0.56
Ranong	523	8.2	0.10
Khlong Marui	348	8.0	0.01

*$M_w = 5.08 + 1.16 \log (L)$ by Wells & Coppersmith (1994).

et al. (2007). They also documented that these main faults are strike-slip faults. In Table 1, the moment magnitude of earthquakes that can be trigger by these major active faults was also estimated by a recognized equation of Wells & Coppersmith (1994). Most of them can approximately activate the earthquake with $M_w \geq 8.0$ in magnitude.

1.4 Peak Ground Acceleration (PGA) at Laem Chabang port

To estimate *PGA* at Laem Chabang port from the nearest major faults in Thailand, the attenuation relationships of this region, proposed by Warnitchai et al. (2001), will be used. The two bands of best-estimate regional attenuation relationship were displayed, by thick yellow curves, in Figure 4. Before the estimation of *PGA* by attenuation models can be performed, the earthquake sources and source-to-site distances must be firstly identified. According to the location of Laem Chabang Port in Figure 2, we can see that Laem Chabang Port situates near two major active faults,

Ground		
SM, SP-SM	N-Value = 23,	ρ = 1900 kg/m³
SM, SP-SM	N-Value = 15,	ρ = 2100 kg/m³
SP-SM	N-Value = 25,	ρ = 2150 kg/m³
SM, SP-SM	N-Value = 28,	ρ = 2150 kg/m³
SM, SP-SM	N-Value = 25,	ρ = 2180 kg/m³
SM	N-Value = 10,	ρ = 2200 kg/m³
SM	N-Value = 5,	ρ = 2200 kg/m³
SM	N-Value = 2,	ρ = 2200 kg/m³
SM	N-Value = 2,	ρ = 2180 kg/m³
SM	N-Value = 10,	ρ = 2200 kg/m³
SC	N-Value = 20,	ρ = 2200 kg/m³
SC	N-Value = 20,	ρ = 2180 kg/m³
SC	N-Value = 22,	ρ = 2100 kg/m³
SC	N-Value = 21,	ρ = 2080 kg/m³
SC	N-Value = 25,	ρ = 2150 kg/m³
SC	N-Value = 36,	ρ = 2170 kg/m³
SC	N-Value = 40,	ρ = 2180 kg/m³
SC	N-Value = 42,	ρ = 2200 kg/m³

Backfill, -5 m., Backfill, -10m., Backfill, -15m., Seabed

Figure 2.   Generalize backfill soil profile at site.

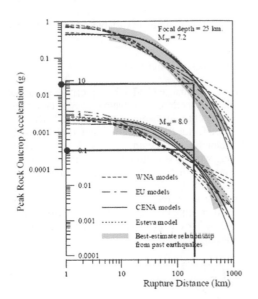

Figure 4. The attenuation relationships of Burma-Thailand-Indochina region (Thick yellow curves) (Warnitchai et al. 2001).

i.e. Three Pagodas fault and Ranong fault. Though their rate of seismic activity is rather low but the source-to-site distance is fairly close, namely about 200 km. The rupture length and corresponding moment magnitude, M_w, calculated from a regression equation of Wells & Coppersmith (1994), of Three Pagodas fault and Ranong fault are 380 km ($M_w = 8.0$) and 523 km ($M_w = 8.2$), respectively. Even if, Ranong fault can cause the strong earthquake up to $M_w = 8.2$, but in Figure 4, PGA can be only estimated for $M_w = 7.2$ and 8.0. Therefore, the maximum PGA at Laem Chabang port from the two nearest faults will be forecasted based on $M_w = 8.0$; the predicted PGA is approximately 0.10 g.

2 LIQUEFACTION POTENTIAL ASSESMENT

2.1 Liquefaction potential assessment by simplified method

Liquefaction is defined as the transformation of a granular material from a solid to a liquefied state as an outcome of increasing in pore water pressure and reducing in effective stress inside granular soils. The evaluation of the liquefaction resistance of soils has evolved for over 30 years. The methodology called "simplified procedure" is generally employed in the liquefaction potential estimation by engineers and researchers. This simplified method though based on empirical but capable to evaluate has been originally developed by Seed & Idriss since 1971 followed from the devastating

earthquakes in Alaska, USA and in Niigata, Japan in 1964. At that time Seed & Idriss correlated blow counts from the standard penetration test (SPT) with a parameter called the cyclic stress ratio that represents the cyclic loading on the soils. After that this procedure has been developed, modified and improved empirically and occasionally relied on field observations and field and laboratory test data.

Calculation of two variables is required for evaluation of liquefaction potential of suspected soils:

1. The level of cyclic loading on the soil caused by earthquake, expressed in term of cyclic stress ratio or *CSR*
2. The capability of the soil to resist liquefaction, express in term of cyclic resistance ratio or *CRR*

In practice, the calculation of *CSR* will follow the derivation of Seed & Idriss (1971) as expressed in (1) with some revisions in a stress reduction coefficient, r_d, which will be discussed later. The evaluation of *CRR* can be retrieved and tested from the undisturbed soil specimens in the laboratory. Unfortunately, sampling techniques and testing granular soil samples in laboratory are too disturbed to obtain the meaningful results. To avoid the difficulties associated with sampling and laboratory testing, field or in-situ tests have become the state-of-practice for routine liquefaction investigations. Several field tests become common usage for evaluation of *CRR*, including the standard penetration test (SPT), the cone penetration test (CPT), shear-wave velocity measurements (V_s). SPT and CPT are generally preferred because of the more extensive databases and past experiences. A promising alternative, or supplement, to the penetration test approaches is provided by in-situ measurements of small-strain shear-wave velocity, V_s. The use of V_s as an index of liquefaction resistance is soundly based because both V_s and liquefaction resistance are similarly influenced by many of the same factors; e.g. void ratio, stress state, stress history and geologic age. A recent update of these simplified methods has been documented in Youd et al. (2001).

2.2 Evaluation of Cyclic Stress Ratio (CSR)

The level of cyclic loading induced by potential earthquake can be expressed in term of Cyclic Stress Ratio (*CSR*):

$$CSR = \frac{\tau_{av}}{\sigma'_{vo}} = 0.65 \left(\frac{a_{max}}{g} \right) \left(\frac{\sigma_{vo}}{\sigma'_{vo}} \right) r_d \qquad (1)$$

where a_{max} = Peak Ground Acceleration (*PGA*) generated by earthquake source; g = gravitational acceleration; σ_{vo} and σ'_{vo} are total and effective vertical stress, respectively; and r_d = stress reduction factor calculated by Equation (2).

$$r_d = \frac{(1.000 - 0.4113z^{0.5} + 0.04052z + 0.001753z^{1.5})}{1.000 - 0.4177z^{0.5} + 0.05729z - 0.006205z^{1.5} + 0.001210z^2} \qquad (2)$$

where z = depth beneath ground surface in meters.

2.3 Evaluation of Cyclic Resistance Ratio (CRR) based on SPT value

The evaluation of Cyclic Resistance Ratio (CRR) of backfill soils based on SPT value requires SPT blow counts and soil density at the depth concerned. The CRR value can be estimated, according to Youd et al. (2001), from CRR curves in Figure 5 and valid only for magnitude 7.5 earthquakes.

For routine engineering calculations in spreadsheets and other techniques, the clean sand base curve according to Figure 5 can be approximated by the following equations;

$$CRR_{7.5} = \frac{1}{34 - (N_1)_{60}} + \frac{(N_1)_{60}}{135}$$
$$+ \frac{50}{[10 \cdot (N_1)_{60} + 45]^2} - \frac{1}{200} \quad (3)$$

where $CRR_{7.5}$ = cyclic resistance ratio for 7.5 of earthquake magnitude; $(N_1)_{60}$ = the corrected SPT value for equipment and effective overburden stress.

Equation (3) valids only for $(N_1)_{60} < 30$. For $(N_1)_{60} > 30$, clean sand soils are too dense to liquefy and are classified as non-liquefiable.

However, many corrections on both SPT and CRR value have to be cautiously performed. First we do the corrections of SPT N-value as follows;

The effect of equipment variables can be found from Equation (4)

$$(N_1)_{60} = N_m C_N C_E C_B C_R C_S \quad (4)$$

where N_m = measured SPT N-value; C_N = normalized factor to reference effective overburden stress; C_E = correction for hammer energy ratio (ER), C_B = correction factor for borehole diameter; C_R = correction factor for rod length; and C_S = correction for samplers with or without liners. The ranged and recommended of these correction values can be found in Youd et al. (2001).

The effect of effective overburden stress calculates from either Equation (5) or Equation (6), recommended by Youd et al. (2001) for routine engineering application.

$$C_N = \left(\frac{P_a}{\sigma'_{vo}}\right)^{0.5} \quad (5)$$

$$C_N = 2.2/\left(1.2 + \frac{\sigma'_{vo}}{P_a}\right) \quad (6)$$

where P_a = atmospheric pressure approximately equal to 100 kPa (1 atm); σ'_{vo} = initial effective vertical stress in the same unit as P_a

The influence of fine contents corrects according to Equation (7)

$$(N_1)_{60cs} = \alpha + \beta(N_1)_{60} \quad (7)$$

where α and β = coefficients relating to fine contents, detailed in Youd et al. (2001).

For simplified analysis, the CRR must also be corrected for Magnitude Scaling Factor (MSF), because the CRR curves in Figure 5 is based on magnitude 7.5 earthquakes. If the estimations depend on other magnitudes, we have to adjust CRR value according to Figure 6.

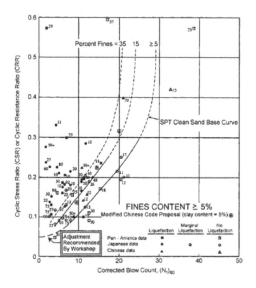

Figure 5. Corrected SPT [$(N_1)_{60}$] curves for Magnitude 7.5 earthquakes (Youd et al. 2001).

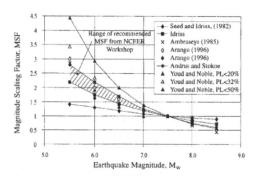

Figure 6. Magnitude Scaling Factor derived from various investigators (Youd et al. 2001).

2.4 Potential for Liquefaction (Factor of Safety, FS)

A common way to quantify the potential for lique-faction is in term of a Factor of Safety (FS). The FS against liquefaction can be defined by;

Liquefaction is predicted to happen when $FS \leq 1$, and liquefaction is predicted not to occur when $FS > 1$.

The suitable value of FS will depend on many factors, including the potential for ground deformation, extent and accuracy of seismic measurements, availability of other site information, and in determination of the design earthquake magnitude and expected value of PGA or a_{max}. The equation for FS against liquefaction can be written in terms of $CRR_{7.5}$, CSR, MSF, K_σ, and K_α, as follows:

$$FS = \left(\frac{CRR_{7.5}}{CSR} \right) MSF \cdot K_\sigma \cdot K_\alpha \qquad (8)$$

where $CRR_{7.5}$ = cyclic resistance ratio for 7.5 of earthquake magnitude; CSR = estimated cyclic stress ratio generated by possible earthquake; MSF = Magnitude Scaling Factor

Because the simplified procedures develop for level surface and for depths less than 15 m, therefore, the effects of high overburden stresses, K_σ, and static shear stresses (sloping ground), K_α, are beyond routine practice and required experiences to explore. Youd et al. (2001) suggested that K_σ and K_α apply mostly to liquefaction hazard analyses of embankment dam and other large structures. Therefore, in this analysis, we do not include the effects of these two parameters.

3 LIQUEFACTION POTENTIAL OF BACKFILL SOILS AT LAEM CHABANG PORT

From the earthquake sources in Thailand and their fault parameters, the data of backfill soil properties and soil profiles and the procedures and corrections of simplified method in Youd et al. (2001), we can practically evaluate the liquefaction potential at Laem Chabang port according to that data and analysis in the previous sections. The maximum PGA using in the evaluation of CSR at Laem Chabang port is 0.1 g; the total and effective vertical stresses vary through the backfill soil depth. The depth of water table is −0.8 m below backfill ground surface and r_d estimates according to Equation (2). The computation of CSR reveals that the CSR values will almost linearly reduce with greater depth below backfill ground surface.

The evaluation of CRR will follow Equation (3) to Equation (6) and the backfill soils are supposed to be clean sand. So, the correction of fine contents will not be carried out in this study. The $CRR_{7.5}$ is estimated by Equation (3) with the corrected SPT N-value, $(N_1)_{60}$. The $(N_1)_{60}$ calculates using Equation (4) with the normalized factor, C_N, varying with depth. Due to lacking of equipment data of SPT field test at the interested

site, other factors, C_E, C_B, C_R, C_S, will be estimated according to Ukritchon & Sangkhawilai (2004). They suggested, for Bangkok clay sampling, the value of approximately 0.7–0.8 including effects of all factors of equipment variables. For this study, the value of 0.75 for the correction of equipment variables will be utilized.

Finally, the potential for liquefaction in term of FS will be calculated according to Equation (8) with the $MSF = 0.75$ ($M_w = 8$ in Figure 6). The outcome of the prediction reveals that some backfill soil layers, e.g. −6.00 to −10.00 m, with low SPT N-value are extremely suspected to liquefaction if subjected to strong earthquake from the nearest fault sources. In addition, we observe the upper bound value of PGA that still make the port secure from liquefaction phenomena. The result exposes that the PGA value of 0.03 g and smaller will not cause any backfill soil layers at Laem Chabang port to be liquefied. The PGA above 0.03 g may liquefy the very low SPT N-value, i.e. less than 5, of backfill soil layers.

4 CONCLUSIONS

The study of the liquefaction potential at Laem Chabang port Thailand by simplified analysis unveils that some backfill soil layers have a potential to be liquefied under strong earthquake activated by the nearest active faults. It is found that the PGA larger than about 0.03 g may causes liquefaction in the very low value of penetration resistance (SPT) of backfill soil layers. The small performance to resist liquefaction of some backfill soil layers at this study may explain by a number of reasons as follows:

1. This study does not include the effect of fine contents, which can apparently increase the value of CRR in soil layers.
2. The high level of water table, i.e. −0.8 m, below backfill surface, can generate excess pore water pressure when subjected to strong shaking.

Some remediation of liquefiable soils to improve the performance to resist liquefaction of some backfill soils layers shall be undertaken by either increasing the liquefaction strength of the soil, i.e. preload, cementation and replacement, or lowering the underground water level to increase effective vertical stress of soils, i.e. drain or replacement by gravel. This remediation can help dropping the damage of structures and facilities around the ports.

REFERENCES

Andrus, R.D. & Stokoe, K.H. 2000. Liquefaction resistance of soils from shear-wave velocity. *Journal of Geotechnical and Geoenvironmental Engineering* 126(11): 1015–1025.

International Navigation Association. 2001. *Seismic Design Guidelines for Port Structures.* Working Group No. 34 of the Maritime Navigation Commission, International Navigation Association. Balkema, the Netherlands.

Ishihara, K. 1993. Liquefaction and flow failure during earthquakes. *Geotechnique* 43(3): 351–415.

Lee, D.H. 2001. Liquefaction performance of soils at the site of a partially completed ground improvement project during the 1999 Chi-Chi earthquake in Taiwan. *Canadian Geotechnical Journal* December 2001. Vol. 38, 2001.

Lee et al. 2002. A study of liquefaction potential for a new reclaimed land in Taiwan. *Proceedings of the twelfth international offshore and polar engineering conference., Kitakyushu, Japan, May 26–31, 2002.*

Leon et al. 2006. Accounting for soil aging when assessing liquefaction potential." *Journal of Geotechnical and Geoenvironmental Engineering* 132(3): 363–377.

Petersen et al. 2007. Documentation for the southeast asia seismic hazard maps. *Administrative report, September 30, 2007.* U.S. Geological Survey.

Robertson, P.K. & Wride, C. E. 1998. Evaluating cyclic lique faction potential using the cone penetration test. *Canadian Geotechnical Journal.* 35: 442–459.

Seed, H.B. & Idriss, I.M. 1971. Simplified procedure for evaluating soil liquefaction potential. *Journal Soil Mech. and Found. Div.* 97(9): 1249–1273.

Seed et al. 1985. The influence of SPT procedures in soil liquefaction resistance evaluations. *Journal of Geotechnical Engineering* 111(12): 1425–1445.

Shuttle, D.A. & Cunning, J. 2007. Liquefaction potential of silts from CPTu. *Canadian Geotechnical Journal* 44: 1–19.

Teachavorasinskun et al. 1994. Effects of the cyclic prestraining on dilatancy characteristics and liquefaction strength of sand. Shibuya et al. (eds), *Pre-failure deformation of geomaterials*: 75–80. Rotterdam: Balkema.

Ukritchon, B. & Sangkhawilai, T. 2004. An analysis of liquefaction potential for Bangkok first sand layer. *Proceeding of the 9th national convention on civil engineering (NCCE9)., Petchaburi, Thailand, May 19–21, 2004.*

Warnitchai et al. 2001. Seismic hazard in Bangkok due to distant earthquakes. *CUS/INCEDE Report 1, Joint workshop on urban safety, AIT, Thailand. September 21–22, 2001.*

Wells, D.L. & Coppersmith, K.J. 1994. New empirical relationships among magnitude, rupture length, rupture width, rupture area, and surface displacement. *Bulletin of the Seismological Society of America* 84(4): 974–1002.

Youd et al. 2001. Liquefaction resistance of soils: Summary report from the 1996 NCEER and 1998 NCEER/NSF workshops on evaluation of liquefaction resistance of soils. *Journal of Geotechnical and Geoenvironmental Engineering* 127(10): 817–833.

Yunmin et al. 2005. Correlation of shear wave velocity wit liquefaction resistance based on laboratory tests. *Soil Dynamics and Earthquake Engineering* 25: 461–469.

Zhou, Y.G. & Chen, Y.M. 2007. Laboratory investigation on assessing liquefaction resistance of sandy soils by shear wave velocity. *Journal of Geotechnical and Geoenvironmental Engineering* 133(8).

Natural dams built by sliding failure of slope during the Iwate-Miyagi Nairiku Earthquake in 2008

K. Tokida

Osaka University, Osaka, Japan

ABSTRACT: In the Iwate-Miyagi Nairiku Earthquake in 2008 in Japan, natural dams were reported to be built along rivers and block the river channels. These dams were afraid to induce secondary disaster after the Earthquake. Because strong inland earthquakes at mountainous area are afraid to occur in the future in Japan, it is necessary and effective to know the fundamental conditions relating to natural dams in the actual earthquakes. In this paper, the fundamental study were carried out to know the properties on 14 natural dams reported in the above Earthquake based on the data obtained just after the Earthquake. Furthermore, the relations between some properties were discussed. As a result, several lessons on the properties of natural dams in the Earthquake can be obtained.

1 INTRODUCTION

The natural dams were afraid to induce secondary disaster such as downstream and/or upstream floods induced by collapse of dams and rise of water level at the reservoir built temporary after the earthquake. Because strong inland earthquakes at mountainous area are afraid to be occur in the future in Japan, it is necessary and effective to know the fundamental conditions concerning to natural dams with use of few and valuable data got in the actual earthquakes.

The fundamental study were carried out to investigate the properties such as length, width and volume of dams etc. observed at 14 natural dams in the above Earthquake, using additional data such as bird's-eye and aerial pictures photographed and topography drawn just after the Earthquake. Furthermore, several relations between length and volume of natural dams were discussed to know the quantitative and engineering relations to apply them for the future earthquakes.

2 EARTHQUAKE AND NATURAL DAMS

The Earthquake occurred around the boundary of Iwate Prefecture and Miyagi Prefecture in Japan on June 14 in 2008. The epicenter located at the mountainous area with the magnitude of 7.2 and the depth of 8 km. In the Earthquake, the fundamental scale such as width, length and volume of 14 natural dams were reported as shown in Table 1 by Tohoku Regional Bureau of the Ministry of Land, Infrastructure and Transport which is named TRB hereinafter (Web site of TRB). The location of these natural dams: 5 dams in Iwate Prefecture and 9 dams in Miyagi Prefecture

Table 1. Natural dams and fundamentals measured.

Prefecture	River	Site	Fundamentals of Natural Dam		
			Width (m)	Length (m)	Volume of Slid Debris (thousand m₃)
Iwate-1	Iwai	KOGAWARA	30	60	20
Iwate-2		ICHINONOBARA	200	700	1,730
Iwate-3		TSUKINOKIDAIRA	60	100	80
Iwate-4		SUKAWA	130	280	390
Iwate-5		UBUSUMEKAWA	200	260	12,600
Miyagi-1	Hazama	SAKANOSHITA	20	80	90
Miyagi-2		AZABU	220	220	300
Miyagi-3		KOGAWARA	200	520	490
Miyagi-4		NUKUYU	80	580	740
Miyagi-5		YUNOKURAONSEN	90	660	810
Miyagi-8		YUBAMA	200	1,000	2,160
Miyagi-10		KAWARAGOYASAWA	170	400	210
Miyagi-7	San-Hazama	NUMAKURA	120	130	270
Miyagi-9		MUMAKURAURASAWA	160	560	1,190

are shown in Figure 1 within the epicentral distance of about 25 km.

3 FUNDAMENTALS OF NATURAL DAMS

3.1 *Objective fundamentals and how to define*

In this study, the fundamentals of natural dams such as the following 7 items were selected and investigated.

- Item 1: Width of natural dam
- Item 2: Length of natural dam
- Item 3: Volume of natural dam
- Item 4: Angle of slid slope
- Item 5: Formation of river channel along slid slope

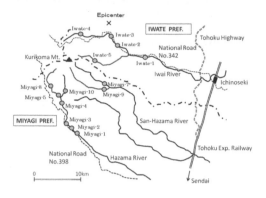

Figure 1. Location of natural dams (drawn based on Web site of TRB).

Photograph 1. Example of Step1: Bird's eye photograph by TRB at Miyagi-5 YUNOKURAONSEN.

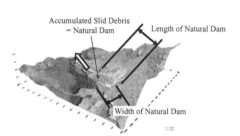

Figure 2. Scale of natural dam identified by TRB.

– Item 6: Topographical form of slid slope along river channel
– Item7: Direction of slid slope against epicenter

As for Items 1, 2 and 3, the fundamentals were already reported as shown in Table 1 where the width, length and volume are defined in Figure 2 and measured.

Other four fundamentals on Items 4, 5, 6 and 7 were selected because they were estimated to relate to the occurrence of slope failure. The four fundamentals were tried to be analyzed using simply available data such as bird's-eye photographs, aerial photographs and topography. The bird's-eye photographs of 14 natural dams were opened to the public by TRB and aerial pictures and topography were done by Geographical Survey Institute of the Ministry of Land, Infrastructure and Transport which is named GSI hereinafter just after the Earthquake (Web site of GSI).

The angle of slid slope on Item 4 was measured by the following working Steps.

– Step 1: Observation of bird's-eye photograph (see Photo 1)
– Step 2: Observation of aerial photograph (see Photo 2)
– Step 3: Observation of topography (see Fig. 3)

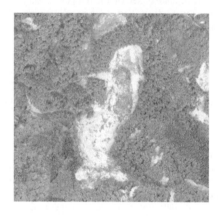

Photograph 2. Example of Step2: Aerial photograph by GSI at Miyagi-5 YUNOKURAONSEN.

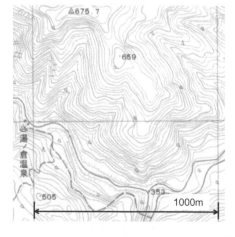

Figure 3. Example of Step3: Topography by GSI at Miyagi-5 YUNOKURAONSEN.

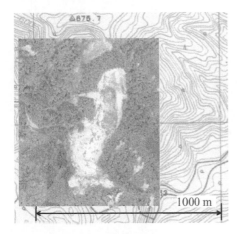

Figure 4. Example of Step4: Putting together aerial photo-graph and topography at Miyagi -5 YUNOKURAONSEN.

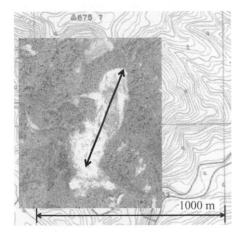

Figure 5. Example of Step5: Drawing horizontal length of slid slope at Miyagi-5 YUNOKURAONSEN.

- Step 4: Putting together aerial photograph and topography (see Fig. 4)
- Step 5: Drawing horizontal length of slid slope (see Fig. 5)
- Step 6: Estimation of height, horizontal length and angle of slid slope (see Fig. 6).

The example on Step 6 is shown in Figure 6 at the one of natural dams: Miyagi-5 YUNOKURAON-SEN where the distance between two vertical grid lines drawn is pointed as 1,000 m. Because the interval of contour lines of height above sea-level is 10 m, the accuracy to estimate the height above sea-level is about ±5 m in this investigation.

The formation of slid slope on Item 5 is classified into 5 factors: concave, slight concave, straight, slight

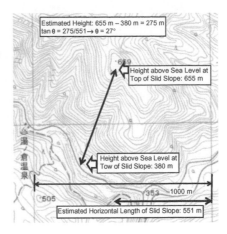

Figure 6. Example of Step6: Estimation of height, horizontal length and angle of slid slope at Miyagi-5 YUNOKURAONSEN.

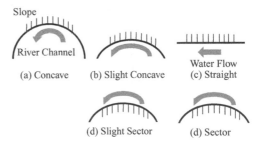

Figure 7. Classification on formation of river channel along slid slope.

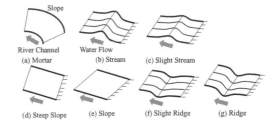

Figure 8. Classification on topographical form of slid slope along river channel.

sector and sector as shown in Figure 7, based on the aerial photograph and topography at natural dams.

The topographical form of slid slope on Item 6 is classified into 7 factors: mortar, stream, slight stream, steep slope, slope, slight ridge and ridge as shown in Figure 8, considering the aerial photograph and topography at natural dams.

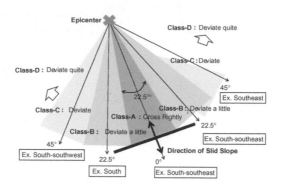

Figure 9.   Classification of directions between epicenter and objective slid slope.

Table 2.   Estimated fundamentals of natural dams.

Prefecture	Site	Average Angle ( )	Form of River Channel	Topographical Form of Slope	Direction of Slope and Epicenter		Class
					Main Direction of Slope	Epicenter Direction	
Iwate-1	KOGAWARA	90	concave	steep slope	South	Southeast	C
Iwate-2	ICHINONOBARA	18	slight sector	slope	East	South-southeast	D
Iwate-3	TSUKINOKIDAIRA	35	concave	slope	North-Northwest	South	B
Iwate-4	[SUKAWA]	16	sector	stream	Northeast	West-southwest	B
Iwate-5	UBUSUMEKAWA	25	concave	mortar	East-southeast	Southwest	D
Miyagi-1	SAKANOSHITA	52	concave	steep slope	South-southwest	South-southeast	A
Miyagi-2	AZABU	28	slight sector	slight ridge	Southwest	South-southeast	B
Miyagi-3	KOGAWARA	24	sector	stream	Southwest	South-southeast	B
Miyagi-4	NUKUYU	42	slight concave	steep slope	West	South-southeast	D
Miyagi-5	YUNOKURAONSEN	27	slight sector	stream	South-southwest	Southwest	B
Miyagi-8	YUBAMA	40	concave	mortar	Southwest	Southwest	A
Miyagi-10	KAWARAGOYASAWA	31	straight	slight ridge	East	Southwest	C
Miyagi-7	NUMAKURA	16	sector	stream	South	South	A
Miyagi-9	MUMAKURAURASAWA	24	slight sector	slight ridge	North-Northwest	South	B

As for the slope failures induced by earthquakes, the relation between the direction of slid slope and the one from the epicenter is very interesting. In this paper, the direction of each slid slope is identified as the right-angle one against the plane of slid slope, and these directions are decided roughly according to classification of 16 directions with an error of ±25/2° (= ±11.25°) shown in Figure 9. According to the deviator of the crossing angle between both directions, the relative relation on Item 7 is classified roughly into 4 classes as follows in this paper (see Fig. 9).

– Class-A: Almost same with deviator less   than ±11.25°
– Class-B: Slight different with deviator from ±11.25° to ±33.75°
– Class-C: Different with deviator from ±33.75° to ±56.25°
– Class-D: Quiet different with deviator more than ±56.25°

3.2   Results of fundamentals

Fundamentals on 7 items can be summarized Table 2 where the height of 10 m and angle of 90° for the site of Iwate-1 KOGAWARA was roughly estimated by author based on the bird's-eye photograph because the measured data were not reported.

The width, length and volume of natural dams can be shown in Figures 10–12, respectively. It can be seen that the width for most of natural dams in the Earthquake ranges from 100 m to 200 m, and the length for most of natural dams ranges from 100 m to 700 m and the maximum one is 1,000 m. Furthermore, the volume for most of natural dams ranges from 200 thousand m³ to 2,000 thousand m³ and as the special case, the volume at Iwate-5 UBUSUMEKAWA is 12,600 thousand m³.

The angle of slid slope can be shown in Figure 13 which is indicated that the angles of the most natural dams can be seen to range from about 10°~50° except the special site with the angle of almost 90° at the

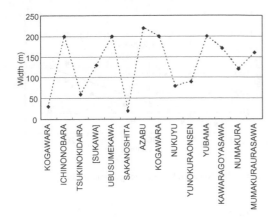

Figure 10.　Width of natural dams.

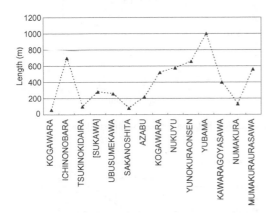

Figure 11.　Length of natural dams.

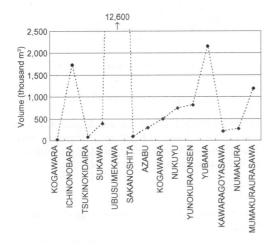

Figure 12.　Volume of natural dams.

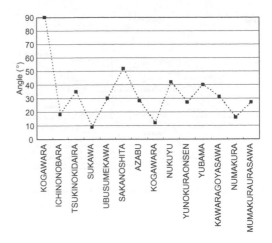

Figure 13.　Estimated angle of slid slopes.

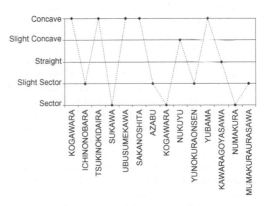

Figure 14.　Formation of river channel along slid slope.

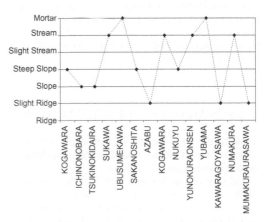

Figure 15.　Topographical form of slid slope along river channel.

587

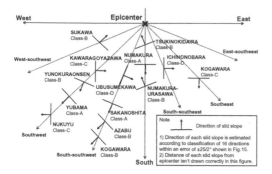

Figure 16. Directions between epicenter and objective slid slope.

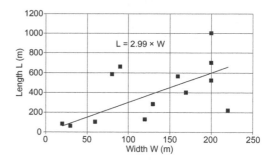

Figure 17. Relation between length and width.

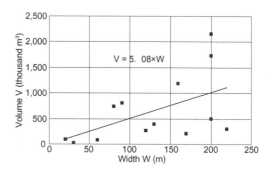

Figure 18. Relation between volume and width.

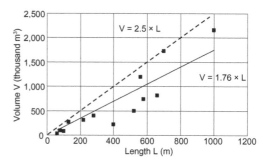

Figure 19. Relation between volume and length.

site of Iwate-1 KOGAWARA. This indicates that the maximum angle of natural slopes in Japan is about 40°~50° in general except for steep slopes.

The formation related to the natural dams can be classified in Figure 14. It can be seen that the concave (5 dams), slight sector (4 dams) and sector (3 dams) can be indicated as the excel formations.

The topographical form related to the natural dams can be classified as shown in Fig. 15. It can be seen that the stream (4 dams), steep slope (3 dams) and slight ridge (3 dams) can be indicated as the excel formations. Comparing the topographical form with the formation of river channel above-mentioned, it can be indicated roughly that mortar and steep slope correspond to concave, and stream corresponds to sector.

The direction of slid slopes at 14 natural dams is summarized in Figure 16 where the distance of each slid slope from the epicenter isn't drawn correctly. As shown in Figure 16, the number of slid slopes corresponding to Class-A, Class-B, Class-C and Class-D are 3 dams, 6 dams, 2 dams and 3 dams, respectively. Especially, because at 9 natural dams against 14 ones, the direction of slid slope corresponds to that of the epicenter with the deviator of angle less than ±33.75°, the direction of slid slope is mostly correspond to that of the epicenter.

4 ESTIMATED SCALE OF NATURAL DAMS

Based on the fundamentals such as width, length and volume of natural dams, 4 kinds of relations between them are investigated as follows.

The relations such as length and width, volume and width, volume and length and volume and width × length can be shown as Figures 17–20, respectively. Herein, the site of Iwate-5 UBUSUME-KAWA shown in Table 2 is deleted on the engineering judgment to estimate the scale of natural dams because the slid

volume is very large comparing with other 13 natural dams.

The relation between width and length of natural dams can be induced as Equation 1. However, their relation is not clear because the correlation coefficient of 0.557 isn't enough.

$$L = 2.99 \times W \tag{1}$$

where L = length of natural dam (m); and W = width of natural dam (m).

The relation between width and volume of natural dams cab de induced as Equation 2. However, their

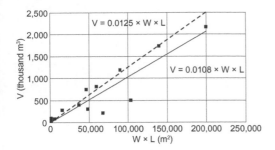

Figure 20.   Relation between volume and width × length.

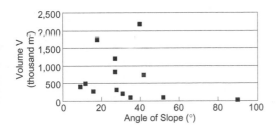

Figure 21.   Relation between volume and angle of slid slope.

relation is not clear because the correlation coefficient of 0.535 isn't enough.

$$V = 5.08 \times W \qquad (2)$$

where V = volume of natural dam (thousand m³).

The relation between length and volume of natural dams can be indicated as Equation 3. They are good relation because the correlation coefficient of 0.906 is enough.

$$V = 1.76 \times L \qquad (3)$$

For improving the correlation, the relation between width × length and volume of natural dams are tried to analyze. Equation 4 can be induced and they have a little better relation than that of Equation 3 because the correlation coefficient is 0.911.

$$V = 0.0108 \times W \times L \qquad (4)$$

While Equations (3) and (4) express the average estimation, the following Equations (5) and (6) to estimate the upper boundary of their relations shown in Figures 19 and 20 can be proposed as the practical ones in this paper.

$$V = 2.5 \times L \qquad (5)$$

$$V = 0.0125 \times W \times L \qquad (6)$$

Additionally, the relation between volume and angle of plane of slid slopes can be shown as Figure 21, where the relation between them isn't clear because the correlation coefficient of −0.224 isn't enough.

5   CONCLUSIONS

In this study, the following lessons can be obtained on the natural dams induced by slid slope along rivers in the Iwate-Miyagi Nairiku Earthquake in 2008.

1. The fundamentals of general natural dams such as width, length, volume of slid debris and slope angle range from 100 m to 200 m, 100 m to 1,000 m, 200 thousand m³ to 2,000 thousand m³ and 10° to 50°, respectively.
2. The formation of the river channel along the slid slope such as concave, slight sector and sector can be indicated as the typical excel formations for natural dams.
3. The original topographical form of the slid slope along the river channel such as stream, steep slope and slight ridge can be indicated as the typical excel formations for natural dams.
4. The direction of slid slope is mostly related to that of epicenter within the deviator of crossing angle less than ±33.75°.
5. The volume (V: thousand m³) of slid debris as for most of natural dams can be estimated roughly with use of length (L: m) and/or width (W: m) of slid debris with the equations of $V = 1.76 \times L$ or $V = 0.0108 \times W \times L$ for the average volume and the equations of $V = 2.5 \times L$ or $V = 0.0125 \times W \times L$ for the maximum one.

Because the examples of natural dams whose detail data could be obtained in the past earthquakes are few at present, further studies are necessary to be conducted to estimate and/or take measures against natural dams in the future earthquakes.

ACKNOWLEGEMENT

The author would like to express sincere gratitude to Mr. N. Kato and Mr. H. Miura of Tohoku Regional Bureau of the Ministry of Land, Infrastructure and Transport for the fruitful data on natural dams built during the Iwate-Miyagi Nairiku Earthquake in 2008.

REFERENCES

Web site of Geographical Survey Institute of the Ministry of Land, Infrastructure and Transport: http://zgate.gsi.go.jp/iwate2008/index.htm

Web site of Tohoku Regional Bureau of the Ministry of Land, Infrastructure and Transport: http://www.thr.go.jp/

Prediction and Simulation Methods for Geohazard Mitigation – Oka, Murakami & Kimoto (eds)
© 2009 Taylor & Francis Group, London, ISBN 978-0-415-80482-0

Site investigation on Toki landslide and its countermeasure on March, 2007

M. Iwata, A. Yashima & K. Sawada
Gifu University, Gifu, Japan

Y. Murata
NPO; Network for the Action against Geo-hazards, Hashima, Japan

T. Suzuki
Gifu Prefectural Government, Gifu, Japan

ABSTRACT: A landslide occurred in Toki city, Gifu, Japan on March, 2007. A heavy rain was not observed before the landslide occurred. Therefore it was difficult to specify the trigger of the landslide. In order to understand the cause and the mechanism of the landslide, intensive literature researches and site investigations were conducted. To stabilize the landslide, the temporary countermeasures were conducted such as the soil removal work, the counterweight fill and so on. In this paper, the details of the site investigations, temporary countermeasures and field observational results are explained. In addition, the permanent countermeasures based on results of investigations and observations are discussed.

1 INTRODUCTION

A landslide with 55 m width and 95 m length occurred in Toki, Gifu, Japan on March 6th, 2007. It was confirmed that the total volume of the landslide was about 40,000 m³. The average speed of the landslide movement was 13 mm/hour at the first time.

The slope was temporarily stabilized by countermeasures such as the soil removal work at the top of the slope, the counterweight fill by large-scaled sandbags at the foot of the slope and the drainage boring work. However, the landslide accelerated again by the influence of rainfall caused by the No. 4 typhoon. Therefore, soil removal work and drainage boring work were additionally carried out. After these countermeasures were finished, the movement of the landslide has not been observed again (Tajimi Civil Engineering Office 2007-a, b).

In this paper, the details of the site investigations, temporary countermeasures and field observational results are explained. In addition, the permanent countermeasures based on the results of investigations and observations are discussed.

2 LANDSLIDE INVESTIGATION

In order to understand the cause and the mechanism of the landslide, the landslide investigations and observations were carried out. Figure 1 shows a flowchart from the initial investigations to the permanent countermeasures.

2.1 History of landslide site

2.1.1 Disaster history and meteorological feature around the landslide site

As shown in Figure 2, there are Yamagami landslide and Tsubakisawa landslide sites around the failure site. The countermeasures against these landslides had been carried out. Additionally, old landslide sites in 50 to 100 m width and 300 to 500 m length are also confirmed around the studied failure site. It was concluded that the landslide site was fragile and weak because the site was located on an extension of the Byobu-yama active fault.

This area experienced Tokai heavy rainfall disaster in 2000; the rainfall was 58 mm/hour, 272 mm/day and 522 mm/month. The feature of weather at the area has been subjected to torrential rains of more than

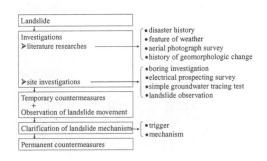

Figure 1. Flowchart from investigations to countermeasures.

100 mm/day and 400 mm/month at least once every 4 years. However, the observed maximum rainfalls were 6 mm/hour, 26 mm/day and 65 mm/month during 2 months before this landslide. It was not observed a heavy rain before the landslide occurred. Therefore, it was difficult to specify the trigger of the landslide.

2.1.2 *History of geomorphologic change by aerial photograph survey*

Past topographical maps and aerial photographs of the failure site were collected and surveyed to confirm whether past landslide took place and the change in terrain as a trigger for the landslide. Figure 3 shows the aerial photographs of the failure site in 1949, 1976, 1987 and 1994, respectively.

From the aerial photograph in 1949, a trace of a landslide was confirmed at the northern side of the slope. Moreover, it was confirmed that a river which was running just below at the failure site was curved toward opposite shore. Therefore, it was estimated that the slope had shown a sign of the old landslide.

From the aerial photograph in 1976, it was confirmed that the construction work had been carried out at the area. A bistered part was confirmed at the top of the slope by the aerial photograph of 1987. There is a possibility that this part is an embankment.

From the results of the aerial photograph survey, it was confirmed that the slope repeatedly slipped several times. Moreover, it was found that a small size of embankment had been constructed at the top of the slope.

2.2 Site investigation

2.2.1 *Boring investigation*

Figure 4 shows the location of measurement equipments for site investigations. A geological cross section along a survey line A of the slope from the boring investigations is illustrated in Figure 5.

It was verified that embankment was distributed near the slope surface and Tokiguchi porcelain clay formation was widely distributed under the embankment. Moreover, it was found that bedrock was Mizunami group and foundation of the landslide consisted of mudstone, sand stone and lignite of Mizunami

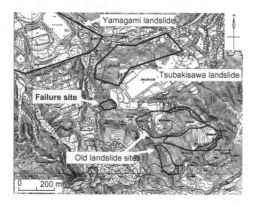

Figure 2.   Topographic map around the failure site.

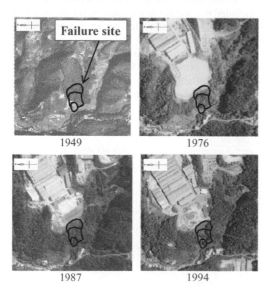

Figure 3.   Aerial photographs around the failure site.

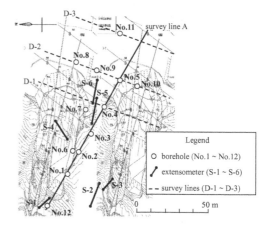

Figure 4.   Measurement equipments for the site investigation.

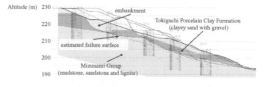

Figure 5.   Geological cross section along the survey line A.

group. Landslide mass consisted of embankment and clayey sand with gravel of Tokiguchi porcelain clay formation. It was assessed that the failure surface was about 15 degrees slope and the layer thicknesses of the landslide at the top and the bottom of the inclination were 8 m and 6 m, respectively. It was ascertained that a boundary between Mizunami group and Tokiguchi porcelain clay formation was dip slope structure.

2.2.2 Electrical prospecting survey

The electrical prospecting surveys were carried out along the survey lines (D-1 to D-3) to understand the groundwater movement with rainfall at the site. An apparent specific resistance value and a change of specific resistance value were measured. In order to observe the influence of rainfalls effectively, the prospecting surveys were conducted before and after No. 4 typhoon attacked the site.

Figure 6 shows the contour figures of the apparent specific resistance value at the each survey line at 0:00 a.m. on July 14th. Figure 7 shows the rate of change of the specific resistance value before and after the typhoon. Rainfalls caused by the typhoon continued from July 14th at 3:00 a.m. to July 15th at 8:00 a.m.

The survey line D-1 was the cross section of the center of upper portion of the landslide. It was confirmed that the specific resistance value of the surface dropped to a lower value on July 14th at 4:00 p.m. It was presumed this phenomenon was due to the percolation of surface stream water. The trench cut, one of the temporary countermeasures, was completed on July 13th. Therefore, due to the effect of the trench cut, the groundwater was drained from the top of the slope and the specific resistance value went up after July 15th at 8:00 a.m.

The survey line D-2 was the cross section just above the top of the landslide. It was confirmed that the water level of borehole No. 5 rose after rainfall caused by the typhoon. It was presumed that the groundwater flowed from the backland to the landslide mass. There is a high specific resistance part at a horizontal distance 42 m

and at an elevation 222 m after July 15th at 8:00 a.m. It was presumed that the large void and high permeability zone, so-called "water passageway", exists locally inside in this slope.

The survey line D-3 was the cross section of the backland of the landslide. It was confirmed that a low specific resistance part expanded from the ground surface after July 15th at 8:00 a.m. The water penetration from a bare field was main cause of this change.

2.2.3 Simple groundwater tracing test

A simple groundwater tracing test was carried out to understand the groundwater flow in the slope. For this test, calcium chloride was used as tracer and was thrown into the borehole No. 11. Figure 8 shows the rate of change of the specific resistance value by the simple groundwater tracing test.

Along the survey line D-1, a low specific resistance part was spread with time from the place at horizontal distance 40 m and an elevation 215 m. The groundwater did not exist inside the landslide mass. It was presumed that this fact was based on the effect of the drainage boring works which were completed before the simple groundwater tracing test. On the other hand, it was confirmed that a great amount of groundwater was submerged under the landslide mass.

About the survey line D-2, a low specific resistance part was spread with time from the place at a horizontal distance 42 m and an elevation 222 m after 72 hours since tracer dropping. The change of the specific resistance value at the survey line D-2 was slower than at D-1. Therefore it was presumed that the tracer flowed in "water passageway" and percolation to periphery was very slow. It was thought that the infiltration capacity of the slope is low and the groundwater from the backland to the inside of the slope flows at a slow speed.

About the survey line D-3, it was apparent that a low specific resistance part slowly spread in the vertical direction from the borehole No. 11. Therefore it was also thought that the infiltration capacity of the slope is low from the result of the survey line D-3.

3 TEMPORARY COUNTERMEASURE AND LANDSLIDE OBSERVATION

The temporary countermeasures and the observations of the landslide movement have been conducted by the estate owner after the landslide event. The followings are details of the temporary countermeasures and results of the landslide observations.

3.1 Temporary countermeasure

Table 1 summarizes the temporary countermeasures and Figure 9 shows positions of the temporary countermeasures. The soil removal works at the top of

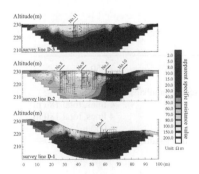

Figure 6. Contour Figure of apparent specific resistance value.

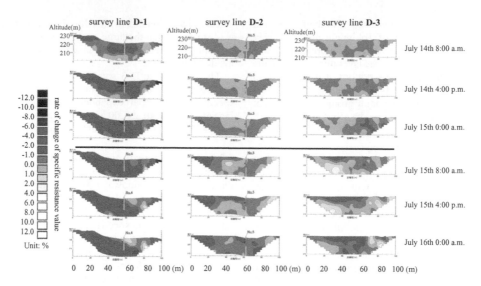

Figure 7. Rate of change of specific resistance value.

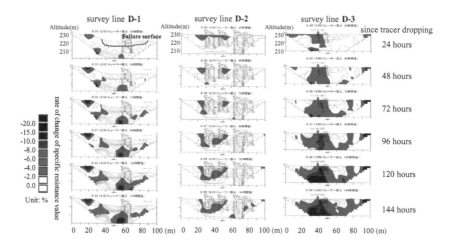

Figure 8. Rate of change of specific resistance value by simple groundwater tracing test.

Table 1. List of temporary countermeasures.

Stage	Date	Countermeasure	
First	Mar. 8th–10th	Soil removal work	7,850 m³
	Mar. 9th–10th	Corrugated pipe	about 75 m
	Mar. 12th–16th	Counterweight fill by large-scaled sandbags	107 pieces
	Mar. 13th–14th	Drainage boring work	4 lines, total 186 m
	Mar. 20th–23rd	Drainage boring work	4 lines, total 180 m
Second	Apr. 4th–23rd	Soil removal work	6,000 m³
	May 7th–10th	Counterweight fill by large-scaled sandbags	61 pieces
Third	Jul. 7th–13th	Trench cut	4 places
	Jul. 18th–20th	Counterweight fill	
Fourth	Sep. 17th–19th	Soil removal work	140 m³
	Sep. 24th–29th	Drainage boring work	8 lines, total 444 m

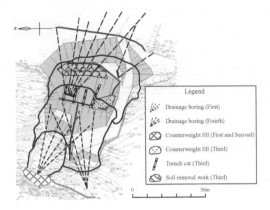

Figure 9. Map of position of temporary countermeasures.

Legend

🐾 Drainage boring (First)

🐾 Drainage boring (Fourth)

⊗ Counterweight fill (First and Second)

⦂ Counterweight fill (Third)

⌿ Trench cut (Third)

⬚ Soil removal work (Third)

0   50m

the slope and the counterweight fills by large-scaled sandbags were carried out to stabilize the slope. The drainage boring works were conducted to remove the groundwater from the landslide mass. Additionally, the landslide observations have been carried out at the same time to evaluate the temporary countermeasures.

3.2 Result of landslide observation

In order to understand the relation between landslide movement and groundwater level, the observations by pipe strain gauges, borehole water level gauges and ground surface extensometers have been carried out. The location of the measurement equipments is also shown in Figure 4. Figure 10 shows the observational results.

3.2.1 Precipitation

According to the observational data by Tsumagi observatory of Ministry of Land, Infrastructure, Transport and Tourism, more than 40 mm continuous precipitation was observed several times between March 10th and September 20th, 2007. Especially, the continuous precipitation recorded 146 mm when the typhoon attacked the site at July 14th and 15th.

3.2.2 Pipe strain gauge

The observations by the pipe strain gauges have been conducted at the boreholes No. 5 to 8, 10 and 12. By the rainfall on June 24th to 25th, a displacement at the upper block was observed and the maximum displacement was recorded at a depth of 4 m of the pipe strain gauge No. 7. The displacements at the center block (No. 6) and at the backland of the landslide (No. 5, 8 and 10) were not observed even when the typhoon attacked the site.

3.2.3 Borehole water level gauge

The observations by the borehole water level gauges have been conducted at the boreholes except No. 12.

Water levels in the boreholes (No. 1 to 3) at the center and the lower block gradually went down since the observations started. The water level had not been rose even when the typhoon attacked the site and always have been kept a low condition. Therefore, it was estimated that the drainage boring works displayed the effectiveness as a temporary countermeasure.

It was also estimated that the trench cut displayed the effectiveness as a temporary countermeasure because water levels in the boreholes (No. 4 and 7) at the upper block were kept low since the trench cut was finished. In addition, the water level also kept low when rainfalls on August 31st and September 11th to 12th.

Due to the rainfall during the typhoon, it was confirmed that the water level in the borehole (No. 5) at the backland of the landslide started to rise slowly. Moreover, the water level in the boreholes No. 9 and 10 rose with rainfalls on August 31st and September 11th to 12th. It was found that the groundwater level at the backland of the landslide depended on the rainfall. Therefore it was concluded that the permanent countermeasures are needed to remove the groundwater from the backland of the landslide against future rainfall.

3.2.4 Ground surface extensometer

By the rainfall on March 24th to 25th, the displacement at the center and the lower block (S-1 to 3) continued to increase and a maximum speed was recorded of 4 mm/hour by the ground surface extensometer S-1. After that time, the soil removal work was conducted at the top of the slope as a temporary work. The displacement of the lower block has not been observed after the second temporary countermeasures were finished.

Due to rainfalls on May 25th to 26th and June 24th to 25th, the displacements of the upper block (S-5 and 6) were observed after the second temporary countermeasures. Therefore, the fourth temporary countermeasures were additionally carried out.

The displacements of the all blocks have not been observed after the typhoon attacked the site. Therefore it was judged that the landslide has been stabilized by the temporary countermeasures.

4 MECHANISM OF THE LANDSLIDE

The mechanism of the landslide is summarized based on investigations as follows.

Mechanical factor:

- The bedrock of the site was fragile and weak because the site was located on an extension of the Byobuyama active fault.

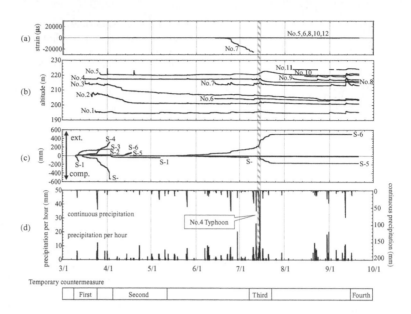

Figure 10. Observational result. (a) Pipe strain gauge (b) Borehole water level gauge (c) Ground surface extensometer (d) Precipitation.

Table 2. List of permanent countermeasures.

Countermeasure	Geometry (Unit: mm)	Quantity	Memo
For safety factor $FS = 1.15$			
Drainage well	ϕ3500	14 m	
Water catchment boring work	VP40, ϕ90	300 m	30 m × 10
Drainage boring work	SGPW90, ϕ135	55 m	55 m × 1
Ground anchor works (3)	5–2, ϕ90	50 m	10 m × 5
Field grating crib work (3)	500 sq., 4000 × 4000	40 m	8 m × 5
Corrugated drainage	350 × 350	700 m	
Counterweight fill		1,500 m³	
Crib retaining wall	Length: 1660, height: 7900	34 m	
For safety factor $FS = 1.20$			
Ground anchor works (1)	5–4, ϕ90	576 m	48 m × 12
Ground anchor works (2)	5–2, ϕ90	42 m	14 m × 3
Field grating crib work (1)	500 sq., 4000 × 4000	192 m	8 m × 24
Field grating crib work (2)	500 sq., 4000 × 4000	24 m	8 m × 3
Soil removal work		2,000 m³	
Pile works	ϕ216.3 × 10	130 m	10 m × 13

- The failure surface consists of Tokiguchi porcelain clay formation and the lignite of Mizunami group with dip slope structure.
- The slope was unstable by the past landslides.

Trigger:

- A small size of embankment was constructed at the top of the slope.
- The groundwater level inside the slope went up by rainfalls.

5 PERMANENT COUNTERMEASURE

As already explained, the landslide was stabilized by the temporary countermeasures. Therefore, it was judged that a safety factor of the slope Fs was equal to 1.05 after the all temporary countermeasures were finished. The permanent countermeasures which satisfied the conditions of a target safety factor of the slope Fs of 1.15 were proposed.

Table 2 summarizes the permanent countermeasures. Figure 11 shows positions of the permanent

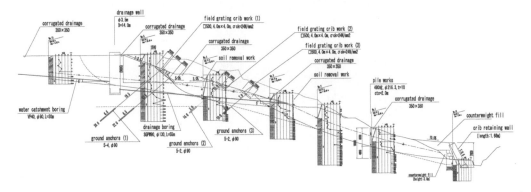

Figure 11. Cross-section view of layout of permanent countermeasures.

countermeasures. Base on the results of the observations, priorities of the permanent countermeasures were decided. The permanent countermeasures are mainly the soil removal work, the counterweight fill with the crib retaining wall, the pile works, the ground anchor works and the drainage well. A plan of the permanent countermeasures is as follows.

In order to satisfy the safety factor Fs of 1.15, the counterweight fill with the crib retaining wall, ground anchor works at lower level and the drainage well are conducted. The constructions of prevention works such as the pile works and the ground anchor works at higher level will be reconsidered based on the future measurements.

The drainage well was adopted because it is necessary to avoid an excess flow of the groundwater from the backland in order to prevent the spread of the landslide. If the further displacement is observed in the future, the ground anchors are planned to be constructed.

6 FUTURE OBSERVING PROGRAM

The observations using the pipe strain gauges, the borehole water level gauges and the ground surface extensometers are planned in the future. Additionally, a drainage volume from the drainage well will be measured.

The observational data will be picked up every other month and the stability will be evaluated. However,

if the rainfalls over 100 mm/day are observed at Tsumagi observatory of Ministry of Land, Infrastructure, Transport and Tourism, the observational data are collected immediately after the rain. These observations are continued for 1 to 3 years after the permanent countermeasures are completed.

7 CONCLUSIONS

A secondary damage due to landslide should be prevented by appropriate temporary countermeasures in a landslide area. Additionally, it is also necessary that the effective and efficient countermeasures have to be selected to prevent the future acceleration of the landslide.

In this case, the landslide investigations and observations were carried out just after the landslide event. The countermeasures from the temporary countermeasures to the permanent countermeasures were safely carried out.

REFERENCES

Tajimi Civil Engineering Office of Gifu Prefecture. 2007-a. First committee's report of Task Force Committee for Toki landslide (in Japanese).
Tajimi Civil Engineering Office of Gifu Prefecture. 2007-b. Second committee's report of Task Force Committee for Toki landslide (in Japanese).

Prediction and Simulation Methods for Geohazard Mitigation – Oka, Murakami & Kimoto (eds)
© 2009 Taylor & Francis Group, London, ISBN 978-0-415-80482-0

Dynamics of geohazards erosive processes at the Prikaspiy region

A.Z. Zhusupbekov
L.N. Gumilyov Eurasian National University, Astana, Kazakhstan

A. Bogomolov
Volgograd State Architectural and Building University, Volgograd, Russia

ABSTRACT: The valley of Volga and Ural Rivers are located within the Northern Prikaspiy region that includes areas of the Russian Federation and the Republic of Kazakhstan. This valley forms the extensive Volgo-Akhtubinskaya flood-plain at the underset current. This region is characterized by active growth of the erosive processes all along the river valley from Volgograd city to Astrakhan city, included also Atyrau city. The most intensive erosive processes are on the right-bank of Volga and Ural Rivers, where we can notice simultaneous landslide on some areas.

1 INTRODUCTION

The valley of Volga and Ural Rivers are located within the Northern Prikaspiy region, that include areas of the Russian Federation and the Republic of Kazakhstan. This valley forms the extensive Volgo-Akhtubinskaya flood-plain at the underset current. This region is characterized by active growth of the erosive processes all along the river valley from Volgograd city to Astrakhan city. The most intensive erosive processes are on the right-bank of Volga River, where we can notice simultaneous landslide on some areas.

We can mark the following reasons that influence on the marginal erosion of Volga and Ural banks. There are morphology, geological structure of slopes, ice and level modes of the river, as well as flowing currents, that define accumulation and deposits moving. Because of the small length of dispersal the wind excitement does not develop and does not render an essential influence on processes of formation of coast owing.

Changes of water level on the Nizhnyaya Volga occur during all over the year. These changes are in direct dependence on a mode of water dump through the dam of the Volga HPS.

2 ABOUT DYNAMICS OF GEOHAZARDS EROSIVE PROCESSES

River bank slopes of the rivers of Volga and Akhtuba are consisting of erosive breeds: friable sandy-clay deposits of khvalynsky terrace and modern alluvial sand, loams and sandy loams. Periodic saturation by water and drainage of such soils at passage of high waters and daily fluctuations of a level substantially promote bank destruction. Influence of current is shown intensively on sites of abrupt turns of a channel of the river where the washout size of alluvial sand can reach 50–60 meters a year.

Dike currents make active processes of erosion due to a river bank slope cutting on various marks at changes of height of a water level, washing out and moving of the fallen breeds. Also an ice mode influences to processes of river erosion. Influence of a water stream on a river bank slope is absent during freeze-up, however the cutting of the bottom part of a slope, its collapsing and an edge recession of a riverside ledge with the following washout of the fallen material by freshet waters occur during an ice drift.

Monitoring of river banks erosion was done for a number of years (1975–1990) on the sites differing by the maximal intensity of process for an estimation of adverse hydro meteorological and technogenical factors and the engineering-geological processes causing significant damage to territories, industrial and economic objects. In 1999–2003 morpho-dynamic researches of Volga-Akhtuba floodplain (Fig. 1) were carried out by scientists from Moscow State Universities together with the RIZA Institute (the Netherlands) in order to decide a problem of restoration of the rivers of the Europe (Bogomolov et al. 2008, Selivanova 2007, Middelkoop et al. 2005, Aytaliyev et al. 2002, Schoor et al. 2001).

The analysis of satellite pictures of the floodplain territory for the period 1985–2001 and field observations of key sites of the floodplain were conducted to perform this work.

In 2007–2008 stationary research of erosive processes was continued, in particular, on Svetly Yar site on the right bank of Volga River and on a number of sites of the left bank of Akhtuba River.

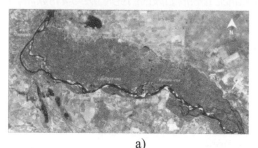

a)

b)

Figure 1. Morphological changes of Volga-Akhtuba floodplain in 1986–1996 (a) and scheme of geographical location (b).

Figure 2. Erosion of a slope on a Svetly Yar site.

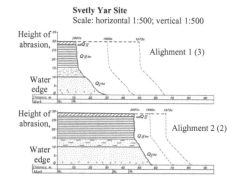

Svetly Yar Site
Scale: horizontal 1:500; vertical 1:500

Figure 3. Structures of a riverside slope of a Svetly Yar site.

soils that creates the best conditions for a filtration of waters. Outputs of subsoil waters on river bank slopes are observed on a site, due to chemical and mechanical suffusion are developing.

Here two types of processing of river bank are marked: erosion-falling and erosive landslide-crumbling.

Riverside ledge on Svetly Yar site is steep, in the top it is vertical and more flat only in the south part of the area. The slope altitude is from 25 up to 30 m above the low-water level in Volga River. Edge is plain, sandy-clay, with a bias to a channel under a corner 7–10°, width is 15–20 m (Fig. 2).

Tetradic loams, layered clay and clay fine-grained sand (Fig. 3) take part in a geological structure of a riverside ledge. Structures of comparison of a relief for the period of supervision from 1972 to 2007 show an average slope erosion speed at a rate of 0.7–1.1 meters a year.

Erosion of river bank occurs as a result of undermining a riverside ledge to formation of collapses and taluses. Soils saturation by water at high levels of the river reduces stability of a slope and leads to occurrence of landslips. Increase and downturn of a level of subsoil waters leads to cracks and times increasing in

3 EROSION-FALLING TYPE

Sites of a riverside slope of this type are incurred to influence of simultaneously two and more processes. At the top of the riverside ledge the landslide phenomena is observed, at the bottom—erosion-falling deformations are observed too. This type of processing is widespread on sites by the general extent up to 40 km.

The most part of Volga River banks concern to this type (120 km of the right and about 100 km of the left bank). For banks of this type it is typical the presence of the high (15–30) abrupt erosive ledges, combined by loams, sandy loam, sand and clay inundated adjournment. Sometimes sandy towpath and sandy-argillaceous in the width 5–20 m are covered by a grass. Outputs of grey clay are noticed on towpath in many places, subsoil waters filter on its housetop.

4 EROSIVE LANDSLIDELY-CRUMBLING TYPE

The maximal intensity of erosive cutting of the riverside ledge is observed on sites with abrupt bends of the river channel due to increase of a role of current, that has greater speed here in a riverside part of the channel than on the rectilinear sites.

Stationary observation of erosion on a riverside slope was conducted here in 1972–1990. Total edge indention of the riverside ledge was from 3.1 m up to 11.90 m, and the mid-annual size of processing of bank was within the limits of 0.5–1.7 m for this period.

Survey of the erosive processes dynamics on the sites of Akhtuba River was fulfilled in 1982–1990 in order to make special recommendations delivery of recommendations for necessities ot protection of economic objects from destruction.

Intensity of erosive processes on Akhtuba River is defined by the same hydro meteorological factors, as on Nizhnyaya Volga including the rise of a water level during the freshet period due to the miss of water of the Volga HPS; cutting of a riverside ledge by ice drift with the subsequent formation of collapses, taluses and sliding deformations; a drain of superficial waters during spring snowmelt and atmospheric precipitation losses.

Levels mode of Akhtuba River depends on an operating mode of the Volga HPS. The schedule of a level change repeats the schedule of water dump through a dam of power station with reduction of amplitude of fluctuation depending on remoteness of a water-measured post from Volgograd City. So seasonal amplitudes of fluctuation of a water level in Akhtuba River in 1988 changed from 7.34 m up to 5.91 m, and duration of the freshet period has made 56 days with peak on May, 1–7st.

Observant sites on Akhtuba River are located on the left bank of the marine khvalynskaya plain. The Volzhsky site (Metallurgist settlement) is the most representative, it is chosen for carrying out of erosion monitoring of Akhtuba River bank in 2007–2008.

Within the limits the Akhtuba River valley has a sharp turn, this causes an activity of erosion on this site. The valley is asymmetric: its left bank is abrupt, and right is flat (Fig. 4).

The width of a channel at horizon of water minus 9–10 m makes about 125–130 m. Depth of the river varies from 0.5 up to 2.5 m, and the greatest depths are dated for the left bank.

Loams, sandy loam, sand (Fig. 5) take part in the geological structure of the riverside ledge. The riverside slope is flat, the ledge is abrupt, steep, and it has height up to 20 m. At a sole of the ledge taluses and collapses, large columnar chars of breeds, traces of falls deformations are noted. The beach is narrow and sandy, its width makes 1–2 m. Washout of bank is connected basically with passage of the high

Figure 4. An erosive slope of Akhtuba River valley.

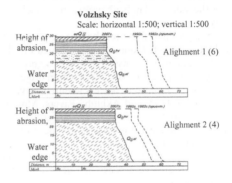

Figure 5. Structures of a riverside slope on Volzhsky site.

water cutting landslide-crumbling accumulations and directly riverside ledge that defines high activity of erosive processes.

Mid-annual speed of the riverside slope processing combined by khvalynsky clays, makes about 3.0 m, and the coast recedes on the average on 1.3 m a year in sand.

On the Figures 3 and 5 are shown following conventional signs: edQ$_{IV}$—recent eluvial-dealluvial deposits—loamy soil; Q$_{III}$hv—highquaternary khvalynsky deposits—clay, in there is clay sand in the bases; Q$_{II}$hz—middlequaternary hazarsky deposits—silica sand, there is gravel in the bases.

Now erosion of the slope represents the greatest danger for some cottages located on distance of 40–50 m from the coast. Besides it is necessary to note, that absence of regulation of a superficial drain leads to khvalynsky clays humidifying and to possible development of fall slope deformations.

The further activization of erosion-falling processes on this site will demand performance of protective actions.

5 CONCLUSIONS

Thus, at the present time the erosive processes of the banks of Volga, Ural, and Akhtuba Rivers are very active both on the right and on the left banks. On the right bank these processes are inseparably linked with landslide deformations. According to the observations an average speed of indent on the right coast is 0.3–0.5 m/year, on some sites reaches the level of 2.5 m/years as a result of erosive cutting.

The erosive processes on the left bank of Volga river are incomparably more intensive (on the left bank there are rest zones, very important agricultural lands and various constructions). Studying of the topographical plans of the left bank of different years has shown that speed of washout can reach 50–60 m/year. The washed away material collects below on the current, forming bars and causing growth of islands in a local part of the valley. Clay bank are washed much more slightly.

Keeping in mind these results it is very important to make the further stationary researches of the characteristics and the intensity of these erosive processes during a development of any construction project and exploitation of the current objects within the riverside zone of Volga, Ural, and Akhtuba Rivers. Moreover it is required to make a forecast of the further development of these erosive processes in order to realize early protection actions.

These investigations are important for the prognosis of geohazards erosive process of river coasts at the Pricaspiy region.

REFERENCES

Bogomolov, A., Shijan, S., Shubin, M., Zhusupbekov, A., Zhapbassov, R. & Kabashev, R. 2008. Dynamics of Erosive Processes in the Nizhnyaya-Volga Prikaspiy Region. *Proceedings of Forth International Conference on Scour and Erosion, 5–7 November, 2008*: 291–294.

Selivanova, T.V. 2007. The Development Dangerous Geological Processes in the Piedmont Region of Primorskiy Krai (Russian Far East). *Proceedings of the International Geotechnical Symposium ≪Geotechnical Engineering for Disaster Prevention & Reduction≫, 24–26 July, 2007*: 240–242.

Middelkoop, H., Schoor, M.M., Babich, D.B., Alabyan, A.M., Shoubin, M.A., Van den Berg, J.H., De Kramer, J. & Dijkstra, J.T. 2005. *Bio-morphodynamics of the Lower Volga river—a reference for river rehabilitation in The Netherlands, Archiv fur Hydrobiologie Supplement 155* (Large Rivers 15 (1–4)): 89–103.

Aytaliyev, S.M., Amanniyazov, K.N. & Baymakhan, R.B. 2002. Geodynamics of Caspian-Round Region with the Elements of Large-Scale Geomechanical Modeling. *Proceedings of the International Conference on Coastal Geotechnical Engineering in Practice, 21–23 May, 2002*: 291–295.

Schoor, M.M., Middelkoop, H., Van de Ven, T., Shoubin, M.A. & Babich, D.B. 2001. *Morphodynamics of the Lower Volga River, Proc. 1st Int. Symp. on Landscape Dynamics in Riverine Corridors, Ascona, Switzerland, 25–30 March*.

Prediction of landslide a few months in advance of its occurrence with chemical sensors for groundwater composition observation

H. Sakai

Railway Technical Research Institute, Tokyo, Japan

ABSTRACT: To provide railroad customers with safe and comfortable travel services, long-term remote sensing technologies have been developed even against abnormal climate changes to suddenly increase precipitation at a time and cause serious ground disasters including landslides. With this technique to use chemical sensors, ground disasters or landslides can be predicted around a few months before they occur.

1   INTRODUCTION

In recent years, the climate has changed throughout the world especially in Asia and Europe resulting in changing precipitation patterns. Thus, large scale short-time rainfalls have been observed frequently in Japan for the last 10 years. Japan Railways provide railroad services all over Japan where over 50% of service lines in mileage are located in mountains or on the sea shore with steep cliffs in back. In the meantime, weather forecast services have been improved to estimate when and where localized torrential rainfalls will occur in advance. In parallel, the relationship between precipitation and risk of ground disaster including landslides has been confirmed. In this situation, however, the information from such sources issues warning to make us be aware of ground disasters taking place on railroad tracks in a short time, mostly right before the incidents caused by rainfalls. Furthermore, it is not always the case that engineers are standing by all the time to watch such disasters closely to the places where the disaster is apprehended. Instead, they are mostly far to take at least one hour to reach the site for a check. Thus, such sort of classical way does not work for arrangements successfully for customers to prepare alternative travel ways. To ensure safe and steady railroad operation by protecting customers from natural disasters, therefore, long-term remote observation systems are eagerly desired to predict the next occurrence of ground disaster.

Basically, current technologies to predict ground disasters are all based on the method to measure the progress with the development of distortion or displacement of the ground, structures or railroad tracks. However, such detection will give information only on the current movement of the ground or conditions only a few days prior to disasters. Since this method is effective just in the case of imminent disasters, it is impossible to organize alternative transportation services in advance to avoid the effect of natural disasters, which would disturb railroad operation. What is more important is to more easily know when such a natural disaster comes up, namely, within how many weeks or months.

To overcome the lack of desired information, which is not available with the conventional technologies for predicting disasters, a noble technique has been developed by installing a chemical sensor, which has never been used for ground disaster detection work. Chemical sensors or ion-selective electrodes can be used to monitor the chemical compositions of the groundwater seeping from the ground where landslides are anticipated. The groundwater contains inorganic ions that originate from soil particles in the ground. The concentrations of the ions do not change while the ground remains stable except when the deep and virgin ground is distorted resulting in the displacement of the surface ground, which leads to a landslide. Some specific ions in the groundwater, therefore, indicate a possibility of landslide by increasing their concentrations. The changes in the concentrations are more easily detected than when displacement comes up on the surface of the ground. This is because the groundwater seeps out from the ground continuously and expeditiously. In contrast, the distortion inside the ground takes a long time to cause displacement of the surface ground. It turns out that the groundwater composition suggests the occurrence of ground displacement like landslides well before it occurs. As a matter of fact, latest results showed that the increases in specific ion concentrations appeared in advance whenever the ground displaced by landslides four times while watched with a prediction system, which was carried out at an inspection site in Japan continuously for 700 days.

2 EXPERIMENTS AND OBSERVATION

2.1 Groundwater collection

To monitor changes in the chemical composition of the groundwater in a batchwise operation, its samples were periodically collected from landslide areas (Sakai et al. 1996, Sakai et al. 2000). The concentrations of sodium, potassium, magnesium, calcium, chloride and sulfate were determined by an ion chromatographic detection system and a conductivity detection method. To observe the changes in the groundwater composition, it is recommended that the groundwater be collected right from sliding surfaces. Unfortunately, however, it is not so easy to successfully obtain such groundwater in general. At any rate, the groundwater available inside landslide areas will pass or originate from adjacent sliding surfaces. Samples can usually be obtainable from drainpipes inserted into the ground, which was carefully chosen to introduce drainage from the spots near a sliding surface. Eventually, groundwater of 100 mL were sampled in a polyethylene bottle every one to two weeks.

2.2 Instrumental measurement

For continuous measurement of chemical composition of groundwater, an ion-selective electrode was employed (Sakai 2008). The electrode was soaked into the flow of the groundwater introduced into a well to temporarily store the drainage from drainpipes for a short time. The electrode worked for more than one year without any special care.

In general, the behavior of landslide is practically measured with such instruments as wire-line extensometers, inclinometers and regular extensometers. This means that the ground behavior is realized when displacement measured by such instruments appears, which exceeds the detectable limit of the measurements. This is to be stressed to exclude ambiguity. In this work, the landslide movement means the ground displacement over the limit by inclinometers that is normally 0.5 mm. The movement with a displacement under the limit does not quantitatively be found in this manner.

2.3 Telecommunications for data transfer

The signal of ion concentrations in the groundwater was converted by a microcomputer and saved in a data logger. The data recorded in the logger was sent to PCs at track work offices and mobile phones held by track maintenance workers in charge through a commercial public telecommunication system. The data was processed to make the workers easily grasp the current situation of the hot landslide with a visual display. To urge them to decide if trains should be stopped, emergency functions were also furnished to make them aware of the event to be apprehended on railroad tracks in the immediate future by flashing lights set on the PC or mobile phones as well as making them vibrate or issue noise.

3 RESULTS AND DISCUSSIONS

The purpose of this paper is to show the possibility to prepare for natural disasters including landslides by using groundwater chemical information. The details of basic experiments are skipped at this opportunity because how and why the groundwater composition changes before soil sample distortion was already confirmed by using soil core samples and groundwater collected from an actual landslide site. As had been expected, the changes of groundwater composition appeared before the core sample was distorted. This phenomenon took place whenever distortion was observed.

3.1 Relationship between landslide movement and groundwater composition

Before the groundwater was automatically and remotely monitored by a chemical sensor at a landslide site, the chemical composition changes and ground displacements had been observed by visual means for 15 years or over in total. At that time, there were no automatic measurements available for groundwater composition and ground displacement. It was enforced, therefore, to collect groundwater samples directly from the site for laboratory analysis. The chemical analysis was also performed manually. Furthermore, the ground movement was checked at intervals of one month or more at a site. Groundwater composition and ground displacement were all measured manually as there were no telecommunication services readily available. This took a long time to confirm the relationship between the changes in the groundwater composition and ground movement by landslides. The investigation was repeated at 30 sites on Japan Railways service lines across Japan. Finally, only six sites were found, where groundwater composition changed before the ground moved. At other sites, there were no changes in the groundwater composition or ground displacement while monitoring those at the same site for at least one year or over. To be clear, there were no cases where groundwater changed even while the ground moved or vice versa.

Japan is covered mostly with mudstone or granite. Thus, there are two typical types where landslides attacked Japan Railways service lines. What ions in groundwater can significantly change their concentrations are dependent upon the layers. The results from the site investigations suggested that sodium, calcium and sulfate are sensitive in the case of mudstone layers. Sodium and calcium work for granite landslides.

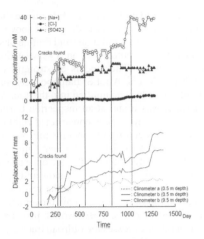

Figure 1. Relationship between changes in groundwater composition inside a landslide area and displacement appearing in inclinometer holes caused by the landslide.

Table 1. The number of days before seeing a landslide occurrence after the change of the groundwater composition recognized as its peak in the case indicated in Fig. 1.

| Number in order | Days after starting the observation | | Time[a] |
	Peak in ion concentrations	Ground movement recorded	
1	85th day	102nd day	17
2	272nd day	342nd–377th day	70
3	300th day	404th–415th day	104
4	553rd day	593rd–622nd day	40
5	837th day	925th–952nd day	88
6	1044th day	1,231st–1,268th day	187
Average	–	–	84.3

[a] Time taken from the ion concentration peak to the ground movement in days.

By using the specific ions in groundwater, a long-term observation of groundwater chemical composition and ground displacement had been performed to make sure of the relationship between the changes in the groundwater composition and ground displacement taking place at a site. For example, Fig. 1 shows the changes and occurrences which were determined in a laboratory in a batch-wise manner after groundwater samples were collected and measured by an inclinometer at the site. Whenever the ground moved, sodium and sulfate ion concentrations increased a few months prior to the event. See Table 1. Thus, the changes in groundwater composition seem to be a symptom to suggest ground movement measurable with equipment like inclinometers following changes in the groundwater composition.

As mentioned above, all the measurement was carried out manually, including groundwater collection at a site, sample analysis and ground displacement measurement with a scale such as inclinometers. Thus, the date recorded were not continuous or given at intervals of one or two weeks. There should be possibilities that groundwater composition changes and ground displacement appeared during the interval between groundwater sample collecting and inclinometer checking times. To overcome this drawback, continuous observation work had keenly been desired.

3.2 Continuous and remote monitoring at a site

By using an ion-selective electrode as a chemical sensor for groundwater composition and telecommunication systems for date transfer, continuous observation was performed at a site to remotely watch changes in the groundwater composition and ground displacement.

Ion-selective electrodes are sensitive to specific ions in water samples. The electrode works without any care at least for half a year even in case it is left in flowing groundwater. An ion-electrode was installed, therefore, at a landslide site to find changes in the groundwater composition. It was also confirmed that the electrode continued to work for one year without any treatment like re-calibration by keeping the groundwater flow always in contact with the surface thereof. This effort has successfully rinsed the electrode surface all the time to keep its clean and fresh condition without bacteria metabolization like bio-films. Some inorganic ion concentrations in the groundwater, its temperature and conductivity were measured by an electrode, thermometers and conductive detection cells, respectively, when required. Figure 2 schematically illustrates how the data were saved in a data logger to transmit to PCs at track work offices and mobile phones of workers responsible for train operation. The system including the electrode, data processors and data loggers were powered by solar batteries. See Fig. 3. Thus, it is not needed to find a place where the commercial power source is available to the system. The groundwater composition can be measured anytime, desirably once a day. When it is required to predict a landslide attack in the immediate future or check the groundwater composition in case a drastic change is observed, it can be checked every five minutes and send the data to the workers every one to 24 hours per day. The frequency of inspection depends on the demand. Workers send commands to the system to set the sequence to monitor the composition and deliver the date to the recipients or the workers. Twenty channels are available to telecommunicate with a recipient at a time, including PCs and mobile phones. The displays equipped on the PC and mobile phones show the changes in the conductivity and turbidity of the groundwater besides its ion

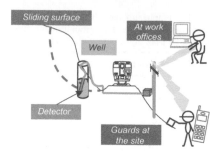

Figure 2. The telecommunication system to successfully send information on the groundwater composition on demand to workers in charge of and responsible for train operation. The ion-selective electrodes soaked into the groundwater in the well periodically determine the concentrations of ions in the groundwater. The signals from the electrodes are processed to save the data in a data logger, which are transmitted to the relevant workers through the public telecommunication service.

Figure 3. Ion-selective electrodes in a well, where the groundwater composition should be monitored. The groundwater was introduced through a pipe to squeeze itself from the landslide site and poured in a pail. The electrodes were left in it. The signal from the electrode was transmitted through a cable to a data logger placed outside the well for processing.

concentrations. For maintenance, the inner temperature of the electrode and the current solar battery voltage are also provided in the display.

To make sure when ground moves after seeing the changes in the groundwater composition, an inclinometer was installed to instrumentally watch the behavior of the landslide, which was set in the ground beside the railroad track. The information given by the inclinometer will confirm that ground displacement was caused as predicted by the changes in groundwater composition. All the data available from the chemical sensors for groundwater composition and the inclinometer for ground displacement were processed by a microprocessor equipped in the main

body to transfer to a PC and a mobile phone through telecommunication systems as described above.

The site where this continuous and remote observation system was installed was on a service line run by Japan Railways. The land is of strongly weathered mudstone layers. A lot of bamboo trees suggest a plenty of groundwater supplied into the site. A high level of water was found at an inspection boring the ground. The railway track is constructed right on the mudstone layer by cutting and opening the low-pitched slope developed by landslides. The landslide ground has repeatedly shifted on a small scale once every five to 20 years. The movement is very slow and intermittent but sturdy in the direction to the railroad track side from the top of mountain. This movement occasionally makes the railroad bed swollen just at a point where the landslide crosses the railroad track resulting in track irregularities.

3.3 Landslide prediction using the remote monitoring system

The results achieved in this work of two years verify that this method to continuously monitor the groundwater composition with chemical sensors satisfactorily predicts the subsequent occurrence of landslide in a simple manner. This is because the phenomena were basically confirmed by long time observation at so many landslide inspection sites, even if the chemical analysis of the groundwater was implemented in a batchwise manner.

Figure 4 indicates that changes in the calcium concentration in the groundwater seeping out from the investigation site and the behavior of the ground displaced by landslides at the same site. While watching two parameters, or the ion concentration and ground displacement for two years, four large-scale changes in the calcium concentration were observed with chemical sensors.

The inclinometer grasped the first small-gap displacement numbered 1 in the Fig, the second large displacement numbered 2, the third small displacement starting creeping numbered 3 and the forth drastic displacement numbered 4 during the observation period. Unfortunately, however, lightning attacked the inclinometer to damage resulting in unavailability of the data on the progress in the ground movement after June 7, 2007. Eventually, no information came up on the ground displacement until September 17, 2007 when the damaged inclinometer was repaired to restart working. However, a significant difference in the absolute position of the spots watched in the inclinometer hole in the ground definitely indicates obvious ground distortion without doubt for about three months from June 7 to September 17, 2008. In the process where the ground movement was measured by an inclinometer as ground displacement,

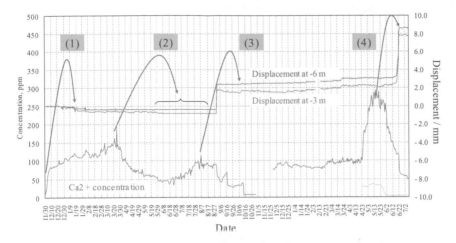

Figure 4.   Changes in the groundwater composition and the ground movement observed November 2006 through July 2008 at a site on a Japan Railways service line by the continuous and remote monitoring system using a chemical sensor for groundwater composition.

increases in the calcium concentration in the groundwater accompanied ground displacement taking place around 60 days in average prior to each event. As just explained, at the second displacement, the exact time when the ground moved is not clear, which should be from June 7 to September 17, 2008. Even in this unexpected situation, however, a highest peak of calcium concentration was observed on March 25, 2007, which is at least 74 days before the displacement.

4   CONCLUSION

For warning to suspend train operation, to let train dispatchers know what is going on in landslides beside railroad tracks in charge, a notice should also be delivered thereto. The ion concentration changes at least two weeks prior to the event including ground distortion and displacement subjected to a landslide. Thus, there is enough time created to prepare for not unexpected but unfortunate situation or landslide attacks coming up in the future by continuously monitoring groundwater composition at a site. This time, for research purposes, an inclinometer was also installed at the site to confirm the following ground displacement after seeing changes in the groundwater composition.

The occurrence of landslide was predicted by monitoring the ion concentrations in the groundwater with chemical sensors. It is not easy by this method to estimate how large the scale of ground displacement caused by landslide is or exactly when it comes, at this momrent. However, it is effortlessly possible to forecast the occurrence normally a few months prior thereto. Furthermore, this equipment calls for less maintenance and no power supply. Thus, the system can simply be installed at any place where such prediction is required even if it is so distant from the nearest observation base to watch landslides.

This technology was developed to successfully protect railroad customers from inconvenience or danger from landslides. What is the most important and indispensable is to know in advance if a ground disaster does or does not come in the immediate future. This is because it would mostly be impossible to hold ground disasters like landslides with a massive ground movement. It is required for us, therefore, to make most of information available to work for the enhancement of safety for human activities, including vehicle traffic on public roads and residences with a back of steep slopes.

REFERENCES

Sakai, H., Murata, O. & Tarumi, T. 1996. A variety of information obtainable from specific chemical contents of groundwater in landslide area. *Proceedings of the 7th International Symposium on Landslide* 2: 867–870, June 17–June 21, 1996.

Sakai, H. & Tarumi, H. 2000. Estimation of the next happening of a landslide by observing the change in the groundwater composition. *Proceedings of the 8th International Symposium on Landslide* 3: 1289–1294, June 26–June 30, 2000.

Sakai, H. 2008. A Warning System Using Chemical Sensors and Telecommunication Technologies to Protect Railroad Operation from Landslide. *Proceedings of the 10th International Symposium on Landslides and Engineered Slopes* 2: 1277–1281, June 30–July 4, 2008.

Author index